离散与计算几何手册

Handbook of Discrete and Computational Geometry, 3e

［美］ 雅各布·E. 古德曼 (Jacob E. Goodman)

［美］ 约瑟夫·奥罗克 (Joseph O'Rourke)

［美］ 乔鲍·D. 托特 (Csaba D. Tóth)

主编

（中册·第三版）

（英文）

哈尔滨工业大学出版社
HARBIN INSTITUTE OF TECHNOLOGY PRESS

U0276502

黑版贸审字 08－2020－202 号

Handbook of Discrete and Computational Geometry, 3e/by Jacob E. Goodman, Joseph O'Rourke, Csaba D. Tóth/ISBN:978-1-4987-1139-5

图书在版编目(CIP)数据

离散与计算几何手册:第三版＝Handbook of Discrete and Computational Geometry, 3e. 中册:英文/(美)雅各布·E.古德曼(Jacob E. Goodman), (美)约瑟夫·奥罗克(Joseph O'Rourke), (美)乔鲍·D.托特(Csaba D. Tóth)主编. —哈尔滨:哈尔滨工业大学出版社,2023.1
ISBN 978-7-5767-0652-9

Ⅰ.①离… Ⅱ.①雅… ②约… ③乔… Ⅲ.①离散数学－计算几何－英文 Ⅳ.①O18

中国国家版本馆 CIP 数据核字(2023)第 030343 号

LISAN YU JISUAN JIHE SHOUCE:DI-SAN BAN(ZHONGCE)

策划编辑　刘培杰　杜莹雪
责任编辑　刘立娟
封面设计　孙茵艾
出版发行　哈尔滨工业大学出版社
社　　址　哈尔滨市南岗区复华四道街 10 号　邮编 150006
传　　真　0451－86414749
网　　址　http://hitpress.hit.edu.cn
印　　刷　哈尔滨市石桥印务有限公司
开　　本　787 mm×1 092 mm　1/16　印张 123　字数 2 371 千字
版　　次　2023 年 1 月第 1 版　2023 年 1 月第 1 次印刷
书　　号　ISBN 978-7-5767-0652-9
定　　价　248.00 元(全 3 册)

(如因印装质量问题影响阅读,我社负责调换)

Part IV

ALGORITHMS AND COMPLEXITY OF FUNDAMENTAL GEOMETRIC OBJECTS

26 CONVEX HULL COMPUTATIONS
Raimund Seidel

INTRODUCTION

The "convex hull problem" is a catch-all phrase for computing various descriptions of a polytope that is either specified as the convex hull of a finite point set in $\mathbb{R}^d$ or as the intersection of a finite number of halfspaces. We first define the various problems and discuss their mutual relationships (Section 26.1). We discuss the very special case of the irredundancy problem in Section 26.2. We consider general dimension d in Section 26.3 and describe the most common general algorithmic approaches along with the best run-time bounds achieved so far. In Section 26.4 we consider separately the case of small dimensions $d = 2, 3, 4, 5$. Finally, Section 26.5 addresses various issues related to the convex hull problem.

26.1 DESCRIBING CONVEX POLYTOPES AND POLYHEDRA

"Computing the convex hull" is a phrase whose meaning varies with the context. Consequently there has been confusion regarding the applicability and efficiency of various "convex hull algorithms." We therefore first discuss the different versions of the "convex hull problem" along with versions of the "halfspace intersection problem" and how they are related via polarity.

CONVEX HULLS

The generic convex hull problem can be stated as follows: Given a finite set $S \subset \mathbb{R}^d$, compute a description of $P = \text{conv} S$, the **polytope** formed by the convex hull of S.

A convex polytope P can be described in many ways. In our context the most important descriptions are those listed below.

GLOSSARY

(See Chapter 15 for basic concepts and results of polytope theory.)

Vertex description: The set of all vertices of P (specified by their coordinates).

Facet description: The set of all facets of P (specified by their defining linear inequalities).

Double description: The set of vertices of P, the set of facets of P, and the incidence relation between the vertices and the facets (specified by an incidence matrix).

Lattice description: The face lattice of P (specified by its *Hasse diagram* (cf.

below), with vertex and facet nodes augmented by coordinates and defining linear inequalities, respectively).

Boundary description: A triangulation of the boundary of P (specified by a simplicial complex, with vertices and maximal simplices augmented by coordinates and defining normalized linear inequalities, respectively).

Hasse diagram: A directed graph of an order relation that joins nodes a to b iff $a \leq b$ and there are no elements between a and b in the sense that if $a \leq c \leq b$ then either $c = a$ or $c = b$. For the face lattice, the order relation is containment.

The five descriptions above assume that P is full-dimensional. If it is not, then a specification of the smallest affine subspace containing P has to be added to all but the vertex description.

These five descriptions make explicit to varying degrees the geometric information carried by polytope P and the combinatorial information of its facial structure. The vertex description and the facet description each carry only rudimentary geometric information about P. We therefore call them **purely geometric descriptions**. The other three descriptions we call **combinatorial** since they also carry more or less complete combinatorial information about the face structure of P. As a matter of fact, these three descriptions are equivalent in the sense that one can be computed from the other by purely combinatorial means, i.e., without the use of arithmetic operations on real numbers.

Which description is to be computed depends on the application at hand. It is important to keep in mind, however, that these descriptions can differ drastically in terms of their sizes (see Section 26.3).

INTERSECTION OF HALFSPACES

Closely related to the convex hull problem is the **halfspace intersection** problem: Given a finite set H of halfspaces in $\mathbb{R}^d$, compute a description of the *polyhedron* $Q = \bigcap H$.

Convex polyhedra are more general objects than convex polytopes in that they need not be bounded. Consequently their descriptions are slightly more complicated. Every polyhedron Q admits a "factorization" $Q = L + C + R$, where L is a linear subspace orthogonal to C and R, the set C is a convex cone, and R is a convex polytope. The "vertex description" of Q then consists of a minimal set of vectors spanning L, the set of extreme rays of C, and the set of vertices of R. Our other four description methods for convex polytopes have to be adjusted accordingly in order to apply to polyhedra. Also, the triangulations appearing in the boundary description need to allow for unbounded simplices (this concept makes sense if one views a k-simplex as an intersection of $k + 1$ halfspaces).

Because polyhedra are more general than polytopes, all statements about the size differences among the various descriptions of the latter apply also to the former.

POLARITY

The relationship between computing convex hulls and computing the intersection of halfspaces arises because of **polarity** (Section 15.1.2). Let S be a finite set in $\mathbb{R}^d$ and let H_S be the set of halfspaces $\{h_p \mid p \in S\}$, with $h_p = \{x \mid \langle x, p \rangle \leq 1\}$. Let

$P = \text{conv}S$ and let $Q = \bigcap H_S$. Polarity yields a 1-1 correspondence between the k-faces of Q and the $(d{-}k)$-faces of P that admit supporting hyperplanes having P and the origin strictly on the same side. In particular, if the origin is contained in the relative interior of P, then the face lattices of P and Q are anti-isomorphic.

It is thus easy to reduce a convex hull problem to a halfspace intersection problem: First translate S by $-\sum_{p \in S} p/|S|$ to insure that the origin is contained in the relative interior of P, and compute $Q = \bigcap H_S$ for the resulting H_S. The polytope Q is then the polar $P^{\triangle}$ of P, and, assuming that P is full-dimensional, we have straightforward correspondences between the vertex description of Q and the facet description of P, between the facet description of Q and the vertex description of P, between the double descriptions of Q and of P (reverse the roles of vertices and facets), and between the lattice descriptions of Q and P (reverse the order of the lattice). Note that there is *no* correspondence between the boundary descriptions. If P has dimension $l < d$ then $Q = Q' \times L$, where polytope Q' has dimension l and L is a linear subspace of dimension $d - l$. The indicated correspondences then hold between P and Q'.

Reducing a halfspace intersection problem to a convex hull problem is more difficult. Polarity assumes all halfspaces to be describable as $\{x \mid \langle a, x \rangle \leq 1\}$, which means they must strictly contain the origin. In general not all halfspaces in a set H will be of such a form. In order to achieve this form the origin must be translated to a point r that is contained in the interior of $Q = \bigcap H$. Determining such a point r requires solving a linear program. Moreover, such an r does not exist if Q is empty, in which case the halfspace intersection problem has a trivial solution, or if Q is not full-dimensional, in which case one has to perform some sort of dimension reduction.

In general, halfspace intersection appears to be a slightly more general and versatile problem, especially in a homogenized formulation, which very elegantly avoids various special cases (see, e.g., [MRTT53]). Nevertheless, we will concentrate exclusively on the convex hull problem. The stated results can be translated *mutatis mutandis* to the halfspace intersection problem. In many cases the algorithms can be "dualized" to apply directly to the halfspace intersection problem, or the algorithms were originally stated for the halfspace intersection problem and were "dualized" to the convex hull problem.

26.2 THE IRREDUNDANCY PROBLEM

GLOSSARY

Irredundancy problem: Given a set S of n points in $\mathbb{R}^d$, compute the vertex description of $P = \text{conv}S$.

$\lambda(n,d)$: The time to solve a linear programming problem in d variables with n constraints. $O(n)$ for fixed d (see Chapter 49).

This problem seeks to compute all points in S that are irredundant, in the sense that they cannot be represented as a convex combination of the remaining points in S. The equivalent polar formulation requires computation of the facet description of $Q = \bigcap H$, given a set H of n halfspaces in $\mathbb{R}^d$. We will follow the primal formulation.

The flavor of this version of the convex hull problem is very different from the other versions. Testing whether a point $p \in S$ is irredundant amounts to solving a linear programming problem in d variables with $n - 1$ constraints. The straightforward method of successively testing points for irredundancy results in an algorithm with running time $O(n\lambda(n-1, d))$, which for fixed dimension d is $O(n^2)$.

Clarkson [Cla94] and independently Ottmann et al. [OSS95] have ingeniously improved this method so that every linear program involves only at most V constraints, where V is the number of vertices of P, i.e., the output size. The resulting running time is $O(n\lambda(V, d))$, which for fixed d is $O(nV)$.

In each of these two methods the n linear programs that occur are closely related to each other. This can be exploited, at least theoretically, by using data structures for so-called linear programming queries [Mat93, Cha96a, Ram00]. This was first done by Matoušek for the naive method [Mat93], and then by Chan for the improved method [Cha96], resulting for fixed $d > 3$ in an asymptotic time bound of

$$O(n \log^{d+2} V + (nV)^{1-1/(\lfloor d/2 \rfloor +1)} \log^{O(1)} n).$$

Finally, note that for the small-dimensional case $d = 2, 3$ there are even algorithms with running time $O(n \log V)$ (see Chapter 42), which can be shown to be asymptotically worst-case optimal [KS86].

26.3 COMPUTING COMBINATORIAL DESCRIPTIONS

GLOSSARY

Facet enumeration problem: Compute the facet description of $P = \text{conv}S$, given S.

Vertex enumeration problem: Compute the vertex description of $Q = \bigcap H$, given H.

The facet and vertex enumeration problems are classical and were already considered as early as 1824 by Fourier (see [Sch86, pp. 209–225] for a survey). Interestingly, no *efficient* algorithm is known that solves these enumeration problems without also computing, besides the desired purely geometric description, some combinatorial description of the polyhedron involved. Consequently we now concentrate on computing combinatorial descriptions.

THE SIZES OF COMBINATORIAL DESCRIPTIONS

It is important to understand how the three combinatorial descriptions differ in terms of their sizes. Let S be a set of n points in $\mathbb{R}^d$ and let $P = \text{conv}S$. Assume that P is a d-polytope and that it has m facets. As a consequence of McMullen's Upper Bound Theorem (Chapter 15) and of polarity, the following inequalities hold between n and m and are tight:

$$n \leq \mu(d, m) \qquad \text{and} \qquad m \leq \mu(d, n),$$

where

$$\mu(d,x) = f_{d-1}(C_d(x)) = \binom{x - \lceil d/2 \rceil}{\lfloor d/2 \rfloor} + \binom{x - 1 - \lceil (d-1)/2 \rceil}{\lfloor (d-1)/2 \rfloor},$$

which is $\Theta(x^{\lfloor d/2 \rfloor})$ for fixed d.

For the sake of definiteness let us define the sizes of the various descriptions as follows. For the double description of P it is the number of vertex-facet incidences, for the lattice description it is the total number of faces (of all dimensions) of P, and for the boundary description it is the number of $(d-1)$-simplices in the boundary triangulation.

Note that for the double and the lattice descriptions the sizes are completely determined by P, whereas the size of a boundary description depends on the boundary triangulation that is actually used. The sizes of those triangulations for a given P can vary quite drastically, even if, as we assume from now on, all vertices of the triangulation must be from S.

These size measures are only crude approximations of the space required to store such descriptions in memory (in particular, in case of the lattice description the edges of the Hasse diagram are completely ignored). However, these approximations suffice to convey the possible similarities and differences between the sizes of the different descriptions.

For such a comparison between the description sizes of $P = \text{conv}\,S$ consider Table 26.3.1, whose columns deal with three cases. The first column lists worst-case upper bounds in terms of n and d. The second column lists upper bounds in terms of m and d under the assumption that S is in nondegenerate position, i.e., no $d + 1$ points in S lie in a common hyperplane, which means that P must be simplicial. Note that in this case there is a unique boundary description. Finally, the third column lists asymptotic bounds (d fixed) for *products of cyclic polytopes* $CC_d(n)$, a certain class of highly degenerate polytopes described in [ABS97]. (See Section 15.1.4 for a discussion of cyclic polytopes.) In this third table column, $\delta = \lfloor \sqrt{d/2} \rfloor$.

TABLE 26.3.1   Polytope description sizes.

DESCRIPTION	WORST CASE	NONDEGENERATE	DEGENERATE CLASS $CC_d(n)$
Double	$d \cdot \mu(d,n)$	$d \cdot m$	$\Theta(n \cdot m^{1-1/\delta})$
Lattice	$2^d \cdot \mu(d,n)$	$2^d \cdot m$	$\Theta((n+m)^\delta)$
Boundary	$\mu(d,n)$	m	$\Omega((n+m)^\delta)$

The bounds in the table are based on the fact that all description sizes are maximized when P is a cyclic polytope, that each facet of a simplicial d-polytope contains 2^d faces, and that the Upper Bound Theorem also applies to simplicial spheres. The lower bound on the size of the boundary description of $CC_d(n)$ applies no matter which triangulation of the boundary is actually used.

The implication of this table is that in the worst case and also in the nondegenerate case all three combinatorial descriptions of P have approximately the same size. If d is considered constant, then the sizes are $\Theta(n^{\lfloor d/2 \rfloor})$ in the worst case, where n is the number of points in S (i.e., n is the input size), and the description sizes are $\Theta(m)$ in the nondegenerate case, where m is the number of facets of P (in a way the output size). The third column of the table, however, shows that in the general case the double description of a polytope P may be substantially more compact than the lattice description or the boundary description.

MAIN RESULTS AND OPEN PROBLEMS

The main positive results are that in the sense of asymptotic worst case complexity the convex hull problem has been solved completely, and that in the case of nondegenerate input, each of the three combinatorial descriptions can be found in time polynomial in the size of the input and the size of the output. In the case of general input this has only been shown for the lattice and for a boundary description, whereas it is unknown whether this is also possible for the double description.

In the following let $P = \text{conv}\,S$ be a d-polytope, and $|S| = n$.

THEOREM 26.3.1   *Chazelle* [Cha93]

If the dimension d is considered constant, then given S, each of the three combinatorial descriptions of $P = \text{conv}\,S$ can be computed in time $O(n \log n + n^{\lfloor d/2 \rfloor})$ using space $O(n^{\lfloor d/2 \rfloor})$. This is asymptotically worst-case optimal.

THEOREM 26.3.2   *Avis-Fukuda* [AF92]

Given S, a boundary description of $P = \text{conv}\,S$ can be computed in time $O(dnM)$ using space $O(dn)$, where M is the size of the boundary description produced.

If S is nondegenerate, then each of the three combinatorial descriptions of P can be computed in time $O(d^{O(1)}nM)$, where M is the size of the respective description.

THEOREM 26.3.3   *Swart* [Swa85] *and Chand-Kapur* [CK93]

Given S, the lattice description of $P = \text{conv}\,S$ can be computed in time and space polynomial in d, n, and the size of the output.

OPEN PROBLEM 26.3.4

Is there an algorithm that, given S, computes the double description of $P = \text{conv}\,S$ in time polynomial in d, n, and the size of the double description?

The algorithm in Chazelle's theorem appears to be of theoretical interest only. The algorithm of Avis-Fukuda is quite practical, the algorithms of Swart and of Chand and Kapur are less so because of the potentially large space requirements. (See Chapters 67 and 68 for descriptions of available code.) The running times of the last two algorithms admit some theoretical improvements, as will be discussed in the following sections.

Almost all algorithms that have been published for solving the different versions of the convex hull problem and the halfspace intersection problem appear to be variations of three general methods: incremental, graph traversal, and divide-and-conquer. We discuss the incremental and the graph traversal methods in the next two subsections. Divide-and-conquer has proven useful only for very small dimension, and we will discuss it in that context in Section 26.4. Methods that fall outside this threefold classification are discussed in Subsection 26.3.3.

26.3.1 THE INCREMENTAL METHOD

The incremental method puts the points in S in some order $p_1, \ldots, p_n$ and then successively computes a description of $P_i = \text{conv} S_i$ from the description of P_{i-1} and p_i, where $S_i = \{p_1, \ldots, p_i\}$.

Before discussing details it should be noted that no matter how the incremental method is implemented, it has a serious shortcoming in that the intermediate polytopes P_i may have many more facets than the final $P_n = P$ (see, e.g., [ABS97]). Thus the description sizes of the intermediate polytopes may be much larger than the size of the description of the final result, and hence this method cannot have running time that depends reasonably on the output size.

This is not necessarily just the result of an unfortunate choice of the insertion order, since Bremner [Bre99] has shown that if S is the vertex set of the aforementioned product of cyclic polytopes $CC_d(n)$, then P_{n-1} has $\Omega(m^{\lfloor \sqrt{d/2} \rfloor - 1})$ facets no matter which insertion order is used, where m is the number of facets of $P_n = P$.

We first present a selection of algorithms implementing the incremental method and list their asymptotic worst-case or expected running times for fixed d (Table 26.3.2). All these algorithms compute boundary descriptions, except for [Sei81] (see also [Ede87, Section 8.4]), which can also be made to compute a lattice description, and [MRTT53], which computes a double description.

TABLE 26.3.2   Sample of incremental algorithms.

ALGORITHM	TIME	BOUND TYPE
Kallay [PS85, Section 3.4.2]	$n^{\lfloor d/2 \rfloor + 1}$	worst-case
Seidel [Sei81]	$n \log n + n^{\lceil d/2 \rceil}$	worst-case
Chazelle [Cha93]	$n \log n + n^{\lfloor d/2 \rfloor}$	worst-case
Clarkson-Shor [CS89]	$n \log n + n^{\lfloor d/2 \rfloor}$	expected
Clarkson et al. [CMS93]	$n \log n + n^{\lfloor d/2 \rfloor}$	expected
Motzkin et al. [MRTT53]	$n^{3 \lfloor d/2 \rfloor + 1}$	worst-case

We now concentrate on how P_{i-1} and P_i differ. For the sake of simplicity we will first assume that S is nondegenerate and hence all involved polytopes are simplicial. Moreover we will ignore how the insertion method starts and assume that P_{i-1} and P_i are full-dimensional. We say that a facet of P_{i-1} is **visible** (from p_i) if its supporting hyperplane separates P_{i-1} and p_i. Otherwise the facet is **obscured**.

The facet set of P_i consists of "old facets," namely all obscured facets of P_{i-1}, and "new facets," namely facets of the form $\text{conv}(R \cup \{p_i\})$, where R is a "horizon" ridge of P_{i-1}, i.e., R is contained in a visible and in an obscured facet of P_{i-1}.

Updating P_{i-1} to P_i thus requires solving three subproblems: finding (and deleting) all visible facets of P_{i-1}; finding all horizon ridges; forming all new facets. The various incremental algorithms only differ in how they solve those subproblems, and they differ in the type of insertion order used.

Visible facets. The simplest way of finding the visible facets is simply to check each facet of P_{i-1}. This is done in Kallay's "beneath-beyond" method [PS85, Section 3.4.2] and in the "double description method" of Motzkin et al. [MRTT53].

Since P_i may have $\Theta(i^{\lfloor d/2 \rfloor})$ facets, such an approach automatically leads to a suboptimal overall running time of $\Omega(n^{\lfloor d/2 \rfloor +1})$ in the worst case.

Another way is to maintain "conflict lists" between facets and not yet inserted points. In the worst case this is no better than the previous method. However, if the insertion order is a random permutation of the points in S, then in expectation this method works in $O(n^{\lfloor d/2 \rfloor})$ time [CS89].

The last method requires the maintenance of a **facet graph**, whose nodes are the facets and whose arcs connect facets if they share a common ridge. The visible facets form a connected subgraph of this facet graph. Thus they can be determined by graph search, such as depth-first search. This takes time proportional to the number of visible facets, which means that in the amortized sense this takes no time since all those visible facets will be deleted. This graph search requires that one starting visible facet be known. Such an initial visible facet can be determined relatively efficiently by a special choice of the insertion order, as in [Sei81], by maintaining "canonical visible facets," as in [CS89] and [CMS93], or by linear programming, as in [Sei91].

Horizon ridges. Determining the horizon ridges is trivial if the facet graph is used, since those ridges correspond to arcs connecting visible and obscured facets. Otherwise one has to use data structuring techniques to determine which of the ridges incident to the visible facets are incident to exactly one visible facet.

New facets. After the horizon ridges are determined, the new facets are easily constructed in time proportional to their number. Keeping this number small is one of the main difficulties of making the insertion method efficient. In the worst case there may be as many as $\mu(d-1, i-1) = \Theta(i^{\lfloor (d-1)/2 \rfloor})$ such new facets. For even d this is $\Theta(i^{\lfloor d/2 \rfloor -1})$, which is the main reason why it was relatively easy to obtain an asymptotically worst-case optimal running time of $O(n^{\lfloor d/2 \rfloor})$ for even d [Sei81]. For general d, using a random insertion order [CS89, CMS93, Sei91] appears to be the only known way to keep this number low, at least in terms of expectation. Chazelle's celebrated deterministic algorithm [Cha93] applies derandomization and thus in effect "simulates" random insertion order so that the number of new facets is not only small in the expected sense but also in the worst case.

Finally, if a facet graph is used, then the arcs corresponding to the ridges between the new facets need to be generated, which can be done via data structuring techniques, as in [Sei91], or by graph traversal techniques, as in [CS89, CMS93]. We should mention that if we remove the nondegeneracy assumption this problem of determining the new ridges seems to become very difficult.

Degenerate input. So far we have assumed that the input set S be nondegenerate. If this is not the case, then this can be simulated using perturbation techniques [Sei96]. This way the algorithms produce a boundary description from which a lattice description or a double description could be computed in $O(n^{\lfloor d/2 \rfloor})$ worst-case time.

The algorithm of Seidel [Sei81] (see also [Ede87, Section 8.4]) also works with degenerate input and then produces a lattice description. Most interesting, though, in the case of degeneracy is the so-called double description algorithm of Motzkin et al. [MRTT53].

THE DOUBLE DESCRIPTION METHOD

Although it is one of the oldest published incremental algorithms, this method has received little attention in the computational geometry community. This method maintains only the double descriptions of the polytopes P_i. It makes no assumptions about nondegeneracy. In fact, despite its poor worst-case complexity, empirically this method works well for degenerate inputs, where all other methods seem to fail, running out of time or space.

The algorithm determines the visible facets by simply checking all facets of P_{i-1}. The interesting point is how it determines the horizon ridges, from which the new facets are then constructed. In contrast to the other methods it does not maintain ridges, since, as we already mentioned, determining the new ridges created during an insertion is difficult. The double description method simply considers each pair of visible and obscured facets of P_{i-1} and checks whether their intersection A forms a horizon ridge. This is achieved by testing whether the vertex set in A is contained in some other facet of P_{i-1}. If it is, then A is not a ridge and hence not a horizon ridge.

A straightforward implementation of this idea will require $\Theta(i^{3\lfloor d/2 \rfloor})$ time in the worst case to discover all horizon ridges of P_{i-1}, resulting in a high worst-case overall running time. Although a number of heuristics have been proposed to speed up this process (see [Zie94, p. 48]), experiments show that this method is unbearably slow in the nondegenerate case when compared to other algorithms. However, in the case of degenerate input it still appears to be the method of choice with the new primal-dual approach (Section 26.3.3) as a possible contender.

Finally, we should mention that convex hull algorithms based on so-called Fourier-Motzkin elimination are nothing but incremental algorithms dressed up in an algebraic formulation.

26.3.2 THE GRAPH TRAVERSAL METHOD

This method attempts to traverse the facet graph of polytope $P = \text{conv} S$ in an organized fashion. The basic step is: given a facet F of P and a ridge R contained in F, find the other facet F' of P that also contains R. Geometrically this amounts to determining the point $p \in S$ such that the hyperplane spanned by R and p maximizes the angle to F. In analogy to a 3D physical realization this operation is therefore known as a "gift-wrapping step," and these algorithms are known as *gift-wrapping algorithms*. In the polar context of intersecting halfspaces, this step corresponds to moving along an edge from one vertex to another and is equivalent to a pivoting step of the simplex algorithm for linear programming. Thus these algorithms are also known as *pivoting algorithms*.

The basic outline of the graph traversal method is as follows: Find some initial facet of $P = \text{conv} S$ and the ridges that it contains. As long as there is an *open ridge* R, i.e., one for which only one containing facet F is known, perform a gift-wrapping step to discover the other facet F' containing R and determine the ridges that F' contains.

This general method faces three problems:

(a) How does one maintain the set of open ridges?

(b) How can the ridges of the new facet F' be quickly discovered?

(c) How can an individual gift-wrapping step be performed quickly?

THE NONDEGENERATE CASE

Let us again first assume that the input set S is in nondegenerate position. This trivializes problem (b) since every facet is a $(d-1)$-simplex and each of the d subsets with $d-1$ of its d vertices will span a ridge.

The most straightforward way to deal with problem (a) is to use some sort of dictionary data structure to store the set of open ridges. The most straightforward way to deal with (c) is to scan through all the points in S to find the best candidate, leading to work proportional to n per discovered facet. This straightforward method has been proposed many times (see [Sch86, p. 224] and [Chv83, p. 282] for references) and has running time $O(d^2nM)$ using $O(d(M+n))$ space, where M is the number of facets of P.

The gift-wrapping steps can be performed faster if a special data structure (for the dual of ray-shooting queries) is used. This was developed by Chan [Cha96], who achieved for fixed $d > 3$ an asymptotic time bound of

$$O(n \log M + (nM)^{1-1/(\lfloor d/2 \rfloor + 1)} \log^{O(1)} n) \,.$$

Avis and Fukuda [AF92] proposed an ingenious way to deal with problem (a) so that no storage space is needed. They pointed out that there is a way of defining a canonical spanning tree T of the facet graph of polytope P so that the arcs of T can be recognized locally. Gift-wrapping steps are then performed only over ridges corresponding to arcs of T. Doing this in the form of a depth-first search traversal of T avoids the use of any extra storage space. Facets can be output as soon as they are discovered. Their algorithm is eminently practical and has a running time of $O(dnM)$ using only $O(dn)$ space.

In theory the gift-wrapping step improvement of Chan also could be applied to the algorithm of Avis and Fukuda. However, this appears to be of little practical relevance.

A completely different way of simultaneously addressing problems (a) and (c) was suggested by Seidel [Sei86a]. He proposed to try to discover the facets in an order corresponding to a straight-line shelling of P. In many cases gift-wrapping steps over several currently open ridges would yield the same new facet F'. However, in that case the entire vertex set of F' is known already and the expensive scan to solve problem (c) is not necessary. The facets of P for which this trick is not applicable can be discovered in advance by linear programming. This "shelling algorithm" has running time $O(n\lambda(n-1, d-1) + d^3 M \log n)$, where $\lambda(n-1, d-1)$ is again the time necessary to solve a linear program with $n-1$ constraints in $d-1$ variables. From the way a shelling proceeds, one can prove that the space requirement for storing the open ridges is somewhat lower than in an ordinary gift-wrapping algorithm.

The linear programs that need to be solved are similar to the ones in the irredundancy problem of Section 26.2. Again, improvements can be achieved by applying linear programming queries ([Mat93]), and the $n\lambda(n-1, d-1)$ factor can be improved to $n^{2-2/(\lfloor d/2 \rfloor + 1)} \log^{O(1)} n)$.

THE GENERAL CASE

There are two ways to approach the general case where P is not simplicial. The first is again to apply perturbations in order to simulate nondegeneracy of S. This way all previously mentioned algorithms still apply, however they now compute a boundary description of P. The parameter M is now the size of the triangulation that happens to be constructed. Moreover, the perturbed computations slow down the running times by a polynomial factor in d.

The second way to deal with the general case is to generalize the algorithms so that they compute the lattice description of P. The main obstacle that must be overcome in the degenerate case is problem (b), the discovery of the ridges of a new facet F'. The obvious way to address this problem is to view the construction of F' as a recursive subproblem one dimension down. Some care must be taken however that in the many recursions small-dimensional faces are not reconstructed too often. This method was proposed by Chand and Kapur [CK93] and their algorithm was later improved and analyzed by Swart [Swa85] who showed a running time of $O(d^2 n K_1 + d^3 K_2 \log K_0)$, where K_i is the number of directed $(i+1)$-vertex paths in the Hasse diagram of the face lattice of P.

Rote [Rot92] generalized the algorithm of Avis and Fukuda to produce the lattice description using little storage space. Its running time is $O(dK_{d+1}n)$ and it appears to be not as relevant in practice as the original algorithm.

Finally, Seidel [Sei86b] generalized his shelling algorithm to produce the lattice description in time $O(n\lambda(n-1, d-1) + K_2(d^2 + \log K_0))$. Because of the recursive nature of straight-line shellings, this generalization avoids reconstruction of small-dimensional faces. Again the improvement via linear programming queries applies.

26.3.3 OTHER METHODS

THE BRUTE-FORCE APPROACH

Let S be a set of n points in $\mathbb{R}^d$ and let $P = \text{conv} S$. Assume w.l.o.g. that the origin is contained in the interior of P (otherwise apply a translation) and assume that S is irredundant in the sense that every point in S is a vertex of P (otherwise apply the results of Section 26.2).

A set $T \subset S$ spans a face of P iff there is a halfspace that has T on its boundary and $S \setminus T$ in its interior. Algebraically this can be tested by determining

$$y_T = \max\{y \in \mathbb{R} | \exists x \in \mathbb{R}^d : \forall p \in T : \langle x, p \rangle = 1 \text{ and } \forall p \in S \setminus T : \langle x, p \rangle + y \le 1\},$$

which can be computed via linear programming, and checking that $y_T > 0$.

This characterization immediately yields a straightforward algorithm with running time $O(2^n \lambda(n, d))$ for generating all faces and also the lattice description of P: Simply test each subset of S whether it spans a face of P. This brute-force approach can be substantially improved by applying backtrack-search techniques ([Bal61],[FLM97]). Fukuda et al. [FLM97] even achieve a running time of $O(nK_0\lambda(n, d))$ this way, using just $O(dn)$ space. Unfortunately this backtrack-search approach does not seem to yield an efficient method to compute the double description of P.

THE PRIMAL-DUAL METHOD

Let S be a set of n points in $\mathbb{R}^d$, let $P = \text{conv} S$, and let $\mathcal{F}$ be the set of facets of P. Determining $\mathcal{F}$ from S is difficult if P is degenerate in the sense that it is not simplicial, i.e., its facets are not all simplices. However, in this case determining S from $\mathcal{F}$ may not be so difficult. The primal-dual method [BFM98] of Bremner, Fukuda, and Marzetta tries to exploit this possibility, despite the fact that $\mathcal{F}$ is unknown and S is the input.

The basic idea of their algorithm is as follows: For a facet $F \in \mathcal{F}$, let H_F be the halfspace that has F on its boundary and contains P, and for $\mathcal{G} \subset \mathcal{F}$ let $H_{\mathcal{G}} = \{H_G | G \in \mathcal{G}\}$. Assume some $\mathcal{G} \subset \mathcal{F}$ is known already. Enumerate the vertices of the polyhedron $P_{\mathcal{G}} = \bigcap H_{\mathcal{G}} \supset P$. If all the vertices found are points in S and if $P_{\mathcal{G}}$ is bounded, then it must be the case that $P_{\mathcal{G}} = P$ and $\mathcal{G} = \mathcal{F}$ and all facets of P have been found, and we are done. If this is not the case (and this can be determined after at most $n+1$ vertices of $P_{\mathcal{G}}$ have been enumerated), then it is easy to find a point $v \in P_{\mathcal{G}} \setminus P$ (either a vertex not in S or a point on an extreme ray of $P_{\mathcal{G}}$). But now clearly $\mathcal{G} \neq \mathcal{F}$. Moreover it is easy to find a facet $G \in \mathcal{F} \setminus \mathcal{G}$ (or rather the halfspace H_G) that separates v from P. This amounts to performing the initial facet finding step of the gift-wrapping algorithm and can be done (without linear programming!) in $O(d^2 n)$ time. Now add G to $\mathcal{G}$ and repeat.

The method suggests that the complexity of computing the facet description of a polytope P from its vertex description is related to the complexity of computing the vertex description from the facet description. It is difficult to make this theoretical statement precise without introducing assumptions about the intermediate polyhedra $P_{\mathcal{G}}$. However, on the practical side, the authors of [BFM98] present experimental evidence showing that the primal-dual method outperforms other algorithms in certain "degenerate" cases.

26.4 THE CASE OF SMALL DIMENSION

Convex hull computations in very small dimension are special. We have strong geometric intuitions about 2D and 3D space (and via Schlegel diagrams even about 4-polytopes). Moreover the situation is simpler in the case $d = 2, 3$ since our five polytope descriptions cannot differ much in terms of their sizes (they are all within a constant factor of each other), which means there is little need for keeping an exact distinction. Algorithmically, small dimensions are special in that besides the incremental and the graph traversal method, divide-and-conquer methods have also been brought to fruition.

THE 2-DIMENSIONAL CASE

The planar convex hull problem has drawn considerable attention and many different algorithmic paradigms have been tried (see textbooks such as [PS85] or [O'R98]). The graph traversal method was rediscovered and is known in the planar case as the **Jarvis march** with running time $O(nM)$, and the incremental method was rediscovered and is known in a rather different guise as the **Graham scan** with running time $O(n \log n)$ (as usual n and M are the sizes of the input and output, respectively). It was easy and natural to apply the divide-and-conquer paradigm to obtain further $O(n \log n)$ time algorithms. By giving this paradigm the extra twist of "marriage-before-conquest" it was possible even to obtain an $O(n \log M)$ algorithm, which was also shown to be worst-case optimal in the algebraic computation tree model of computation [KS86]. This algorithm required the use of 2D linear programming. Much later Chan, Snoeyink, and Yap [CSY97] showed how to avoid this and substantially simplified the algorithm in way that allowed its generalization to higher dimensions. Later, Chan [Cha96] showed quite surprisingly that by using simple data structures and the method of guessing the output size by repeated squaring, the Jarvis march algorithm can be sped up to also run in time $O(n \log M)$.

THE 3-DIMENSIONAL CASE

In 3 dimensions the output size M is $O(n)$ in the worst case. However, the straight-forward implementations of the standard incremental and the graph traversal methods only yield algorithms with worst-case running time $O(n^2)$. In this context the use of the divide-and-conquer paradigm was decisive in obtaining $O(n \log n)$ running time, which was achieved by Preparata and Hong (see [PS85, Section 3.4.4]; for a more detailed account, [Ede87, Section 8.5]). This running time was later matched in the expected sense by the randomized incremental algorithm of Clarkson and Shor [CS89], who also gave another randomized algorithm with expected performance $O(n \log M)$.

The question whether this optimal output-size sensitive bound could also be achieved deterministically was open for a long time. Edelsbrunner and Shi [ES91] first generalized the "marriage-before-conquest" method of [KS86] but achieved only a running time of $O(n \log^2 M)$. Eventually Chazelle and Matoušek [CM95] succeeded in derandomizing the randomized algorithm of Clarkson and Shor and obtained, at least theoretically, this optimal $O(n \log M)$ time bound. Later, Chan [Cha96] showed that there is a relatively simple algorithm for achieving this bound, again by the method of speeding up the gift-wrapping method using data structures and guessing the output size by repeated squaring.

THE CASE $d = 4, 5$

In this case the sizes of the combinatorial descriptions may be as large as $\Theta(n^2)$. All the methods and bounds mentioned in Section 26.3 apply. In addition there are methods for computing a boundary description based on sophisticated divide-and-conquer and some additional pruning mechanisms. Worst-case time bounds of $O((n + M) \log^{d-2} M)$ were achieved by Chan, Snoeyink, and Yap [CSY97] for $d = 4$, and by Amato and Ramos [AR96] for $d = 4, 5$. The latter paper also states that their bound applies to computing the lattice description in the case $d = 4$.

26.5 RELATED TOPICS

There has been some work on determining the intrinsic computational complexity of versions of the convex hull problem. The strongest results at this point are:

1. For fixed $d \geq 2$ the time necessary to determine whether exactly V of n points in $\mathbb{R}^d$ are extreme is $\Omega(n \log V)$ in the algebraic computation tree model [KS86]. This is asymptotically best possible for $d = 2$.

2. For fixed $d \geq 2$ the time necessary to determine whether the convex hull of n points in $\mathbb{R}^d$ has exactly M facets is $\Omega(n^{\lceil d/2 \rceil - 1} + n \log n)$ in a specialized but realistic model of computation [Eri99]. This is asymptotically best possible for odd $d > 1$.

The expected sizes of convex hulls of point sets drawn according to some statistical distribution are typically much smaller than the worst-case sizes. Constructing such convex hulls has been explicitly studied by several authors (see, e.g.,[DT81, Dwy91, BGJR91]). One should also mention in this context the randomized incremental algorithm [CS89]. With input set $S \subset \mathbb{R}^d$ its expected running time for constructing a boundary description is

$$O \left(\sum_{d+1 < r \leq n} df_r(S)/r + \sum_{d+1 \leq r < n} d^2 n f_r(S)/r^2 \right),$$

where $f_r(S)$ is the expected size of the boundary description of the convex hull of a random subset of S of size r. For many distributions f_r is sufficiently sublinear so that this randomized incremental algorithm has $O(n)$ expected running time.

The problem of maintaining convex hulls under insertions and deletions of points has been addressed also. In higher dimensions randomized incremental algorithms have been adapted by several authors to process updates [Mul94, Sch91, CMS93]. However, the analyses are all based on some probabilistic model of which updates actually occur. More satisfactory solutions have only been obtained in the planar case. Solutions with $O(\log n)$ update time were obtained for the insertions-only case (see [PS85, Section 3.3.6]) and also for the deletions-only case [HS92]. For the general dynamic case $O(\log^2 n)$ update times were achieved early on [OL81, Gow80], and only very recently they were improved to $O(\log n)$ in [Cha01, BJ02].

For some time there was hope that additional input information might help compute convex hulls. Although this is true in the planar case, where having points presorted or having them given along a nonintersecting polygonal line [Mel87] leads to linear-time algorithms, it has been shown [Sei85] that for dimension $d \geq 3$ such additional information does not help. Having a 3D set S presorted or even knowing a nonself-intersecting polyhedral surface whose vertex set is S does not in general make it easier to find the convex hull of S.

There have been some attempts to generalize the convex hull construction problem so that the input S does not consist of points but of more general objects such as algebraically described regions in the plane [BK91, NY98], balls in $\mathbb{R}^d$ [BCD+96], ellipsoids in $\mathbb{R}3$ [Wol02], or sets of polyhedra [FLL01].

Finally, parallel algorithms for the convex hull problem have been developed; see Chapter 46.

26.6 SOURCES AND RELATED MATERIALS

FURTHER READING

[Zie94]: A modern account of polytope theory.

[MR80]: A survey of vertex enumeration methods from the dual standpoint.

RELATED CHAPTERS

Chapter 15: Basic properties of convex polytopes
Chapter 17: Face numbers of polytopes and complexes
Chapter 27: Voronoi diagrams and Delaunay triangulations
Chapter 49: Linear programming

REFERENCES

[ABS97]   D. Avis, D. Bremner, and R. Seidel. How good are convex hull algorithms? *Comput. Geom.*, 7:265–301, 1997.

[AF92]   D. Avis and K. Fukuda. A pivoting algorithm for convex hulls and vertex enumeration of arrangements and polyhedra. *Discrete Comput. Geom.*, 8:295–313, 1992.

[AR96]   N.M. Amato and E.A. Ramos. On computing Voronoi diagrams by divide-prune-and-conquer. In *Proc. 12th Sympos. Comput. Geom.*, pages 166–175, ACM Press, 1996.

[BK91]   C.L. Bajaj and M.-S. Kim. Convex hulls of objects bounded by algebraic curves. *Algorithmica*, 6:533–553, 1991.

[Bal61]   M.L. Balinski. An algorithm for finding all vertices of convex polyhedral sets. *SIAM J. Appl. Math.*, 9:72–81, 1961.

[BCD⁺96]   J.-D. Boissonnat, A. Cérézo, O. Devillers, J. Duquesne, and M. Yvinec. An algorithm for constructing the convex hull of a set of spheres in dimension d. *Comput. Geom.*, 6:123–130, 1996.

[BGJR91]   K.H. Borgwardt, N. Gaffke, M. Jünger, and G. Reinelt. Computing the convex hull in the Euclidean plane in linear expected time. In P. Gritzmann and B. Sturmfels, editors, *Applied Geometry and Discrete Mathematics: The Victor Klee Festschrift*, vol. 4 of *DIMACS Ser. Discrete Math. Theoret. Comput. Sci.*, pages 91–107, AMS, Providence, 1991.

[Bre99]   D. Bremner. Incremental convex hull algorithms are not output sensitive. *Discrete Comput. Geom.*, 21:57–68, 1999.

[BFM98]   D. Bremner, K. Fukuda, and A. Marzetta. Primal-dual methods for vertex and facet enumeration. *Discrete Comput. Geom.*, 20:333–357, 1998.

[BJ02]   G.S. Brodal and R. Jacob. Dynamic planar convex hull. *Proc. 43rd IEEE Sympos. Found. Comput. Sci.*, pages 617–626, 2002.

[Cha93]   B. Chazelle. An optimal convex hull algorithm in any fixed dimension. *Discrete Comput. Geom.*, 10:377–409, 1993.

[Cha96]   T.M. Chan. Output-sensitive results on convex hulls, extreme points, and related problems. *Discrete Comput. Geom.*, 16:369–387, 1996.

[Cha96a]   T.M. Chan. Fixed-dimensional linear programming queries made easy. In *Proc. 12th Sympos. Comput. Geom.*, pages 284–290, ACM Press, 1996.

[Cha01]   T.M. Chan. Dynamic planar convex hull operations in near-logarithmic time. *J. ACM*, 48:1–12, 2001.

[Chv83]   V. Chvátal. *Linear Programming*. W.H. Freeman, New York, 1983.

[CK70]   D.R. Chand and S.S. Kapur. An algorithm for convex polytopes. *J. ACM*, 17:78–86, 1970.

[Cla94]   K.L. Clarkson. More output-sensitive geometric algorithms. In *Proc. 35th IEEE Sympos. Found. Comput. Sci.*, pages 695–702, 1994.

[CM95]   B. Chazelle and J. Matoušek. Derandomizing an output-sensitive convex hull algorithm in three dimensions. *Comput. Geom.*, 5:27–32, 1995.

[CMS93]   K.L. Clarkson, K. Mehlhorn, and R. Seidel. Four results on randomized incremental constructions. *Comput. Geom.*, 3:185–212, 1993.

[CS89]   K.L. Clarkson and P.W. Shor. Applications of random sampling in computational geometry, II. *Discrete Comput. Geom.*, 4:387–421, 1989.

[CSY97]   T.M. Chan, J. Snoeyink, and C.K. Yap. Primal dividing and dual pruning: Output-sensitive construction of four-dimensional polytopes and three-dimensional Voronoi diagrams. *Discrete Comput. Geom.*, 18:433–454, 1997.

[DT81]   L. Devroye and G.T. Toussaint. A note on linear expected time algorithms for finding convex hulls. *Computing*, 26:361–366, 1981.

[Dwy91]   R.A. Dwyer. Rex A. Dwyer Convex hulls of samples from spherically symmetric distributions. *Discrete Appl. Math.*, 31:113–132, 1991.

[Ede87]   H. Edelsbrunner. *Algorithms in Combinatorial Geometry*, vol. 10 of *EATCS Monogr. Theoret. Comput. Sci.* Springer-Verlag, Heidelberg, 1987.

[ES91]   H. Edelsbrunner and W. Shi. An $O(n \log^2 h)$ time algorithm for the three-dimensional convex hull problem. *SIAM J. Comput.*, 20:259–277, 1991.

[Eri99]   J. Erickson. New lower bounds for convex hull problems in odd dimensions. *SIAM J. Comput.*, 28:1198–1214, 1999.

[FLL01]   K. Fukuda, T.M. Liebling, and C. Lütolf. Extended convex hull. *Comput. Geom.*, 20:13–23, 2001.

[FLM97]   K. Fukuda, T.M. Liebling, and F. Margot. Analysis of backtrack algorithms for listing all vertices and all faces of a convex polyhedron. *Comput. Geom.*, 8:1–12, 1997.

[Gow80]   I.G. Gowda. *Dynamic problems in computational geometry*. M.Sc. thesis, Dept. Comput. Sci., Univ. British Columbia, Vancouver, 1980.

[HS92]   J. Hershberger and S. Suri. Applications of a semi-dynamic convex hull algorithm. *BIT*, 32:249–267, 1992.

[KS86]   D.G. Kirkpatrick and R. Seidel. The ultimate planar convex hull algorithm? *SIAM J. Comput.*, 15:287–299, 1986.

[Mat93]   J. Matoušek. Linear optimization queries. *J. Algorithms*, 14:432–448, 1993.

[Mel87]   A. Melkman. On-line construction of the convex hull of a simple polyline. *Inform. Process. Lett.*, 25:11–12, 1987.

[MR80]   T.H. Mattheiss and D. Rubin. A survey and comparison of methods for finding all vertices of convex polyhedral sets. *Math. Oper. Res.*, 5:167–185, 1980.

[MRTT53]   T.S. Motzkin, H. Raiffa, G.L. Thompson, and R.M. Thrall. The double description method. In H.W. Kuhn and A.W. Tucker, editors, *Contributions to the Theory of Games II*, vol. 8 of *Ann. Math. Stud.*, pages 51–73. Princeton University Press, 1953.

[Mul94]   K. Mulmuley. *Computational Geometry: An Introduction through Randomized Algorithms*. Prentice Hall, Englewood Cliffs, 1994.

[NY98]   F. Nielsen and M. Yvinec. Output-sensitive convex hull algorithms of planar convex objects. *Comput. Geom.* 8:39–66, 1998.

[OL81]   M.H. Overmars and J. van Leeuwen. Maintenance of configurations in the plane. *J. Comput. Syst. Sci.*, 23:166–204, 1981.

[O'R98]   J. O'Rourke. *Computational Geometry in C*, second edition. Cambridge University Press, 1998.

[OSS95]   T.A. Ottmann, S. Schuierer, and S. Soundaralakshmi. Enumerating extreme points in higher dimensions. In *Proc. 12th Sympos. Theoret. Aspects Comp. Sci.*, vol. 900 of LNCS, pages 562–570, Springer, Berlin, 1995.

[PS85]   F.P. Preparata and M.I. Shamos. *Computational Geometry: An Introduction.* Springer-Verlag, New York, 1985.

[Ram00]   E.A. Ramos. Linear optimization queries revisited. In *Proc. 16th Sympos. Comput. Geom.*, pages 176–181, ACM Press, 2000.

[Rot92]   G. Rote. Degenerate convex hulls in high dimensions without extra storage. In *Proc. 8th Sympos. Comput. Geom.*, pages 26–32, ACM Press, 1992.

[Sch86]   A. Schrijver. *Theory of Linear and Integer Programming.* Wiley-Interscience, New York, 1986.

[Sch91]   O. Schwarzkopf. Dynamic maintenance of geometric structures made easy. In *Proc. 32nd IEEE Sympos. Found. Comput. Sci.*, pages 197–206, 1991.

[Sei81]   R. Seidel. *A convex hull algorithm optimal for point sets in even dimensions.* M.Sc. thesis, Dept. Comput. Sci., Univ. British Columbia, Vancouver, 1981.

[Sei85]   R. Seidel. A method for proving lower bounds for certain geometric problems. In G.T. Toussaint, editor, *Computational Geometry*, pages 319–334, North-Holland, Amsterdam, 1985.

[Sei86a]   R. Seidel. Constructing higher-dimensional convex hulls at logarithmic cost per face. In *Proc. 18th Sympos. Theory Comput.*, pages 404–413, ACM press, 1986.

[Sei86b]   R. Seidel. *Output-size sensitive algorithms for constructive problems in computational geometry.* Ph.D. thesis, Dept. Comput. Sci., Cornell Univ., Ithaca, 1986.

[Sei91]   R. Seidel. Small-dimensional linear programming and convex hulls made easy. *Discrete Comput. Geom.*, 6:423–434, 1991.

[Sei96]   R. Seidel. The meaning and nature of perturbations in geometric computing. *Discrete Comput. Geom.*, 19:1–17, 1996.

[Swa85]   G.F. Swart. Finding the convex hull facet by facet. *J. Algorithms*, 6:17–48, 1985.

[Wol02]   N. Wolpert. *An exact and efficient approach for computing a cell in an arrangement of quadrics.* Ph.D. thesis, FR Informatik, Univ. des Saarlandes, Saarbrücken, 2002.

[Zie94]   G.M. Ziegler. *Lectures on Polytopes*, vol. 152 of *Graduate Texts in Math.* Springer-Verlag, New York, 1994.

27 VORONOI DIAGRAMS
AND DELAUNAY TRIANGULATIONS
Steven Fortune

INTRODUCTION

The Voronoi diagram of a set of sites partitions space into regions, one per site; the region for a site s consists of all points closer to s than to any other site. The dual of the Voronoi diagram, the Delaunay triangulation, is the unique triangulation such that the circumsphere of every simplex contains no sites in its interior. Voronoi diagrams and Delaunay triangulations have been rediscovered or applied in many areas of mathematics and the natural sciences; they are central topics in computational geometry, with hundreds of papers discussing algorithms and extensions.

Section 27.1 discusses the definition and basic properties in the usual case of point sites in $\mathbb{R}^d$ with the Euclidean metric, while Section 27.2 gives basic algorithms. Some of the many extensions obtained by varying metric, sites, environment, and constraints are discussed in Section 27.3. Section 27.4 finishes with some interesting and nonobvious structural properties of Voronoi diagrams and Delaunay triangulations.

GLOSSARY

Site: A defining object for a Voronoi diagram or Delaunay triangulation. Also generator, source, Voronoi point.

Voronoi face: The set of points for which a single site is closest (or more generally a set of sites is closest). Also Voronoi region, Voronoi cell.

Voronoi diagram: The set of all Voronoi faces. Also Thiessen diagram, Wigner-Seitz diagram, Blum transform, Dirichlet tessellation.

Delaunay triangulation: The unique triangulation of a set of sites such that the circumsphere of each full-dimensional simplex has no sites in its interior.

27.1 POINT SITES IN THE EUCLIDEAN METRIC

See [Aur91, Ede87, For95] for more details and proofs of material in this section.

GLOSSARY

Sites: Points in a finite set S in $\mathbb{R}^d$.

Voronoi face of a site s: The set of all points of $\mathbb{R}^d$ strictly closer to the

FIGURE 27.1.1

Voronoi diagram and Delaunay triangulation of the same set of sites in two dimensions (a,b) and three dimensions (c,d).

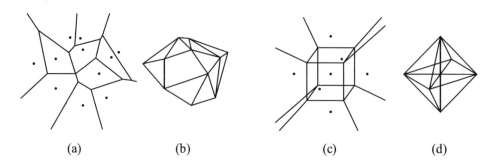

(a)   (b)   (c)   (d)

    site $s \in S$ than to any other site in S. The Voronoi face of a site is always a nonempty, open, convex, full-dimensional subset of $\mathbb{R}^d$.

Voronoi face $V(T)$ of a subset T: For T a nonempty subset of S, the set of points of $\mathbb{R}^d$ equidistant from all members of T and closer to any member of T than to any member of $S \backslash T$.

Voronoi diagram of S: The collection of all nonempty Voronoi faces $V(T)$, for $T \subseteq S$. The Voronoi diagram forms a cell complex partitioning $\mathbb{R}^d$. In two dimensions (Figure 27.1.1(a)), the Voronoi face of a site is the interior of a convex, possibly infinite polygon; its boundary consists of **Voronoi edges** (1-dimensional faces) equidistant from two sites and **Voronoi vertices** (0-dimensional faces) equidistant from at least three sites. Figure 27.1.1(c) shows a Voronoi diagram in three dimensions.

Delaunay face $D(T)$ of a subset T: The Delaunay face $D(T)$ is defined for a subset T of S whenever there is a sphere through all the sites of T with all other sites exterior (equivalently, whenever $V(T)$ is not empty). Then $D(T)$ is the (relative) interior of the convex hull of T. For example, in two dimensions (Figure 27.1.1(b)), a **Delaunay triangle** is formed by three sites whose circumcircle is empty and a **Delaunay edge** connects two sites that have an empty circumcircle (in fact, infinitely many empty circumcircles).

Delaunay triangulation of S: The collection of all Delaunay faces. The Delaunay triangulation forms a cell complex partitioning the convex hull of S.

There is an obvious one-one correspondence between the Voronoi diagram and the Delaunay triangulation; it maps the Voronoi face $V(T)$ to the Delaunay face $D(T)$. This correspondence has the property that the sum of the dimensions of $V(T)$ and $D(T)$ is always d. Thus, in two dimensions, $V(T)$ is a Voronoi vertex iff $D(T)$ is an open polygonal region; $V(T)$ is an edge iff $D(T)$ is; $V(T)$ is an open polygonal region iff $D(T)$ is a vertex, i.e., a site. In fact, the 1–1 correspondence is a duality between cell complexes, reversing face ordering: for subsets $T, T' \subseteq S$, $V(T')$ is a face of $V(T)$ iff $D(T)$ is a face of $D(T')$.

    The set of sites $S \subset \mathbb{R}^d$ is in **general position** (or is **nondegenerate**) if no $d+2$ points lie on a common d-sphere and no $k+2$ points lie on a common k-flat, for $k < d$. If S is in general position, then the Delaunay triangulation of S is a simplicial complex, and every vertex of the Voronoi diagram is incident to $d+1$ edges in the Delaunay triangulation. If S is not in general position, then Delaunay faces need not be simplices; for example, the four cocircular sites in Figure 27.1.1(b) form a Delaunay quadrilateral. A ***completion*** of a Delaunay triangulation is obtained by splitting nonsimplicial faces into simplices without adding new vertices.

RELATION TO CONVEXITY

There is an intimate connection between Delaunay triangulations in $\mathbb{R}^d$ and convex hulls in $\mathbb{R}^{d+1}$, and between Voronoi diagrams in $\mathbb{R}^d$ and halfspace intersections in $\mathbb{R}^{d+1}$. To see the connections, consider the special case of $d = 2$. Identify $\mathbb{R}^2$ with the plane spanned by the first two coordinate axes of $\mathbb{R}^3$, and call the third coordinate direction the *vertical* direction.

The ***lifting map*** $\lambda : \mathbb{R}^2 \to \mathbb{R}^3$ is defined by $\lambda(x_1, x_2) = (x_1, x_2, x_1^2 + x_2^2)$; $\Lambda = \lambda(\mathbb{R}^2)$ is a paraboloid of revolution about the vertical axis. See Figure 27.1.2(a). Let H be the convex hull of the lifted sites $\lambda(S)$.

The Delaunay triangulation of S is exactly the orthogonal projection into $\mathbb{R}^2$ of the lower faces of H (a face is ***lower*** if it has a supporting plane with inward normal having positive vertical coordinate). To see this informally, suppose that triangle $\lambda(s)\lambda(t)\lambda(u)$ is a lower facet of H, and that plane P passes through $\lambda(s)\lambda(t)\lambda(u)$. The intersection of P with Λ is an ellipse that projects orthogonally to a circle in $\mathbb{R}^2$ (Figure 27.1.2(a)). Since all other lifted sites are above the plane, all other unlifted sites are outside the circle, and stu is a Delaunay triangle. The opposite direction, that a Delaunay triangle is a lower facet, is similar.

For Voronoi diagrams, assign to each site $s = (s_1, s_2)$ the plane

$$P_s = \{(x_1, x_2, x_3) : x_3 = -2x_1 s_1 + s_1^2 - 2x_2 s_2 + s_2^2\}.$$

Let I be the intersection of the lower halfspaces of the planes P_s. The Voronoi diagram is exactly the orthogonal projection into $\mathbb{R}^2$ of the upper faces of I. To

FIGURE 27.1.2
(a) *The intersection of a plane with Λ is an ellipse that projects to a circle;* (b) *on any vertical line, the surfaces $\{D_s\}$ appear in the same order as the planes $\{P_s\}$.*

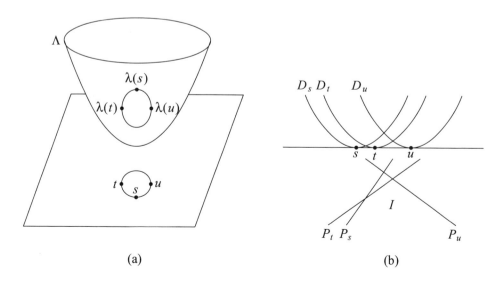

(a)   (b)

see this informally, consider the surfaces

$$D_s = \{(x_1, x_2, x_3) : x_3 = ((x_1 - s_1)^2 + (x_2 - s_2)^2\}$$

(see Figure 27.1.2). Viewed as a function from $\mathbb{R}^2$ into R, D_s gives the squared distance to site s. Furthermore, P_s and D_s differ only by the quadratic term $x_1^2 + x_2^2$, which is independent of s. Hence a point $x \in \mathbb{R}^2$ is in the Voronoi cell of site t iff on the vertical line through x, D_t is lowest among all surfaces $\{D_s\}$. This happens exactly if, on the same line, P_t is lowest among all planes $\{P_s\}$, i.e., x is in the projection of the upper face of I formed by P_t.

COMBINATORIAL COMPLEXITY

In dimension 2, a Voronoi diagram of $n \geq 3$ sites has at most $2n - 5$ vertices and $3n - 6$ edges (and the Delaunay triangulation has at most as many triangles and edges, respectively).

In dimension $d \geq 3$ the Voronoi diagram and Delaunay triangulation can have $\Theta(n^{\lceil d/2 \rceil})$ faces. Exact bounds can be given using results from convex polytope theory (Chapter 15). For n sites in d dimensions, the maximum number of Voronoi k-dimensional faces, $k < d$, is $f_{n-k}(C_{d+1}(n)) - \delta_{0k}$, where $C_{d+1}(n)$ is the $d+1$-dimensional cyclic polytope, f_{n-k} gives the number of $n-k$ dimensional faces (see Section 17.3 and Theorem 17.3.4), and $\delta_{0k} = 1$ if $k = 0$ and 0 otherwise.

For a simple lower bound example in dimension 3, choose $n/2$ distinct point sites on each of two noncoplanar line segments l and l'. Then there is an empty sphere through each quadruple of sites (a, a', b, b') with a, a' adjacent on l and b, b' adjacent on l'. Since there are $\Omega(n^2)$ such quadruples, there are as many Delaunay tetrahedra (and Voronoi vertices).

If point sites are chosen uniformly at random from inside a sphere, then the expected number of faces is linear in the number of sites. In dimension 2, the expected number of Delaunay triangles is $2n$; in dimension 3, the expected number of Delaunay tetrahedra is $\sim 6.77n$; in dimension 4, the expected number of Delaunay 4-simplices is $\sim 31.78n$ [Dwy91]. Similar bounds probably hold for other distributions, but proofs are lacking.

Subquadratic bounds on the complexity of the Delaunay triangulation of point sites in $\mathbb{R}^3$ can be obtained in a few cases. The **spread (of points)** of a set of points is the ratio between largest and smallest interpoint distances. A point set in $\mathbb{R}^3$ of size n with spread Δ can have at most $O(\Delta^3)$ Delaunay tetrahedra, for all $\Delta = O(\sqrt{n})$ [Eri05]. Thus if the point set is dense, i.e., has spread $O(n^{1/3})$, there are only $O(n)$ tetrahedra. Points chosen on a surface can also have subquadratic complexity. For example, if the surface is sufficiently continuous and satisfies mild genericity conditions, then any (ϵ, κ)-sample of n points on the surface has complexity $O(n \log n)$ [ABL03]. A set of points is an (ϵ, κ)-*sample* if for any point in the set, there is at least one and at most κ other points in the set within geodesic distance ϵ measured on the surface.

27.2 BASIC ALGORITHMS

Table 27.2.1 lists basic algorithms that compute the Delaunay triangulation of n point sites in $\mathbb{R}^d$ using the Euclidean metric. Using the connection with convexity, any $(d+1)$-dimensional convex hull algorithm can be used to compute a d-dimensional Delaunay triangulation; in fact the divide-and-conquer, incremental, and gift-wrapping algorithms are specialized convex hull algorithms. Running times are given both for worst-case inputs, and for inputs chosen uniformly at random inside a sphere, with expectation taken over input distribution. The Voronoi diagram can be obtained in linear time from the Delaunay triangulation, using the one-one correspondence between their faces. See [Aur91, Ede87, For95, AK00, AKL13] for more references. Chapter 67 lists available implementations of Voronoi diagram algorithms.

TABLE 27.2.1   Delaunay Triangulation algorithms in the Euclidean metric for point sites.

ALGORITHM	DIM	WORST CASE	UNIFORM
Flipping	2	$O(n^2)$	
Plane sweep	2	$O(n \log n)$	
Divide-and-conquer	2	$O(n \log n)$	$O(n)$
Randomized incremental	2	$O(n \log n)$	
Randomized incremental	≥ 3	$O(n^{\lceil d/2 \rceil})$	$O(n \log n)$
Gift-wrapping	≥ 2	$O(n^{\lceil d/2 \rceil + 1})$	$O(n)$

THE RANDOMIZED INCREMENTAL ALGORITHM

The incremental algorithm adds sites one by one, updating the Delaunay triangulation after each addition. The update consists of discovering all Delaunay faces whose circumspheres contain the new site. These faces are deleted and the empty region is partitioned into new faces, each of which has the new site as a vertex. See Figure 27.2.1. An efficient algorithm requires a good data structure for finding the faces to be deleted. Then the running time is determined by the total number of face updates, which depends upon site insertion order. The bounds given in Table 27.2.1 are the expected running time of an algorithm that makes a random choice of insertion order, with each insertion permutation equally likely; the bounds for the worst-case insertion order are about a factor of n worse. For uniform data there is a double expectation, over both insertion order and input distribution. With additional algorithmic complexity, it is possible to obtain deterministic algorithms with the same worst-case running times [Cha93].

THE PLANE SWEEP ALGORITHM

The plane sweep algorithm computes a planar Delaunay triangulation using a horizontal line that sweeps upward across the plane. The algorithm discovers a Delaunay triangle when the sweepline passes through the topmost point of its circumcircle; in Figure 27.2.2, the Delaunay triangles shown have already been discovered. A sweepline data structure stores an ordered list of sites; the entry for site s corresponds to an interval I_s on the sweepline where each maximal empty circle with

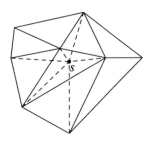

FIGURE 27.2.1

The addition of site s deletes four triangles and adds six (shown dashed).

topmost point in I_s touches site s. The sweepline moves in discrete steps only when the ordered list changes. This happens when a new site is encountered or when a new Delaunay triangle is discovered (at the topmost point of the circumcircle of three sites that are consecutive on the sweepline list). A priority queue is needed to determine the next sweepline move. The running time of the algorithm is $O(n \log n)$ since the sweepline moves $O(n)$ times—once per site and once per triangle—and it costs time $O(\log n)$ per move to maintain the priority queue and sweepline data structure.

FIGURE 27.2.2

The sweepline list is x, s, t, u, v, w, x. The next Delaunay triangle is tuv.

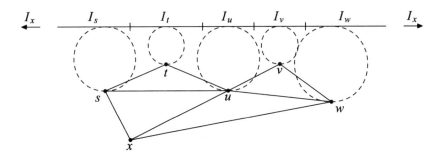

OTHER ALGORITHMS

The planar divide-and-conquer algorithm uses a splitting line to partition the point set into two equal halves, recursively computes the Delaunay triangulation of each half, and then merges the two subtriangulations in linear time. If the sites form the vertices of a convex polygon, then the Voronoi diagram can be computed in linear time [AGSS89]. In any dimension, the gift-wrapping algorithm is a specialization of the convex-hull gift-wrapping algorithm (Chapter 26) to Delaunay triangulations. There is an approximately output-sensitive algorithm, with running time $O(f \log n \log \Delta)$ where f is the number of output simplices, n is the number of points and Δ is the spread of the point set [MS14].

Graphics hardware, in particular Z-buffers, allow efficient practical computation of fixed-resolution approximate Voronoi diagrams for quite general sites and distance functions [HCK+99].

27.3 EXTENSIONS

GLOSSARY

Order-*k* Voronoi diagram: The order-*k* Voronoi diagram partitions $\mathbb{R}^d$ on the basis of the first *k* closest sites (without distinguishing order among them).

Farthest site Voronoi diagram: The farthest site Voronoi diagram partitions $\mathbb{R}^d$ on the basis of the farthest site, or equivalently, the closest *n*−1 of *n* sites.

Constrained Delaunay triangulation: Constrained Delaunay triangulations are defined relative to a set of **constraint facets** that restrict visibility. The constrained Delaunay triangulation of a set of sites has the property that for every simplex, the interior of the simplex circumsphere contains no site visible from the interior of the simplex.

Conforming Delaunay triangulation: Fix a set of noncrossing **constraint facets** *E* and a set of point sites *S*. A conforming Delaunay triangulation is the Delaunay triangulation of a set of sites $S' \supseteq S$ so that every facet in *E* is the union of Delaunay faces of S'.

Power or Laguerre diagram: A Voronoi diagram for sites s_i with weights w_i where the distance from a point *x* is measured along a tangent to the sphere of radius $\sqrt{w_i}$ centered on s_i.

HIGHER-ORDER VORONOI DIAGRAMS

The order-*k* Voronoi diagram can be obtained as an appropriate projection of the *k*-level of an arrangement of hyperplanes (see [Ede87, For93] and Section 28.2 of this Handbook); it can also be obtained as the orthogonal projection of an intersection polytope [AS92]. In dimension 2, the order-*k* Voronoi diagram has $O(k(n - k))$ faces. In dimensions $d \geq 3$, the sum of the number of faces of the order-*j* diagrams, $j \leq k$, is $O(n^{\lceil d/2 \rceil} k^{\lfloor d/2 \rfloor + 1})$ [CS89]; finding good bounds for fixed *k* remains an open problem. See Table 27.3.1 for running time bounds.

TABLE 27.3.1   Algorithms for order-*k* Voronoi diagrams of point sites in the Euclidean metric.

PROBLEM	DIM	TIME
Farthest site	2	$O(n \log n)$
Farthest site	≥ 3	$O(n^{\lceil d/2 \rceil})$
Order-*k*	2	$O(k(n - k) \log n + n \log^3 n)$
Order-*j*, $1 \leq j \leq k$	≥ 3	$O(n^{\lceil d/2 \rceil} k^{\lfloor d/2 \rfloor + 1})$

CONSTRAINED DELAUNAY TRIANGULATIONS

Let S be a set of n point sites in $\mathbb{R}^2$ and E a set of noncrossing **constraint** edges with endpoints in S. A point $p \in \mathbb{R}^2$ is **visible** from a site s if the open segment ps does not intersect any edge of E. The **constrained Delaunay triangulation** (of S with respect to E) is a triangulation of S extending the edges in E so that the circumcircle of every triangle contains no site that is visible from the interior of the triangle. In $\mathbb{R}^2$ the constrained Delaunay triangulation always exists; it is as close as possible to the true Delaunay triangulation, subject to the constraint that the edges in E must be used. See also Section 28.2.

The **bounded distance** from a site to a point is Euclidean distance if the point is visible, and infinite otherwise; the **bounded Voronoi diagram** of S using E is defined using bounded distance. The bounded Voronoi diagram is dual to a subgraph of the constrained Delaunay triangulation.

Both the constrained Delaunay triangulation and the bounded Voronoi diagram can be computed in time $O(n \log n)$ using either divide-and-conquer or the sweepline paradigm. If the sites and constraint edges are the vertices and edges of a simple polygon, respectively, then the constrained Delaunay triangulation can be computed in linear time [KL96].

The constrained Delaunay triangulation can be generalized to dimension $d > 2$. Let S be a set of point sites in $\mathbb{R}^d$ and E a set of $d-1$-dimensional closed simplicial **constraint facets** with vertices in S that are noncrossing (the intersection of two constraint simplices is either empty or a face of both). A **constrained Delaunay triangulation** (of S with respect to E) is a triangulation such that constraint facets are triangulation facets and such that, for every d-simplex, the interior of the circumsphere of the simplex contains no site visible from the interior of the simplex. In dimension $d > 2$, a constrained Delaunay triangulation does not always exist; for example, the Schönhardt polyhedron in dimension 3 cannot be triangulated without extra Steiner points (see Section 29.5). Shewchuk [She08] gives a sufficient condition for the existence of constrained Delaunay triangulations: the constraint simplices must be **ridge-protected**, that is, each j-face of a constraint simplex, $j \leq d-2$, must have a closed circumsphere not containing any sites. He also gives two algorithms that construct the constrained Delaunay triangulation when it exists, one a sweep algorithm [She00] and the second a flipping algorithm [She03].

CONFORMING DELAUNAY TRIANGULATIONS

Let S be a set of point sites in $\mathbb{R}^d$ and E a set of noncrossing j-dimensional **constraint simplices**, $j < d$. A **conforming Delaunay triangulation** of E is the Delaunay triangulation of a set of sites $S' \supseteq S$ so that every simplex in E is the union of faces of Delaunay simplices of S'. In $\mathbb{R}^2$, Edelsbrunner and Tan [ET93] give an algorithm for conforming Delaunay triangulations, where the cardinality of S' is $O(n^3)$, n the cardinality of S. See [CP06] for an approach in $\mathbb{R}^3$.

OTHER DISTANCE MEASURES

Table 27.3.2 lists Voronoi diagram algorithms where "distance" is altered. The distance from a site s_i to a point x can be a function of the Euclidean distance $e(s_i, x)$ and a site-specific real weight w_i.

TABLE 27.3.2   Algorithms for point sites in $\mathbb{R}^2$, other distance measures.

PROBLEM	DISTANCE TO x	TIME
Additive weights	$w_i + e(s_i, x)$	$O(n \log n)$
Multiplicative weights	$w_i e(s_i, x)$	$O(n^2)$
Laguerre or power	$\sqrt{e(s_i, x)^2 - w_i}$	$O(n \log n)$
L_p	$\|s_i - x\|_p$	$O(n \log n)$
Skew	$e(s_i, x) + \kappa \, \Delta_y(s_i, x)$	$O(n \log n)$
Convex distance function		$O(n \log n)$
Abstract	axiomatic	$O(n \log n)$
Simple polygon	geodesic	$O(n \log^2 n)$
Crystal growth	$w_i \cdot SP(s_i, x)$	$O(n^3 + nS \log S)$
Anisotropic	local metric tensor	$O(n^{2+\epsilon})$

The seemingly peculiar ***power distance*** [Aur87] is the distance from x to the sphere of radius $\sqrt{w_i}$ about s_i along a line tangent to the sphere. Many of the basic Voronoi diagram algorithms extend immediately to the power distance, even in higher dimension.

A ***(polygonal) convex distance function*** [CD85] is defined by a convex polygon C with the origin in its interior. The distance from x to y is the real $r \geq 0$ so that the boundary of $rC + x$ contains y. Polygonal convex distance functions generalize the L_1 and L_∞ metrics (C is a diamond or square, respectively); a polygonal convex distance function is a metric exactly if C is symmetric about the origin.

The ***(skew) distance*** [AAC+99] between two points is the Euclidean distance plus a constant times the difference in y-coordinate. It can be viewed as a measure of the difficulty of motion on a plane that has been rotated in three dimensions about the x-axis.

An ***abstract*** Voronoi diagram [KMM93] is defined by the "bisectors" between pairs of sites, which must satisfy special properties.

The ***geodesic*** distance inside an environment of polygonal obstacles is the length of the shortest path that avoids obstacle interiors. Some progress using the geodesic metric appears in [HS99].

The ***crystal growth*** Voronoi diagram [SD91] models crystal growth where each crystal has a different growth rate. The distance from a site s_i to a point x in the Voronoi face of s_i is $w_i \cdot SP(s_i, x)$, where w_i is a weight and $SP(s_i, x)$ is the shortest path distance lying entirely within the Voronoi face of s_i. The parameter S in the running time measures the time to approximate bisectors numerically.

An ***anisotropic*** Voronoi diagram [LS03] requires a metric tensor at each site to specify how distance is measured from that site. The anisotropic Voronoi diagram generalizes the multiplicatively weighted diagram; both have the property that the region of a site may be disconnected or not simply-connected.

The ***Bregman divergence*** from a point p to a point q is defined relative to a convex function F and roughly measures how $F(q)$ differs from the first-order Taylor approximation obtained from F at p. The Bregman divergence generalizes various functions from statistics and machine learning, e.g., the Kullback-Leibler and Itakura-Saito divergences. Boissonnat *et al.* [BNN10] define various versions and give algorithms for the Voronoi diagram and Delaunay triangulation defined using Bregman divergence.

OTHER SITES

Many classes of sites besides points have been used to define Voronoi diagram and Voronoi-diagram-like objects. For example, the Voronoi diagram of a set of disjoint circles in the plane is just an additively-weighted point-site Voronoi diagram.

The Voronoi diagram of a set of n line segment sites in $\mathbb{R}^2$ can be computed in time $O(n \log n)$ using the sweepline method or the divide-and-conquer method. The well-known medial axis of a polygon or polygonal region can be obtained from the Voronoi diagram of its constituent line segments. The medial axis of a simple polygon can be found in linear time, using the linear-time triangulation algorithm [AGSS89].

Aichholzer *et al.* [AAA+10] describe a randomized algorithm that handles segments and circular arcs; more general free-form curves can be handled by approximation using polygonal and curved arcs. The algorithm uses an unusual divide-and-conquer paradigm and runs in expected time $O(n \log n)$ under some assumptions.

The *straight skeleton* of a simple polygon [AAAG95] is structurally similar to the medial axis, though it is not strictly a Voronoi diagram. It is defined as the trace of the vertices of the polygon, as the polygon is shrunk by translating each edge inward at a constant rate. Unlike the medial axis, it has only polygonal edges. Several algorithms achieve time and space bounds of roughly $O(nr)$, r the number of reflex vertices; a subquadratic worst-case bound is known [EE99]. Some progress on extension to three-dimensional polyhedra is available [BEGV08].

The worst-case combinatorial and algorithmic complexity of Voronoi diagrams of general sites in three dimensions is not well understood [KS03, KS04]. For many sites and metrics in $\mathbb{R}^3$, roughly cubic upper bounds on the combinatorial complexity of the Voronoi diagram can be obtained using the general theory of lower envelopes of trivariate functions (see Chapter 28 on arrangements). Known lower bounds are roughly quadratic, and upper bounds are conjectured to be quadratic.

A specific long-standing open problem is to give tight bounds on the combinatorial complexity of the Voronoi diagram of a set of n lines in $\mathbb{R}^3$ using the Euclidean metric. Roughly quadratic upper bounds are known if the lines have only a constant number of orientations [KS03]. The boundary of the union of infinite cylinders of fixed radius is also known to have roughly quadratic complexity [AS00]; this boundary can be viewed as the level set of the line Voronoi diagram at fixed distance. Everett *et al.* [EGL+09, ELLD09] give a complete combinatorial analysis of all possible Voronoi diagrams of three lines; Hemmer *et al.* [HSH10] describe a nearly-cubic exact-arithmetic algorithm that computes the Voronoi diagram of an arbitrary set of lines in $\mathbb{R}^3$.

Dwyer [Dwy97] shows that the expected complexity of the Euclidean Voronoi diagram of n k-flats in $\mathbb{R}^d$ is $\theta(n^{d/(d-k)})$, as long as $d \geq 3$ and $0 \leq k < d$. The flats are assumed to be drawn independently from the uniform distribution on k-flats intersecting the unit ball. Thus the expected complexity of the Voronoi diagram of a set of n uniformly random lines in $\mathbb{R}^3$ is $O(n^{3/2})$.

Voronoi diagrams in $\mathbb{R}^3$ can be defined by convex distance functions, as in the plane. If the distance function is determined by a convex polytope with a constant number of facets, then the Voronoi diagram of a set of disjoint polyhedra has combinatorial complexity roughly quadratic in the total number of vertices of all polytopes [KS04].

KINETIC VORONOI DIAGRAMS

Consider a set of n moving point sites in $\mathbb{R}^d$, where the position of each site is a continuous function of a real parameter t, representing time. In general the Voronoi diagram of the points will vary continuously with t, without any change to its combinatorial structure; however at certain discrete values t_i, $i = 1, \ldots$, the combinatorial structure will change. A ***kinetic Voronoi diagram algorithm*** determines times that the structure changes and at each change updates a data structure representation of the Voronoi diagram. See a survey in [DC08] and also Chapter 53.

Rubin [Rub15] recently showed a nearly-quadratic upper bound on the number of combinatorial changes to a planar Delaunay triangulation, provided that the points are each moving at unit speed along a straight line.

OTHER SURFACES

The Delaunay triangulation of a set of points on the surface of a sphere S^d has the same combinatorial structure as the convex hull of the set of points, viewed as sitting in $\mathbb{R}^{d+1}$. On a closed Riemannian manifold, the Delaunay triangulation of a set of sites exists and has properties similar to the Euclidean case, as long as the set of sites is sufficiently dense [LL00, GM01].

MOTION PLANNING

The motion planning problem is to find a collision-free path for a robot in an environment filled with obstacles. The Voronoi diagram of the obstacles is quite useful, since it gives a lower-dimensional skeleton of maximal clearance from the obstacles. In many cases the shape of the robot can be used to define an appropriate metric for the Voronoi diagram. See Section 50.2 for more on the use of Voronoi diagrams in motion planning.

SURFACE RECONSTRUCTION

The *surface reconstruction problem* is to construct an approximation to a two-dimensional surface embedded in $\mathbb{R}^3$, given a set of points sampled from the surface. A whole class of surface reconstruction methods are based on the computation of Voronoi diagrams [AB99, Dey07] (see Chapters 35 on surface reconstruction).

DELAUNAY REFINEMENT

Delaunay refinement builds a mesh of "fat" triangles, i.e., with no small or large angles, by maintaining a Delaunay triangulation while judiciously adding new sites. The original ideas come from Chew [Che89], who showed that skinny triangles can be eliminated by repeatedly adding as new site the circumcenter of any skinny triangle, and Ruppert [Rup95], who with an additional update rule gave an algorithm that produces meshes that have the minimum possible number of triangles, up to a constant factor. See [CDS13] for the now-extensive theory including extensions to higher dimensions.

IMPLEMENTATIONS

There are a number of available high-quality implementations of algorithms that compute Delaunay triangulations and Voronoi diagrams of point sites in the Euclidean metric. These can be obtained from the web and from the algorithms libraries CGAL and LEDA (see Chapters 67 and 68 on implementations). It is typically challenging to implement algorithms for sites other than points or metrics other than the Euclidean metric, largely because of issues of numerical robustness. See [Bur96, Hel01, SIII00] for approaches for line segment sites in the plane.

27.4 IMPORTANT PROPERTIES

ROUNDNESS

The Delaunay triangulation is "round," that is, skinny simplices are avoided. This can be formalized in two dimensions by Lawson's classic result: over all possible triangulations, the Delaunay triangulation maximizes the minimum angle of any triangle. No generalization using angles is known in higher dimension. However, define the **enclosing radius** of a simplex as the minimum radius of an enclosing sphere. In any dimension and over all possible triangulations of a point set, the Delaunay triangulation minimizes the maximum enclosing radius of any simplex [Raj94]. Also see Section 29.4 on mesh generation.

OPTIMALITY

Fix a set S of sites in $\mathbb{R}^d$. For a triangulation T of S with simplices $t_1, \ldots, t_n$, define

$$
\begin{aligned}
v_i &= \text{sum of squared vertex norms of } t_i \\
c_i &= \text{squared norm of barycenter of } t_i \\
a_i &= \text{volume of } t_i \\
s_i &= \text{sum of squared edge lengths of } t_i \\
r_i &= \text{circumradius of } t_i.
\end{aligned}
$$

Over all triangulations T of S, the Delaunay triangulation attains the unique minimum of the following functions, where κ is any positive real [Mus97]:

$$V(T) = \sum_i v_i a_i$$

$$C(T) = \sum_i c_i a_i$$

$$H(T) = \sum_i s_i/a_i \qquad d = 2 \text{ only}$$

$$R(T, \kappa) = \sum_i r_i^\kappa \qquad d = 2 \text{ only.}$$

VISIBILITY DEPTH ORDERING

Choose a viewpoint v and a family of disjoint convex objects in $\mathbb{R}^d$. Object A is *in front of* object B from v if there is a ray starting at v that intersects A and then B in that order. Though an arbitrary family can have cycles in the "in front of" relation, the relation is acyclic for the faces of the Delaunay triangulation, for any viewpoint and any dimension [Ede90].

An application comes from computer graphics. The ***painter's algorithm*** renders 3D objects in back to front order, with later objects simply overpainting the image space occupied by earlier objects. A valid rendering order always exists if the "in front of" relation is acyclic, as is the case if the objects are Delaunay tetrahedra, or a subset of a set of Delaunay tetrahedra.

SUBGRAPH RELATIONSHIPS

The edges of a Delaunay triangulation form a graph DT whose vertices are the sites. In any dimension, the following subgraph relations hold:

$$NNG \subseteq EMST \subseteq RNG \subseteq GG \subseteq DT$$

where EMST is the Euclidean minimum spanning tree, RNG is the relative neighborhood graph, GG is the Gabriel graph, and NNG is the nearest neighbor graph with edges viewed as undirected; the containment $NNG \subseteq EMST$ requires that the sites be in general position. See Section 32.1 on proximity graphs.

DILATION

A geometrically embedded graph G has ***dilation*** c if for any two vertices, the shortest path distance along the edges of G is at most c times the Euclidean distance between the vertices. In $\mathbb{R}^2$, the Delaunay triangulation has dilation at most ~ 1.998 [Xia13] and examples exist with dilation at least ~ 1.5932. With an equilateral-triangle convex distance function, the dilation is at most 2.

INTERPOLATION

Suppose each point site $s_i \in S \subset \mathbb{R}^d$ has an associated function value f_i. For $p \in \mathbb{R}^d$ define $\lambda_i(p)$ as the proportion of the area of s_i's Voronoi cell that would be removed if p were added as a site. Then the **natural neighbor** interpolant $f(p) = \sum \lambda_i(p) f_i$ is C^0, and C^1 except at sites. This construction can be generalized to give a C^k interpolant for any fixed k [HS02].

Alternatively, for a triangulation of S in $\mathbb{R}^2$, consider the piecewise linear surface defined by linear interpolation over each triangle. Over all possible triangulations, the Delaunay triangulation minimizes the roughness of the resulting surface, where **roughness** is the square of the L_2 norm of the gradient of the surface, integrated over the triangulation [Rip90].

27.5 SOURCES AND RELATED MATERIAL

FURTHER READING

[Aur91, For95, AK00]: Survey papers that cover many aspects of Delaunay triangulations and Voronoi diagrams.

[AKL13, OBSC00]: General reference books on Voronoi diagrams and Delaunay triangulations; the latter has an extensive discussion of applications.

[Ede87, PS85, BKOS97]: Basic references for geometric algorithms.

[CDS13]: Extensive discussion of Delaunay mesh generation.

`www.voronoi.com`, `www.ics.uci.edu/~eppstein/gina/voronoi.html`: Web sites devoted to Voronoi diagrams.

RELATED CHAPTERS

Chapter 26: Convex hull computations
Chapter 28: Arrangements
Chapter 29: Triangulations and mesh generation
Chapter 32: Proximity algorithms
Chapter 36: Computational convexity
Chapter 50: Algorithmic motion planning

REFERENCES

[AS00]   P.K. Agarwal and M. Sharir. Pipes, cigars, and kreplach: the union of Minkowski sums in three dimensions. *Discrete Comput. Geom.*, 24:645–685, 2000.

[AGSS89] A. Aggarwal, L.J. Guibas, J.B. Saxe, and P.W. Shor. A linear-time algorithm for computing the Voronoi diagram of a convex polygon. *Discrete Comput. Geom.*, 4:591–604, 1989.

[AAA⁺10]   O. Aichholzer, W. Aigner, F. Aurenhammer, T. Hackl, B. Jüttler, E. Pilgerstorfer, and M. Rabl. Divide-and-conquer for Voronoi diagrams revisited. *Comput. Geom.*, 43:688–699, 2010.

[AAAG95]   O. Aichholzer, F. Aurenhammer, D. Alberts, and B. Gärtner. A novel type of skeleton for polygons. *J. Univer. Comput. Sci.*, 1:752–761, 1995.

[AAC⁺99]   O. Aichholzer, F. Aurenhammer, D.Z. Chen, D.T. Lee, and E. Papadopoulou. Skew Voronoi diagrams. *Internat. J. Comput. Geom. Appl.*, 9:235–247, 1999.

[AB99]   N. Amenta and M. Bern. Surface reconstruction by Voronoi filtering. *Discrete Comput. Geom.*, 22:481–504, 1999.

[ABL03]   D. Attali, J.-D. Boissonnat, and A. Lieutier. Complexity of the Delaunay triangulation of points on surfaces: The smooth case. *Proc. 19th. Sympos. Comput. Geom.*, pages 201–210, ACM Press, 2003.

[AS92]   F. Aurenhammer and O. Schwarzkopf. A simple randomized incremental algorithm for computing higher order Voronoi diagrams. *Internat. J. Comput. Geom. Appl.*, 2:363–381, 1992.

[Aur87]   F. Aurenhammer. Power diagrams: properties, algorithms, and applications. *SIAM J. Comput.*, 16:78–96, 1987.

[Aur91]   F. Aurenhammer. Voronoi diagrams—a survey of a fundamental geometric data structure. *ACM Comput. Surv.*, 23:345–405, 1991.

[AK00]   F. Aurenhammer and R. Klein. Voronoi diagrams. In J. Sack and J. Urrutia, editors, *Handbook of Computational Geometry*, pages 201–290, Elsevier, Amsterdam, 2000.

[AKL13]   F. Aurenhammer, R. Klein, and D.T. Lee. *Voronoi Diagrams and Delaunay Triangulations* World Scientific, Singapore, 2013.

[BEGV08]   G. Barequet, D. Eppstein, M.T. Goodrich, and A. Vaxman. Straight skeletons of three-dimensional polyhedra. *Proc. 16th Europ. Sympos. Algorithms.*, vol. 5193 of *LNCS*, pages 148–160, Springer, Berlin, 2008.

[BKOS97]   M. de Berg, M. van Kreveld, M.H. Overmars, and O. Schwarzkopf. *Computational Geometry: Algorithms and Applications.* Springer, Berlin, 1997; 3rd edition, 2008.

[BNN10]   J.-D. Boissonnat, F. Nielsen, and R. Nock. Bregman Voronoi diagrams. *Discrete Comput. Geom.*, 44:281–307, 2010.

[Bur96]   C. Burnikel. *Exact computation of Voronoi diagrams and line segment intersections.* Ph.D. Thesis, Universität des Saarlandes, 1996.

[CD85]   L.P. Chew and R.L. Drysdale. Voronoi diagrams based on convex distance functions. In *Proc. 1st Sympos. Comput. Geom.*, pages 234–244, ACM Press, 1985.

[CDS13]   S.-W. Cheng, T.K. Dey, and J.R. Shewchuk. *Delaunay Mesh Generation*, CRC Press, Boca Raton, 2013.

[Cha93]   B. Chazelle. An optimal convex hull algorithm in any fixed dimension. *Discrete Comput. Geom.*, 10:377–409, 1993.

[Che89]   L.P. Chew. Guaranteed quality triangular meshes. Technical Report TR-89-983, Dept. Comp. Sci., Cornell University, 1989.

[CP06]   S.-W. Cheng and S.-H. Poon. Three-dimensional Delaunay mesh generation. *Discrete Comput. Geom.* 36:419–456, 2006.

[CS89]   K.L. Clarkson and P.W. Shor. Applications of random sampling in computational geometry, II. *Discrete Comput. Geom.*, 4:387–421, 1989.

[DC08]   O. Devillers and P.M.M. de Castro. State of the Art: Updating Delaunay triangulations for moving points. Research Report RR-6665, INRIA, 2008.

[Dey07]   T.K. Dey *Curve and Surface Reconstruction: Algorithms with Mathematical Analysis.* Cambridge Monog. Appl. Comput. Math., Cambridge University Press, 2006.

[Dwy91]   R.A. Dwyer. Higher-dimensional Voronoi diagrams in linear expected time. *Discrete Comput. Geom.*, 6:343–367, 1991.

[Dwy97]   R. Dwyer. Voronoi diagrams of random lines and flats. *Discrete Comput. Geom.*, 17:123–136, 1997.

[Ede87]   H. Edelsbrunner. *Algorithms in Combinatorial Geometry.* Springer, Berlin, 1987.

[Ede90]   H. Edelsbrunner. An acyclicity theorem for cell complexes in *d* dimensions. *Combinatorica*, 10:251–260, 1990.

[EE99]   D. Eppstein and J. Erickson. Raising roofs, crashing cycles, and playing pool: applications of a data structure for finding pairwise interactions. *Discrete Comput. Geom.*, 22:569–592, 1999.

[EGL⁺09]   H. Everett, C. Gillot, D. Lazard, S. Lazard, and M. Pouget. The Voronoi diagram of three arbitrary lines in $\mathbb{R}^3$. *Abstracts of 25th Europ. Workshop. Comput. Geom.*, 2009.

[ELLD09]   H. Everett, D. Lazard, S. Lazard, M.S. El Din. The Voronoi diagram of three lines. *Discrete Comput. Geom.*, 42:94–130, 2009

[Eri05]   J. Erickson. Dense point sets have sparse Delaunay triangulations. *Discrete Comput. Geom.* 33:83–15, 2005.

[ET93]   H. Edelsbrunner and T.-S. Tan. An upper bound for conforming Delaunay triangulations. *Discrete Comput. Geom.*, 10:197–213, 1993.

[For93]   S.J. Fortune. Progress in computational geometry. In R. Martin, editor, *Directions in Geometric Computing*, pages 81–128, Information Geometers, Winchester, 1993.

[For95]   S.J. Fortune. Voronoi diagrams and Delaunay triangulations. In F. Hwang and D.Z. Du, editors, *Computing in Euclidean Geometry*, 2nd edition, pages 225–265, World Scientific, Singapore, 1995.

[GM01]   C.I. Grima and A. Márquez. *Computational Geometry on Surfaces.* Kluwer, Dordrecht, 2001.

[HCK⁺99]   K.E. Hoff III, T. Culver, J. Keyser, M.C. Lin, and D. Manocha. Fast computation of generalized Voronoi diagrams using graphics hardware. *Proc. ACM Conf. SIGGRAPH*, pages 277–286, 1999.

[Hel01]   M. Held. VRONI: An engineering approach to the reliable and efficient computation of Voronoi diagrams of points and line segments. *Comput. Geom.*, 18:95–123, 2001.

[HS99]   J. Hershberger and S. Suri. An optimal algorithm for Euclidean shortest paths in the plane. *SIAM J. Comput.*, 26:2215–2256, 1999.

[HS02]   H. Hiyoshi and K. Sugihara. Improving continuity of Voronoi-based interpolation over Delaunay spheres. *Comput. Geom.*, 22:167–183, 2002.

[HSH10]   M. Hemmer, O. Setter, and D. Halperin. Constructing the exact Voronoi diagram of arbitrary lines in space. In *Proc. 18th Europ. Sympos. Algorithms*, vol. 6346 of *LNCS*, pages 398–409, Springer, Berlin, 2010.

[KL96]   R. Klein and A. Lingas. A linear-time randomized algorithm for the bounded Voronoi diagram of a simple polygon. *Internat. J. Comput. Geom. Appl.*, 6:263–278, 1996.

[KMM93]   R. Klein, K. Mehlhorn, and S. Meiser. Randomized incremental construction of abstract Voronoi diagrams. *Comput. Geom.*, 3:157–184, 1993.

[KS03]   V. Koltun and M. Sharir. Three dimensional Euclidean Voronoi diagrams of lines with a fixed number of orientations. *SIAM J. Comput.*, 32:616–642, 2003.

[KS04]   V. Koltun and M. Sharir. Polyhedral Voronoi diagrams of polyhedra in three dimensions. *Discrete Comput. Geom.*, 31:83–124, 2004.

[LL00]   G. Leibon and D. Letscher. Delaunay triangulations and Voronoi diagrams for Riemannian manifolds. *Proc. 16th Sympos. Comput. Geom.*, pages 341–349, ACM Press, 2000.

[LS03]   F. Labelle and J.R. Shewchuk. Anisotropic Voronoi diagrams and guaranteed-quality anisotropic mesh generation. *Proc. 19th Sympos. Comput. Geom.*, pages 191–200, ACM Press, 2003.

[MS14]   G.L. Miller and D.R. Sheehy. A new approach to output-sensitive Voronoi diagrams and Delaunay triangulations. *Discrete Comput. Geom.*, 52:476–491, 2014.

[Mus97]   O.R. Musin. Properties of the Delaunay triangulation. *Proc. 13th Sympos. Comput. Geom.*, pages 424–426, ACM Press, 1997.

[OBSC00]   A. Okabe, B. Boots, K. Sugihara, and S.N. Chio. *Spatial Tessellations: Concepts and Applications of Voronoi Diagrams*, second edition. Wiley, Chichester, 2000.

[PS85]   F.P. Preparata and M.I. Shamos. *Computational Geometry*. Springer-Verlag, New York, 1985.

[Raj94]   V.T. Rajan. Optimality of the Delaunay triangulation in $\mathbb{R}^d$. *Discrete Comput. Geom.*, 12:189–202, 1994.

[Rip90]   S. Rippa. Minimal roughness property of the Delaunay triangulation. *Comput. Aided Design*, 7:489–497, 1990.

[Rub15]   N. Rubin. On kinetic Delaunay triangulations: a near-quadratic bound for unit speed motions. *J. ACM*, 62:25, 2015.

[Rup95]   J. Ruppert. A Delaunay refinement algorithm for quality 2-dimensional mesh generation. *J. Algorithms*, 18:548–585, 1995.

[SD91]   B. Schaudt and R.L. Drysdale. Multiplicatively weighted crystal growth Voronoi diagram. In *Proc. 7th Sympos. Comput. Geom.*, pages 214–223, ACM Press, 1991.

[She00]   J.R. Shewchuk. Sweep algorithms for constructing higher-dimensional constrained Delaunay triangulations. In *Proc. 16th Sympos. Comput. Geom.*, pages 350–359, ACM Press, 2000.

[She03]   J.R. Shewchuk. Updating and constructing constrained Delaunay and constrained regular triangulations by flips. In *Proc. 19th Sympos. Comput. Geom.*, pages 181–190, ACM Press, 2003.

[She08]   J.R. Shewchuk. General-dimensional constrained Delaunay and constrained regular triangulations, I: Combinatorial Properties. *Discrete Comput. Geom.*, 39:580–637, 2008

[SIII00]   K. Sugihara, M. Iri, H. Inagaki, and T. Imai. Topology-oriented implementation—An approach to robust geometric algorithms. *Algorithmica*, 27:5–20, 2000.

[Xia13]   G. Xia. The stretch factor of the Delaunay triangulation is less than 1.998. *SIAM J. Comput.*, 42:1620–1659, 2013.

28 ARRANGEMENTS

Dan Halperin and Micha Sharir

INTRODUCTION

Given a finite collection S of geometric objects such as hyperplanes or spheres in $\mathbb{R}^d$, the *arrangement* $\mathcal{A}(S)$ is the decomposition of $\mathbb{R}^d$ into connected open cells of dimensions $0, 1, \ldots, d$ induced by S. Besides being interesting in their own right, arrangements of hyperplanes have served as a unifying structure for many problems in discrete and computational geometry. With the recent advances in the study of arrangements of curved (algebraic) surfaces, arrangements have emerged as the underlying structure of geometric problems in a variety of "physical world" application domains such as robot motion planning and computer vision. This chapter is devoted to arrangements of hyperplanes and of curved surfaces in low-dimensional Euclidean space, with an emphasis on combinatorics and algorithms.

In the first section we introduce basic terminology and combinatorics of arrangements. In Section 28.2 we describe substructures in arrangements and their combinatorial complexity. Section 28.3 deals with data structures for representing arrangements and with special refinements of arrangements. The following two sections focus on algorithms: algorithms for constructing full arrangements are described in Section 28.4, and algorithms for constructing substructures in Section 28.5. In Section 28.6 we discuss the relation between arrangements and other structures. Several applications of arrangements are reviewed in Section 28.7. Section 28.8 deals with robustness issues when implementing algorithms and data structures for arrangements and Section 28.9 surveys software implementations. We conclude in Section 28.10 with a brief review of Davenport-Schinzel sequences, a combinatorial structure that plays an important role in the analysis of arrangements.

28.1 BASICS

In this section we review basic terminology and combinatorics of arrangements, first for arrangements of hyperplanes and then for arrangements of curves and surfaces.

28.1.1 ARRANGEMENTS OF HYPERPLANES

GLOSSARY

Arrangement of hyperplanes: Let $\mathcal{H}$ be a finite set of hyperplanes in $\mathbb{R}^d$. The hyperplanes in $\mathcal{H}$ induce a decomposition of $\mathbb{R}^d$ (into connected open cells), the arrangement $\mathcal{A}(\mathcal{H})$. A d-dimensional cell in $\mathcal{A}(\mathcal{H})$ is a maximal connected region of $\mathbb{R}^d$ not intersected by any hyperplane in $\mathcal{H}$; any k-dimensional cell in $\mathcal{A}(\mathcal{H})$, for $0 \leq k \leq d - 1$, is a maximal connected region of dimension k in the intersection of a subset of the hyperplanes in $\mathcal{H}$ that is not intersected by any other hyperplane in $\mathcal{H}$. It follows that any cell in an arrangement of hyperplanes is convex.

Simple arrangement: An arrangement $\mathcal{A}(\mathcal{H})$ of a set $\mathcal{H}$ of n hyperplanes in $\mathbb{R}^d$, with $n \geq d$, is called simple if every d hyperplanes in H meet in a single point and if any $d + 1$ hyperplanes have no point in common.

Vertex, edge, face, facet: $0, 1, 2$, and $(d-1)$-dimensional cell of the arrangement, respectively. (What we call *cells* here are in some texts referred to as *faces*.)

k-cell : A k-dimensional cell in the arrangement.

Combinatorial complexity of an arrangement: The overall number of cells of all dimensions in the arrangement.

EXAMPLE: AN ARRANGEMENT OF LINES

Let $\mathcal{L}$ be a finite set of lines in the plane, let $\mathcal{A}(\mathcal{L})$ be the arrangement induced by $\mathcal{L}$, and assume $\mathcal{A}(\mathcal{L})$ to be simple. A 0-dimensional cell (a vertex) is the intersection point of two lines in $\mathcal{L}$; a 1-dimensional cell (an edge) is a maximal connected portion of a line in $\mathcal{L}$ that is not intersected by any other line in $\mathcal{L}$; and a 2-dimensional cell (a face) is a maximal connected region of $\mathbb{R}^2$ not intersected by any line in $\mathcal{L}$. See Figure 28.1.1.

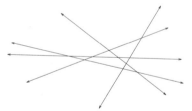

FIGURE 28.1.1

A simple arrangement of 5 lines.
It has 10 vertices, 25 edges (10 of which are unbounded),
and 16 faces (10 of which are unbounded).

COUNTING CELLS

A fundamental question in the study of arrangements is how complex a certain arrangement (or portion of it) can be. Answering this question is often a prerequisite to the analysis of algorithms on arrangements.

THEOREM 28.1.1

Let $\mathcal{H}$ be a set of hyperplanes in $\mathbb{R}^d$. The maximum number of k-dimensional cells in the arrangement $\mathcal{A}(\mathcal{H})$, for $0 \leq k \leq d$, is

$$\sum_{i=0}^{k} \binom{d-i}{k-i} \binom{n}{d-i}.$$

The maximum is attained exactly when $\mathcal{A}(\mathcal{H})$ is simple.

FIGURE 28.1.2

A simple arrangement of 5 x-monotone bounded arcs, where s = 2.

It has 17 vertices (10 of which are arc endpoints), 19 edges, and 4 faces (one of which is unbounded).

We assume henceforth that the dimension d is a (small) constant. With few exceptions, we will not discuss *exact* combinatorial complexity bounds, as in the theorem above, but rather use the big-O notation. Theorem 28.1.1 implies the following:

COROLLARY 28.1.2

The maximum combinatorial complexity of an arrangement of n hyperplanes in $\mathbb{R}^d$ is $O(n^d)$. If the arrangement is simple its complexity is $\Theta(n^d)$. In these bounds the constant of proportionality depends on d.

28.1.2 ARRANGEMENTS OF CURVES AND SURFACES

We now introduce more general arrangements, allowing for objects that are non-linear and/or bounded. We distinguish between planar arrangements and arrangements in three or higher dimensions. For planar arrangements we require only that the objects defining the arrangement be x-monotone Jordan arcs with a constant maximum number of intersections per pair. For arrangements of surfaces in three or higher dimensions we require that the surfaces be algebraic of constant maximum degree, or suitable semi-algebraic portions ("patches") of such surfaces (a more precise definition is given below). This requirement simplifies the analysis and computation of such arrangements, and it does not seem to be too restrictive, as in most applications the arrangements that arise are of low-degree algebraic surfaces or surface patches.

In both cases we typically assume that the objects (curves or surfaces) are in general position. This is a generalization to the current setting of the simplicity assumption for hyperplanes made above. (This assumption is reconsidered in Section 28.8.) All the other definitions in the Glossary carry over to arrangements of curves and surfaces.

PLANAR ARRANGEMENTS

Let $\mathcal{C} = \{c_1, c_2, \ldots, c_n\}$ be a collection of (bounded or unbounded) Jordan arcs in the xy-plane, such that each arc is **x-monotone** (i.e., every line parallel to the y-axis intersects an arc in at most one point), and each pair of arcs in $\mathcal{C}$ intersect in at most s points for some fixed constant s. The arrangement $\mathcal{A}(\mathcal{C})$ is the decomposition of the plane into open cells of dimensions $0, 1$, and 2 induced by the arcs in $\mathcal{C}$. Here, a 0-dimensional cell (a vertex) is either an endpoint of one arc or an intersection point of two arcs. See Figure 28.1.2.

We assume that the arcs in $\mathcal{C}$ are in **general position**; here this means that each intersection of a pair of arcs in $\mathcal{C}$ is either a common endpoint or a transversal intersection at a point in the relative interior of both arcs, and that no three arcs intersect at a common point.

THEOREM 28.1.3

If $\mathcal{C}$ is a collection of n Jordan arcs as defined above, then the maximum combinatorial complexity of the arrangement $\mathcal{A}(\mathcal{C})$ is $O(n^2)$. There are such arrangements whose complexity is $\Theta(n^2)$. In these bounds the constant of proportionality depends linearly on s.

PSEUDO LINES, SEGMENTS, OR CIRCLES

Several special classes of curves have arrangements with favorable properties.

GLOSSARY

A collection of pseudo-lines: A set Γ of unbounded x-monotone connected curves (which can be regarded as graphs of totally-defined continuous functions), each pair of which intersect (transversally) at most once.

A collection of pseudo-segments: Same as above, but the curves of Γ are bounded (graphs of functions defined over bounded intervals).

A collection of pseudo-circles: A set C of simple closed curves, every pair of which intersect at most twice. If the curves are unbounded, we call C **a collection of pseudo-parabolas**.

As it turns out, arrangements of such families of curves, defined in a purely topological manner, share many properties with arrangements of their standard counterparts—lines, segments, and circles (or parabolas); some of these properties will be noted later in this chapter. For example, a collection of n pseudo-circles has the useful property that the complexity of the union of n regions bounded by pseudo-circles or pseudo-parabolas is at most $6n - 12$ [KLPS86]. In certain applications, it is desirable to cut the curves into subarcs, so that each pair of them intersect at most once (i.e., cut the curves into pseudo-segments), and then solve a variety of combinatorial and algorithmic problems on the resulting pseudo-segments. An extensive work on this problem, starting with Tamaki and Tokuyama [TT98], has culminated in works by Agarwal et al. [ANP⁺04] and by Marcus and Tardos [MT06], showing that n pseudo-circles can be cut into $O(n^{3/2} \log n)$ pseudo-segments. See [ANP⁺04, AS05] for several combinatorial and algorithmic applications of this result.

THREE AND HIGHER DIMENSIONS

We denote the coordinates of $\mathbb{R}^d$ by $x_1, x_2, \ldots, x_d$. For a collection $\mathcal{S} = \{s_1, \ldots, s_n\}$ of (hyper)surface patches in $\mathbb{R}^d$ we make the following assumptions:

1. Each surface patch is contained in an algebraic surface of constant maximum degree.
2. The boundary of each surface patch is determined by at most some constant number of algebraic surface patches of constant maximum degree each. (Formally, each surface patch is a semialgebraic set of $\mathbb{R}^d$ defined by a Boolean combination of a constant number of d-variate polynomial equalities or inequalities of constant maximum degree each.)

3. Every d surface patches in $\mathcal{S}$ meet in at most s points.

4. Each surface patch is **monotone** in $x_1, \ldots, x_{d-1}$, namely every line parallel to the x_d-axis intersects the surface patch in at most one point.

5. The surface patches in $\mathcal{S}$ are in *general position*.

We use the simplified term **arrangement of surfaces** to refer to arrangements whose defining objects satisfy the assumptions above. A few remarks regarding these assumptions (see [AS00a, Section 2], [Mat02, Section 7.7], [Sha94], for detailed discussions of the required assumptions):

- Assumptions (1) and (2), together with the general position assumption (5), imply that every d-tuple of surfaces meet in at most some constant number of points. One can bound this number using Bézout's Theorem (see Chapter 37). The bound s on the number of d-tuple intersection points turns out to be a crucial parameter in the combinatorial analysis of substructures in arrangements. Often, one can get a better estimate for s than the bound implied by Bézout's theorem.

- Assumption (4) is used in results cited below. It can however be easily relaxed without affecting these results: If a surface patch does not satisfy this assumption, it can be decomposed into pieces that satisfy the assumption, and by assumptions (1) and (2) the number of these pieces will be bounded by a constant and their boundaries will satisfy assumption (2).

- Assumption (5), which is a generalization of the simplicity assumption for hyperplanes (and is discussed in detail in [AS00a, Section 2]), often does not affect the worst-case combinatorial bounds obtained for arrangements or their substructures, because it can be shown that the asymptotically highest complexity is obtained when the surfaces are in general position [Sha94]. For algorithms, this assumption is more problematic. There are general relaxation methods but these seem to introduce new difficulties [Sei98] (see also Section 28.8).

THEOREM 28.1.4

Given a collection $\mathcal{S}$ of n surfaces in $\mathbb{R}^d$, as defined above, the maximum combinatorial complexity of the arrangement $\mathcal{A}(\mathcal{S})$ is $O(n^d)$. There are such arrangements whose complexity is $\Theta(n^d)$. The constant of proportionality in these bounds depends on d and on the maximum algebraic degree of the surfaces and of the polynomials defining their boundaries.

ARRANGEMENTS ON CURVED SURFACES

Although we do not discuss such arrangements directly in this chapter, many of the combinatorial and algorithmic results that we survey carry over to arrangements on curved surfaces (which are assumed to be algebraic of constant degree) with only slight adjustments. Arrangements on spheres are especially prevalent in applications. The ability to analyze or construct arrangements on curved surfaces is implicitly assumed and exploited in the results for arrangements of surfaces in Euclidean space, since we often need to consider the lower-dimensional arrangement induced on a surface by its intersections with all the other surfaces that define the arrangement.

ADDITIONAL TOPICS

We focus in this chapter on simple arrangements. We note, however, that non-simple arrangements raise interesting questions; see, for example, [Szé97]. Another noteworthy topic that we will not cover here is **combinatorial equivalence** of arrangements; see Chapter 6 and [BLW+93].

28.2 SUBSTRUCTURES IN ARRANGEMENTS

A substructure in an arrangement (i.e., a portion of an arrangement), rather than the entire arrangement, may be sufficient to solve a problem at hand. Also, the analysis of several algorithms for constructing arrangements relies on combinatorial bounds for substructures. We survey substructures that are known in general to have significantly smaller complexity than that of the entire arrangement. For simplicity, some of the substructures are defined below only for the planar case.

GLOSSARY

Let $\mathcal{C}$ be a collection of n x-monotone Jordan arcs as defined in Section 28.1.

Lower (upper) envelope: For this definition we regard each curve c_i in $\mathcal{C}$ as the graph of a continuous univariate function $c_i(x)$ defined on an interval. The lower envelope Ψ of the collection $\mathcal{C}$ is the pointwise minimum of these functions: $\Psi(x) = \min c_i(x)$, where the minimum is taken over all functions defined at x. (The lower envelope is the 0-level of the arrangement $\mathcal{A}(\mathcal{C})$; see below.) Similarly, the upper envelope of the collection $\mathcal{C}$ is defined as the pointwise maximum of these functions. Lower and upper envelopes are completely symmetric structures, and from this point on we will discuss only lower envelopes.

Minimization diagram of $\mathcal{C}$: The subdivision of the x-axis into maximal intervals so that on each interval the same subset of functions attains the minimum. In $\mathbb{R}^d$ we regard the surface patches in $\mathcal{S}$ as graphs of functions in the variables $x_1, \ldots, x_{d-1}$, the lower envelope is the pointwise minimum of these functions, and the minimization diagram is the subdivision of $\mathbb{R}^{d-1}$ into maximal connected relatively-open cells such that over each cell the lower envelope is attained by a fixed subset of $\mathcal{S}$.

Zone: For an additional curve γ, the collection of faces of the arrangement $\mathcal{A}(\mathcal{C})$ intersected by γ. See Figure 28.4.1. In earlier works, the zone is sometimes called the **horizon**.

Single cell: In this section, a d-cell in an arrangement in $\mathbb{R}^d$.

Sandwich region: Given two sets of surfaces, this is the closure of the intersection of the cell below the lower envelope of one set and the cell above the upper envelope of the other set.

Many cells (m cells): Any m distinct d-cells in an arrangement in $\mathbb{R}^d$.

Sides and borders: Let e be an edge in an arrangement of lines, and let l be the line containing e. The line l divides the plane into two halfplanes h_1, h_2. We regard e as two-sided, and denote the two sides by (e, h_1) and (e, h_2). The edge e is on the boundary of two faces f_1 and f_2 in the arrangement. e is said to be a

1-*border* of either face, marked (e, f_1) and (e, f_2), respectively (more precisely, (e, f_1) corresponds to the side (e, h_1), in the sense that f_1 lies in h_1, near e, and (e, f_2) corresponds to (e, h_2)). Similarly a vertex in a simple arrangement of lines has four sides, and it is a **0-*border*** of four faces. The definition extends in an obvious way to arrangements of hyperplanes in higher dimensions and to arrangements of curved surfaces.

k-level: We assume here, for simplicity, that the curves are unbounded; the definition can be extended to the case of bounded curves. A point p in the plane is said to be at level k, if there are exactly k curves in $\mathcal{C}$ lying strictly below p (i.e., a relatively open ray emanating from p in the negative y direction intersects exactly k curves in $\mathcal{C}$). The level of an (open) edge e in $\mathcal{A}(\mathcal{C})$ is the level of any point of e; the level is not necessarily fixed on an edge when the arcs are bounded. The k-level of $\mathcal{A}(\mathcal{C})$ is the closure of the union of edges of $\mathcal{A}(\mathcal{C})$ that are at level k; see Figure 28.2.1. The ***at-most-k-level*** of $\mathcal{A}(\mathcal{C})$, denoted $(\leq k)$-***level***, is the union of points in the plane at level j, for $0 \leq j \leq k$. Different texts use slight variations of the above definitions. In particular, in some texts the ray is directed upwards thus counting the levels from top to bottom. k-levels in arrangements of hyperplanes are closely related (through *duality*, see Section 28.6) to k-sets in point configurations; see Chapter 1.

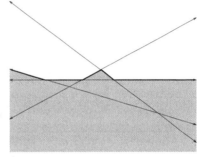

FIGURE 28.2.1
The bold polygonal line is the 2-level of the arrangement of four lines.
The shaded region is the (≤ 2)-level of the arrangement.

Union boundary: If each surface s in an arrangement in $\mathbb{R}^d$ is the boundary of a d-dimensional object (here we no longer assume monotonicity of the surfaces), then the boundary of the union of the objects is another interesting substructure. The study of the union boundary has largely been motivated by robot motion planning problems; for details see Chapter 50.

$\boldsymbol{\alpha(n)}$**:** The extremely slowly growing functional inverse of Ackermann's function.

MEASURING THE COMPLEXITY OF A SUBSTRUCTURE

For an arrangement in $\mathbb{R}^d$, if a substructure consists of a collection C of d-cells, its combinatorial complexity is defined to be the overall number of cells of any dimension on the boundary of each of the d-cells in C. This means that we count certain cells of the arrangement with multiplicity (as *borders* of the corresponding d-cells). For example, for the zone of a line l in an arrangement of lines, each edge of the arrangement that intersects l will be counted twice. However, since we assume that our arrangements reside in a space of a fixed (low) dimension, this only implies a constant multiplicative factor in our count.

The complexity of the lower envelope of an arrangement is defined to be the complexity of its minimization diagram. In three or higher dimensions, this means that we count features that do not appear in the original arrangement. For example, in the lower envelope of a collection of triangles in 3-space, the projection of the edges of two distinct triangles may intersect in the minimization diagram although the two triangles are disjoint in 3-space.

The complexity of a k-level in an arrangement is defined in a similar way to the complexity of an envelope. The complexity of the ($\leq k$)-level is defined as the overall number of cells of the arrangement that lie in the region of space whose points are at level at-most-k.

COMBINATORIAL COMPLEXITY BOUNDS FOR SUBSTRUCTURES

In the rest of this section we list bounds on the maximum combinatorial complexity of substructures. For lines, hyperplanes, Jordan arcs, and surfaces, these are arranged in Tables 28.2.1, 28.2.2, 28.2.3, and 28.2.4, respectively. A bound of the form $(n^{k+\epsilon})$ means a bound $A_\epsilon n^{k+\epsilon}$ for every $\epsilon > 0$, where the coefficient A_ϵ depends on ϵ. In the bounds for k-levels and ($\leq k$)-levels we assume that $k \geq 1$ (otherwise one should use $k + 1$ instead of k). For each substructure, many special cases of arrangements have been considered and the results are too numerous to cover here. For an extensive review of results for k-levels see [Mat02, Chapter 11], for other substructures see [AS00a], [Mat02, Chapter 7].

TABLE 28.2.1   Substructures in arrangements of n lines or pseudo-lines in the plane.

SUBSTRUCTURE	BOUND	NOTES
Envelope	n edges	
Single face	n edges	
Zone of a line	$\Theta(n)$	See [Ede87] for an exact bound on the number of 0- and 1-borders
m faces	$\Theta(m^{2/3}n^{2/3} + m + n)$	Upper bound [CEG$^+$90]; lower bound [Ede87]
k-level	$O(nk^{1/3})$	[Dey98]
	$n2^{\Omega(\sqrt{\log k})}$	[Tót01], [Niv08]
($\leq k$)-level	$\Theta(nk)$	[AG86]

All the results in Table 28.2.1 also hold for arrangements of pseudo-lines; in the case of a zone, we assume that the curve defining the zone is another pseudo-line.

CURVES

For a collection $\mathcal{C}$ of n well-behaved curves as defined in Section 28.1, the complexity bounds for certain substructures involve functions related to *Davenport-Schinzel sequences*. The function $\lambda_s(n)$ is defined as the maximum length of a Davenport-Schinzel sequence of order s on n symbols, and it is almost linear in n for any

TABLE 28.2.2   Substructures in arrangements of n hyperplanes in $\mathbb{R}^d$.

SUBSTRUCTURE	BOUND	NOTES
Envelope	$\Theta(n^{\lfloor \frac{d}{2} \rfloor})$	Upper bound theorem [McM70]
Single cell	$\Theta(n^{\lfloor \frac{d}{2} \rfloor})$	Upper bound theorem [McM70]
Zone of a hyperplane	$\Theta(n^{d-1})$	[ESS93]
Zone of p-dimensional algebraic surface of constant degree	$O(n^{\lfloor (d+p)/2 \rfloor} \log^\gamma n)$	$\gamma = d + p \pmod 2$ [APS93], the bound is almost tight in the worst case
m cells	$O(m^{\frac{1}{2}} n^{\frac{d}{2}} \log^{(\lfloor \frac{d}{2} \rfloor - 2)/2} n)$	Bound is almost tight [AS04], [AMS94]; see [AA92] for bounds on no. of facets
k-level, $d = 3$	$O(nk^{3/2})$	[SST01]
k-level, $d = 4$	$O(n^{4-1/18})$	[Sha11]
k-level, $d \geq 5$	$O(n^{\lfloor d/2 \rfloor} k^{\lceil d/2 \rceil - \epsilon_d})$	[AACS98], constant $\epsilon_d > 0$
($\leq k$)-level	$\Theta(n^{\lfloor d/2 \rfloor} k^{\lceil d/2 \rceil})$	[CS89]

fixed s. Davenport-Schinzel sequences play a central role in the analysis of substructures of arrangements of curves and surfaces, and are reviewed in more detail in Section 28.10 below.

THEOREM 28.2.1

For a set $\mathcal{C}$ of n x-monotone Jordan arcs such that each pair intersects in at most s points, the maximum number of intervals in the minimization diagram of $\mathcal{C}$ is $\lambda_{s+2}(n)$. If the curves are unbounded, then the maximum number of intervals is $\lambda_s(n)$.

The connection between a zone and a single cell. As observed in [EGP+92], a bound on the complexity of a single cell in general arrangements of arcs implies the same asymptotic bound on the complexity of the zone of an additional well-behaved curve γ in the arrangement; "well-behaved" meaning that γ does not intersect any curve in $\mathcal{C}$ more than some constant number of times. This observation extends to higher dimensions and is exploited in the result for zones in arrangements of surfaces [HS95a].

The results in Table 28.2.3 are for Jordan arcs (bounded curves). There are slightly better bounds in the case of unbounded curves. For subquadratic bounds on k-levels in special arrangements of curves see [TT98], [Cha03], [ANP+04],[MT06]. Improved bounds on the complexity of m faces in special arrangements of curves are given in [AEGS92] for segments, [AAS03] for pseudo-segments and for circles, and [ANP+04],[MT06] for pseudo circles and some other types of curves.

Inner vs. outer zone. If γ is a Jordan curve, namely a simple closed curve, we distinguish between the portion of the zone in the interior region bounded by γ and the portion in the exterior region, which we call the **inner zone** and **outer zone** respectively. If γ is the boundary of a convex region then the complexity of the *outer zone* of γ in an arrangement of n lines is $\Theta(n)$ [AD11]. Similarly in higher dimensions, the complexity of the outer zone of the boundary of a convex shape in an arrangement of n hyperplanes in $\mathbb{R}^d$ is $\Theta(n^{d-1})$ [Raz15]. In either case the bound on the complexity of the corresponding inner zone is a tad larger (e.g., $O(n\alpha(n))$ in the planar case), and it is not known whether it is tight. See [Niv15] for recent progress on this problem.

TABLE 28.2.3   Substructures in arrangements of n Jordan arcs.

SUBSTRUCTURE	BOUND	NOTES
Envelope	$\Theta(\lambda_{s+2}(n))$	See Theorem 28.2.1
Single face, zone	$\Theta(\lambda_{s+2}(n))$	[GSS89]
m cells, general	$O(m^{1/2}\lambda_{s+2}(n))$	[EGP^{+}92]
m cells, pseudo-segments	$O(m^{2/3}n^{2/3} + n\log^2 n)$	[AAS03]
m cells, circles	$O(m^{6/11+\epsilon}n^{9/11} + n\log n)$	[AAS03]
m cells	$\Omega(m^{2/3}n^{2/3})$	Lower bound for lines
$(\leq k)$-level	$\Theta(k^2\lambda_{s+2}(\lfloor\frac{n}{k}\rfloor))$	[Sha91]

Notice that the maximum number of edges in envelopes of n Jordan arcs as above is exactly $\lambda_{s+2}(n)$.

TABLE 28.2.4   Substructures in arrangements of n surfaces.

OBJECTS	SUBSTRUCTURE	BOUND	NOTES
Surfaces in $\mathbb{R}^d$	Lower envelope	$O(n^{d-1+\epsilon})$	[HS94],[Sha94]
	Single cell, zone	$O(n^{d-1+\epsilon})$	[Bas03],[HS95a]
	$(\leq k)$-level	$O(n^{d-1+\epsilon}k^{1-\epsilon})$	Combining [CS89] and Lower envelopes bound
$(d-1)$-simplices in $\mathbb{R}^d$	Lower envelope	$\Theta(n^{d-1}\alpha(n))$	[PS89], [Ede89]
	Single cell, zone	$O(n^{d-1}\log n)$	[AS94]
$(d-1)$-spheres in $\mathbb{R}^d$	Lower envelope, single cell	$\Theta(n^{\lceil\frac{d}{2}\rceil})$	Linearization

UNION BOUNDARY

For a collection of n pseudo-disks (regions bounded by pseudo-circles), there are at most $6n - 12$ intersection points (for $n \geq 3$) between curves on the union boundary [KLPS86]. This bound is tight in the worst case. For variants and extensions of this result see [EGH^{+}89], [PS99], [AEHS01].

Many of the interesting results in this area are for Minkowski sums where one of the operands is convex, motivated primarily by motion planning problems. In the plane this reduces to the union of pseudo-disks; see [KLPS86]. These results are reviewed in Chapter 50. We mention one exemplary result in three dimensions that (almost) settles a long-standing open problem: the complexity of the union boundary of n congruent infinite cylinders (namely, each cylinder is the Minkowski sum of a line in 3-space and a unit ball) is $O(n^{2+\epsilon})$ [AS00]. The combinatorial complexity of the union of n infinite cylinders in $\mathbb{R}^3$, having arbitrary radii, is $O(n^{2+\epsilon})$, for any $\epsilon > 0$ where the bound is almost tight in the worst case [Ezr11].

Another family of results is for so-called *fat* objects. For example, a triangle is considered fat if all its angles are at least some fixed constant $\delta > 0$. For such triangles it is shown [MPS^{+}94] that they determine at most a linear number of *holes* (namely connected components of the complement of the union) and that their union boundary has near-linear complexity (see below). Several works establish near-linear bounds for other classes of objects in $\mathbb{R}^2$; they are summarized in

[APS08]. Perhaps the most comprehensive result is due to de Berg et al. [ABES14] who show that the complexity of the union of n *locally γ-fat* objects of constant descriptive complexity is $\frac{n}{\gamma^4} 2^{O(\log^* n)}$, where an object K is locally γ-fat if, for any disk D whose center lies in K and that does not fully contain K, we have $\text{area}(D \sqcap K) \geq \gamma \cdot \text{area}(D)$, where $D \sqcap K$ is the connected component of $D \cap K$ that contains the center of D. This is the most general class of fat objects. For γ-fat triangles (a special case), the union complexity improves to $O(n \log^8 n + \frac{n}{\gamma} \log^2 \frac{1}{\gamma})$. Typically (but not always) fatness precludes constructions with high union complexity, such as grid-like patterns with complexity $\Omega(n^d)$ in $\mathbb{R}^d$. See [ABES14] for references to many previous results, with slightly inferior bounds, for the union complexity of other classes of fat objects.

In three and higher dimensions, the following results are known: (i) The complexity of the union of n balls in $\mathbb{R}^d$ is $O(n^{\lceil d/2 \rceil})$ (easily established by lifting the balls into halfspaces in $\mathbb{R}^{d+1}$). (ii) The complexity of the union of n axis-parallel cubes in $\mathbb{R}^d$ is also $O(n^{\lceil d/2 \rceil})$, and it drops (in odd dimensions) to $O(n^{\lfloor d/2 \rfloor})$ [BSTY98]. (iii) In three dimensions, the complexity of the union of k convex polytopes with a total of n facets is $O(k^3 + nk \log k)$, and can be $\Omega(k^3 + nk\alpha(k))$ in the worst case [AST97]. (iv) The preceding bound improves to $O(nk \log k)$ (and the lower bound is $\Omega(nk\alpha(k))$) where the polytopes are Minkowski sums of a fixed convex polytope with a collection of k pairwise disjoint convex polytopes [AS97]. If the fixed polytope in the sum is a box, the bound further improves to $O((n^2\alpha(n))$ [HY98]. (v) The complexity of the union of n arbitrary fat tetrahedra in $\mathbb{R}^3$ is $O(n^{2+\epsilon})$, for any $\epsilon > 0$ [ES09]. In particular, the complexity of the union of n arbitrarily aligned cubes in 3-space is $O(n^{2+\epsilon})$, for any $\epsilon > 0$ (see also [PSS03] for a different proof for the case of nearly congruent cubes). (vi) The complexity of the union of n κ-round (not necessarily convex) objects in $\mathbb{R}^3$ (resp., in $\mathbb{R}^4$) of constant descriptive complexity is $O(n^{2+\epsilon})$ (resp., $O(n^{3+\epsilon})$), for any $\epsilon > 0$; an object c is called κ-round if for every $p \in \partial c$ there exists a ball that contains p, is contained in c, and has radius $\kappa \cdot \text{diam}(c)$ [AEKS06]. (vii) The maximum number of holes in the union of n translates of a convex set in $\mathbb{R}^3$ is $\Theta(n^3)$ [ACDG17].

ADDITIONAL COMBINATORIAL BOUNDS

The following bounds, while not bounds on the complexity of substructures, are useful in the analysis of algorithms for computing substructures and in obtaining other combinatorial bounds on arrangements.

Sum of squares of cell complexities. Let $\mathcal{H}$ be a collection of n hyperplanes in $\mathbb{R}^d$. For each d-cell c of the arrangement $\mathcal{A}(\mathcal{H})$, let $f(c)$ denote the number of cells of any dimension on the boundary of c. Aronov et al. [AMS94] show that $\sum_c f^2(c) = O(n^d \log^{\lfloor \frac{d}{2} \rfloor - 1} n)$, where the sum extends over all d-cells of the arrangement; see also [AS04] for a simpler proof. They use it to obtain bounds on the complexity of m cells in the arrangement. An application of the zone theorem [ESS93] implies a related bound: If we denote the number of hyperplanes appearing on the boundary of the cell c by $f_{d-1}(c)$ (this is equal to the number of facets of ∂c), then $\sum_c f(c) f_{d-1}(c) = O(n^d)$, where the sum extends over all d-cells of the arrangement.

Overlay of envelopes. For two sets A and B of objects in $\mathbb{R}^d$, the complexity of the *overlay of envelopes* is defined as the complexity of the subdivision of $\mathbb{R}^{d-1}$ induced by superposing the minimization diagram of A on that of B. Given two sets

$\mathcal{C}_1$ and $\mathcal{C}_2$, each of n x-monotone Jordan arcs, such that no pair of (the collection of $2n$) arcs intersects more than s times, the complexity of the overlay is easily seen to be $\Theta(\lambda_{s+2}(n))$. In 3-space, given two sets each of n well-behaved surfaces, the complexity of the overlay is $O(n^{2+\epsilon})$ [ASS96] (a simpler proof of the bound appears in [KS03]). The bound is applied to obtain a simple divide-and-conquer algorithm for computing the envelope in 3-space, and for obtaining bounds on the complexity of *transversals* (see Chapter 4). The bound in $\mathbb{R}^4$ is $O(n^{3+\epsilon})$ [KS03], but analogously sharp bounds are not known in higher dimensions; see [KS09] for some progress in this direction. An interesting variant of this theme is presented by Kaplan et al. [KRS11], who show that the expected complexity of the overlay of all the minimization diagrams obtained during a randomized incremental construction of the lower envelope of n planes in $\mathbb{R}^3$ is $O(n \log n)$.

Sandwich region. As an immediate application of the bound on the overlay of envelopes, one can derive the same asymptotic bound for the complexity of the sandwich region for two families of n surfaces in total. That is, the complexity of the sandwich region is $O(n^{2+\epsilon})$ in $\mathbb{R}^3$ [ASS96] and $O(n^{3+\epsilon})$ in $\mathbb{R}^4$ [KS03]

OPEN PROBLEMS

1. What is the complexity of the k-level in an arrangement of lines in the plane? For the gap between the known lower and upper bounds see Table 28.2.1. This is a long-standing open problem in combinatorial geometry. The analogous questions for arrangements of planes or hyperplanes in higher dimensions are equally challenging with wider gaps between the known bounds.

2. What is the complexity of m faces in an arrangement of well-behaved Jordan arcs? For lines and pseudo-lines a tight bound is known, as well as an almost tight bound for segments and a sharp bound for circles, whereas for more general curves a considerable gap still exists—see Table 28.2.3.

28.3 REPRESENTATIONS AND DECOMPOSITIONS

Before describing algorithms for arrangements in the next sections, we discuss how to represent an arrangement. The appropriate data structure for representing an arrangement depends on its intended use. Two typical ways of using arrangements are: (i) traversing the entire arrangement cell by cell; and (ii) directly accessing certain cells of the arrangement. We will present three structures, each providing a method for traversing the entire arrangement: the *incidence graph*, the *cell-tuple structure*, and the *complete skeleton*. We will then discuss refined representations that further subdivide an arrangement into subcells. These refinements are essential to allow for efficient access to cells of the arrangement. For algebraic geometry-oriented representations and decompositions see Chapters 37 and 50.

GLOSSARY

Let $\mathcal{S}$ be a collection of surfaces in $\mathbb{R}^d$ (or curves in $\mathbb{R}^2$) as defined in Section 28.1, and $\mathcal{A}(\mathcal{S})$ the arrangement induced by $\mathcal{S}$. Let c_1 be a k_1-dimensional cell of $\mathcal{A}(\mathcal{S})$ and c_2 a k_2-dimensional cell of $\mathcal{A}(\mathcal{S})$.

Subcell, supercell: If $k_2 = k_1 + 1$ and c_1 is on the boundary of c_2, then c_1 is a subcell of c_2, and c_2 is a supercell of c_1.

(-1)-dimensional cell, $(d+1)$-dimensional cell: Some representations assume the existence of two additional cells in an arrangement. The unique (-1)-dimensional cell is a subcell of every vertex (0-dimensional cell) in the arrangement, and the unique $(d+1)$-dimensional cell is a supercell of all the d-dimensional cells in the arrangement.

Incidence: If c_1 is a subcell of c_2, then c_1 and c_2 are *incident* to one another. We say that c_1 and c_2 define an **incidence**.

28.3.1 REPRESENTATIONS

INCIDENCE GRAPH

The incidence graph (sometimes called the **facial lattice**) of the arrangement $\mathcal{A}(\mathcal{S})$ is a graph $G = (V, E)$ where there is a node in V for every k-cell of $\mathcal{A}(\mathcal{S})$, $-1 \leq k \leq d + 1$, and an edge between two nodes if the corresponding cells are incident to one another (cf. Figure 16.1.3). For an arrangement of n surfaces in $\mathbb{R}^d$ the number of nodes in V is $O(n^d)$ by Theorem 28.1.4. This is also a bound on the number of edges in E: every cell (besides the (-1)-dimensional cell) in an arrangement $\mathcal{A}(\mathcal{S})$ in general position has at most a constant number of supercells. For an exact bound in the case of hyperplanes, see [Ede87, Section 1.2].

CELL-TUPLE STRUCTURE

While the incidence graph captures all the cells in an arrangement and (as its name implies) their incidence relation, it misses *order* information between cells. For example, there is a natural order among the edges that appear along the boundary of a face in a planar arrangement. This leads to the **cell-tuple structure** [Bri93] which is a generalization to any dimension of the two-dimensional doubly-connected-edge-list (DCEL) [BCKO08] or the similar **quad-edge structure** [GS85] and the 3D **facet-edge structure** [DL89]. The cell-tuple structure gives a simple and uniform representation of the adjacency and ordering information in the arrangement.

SKELETON

Let $\mathcal{H}$ be a finite set of hyperplanes in $\mathbb{R}^d$. A **skeleton** in the arrangement $\mathcal{A}(\mathcal{H})$ is a connected subset of edges and vertices of the arrangement. The **complete skeleton** is the union of all the edges and vertices of the arrangement. Edelsbrunner [Ede87] proposes a representation of the skeleton as a digraph, which allows for a systematic traversal of the entire arrangement (in the case of a complete skeleton) or a substructure of the arrangement. Using a one-dimensional skeleton to represent an arrangement in an arbitrary-dimensional space is a notion that appears also in algebro-geometric representations. There, however, the skeleton, or ***roadmap***, is far more complicated (indeed it represents more general arrangements); see [BPR06] and Chapter 50.

28.3.2 DECOMPOSITIONS

A raw arrangement may still be an unwieldy structure as cells may have complicated shapes and many bounding subcells. It is often desirable to decompose the cells of the arrangement into subcomponents so that each subcomponent has a constant descriptive complexity and is homeomorphic to a ball. Besides the obvious convenience that such a decomposition offers (just like a triangulation of a simple polygon), it turns out to be crucial to the design and analysis of randomized algorithms for arrangements, as well as to combinatorial analysis of arrangements.

For a decomposition to be useful, we aim to add as few extra features as possible. The three decompositions described in this section have the property that the complexity of the decomposed arrangement is asymptotically close to (sometimes the same as) that of the original arrangement. (This is still not known for the *vertical decomposition* in higher dimensions—see the open problem below.)

BOTTOM VERTEX DECOMPOSITION OF HYPERPLANE ARRANGEMENTS

Consider an arrangement of lines $\mathcal{A}(\mathcal{L})$ in the plane. For a face f let $v_b = v_b(f)$ be the bottommost vertex of f (the vertex with lowest y coordinate, ties can be broken by the lexicographic ordering of the coordinate vectors of the vertices). Extend an edge from v_b to each vertex on the boundary of f that is not incident to an edge incident to v_b; see Figure 28.3.1. Repeat for all faces of $\mathcal{A}(\mathcal{L})$ (unbounded faces require special care). The original arrangement, together with the added edges, constitutes the ***bottom vertex decomposition*** of $\mathcal{A}(\mathcal{L})$, which is a decomposition of $\mathcal{A}(\mathcal{L})$ into triangles. The notion extends to arrangements of hyperplanes in higher dimensions, and it is carried out recursively [Cla88]. The combinatorial complexity of the decomposition is asymptotically the same as that of the original arrangement.

FIGURE 28.3.1
The bottom vertex decomposition of a face in an arrangement of lines.

VERTICAL DECOMPOSITION

The bottom vertex decomposition does not in general extend to arrangements of nonlinear objects, or even of line segments. Fortunately there is an alternative, rather simple, decomposition method that applies to almost any reasonable arrangement. This is the ***vertical decomposition*** or ***trapezoidal decomposition***.

See Figure 28.3.2. It is optimal for two-dimensional arrangements, namely its complexity is asymptotically the same as that of the underlying arrangement. It is near-optimal in three and four dimensions. In higher dimensions it is still the general decomposition method that is known to have the best (lowest) complexity.

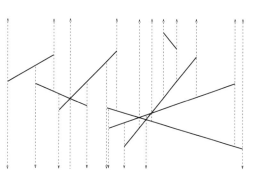

FIGURE 28.3.2

The vertical decomposition of an arrangement of segments: a vertical line segment is extended upwards and downwards from each vertex of the arrangement until it either hits another segment or extends to infinity. These segments decompose the arrangement into trapezoids, triangles, and degenerate variants thereof.

The extension to higher dimensions is defined recursively and is presented in full generality in [CEGS91]. For details of the extension to three dimensions, see [CEG+90] for the case of spheres, and [BGH96] for the case of triangles. The four-dimensional case is studied in [Kol04a], [Kol04b]. Table 28.3.1 summarizes the bounds on the maximum combinatorial complexity of the vertical decomposition for several types of arrangements and substructures. Certain assumptions that the input curves and surfaces are "well-behaved" are not detailed.

TABLE 28.3.1   Combinatorial bounds on the maximum complexity of the vertical decomposition of n objects.

OBJECTS	BOUND	NOTES
Curves in $\mathbb{R}^2$	$\Theta(K)$	K is the complexity of $\mathcal{A}$
Surfaces in $\mathbb{R}^3$	$O(n^2\lambda_t(n))$	[CEGS91], t depends on the algebraic complexity
Surfaces in $\mathbb{R}^d, d \geq 4$	$O(n^{2d-4+\epsilon})$	[CEGS91], [Kol04a]
Triangles in $\mathbb{R}^3$	$\Theta(n^3)$	[BGH96]
Triangles in $\mathbb{R}^3$	$O(n^2\alpha(n)\log n + K)$	K is the complexity of $\mathcal{A}$ [Tag96]
Surfaces in $\mathbb{R}^3$, single cell	$O(n^{2+\epsilon})$	[SS97]
Surfaces in $\mathbb{R}^3$, $(\leq k)$-level	$O(n^{2+\epsilon}k)$	See [AES99] for refined bounds
Hyperplanes in $\mathbb{R}^4$	$\Theta(n^4)$	[Kol04b]
Simplices in $\mathbb{R}^4$	$O(n^4\alpha(n)\log n)$	[Kol04b]

CUTTINGS

All the decompositions described so far have the property that each cell of the decomposition lies fully in a single cell of the arrangement. In various applications this property is not required and other decomposition schemes may be applied, such as *cuttings* (Chapter 44). Cuttings are the basis of efficient divide-and-conquer algorithms for numerous geometric problems on arrangements and otherwise.

POLYNOMIAL PARTITIONING

A novel approach to decomposing arrangements is due to Guth and Katz [GK15]; see also [Gut16]. Using algebraic techniques, combined with the polynomial Ham-Sandwich theorem of Stone and Tukey, they have obtained the following result.

THEOREM 28.3.1

Let P be a set of n points in $\mathbb{R}^d$, and let $r < n$ be a given parameter. There exists a real d-variate polynomial f, of degree $O(r^{1/d})$, such that each of the $O(r)$ connected components of $\mathbb{R}^d \setminus Z(f)$ (where $Z(f)$ denotes the zero set of f) contains at most n/r points of P.

Note several features of this *polynomial partitioning* technique. First, it offers no guarantee about the size of $P \cap Z(f)$. In principle, all the points of P could lie on $Z(f)$. Second, this is a technique for partitioning a set of points and not an arrangement of surfaces. Nevertheless, using standard techniques from real algebraic geometry (for which see, e.g., [BPR06]), any algebraic surface of constant degree and of dimension k intersects only $O(r^{k/d})$ cells of the partition (i.e., components of $\mathbb{R}^d \setminus Z(f)$). Hence, given a collection of n k-dimensional algebraic surfaces of constant descriptive complexity, each cell of the partition is crossed, on average, by $O(n/r^{(d-k)/d})$ surfaces. A more recent construction of Guth [Gut14] provides an alternative similar construction, where each cell of the partition is guaranteed to be crossed by at most $O(n/r^{(d-k)/d})$ of the surfaces.

This new approach strengthens considerably the earlier decomposition techniques mentioned above: (i) It applies to surfaces of any dimension (e.g., it provides a decomposition scheme for a set of lines in 3-space), which the older techniques could not do (for an exception, see Koltun and Sharir [KS05]). (ii) It provides sharp bounds for the size of the subproblems within the partition cells, which the older schemes (based on vertical decomposition) failed so far to do in dimension greater than 4.

A weak aspect of the new technique is that it does not provide a general scheme for handling the points of P that lie on the zero set $Z(f)$; as mentioned, there might be many such points. Many of the recent applications of polynomial partitioning had to provide ad-hoc solutions for this part of the problem, and a significant portion of the current research aims to provide general-purpose techniques for further partitioning $P \cap Z(f)$.

Yet another handicap is that the technique does not offer an efficient procedure for constructing the partition, mainly because there are no known efficient algorithms for constructing polynomial Ham-Sandwich cuts in higher dimensions. See Agarwal et al. [AMS13] for an efficient scheme for constructing approximate polynomial partitionings, with algorithmic applications.

In spite of these weaknesses, polynomial partitioning had a tremendous impact on combinatorial geometry, and has led to many new results on incidences between points and curves or surfaces in higher dimensions, distinct distances between points in a given set, repeated distances and other repeated patterns, efficient range searching with semi-algebraic sets, and more. The most dramatic achievement of polynomial partitioning, in Guth and Katz's original paper [GK15], was to obtain the almost tight lower bound $\Omega(n/\log n)$ on the number of distinct distances determined by any set of n points in the plane, a classical problem posed by Erdős in 1946; see also Chapter 1.

OPEN PROBLEMS

We summarize parts of the preceding discussion into two major open problems:

1. How fast can a partitioning polynomial be constructed? In contrast with cuttings and other earlier decomposition techniques, where optimal or near-optimal algorithms are known for many cases, here it is not known whether a partitioning polynomial can be constructed in polynomial time. This is because a key step in the construction is the polynomial Ham Sandwich cut of Stone and Tukey, whose original proof of existence is nonconstructive, and the only know constructive proof, for discrete sets of points, takes exponential time. A partial solution of this problem is given in Agarwal et al. [AMS13].

2. Another challenge concerning partitioning polynomials has to do with the issue that, while the zero set $Z(f)$ of the partitioning polynomial f distributes the given points evenly among the cells of $\mathbb{R}^d \setminus Z(f)$, it may leave an uncontrolled number of points on $Z(f)$ itself. In many problems, handling these points becomes a nontrivial issue. Several methods have been proposed, such as the construction of a second polynomial g, which partitions evenly the points on $Z(f)$ among the cells of $Z(f) \setminus Z(g)$ (see [KMSS12, Zah11]), and even a third partitioning polynomial (see [BS16]), but the general problem, especially the variants that involve higher-dimensional surfaces that interact with the given points in higher dimensions, is still open.

28.4 ALGORITHMS FOR ARRANGEMENTS

This section covers the algorithmic problem of constructing an arrangement: producing a representation of an arrangement in one of the forms described in the previous section (or in a similar form). We distinguish between algorithms for the construction of the entire arrangement (surveyed in this section), and algorithms for constructing substructures of an arrangement (in the next section). We start with deterministic algorithms and then describe randomized ones.

MODEL OF COMPUTATION

We assume the standard model in computational geometry: infinite precision real arithmetic [PS85]. For algorithms computing arrangements of curves or surfaces, we further assume that certain operations on a small number of curves or surfaces take unit time each. For algebraic curves or surfaces, the unit cost assumption for these operations is theoretically justified by results on the solution of sets of polynomial equations, or more generally, on construction and manipulation of semi-algebraic sets; see Chapter 37. When implementing algorithms for arrangements some of these assumptions need to be reconsidered from the practical point of view; see Sections 28.8 and 28.9.

28.4.1 DETERMINISTIC ALGORITHMS

Incremental construction. The incremental algorithm proceeds by adding one object after the other to the arrangement while maintaining (a representation of) the arrangement of the objects added so far. This approach yields an optimal-time algorithm for arrangements of hyperplanes. The analysis of the running time is based on the zone result [ESS93] (Section 28.2).

We present the algorithm for a collection $\mathcal{L} = \{l_1, \ldots, l_n\}$ of n lines in the plane, assuming that the arrangement $\mathcal{A}(\mathcal{L})$ is simple. Let $\mathcal{L}_i$ denote the set $\{l_1, \ldots, l_i\}$. At stage $i+1$ we add l_{i+1} to the arrangement $\mathcal{A}(\mathcal{L}_i)$. We maintain the DCEL representation [BCKO08] for $\mathcal{A}(\mathcal{L}_i)$, so that in addition to the incidence information, we also have the order of edges along the boundary of each face. The addition of l_{i+1} is carried out in two steps: (i) we find a point p of intersection between l_{i+1} and an edge of $\mathcal{A}(\mathcal{L}_i)$ and split that edge into two, and (ii) we walk along l_{i+1} from p to the left (assuming l_{i+1} is not vertical) updating $\mathcal{A}(\mathcal{L}_i)$ as we go; we then walk along l_{i+1} from p to the right completing the construction of $\mathcal{A}(\mathcal{L}_{i+1})$. See Figure 28.4.1.

FIGURE 28.4.1

Adding the line l_{i+1} to the arrangement $\mathcal{A}(\mathcal{L}_i)$.
The shaded region is the zone of l_{i+1} in the arrangement of the other four lines.

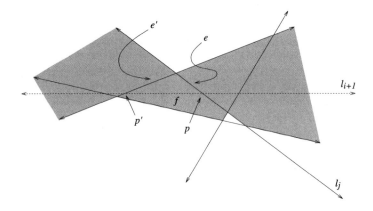

Finding an edge of $\mathcal{A}(\mathcal{L}_i)$ that l_{i+1} intersects can be done in $O(i)$ time by choosing one line l_j from $\mathcal{L}_i$ and checking all the edges of $\mathcal{A}(\mathcal{L}_i)$ that lie on l_j for intersection with l_{i+1}. This intersection point p lies on an edge e that borders two faces of $\mathcal{A}(\mathcal{L}_i)$. We split e into two edges at p. Next, consider the face f intersected by the part of l_{i+1} to the left of p. Using the order information, we walk along the edges of f away from p and we check for another intersection p' of l_{i+1} with an edge e' on the boundary of f. At the intersection we split e' into two edges, we add an edge to the arrangement for the portion $\overline{pp'}$ of l_{i+1}, and we move to the face on the other (left) side of e'. Once we are done with the faces of $\mathcal{A}(\mathcal{L}_i)$ crossed by l_{i+1} to the left of p, we go back to p and walk to the other side. This way we visit all the faces of the zone of l_{i+1} in $\mathcal{A}(\mathcal{L}_i)$, as well as some of its edges. Updating the DCEL structure due to the splitting or addition of edges is straightforward. The amount of time spent is proportional to the number of edges we visit, and hence bounded by the complexity of the zone of l_{i+1} in $\mathcal{A}(\mathcal{L}_i)$, which is $O(i)$. The total time, over all insertions steps, is thus $O(n^2)$. The space required for the algorithm is the space to maintain the DCEL structure. The same approach extends to higher dimensions; for details see [Ede87, Chapter 7].

THEOREM 28.4.1

If $\mathcal{H}$ is a set of n hyperplanes in $\mathbb{R}^d$ such that $\mathcal{A}(\mathcal{H})$ is a simple arrangement, then $\mathcal{A}(\mathcal{H})$ can be constructed in $\Theta(n^d)$ time and space.

The time and space required by the algorithm are clearly optimal. However, it turns out that for arrangements of lines one can do better in terms of *working space*. This is explained below in the subsection *topological sweep*. See [Goo93], [HJW90] for parallel algorithms for arrangements of hyperplanes.

The incremental approach can be applied to constructing planar arrangements of curves, using the vertical decomposition of the arrangement [EGP+92]:

THEOREM 28.4.2

Let $\mathcal{C}$ be a set of n Jordan arcs as defined in Section 28.1. The arrangement $\mathcal{A}(\mathcal{C})$ can be constructed in $O(n\lambda_{s+2}(n))$ time using $O(n^2)$ space.

Sweeping over the arrangement. The sweep paradigm, a fundamental paradigm in computational geometry, is also applicable to constructing arrangements. For planar arrangements, its worst-case running time is slightly inferior to that of the incremental construction described above. It is, however, output sensitive.

THEOREM 28.4.3

Let $\mathcal{C}$ be a set of n Jordan arcs as defined in Section 28.1. The arrangement $\mathcal{A}(\mathcal{C})$ can be constructed in $O((n + k) \log n)$ time and $O(n + k)$ space, where k is the number of intersection points in the arrangement.

One can similarly sweep a plane over an arrangement of surfaces in $\mathbb{R}^3$. There is an output-sensitive algorithm for constructing the vertical decomposition of an arrangement of n surfaces that runs in time $O(n \log^2 n + V \log n)$, where V is the combinatorial complexity of the vertical decomposition. For details see [SH02].

Topological sweep. Edelsbrunner and Guibas [EG89] devised an algorithm for constructing an arrangement of lines that requires only linear working storage and runs in optimal $O(n^2)$ time. Instead of sweeping the arrangement with a straight line, they sweep it with a pseudoline that serves as a "topological wavefront."

The most efficient deterministic algorithm for computing the intersections in a collection of well-behaved curves is due to Balaban [Bal95]. It runs in $O(n \log n + k)$ time and requires $O(n)$ working storage.

28.4.2 RANDOMIZED ALGORITHMS

Most randomized algorithms for arrangements follow one of two paradigms: (i) incremental construction or (ii) divide-and-conquer using random sampling. The randomization in these algorithms is in choices made by the algorithm; for example, the order in which the objects are handled in an incremental construction. In the expected performance bounds, the expectation is with respect to the random choices made by the algorithm. We do not make any assumptions about the distribution of the objects in space. See also Chapter 44.

In constructing a full arrangement, these two paradigms are rather straightforward to apply. Most of these algorithms use an efficient decomposition as discussed in Section 28.3.

Incremental construction.   Here the randomization is in the order that the objects defining the arrangement are inserted. For the construction of an arrangement of curves, the algorithm is similar to the deterministic construction mentioned above.

THEOREM 28.4.4   [Mul93]

Let C be a set of n Jordan arcs as defined in Section 28.1. The arrangement $\mathcal{A}(C)$ can be constructed by a randomized incremental algorithm in $O(n \log n + k)$ expected time and $O(n + k)$ expected space, where k is the number of intersection points in the arrangement.

Divide-and-conquer by random sampling.   For a set $\mathcal{V}$ of n objects in $\mathbb{R}^d$ the paradigm is: choose a subset $\mathcal{R}$ of the objects at random, construct the arrangement $\mathcal{A}(\mathcal{R})$, decompose it further into constant complexity components (using, for example, one of the methods described in Section 28.3), and recursively construct the portion of the arrangement in each of the resulting components. Then glue all the substructures together into the full arrangement. The theory of random sampling is then used to show that with high probability the size of each subproblem is considerably smaller than that of the original problem, and thus efficient resource bounds can be proved. See Chapters 44, 47, and 48 for further application.

The divide-and-conquer counterpart of Theorem 28.4.4 is due to Amato et al. [AGR00]. It has the same running time, and uses slightly more space (or exactly the same space for the case of segments).

The result stated in the following theorem is obtained by applying this paradigm to arrangements of algebraic surfaces and it is based on the vertical decomposition of the arrangement.

THEOREM 28.4.5   [CEGS91], [Kol04a]

Given a collection S of n algebraic surfaces in $\mathbb{R}^d$ as defined in Section 28.1, a data structure of size $O(n^{2d-4+\epsilon})$ for the arrangement $\mathcal{A}(S)$ can be constructed in $O(n^{2d-4+\epsilon})$ time, for any $\epsilon > 0$, so that a point-location query can be answered in $O(\log n)$ time. In these bounds the constant of proportionality depends on ϵ, the dimension d, and the maximum algebraic degree of the surfaces and their boundaries.

If only traversal of the entire arrangement is needed, it is plausible that a simpler structure such as the incidence graph could be constructed using less time and storage space, close to $O(n^d)$ for both. See [Can93], [BPR06] for algebro-geometric methods.

Derandomization.   Techniques have been proposed to derandomize many randomized geometric algorithms, often without increase in their asymptotic running time; see Chapter 44. However, in most cases the randomized versions are conceptually much simpler and hence may be better candidates for efficient implementation.

28.4.3 OTHER ALGORITHMIC ISSUES

For algebro-geometric tools, see Chapter 37. See Chapter 45 (and Section 28.8) for a discussion of precision and degeneracies. Parallel algorithms are discussed in Chapter 46.

28.5 CONSTRUCTING SUBSTRUCTURES

ENVELOPE AND SINGLE CELL IN ARRANGEMENTS OF HYPERPLANES

Computing a single cell or an envelope in an arrangement of hyperplanes is equivalent (through duality) to computing the convex hull of a set of points in $\mathbb{R}^d$ (Chapter 26). For the case of a single cell, one also needs to find a point inside the cell, to facilitate the duality, which can be done by linear programming (Chapter 49).

Using linearization [AM94], we can solve these problems for arrangements of spheres in $\mathbb{R}^d$. We first transform the spheres into hyperplanes in $\mathbb{R}^{d+1}$, and then solve the corresponding problems in $\mathbb{R}^{d+1}$. (transporting the solution back to $\mathbb{R}^d$ requires some care, as the single cell in $\mathbb{R}^{d+1}$ might be split into several subcells in $\mathbb{R}^d$.

LOWER ENVELOPE

The lower envelope of a collection of n well-behaved curves (where each pair intersect in at most s points) can be computed, in a suitable model of computation, by a simple divide-and-conquer algorithm that runs in time $O(\lambda_{s+2}(n) \log n)$ and requires $O(\lambda_{s+2}(n))$ storage. Hershberger [Her89] devised an improved algorithm that runs in time $O(\lambda_{s+1}(n) \log n)$; in particular, for the case of line segments, it runs in optimal $O(n \log n)$ time. In 3-space, Agarwal et al. [ASS96] showed that a simple divide-and-conquer scheme can be used to compute the envelope of n surfaces in time $O(n^{2+\epsilon})$. This is an application of the bound on the complexity of the overlay of envelopes cited in Section 28.2. Boissonnat and Dobrindt give a randomized incremental algorithm for computing the envelope [BD96]. There are efficient algorithms for computing the envelope of $(d-1)$-simplices in $\mathbb{R}^d$ (see [EGS89] for the algorithm in 3D which can be efficiently extended to higher dimensions), and an efficient data structure for point location in the minimization diagram of surfaces in $\mathbb{R}^4$ [AAS97]. Output-sensitive construction of the envelope of triangles in $\mathbb{R}^3$ has mainly been studied in relation to hidden-surface removal (see [Ber93]). Partial information of the minimization diagram (vertices, edges and 2-cells) can be computed efficiently for arrangements of surfaces in any fixed dimension [AAS97]. See also [KRS11, AES99].

SINGLE CELL AND ZONE

All the results cited below for a single cell in arrangements of bounded objects hold for the zone problem as well (see the remark in Section 28.2 on the connection between the problems).

Computing a single face in an arrangement of n Jordan arcs as defined in Section 28.1 can be accomplished in worst-case near-optimal time: deterministically in $O(\lambda_{s+2}(n) \log^2 n)$ time, and using randomization in $O(\lambda_{s+2}(n) \log n)$ time [SA95].

In three dimensions, Schwarzkopf and Sharir [SS97] give an algorithm with running time $O(n^{2+\epsilon})$ for any $\epsilon > 0$ to compute a single cell in an arrangement of n well-behaved surfaces. Algorithms with improved running time to compute a single cell in 3D arrangements are known for arrangements of surfaces induced by certain motion planning problems [Hal92], [Hal94], and for arrangements of triangles [BDS95].

It is still not known how to compute a single cell in arrangements of surfaces in dimension $d > 4$ in time $O(n^{d-1+\epsilon})$, similar to the bound on the complexity of a cell. However, less efficient algorithms from real algebraic geometry are known (see Basu et al. [BPR06]).

LEVELS

In an arrangement of n lines in the plane, the k-level can be computed in $O((n + f)\log n)$ time, where f is the combinatorial complexity of the k-level—the bound is for the algorithm described in [EW86] while using the data structure in [BJ02] which in turn builds on ideas in [Cha01]. For computing the k-level in an arrangement of hyperplanes in $\mathbb{R}^d$ see [AM95], [Cha96].

The $(\leq k)$-level in arrangements of lines can be computed in worst-case optimal time $O(n\log n + kn)$ [ERK96]. Algorithms for computing the $(\leq k)$-level in arrangements of Jordan arcs are described in [ABMS98], the $(\leq k)$-level in arrangements of planes in $\mathbb{R}^3$ (in optimal $O(n\log n + k^2n)$ expected time) in [Cha00], and in arrangements of surfaces in $\mathbb{R}^3$ in $O(n^{2+\epsilon})$ time [AES99].

The $(\leq k)$-level of an arrangement of n surfaces in $\mathbb{R}^d$ is closely related to the notion of k-shallow $(1/r)$-cuttings, where we wish to partition the portion of space lying at or below the k-level into a small number of cells of constant descriptive complexity, each crossed by at most n/r of the surfaces. The fact that the complexity of the $(\leq k)$-level is smaller than that of the whole arrangement leads to improved bounds on the size of shallow cuttings, and, subsequently, to a variety of applications, most notably halfspace range reporting in arrangements of hyperplanes [Mat92, Ram99, AC09, CT16, HKS16] and of more general surfaces [AES99].

UNION BOUNDARY

For a given family of planar regions bounded by well-behaved curves, let $f(m)$ be the maximum complexity of the union boundary of a collection of m objects of the family (the interesting case is when $f(m)$ is linear or close to linear in $\mathbb{R}^2$). Then the union of n such objects can be constructed deterministically in $O(f(n)\log^2 n)$ time or by a randomized incremental algorithm in expected $O(f(n)\log n)$ time [BDS95]. A slightly faster algorithm for the case of fat triangles is given in [MMP+91]. A practically efficient algorithm is described in [EHS04]. An efficient randomized algorithm for computing the union of convex polytopes in $\mathbb{R}^3$ is given in [AST97]. The case of the boundary of the union of Minkowski sums (all having one summand in common) is covered in detailed in Chapter 50.

MANY CELLS

There are efficient algorithms (deterministic and randomized) for computing a set of selected faces in arrangements of lines or segments in the plane. These algorithms are nearly worst-case optimal [AMS98]. Algorithms for arrangements of planes are described in [EGS90], and for arrangements of triangles in 3-space in [AS90].

The related issue of computing the incidences between a set of objects (lines, unit circles) and a set of points is dealt with in [Mat93], with results that extend to higher dimensions [AS00a]. Generally, the bounds for the running time are roughly the same as those for the number of incidences. For lower bounds for the related *Hopcroft's problem* see [Eri96], [BK03].

OPEN PROBLEMS

Devise efficient algorithms for computing:

1. The lower envelope of an arrangement of surfaces in five and higher dimensions; for an algorithm that computes partial information see [AAS97].

2. A single cell in an arrangement of surfaces in four and higher dimensions; for a worst-case near-optimal algorithm in three dimensions see [SS97].

28.6 RELATION TO OTHER STRUCTURES

Arrangements relate to a variety of additional structures. Since the machinery for analyzing and computing arrangements is rather well developed, problems on related structures are often solved by first constructing (or reasoning about) the corresponding arrangement.

Using *duality* one can transform a set (or *configuration*) of points in $\mathbb{R}^d$ (the primal space) into a set of hyperplanes in $\mathbb{R}^d$ (the dual space) and vice versa. Different duality transforms are advantageous in different situations [O'R98].

Edelsbrunner [Ede87, Chapter 12] describes a collection of problems stated for point configurations and solved by operating on their corresponding dual arrangements. An example is given in the next section. See also Chapter 1. Another example is computing incidences between m points and n lines in the plane, or constructing a set of m marked faces in an arrangement of n lines, or computing the number of intersections between n line segments. In these problems one first constructs a decomposition of the plane of the sorts mentioned in Section 28.3.2, obtains subproblems within the cells of the decomposition, and solves each subproblem by passing to the dual plane. See, e.g., [Aga90].

Since many properties of line arrangements extend to pseudo-line arrangements (i.e., unbounded x-monotone curves, each pair of which intersect at most (or exactly) once), it is desirable to apply duality in pseudo-line arrangements too (e.g., for solving variants of the aforementioned problems). Such an effective (albeit fairly involved) scheme is presented in Agarwal and Sharir [AS05].

Plücker coordinates are a tool that enables one to treat k-flats in $\mathbb{R}^d$ as points or hyperplanes in a possibly different higher-dimensional space. This has been taken advantage of in the study of families of lines in 3-space; see Chapter 41.

Lower envelopes (or more generally k-levels in arrangements) relate to Voronoi diagrams; see Chapter 27.

For the connection of arrangements to polytopes and zonotopes see [Ede87] and Section 17.5 of this Handbook. For the connection to oriented matroids see Chapter 6.

28.7 APPLICATIONS

A typical application of arrangements is for solving a problem on related structures. We first transform the original structure (e.g., a point configuration) into an arrangement and then solve the problem on the resulting arrangement. See Section 28.6 above and Chapters 1, 27, 41, and 50.

EXAMPLE: MINIMUM AREA TRIANGLE

Let P be a set of n points in the plane. We wish to find three points of P such that the triangle that they define has minimum area. We use the duality transform that maps a point $p := (a, b)$ to the line $p^* := (y = ax - b)$, and maps a line $l := (y = cx + d)$ to the point $l^* := (c, -d)$. One can show that if we fix two points $p_i, p_j \in P$, and the line p_k^* has the smallest vertical distance to the intersection point $p_i^* \cap p_j^*$ among all other lines in $P^* = \{p^* | p \in P\}$, then the point p_k defines the minimum area triangle with the fixed points p_i, p_j over all points in $P \setminus \{p_i, p_j\}$. Finding the triple of lines as above (an intersecting pair and the other line closest to the intersection) is easy after constructing the arrangement $\mathcal{A}(P^*)$ (Section 28.4), and can be done in $\Theta(n^2)$ time in total. As a special case, we can determine whether P contains three collinear points (a zero-area triangle) in $\Theta(n^2)$ time. This is the most efficient algorithm known for the minimum-area problem [GO95], which for now survives the recent successful attacks on the related 3-SUM problem and its relatives [JP14]. The minimum volume simplex defined by $d + 1$ points in a set of n points in $\mathbb{R}^d$ can be found using arrangements of hyperplanes in $\Theta(n^d)$ time.

OTHER APPLICATIONS

Another strand of applications consists of "robotic" or "physical world" applications [HS95b]. In these problems a continuous space is decomposed into a finite number of cells so that in each cell a certain invariant is maintained. Here, arrangements are used to discretize a continuous space without giving up the completeness or exactness of the solution. An example of an application of this kind solves the following problem: Given a convex polyhedron in 3-space, determine how many combinatorially distinct orthographic and perspective views it induces; see Table 33.8.4. The answer is given using an arrangement of circles on the sphere (for orthographic views) and an arrangement of planes in 3-space (for perspective views) [BD90].

Many developments in the study of arrangements of curves and surfaces have been primarily motivated by problems in robot motion planning (Chapter 50) and several of its variants (Chapter 51). For example, the most efficient algorithm known for computing a collision-free path for an arbitrary polygonal robot (not necessarily convex) moving by translation and rotation among polygonal obstacles in the plane is based on computing a single connected component in an arrangement of surfaces in 3-space. The problem of planning a collision-free motion for a robot among obstacles is typically studied in the *configuration space* where every point represents a possible configuration of the robot. The related arrangements are of surfaces that represent all the contact configurations between the boundary of the robot and the boundaries of obstacles and thus partition configuration space into free cells (describing configurations where the robot does not intersect any obstacle) and forbidden cells. Given the initial (free placement) of the robot, we need only explore the cell that contains this initial configuration in the arrangement.

A concept similar to configuration space of motion planning has been applied in assembly planning (Section 51.3). The assembly planning problem is converted into a problem in *motion space* where every point represents an allowed path (motion) of a subcollection of the assembly relative the rest of the assembly [HLW00]. The motion space is partitioned by a collection of constraint surfaces such that for all possible motions inside a cell of the arrangement, the collection of movable subsets of the assembly is invariant.

As mentioned earlier, arrangements on spheres are prevalent in applications. Aside from vision applications, they also occur in: computer-assisted radio-surgery [SAL93], molecular modeling [HS98], assembly planning (Section 51.3), manufacturing [ABB+02], and more.

Arrangements have been used to solve problems in many other areas including geometric optimization [AS98], range searching (Chapter 40), statistical analysis (Chapter 58), and micro robotics [BDH99], to name a few. More applications can be found in the sources cited below and in several other chapters in this book.

28.8 ROBUSTNESS

Transforming the data structures and algorithms described above into effective computer programs is a difficult task. The typical assumptions of (i) the real RAM model of computation and (ii) general position, are not realistic in practice. This is not only a problem for implementing software for arrangements but rather a general problem in computational geometry (see Chapter 68). However, it is especially acute in the case of arrangements since here one needs to compute *intersection points* of curves and surfaces and use the computed values in further operations (to distinguish from say convex hull algorithms that only select a subset of the input points).

EXACT COMPUTING

A general paradigm to overcome robustness problems is to compute exactly. For arrangements of linear objects, namely, arrangements of hyperplanes or of simplices, there is a fairly straightforward solution: using arbitrary precision rational arithmetic. This is regularly done by keeping arbitrary long integers for the enumerator and denominator of each number. Of course the basic numerical operations now become costly, and methods were devised to reduce the cost of rational arithmetic predicates through the use of *floating point filters* (Chapter 45) which turn out to be very effective in practice, especially when the input is nondegenerate.

Matters are more complicated when the objects are not linear. First, there is the issue of representation. Consider the following simplest planar arrangement of the line $y = x$ and the circle $x^2 + y^2 = 1$ (both described by equations with integer coefficients). The upper vertex (intersection point) v_1 has coordinates $(\sqrt{2}/2, \sqrt{2}/2)$. This means that we cannot have a simple numerical representation of the vertices of the arrangement. An elegant solution to this problem is provided by special number types, so-called *algebraic number types*. The approach is transparent to the user who just has to substitute the standard machine type (e.g., double) for the corresponding novel number type (which is a C++ class). Two software libraries support such number types (called *real* in both): LEDA [MN00] (Chapter 68) and Core [KLPY99] (Chapter 45). The ideas behind the solution proposed by both are similar and rely on separation bounds. In terms of arrangements the power that these number types provide is that we can determine the exact topology of the arrangement in all cases including degenerate cases.

While exact computing may seem to be the solution to all problems, the situation is far from being satisfactory for several reasons: (i) The existing number types considerably slow down the computation compared with standard machine arithmetic. (ii) It is difficult to implement the full-fledged number types required for arrangements of curves and surfaces. Significant progress has been made for planar algebraic curves of arbitrary degree [EK08]; a detailed review of support for special types of curves appears in [FHW12, Section 5.7]. (iii) It still leaves open the question of handling degeneracies (see the section Perturbation below).

The high cost of exact predicates has led researchers to look for alternative algorithmic solutions (for problems where good solutions, in the standard measures of computational geometry, have been known), solutions that use less costly predicates; see, e.g., [BP00].

ROUNDING

In rounding we transform an arbitrary precision arrangement into fixed precision representation. The most intensively studied case is that of planar arrangements of segments. A solution proposed independently by Hobby [Hob99] and by Greene (improving on an earlier method in [GY86]), snaps vertices of the arrangement to centers of pixels in a prespecified grid. The method preserves several topological properties of the original arrangement and indeed expresses the vertices of the arrangement with limited precision numbers (say bounded bit-length integers). A dynamic algorithm is described in [GM98], and an improved algorithm for the case where there are many intersections within a pixel is given in [GGHT97]. Snap rounding has several drawbacks though: a line is substituted by a polyline possibly with many links (a "shortest-path" rounding scheme is proposed in [Mil00] that sometimes introduces fewer links than snap rounding), and a vertex of the arrangement can become very close to a nonincident edge. The latter problem has been overcome in an alternative scheme *iterated snap rounding* which guarantees a large separation between such features of the arrangement but pays in the quality of approximation [HP02, Pac08]. Several more efficient algorithms and variants have also been proposed (see, e.g. [BHO07, Her13]). Notice that in the snap-rounded arrangement the rounded versions of a pair of input segments may intersect an arbitrarily large number of times. Finally, the 3D version seems to produce a huge number of extra features [For99]: a polyhedral subdivision of complexity n turns into a snapped subdivision of complexity $O(n^4)$; in addition the rounding precision depends on the combinatorial complexity of the input.

Effective and consistent rounding of arrangements remains an important and largely open problem. The importance of rounding arrangements stems not only from its being a means to overcome robustness issues, but, not less significantly, from being a way to express the arrangement numerically with reasonable bit-size numbers.

APPROXIMATE ARITHMETIC IN PREDICATE EVALUATION

The behavior of fundamental algorithms for computing line arrangements (both sweep line and incremental) while using limited precision arithmetic is studied in [FM91]. It is shown that the two algorithms can be implemented such that for n lines the maximum error of the coordinates of vertices is $O(n\epsilon)$ where ϵ is the relative error of the approximate arithmetic used (e.g., floating point). An approximate algorithm for constructing curve arrangements is presented in [MS07].

PERTURBATION

An arrangement of lines is considered degenerate if it is not simple (Section 28.1). A degeneracy occurs for example when three lines meet at a common point. Intuitively this is a degeneracy since moving the lines slightly will result in a topologically different arrangement. Degeneracies in arrangements pose difficulties for two reasons. First and foremost they incredibly complicate programming. Although it has been proposed that handling degeneracies could be the solution in practice to relax the general position assumption [BMS94b], in three and higher dimensions handling all degeneracies in arrangements seems an extremely difficult task. The second difficulty posed by degeneracies is that the numerical computation at or near degeneracies typically requires higher precision and will for example cause floating point filters to fail and resort to exact computing resulting in longer running time.

To overcome the first difficulty, symbolic perturbation schemes have been proposed. They enable a consistent perturbation of the input objects so that all degeneracies are removed. These schemes modify the objects only symbolically and a limiting process is used to define the perturbed objects (corresponding to infinitesimal perturbations) such that all predicates will have nonzero results. They require the usage of exact arithmetic, and a postprocessing stage to determine the structure of the output. The case of arrangements of hyperplanes can be approached by *simulation of simplicity* [EM90] via point-hyperplane duality. For a unifying view of these schemes and a discussion of their properties, see [Sei98].

An alternative approach is to *actually* perturb the objects from their original placement. One would like to perturb the input objects as little as possible so that precision problems are resolved. This approach is viable in situations where the exact placement of the input can be compromised, as is the case in many engineering and scientific applications where the input is inexact due to measurement or modeling errors. An efficient such scheme for arrangements of spheres that model molecules is described in [HS98]; it has been adapted and extended to handle arrangements of line segments [Pac11], circles [HL04], polyhedral surfaces [Raa99], as well as Delaunay triangulations [FKMS05]. It is referred to as *controlled* perturbation since it guarantees that the final arrangement is degeneracy free (and predicates can be safely computed with limited precision arithmetic), to distinguish from heuristic perturbation methods. A general analysis methodology for controlled-perturbation algorithms is presented in [MOS11]. A variant called controlled *linear* perturbation has been devised and used to robustly compute three-dimensional Minkowski sums [SMK11].

OPEN PROBLEM

Devise efficient and consistent rounding schemes for arrangements of curves in the plane and for arrangements in three and higher dimensions.

28.9 SOFTWARE

In spite of the numerous applications of arrangements, robust software for computing and manipulating arrangements has barely been available until about a decade ago. The situation has changed significantly over the past decade, with the increased understanding of the underlying difficulties, the research on overcoming these difficulties that has intensified during the last several years (Chapter 45), and the appearance of infrastructure for developing such software in the form of computational geometry libraries that emphasize robustness (Chapter 68).

28.9.1 2D ARRANGEMENTS

LEDA enables the construction of arrangements of segments via a sweep line algorithm. The resulting subdivision is represented as a LEDA graph. Point location based on persistent search trees is supported. The construction is robust through the use of arbitrary precision rationals.

Arrangements of general types of curves are supported by CGAL as we describe next, not limited to the planar case but rather supporting arrangement on surfaces.

2D ARRANGEMENTS IN CGAL

The most generic arrangement package at the time of the writing is the CGAL arrangements package [FHW12]. The genericity is obtained through the separation of the combinatorial part of the algorithms and the numerical part [FHH+00], [WFZH07]. (The overall design follows [Ket99].) The combinatorial algorithms are coded assuming that a small set of numerical/geometric operations (predicates and constructions) is supplied by the user for the desired type of curves. These operations are packed in a traits class (Chapter 68) that is passed as a parameter to the algorithms. The algorithms include the dynamic construction of the arrangement, represented as a doubly-connected-edge-list (DCEL), allowing for insertion and deletion of curves. Alternatively one can construct the arrangement using a sweep line algorithm. Several algorithms for point location are supported [HH08], most notably a complete implementation of random incremental construction of a trapeziodal-map based structure for arbitrary curves [FHH+00, HKH12]. All algorithms handle arbitrary input, namely they do not assume general position. Several traits classes are supplied with the package for: line segments, circular arcs, canonical parabolas, polylines, and planar algebraic curves of arbitrary degree [EK08]; for a list of supported types of curves, see [FHW12, Section 5.7].

Several tools were built on top of the CGAL arrangements package for computing: Envelopes of surfaces [Mey06], which in turn have paved the way to computing general Voronoi diagrams [SSH10], Boolean operations, and Minkowski sums.

The CGAL arrangement package has been used to compute Voronoi diagrams of lines in space [HSH10], to implement motion planning algorithms, [HH02],[SHRH13], several versions of snap rounding [HP02], art gallery optimization [KBFS12], NMR analysis tools [MYB+11], and many more applications.

A major recent development is the extension of CGAL's arrangement package from planar arrangements to arrangements on parametric surfaces [BFH+10b, BFH+10a]. The extended framework can handle planes, cylinders, spheres, tori, and surfaces homeomorphic to them. This extension has already been applied to computing Voronoi diagrams on the sphere, Minkowski sums of convex polytopes [BFH+10a] and to plan disassembly with infinite translations [FH13], among others.

28.9.2 3D ARRANGEMENTS

Software to construct arrangements of triangles in 3-space exactly, assuming general position, is described in [SH02]. The implementation uses a space sweep algorithm and exact rational arithmetic. The arrangement is represented by its vertical decomposition or a sparser variant called the *partial vertical decomposition*. Arrangements of algebraic surfaces in 3-space pose a much bigger challenge. Steps in this direction, including the handling of degeneracies, based on an efficient variant of Collins decomposition are described in [BS08], [BKS10].

OPEN PROBLEMS

1. Devise a systematic method to directly handle degeneracies in arrangements in three and higher dimensions (that is, to compute and represent degeneracies without removing them).

2. Extend the full-fledged support for 2D arrangements of curves to 3D arrangements of surfaces.

28.10 DAVENPORT-SCHINZEL SEQUENCES

Davenport-Schinzel sequences are interesting and powerful combinatorial structures that arise in the analysis and calculation of the lower or upper envelope of collections of functions, and therefore have applications in many geometric problems, including numerous motion planning problems, which can be reduced to the calculation of such an envelope. A comprehensive survey of Davenport-Schinzel sequences and their geometric applications can be found in [SA95].

An (n, s) **Davenport-Schinzel sequence**, where n and s are positive integers, is a sequence $U = (u_1, \ldots, u_m)$ composed of n symbols with the properties:

(i) No two adjacent elements of U are equal: $u_i \neq u_{i+1}$ for $i = 1, \ldots, m-1$.

(ii) U does not contain as a subsequence any alternation of length $s + 2$ between two distinct symbols: there do not exist $s + 2$ indices $i_1 < i_2 < \cdots < i_{s+2}$ so that $u_{i_1} = u_{i_3} = u_{i_5} = \cdots = a$ and $u_{i_2} = u_{i_4} = u_{i_6} = \cdots = b$, for two distinct symbols a and b.

Thus, for example, an $(n,3)$ sequence is not allowed to contain any subsequence of the form $(a \cdots b \cdots a \cdots b \cdots a)$. Let $\lambda_s(n)$ denote the maximum possible length of an (n,s) Davenport-Schinzel sequence.

The importance of Davenport-Schinzel sequences lies in their relationship to the combinatorial structure of the lower (or upper) envelope of a collection of functions (Section 28.2). Specifically, for any collection of n real-valued continuous functions $f_1, \ldots, f_n$ defined on the real line, having the property that each pair of them intersect in at most s points, one can show that the sequence of function indices i in the order in which these functions attain their lower envelope (i.e., their pointwise minimum $f = \min_i f_i$) from left to right is an (n,s) Davenport-Schinzel sequence. Conversely, any (n,s) Davenport-Schinzel sequence can be realized in this way for an appropriate collection of n continuous univariate functions, each pair of which intersect in at most s points.

The crucial and surprising property of Davenport-Schinzel sequences is that, for any fixed s, the maximal length $\lambda_s(n)$ is nearly linear in n, although for $s \geq 3$ it is slightly super-linear.

The best bounds on $\lambda_s(n)$, for every s, are due to Pettie [Pet15], and they are all asymptotically tight, or nearly tight. They are

$$
\lambda_s(n) = \begin{cases}
n & s = 1 \\
2n - 1 & s = 2 \\
2n\alpha(n) + O(n) & s = 3 \\
\Theta\left(n2^{\alpha(n)}\right) & s = 4 \\
\Theta\left(n\alpha(n)2^{\alpha(n)}\right) & s = 5 \\
n \cdot 2^{(1+o(1))\alpha^t(n)/t!} & \text{for both even and odd } s \geq 6; \ t = \lfloor \frac{s-2}{2} \rfloor
\end{cases}
$$

where $\alpha(n)$ is the inverse of Ackermann's function. Ackermann's function $A(n)$ grows extremely quickly, with $A(4)$ equal to an exponential "tower" of 65636 2's. Thus $\alpha(n) \leq 4$ for all practical values of n. See [SA95].

If one considers the lower envelope of n continuous, but only partially defined, functions, then the complexity of the envelope is at most $\lambda_{s+2}(n)$, where s is the maximum number of intersections between any pair of functions [SA95]. Thus for a collection of n line segments (for which $s = 1$), the lower envelope consists of at most $O(n\alpha(n))$ subsegments. A surprising result is that this bound is tight in the worst case: there are collections of n segments, for arbitrarily large n, whose lower envelope does consist of $\Omega(n\alpha(n))$ subsegments. This is perhaps the most natural example of a combinatorial structure defined in terms of n simple objects, whose complexity involves the inverse Ackermann's function; see [SA95, WS88].

28.11 SOURCES AND RELATED MATERIAL

FURTHER READING

The study of arrangements through the early 1970s is covered by Grünbaum in [Grü67, Chapter 18], [Grü71], and [Grü72]. See also the monograph by Zaslavsky [Zas75].

In this chapter we have concentrated on more recent results. Details of many of these results can be found in the following books. The book by Edelsbrunner [Ede87] takes the view of "arrangements of hyperplanes" as a unifying theme for a large part of discrete and computational geometry until 1987. Sharir and Agarwal's book [SA95] is an extensive report on results for arrangements of curves and surfaces. See also the more recent survey [AS00a] and book [PS09]. Chapters dedicated to arrangements of hyperplanes in books: Mulmuley emphasizes randomized algorithms [Mul93], O'Rourke discusses basic combinatorics, relations to other structures and applications [O'R98], de Berg et al. discuss planar arrangements of lines with application to discrepancy [BCKO08], and Pach and Agarwal [PA95] discuss problems involving arrangements in discrete geometry. Boissonnat and Yvinec [BY98] discuss, in addition to arrangements of hyperplanes, arrangements of segments and of triangles. Arrangements of hyperplanes and of surfaces are also the topics of chapters in Matoušek's book [Mat02].

RELATED CHAPTERS

REFERENCES

[AA92]   P.K. Agarwal and B. Aronov. Counting facets and incidences. *Discrete Comput. Geom.*, 7:359–369, 1992.

[AACS98]   P.K. Agarwal, B. Aronov, T.M. Chan, and M. Sharir. On levels in arrangements of lines, segments, planes, and triangles. *Discrete Comput. Geom.*, 19:315–331, 1998.

[AAS97]   P.K. Agarwal, B. Aronov, and M. Sharir. Computing envelopes in four dimensions with applications. *SIAM J. Comput.*, 26:1714–1732, 1997.

[AAS03]   P.K. Agarwal, B. Aronov, and M. Sharir. On the complexity of many faces in arrangements of pseudo-segments and of circles. In B. Aronov et al., editors, *Discrete and Computational Geometry—The Goodman-Pollack Festschrift*, pages 1–23. Springer, Berlin, 2003.

[AC09]   P. Afshani and T.M. Chan. Optimal halfspace range reporting in three dimensions. In *Proc. 20th ACM-SIAM Sympos. Discrete Algorithms*, pages 180–186, 2009.

[ACDG17]   B. Aronov, O. Cheong, M.G. Dobbins, and X. Goaoc. The number of holes in the union of translates of a convex set in three dimensions. *Discrete Comput. Geom.*, 57:104–124, 2017.

[AD11]   B. Aronov and D. Drusvyatskiy. Complexity of a single face in an arrangement of *s*-intersecting curves. Preprint, arXiv:1108.4336, 2011.

[ABB⁺02]   H.-K. Ahn, M. de Berg, P. Bose, S.-W. Cheng, D. Halperin, J. Matoušek, and O. Schwarzkopf. Separating an object from its cast. *Computer-Aided Design*, 34:547–559, 2002.

[ABES14]   B. Aronov, M. de Berg, E. Ezra, and M. Sharir. Improved bounds for the union of locally fat objects in the plane. *SIAM J. Comput.*, 43:543–572, 2014.

[ABMS98]   P.K. Agarwal, M. de Berg, J. Matoušek, and O. Schwarzkopf. Constructing levels in arrangements and higher order Voronoi diagrams. *SIAM J. Comput.*, 27:654–667, 1998.

[AEGS92]   B. Aronov, H. Edelsbrunner, L.J. Guibas, and M. Sharir. The number of edges of many faces in a line segment arrangement. *Combinatorica*, 12:261–274, 1992.

[AEHS01]   B. Aronov, A. Efrat, D. Halperin, and M. Sharir. On the number of regular vertices of the union of Jordan regions. *Discrete Comput. Geom.*, 25:203–220, 2001.

[AEKS06]   B. Aronov, A. Efrat, V. Koltun, and M. Sharir. On the union of κ-round objects in three and four dimensions. *Discrete Comput. Geom.*, 36:511–526, 2006.

[AES99]   P.K. Agarwal, A. Efrat, and M. Sharir. Vertical decomposition of shallow levels in 3-dimensional arrangements and its applications. *SIAM J. Comput.*, 29:912–953, 1999.

[AG86]   N. Alon and E. Győri. The number of small semispaces of a finite set of points in the plane. *J. Combin. Theory Ser. A*, 41:154–157, 1986.

[Aga90]   P.K. Agarwal. Partitioning arrangements of lines II: applications. *Discrete Comput. Geom.*, 5:533–573, 1990.

[AGR00]   N.M. Amato, M.T. Goodrich, and E.A. Ramos. Computing the arrangement of curve segments: divide-and-conquer algorithms via sampling. In *Proc. 11th ACM-SIAM Sympos. Discrete Algorithms*, pages 705–706, 2000.

[AM94]   P.K. Agarwal and J. Matoušek. On range searching with semialgebraic sets. *Discrete Comput. Geom.*, 11:393–418, 1994.

[AM95]   P.K. Agarwal and J. Matoušek. Dynamic half-space range reporting and its applications. *Algorithmica*, 13:325–345, 1995.

[AMS94]   B. Aronov, J. Matoušek, and M. Sharir. On the sum of squares of cell complexities in hyperplane arrangements. *J. Combin. Theory Ser. A*, 65:311–321, 1994.

[AMS98]   P.K. Agarwal, J. Matoušek, and O. Schwarzkopf. Computing many faces in arrangements of lines and segments. *SIAM J. Comput.*, 27:491–505, 1998.

[AMS13]   P.K. Agarwal, J. Matoušek, and M. Sharir. On range searching with semialgebraic sets, II. *SIAM J. Comput.*, 42:2039–2062, 2013.

[ANP⁺04]   P.K. Agarwal, E. Nevo, J. Pach, R. Pinchasi, M. Sharir, and S. Smorodinsky. Lenses in arrangements of pseudo-circles and their applications. *J. ACM*, 51:139–186, 2004.

[APS93]   B. Aronov, M. Pellegrini, and M. Sharir. On the zone of a surface in a hyperplane arrangement. *Discrete Comput. Geom.*, 9:177–186, 1993.

[APS08]   P.K. Agarwal, J. Pach, and M. Sharir. State of the union (of geometric objects). In J.E. Goodman, J. Pach, and R. Pollack, editors, *Surveys on Discrete and Computational Geometry: Twenty Years Later*, vol. 453 of *Contemp. Math.*, pages 9–48, AMS, Providence, 2008.

[AS90]   B. Aronov and M. Sharir. Triangles in space or building (and analyzing) castles in the air. *Combinatorica*, 10:137–173, 1990.

[AS94]   B. Aronov and M. Sharir. Castles in the air revisited. *Discrete Comput. Geom.*, 12:119–150, 1994.

[AS97]   B. Aronov and M. Sharir. On translational motion planning of a convex polyhedron in 3-space. *SIAM J. Comput.*, 26:1785–1803, 1997.

[AS98]   P.K. Agarwal and M. Sharir. Efficient algorithms for geometric optimization. *ACM Comput. Surv.*, 30:412–458, 1998.

[AS00a]   P.K. Agarwal and M. Sharir. Arrangements and their applications. In J.-R. Sack and J. Urrutia, editors, *Handbook of Computational Geometry*, pages 49–119. Elsevier, Amsterdam, 2000.

[AS00b]   P.K. Agarwal and M. Sharir. Pipes, cigars, and kreplach: The union of Minkowski sums in three dimensions. *Discrete Comput. Geom.*, 24:645–685, 2000.

[AS04]   B. Aronov and M. Sharir. Cell complexities in hyperplane arrangements. *Discrete Comput. Geom.*, 32:107–115, 2004.

[AS05]   P.K. Agarwal and M. Sharir. Pseudo-line arrangements: Duality, algorithms, and applications. *SIAM J. Comput.*, 34:526–552, 2005.

[ASS96]   P.K. Agarwal, O. Schwarzkopf, and M. Sharir. The overlay of lower envelopes and its applications. *Discrete Comput. Geom.*, 15:1–13, 1996.

[AST97]   B. Aronov, M. Sharir, and B. Tagansky. The union of convex polyhedra in three dimensions. *SIAM J. Comput.*, 26:1670–1688, 1997.

[Bal95]   I.J. Balaban. An optimal algorithm for finding segment intersections. In *Proc. 11th Sympos. Comput. Geom.*, pages 211–219, ACM Press, 1995.

[Bas03]   S. Basu. The combinatorial and topological complexity of a single cell. *Discrete Comput. Geom.*, 29:41–59, 2003.

[BCKO08] M. de Berg, O. Cheong, M. van Kreveld, and M. Overmars. *Computational Geometry: Algorithms and Applications*. Springer, Berlin, 3rd edition, 2008.

[BD90]   K.W. Bowyer and C.R. Dyer. Aspect graphs: An introduction and survey of recent results. *Internat. J. Imaging Syst. Tech.*, 2:315–328, 1990.

[BD96]   J.-D. Boissonnat and K. Dobrindt. On-line construction of the upper envelope of triangles and surface patches in three dimensions. *Comput. Geom.* 5:303–320, 1996.

[BDH99]   K.-F. Böhringer, B.R. Donald, and D. Halperin. On the area bisectors of a polygon. *Discrete Comput. Geom.*, 22:269–285, 1999.

[BDS95]   M. de Berg, K. Dobrindt, and O. Schwarzkopf. On lazy randomized incremental construction. *Discrete Comput. Geom.*, 14:261–286, 1995.

[Ber93]   M. de Berg. *Ray Shooting, Depth Orders and Hidden Surface Removal.* Vol. 703 of *Lecture Notes Comp. Sci.*, Springer, Berlin, 1993.

[BFH⁺10a] E. Berberich, E. Fogel, D. Halperin, M. Kerber, and O. Setter. Arrangements on parametric surfaces II: concretizations and applications. *Math. Comput. Sci.*, 4:67–91, 2010.

[BFH⁺10b] E. Berberich, E. Fogel, D. Halperin, K. Mehlhorn, and R. Wein. Arrangements on parametric surfaces I: general framework and infrastructure. *Math. Comput. Sci.*, 4:45–66, 2010.

[BGH96]   M. de Berg, L.J. Guibas, and D. Halperin. Vertical decompositions for triangles in 3-space. *Discrete Comput. Geom.*, 15:35–61, 1996.

[BHO07]   M. de Berg, D. Halperin, and M.H. Overmars. An intersection-sensitive algorithm for snap rounding. *Comput. Geom.*, 36:159–165, 2007.

[BJ02]   G.S. Brodal and R. Jacob. Dynamic planar convex hull. In *Proc. 43rd IEEE Sympos. Found. Comp. Sci.*, pages 617–626, 2002.

[BK03]   P. Braß and C. Knauer. On counting point-hyperplane incidences. *Comput. Geom.*, 25:13–20, 2003.

[BKS10]   E. Berberich, M. Kerber, and M. Sagraloff. An efficient algorithm for the stratification and triangulation of an algebraic surface. *Comput. Geom.*, 43:257–278, 2010.

[BLW+93] A. Björner, M. Las Vergnas, N. White, B. Sturmfels, and G.M. Ziegler. *Oriented Matroids*. Vol. 46 of *Encycl. Math.*, Cambridge Univ. Press, 1993.

[BMS94]   C. Burnikel, K. Mehlhorn, and S. Schirra. On degeneracy in geometric computations. In *Proc. ACM-SIAM Sympos. Discrete Algorithms*, pages 16–23, 1994.

[BP00]   J.-D. Boissonnat and F.P. Preparata. Robust plane sweep for intersecting segments. *SIAM J. Comput.*, 29:1401–1421, 2000.

[BPR06]   S. Basu, R. Pollack, and M.-F. Roy. *Algorithms in Real Algebraic Geometry (Algorithms and Computation in Mathematics)*. Springer, New York, 2006.

[Bri93]   E. Brisson. Representing geometric structures in d dimensions: Topology and order. *Discrete Comput. Geom.*, 9:387–426, 1993.

[BS08]   E. Berberich and M. Sagraloff. A generic and flexible framework for the geometrical and topological analysis of (algebraic) surfaces. In *Proc. ACM Sympos. Solid Physical Modeling*, pages 171–182, 2008.

[BS16]   S. Basu and M. Sombra. Polynomial partitioning on varieties of codimension two and point-hypersurface incidences in four dimensions. *Discrete Comput. Geom.* 55:158–184, 2016.

[BSTY98] J.-D. Boissonnat, M. Sharir, B. Tagansky, and M. Yvinec. Voronoi diagrams in higher dimensions under certain polyhedral distance functions. *Discrete Comput. Geom.*, 19:485–519, 1998.

[BY98]   J.-D. Boissonnat and M. Yvinec. *Algorithmic Geometry*. Cambridge University Press, 1998. Translated by H. Brönnimann.

[Can93]   J.F. Canny. Computing roadmaps in general semialgebraic sets. *Comput. J.*, 36:504–514, 1993.

[CEG+90] K.L. Clarkson, H. Edelsbrunner, L.J. Guibas, M. Sharir, and E. Welzl. Combinatorial complexity bounds for arrangements of curves and spheres. *Discrete Comput. Geom.*, 5:99–160, 1990.

[CEGS91] B. Chazelle, H. Edelsbrunner, L.J. Guibas, and M. Sharir. A singly-exponential stratification scheme for real semi-algebraic varieties and its applications. *Theoret. Comput. Sci.*, 84:77–105, 1991. An improved bound appears in the proceedings version: *Proc. ICALP*, vol. 372 of *LNCS*, pages 179–193, Springer, Berlin, 1989.

[Cha96]   T.M. Chan. Output-sensitive results on convex hulls, extreme points, and related problems. *Discrete Comput. Geom.*, 16:369–387, 1996.

[Cha00]   T.M. Chan. Random sampling, halfspace range reporting, and construction of ($\leq k$)-levels in three dimensions. *SIAM J. Comput.*, 30:561–575, 2000.

[Cha01]   T.M. Chan. Dynamic planar convex hull operations in near-logarithmic amortized time. *J. ACM*, 48:1–12, 2001.

[Cha03]   T.M. Chan. On levels in arrangements of curves. *Discrete Comput. Geom.*, 29:375–393, 2003.

[Cla88]   K.L. Clarkson. A randomized algorithm for closest-point queries. *SIAM J. Comput.*, 17:830–847, 1988.

[CS89]   K.L. Clarkson and P.W. Shor. Applications of random sampling in computational geometry, II. *Discrete Comput. Geom.*, 4:387–421, 1989.

[CT16]   T.M. Chan and K. Tsakalidis. Optimal deterministic algorithms for 2-d and 3-d shallow cuttings. *Discrete Comput. Geom.*, 56:866–881, 2016.

[Dey98]   T.K. Dey. Improved bounds on planar k-sets and related problems. *Discrete Comput. Geom.*, 19:373–382, 1998.

[DL89]   D.P. Dobkin and M.J. Laszlo. Primitives for the manipulation of three-dimensional subdivisions. *Algorithmica*, 4:3–32, 1989.

[Ede87]   H. Edelsbrunner. *Algorithms in Combinatorial Geometry*. Springer, Berlin, 1987.

[Ede89]   H. Edelsbrunner. The upper envelope of piecewise linear functions: Tight complexity bounds in higher dimensions. *Discrete Comput. Geom.*, 4:337–343, 1989.

[EG89]   H. Edelsbrunner and L.J. Guibas. Topologically sweeping an arrangement. *J. Comput. Syst. Sci.*, 38:165–194, 1989. Corrigendum in 42:249–251, 1991.

[EGH⁺89]   H. Edelsbrunner, L.J. Guibas, J. Hershberger, J. Pach, R. Pollack, R. Seidel, M. Sharir, and J. Snoeyink. Arrangements of Jordan arcs with three intersections per pair. *Discrete Comput. Geom.*, 4:523–539, 1989.

[EGP⁺92]   H. Edelsbrunner, L.J. Guibas, J. Pach, R. Pollack, R. Seidel, and M. Sharir. Arrangements of curves in the plane: Topology, combinatorics, and algorithms. *Theoret. Comput. Sci.*, 92:319–336, 1992.

[EGS89]   H. Edelsbrunner, L.J. Guibas, and M. Sharir. The upper envelope of piecewise linear functions: algorithms and applications. *Discrete Comput. Geom.*, 4:311–336, 1989.

[EGS90]   H. Edelsbrunner, L.J. Guibas, and M. Sharir. The complexity of many cells in arrangements of planes and related problems. *Discrete Comput. Geom.*, 5:197–216, 1990.

[EHS04]   E. Ezra, D. Halperin, and M. Sharir. Speeding up the incremental construction of the union of geometric objects in practice. *Comput. Geom.*, 27:63–85, 2004.

[EK08]   A. Eigenwillig and M. Kerber. Exact and efficient 2d-arrangements of arbitrary algebraic curves. In *Proc. 19th ACM-SIAM Sympos. Discrete Algorithms*, pages 122–131, 2008.

[EM90]   H. Edelsbrunner and E.P. Mücke. Simulation of simplicity: A technique to cope with degenerate cases in geometric algorithms. *ACM Trans. Graph.*, 9:66–104, 1990.

[Eri96]   J. Erickson. New lower bounds for Hopcroft's problem. *Discrete Comput. Geom.*, 16:389–418, 1996.

[ERK96]   H. Everett, J.-M. Robert, and M. van Kreveld. An optimal algorithm for the $(\leq k)$-levels, with applications to separation and transversal problems. *Internat. J. Comput. Geom. Appl.*, 6:247–261, 1996.

[ES09]   E. Ezra and M. Sharir. On the union of fat tetrahedra in three dimensions. *J. ACM*, 57(1), 2009.

[ESS93]   H. Edelsbrunner, R. Seidel, and M. Sharir. On the zone theorem for hyperplane arrangements. *SIAM J. Comput.*, 22:418–429, 1993.

[EW86]   H. Edelsbrunner and E. Welzl. Constructing belts in two-dimensional arrangements with applications. *SIAM J. Comput.*, 15:271–284, 1986.

[Ezr11]   E. Ezra. On the union of cylinders in three dimensions. *Discrete Comput. Geom.*, 45:45–64, 2011.

[FH13]   E. Fogel and D. Halperin. Polyhedral assembly partitioning with infinite translations or the importance of being exact. *IEEE Trans. Autom. Sci. Eng.*, 10:227–241, 2013.

[FHH⁺00]   E. Flato, D. Halperin, I. Hanniel, O. Nechushtan, and E. Ezra. The design and implementation of planar maps in CGAL. *ACM J. Exper. Algorithmics*, 5:13, 2000.

[FHW12]   E. Fogel, D. Halperin, and R. Wein. *CGAL Arrangements and Their Applications—A Step-by-Step Guide.* Vol. 7 of *Geometry and Computing*, Springer, Berlin, 2012.

[FKMS05]   S. Funke, C. Klein, K. Mehlhorn, and S. Schmitt. Controlled perturbation for Delaunay triangulations. In *Proc. 16th ACM-SIAM Sympos. Discrete Algorithms*, pages 1047–1056, 2005.

[FM91]   S. Fortune and V. Milenkovic. Numerical stability of algorithms for line arrangements. In *Proc. 17th Sympos. Comput. Geom.*, pages 334–341, ACM Press, 1991.

[For99]   S. Fortune. Vertex-rounding a three-dimensional polyhedral subdivision. *Discrete Comput. Geom.*, 22:593–618, 1999.

[GGHT97]   M.T. Goodrich, L.J. Guibas, J. Hershberger, and P. Tanenbaum. Snap rounding line segments efficiently in two and three dimensions. *Proc. 13th Sympos. Comput. Geom.*, pages 284–293, ACM Press, 1997.

[GK15]   L. Guth and N.H. Katz. On the Erdős distinct distances problem in the plane. *Ann. of Math.*, 181:155–190, 2015.

[GM98]   L.J. Guibas and D. Marimont. Rounding arrangements dynamically. *Internat. J. Comput. Geom. Appl.*, 8:157–176, 1998.

[GO95]   A. Gajentaan and M.H. Overmars. On a class of $O(n^2)$ problems in computational geometry. *Comput. Geom.*, 5:165–185, 1995.

[Goo93]   M.T. Goodrich. Constructing arrangements optimally in parallel. *Discrete Comput. Geom.*, 9:371–385, 1993.

[Grü67]   B. Grünbaum. *Convex Polytopes.* John Wiley & Sons, New York, 1967.

[Grü71]   B. Grünbaum. Arrangements of hyperplanes. *Congr. Numer.*, 3:41–106, 1971.

[Grü72]   B. Grünbaum. *Arrangements and Spreads.* Regional Conf. Ser. Math., AMS, Providence, 1972.

[GS85]   L.J. Guibas and J. Stolfi. Primitives for the manipulation of general subdivisions and computation of Voronoi diagrams. *ACM Trans. Graph.*, 4:74–123, 1985.

[GSS89]   L.J. Guibas, M. Sharir, and S. Sifrony. On the general motion planning problem with two degrees of freedom. *Discrete Comput. Geom.*, 4:491–521, 1989.

[Gut14]   L. Guth. Polynomial partitioning for a set of varieties. *Math. Proc. Cambridge Phil. Soc.*, 159:459–469, 2015.

[Gut16]   L. Guth. *Polynomial Methods in Combinatorics.* AMS, Providence, 2016.

[GY86]   D.H. Greene and F.F. Yao. Finite-resolution computational geometry. *Proc. 27th IEEE Sympos. Found. Comp. Sci.*, pages 143–152, 1986.

[Hal92]   D. Halperin. *Algorithmic Motion Planning via Arrangements of Curves and of Surfaces.* Ph.D. thesis, Computer Science Department, Tel-Aviv University, 1992.

[Hal94]   D. Halperin. On the complexity of a single cell in certain arrangements of surfaces related to motion planning. *Discrete Comput. Geom.*, 11:1–34, 1994.

[Her89]   J. Hershberger. Finding the upper envelope of n line segments in $O(n \log n)$ time. *Inform. Process. Lett.*, 33:169–174, 1989.

[Her13]   J. Hershberger. Stable snap rounding. *Comput. Geom.*, 46:403–416, 2013.

[HH02]   S. Hirsch and D. Halperin. Hybrid motion planning: Coordinating two discs moving among polygonal obstacles in the plane. In *Algorithmic Foundations of Robotics V.* Vol. 7 of *Springer Tracts in Advanced Robotics*, pages 239–256, 2002.

[HH08]   I. Haran and D. Halperin. An experimental study of point location in planar arrangements in CGAL. *ACM J. Exper. Algorithmics*, 13, 2008.

[HJW90]   T. Hagerup, H. Jung, and E. Welzl. Efficient parallel computation of arrangements of hyperplanes in d dimensions. In *Proc. 2nd ACM Sympos. Parallel Algorithms Architect.*, pages 290–297, 1990.

[HKH12]   M. Hemmer, M. Kleinbort, and D. Halperin. Improved implementation of point location in general two-dimensional subdivisions. In *Proc. 20th European Sympos. Algorithms*, pages 611–623, vol. 7501 of *LNCS*, Springer, Berlin, 2012.

[HKS16]   S. Har-Peled, H. Kaplan, and M. Sharir. Approximating the k-level in three-dimensional plane arrangements. In *Proc. 27th ACM-SIAM Sympos. Discrete Algorithms*, pages 1193–1212, 2016.

[HL04]   D. Halperin and E. Leiserowitz. Controlled perturbation for arrangements of circles. *Internat. J. Comput. Geom. Appl.*, 14:277–310, 2004.

[HLW00]   D. Halperin, J.-C. Latombe, and R.H. Wilson. A general framework for assembly planning: The motion space approach. *Algorithmica*, 26:577–601, 2000.

[Hob99]   J.D. Hobby. Practical segment intersection with finite precision output. *Comput. Geom.*, 13:199–214, 1999.

[HP02]   D. Halperin and E. Packer. Iterated snap rounding. *Comput. Geom.*, 23:209–225, 2002.

[HS94]   D. Halperin and M. Sharir. New bounds for lower envelopes in three dimensions, with applications to visibility in terrains. *Discrete Comput. Geom.*, 12:313–326, 1994.

[HS95a]   D. Halperin and M. Sharir. Almost tight upper bounds for the single cell and zone problems in three dimensions. *Discrete Comput. Geom.*, 14:385–410, 1995.

[HS95b]   D. Halperin and M. Sharir. Arrangements and their applications in robotics: Recent developments. In K. Goldbergs et al., editors, *Algorithmic Foundations of Robotics*, pages 495–511. A.K. Peters, Boston, 1995.

[HS98]   D. Halperin and C.R. Shelton. A perturbation scheme for spherical arrangements with application to molecular modeling. *Comput. Geom.*, 10:273–287, 1998.

[HSH10]   M. Hemmer, O. Setter, and D. Halperin. Constructing the exact Voronoi diagram of arbitrary lines in three-dimensional space—with fast point-location. In *Proc. 18th European Sympos. Algorithms*, pages 398–409, vol. 6346 of *LNCS*, Springer, Berlin, 2010.

[HY98]   D. Halperin and C.K. Yap. Combinatorial complexity of translating a box in polyhedral 3-space. *Comput. Geom.*, 9:181–196, 1998.

[JP14]   A.G. Jørgensen and S. Pettie. Threesomes, degenerates, and love triangles. In *55th IEEE Sympos. Found. Comp. Sci.*, pages 621–630, 2014.

[KBFS12]   A. Kröller, T. Baumgartner, S.P. Fekete, and C. Schmidt. Exact solutions and bounds for general art gallery problems. *ACM J. Exper. Algorithmics*, 17, 2012.

[Ket99]   L. Kettner. Using generic programming for designing a data structure for polyhedral surfaces. *Comput. Geom.*, 13:65–90, 1999.

[KLPS86]   K. Kedem, R. Livne, J. Pach, and M. Sharir. On the union of Jordan regions and collision-free translational motion amidst polygonal obstacles. *Discrete Comput. Geom.*, 1:59–71, 1986.

[KLPY99]   V. Karamcheti, C. Li, I. Pechtchanski, and C. Yap. *The CORE Library Project*, 1.2 edition, 1999. http://www.cs.nyu.edu/exact/core/.

[KMSS12]   H. Kaplan, J. Matoušek, M. Sharir, and Z. Safernová. Unit distances in three dimensions. *Combin. Probab. Comput.* 21:597–610, 2012.

[Kol04a]   V. Koltun. Almost tight upper bounds for vertical decompositions in four dimensions. *J. ACM*, 51:699–730, 2004.

[Kol04b]   V. Koltun. Sharp bounds for vertical decompositions of linear arrangements in four dimensions. *Discrete Comput. Geom.*, 31:435–460, 2004.

[KRS11]   H. Kaplan, E. Ramos, and M. Sharir. The overlay of minimization diagrams in a randomized incremental construction. *Discrete Comput. Geom.*, 45:371–382, 2011.

[KS03]   V. Koltun and M. Sharir. The partition technique for overlays of envelopes. *SIAM J. Comput.*, 32:841–863, 2003.

[KS05]   V. Koltun and M. Sharir. Curve-sensitive cuttings. *SIAM J. Comput.*, 34:863–878, 2005.

[KS09]   V. Koltun and M. Sharir. On overlays and minimization diagrams. *Discrete Comput. Geom.*, 41:385–397, 2009.

[Mat92]   J. Matoušek. Reporting points in halfspaces. *Comput. Geom.*, 2:169–186, 1992.

[Mat93]   J. Matoušek. Range searching with efficient hierarchical cuttings. *Discrete Comput. Geom.*, 10:157–182, 1993.

[Mat02]   J. Matoušek. *Lectures on Disrete Geometry*. Vol. 212 of *Graduate Texts in Mathematics*, Springer, New York, 2002.

[McM70]   P. McMullen. The maximal number of faces of a convex polytope. *Mathematika*, 17:179–184, 1970.

[Mey06]   M. Meyerovitch. Robust, generic and efficient construction of envelopes of surfaces in three-dimensional spaces. In *Proc. 14th European Sympos. Algorithms*, pages 792–803, vol. 4168 of *LNCS*, Springer, Berlin, 2006.

[Mil00]   V.J. Milenkovic. Shortest path geometric rounding. *Algorithmica*, 27:57–86, 2000.

[MMP+91]   J. Matoušek, N. Miller, J. Pach, M. Sharir, S. Sifrony, and E. Welzl. Fat triangles determine linearly many holes. In *Proc. 32nd IEEE Sympos. Found. Comp. Sci.* pages 49–58, 1991.

[MN00]   K. Mehlhorn and S. Näher. *LEDA: A Platform for Combinatorial and Geometric Computing*. Cambridge University Press, 2000.

[MOS11]   K. Mehlhorn, R. Osbild, and M. Sagraloff. A general approach to the analysis of controlled perturbation algorithms. *Comput. Geom.*, 44:507–528, 2011.

[MPS+94]   J. Matoušek, J. Pach, M. Sharir, S. Sifrony, and E. Welzl. Fat triangles determine linearly many holes. *SIAM J. Comput.*, 23:154–169, 1994.

[MS07]   V. Milenkovic and E. Sacks. An approximate arrangement algorithm for semi-algebraic curves. *Internat. J. Comput. Geom. Appl.*, 17:175–198, 2007.

[MT06]   A. Marcus and G. Tardos. Intersection reverse sequences and geometric applications. *J. Combin. Theory Ser. A*, 113:675–691, 2006.

[Mul93]   K. Mulmuley. *Computational Geometry: An Introduction through Randomized Algorithms*. Prentice Hall, Englewood Cliffs, 1993.

[MYB+11]   J.W. Martin, A.K. Yan, C. Bailey-Kellogg, P. Zhou, and B.R. Donald. A geometric arrangement algorithm for structure determination of symmetric protein homo-oligomers from NOEs and RDCs. *J. Comput. Bio.*, 18:1507–1523, 2011.

[Niv08]   G. Nivasch. An improved, simple construction of many halving edges. In J.E. Goodman and J. Pach and R. Pollack, editor, *Surveys on Discrete and Computational Geometry: Twenty Years Later*, vol. 453 of *Contemp. Math.*, pages 299–305, AMS, Providence, 2008.

[Niv15]   G. Nivasch. On the zone of a circle in an arrangement of lines. *Elect. Notes Discrete Math.*, 49:221–231, 2015.

[O'R98]   J. O'Rourke. *Computational Geometry in C*, second edition. Cambridge University Press, 1998.

[PA95]   J. Pach and P.K. Agarwal. *Combinatorial Geometry*. John Wiley & Sons, New York, 1995.

[Pac08]   E. Packer. Iterated snap rounding with bounded drift. *Comput. Geom.*, 40:231–251, 2008.

[Pac11]   E. Packer. Controlled perturbation of sets of line segments in $\mathbb{R}^2$ with smart processing order. *Comput. Geom.*, 44:265–285, 2011.

[Pet15]   S. Pettie. Sharp bounds on Davenport-Schinzel sequences of every order. *J. ACM*, 62:36, 2015.

[PS85]   F.P. Preparata and M.I. Shamos. *Computational Geometry: An Introduction*. Springer-Verlag, New York, 1985.

[PS89]   J. Pach and M. Sharir. The upper envelope of piecewise linear functions and the boundary of a region enclosed by convex plates: Combinatorial analysis. *Discrete Comput. Geom.*, 4:291–309, 1989.

[PS99]   J. Pach and M. Sharir. On the boundary of the union of planar convex sets. *Discrete Comput. Geom.*, 21:321–328, 1999.

[PS09]   J. Pach and M. Sharir. *Combinatorial Geometry and its Algorithmic Applications: The Alcalá Lecturess*. AMS, Providence, 2009.

[PSS03]   J. Pach, I. Safruti, and M. Sharir. The union of congruent cubes in three dimensions. *Discrete Comput. Geom.*, 30:133–160, 2003.

[Raa99]   S. Raab. Controlled perturbation for arrangements of polyhedral surfaces with application to swept volumes. In *Proc. 15th Sympos Comput. Geom.*, pages 163–172, ACM Press, 1999.

[Ram99]   E.A. Ramos. On range reporting, ray shooting and k-level construction. In *Proc. 15th Sympos. Comput. Geom.*, pages 390–399, ACM Press, 1999.

[Raz15]   O.E. Raz. On the zone of the boundary of a convex body. *Comput. Geom.*, 48:333–341, 2015.

[SA95]   M. Sharir and P.K. Agarwal. *Davenport-Schinzel Sequences and Their Geometric Applications*. Cambridge University Press, 1995.

[SAL93]   A. Schweikard, J.E. Adler, and J.-C. Latombe. Motion planning in stereotaxic radiosurgery. In *Proc. IEEE Int. Conf. Robot. Autom.*, pages 764–774, 1993.

[Sei98]   R. Seidel. The nature and meaning of perturbations in geometric computing. *Discrete Comput. Geom.*, 19:1–17, 1998.

[SH02]   H. Shaul and D. Halperin. Improved construction of vertical decompositions of three-dimensional arrangements. In *Proc. 18th Sympos. Comput. Geom.*, pages 283–292, ACM Press, 2002.

[Sha91]   M. Sharir. On k-sets in arrangements of curves and surfaces. *Discrete Comput. Geom.*, 6:593–613, 1991.

[Sha94]   M. Sharir. Almost tight upper bounds for lower envelopes in higher dimensions. *Discrete Comput. Geom.*, 12:327–345, 1994.

[Sha11]   M. Sharir. An improved bound for k-sets in four dimensions. *Combin. Probab. Comput.*, 20:119–129, 2011.

[SHRH13]   O. Salzman, M. Hemmer, B. Raveh, and D. Halperin. Motion planning via manifold samples. *Algorithmica*, 67:547–565, 2013.

[SMK11]   E. Sacks, V. Milenkovic, and M.-H. Kyung. Controlled linear perturbation. *Computer-Aided Design*, 43:1250–1257, 2011.

[SS97]   O. Schwarzkopf and M. Sharir. Vertical decomposition of a single cell in a three-dimensional arrangement of surfaces and its applications. *Discrete Comput. Geom.*, 18:269–288, 1997.

[SSH10]   O. Setter, M. Sharir, and D. Halperin. Constructing two-dimensional Voronoi diagrams via divide-and-conquer of envelopes in space. In *Transactions on Computational Science IX*, pages 1–27, vol. 6290 of *LNCS*, Springer, 2010.

[SST01]   M. Sharir, S. Smorodinsky, and G. Tardos. An improved bound for k-sets in three dimensions. *Discrete Comput. Geom.*, 26:195–204, 2001.

[Szé97]   L.A. Székely. Crossing numbers and hard Erdős problems in discrete geometry. *Combin. Probab. Comput.*, 6:353–358, 1997.

[Tag96]   B. Tagansky. A new technique for analyzing substructures in arrangements of piecewise linear surfaces. *Discrete Comput. Geom.*, 16:455–479, 1996.

[Tót01]   G. Tóth. Point sets with many k-sets. *Discrete Comput. Geom.*, 26:187–194, 2001.

[TT98]   H. Tamaki and T. Tokuyama. How to cut pseudoparabolas into segments. *Discrete Comput. Geom.*, 19:265–290, 1998.

[WFZH07]   R. Wein, E. Fogel, B. Zukerman, and D. Halperin. Advanced programming techniques applied to CGAL's arrangement package. *Comput. Geom.*, 38:37–63, 2007.

[WS88]   A. Wiernik and M. Sharir. Planar realizations of nonlinear Davenport-Schinzel sequences by segments. *Discrete Comput. Geom.*, 3:15–47, 1988.

[Zah11]   J. Zahl. An improved bound on the number of point-surface incidences in three dimensions. *Contrib. Discrete Math.* 8:100–121, 2013.

[Zas75]   T. Zaslavsky. *Facing up to Arrangements: Face-Count Formulas for Partitions of Space by Hyperplanes*. Vol. 154 of *Memoirs of the AMS*, AMS, Providence, 1975.

29 TRIANGULATIONS AND MESH GENERATION
Marshall Bern, Jonathan R. Shewchuk, and Nina Amenta

INTRODUCTION

A triangulation is a partition of a geometric domain, such as a point set, polygon, or polyhedron, into simplices that meet only at shared faces. (For a point set, the triangulation covers the convex hull.) Triangulations are important for representing complicated geometry by piecewise simple geometry and for interpolating numerical fields. The first four sections of this chapter discuss two-dimensional triangulations: Delaunay triangulation of point sets (Section 29.1); triangulations of polygons, including constrained Delaunay triangulations (Section 29.2); other optimal triangulations (Section 29.3); and mesh generation (Section 29.4). The last three sections treat triangulations of surfaces embedded in $\mathbb{R}^3$ (Section 29.5), triangulations (composed of tetrahedra) of polyhedra in $\mathbb{R}^3$ (Section 29.6), and triangulations in arbitrary dimension $\mathbb{R}^d$ (Section 29.7).

FIGURE 29.0.1
Triangulations of a point set, a simple polygon, and a polyhedron.

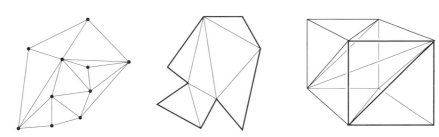

29.1 DELAUNAY TRIANGULATION

The Delaunay triangulation is the most famous and useful triangulation of a point set. Chapter 27 discusses this construction in conjunction with the Voronoi diagram.

GLOSSARY

Empty circle: No input points strictly inside the circle.
Delaunay triangulation (DT):   All triangles have empty circumcircles.
Completion: Adding edges to a polyhedral subdivision to make a triangulation.
Edge flipping: Replacing an edge by a crossing edge; used to compute a DT.

BASIC FACTS

Let $S = \{s_1, s_2, \ldots, s_n\}$ (for "sites") be a set of points in the Euclidean plane $\mathbb{R}^2$. The Delaunay triangulation (DT) is a triangulation of S defined by the **empty circle condition**: a triangle $s_i s_j s_k$ appears in the DT only if its circumscribing circle (**circumcircle**) has no point of S strictly inside it.

The **Delaunay subdivision** is a subdivision of the convex hull of S into polygons with cocircular vertices and empty circumcircles. The Delaunay subdivision is the planar dual of the Voronoi diagram, meaning that an edge $s_i s_j$ appears in the subdivision if and only if the Voronoi cells of s_i and s_j share a boundary edge. If no four points in S are cocircular, the Delaunay subdivision is a triangulation of S. If four or more points in S lie on a common empty circle, the Delaunay subdivision has one or more faces with more than three sides. These can be triangulated to **complete** a Delaunay triangulation of S. The triangulations that complete these faces can be chosen arbitrarily, so the DT is not always unique.

There is a connection between a Delaunay subdivision in $\mathbb{R}^2$ and a convex polytope in $\mathbb{R}^3$. If we **lift** S onto the paraboloid with equation $z = x^2 + y^2$ by mapping $s_i = (x_i, y_i)$ to $(x_i, y_i, x_i^2 + y_i^2)$, then the Delaunay subdivision turns out to be the projection of the lower convex hull of the lifted points. See Figure 27.1.2.

ALGORITHMS

There are a number of practical planar DT algorithms [For95], including edge flipping, incremental construction, plane sweep, and divide and conquer. The last three algorithms can be implemented to run in $O(n \log n)$ time. We describe only the edge flipping algorithm, even though its worst-case running time of $O(n^2)$ is not optimal, because it is most relevant to our subsequent discussion.

The edge flipping algorithm starts from any triangulation of S and then locally optimizes each edge. Let e be an internal edge (not on the boundary of the convex hull of S) and Q_e be the triangulated quadrilateral formed by the two triangles sharing e. Q_e is *reversed* if the two angles opposite the diagonal sum to more than $180°$, or equivalently, if each triangle's circumcircle encloses the opposite vertex. If Q_e is reversed, we "flip" it by exchanging e for the other diagonal of Q_e.

> Compute an initial triangulation of S
> Place all internal edges into a queue
> **while** the queue is not empty **do**
>   Remove an edge e from the queue
>   **if** quadrilateral Q_e is reversed **then**
>   Flip e; add the four outside edges of Q_e to the queue **fi od**

An initial triangulation can be computed by a plane-sweep algorithm that adds the points of S by x-coordinate order (breaking ties by y-coordinate), as shown in Figure 29.1.1. Upon each addition, the algorithm walks around the convex hull of the already-added points, connecting the new vertex to the vertices it can see.

The following theorem guarantees the success of edge flipping: a triangulation in which no quadrilateral is reversed must be a DT. This theorem can be proved with the lifting map: a reversed quadrilateral lifts to a reflex edge, and a surface without reflex edges must be the lower convex hull.

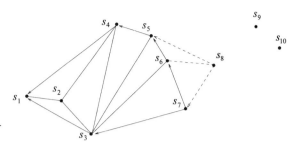

FIGURE 29.1.1
A generic step in computing an initial triangulation.

OPTIMALITY PROPERTIES

Certain measures of quality are improved by flipping a reversed quadrilateral [BE95]. For example, the minimum angle in a triangle of Q_e must increase. Hence, a triangulation that maximizes the minimum angle cannot have a reversed quadrilateral, implying that it is a DT. Among all triangulations of the input points, some DT:

- maximizes the minimum angle (moreover, lexicographically maximizes the list of angles ordered from smallest to largest);
- minimizes the maximum radius of the triangles' circumcircles;
- minimizes the maximum radius of the triangles' smallest enclosing circles;
- maximizes the sum of the radii of the triangles' inscribed circles;
- minimizes the "potential energy" (sum of area-weighted squared gradients) of an interpolated piecewise-linear surface; and
- minimizes the surface area of a piecewise-linear surface for elevations scaled sufficiently small.

Two additional properties of the DT: Delaunay triangles are acyclically ordered by distance from any fixed reference point. The distance between any pair of vertices, measured along edges of the DT, is at most a constant (less than 1.998) times the Euclidean distance between them [Xia13].

WEIGHTED DELAUNAY TRIANGULATIONS

Voronoi diagrams can be defined for various distance measures (Section 27.3), and some of them induce Delaunay-like triangulations by duality. Here we mention one generalization that retains most of the rich mathematical structure. Assign each point $s_i = (x_i, y_i)$ in S a real weight w_i. The **weighted Delaunay triangulation** of S is the projection of the lower convex hull of the points $(x_i, y_i, x_i^2 + y_i^2 - w_i)$. With a small (perhaps negative) weight, a site can fail to appear in the weighted Delaunay triangulation, because the corresponding point in $\mathbb{R}^3$ lies inside the convex hull. Hence, in general the weighted Delaunay triangulation is a graph on a subset of the sites S. In the special case that the weights are all zero (or all equal), the weighted Delaunay triangulation is the DT.

A **regular triangulation** is a triangulation in $\mathbb{R}^2$ found as a projection of the lower surface of a polytope in $\mathbb{R}^3$. Not all triangulations are regular; see Section 16.3 for a counterexample. Every regular triangulation can be expressed as a weighted Delaunay triangulation, because the w_i weights are arbitrary. (The problem of finding suitable weights can be expressed as a linear program.)

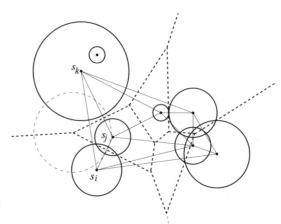

FIGURE 29.1.2
Power diagram (dashed) and weighted Delaunay triangulation. The dashed circle is the orthogonal circle for triangle $s_i s_j s_k$.

The planar dual of the weighted Delaunay triangulation is the **power diagram**, a Voronoi diagram in which the distance from a site $s_i \in S$ to a point in $\mathbb{R}^2$ is the square of the Euclidean distance minus w_i. We can regard the sites in a power diagram as circles, with the radius of site i being $\sqrt{w_i}$. See Figure 29.1.2. In weighted Delaunay triangulations, the analogue of the empty circle condition is the **orthogonal circle condition**: a triangle $s_i s_j s_k$ appears in the triangulation only if the circle that crosses circles i, j, and k at right angles penetrates no other site's circle more deeply.

29.2 TRIANGULATIONS OF POLYGONS

We now discuss triangulations of more complicated inputs: polygons and planar straight-line graphs. We start with the problem of computing any triangulation at all; then we progress to constrained Delaunay triangulations.

GLOSSARY

Simple polygon: Boundary is a loop made of edges without self-intersections.

Monotone polygon: Intersection with any vertical line is one segment.

Constrained Delaunay triangulation: Allows input edges as well as vertices. Triangles have empty circumcircles, meaning no *visible* input vertices.

TRIANGULATIONS OF SIMPLE POLYGONS

Triangulating a simple polygon is both an interesting problem in its own right and an important preprocessing step in other computations. For example, the following problems are known to be solvable in linear time once the input polygon P is triangulated: computing link distances from a given source, finding a monotone path within P connecting two given points, and computing the portion of P illuminated by a given line segment.

How much time does it take to triangulate a simple polygon? For practical purposes, one should use either an $O(n \log n)$ deterministic algorithm (such as the one given below for the more general case of planar straight-line graphs) or a slightly faster randomized algorithm (such as one with running time $O(n \log^* n)$ described by Mulmuley [Mul94]).

However, for theoretical purposes, achieving the ultimate running time was for several years an outstanding open problem. After a sequence of interim results, Chazelle [Cha91] devised a linear-time algorithm. Chazelle's algorithm, like previous algorithms, reduces the problem to that of computing the ***horizontal visibility map*** of P—the partition obtained by shooting horizontal rays left and right from each of the vertices. The "up-phase" of this algorithm recursively merges coarse visibility maps for halves of the polygon (polygonal chains); the "down-phase" refines the coarse map into the complete horizontal visibility map.

TRIANGULATIONS OF PLANAR STRAIGHT-LINE GRAPHS

Let G be a planar straight-line graph (PSLG). We describe an $O(n \log n)$ algorithm [PS85] that triangulates G in two stages, called regularization and triangulation. Regularization adds edges to G so that each vertex, except the first and last, has at least one edge extending to the left and one extending to the right. Conceptually, we sweep a vertical line ℓ from left to right across G while maintaining the list of intervals of ℓ between successive edges of G. For each vertical interval I in ℓ, we remember a vertex $v(I)$ visible to all points of I: this vertex is either an endpoint of one of the two edges bounding I or a vertex between these edges, lacking a right edge. When we hit a vertex u with no left edge, we add the edge $\{u, v(I)\}$, where I is the interval containing u, as shown in Figure 29.2.1(a). After the left-to-right sweep, we sweep from right to left, adding right edges to vertices lacking them.

> Start vertical line ℓ to the left of all vertices in G
> **for** each vertex u of G from left to right **do**
> > **if** u has no left edges **and** u isn't the first vertex **then**
> > > Add edge $\{u, v(I)\}$ where I is the interval containing u **fi**
> >
> > Delete u's left edges from interval list
> > Insert u's right edges with $v(I) \leftarrow u$ for each new vertical interval I **od**
> > **if** u has no right edges **then**
> > > Set $v(I) \leftarrow u$ for the interval I that contains u **fi**
> >
> Repeat the steps above for vertices from right to left

After the regularization stage, each bounded face of G is a ***monotone polygon***, meaning that every vertical line intersects the face in at most one interval. We consider the vertices $u_1, u_2, \ldots, u_n$ of a face in left-to-right order, using a stack to store the not-yet-triangulated vertices (a reflex chain) to the left of the current vertex u_i. If u_i is adjacent to u_{i-1}, the topmost vertex on the stack, as shown in the upper illustration of Figure 29.2.1(b), then we pop vertices off the stack and add diagonals from these vertices to u_i, until the vertices on the stack—u_i on top—again form a reflex chain. If u_i is instead adjacent to the leftmost vertex on the stack, as shown in the lower picture, then we can add a diagonal from each vertex on the stack, and clear the stack of all vertices except u_i and u_{i-1}.

FIGURE 29.2.1
(a) *Sweep-line algorithm for regularization.* (b) *Stack-based triangulation algorithm.*

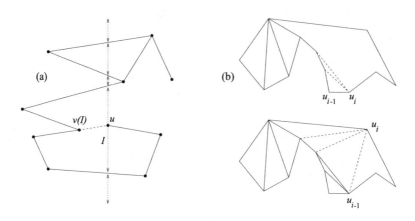

CONSTRAINED DELAUNAY TRIANGULATIONS

The constrained Delaunay triangulation [LL86] provides a way to force the edges of a planar straight-line graph G into the DT. A point p is **visible** to point q if line segment pq does not intersect any edge or vertex in G, except maybe at its endpoints. A triangle abc with vertices from G appears in the **constrained Delaunay triangulation** (CDT) if its circumcircle encloses no vertex of G visible to some point in the interior of abc, and moreover, no edge of G intersects the interior of abc. If G is a graph with vertices but not edges, then the CDT is the ordinary, unconstrained Delaunay triangulation. If G is a polygon or polygon with holes, as in Figure 29.2.2(b), then the CDT retains only the triangles interior to G.

FIGURE 29.2.2
Constrained Delaunay triangulations of (a) *a PSLG and* (b) *a polygon with a hole.*

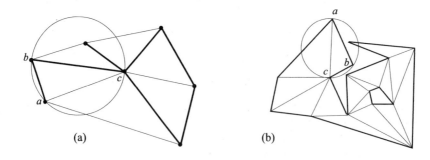

The edge flipping algorithm works for CDTs, with the modification that edges of G are never placed on the queue or flipped. There are also $O(n \log n)$-time algorithms for the CDT [Sei88, Che89], and even $O(n)$-time algorithms for the case that G is just a simple polygon [KL93, CW98]. A slightly slower incremental edge insertion algorithm is usually used in practice [SB15]. See Section 67.2 for pointers to software for computing the constrained Delaunay triangulation.

29.3 OPTIMAL TRIANGULATIONS

We have already seen two types of optimal triangulations: the DT and the CDT. Some applications, however, demand triangulations with properties other than those optimized by DTs and CDTs. Table 29.3.1 summarizes some results.

GLOSSARY

Edge insertion: Local improvement operation, more general than edge flipping.

Local optimum: A solution that cannot be improved by local moves.

Greedy triangulation: Repeatedly add the shortest non-crossing edge.

Steiner triangulation: Extra vertices, not in the input, are allowed.

TABLE 29.3.1   Optimal triangulation results.   (Constrained) Delaunay triangulations maximize the minimum angle and optimize many other objectives.

PROPERTY	INPUTS	ALGORITHMS	TIME
various properties	polygons	dynamic programming [Kli80]	$O(n^3)$
constrained Delaunay	polygons	divide and conquer [KL93, CW98]	$O(n)$
minimize total edge length	polygons	approximation algorithms [Epp94, LK98]	$O(n \log n)$
Delaunay	point sets	various algorithms [For95]	$O(n \log n)$
minimize maximum angle	point sets	fast edge insertion [ETW92]	$O(n^2 \log n)$
minmax slope terrain	point sets	edge insertion [BEE$^+$93]	$O(n^3)$
minmax edge length	point sets	MST induces polygons [ET91]	$O(n^2)$
greedy edge length	point sets	dynamic Voronoi diagram [LL92]	$O(n^2)$
constrained Delaunay	PSLGs	various algorithms [Sei88, Che89, SB15]	$O(n \log n)$

OPTIMAL TRIANGULATIONS OF SIMPLE POLYGONS

Many problems in finding an optimal triangulation of a simple polygon can be solved in $O(n^3)$ time by a dynamic programming algorithm of Klincsek [Kli80]. For example, the algorithm solves any problem that assigns a weight to each possible triangle and/or edge and asks to find the triangulation that minimizes or maximizes the minimum or maximum weight or the sum of the weights. Examples include the minimum/maximum angle in the triangulation, the minimum/maximum length of an edge in the triangulation, and the sum of edge lengths in the triangulation. The triangulation that minimizes the sum of edge lengths is called the *minimum weight triangulation* and is discussed further below.

The running time can be improved for some nonconvex polygons. Let p be the number of pairs of vertices of the polygon that can "see" each other; that is, the line segment connecting them is in the polygon. The optimal triangulation can be found in $O(n^2 + p^{3/2})$ time by first computing a *visibility graph* [BE95].

EDGE FLIPPING AND EDGE INSERTION

The edge flipping DT algorithm can be modified to compute other locally optimal triangulations of point sets. For example, if we redefine "reversed" to mean a quadrilateral triangulated with the diagonal that forms the larger maximum angle, then edge flipping can be used to minimize the maximum angle. For the minmax angle criterion, however, edge flipping computes only a local optimum, not necessarily the true global optimum.

Although edge flipping seems to work well in practice [ETW92], its theoretical guarantees are very weak: the running time is not known to be polynomially bounded and the local optimum it finds may be greatly inferior to the true optimum.

A more general local improvement method, called **edge insertion** [BEE⁺93, ETW92] exactly solves certain minmax optimization problems, including minmax angle and minmax slope of a piecewise-linear interpolating surface.

Assume that the input is a planar straight-line graph G, and we are trying to minimize the maximum angle. Starting from some initial triangulation of G, edge insertion repeatedly adds a candidate edge e that subdivides the maximum angle. (In general, edge insertion always breaks up a worst triangle by adding an edge incident to its "worst vertex.") The algorithm then removes the edges that are crossed by e, forming two polygonal holes alongside e. Holes are retriangulated by repeatedly removing **ears** (triangles with two sides on the boundary, as shown in Figure 29.3.1) with maximum angle smaller than the old worst angle $\angle cab$. If retriangulation succeeds, then the overall triangulation improves and edge bc is eliminated as a future candidate. If retriangulation fails, then the overall triangulation is returned to its state before the insertion of e, and e is eliminated as a future candidate. Each candidate insertion takes time $O(n)$, giving a total running time of $O(n^3)$.

> Compute an initial triangulation with all $\binom{n}{2}$ edge slots unmarked
> **while** $\exists$ an unmarked edge e cutting the worst vertex of worst triangle abc **do**
>   Add e and remove all edges crossed by e
>   Try to retriangulate the polygonal holes by removing ears better than abc
>   **if** retriangulation succeeds **then**
>   mark bc
>   **else** mark e and undo e's insertion **fi od**

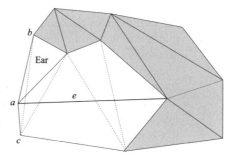

FIGURE 29.3.1

The maximum angle $\angle cab$ is subdivided by the insertion of an edge e, causing the deletion of the dashed edges. The algorithm retriangulates the two polygonal holes alongside e by removing sufficiently good ears. (From [BE95], with permission.)

Edge insertion can compute the minmax "eccentricity" triangulation or the minmax slope surface [BEE+93] in time $O(n^3)$. By inserting candidate edges in a certain order, one can improve the running time to $O(n^2 \log n)$ for minmax angle [ETW92] and maxmin triangle height.

MINIMUM WEIGHT TRIANGULATIONS

Several natural optimization criteria can be defined using edge lengths [BE95]. The most famous such criterion—called ***minimum weight triangulation***—asks for a triangulation of a planar point set minimizing the total edge length. We have already seen that for simple polygons it can be computed in $O(n^3)$ time, but the dynamic programming algorithm does not apply to point sets. Since the problem was posed in 1970, the suspicion that it is NP-hard grew but was only confirmed in 2006, by Mulzer and Rote [MR08]. (It is still not known whether the problem is in NP, because of technicalities related to computer arithmetic with radicals: it is not known how to compare sums of Euclidean lengths in polynomial time.)

However, there is an algorithm that is quite fast in practice for most point sets [DKM97]. This algorithm uses a local criterion to find edges sure to be in the minimum weight triangulation. These edges break the convex hull of the point set into regions, such as simple polygons or polygons with one or two disconnected interior points, that can be triangulated optimally using dynamic programming.

The best polynomial-time approximation algorithm for minimum weight triangulation, by Levcopoulos and Krznaric [LK98], gives a solution within a constant multiplicative factor of the optimal length. Remy and Steger [RS09] give an approximation scheme that finds a $(1 + \epsilon)$-approximation for any fixed $\epsilon > 0$, but runs in quasi-polynomial $n^{O(\log^8 n)}$ time. Eppstein [Epp94] gives a constant-factor approximation ratio for minimum weight Steiner triangulation, in which extra vertices are allowed.

A commonly used heuristic for minimum weight triangulation is ***greedy triangulation***. This algorithm adds edges one at a time, each time choosing the shortest edge that is not already crossed. The greedy triangulation can be viewed as an optimal triangulation in its own right, because it lexicographically minimizes the sorted vector of edge lengths. For planar point sets, the greedy triangulation approximates the minimum weight triangulation by a factor of $O(\sqrt{n})$, and it can be computed in time $O(n^2)$ by dynamic maintenance of a bounded Voronoi diagram [LL92].

Another natural criterion asks for a triangulation minimizing the maximum edge length. Edelsbrunner and Tan [ET91] show that such a triangulation—like the DT—must contain the edges of the minimum spanning tree (MST). The MST, together with the edges of the convex hull, divides the polygon into regions that are weakly-simple polygons (each interior is a topological disk, but an edge might appear multiple times on the boundary). This geometric lemma gives the following $O(n^2)$-time algorithm: compute the MST, then triangulate the resulting weakly-simple polygons optimally using dynamic programming.

OPEN PROBLEMS

1. Explain the empirical success and limits of edge flipping for non-Delaunay optimization criteria—both solution quality and running time.
2. Can the minimum weight triangulation of a convex polygon be computed in $o(n^3)$ time?
3. Find a polynomial-time approximation scheme for the minimum weight triangulation.
4. Show that the minimum weight Steiner triangulation exists; that is, rule out the possibility that more and more Steiner points decrease the total edge length forever.

29.4 PLANAR MESH GENERATION

A *mesh* is a decomposition of a geometric domain into *elements*, usually triangles or quadrilaterals in $\mathbb{R}^2$. (For brevity, we ignore a large literature on quadrilateral mesh generation.) Meshes are used to discretize functions, especially solutions to partial differential equations. Piecewise linear discretizations are by far the most popular. Practical mesh generation problems tend to be application-specific: one desires small elements where a function changes rapidly and larger elements elsewhere. However, certain goals apply fairly generally, and computational geometers have formulated problems incorporating these considerations. Table 29.4.1 summarizes these results; below we discuss some of them in detail.

GLOSSARY

Steiner point: An added vertex that is not an input point.

Conforming mesh: Elements exactly cover the input domain.

Quadtree: A recursive subdivision of a bounding square into smaller squares.

TABLE 29.4.1   Some mesh generation results.

PROPERTY	INPUTS	ALGORITHMS	SIZE
no small or obtuse angles	polygons	grid [BGR88], quadtree [BEG94]	$O(1) \cdot$ optimal
no obtuse angles	polygons	disk packing [BMR94]	$O(n)$
no small angles	most PSLGs	Delaunay [Rup95]	$O(1) \cdot$ optimal
no obtuse angles	PSLGs	disk packing & propagation [Bis16]	$O(n^{2.5})$
no large angles	PSLGs	propagating horns [Mit93, Tan94, Bis16]	$O(n^2)$
no extreme dihedral angles	polyhedra	octree [MV92]	$O(1) \cdot$ optimal
no extreme dihedral angles	smooth 3D	octree [LS07]	no bound

NO SMALL ANGLES

Sharp triangles can degrade appearance and accuracy, so most mesh generation methods attempt to avoid small and large angles. (There are exceptions: properly aligned sharp triangles prove quite useful in simulations of fluid flow.) In finite element methods, elements with small angles sometimes cause the associated stiffness matrices to be badly conditioned, and elements with large angles can lead to large discretization errors.

Baker et al. [BGR88] give a grid-based algorithm for triangulating a polygon so the mesh has no obtuse angles and all the *new* angles—a sharp angle in the input cannot be erased—measure at least 14°. Bern et al. [BEG94] use quadtrees instead of a uniform grid and prove the following *size optimality* guarantee: the number of triangles is $O(1)$ times the minimum number in any no-small-angle triangulation of the input. The minimum number of triangles required depends not just on the number of input vertices n, but also on the geometry of the input. The simple example where the input is a long skinny rectangle shows why the number of output triangles depends upon the geometry.

Ruppert [Rup95], building on work of Chew, devised a *Delaunay refinement* algorithm with the same guarantee. The main loop of Ruppert's algorithm attempts to add the circumcenter (the center of the circumcircle) of a too-sharp triangle as a new vertex. If the circumcenter "encroaches" upon a boundary edge, meaning that it falls within the boundary edge's diameter circle, then the algorithm subdivides the boundary edge instead of adding the triangle circumcenter. Ruppert's algorithm accepts PSLGs as input, not merely polygons, but it is guaranteed to work only if the input PSLG has no acute angles. For PSLG inputs with small angles, unlike for polygon inputs, it is impossible to devise a mesh generation algorithm guaranteed to create no *new* small angles.

The size optimality guarantees for Delaunay refinement algorithms follow from a stronger guarantee: at each mesh vertex v, every adjoining mesh edge has length within a constant factor of the "local feature size" at v, which is a local measure of the distance between PSLG features (see also Chapter 35).

NO LARGE ANGLES

A lower bound on each triangle's smallest angle implies an upper bound on its largest angle. A weaker restriction is to prohibit large angles (close to 180°) only. The strictest bound on large angles that does not also imply a bound on small angles is to ask for no obtuse angles, that is, all angles are at most 90°. Surprisingly, it is possible to triangulate any polygon (possibly with holes) with only $O(n)$ nonobtuse triangles. Figure 29.4.1 illustrates an algorithm of Bern, Mitchell, and Ruppert [BMR94]: the domain is packed with nonoverlapping disks until each uncovered region has either 3 or 4 sides; radii to tangencies are added to split the domain into small polygons; and finally these polygons are triangulated with right triangles.

The problem is substantially harder for PSLGs than for polygons. There are PSLGs for which every nonobtuse triangulation has $\Omega(n^2)$ triangles. An algorithm of Bishop [Bis16] constructs a nonobtuse triangulation of a PSLG with $O(n^{2.5})$ triangles. Closing the gap between the $\Omega(n^2)$ lower bound and the $O(n^{2.5})$ upper

FIGURE 29.4.1
Nonobtuse triangulation steps. (From [BMR94], [BE95], *with permission.)*

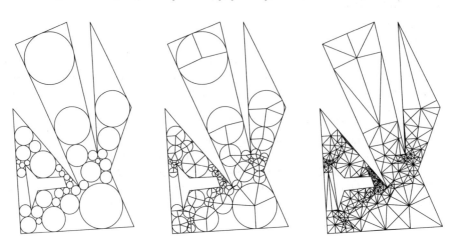

bound is an open problem.

By relaxing the bound on the largest angle from 90° to something larger, one can improve the worst-case complexity of the mesh. Mitchell [Mit93] gives an algorithm that uses $O(n^2 \log n)$ triangles to guarantee that all angles measure less than 157.5°. The algorithm traces a cone of possible angle-breaking edges, called a *horn*, from each vertex (including Steiner points introduced on input edges) with a larger angle. Horns propagate around the PSLG until meeting an exterior edge or another horn. By adding some more horn-stopping "traps," Tan [Tan94] improves the angle bound to 132° and the complexity bound to $O(n^2)$. This complexity bound is tight; there are PSLGs for which a smaller complexity is not possible. Bishop [Bis16] combines the disk-packing algorithm of Bern et al. with propagation of Steiner points, yielding an angle bound of $90° + \epsilon$ and a complexity bound of $O(n^2/\epsilon^2)$ for any fixed $\epsilon > 0$.

CONFORMING DELAUNAY TRIANGULATIONS

For some applications, it suffices to have a mesh that is Delaunay and respects the input edges. A ***conforming Delaunay triangulation*** of a PSLG is a triangulation in which extra vertices—***Steiner points***—are added to the input, until the Delaunay triangulation of the vertices "conforms" to the input, meaning that each input edge is a union of Delaunay edges.

It is easy to verify that every nonobtuse triangulation is Delaunay. The nonobtuse triangulation algorithms discussed above currently offer the best size complexity of any known conforming Delaunay triangulation algorithms—namely, $O(n)$ for polygon inputs and $O(n^{2.5})$ for PSLG inputs. In principle, finding a conforming Delaunay triangulation of a PSLG might be easier than finding a nonobtuse triangulation, but the $\Omega(n^2)$ lower bound on triangulation complexity applies to conforming Delaunay triangulations as well (for worst-case PSLGs).

OPEN PROBLEMS

1. Does every PSLG have a conforming Delaunay triangulation of size $O(n^2)$? Or, more strongly, a nonobtuse triangulation of size $O(n^2)$?

2. Can the algorithms for triangulations with no large (or obtuse) angles be generalized to inputs with curved boundaries?

29.5 SURFACE MESHES

Sitting between two- and three-dimensional triangulations are triangulated surface meshes, which typically enclose 3D solids. Surface meshes are used heavily in computer graphics and in applications such as boundary element methods.

RESTRICTED DELAUNAY TRIANGULATIONS

Many ideas in guaranteed-quality mesh generation extend to surface meshes with the help of the ***restricted Delaunay triangulation*** (RDT) of Edelsbrunner and Shah [ES97], a geometric structure that extends the Delaunay triangulation to curved surfaces. The RDT is always a subcomplex of the three-dimensional Delaunay triangulation. It has proven itself as a mathematically powerful tool for surface meshing and surface reconstruction.

The RDT is defined by dualizing the restricted Voronoi diagram. The ***restricted Voronoi diagram*** of a point set $S \subset \Sigma$ with respect to a surface $\Sigma \subset \mathbb{R}^3$ is a cell complex much like the standard Voronoi diagram, but the restricted Voronoi cells contain only points on the surface Σ. The restricted Delaunay triangulation of S with respect to Σ is the subcomplex of S's 3D Delaunay triangulation containing every Delaunay face whose Voronoi dual face intersects Σ. Typically, the RDT is a subset only of the Delaunay triangles and their edges and vertices; the RDT contains no tetrahedra except in "degenerate" circumstances that can be prevented by infinitesimally perturbing Σ. Another way to characterize the RDT is by observing that its triangles have empty circumscribing spheres whose centers lie on Σ.

If Σ is a smooth surface and the point set S is sampled sufficiently densely from Σ, then the RDT is a triangulation of Σ. In particular, it can be shown that the RDT is homeomorphic to Σ and is a geometrically good approximation of Σ. (Note that if the point set S is not dense enough, the RDT may be a mess that fails to approximate Σ or even to be a manifold.) Partly for these reasons, RDTs have become a standard tool in provably good surface reconstruction and surface meshing algorithms [CDS12]. A surface meshing algorithm of Boissonnat and Oudot [BO05] offers guarantees similar to those of Ruppert's algorithm in the plane. The algorithm can be extended to generate tetrahedral meshes in a volume enclosed by a smooth surface [ORY05]. These algorithms have a simple interaction with the representation of Σ: they require an oracle that, given an arbitrary line segment, returns a point where the line segment intersects Σ. This oracle can be implemented efficiently in practice for many different surface representations.

INTRINSIC DELAUNAY AND SELF-DELAUNAY MESHES

Another way to define a Delaunay-like surface triangulation is to dualize the Voronoi diagram induced by the *intrinsic distance metric* in Σ, defined to be the Euclidean length of the shortest path between each pair of points in Σ, where the paths are restricted to lie on Σ. (This metric is sometimes called the *geodesic distance*, but when Σ is not smooth, the intrinsic shortest path is not necessarily a geodesic.) If we dualize the *intrinsic Voronoi diagram* of a point set $S \subset \Sigma$, we obtain an *intrinsic Delaunay triangulation* in which each dual triangle is *intrinsic Delaunay*, which implies that there is an empty geodesic ball on the surface Σ whose boundary passes through the triangle's vertices. The intrinsic Delaunay triangulation is not always a "proper" triangulation—it might include loop edges or multi-edges.

If Σ is itself a triangulated surface and S is its vertex set, the intrinsic Delaunay triangulation of S might or might not coincide with the triangles that make up Σ (even if all of the triangles of Σ are Delaunay in the usual 3D sense). If they do coincide, so the triangles comprising Σ are all intrinsic Delaunay, we say that Σ is *self-Delaunay*.

Bobenko and Springborn [BS07] show that the intrinsic Delaunay triangulation on a piecewise linear surface Σ in $\mathbb{R}^3$ is unique, and that the corresponding Laplace–Beltrami operator always has positive weights, even if the intrinsic triangulation is not proper; this is useful in geometry processing. Dyer, Zhang and Möller [DZM07] give an algorithm that takes a triangulated surface Σ in $\mathbb{R}^3$ and produces a geometrically similar self-Delaunay mesh, using a combination of flipping and vertex insertion. A recent preprint by Boissonnat, Dyer and Ghosh [BDG13] describes an algorithm that takes a smooth k-dimensional manifold Σ in $\mathbb{R}^d$ as input and produces a self-Delaunay surface triangulation, for any integers $0 < k < d$.

29.6 THREE-DIMENSIONAL POLYHEDRA

In this section we discuss the triangulation (or *tetrahedralization*) of 3D polyhedra. A polyhedron P is a flat-sided solid, usually assumed to be connected and to satisfy the following nondegeneracy condition: around any point on the boundary of P, a sufficiently small ball contains one connected component of each of the interior and exterior of P. With this assumption, the numbers of vertices, edges, and faces (facets) of P are linearly related by Euler's polyhedron formula (adjusted for genus if appropriate).

GLOSSARY

Dihedral angle: The angle separating two polyhedral faces meeting at a shared edge, measured on a plane normal to the shared edge.

Reflex edge: An edge with interior dihedral angle greater than $180°$.

Convex polyhedron: A polyhedron without reflex edges.

General polyhedron: Multiple components, cavities, and handles (higher genus) are permitted.

Circumsphere: The sphere through the vertices of a tetrahedron.

BAD EXAMPLES

Three dimensions is not as nice as two. Triangulations of the same input may contain different numbers of tetrahedra. For example, a triangulation of an n-vertex convex polyhedron may have as few as $n - 3$ or as many as $\binom{n-2}{2}$ tetrahedra. Below et al. [BLR00] proved that finding the minimum number of tetrahedra needed to triangulate (without Steiner points) a convex polyhedron is NP-complete. And when we move to nonconvex polyhedra, we get an even worse surprise: some cannot even be triangulated without Steiner points.

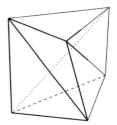

FIGURE 29.6.1
A twisted prism cannot be triangulated without Steiner points.

Schönhardt's polyhedron, shown in Figure 29.6.1, is the simplest example of a polyhedron that cannot be triangulated. Ruppert and Seidel [RS92] prove that it is NP-complete to determine whether a polyhedron can be triangulated without Steiner points, or to test whether k Steiner points suffice.

Chazelle [Cha84] gives an n-vertex polyhedron that requires $\Omega(n^2)$ Steiner points. This polyhedron is a box with thin wedges removed from the top and bottom faces (Figure 29.6.2). The tips of the wedges nearly meet at the hyperbolic surface $z = xy$. From a top view, the wedges appear to subdivide the polyhedron into $\Theta(n^2)$ small squares. We can exhibit $\Theta(n^2)$ points (one in each square) such that no two can see each other within the polyhedron, implying that every subdivision of the polyhedron into convex cells has $\Omega(n^2)$ cells.

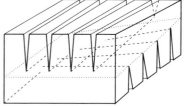

FIGURE 29.6.2
A polyhedron that requires $\Omega(n^2)$ tetrahedra. (From [BE95]*, with permission.)*

TRIANGULATIONS OF POLYHEDRA

Any polyhedron can be triangulated with $O(n^2)$ tetrahedra, matching the lower bound. An algorithm called ***vertical decomposition*** shoots vertical walls up and down from each edge of the polyhedron boundary; walls stop when they reach some other part of the boundary. The tops and bottoms of the resulting "cylinders" are then triangulated to produce $O(n^2)$ triangular prisms, which can each be triangulated with a single interior Steiner point. A better algorithm first plucks off "pointed vertices" with unhindered "caps" [CP90]. Such a vertex, together with its incident faces, forms an empty convex cone. This algorithm uses $O(n + r^2)$ tetrahedra, where r is the number of reflex edges of the original polyhedron.

An alternative algorithm [Cha84] divides the polyhedron into convex solids by incrementally bisecting each reflex angle with a plane that extends away from the reflex angle in all directions until it first contacts the polyhedron boundary. This algorithm produces at most $O(nr + r^{7/3})$ tetrahedra [HS92].

SPECIAL POLYHEDRA

Any n-vertex convex polyhedron can be triangulated with at most $2n - 7$ tetrahedra by "starring" from a vertex. The region between two convex polyhedra (the convex hull of the union, minus the polyhedra), with a total of n vertices, can be triangulated without any Steiner points. If Steiner points are allowed, $O(n)$ tetrahedra suffice. The union of three convex polyhedra can also be tetrahedralized without Steiner points. The region between a convex polyhedron and a terrain can be triangulated with $O(n \log n)$ tetrahedra, and in fact, some such regions require $\Omega(n \log n)$ tetrahedra [CS94].

THREE-DIMENSIONAL MESH GENERATION

Mesh generation for three-dimensional solids is an important practical problem. Current approaches include advancing front methods, Delaunay refinement, octrees (the 3D generalization of quadtrees), bubble meshing, and mesh improvement methods, with none of them being clearly dominant. Some Delaunay and octree methods offer theoretical guarantees. A typical goal is to take a polyhedron input and generate a mesh of tetrahedra that are nicely shaped by some criterion—loosely speaking, they are not "skinny"—in which the number of tetrahedra is small.

Unfortunately, current state-of-the-art algorithms cannot guarantee that all the tetrahedra will have good angles, at least not in theory. (Often it is not difficult to produce good tetrahedra in practice, but there are no guarantees.) Most applications of tetrahedral meshes desire tetrahedra that have no dihedral angle close to 0° or 180° (which implies no solid angle is close to 0° or 360°). Unfortunately, known algorithms for meshing polyhedra can guarantee only very weak bounds on dihedral angles—bounds so weak they are typically not even stated, only proven to exist. Quite a few such algorithms have been published.

Many other algorithms offer no dihedral angle bounds, but instead guarantee that no tetrahedron will have a large *radius-edge ratio*, which is the radius of the tetrahedron's circumscribing sphere divided by the length of its shortest edge. This guarantee rules out most skinny tetrahedron shapes, but *slivers* may remain. A sliver is a tetrahedron whose four vertices lie nearly on a common circle (although the vertices are not exactly coplanar), so that two of its dihedral angles are close to 180° and four are close to 0°. Although bounds on radius-edge ratios may seem disappointing compared to bounds on dihedral angles, they lead to algorithms that are popular in practice because the bounds are tight enough to eliminate most bad tetrahedra and slivers are easy to eliminate in practice, albeit not in theory.

The earliest example of an algorithm offering weak dihedral angle bounds remains important as a theoretical exemplar. Mitchell and Vavasis [MV92] give an octree method that generalizes the quadtree mesh generation algorithm of Bern et al. [BEG94] to polyhedra. In addition to proving that their algorithm generates tetrahedra whose dihedral angles are bounded, Mitchell and Vavasis guarantee that the number of tetrahedra in the mesh is asymptotically "optimal," in the sense that it is within a constant factor of the best possible number for any mesh satisfying the same angle bounds.

Most (but not all) published algorithms that offer bounds on radius-edge ratios use Delaunay triangulations. Shewchuk [She98, CDS12] generalizes Ruppert's 2D Delaunay refinement algorithm to take a polyhedron (with no dihedral angle smaller than 90°) and generate a mesh whose tetrahedra's radius-edge ratios do not exceed 2. Variants of this algorithm have been implemented and found to work well in practice, for instance in Si's software *TetGen* [Si15].

Some Delaunay-based algorithms also offer weak dihedral angle bounds. Slivers can be removed by carefully perturbing the point set, either geometrically [Che97, ELM$^+$00] or by adding small weights to the points and using a weighted Delaunay triangulation, a technique called ***sliver exudation*** [Ede01, LT01, CDS12]. In all algorithms for meshing polyhedra that offer dihedral angle bounds, the bounds are too weak to be reassuring to practitioners, but sliver exudation often performs substantially better in practice than the theory suggests [EG02]. Guaranteed sliver elimination with bounds strong enough to be meaningful in practice remains an important unsolved problem.

There is a special case where it is possible to obtain strong bounds on dihedral angles: when the input domain is not a polyhedron, but a region bounded by a smooth surface. By deforming a body-centered cubic lattice or octree, the ***iso-surface stuffing*** algorithm of Labelle and Shewchuk [LS07] generates a mesh of tetrahedra whose dihedral angles are bounded between 10.7° and 164.8°.

Constrained Delaunay triangulations extend uneasily to $\mathbb{R}^3$, because as Schön-hardt's and Chazelle's polyhedra remind us, not every polyhedron even has a triangulation without Steiner points. Shewchuk [She08] considers adding vertices on the polyhedron's edges. Given a polyhedron X, call a tetrahedron t *constrained Delaunay* if $t \subseteq X$, t's vertices are vertices of X, and t has a circumsphere that encloses no vertex of X visible from any point in the relative interior of t. A CDT of X is a triangulation of X whose tetrahedra are all constrained Delaunay. Call an edge of X *strongly Delaunay* if there exists a closed ball that contains the edge's two vertices but contains no other vertex of X. Shewchuk's *CDT Theorem* states that if every edge of X is strongly Delaunay, then X has a CDT. If a polyhedron does not have a CDT, one can subdivide its edges with new vertices until it has one. Shewchuk and Si [SS14] use the CDT Theorem and this notion of constrained Delaunay triangulation as part of a Delaunay refinement algorithm (implemented in *TetGen*) that is particularly effective for polyhedra that have small input angles.

OPEN PROBLEMS

1. Can the region between k convex polytopes, with n vertices in total, be (Steiner) triangulated with $O(n + k^2)$ tetrahedra?

2. Give an *input-sensitive* tetrahedralization algorithm for polyhedra, for example, one that uses only $O(1)$ times the smallest possible number of tetrahedra.

3. Give a polynomial bound (or even a simple-to-state bound depending upon geometry) on the number of Steiner points needed to make all edges of a polyhedron strongly Delaunay.

4. Give an algorithm for computing tetrahedralizations of point sets or polyhedra, such that each tetrahedron contains its own circumcenter. This condition guarantees a desirable matrix property for a finite-volume formulation of an elliptic partial differential equation [Ber02].

29.7 ARBITRARY DIMENSION

We now discuss triangulation algorithms for arbitrary dimension $\mathbb{R}^d$. In our big-O expressions, we consider the dimension d to be fixed.

GLOSSARY

Polytope: A bounded intersection of halfspaces in $\mathbb{R}^d$.

Face: A subpolytope of any dimension—a vertex, edge, 2D face, 3D face, etc.

Simplex: The convex hull of $d + 1$ affinely independent points in $\mathbb{R}^d$.

Circumsphere: The hypersphere through the vertices of a simplex.

Flip: A local operation, sometimes called a geometric bistellar operation, that exchanges two different triangulations of $d + 2$ points in $\mathbb{R}^d$.

TRIANGULATIONS OF POINT SETS

Delaunay triangulations and weighted Delaunay triangulations generalize to $\mathbb{R}^d$. Every simplex in the DT has a circumsphere (a circumscribing hypersphere) that encloses no input points. The lifting map generalizes as well, so any convex hull algorithm in dimension $d + 1$ can be used to compute d-dimensional DTs. The incremental insertion algorithm also generalizes to $\mathbb{R}^d$; however, the edge flipping algorithm does not. Flips themselves do generalize to $\mathbb{R}^d$: a flip in $\mathbb{R}^d$ replaces a triangulation of $d + 2$ points in convex position with another triangulation of the same points. For example, 5 points in convex position in $\mathbb{R}^3$ can be triangulated by two tetrahedra sharing a face or by three tetrahedra sharing an edge. Unfortunately, the natural generalization of the flip algorithm, which starts from an arbitrary triangulation in $\mathbb{R}^3$ and performs appropriate flips, can get stuck before reaching the DT [Joe89]. The most popular algorithm for computing DTs in three or more dimensions is randomized incremental insertion.

The lifting map can be used to show (from the Upper Bound Theorem for Polytopes) that an n-vertex DT in $\mathbb{R}^d$ contains at most $O(n^{\lceil d/2 \rceil})$ simplices. For practical applications such as interpolation, surface reconstruction, and mesh generation, however, the DT rarely attains its worst-case complexity. The DT of random points within a volume or on a convex surface in $\mathbb{R}^3$ has linear expected complexity, but on a nonconvex surface can have near-quadratic complexity [Eri03]. DT complexity can also be bounded by geometric parameters such as the ratio between longest and shortest pairwise distances [Eri03].

Most 2D DT optimality properties do not generalize to higher dimensions. One exception: the DT minimizes the maximum radius of the simplices' smallest enclosing spheres. The **smallest enclosing sphere** of a simplex is always either the circumscribing sphere of the simplex or the smallest circumscribing sphere of some face of the simplex.

Of interest in algebraic geometry as well as computational geometry is the **flip graph** or **triangulation space**, which has a vertex for each distinct triangulation of a point set and an edge for each flip. The flip graph is sometimes defined so that flips that remove or insert input vertices are permitted; for example, a tetrahedron in $\mathbb{R}^3$ can be split into four tetrahedra by the insertion of a vertex in its interior. If the flip graph includes these flips, then the flip graph of the regular triangulations of a point set has the structure of the skeleton of a high-dimensional polytope called the **secondary polytope** [BFS90, GKZ90]. Therefore, the flip graph is connected.

Unfortunately, the flip graph of *all* triangulations of a point set (including nonregular triangulations) is not well understood. Santos [San00] showed that for some point sets in $\mathbb{R}^5$ this flip graph is not connected, even if the points are in convex position (making it moot whether flips that remove or insert input vertices are permitted, as every triangulation includes all the input vertices). Moreover, in $\mathbb{R}^6$ the flip graph may even have an isolated vertex. The question remains open in $\mathbb{R}^3$ and $\mathbb{R}^4$.

The following is known about Steiner triangulations of point sets in $\mathbb{R}^d$. It is always possible to add $O(n)$ Steiner points, so that the DT of the augmented point set has size only $O(n)$. Moreover, there is always a nonobtuse Steiner triangulation containing at most $O(n^{\lceil d/2 \rceil})$ simplices, all of which are **path simplices**: each includes a path of d pairwise orthogonal edges [BCER95].

TRIANGULATIONS OF POLYTOPES

Triangulations of polytopes in $\mathbb{R}^d$ arise in combinatorics and algebra [GKZ90, Sta80]. Several algorithms are known for triangulating the hypercube, but there is a gap between the algorithm that produces the least number of simplices and the best lower bound on the number of simplices [OS03]; see Section 16.7.2. It is known that the region between two convex polytopes—a nonconvex polytope—can always be triangulated without Steiner points [GP88]; see Section 16.3.2. Below et al. [BBLR00] have shown that there can be a significant difference (linear in the number of vertices) in the minimum numbers of simplices in a triangulation and a **dissection** of a convex polytope (of dimension 3 or greater), which is a partition of a polytope into simplices whose faces may meet only partially (for example, a triangle bordering two other triangles along one of its sides).

OPEN PROBLEMS

1. Is the flip graph of the triangulations of a point set or polytope in $\mathbb{R}^3$ or $\mathbb{R}^4$ always connected?

2. What is the maximum number of triangulations of a set of n points in $\mathbb{R}^d$?

For bounds on this maximum in $\mathbb{R}^2$, see Sharir and Sheffer [SS11] for an upper bound and Dumitrescu et al. [DSST13] for a lower bound.

3. Narrow the gap between the upper and lower bounds on the minimum number of simplices in a triangulation of the d-cube.

29.8 SOURCES AND RELATED MATERIAL

SURVEYS

For more complete descriptions and references, consult the following sources.

[Aur91]: Generalizations of the Voronoi diagram and Delaunay triangulation.

[Ber02]: A survey of mesh generation algorithms.

[CDS12]: A book on Delaunay triangulations and Delaunay refinement algorithms for mesh generation.

[DRS10]: A book on the combinatorics of triangulations, especially regular triangulations, in arbitrary dimensions.

[Ede01]: A book on geometry and topology relevant to triangular and tetrahedral mesh generation.

The web is a rich source on mesh generation and triangulation; see Chapter 67.

RELATED CHAPTERS

Chapter 16: Subdivisions and triangulations of polytopes
Chapter 23: Computational topology of graphs on surfaces
Chapter 27: Voronoi diagrams and Delaunay triangulations
Chapter 30: Polygons
Chapter 35: Curve and surface reconstruction

REFERENCES

[Aur91]   F. Aurenhammer. Voronoi diagrams—A survey of a fundamental geometric data structure. *ACM Comput. Surv.*, 23:345–405, 1991.

[BBLR00]   A. Below, U. Brehm, J.A. De Loera, J. Richter-Gebert. Minimal simplicial dissections and triangulations of convex 3-polytopes. *Discrete Comput. Geom.*, 24:35–48, 2000.

[BCER95]   M. Bern, L.P. Chew, D. Eppstein, and J. Ruppert. Dihedral bounds for mesh generation in high dimensions. In *Proc. 6th ACM-SIAM Sympos. Discrete Algorithms*, pages 189–196, 1995.

[BDG13]   J.-D. Boissonnat, R. Dyer, and A. Ghosh. Constructing intrinsic Delaunay triangulations of submanifolds. Technical Report RR-8273, INRIA, Sophia Antipolis, 2013.

[BE95]   M. Bern and D. Eppstein. Mesh generation and optimal triangulation. In D.-Z. Du and F.K. Hwang, editors, *Computing in Euclidean Geometry*, 2nd edition, pages 47–123, World Scientific, Singapore, 1995.

[BEE+93] M. Bern, H. Edelsbrunner, D. Eppstein, S.A. Mitchell, and T.-S. Tan. Edge-insertion for optimal triangulations. *Discrete Comput. Geom.*, 10:47–65, 1993.

[BEG94] M. Bern, D. Eppstein, and J.R. Gilbert. Provably good mesh generation. *J. Comput. Syst. Sci.*, 48:384–409, 1994.

[Ber02] M. Bern. Adaptive mesh generation. In T. Barth and H. Deconinck, editors, *Error Estimation and Adaptive Discretization Methods in Computational Fluid Dynamics*, pages 1–56, Springer-Verlag, Berlin, 2002.

[BFS90] L. Billera, P. Filliman, and B. Sturmfels. Constructions and complexity of secondary polytopes. *Adv. Math.*, 83:155–179, 1990.

[BGR88] B.S. Baker, E. Grosse, and C.S. Rafferty. Nonobtuse triangulation of polygons. *Discrete Comput. Geom.*, 3:147–168, 1988.

[Bis16] C.J. Bishop. Nonobtuse triangulations of PSLGs. *Discrete Comput. Geom.*, 56:43–92, 2016.

[BLR00] A. Below, J.A. De Loera, and J. Richter-Gebert. Finding minimal triangulations of convex 3-polytopes is NP-hard. In *Proc. 11th ACM-SIAM Sympos. Discrete Algorithms*, pages 65–66, 2000.

[BMR94] M. Bern, S.A. Mitchell, and J. Ruppert. Linear-size nonobtuse triangulation of polygons. In *Proc. 10th Sympos. Comput. Geom.*, pages 221–230, ACM Press, 1994.

[BO05] J.-D. Boissonnat and S. Oudot. Provably good sampling and meshing of surfaces. *Graphical Models*, 67:405–451, 2005.

[BS07] A.I. Bobenko and B.A. Springborn. A discrete Laplace–Beltrami operator for simplicial surfaces. *Discrete Comput. Geom.*, 38:740–756, 2007.

[CDS12] S.-W. Cheng, T.K. Dey, and J.R. Shewchuk. *Delaunay Mesh Generation*. CRC Press, Boca Raton, 2012.

[Cha84] B. Chazelle. Convex partitions of polyhedra: A lower bound and worst-case optimal algorithm. *SIAM J. Comput.*, 13:488–507, 1984.

[Cha91] B. Chazelle. Triangulating a simple polygon in linear time. *Discrete Comput. Geom.*, 6:485–524, 1991.

[Che89] L.P. Chew. Constrained Delaunay triangulations. *Algorithmica*, 4:97–108, 1989.

[Che97] L.P. Chew. Guaranteed-quality Delaunay meshing in 3D. In *Proc. 13th Sympos. Comput. Geom.*, pages 391–393, ACM Press, 1997.

[CP90] B. Chazelle and L. Palios. Triangulating a nonconvex polytope. *Discrete Comput. Geom.*, 5:505–526, 1990.

[CS94] B. Chazelle and N. Shouraboura. Bounds on the size of tetrahedralizations. In *Proc. 10th Sympos. Comput. Geom.*, pages 231–239, ACM Press, 1994.

[CW98] F.Y.L. Chin and C.A. Wang. Finding the constrained Delaunay triangulation and constrained Voronoi diagram of a simple polygon in linear time. *SIAM J. Computing*, 28:471–486, 1998.

[DKM97] M.T. Dickerson, J.M. Keil, and M.H. Montague. A large subgraph of the minimum weight triangulation. *Discrete Comput. Geom.*, 18:289–304, 1997.

[DRS10] J.A. De Loera, J. Rambau, and F. Santos. *Triangulations: Structures for Algorithms and Applications*. Springer-Verlag, Berlin, 2010.

[DSST13] A. Dumitrescu, A. Schulz, A. Sheffer, and C. D. Tóth. Bounds on the maximum multiplicity of some common geometric graphs. *SIAM J. Discrete Math.*, 27:802–826, 2013.

[DZM07]   R. Dyer, H. Zhang, and T. Möller. Delaunay mesh construction. In *Proc. 5th Sympos. Geom. Processing*, pages 273–282, Eurographics, 2007.

[Ede01]   H. Edelsbrunner. *Geometry and Topology for Mesh Generation*. Cambridge University Press, 2001.

[EG02]   H. Edelsbrunner and D. Guoy. An experimental study of sliver exudation. *Engineering with Computers*, 18:229–240, 2002.

[ELM⁺00] H. Edelsbrunner, X.-Y. Li, G. Miller, A. Stathopoulos, D. Talmor, S.-H. Teng, A. Üngör, and N.J. Walkington. Smoothing and cleaning up slivers. In *Proc. 32nd ACM Sympos. Theory Comput.*, pages 273–278, 2000.

[Epp94]   D. Eppstein. Approximating the minimum weight triangulation. *Discrete Comput. Geom.*, 11:163–191, 1994.

[Eri03]   J. Erickson. Nice point sets can have nasty Delaunay triangulations. *Discrete Comput. Geom.*, 30:109–132, 2003.

[ES97]   H. Edelsbrunner and N.R. Shah. Triangulating topological spaces. *Internat. J. Comput. Geom. Appl.*, 7:365–378, 1997.

[ET91]   H. Edelsbrunner and T.-S. Tan. A quadratic time algorithm for the minmax length triangulation. In *Proc. 32nd IEEE Sympos. Found. Comput. Sci.*, pages 414–423, 1991.

[ETW92]   H. Edelsbrunner, T.-S. Tan, and R. Waupotitsch. A polynomial time algorithm for the minmax angle triangulation. *SIAM J. Sci. Statist. Comput.*, 13:994–1008, 1992.

[For95]   S.J. Fortune. Voronoi diagrams and Delaunay triangulations. In F.K. Hwang and D.-Z. Du, editors, *Computing in Euclidean Geometry*, 2nd edition, pages 225–265, World Scientific, Singapore, 1995.

[GKZ90]   I.M. Gelfand, M.M. Kapranov, and A.V. Zelevinsky. Newton polytopes of the classical discriminant and resultant. *Adv. Math.*, 84:237–254, 1990.

[GP88]   J.E. Goodman and J. Pach. Cell decomposition of polytopes by bending. *Israel J. Math.*, 64:129–138, 1988.

[HS92]   J. Hershberger and J. Snoeyink. Convex polygons made from few lines and convex decompositions of polyhedra. In *Proc. 3rd Scand. Workshop Algorithm Theory*, vol. 621 of *LNCS*, pages 376–387, Springer-Verlag, Berlin, 1992.

[Joe89]   B. Joe. Three-dimensional triangulations from local transformations. *SIAM J. Sci. Stat. Comput.*, 10:718–741, 1989.

[KL93]   R. Klein and A. Lingas. A linear-time randomized algorithm for the bounded Voronoi diagram of a simple polygon. *Internat. J. Comput. Geom. Appl.*, 6:263–278, 1996.

[Kli80]   G.T. Klincsek. Minimal triangulations of polygonal domains. *Annals Discrete Math.*, 9:121–123, 1980.

[LK98]   C. Levcopoulos and D. Krznaric. Quasi-greedy triangulations approximating the minimum weight triangulation. *J. Algorithms*, 27:303–338, 1998.

[LL86]   D.T. Lee and A.K. Lin. Generalized Delaunay triangulation for planar graphs. *Discrete Comput. Geom.*, 1:201–217, 1986.

[LL92]   C. Levcopoulos and A. Lingas. Fast algorithms for greedy triangulation. *BIT*, 32:280–296, 1992.

[LS07]   F. Labelle and J.R. Shewchuk. Isosurface stuffing: Fast tetrahedral meshes with good dihedral angles. *ACM Trans. Graph.*, 26:57, 2007.

[LT01]   X.-Y. Li and S.-H. Teng. Generating well-shaped Delaunay meshes in 3D. In *Proc. 12th ACM-SIAM Sympos. Discrete Algorithms*, pages 28–37, 2001.

[Mit93]   S.A. Mitchell. Refining a triangulation of a planar straight-line graph to eliminate large angles. In *Proc. 34th IEEE Sympos. Found. Comput. Sci.*, pages 583–591, 1993.

[MR08]   W. Mulzer and G. Rote. Minimum weight triangulation is NP-hard. *J. ACM*, 55:11, 2008.

[Mul94]   K. Mulmuley. *Computational Geometry: An Introduction through Randomized Algorithms*. Prentice-Hall, Englewood Cliffs, 1994.

[MV92]   S.A. Mitchell and S.A. Vavasis Quality mesh generation in three dimensions. In *Proc. 8th Sympos. Comput. Geom.*, pages 212–221, ACM Press, 1992.

[ORY05]   S. Oudot, L. Rineau, and M. Yvinec. Meshing volumes bounded by smooth surfaces. In *Proc. 14th International Meshing Roundtable*, pages 203–219, Springer, Berlin, 2005.

[OS03]   D. Orden and F. Santos. Asymptotically efficient triangulations of the d-cube. *Discrete Comput. Geom.*, 30:509–528, 2003.

[PS85]   F.P. Preparata and M.I. Shamos. *Computational Geometry: An Introduction*. Springer-Verlag, New York, 1985.

[RS92]   J. Ruppert and R. Seidel. On the difficulty of tetrahedralizing 3-dimensional non-convex polyhedra. *Discrete Comput. Geom.*, 7:227–253, 1992.

[RS09]   J. Remy and A. Steger. A quasi-polynomial time approximation scheme for minimum weight triangulation. *J. ACM*, 56:15, 2009.

[Rup95]   J. Ruppert. A Delaunay refinement algorithm for quality 2-dimensional mesh generation. *J. Algorithms*, 18:548–585, 1995.

[San00]   F. Santos. A point set whose space of triangulations is disconnected. *J. Amer. Math. Soc.*, 13:611–637, 2000.

[SB15]   J.R. Shewchuk and B.C. Brown. Fast segment insertion and incremental construction of constrained Delaunay triangulations. *Comput. Geom.*, 48:554–574, 2015.

[Sei88]   R. Seidel. Constrained Delaunay triangulations and Voronoi diagrams with obstacles. In H. S. Poingratz and W. Schinnerl, editors, *1978–1988 Ten Years IIG*, pages 178–191, Institute for Information Processing, Graz University of Technology, 1988.

[She98]   J.R. Shewchuk. Tetrahedral mesh generation by Delaunay refinement. In *Proc. 14th Sympos. Comput. Geom.*, pages 86–95, ACM Press, 1998.

[She08]   J.R. Shewchuk. General-dimensional constrained Delaunay triangulations and constrained regular triangulations, I: Combinatorial properties. *Discrete Comput. Geom.*, 39:580–637, 2008.

[Si15]   H. Si. TetGen, a Delaunay-based quality tetrahedral mesh generator. *ACM Trans. Math. Software*, 41:11, 2015.

[SS11]   M. Sharir and A. Sheffer. Counting triangulations of planar point sets. *Electr. J. Comb.*, 18(1), 2011.

[SS14]   J.R. Shewchuk and H. Si. Higher-quality tetrahedral mesh generation for domains with small angles by constrained Delaunay refinement. In *Proc. 13th Sympos. Comput. Geom.*, pages 290–299, ACM Press, 2014.

[Sta80]   R.P. Stanley. Decompositions of rational convex polytopes. *Annals Discrete Math.*, 6:333–342, 1980.

[Tan94]   T.-S. Tan. An optimal bound for conforming quality triangulations. In *Proc. 10h Sympos. Comput. Geom.*, pages 240–249, ACM Press, 1994.

[Xia13]   G. Xia. The stretch factor of the Delaunay triangulation is less than 1.998. *SIAM J. Comput.*, 42:1620–1659, 2013.

30 POLYGONS

Joseph O'Rourke, Subhash Suri, and Csaba D. Tóth

INTRODUCTION

Polygons are one of the fundamental building blocks in geometric modeling, and they are used to represent a wide variety of shapes and figures in computer graphics, vision, pattern recognition, robotics, and other computational fields. By a polygon we mean a region of the plane enclosed by a simple cycle of straight line segments; a *simple cycle* means that nonadjacent segments do not intersect and two adjacent segments intersect only at their common endpoint. This chapter describes a collection of results on polygons with both combinatorial and algorithmic flavors. After classifying polygons in the opening section, Section 30.2 looks at simple polygonizations, Section 30.3 covers polygon decomposition, and Section 30.4 polygon intersection. Sections 30.5 addresses polygon containment problems and Section 30.6 touches upon a few miscellaneous problems and results.

30.1 POLYGON CLASSIFICATION

Polygons can be classified in several different ways depending on their domain of application. In chip-masking applications, for instance, the most commonly used polygons have their sides parallel to the coordinate axes.

GLOSSARY

Simple polygon: A closed region of the plane enclosed by a simple cycle of straight line segments.

Convex polygon: The line segment joining any two points of the polygon lies within the polygon.

Monotone polygon: Any line orthogonal to the direction of monotonicity intersects the polygon in a single connected piece.

Star-shaped polygon: The entire polygon is visible from some point inside the polygon.

Orthogonal polygon: A polygon with sides parallel to the (orthogonal) coordinate axes. Sometimes called a *rectilinear polygon*.

Orthogonally convex polygon: An orthogonal polygon that is both x- and y-monotone.

Polygonal chain: A sequence of connected, non-self-intersecting line segments forming a subportion of a simple polygon's boundary. A chain is convex or reflex if all internal angles are convex or reflex respectively.

Spiral polygon: A polygon bounded by one convex chain and one reflex chain.

Crescent polygon: A monotone spiral polygon.

Pseudotriangle: A polygon with exactly three convex angles. Each is pair of convex vertices is connected either by a single segment or a reflex chain.

Histogram polygon: An orthogonal polygon bounded by an x-monotone polygonal chain and a single horizontal line segment.

POLYGON TYPES

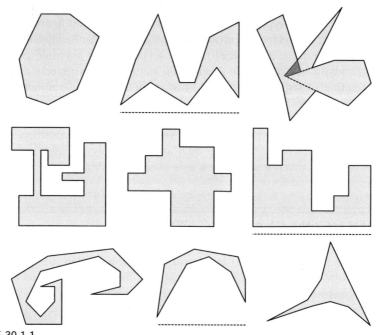

FIGURE 30.1.1

Various varieties of polygons: convex, monotone, star-shaped (with kernel), orthogonal, orthogonally convex, histogram, spiral, crescent, and pseudotriangle.

Before starting our discussion on problems and results concerning polygons, we clarify a few technical issues. The qualifier "simple" in the definition of a simple polygon states a *topological* property, meaning "nonself-intersection." Not to be confused with "uncomplicated polygons," in fact, these polygons include the most complex among polygons that are topologically equivalent to a disk (see the classification below). Finally, we will make a standard ***general position*** assumption throughout this chapter that no three vertices of a polygon are collinear.

   The relationship between several classes of polygons can be understood using the concept of visibility (see Chapter 33). We say that two points x and y in a polygon P are mutually ***visible*** if the line segment $\overline{xy}$ does not intersect the complement of P; thus the segment $\overline{xy}$ is allowed to graze the polygon boundary but not cross it. We call a set of points $K \subset P$ the ***kernel*** of P if all points of P are visible from every point in the kernel. Then, a polygon P is convex if $K = P$; the polygon is star-shaped if $K \neq \emptyset$ (see the star-shaped polygon in Fig. 30.1.1);

FIGURE 30.1.2
A polygon with five holes.

otherwise, the polygon is merely a simple polygon. Speaking somewhat loosely, a monotone polygon can be viewed as a special case of a star-shaped polygon with the exterior kernel at infinity—that is, a monotone polygon can be decomposed into two polygonal chains, each of which is entirely visible from the (same) point at infinity in the extended plane. A pseudotriangle is often but not always star-shaped.

The kernel has been generalized to "left-" and "right-kernels," whose definition we leave to [TTW11].

By definition, a simple polygon P is a polygon *without holes*—that is, the interior of the polygon is topologically equivalent to a disk. A ***polygon with holes*** is a higher-genus variant of a simple polygon, obtained by removing a nonoverlapping set of strictly interior, simple subpolygons from P. Figure 30.1.2 illustrates the distinction between a simple polygon and a polygon with holes.

An important class of polygons are the *orthogonal polygons*, where all edges are parallel to the coordinate axes. These polygons arise quite naturally in industrial applications, and often algorithms are faster on these more structured polygons.

30.2 SIMPLE POLYGONIZATIONS

It would be useful to have a clear notion of a "random polygon" so that algorithms could be tested for typical rather than worst-case behavior. This leads to the issue of generating the ***simple polygonalizations*** of a fixed point set S, a simple polygon whose vertices are precisely the points of S. That every set S of n points in general position has a simple polygonization has been known since Steinhaus in 1964. In fact, every such S has a star-shaped polygonization (Graham), a monotone polygonization (Grünbaum), and a spiral polygonization. See [IM11] for the latter result and a survey of earlier work.

The "space" of all polygonizations of a fixed point set S can be explored through two elementary moves, which is one approach to generating random polygons: generate a random S, then walk between polygonizations to a "random" polygonization in the space [DFOR10]. The size of this space can be exponential.

Considerable research has focused on quantifying this exponential, that is, counting the maximum number of polygonizations over all n-point sets S. (Of course, the minimum is 1 because vertices of a convex polygon can form only one simple polygon.) Current best bounds for the maximum are $\Omega(4.64^n)$ [GNT00] and $O(54.55^n)$ [SSW13].

Subsets of the full space have been explored, in particular, finding polygonizations with special properties. The shortest perimeter polygonization is the Euclidean TSP; see Chapter 31. There is a 2π approximation (under certain conditions) for the longest perimeter polygonization [DT10]. Finding the minimum or maximum area polygonization is NP-hard [Fek00]. Sufficient conditions have been found for the existence of a polygonization of n red points while enclosing [HMO+09] or excluding [FKMU13] a set of blue points.

Another goal has been to minimize the number of reflex vertices in a polygonization of a point set S. This minimum is known as the *reflexivity* of S. A tight upper bound is known as a function of n_I, the number of points of S strictly interior to the convex hull: the worst case is $\lceil n_I/2 \rceil$, and this can be achieved by a class of examples [AFH+03]. As a function of $n = |S|$, the best bound is $\frac{5}{12}n + O(1) \approx 0.4167n$ [AAK09]. Another criterion in the same spirit is to minimize the sum of the "turn angles" at reflex interior angles [Ror14]. The worst case is conjectured to be $2\pi(1 - 1/(n-1))$, but only $2\pi - \pi/((n-1)(n-3))$ is proved.

OPEN PROBLEMS

1. *Simple polygonalization*: Can the number of simple polygonalizations of a set of n points in the plane be computed in polynomial time?

2. *Partial polygons*: Is there a polynomial time algorithm that can decide, for a set of n points and a set of edges among them, whether there is a polygonization that uses all the given edges?

30.3 POLYGON DECOMPOSITION

Many computational geometry algorithms that operate on polygons first decompose them into more elementary pieces, such as triangles or quadrilaterals. There is a substantial body of literature in computational geometry on this subject.

GLOSSARY

Steiner point: A vertex not part of the input set.

Diagonal: A line segment connecting two polygon nonadjacent vertices and contained in the polygon. An **edge** connects adjacent vertices.

Polygon cover: A collection of subpolygons whose union is exactly the input polygon.

Polygon partition: A collection of subpolygons with *pairwise disjoint* interiors whose union is exactly the input polygon.

Dissection: A dissection of one polygon P to another Q is a partition of P into a finite number of pieces that may be reassembled to form Q.

Decompositions may be classified along two primary dimensions: covers or partitions, and with or without Steiner points. A cover permits a polygon in the shape of the symbol "+" to be represented as the union of two rectangles, whereas a minimal partition requires three rectangles, a less natural decomposition. Decompositions without Steiner points use diagonals, and are in general easier to find but less parsimonious. For each of the four types of decomposition, different primitives may be considered. The ones most commonly used are rectangles, convex polygons, star-shaped polygons, spiral polygons, and trapezoids. Restrictions on the shape of the piece being decomposed are often available; for example, orthogonal polygons for rectangle covers. Lastly, the distinction between simple polygons and polygons with holes is often relevant for algorithms. The most celebrated polygon partition problem is the "polygon triangulation problem."

TRIANGULATION ALGORITHMS

The polygon triangulation problem is to dissect a polygon into triangles by drawing a maximal number of noncrossing diagonals. Only the vertices of the polygon are used as triangle vertices, and no additional interior (Steiner) points are allowed. It is an easy and well-known result that every simple polygon can be triangulated, and that the number of triangles is invariant over all triangulations. More precisely:

THEOREM 30.3.1

Every simple polygon admits a triangulation, and every triangulation of an n-vertex polygon has $n - 3$ diagonals and $n - 2$ triangles.

(But note that the natural generalization to $\mathbb{R}^3$—that every polyhedron admits a tetrahedralization—is false: see Chapter 29.) The number of possible diagonals in a polygon may vary from linear (e.g., a spiral polygon) to quadratic (e.g., a convex polygon). A diagonal that breaks the polygon into two roughly equal halves is called a ***balanced*** diagonal. In designing his $O(n \log n)$ time algorithm for triangulating a polygon, Chazelle [Cha82] proved the following fact, which has found numerous applications in divide-and-conquer based algorithms for polygons, e.g., to ray-shooting [HS95]:

THEOREM 30.3.2

Every n-vertex simple polygon admits a diagonal that breaks the polygon into two subpolygons, neither one with more than $\lceil 2n/3 \rceil + 1$ vertices.

By recursively dividing the polygon using balanced diagonals, we get a balanced decomposition of P, which can be modeled by a tree of height $O(\log n)$. The existence of a balanced diagonal follows easily once we consider the graph-theoretic dual of a triangulation. This dual graph of a polygon triangulation is a tree, with maximum node degree three. Diagonals of the triangulation correspond to the edges of the dual tree, and thus a balanced diagonal corresponds to an edge whose removal breaks the tree into two subtrees, each with at most $\lceil 2n/3 \rceil + 1$ nodes.

The problem of computing a triangulation of a polygon has had a long and distinguished history [O'R87], culminating in Chazelle's linear-time algorithm [Cha91]. Table 30.3.1 lists some of the best-known algorithms for this problem. The algorithm in [Sei91] is a randomized Las Vegas algorithm (see Chapter 44). All others are deterministic algorithms, with worst-case time bounds as shown.

TABLE 30.3.1   Results on triangulating a simple polygon.

TIME COMPLEXITY	ALGORITHM	SOURCE
$O(n \log n)$	monotone pieces	[GJPT78]
$O(n \log n)$	divide-and-conquer	[Cha82]
$O(n \log n)$	plane sweep	[HM85]
$O(n \log^\star n)$	randomized	[Sei91]
$O(n)$	polygon cutting	[Cha91]

Chazelle's deterministic linear-time algorithm is formidably complex, but has led to a simpler randomized algorithm that runs in linear expected time [AGR01].

If the polygon contains holes (Figure 30.1.2), then $\Theta(n \log n)$ time is both necessary and sufficient for triangulating the region [HM85]. See Table 30.3.2.

TABLE 30.3.2   Results on triangulating a polygon with holes.

TIME COMPLEXITY	ALGORITHM	SOURCES
$O(n \log n)$	plane sweep	[HM85]
$O(n \log n)$	local sweep	[RR94]

COUNTING TRIANGULATIONS

The number of triangulations of a simple polygon with n vertices is at least 1 (every polygon can be triangulated) and at most $C_{n-2} = \frac{1}{n-1}\binom{2n-4}{n-2}$, the $(n-2)$-th Catalan number (for convex polygons). The number of triangulations of a given polygon of n vertices can be determined in $O(n^3)$ time by dynamic programming [ES94]; or in $O(e^{3/2})$ time, where $e = O(n^2)$ is the number of diagonals of the polygon [DFH+99], if the diagonals are given. In $O(n \log n)$ additional time, random triangulations (with uniform distribution) can be generated [DQTW05], improving on an earlier $O(n^4)$ algorithm [DFH+99].

SPECIAL TRIANGULATIONS

Polygon triangulations with either minimum or maximum edge length can be found in $O(n^2)$ time via dynamic programming [Kli80].

Pseudotriangulations have many applications, for example, to motion planning, and to kinetic data structures. Every simple polygon admits a pseudotriangulation in which the maximum vertex-degree is at most 5, and this bound is the best possible [AHST03]. However, the dual graph of a minimum pseudotriangulation may have degree $\Omega(\log n)$.

COVERS AND PARTITIONS

The problem of decomposing polygons into different types of simpler polygons has numerous applications within and outside computational geometry (see, e.g., Chapter 57). Unlike the triangulation problem, most variants of the covering and partitioning problems turn out to be provably hard. In a covering problem, the goal is to cover the interior of the polygon with the smallest number of subpolygons of a particular type, for instance, convex or star-shaped polygons. Table 30.3.3 lists results for various polygon covering problems. In this table, "cover type" refers to the family of polygons allowed in the cover, while "domain" refers to the polygonal region that needs to be covered. For the most part, we consider only four types of domains: simple polygons, with and without holes, and orthogonal polygons, with and without (orthogonal) holes. In all of these problems, the cover or partition pieces are allowed to use Steiner points for their vertices. Almost all variations of the covering problem are intractable. Even the minimum convex cover problem without Steiner points is NP-complete [Chr11].

TABLE 30.3.3   Results on polygon covering problems.

COVER TYPE	DOMAIN	HOLES	COMPLEXITY	SOURCE
Rectangles	orthogonal	Y	NP-complete	[Mas78]
Convex–star	polygons	Y	NP-hard	[OS83]
Star	polygons	N	NP-hard	[Agg84]
			APX-hard	[ESW01]
Rectangles	orthogonal	N	NP-hard	[CR94]
Squares	orthogonal	N	$O(n^{3/2})$	[ACKO88]
		Y	NP-complete	[ACKO88]
Convex	polygons	N	NP-hard	[CR94]
			APX-hard	[EW03]

The polygon-partitioning problems are similar to the covering problem, except that the tessellating pieces are not allowed to overlap: they have pairwise disjoint interiors. Table 30.3.4 collects results on polygon partitioning problems permitting Steiner points. Polynomial-time algorithms can be achieved for simple polygons using dynamic programming. The same problems, however, turn out to be intractable when the polygon has holes. Disallowing Steiner points also leads to polynomial-time algorithms. For example, partitioning a polygon without holes into the fewest convex pieces, not employing Steiner points, is achievable in $O(n^3 \log n)$ time [Kei85, KS02].

CONVEX COVER APPROXIMATIONS

The intractability of most covering and partitioning problems naturally leads to the question of approximability—how well can we approximate the size of an optimal cover or partition in polynomial time. In many cases, there are only a polynomial number of covering candidates—for instance, rectangle covers or convex poly-

TABLE 30.3.4   Results on polygon partitioning problems.

PARTITION	DOMAIN	HOLES	COMPLEXITY	SOURCE
Convex	polygons	N	$O(n^3)$	[CD79]
Convex	polygons	Y	NP-hard	[CD79]
Trapezoids	polygons	N	$O(n^2)$	[Kei85]
Trapezoids	polygons	Y	NP-complete	[AAI86]
Rectangles	orthogonal	N	$O(n^{3/2}\log n)$	[ACKO88]
		Y	NP-complete	[ACKO88]

gon covers. An easy 4-approximation for minimum convex cover has been known for some time: shoot a ray from each reflex vertex, splitting into two convex angles [HM85]. Without Steiner points, balanced geometric separators have led to a quasi-PTAS [BBV15]. Building on [CR94], a polynomial-time approximation algorithm can achieve an approximation ratio of $O(\log n)$ [EW03].

STAR COVERS: ART GALLERY COVERAGE

Star-shaped covers of polygons corresponds to coverage of an "art gallery" by point-guards (Chapter 33). Vertex-guards is a further, useful restriction to place the guards at vertices. Because minimum star cover is APX-hard (Table 30.3.3) for both point- and vertex-guards [ESW01], the focus has been on approximation algorithms [Gho10]. An algorithm that runs in pseudopolynomial time achieves an approximation factor of $O(\log \text{OPT})$ for point or perimeter guards [DKDS07]. The same approximation ratio was achieved via a randomized algorithm that runs in fully polynomial expected-time [EHP06]. Among the strongest results in this direction are an $O(\log \log \text{OPT})$-approximation for vertex guards, running in $O(n^3)$, and an $O(\log h \log \text{OPT})$-approximation that runs in $O(n^3 h^2)$ for a polygon with h holes [Kin13].

FAT PARTITIONS

Because many algorithms work faster on "fat" shapes, partitioning polygons into fat pieces has become a recent focus. One notion of fatness asks for a partition into convex polygons that minimizes the largest aspect ratio of any piece of the partition. The **aspect ratio** of a polygon P is the ratio of the diameters of the smallest circumscribing circle to the largest inscribed circle. Thus, the fatness corresponds to circularity. If Steiner points are disallowed, i.e., if the pieces of the partition must have their vertices chosen among P's vertices, then a polynomial-time algorithm is known [DI04]. Permitting Steiner points leads to considerable complexity. For example, the optimal partition of an equilateral triangle needs an infinite number of pieces, and the optimal partition for a square is not yet known [DO03]. See Figure 30.3.1.

A variation on the Treemap algorithm, for a tree of n nodes and height h, can construct a convex partition into convex polygons with aspect ratio bounded by $O(\text{poly}(h, \log n))$ [BOS13]. This then leads to a tree partition into fat rectangles with similarly bounded aspect ratio.

A chip manufacturing application has led to another type of "fat" rectangle partition, where the goal is to maximize the shortest rectangle side over all rect-

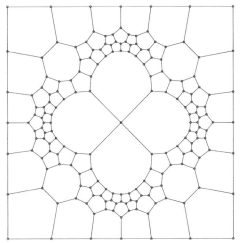

FIGURE 30.3.1
A 92-piece partition achieving an aspect ratio of 1.29950.

angles in the partition, i.e., avoid thin rectangles. The challenge is that the edges of the optimal partition need not be "anchored" to a point on the boundary of the polygon, but may instead float freely inside, as in Fig. 30.3.2. Nevertheless, for simple orthogonal polygons, a polynomial-time algorithm can be achieved, albeit with high complexity [OT04].

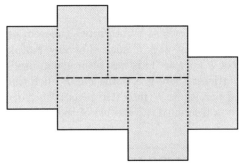

FIGURE 30.3.2
Not every cut of an optimal rectangle partition is "anchored" on the boundary.

ORTHOGONAL POLYGONS

Partitions and covers of orthogonal polygons into rectangles were mentioned above. With the goal of achieving the fewest number of rectangles, finding optimal covers is NP-complete, whereas finding optimal partitions is polynomial, $O(n^{3/2} \log n)$. If the goal is to minimize the total length of the "cuts" between the rectangles (minimum "ink"), then an optimum partition can be found in $O(n^4)$ time for polygons without holes, but is NP-complete with holes [LTL89]. Approximations are available; for example, one that guarantees a solution within a factor of 3 of the minimum length [GZ90]. Another variation on ink minimization seeks a convex partition of a polygon P with the total perimeter of the convex pieces a factor of the perimeter of P. Allowing Steiner points, a total perimeter of $O((\log n / \log \log n) \text{perim}(P))$ can be achieved [DT11]. Covering orthogonal polygons without holes with the fewest squares is polynomial, $O(n^{3/2})$, but NP-complete for polygons with holes [ACKO88].

AREA BISECTION

A particularly useful partition of a polygon P is an **area bisection**: a line determining a halfplane H such that $H \cap P$ and $\bar{H} \cap P$ have the same area. In [DO90] an $O(n \log n)$ algorithm for area bisection was developed, and then used to "ham-sandwich section" a pair of polygons. This result was subsequently improved to $O(n)$-time [She92]. Motivated by positioning parts in industrial part-feeding systems, Böhringer et al. [BDH99] developed an output-size sensitive algorithm for computing the complete set of combinatorially distinct area bisectors, which they show can have size $\Omega(n^2)$.

A *ham-sandwich geodesic* in a polygon P enclosing points is a shortest path connecting two boundary points that simultaneously bisects red points and blue points in the polygon. If n is the number of vertices of P plus the number of interior points, such a geodesic can be found in $O(n \log r)$-time, where r is the number of reflex vertices of P [BDH+07]. This result has been generalized to the situation where there are kn red points and km blue points, and the task is to partition P into k *relatively-convex regions* (closed under shortest paths), each containing n red and m blue points. Then a $O(kn^2 \log^2 n)$-time algorithm is available [BBK06]. A related result partitions a polygon of n vertices containing k points into equal-area convex regions, each containing exactly one point. $O(kn + k^2 \log n)$-time can be achieved [AP10].

A classic result known as Winternitz's theorem says that in every convex polygon P, there is a point $x \in P$ such that any halfplane that contains x contains at least 4/9's of P's area. This has been generalized to nonconvex polygons P of $r \geq 1$ reflex vertices: there is a point $x \in P$ such that any boundary-to-boundary segment chord partitions P into two pieces, such that the piece that contains x contains nearly a fraction $1/(2(r+1))$ of the area of P [BCHM11].

SUM-DIFFERENCE DECOMPOSITIONS

Permitting set subtraction as well as set union leads to natural shape decompositions. This is evident from the field of Constructive Solid Geometry, where shapes are described with CSG trees whose nodes are union or difference operators, and whose leaves are primitive shapes (Section 57.1). Batchelor developed a similar concept for shape description, the **convex deficiency tree** [Bat80]. For a shape P, the root of this tree is its hull conv(P), the children of the root the hulls of the convex deficiencies conv(P) \ P, and so on [O'R98, p. 98].

Chazelle suggested [Cha79] representing a shape by the difference of convex sets: $A \setminus B$ where A and B are unions of convex polygons. It has been established that finding the minimum number of convex pieces in such a sum-difference decomposition of a multiply connected polygonal region is NP-hard [Con90].

DISSECTIONS

A *dissection* of one polygon P to another Q is a partition of P into a finite number of pieces that may be reassembled to form Q. P and Q are then said to be *equidecomposable*. Dissections have been studied as puzzles for centuries. A typical example is shown in Figure 30.3.3 [Fre97, p. 66]. It has been known since the early

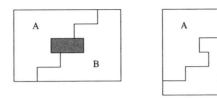

FIGURE 30.3.3
Sam Loyd's "A&P Baking Powder" puzzle reassembles a rectangle with a hole to a rectangle without a hole via a two-piece dissection.

19th century that any two polygons of equal area are equidecomposable [Fre97, p. 221]. The same question for the more constrained *hinged dissections* remained open (implicitly) for years. See Fig. 30.3.4 for the famous Dudeney-McElroy hinged dissection between a square and an equilateral triangle [Fre02]. Now the question is settled positively: any finite collection of polygons of equal area has a common hinged dissection [AAC+12]. For two polygons with vertices on the integer lattice, both the number of pieces of the hinged dissection and the running time of the construction algorithm are pseudopolynomial.

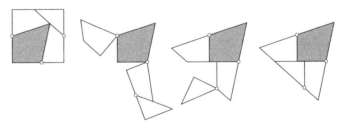

FIGURE 30.3.4
A four-piece hinged dissection between a square and an equilateral triangle.

OPEN PROBLEMS

1. *Approximating the number of art gallery guards:* Give a polynomial-time algorithm for computing a constant-factor approximation of the minimum number of point guards needed to cover a simple polygon.

2. *Fat partition of a square:* What is the optimal partition of a square into "fat" convex polygons?

30.4 POLYGON INTERSECTION

Polygon intersection problems deal with issues of detection and computation of the collision between two polygonal shapes. In the detection problem, one is only interested in deciding *whether* the two polygons have a point in common. In the intersection computation problem, the algorithm is asked to report the overlapping parts of the two polygons. Such problems arise naturally in robotics and computer games; see Chapter 39 for additional material.

The maximum *number* of points at which the boundaries of two polygons may cross each other depends on the type of polygons. If p and q, respectively, denote the number of vertices of the two polygons, then the maximum number of intersections is $\min(2p, 2q)$ if both polygons are convex, $\max(2p, 2q)$ if one is convex, and pq otherwise.

Algorithmically, intersection-detection between convex polygons can be done significantly faster than intersection computation, if we allow reasonable preprocessing of polygons. By a reasonable preprocessing, we mean that the preprocessing algorithm takes into account the *structure* of the polygons but *not their positions*. In Table 30.4.1, n denotes the total number of vertices in the two polygons; that is, $n = p + q$.

TABLE 30.4.1   Intersecting polygons.

POLYGON TYPES	PREPROCESSING	QUERY	SOURCE
Convex-convex	$O(1)$	$O(n)$	[CD80]
Convex-convex	$O(n)$	$O(\log n)$	[CD80]
Simple-simple	$O(1)$	$O(n)$	[Cha91]
Simple-simple	$O(n \log n)$	$O(m \log^2 n)$	[Mou92]

The parameter m in the query time for intersections of two simple polygons is the complexity of a *minimum link witness* for the intersection or disjointness of the two polygons, and we always have $m \leq n$. The preprocessing space requirement is linear when the polygons are preprocessed.

30.5 POLYGON CONTAINMENT

Polygon containment refers to a class of problems that deals with the placement of one polygonal figure inside another. Polygon inscription, polygon circumscription, and polygon nesting are other variants of this type of problem.

GLOSSARY

Inscribed polygon: We will say that a polygon Q is inscribed in polygon P if $Q \subset P$. P is then called a *circumscribing polygon.*

Polygon nesting: P, Q is a nested pair if $Q \subset P$ or vice versa.

CONTAINMENT OF POLYGONS

Let P, Q be two simple polygons with p and q vertices, respectively. The polygon containment problem asks for the largest copy of Q that can be contained in P using rotations, translations, and scaling. (In this section, all scalings are assumed to be *uniform*; thus "shearing" is not permitted.) The containment problems can be solved using the parameter space of all translated or rotated copies of P, by computing the *free space* of placements that lie in Q; see Section 50.2 for further details). The largest scale factor of P for which such a placement exists can be found with parametric search. Table 30.5.1 collects the best results known for the most important cases. See Section 28.10 for a description of the near-linear λ_s function.

TABLE 30.5.1   Results for the polygon containment problem.

P	Q	TRANSFORMS	RESULTS	SOURCE
Convex	convex	translate, scale	$O(p + q \log q)$	[ST94, GP13]
Convex	polygon w holes	translate, scale	$O(pq \log(pq))$	[For85, LS87]
Orthogonal	orthogonal	translate, scale	$O(pq \log(pq))$	[Bar96]
Simple	polygon w holes	translate, scale	$O(p^2 q^2 \log(pq))$	[AFH02]
Convex	convex	translate, rotate, scale	$O(pq^2 \log q)$	[AAS98]
Convex	polygon w holes	translate, rotate, scale	$O(pq\lambda_6(pq) \log^c(pq))$	[AAS99]
Simple	polygon w holes	translate, rotate, scale	$O(p^3 q^3 \log pq)$	[AB88]

It has been shown recently that the decision problem—whether there exists a transformation of Q that permits it to be contained in P—is 3SUM-hard for simple polygons under homotheties and for convex polygons under similarities [BHP01]. Despite recent breakthroughs on the computational complexity of 3SUM [Fre15], it is unlikely the above bounds can be pushed much below quadratic [AVWY15].

When only rigid motions are allowed and P does not contain any copy of Q, one can seek to maximize the area of the intersection of P and a copy of Q. Under translations, this can be done in $O((p+q) \log(pq))$ time for two convex polygons [BCD+98], and in $O(p^2 q^2)$ time for two simple polygons [MSW96]; an $(1-\varepsilon)$-approximation is available in $O(p+q)$ time for every $\varepsilon > 0$ [HPR14]. Under translations and rotations, only approximation algorithms are known [CL13, HPR14].

INSCRIBING/CIRCUMSCRIBING POLYGONS

We now consider problems related to inscribing and circumscribing polygons. In these problems, a polygon P is given, and the task is to find a polygon Q of some specified number of vertices k that is inscribed in (resp. circumscribes) P while maximizing (resp. minimizing) certain measures of Q. The common measures include area and perimeter. See Table 30.5.2 for results concerning this class of problems; n denotes the number of vertices of P.

TABLE 30.5.2   Inscribing and circumscribing polygons.

TYPE	k	P	MEASURE	RESULTS	SOURCE
Inscribe	3	convex	max area	$O(n \log n)$	[KLU$^+$17]
Inscribe	k	convex	max area/perimeter	$O(kn + n \log n)$	[AKM$^+$87, KLU$^+$17]
Inscribe	convex	simple	max area, perimeter	$O(n^7)$, $O(n^6)$	[CY86]
Inscribe	3	simple	max area/perimeter	$O(n^4)$	[MS90]
Circumscribe	3	convex	min area	$O(n)$	[OAMB86]
Circumscribe	3	convex	min perimeter	$O(n)$	[BM02]
Circumscribe	k	convex	min area	$O(kn + n \log n)$	[AP88]

NESTING POLYGONS

The nested polygon problem asks for a polygon with the smallest number of vertices that fits between two nested polygons. More precisely, given two nested polygons P and Q, where $Q \subset P$, find a polygon K of the least number of vertices such that $Q \subset K \subset P$. Generalizing the notion of nested polygons, one can also pose the problem of determining a polygonal subdivision of the least number of *edges* that "separates" a family of polygons. Table 30.5.3 lists the results on these problems. In this table, n is the total number of vertices in the input polygons, while k is the number of vertices in the output polygon (or subdivision).

TABLE 30.5.3   Results for polygon nesting.

TYPES OF P, Q	TYPE OF K	RESULTS	SOURCE
Convex-convex	convex	$O(n \log k)$	[ABO$^+$89]
Simple-simple	simple	$O(n \log k)$	[Gho91]
Polygonal family	subdivision	NP-complete	[Das90]
Polygonal family	subdivision	$O(1)$-Opt in $O(n \log n)$	[MS95b]

Several other results on polygon nesting have been obtained. In particular, if the minimum-vertex nested polygon is nonconvex, then it can be found in $O(n)$ time [GM90]. There is also a relation here to *offset polygons* [BG14] (Chapter 57), and *minimum-link separators* (Chapter 53).

OPEN PROBLEMS

1. *Large empty convex polygons:* Danzer conjectured that for every set S of n points in the unit square $[0, 1]^2$, there is a convex polygon in $[0, 1]^2 \setminus S$ of area $\Omega(1/ \log n)$. There are $(1 - \varepsilon)$-approximation algorithms [DHPT14] for finding the maximum area of a convex polygon in $[0, 1]^2 \setminus S$.

2. *Square Peg Problem:* Toeplitz [Toe11] conjectured that every Jordan curve C in the plane contains four points that are the vertices of a square. The

conjecture has been confirmed in many important special cases, such as piecewise linear curves (e.g., the boundaries of polygons) [Pak10] or smooth curves [CDM14], but remains open in general; see [Mat14] for a survey.

30.6 MISCELLANEOUS

There is a rather large number of results pertaining to polygons, and it would be impossible to cover them all in a single chapter. Having focused on a selected list of topics so far, we now provide below an unorganized collection of some miscellaneous results.

POLYGON MORPHING

To *morph* one polygon into another is to find a continuous deformation from the source polygon to the target polygon. In a *parallel morph*, the deformation maintains the orientation of every edge of the polygon. A parallel morph exists between any two simple n-gons if their edges, taken in counterclockwise order, are parallel and oriented the same way [GCK91]. Hershberger and Suri [GHS00] show that a sequence of $O(n \log n)$ morphing steps suffice where each step consists of a uniform scaling or translation of a part of the polygon.

Between two arbitrary simple n-gons in the plane, there is always a morph in $O(n)$ steps such that in each step all vertices move at constant speed along parallel lines [AAB+17]. Such a morph exists, in general, between any two isotopic straight-line embeddings of a planar graph G; the case of two simple polygons with n vertices corresponds to $G = C_n$. The morph can be computed in $O(n^3)$ time; it crucially relies on the case of morphing triangulations. Aronov et al. [ASS93] showed that for any two simple n-gons, P and Q, there exists straight-line isotopic triangulations (so-called *compatible triangulations*) using $O(n^2)$ Steiner points, and this bound cannot be improved.

Morphing between simple polygons of fixed edge lengths (i.e., linkages) is discussed in Chapter 9.

CSG REPRESENTATION

Peterson proved that every simple polygon in two dimensions admits a representation by a Boolean formula on the halfplanes supporting the edges of the polygon. Furthermore, the resulting formula is *monotone*; that is, there is no negation and each halfplane appears exactly once. A ***Peterson-style formula*** is a "constructive solid geometry" representation, in which the polygon is presented as a set of Boolean operations; see Chapter 57. Interestingly, it turns out that not all 3D polyhedra admit a Peterson-style formula [DGHS93].

Dobkin et al. [DGHS93] give an $O(n \log n)$ time algorithm for computing a Petersen-style formula for a simple n-gon. Chirst et al. [CHOU10] show that every n-gon can be expressed as a monotone Boolean formula of at most $\lfloor (4n2)/5 \rfloor$ wedges (where a wedge is the intersection or the union of two halfplanes), and $\lceil (3n4)/5 \rceil$ wedges are sometimes necessary.

POLYGON SEARCHING AND PURSUIT-EVASION

In these problems, the goal is to design search strategies for an (identifiable) object (intruder) in a polygon. The motivation often comes from surveillance applications in robotics. The "polygon searching" line of research typically assumes that the object of search is *stationary*, the searcher "discovers" the geometry of the polygon during its navigation (*on-line* model), and the goal is to minimize the search cost (distance traveled), measured by its competitive ratio. Table 30.6.1 summarizes some basic results on the polygon searching problems. (The parameter k in the second to last line denotes the number of distinct initial placements of the robot having the same visibility polygon.) The survey article [GK10] is a good starting point for this topic.

TABLE 30.6.1   Results for polygon searching.

ENVIRONMENT	GOAL	COMPETITIVE RATIO	SOURCE
n oriented rectangles	shortest path	$\Theta(\sqrt{n})$	[BRS97]
"Street" polygon	shortest path	$\sqrt{2}$	[IKLS04]
Star-shaped polygon	reach kernel	≈ 3.12	[Pal00]
Orthogonal polygon	exploration	randomized 5/4	[Kle94]
Simple polygon	localization with min travel	$(k-1)$-Opt	[DRW98]
Simple polygon	shortest watchman tour	26.5-Opt	[HIKK01]

In the "pursuit-evasion" line of research, one or more searchers (pursuers) coordinate to locate and capture a *mobile* object (intruder), and the goal is to establish necessary and sufficient conditions for a successful pursuit. The survey article [CHI11] is a good starting point for this topic. The origin of pursuit-evasion goes back to the celebrated "Lion-and-Man" problem, attributed to Rado in 1930s: if a man and a lion are confined to a closed arena, and both have equal maximum speeds, can the lion catch the man? Surprisingly, the man can evade the lion indefinitely as shown by Besicovitch [Lit86]—the lion fails to reach the man in any finite time although it can get arbitrarily close to him. In computational geometry, a primary focus of research is to bound the minimum number of pursuers needed to locate or capture the evader, as a function of environment's complexity. The model assumes that pursuers and the evader can move with the same maximum speed, the geometry of the environment (polygon) is known to all players, and the players move taking alternating turns. Two models of visibility are considered: in the *full visibility* model, each player knows the position of all other players at all times (following the convention of the cops-and-robber game in graphs) while in the *LoS visibility* model, each player is limited to its line-of-sight visibility. Table 30.6.2 below summarizes the current state of the art. (The *minimum feature size* (MFS) condition requires that the minimum (geodesic) distance between two vertices is lower bounded by the distance each player can move in one step.) Open problems include closing the gap between upper and lower bounds as well as extending the pursuit to three-dimensional environments.

TABLE 30.6.2   Results for pursuit-evasion in polygons.

ENVIRONMENT	GOAL	VISIBILITY MODEL	NUMBER OF PURSUERS	SOURCE
Simple polygon	Locate	LoS	$\Theta(\log n)$	[GLL$^+$99]
Polygon with h holes	Locate	LoS	$\Theta(\log n + \sqrt{h})$	[GLL$^+$99]
Polygon with h holes	Capture	Full	3	[BKIS12]
Polygon with holes	Capture	LoS without MFS	$\Omega(n^{2/3})$, $O(n^{5/6})$	[KS15a]
Polygon with h holes	Capture	LoS with MFS	$O(\log n + \sqrt{h})$	[KS15a]
3D polyhedral surface	Capture	Full	3 nec., 4 suff.	[KS15b]

THREE-DIMENSIONAL POLYGONS

A 3D polygon is an unknotted closed chain of segments in $\mathbb{R}^3$ such that adjacent segments share an endpoint, and nonadjacent segments do not intersect. A triangulation of a 3D polygon has the same combinatorial structure as a triangulation of a planar polygon—all triangle vertices are polygon vertices, each polygon edge is a side of one triangle, each diagonal is shared by exactly two triangles—with the surface they define a nonself-intersecting topological disk. This disk is said to *span* the polygon. Barequet et al. [BDE98] proved that determining whether a 3D polygon has a triangulation in this sense is NP-complete. Another negative result along the same lines is that there exist 3D polygons of n vertices that can only be spanned by nonself-intersecting piecewise-linear disks which, when triangulated, need $2^{\Omega(n)}$ triangles [HST03]. Note that here the triangle vertices are not necessarily polygon vertices, i.e., Steiner points are (necessarily) used. This exponential lower bound shows that ***knot triviality*** algorithms (which check whether a closed chain is the trivial "unknot") that search for such spanning disks necessarily lead to exponential-time algorithms [Bur04, Lac15]. This unknotting problem is known to be in NP [HLP99] and co-NP [Lac16].

OPEN PROBLEMS

1. *Random Polygonizations.* Is there an efficient algorithm to generate a random polygonization for n given points in the plane?

2. *3D Peterson formulas:* Characterize the 3D polyhedra that can be represented by Peterson-style formulas.

3. *Computational complexity of unknot recognition:* Is there a polynomial-time algorithm for deciding whether a 3D polygon is a trivial knot.

30.7 SOURCES AND RELATED MATERIAL

SURVEYS

The survey article by Mitchell and Suri [MS95a] addresses optimization problems in computational geometry, many involving polygons. Keil surveys polygon decomposition algorithms in [Kei00]. Link distance problems are surveyed in [MSD00].

RELATED CHAPTERS

REFERENCES

[AAB⁺17]   S. Alamdari, P. Angelini, F. Barrera-Cruz, T.M. Chan, G. Da Lozzo, G. Di Battista, F. Frati, P. Haxell, A. Lubiw, M. Patrignani, V. Roselli, S. Singla, and B.T. Wilkinson. How to morph planar graph drawings. *SIAM J. Comput.*, 46:824–852, 2017.

[AAC⁺12]   T.G. Abbott, Z. Abel, D. Charlton, E.D. Demaine, M.L. Demaine, and S.D. Kominers. Hinged dissections exist. *Discrete Comput. Geom.*, 47:150–186, 2012.

[AAI86]   Ta. Asano, Te. Asano, and H. Imai. Partitioning a polygonal region into trapezoids. *J. ACM*, 33:290–312, 1986.

[AAK09]   E. Ackerman, O. Aichholzer, and B. Keszegh. Improved upper bounds on the reflexivity of point sets. *Comput. Geom.*, 42:241–249, 2009.

[AAS98]   P.K. Agarwal, N. Amenta, and M. Sharir. Largest placement of one convex polygon inside another. *Discrete Comput. Geom.* 19:95–104, 1998.

[AAS99]   P.K. Agarwal, B. Aronov, and M. Sharir. Motion planning for a convex polygon in a polygonal environment. *Discrete Comput. Geom.*, 22:201–221, 1999.

[AB88]   F. Avnaim and J.-D. Boissonnat. Polygon placement under translation and rotation. *Proc. 5th Sympos. Theoret. Aspects Comput. Sci.*, vol. 294 of *LNCS*, pages 322–333, Springer, Berlin, 1988.

[ABO⁺89]   A. Aggarwal, H. Booth, J. O'Rourke, S. Suri, and C.K. Yap. Finding minimal convex nested polygons. *Inform. Comput.*, 83:98–110, 1989.

[ACKO88]   L.J. Aupperle, H.E. Conn, J.M. Keil, and J. O'Rourke. Covering orthogonal polygons with squares. In *Proc. 26th Allerton Conf. Commun. Control Comput.*, pages 97–106, 1988.

[AFH02]   P.K. Agarwal, E. Flato, and D. Halperin. Polygon decomposition for efficient construction of Minkowski sums. *Comput. Geom.*, 21:3961, 2002.

[AFH⁺03]   E.M. Arkin, S.P. Fekete, F. Hurtado, J.S.B. Mitchell, M. Noy, V. Sacristán, and S. Sethia. On the reflexivity of point sets. In *Discrete Comput. Geom.*, pages 139–156, 2003.

[Agg84]   A. Aggarwal. *The art gallery problem: Its variations, applications, and algorithmic aspects.* Ph.D. thesis, Dept. Comput. Sci., Johns Hopkins Univ., Baltimore, 1984.

[AGR01]   N.M. Amato, M.T. Goodrich, and E.A. Ramos. A randomized algorithm for triangulating a simple polygon in linear time. *Discrete Comput. Geom.*, 26:245–265, 2001.

[AHST03]   O. Aichholzer, M. Hoffmann, B. Speckmann, and C.D. Tóth. Degree bounds for constrained pseudo-triangulations. In *Canad. Conf. Comput. Geom.*, pages 155–158, 2003.

[AKM⁺87]   A. Aggarwal, M.M. Klawe, S. Moran, P.W. Shor, and R. Wilber. Geometric applications of a matrix-searching algorithm. *Algorithmica*, 2:195–208, 1987.

[AP88]   A. Aggarwal and J.K. Park. Notes on searching in multidimensional monotone arrays. In *Proc. 29th IEEE Sympos. Found. Comput. Sci.*, pages 497–512, 1988.

[AP10]   D. Adjiashvili and D. Peleg. Equal-area locus-based convex polygon decomposition. *Theoret. Comput. Sci.*, 411:1648–1667, 2010

[ASS93]   B. Aronov, R. Seidel, and D.L. Souvaine. On compatible triangulations of simple polygons. *Comput. Geom.*, 3:27–35, 1993.

[AVWY15]   A. Abboud, V. Vassilevska Williams, and H. Yu. Matching triangles and basing hardness on an extremely popular conjecture. In *Proc. 47th ACM Sympos. Theory Comput.*, pages 41–50, 2015.

[Bar96]   A.H. Barrera. Algorithms for deciding the containment of polygons. *Inform. Process. Lett.*, 59:261–265, 2996.

[Bat80]   B.G. Batchelor. Hierarchical shape description based upon convex hulls of concavities. *J. Cybern.*, 10:205–210, 1980.

[BBK06]   S. Bereg, P. Bose, and D. Kirkpatrick. Equitable subdivisions within polygonal regions. *Comput. Geom.*, 34:20–27, 2006.

[BBV15]   S. Bandyapadhyay, S. Bhowmick, and K. Varadarajan. Approximation schemes for partitioning: Convex decomposition and surface approximation. In *Proc. 26th ACM-SIAM Symp. Discrete Algorithms*, pages 1457–1470, 2015.

[BCD⁺98]   M. de Berg, O. Cheong, O. Devillers, M. van Kreveld, and M. Teillaud. Computing the maximum overlap of two convex polygons under translations. *Theory Comput. Syst.*, 31:613–628, 1998.

[BCHM11]   P. Bose, P. Carmi, F. Hurtado, and P. Morin. A generalized Winternitz theorem. *J. Geometry*, 100:29–35, 2011.

[BDE98]   G. Barequet, M.T. Dickerson, and D. Eppstein. On triangulating three-dimensional polygons. *Comput. Geom.*, 10:155–170, 1998.

[BDH99]   K.-F. Böhringer, B.R. Donald, and D. Halperin. The area bisectors of a polygon. *Discrete Comput. Geom.*, 22:269–285, 1999.

[BDH⁺07]   P. Bose, E.D. Demaine, F. Hurtado, J. Iacono, S. Langerman, and P. Morin. Geodesic ham-sandwich cuts. *Discrete Comput. Geom.*, 37:325–339, 2007.

[BG14]   G. Barequet and A. Goryachev. Offset polygon and annulus placement problems. *Comput. Geom.*, 47:407–434, 2014.

[BHP01]   G. Barequet and S. Har-Peled. Polygon containment and translational min-Hausdorff-distance between segment sets are 3SUM-hard. *Internat. J. Comput. Geom. Appl.*, 11:465–474, 2001.

[BKIS12]   D. Bhadauria, K. Klein, V. Isler, and S. Suri. Capturing an evader in polygonal environments with obstacles: The full visibility case. *I. J. Robot. Res.*, 31:1176–1189, 2012.

[BM02]   B. Bhattacharya and A. Mukhopadhyay. On the minimum perimeter triangle enclosing a convex polygon. In J. Akiyama and M. Kano, editors, *Discrete and Computational Geometry*, vol. 2866 of *LNCS*, pages 84–96, Springer, Berlin, 2002.

[BOS13]   M. de Berg, K. Onak, and A. Sidiropoulos. Fat polygonal partitions with applications to visualization and embeddings. *J. Comput. Geom.*, 4:212–239, 2013.

[BRS97]   A. Blum, P. Raghavan, and B. Schieber. Navigating in unfamiliar geometric terrain. *SIAM J. Comput.*, 26:110–137, 1997.

[Bur04]   B.A. Burton. Introducing Regina, the 3-manifold topology software. *Exp. Math.*, 13:267–272, 2004.

[CD79]   B. Chazelle and D.P. Dobkin. Decomposing a polygon into its convex parts. In *Proc. 11th ACM Sympos. Theory Comput.*, pages 38–48, 1979.

[CD80]   B. Chazelle and D.P. Dobkin. Detection is easier than computation. In *Proc. 12th ACM Sympos. Theory Comput.*, pages 146–153, 1980.

[CDM14]   J. Cantarella, E. Denne, and J. McCleary. Transversality for configuration spaces and the "Square-Peg" Theorem. Preprint, `arXiv:1402.6174`, 2014.

[Cha79]   B. Chazelle. *Computational geometry and convexity.* Ph.D. thesis, Dept. Comput. Sci., Yale Univ., New Haven, 1979. Carnegie-Mellon Univ. Report CS-80-150.

[Cha82]   B. Chazelle. A theorem on polygon cutting with applications. In *Proc. 23rd IEEE Sympos. Found. Comput. Sci.*, pages 339–349, 1982.

[Cha91]   B. Chazelle. Triangulating a simple polygon in linear time. *Discrete Comput. Geom.*, 6:485–524, 1991.

[CHI11]   T.H. Chung, G.A. Hollinger, and V. Isler. Search and pursuit-evasion in mobile robotics. *Autonomous Robots*, 31:299–316, 2011.

[CHOU10]   T. Christ, M. Hoffmann, Y. Okamoto, and T. Uno. Improved bounds for wireless localization. *Algorithmica*, 57:499–516, 2010.

[Chr11]   T. Christ. Beyond triangulation: covering polygons with triangles. In *Proc. 12th Workshop Algorithms Data Structures*, vol. 6844 of *LNCS*, pages 231–242, Springer, Berlin, 2011.

[CL13]   S.-W. Cheng and C.-K. Lam. Shape matching under rigid motion. *Comput. Geom.*, 46:591–603, 2013.

[Con90]   H. Conn. *Some Polygon Decomposition Problems.* Ph.D. thesis, Dept. Comput. Sci., John Hopkins Univ., Baltimore, 1990.

[CR94]   J.C. Culberson and R.A. Reckhow. Covering polygons is hard. *J. Algorithms*, 17:2–24, 1994.

[CY86]   J.S. Chang and C.K. Yap. A polynomial solution for the potato-peeling problem. *Discrete Comput. Geom.*, 1:155–182, 1986.

[Das90]   G. Das. *Approximation schemes in computational geometry.* Ph.D. thesis, Univ. of Wisconsin, 1990.

[DFH+99]   L. Devroye, P. Flajolet, F. Hurtado, M. Noy, and W. Steiger. Properties of random triangulations and trees. *Discrete Comput. Geom.*, 22:105–117, 1999.

[DFOR10]   M. Damian, R. Flatland, J. ORourke, and S. Ramaswami. Connecting polygonizations via stretches and twangs. *Theory Comput. Syst.*, 47:674–695, 2010.

[DGHS93]   D.P. Dobkin, L.J. Guibas, J. Hershberger, and J. Snoeyink. An efficient algorithm for finding the CSG representation of a simple polygon. *Algorithmica*, 10:1–23, 1993.

[DHPT14]   A. Dumitrescu, S. Har-Peled, and C.D. Tóth. Minimum convex partitions and maximum empty polytopes. *J. Comput. Geom.*, 5:86–103, 2014.

[DI04]   M. Damian. Exact and approximation algorithms for computing optimal fat decompositions. *Comput. Geom.*, 28:19–27, 2004.

[DKDS07]   A. Deshpande, T. Kim, E.D. Demaine, and S.E. Sarma. A pseudopolynomial time $o(\log n)$-approximation algorithm for art gallery problems. In *Proc. 10th Workshop Algorithms Data Structs.*, vol. 4619 of *LNCS*, pages 163–174, Springer, Berlin, 2007.

[DO90]   M. Díaz and J. O'Rourke. Ham-sandwich sectioning of polygons. In *Proc. 2nd Canad. Conf. Comput. Geom.*, pages 282–286, 1990.

[DO03]   M. Damian and J. O'Rourke. Partitioning regular polygons into circular pieces I: Convex partitions. In *Proc. 15th Canad. Conf. Comput. Geom.*, pages 43–46, 2003.

[DRW98]   G. Dudek, K. Romanik, and S. Whitesides. Localizing a robot with minimum travel. *SIAM J. Comput.*, 27:583–604, 1998.

[DQTW05] Q. Ding, J. Qian, W. Tsang, and C. Wang. Randomly generating triangulations of a simple polygon. In *Proc. 11th Computing Combin. Conf.*, vol. 3595 of *LNCS*, pages 471–480, Springer, Berlin, 2005.

[DT10]   A. Dumitrescu and C.D. Tóth. Long non-crossing configurations in the plane. *Discrete Comput. Geom.*, 44:727–752, 2010.

[DT11]   A. Dumitrescu and C.D. Tóth. Minimum weight convex Steiner partitions. *Algorithmica*, 60:627–652, 2011.

[EHP06]   A. Efrat and S. Har-Peled. Guarding galleries and terrains. *Inform. Process. Lett.*, 100:238–245, 2006.

[ES94]   P. Epstein and J.-R. Sack. Generating triangulations at random. *ACM Trans. Model. Comput. Simulation*, 4:267–278, 1994.

[ESW01]   S.J. Eidenbenz, C. Stamm, and P. Widmayer. Inapproximability results for guarding polygons and terrains. *Algorithmica*, 31:79–113, 2001.

[EW03]   S.J. Eidenbenz and P. Widmayer. An approximation algorithm for minimum convex cover with logarithmic performance guarantee. *SIAM J. Comput.*, 32:654–670, 2003.

[Fek00]   S.P. Fekete. On simple polygonalizations with optimal area. *Discrete Comput. Geom.*, 23:73–110, 2000.

[FKMU13] R. Fulek, B. Keszegh, F. Morić, and I. Uljarević. On polygons excluding point sets. *Graphs Combin.*, 29:1741–1753, 2013.

[For85]   S.J. Fortune. A fast algorithm for polygon containment by translation. In *Proc. 12th Internat. Colloq. Automata Lang. Program.*, vol. 194 of *LNCS*, pages 189–198, Springer, Berlin, 1985.

[Fre97]   G.N. Frederickson. *Dissections: Plane and Fancy.* Cambridge University Press, 1997.

[Fre02]   G.N. Frederickson. *Hinged Dissections: Swinging & Twisting.* Cambridge University Press, 2002.

[Fre15]   A. Freund. Improved subquadratic 3SUM. *Algorithmica*, online first, 2015.

[GCK91]   U. Grenander, Y. Chow, and D.M. Keenan. *Hands: A Pattern Theoretic Study of Biological Shapes.* Springer-Verlag, Berlin, 1991.

[Gho91]   S.K. Ghosh. Computing visibility polygon from a convex set and related problems. *J. Algorithms*, 12:75–95, 1991.

[Gho10]   S.K. Ghosh. Approximation algorithms for art gallery problems in polygons. *Discrete Applied Math.*, 158:718–722, 2010.

[GHS00]   L. Guibas, J. Hershberger, and S. Suri. Morphing simple polygons. *Discrete Comput. Geom.*, 24:1–3, 2000.

[GJPT78] M.R. Garey, D.S. Johnson, F.P. Preparata, and R.E. Tarjan. Triangulating a simple polygon. *Inform. Process. Lett.*, 7:175–179, 1978.

[GK10]   S.K. Ghosh, and R. Klein. Survey: online algorithms for searching and exploration in the plane. *Computer Science Review*, 4:189–201, 2010.

[GLL+99]   L.J. Guibas, J.-C. Latombe, S.M. LaValle, D. Lin, and R. Motwani. Visibility-based pursuit-evasion in a polygonal environment. *Internat. J. Comput. Geom. Appl.*, 9:471–494, 1999.

[GM90]   S.K. Ghosh and A. Maheshwari. An optimal algorithm for computing a minimum nested nonconvex polygon. *Inform. Process. Lett.*, 36:277–280, 1990.

[GNT00]   A. Garcı, M. Noy, and J. Tejel. Lower bounds on the number of crossing-free subgraphs of K_n. *Comput. Geom.*, 16:211–221, 2000.

[GP13]   M.T. Goodrich and P. Pszona. Cole's parametric search technique made practical. In *Proc. 25th Canad. Conf. Comput. Geom.*, Waterloo, ON, 2013.

[GZ90]   T.F. Gonzalez and S.-Q. Zheng. Approximation algorithms for partitioning a rectangle with interior points. *Algorithmica*, 5:11–42, 1990.

[HIKK01]   F. Hoffmann, C. Icking, R. Klein, and K. Kriegel. The polygon exploration problem. *SIAM J. Comput.*, 31:577–600, 2001.

[HLP99]   J. Hass, J.C. Lagarias, and N. Pippenger. The computational complexity of knot and link problems. *J. ACM*, 46:185–211, 1999.

[HM85]   S. Hertel and K. Mehlhorn. Fast triangulation of the plane with respect to simple polygons. *Inform. Control*, 64:52–76, 1985.

[HMO+09]   F. Hurtado, C. Merino, D. Oliveros, T. Sakai, J. Urrutia, and I. Ventura. On polygons enclosing point sets II. *Graphs Combin.*, 25:327–339, 2009.

[HPR14]   S. Har-Peled and S. Roy. Approximating the maximum overlap of polygons under translation. *Algorithmica*, online first, 2016.

[HS95]   J. Hershberger and S. Suri. A pedestrian approach to ray shooting: Shoot a ray, take a walk. *J. Algorithms*, 18:403–431, 1995.

[HST03]   J. Hass, J. Snoeyink, and W.P. Thurston. The size of spanning disks for polygonal curves. *Discrete Comput. Geom.*, 29:1–18, 2003.

[IKLS04]   C. Icking, R. Klein, E. Langetepe, S. Schuierer, and I. Semrau. An optimal competitive strategy for walking in streets. *SIAM J. Comput.*, 33:462–486, 2004.

[IM11]   J. Iwerks and J.S.B. Mitchell. Spiral serpentine polygonization of a planar point set. In *Proc. XIV Spanish Meeting Comput. Geom.*, vol. 7579 of *LNCS*, pages 146–154, Springer, Berlin, 2011.

[Kei85]   J.M. Keil. Decomposing a polygon into simpler components. *SIAM J. Comput.*, 14:799–817, 1985.

[Kei00]   J.M. Keil. Polygon decomposition. In J.-R. Sack and J. Urrutia, editors, *Handbook of Computational Geometry*, pages 491–518, Elsevier, Amsterdam, 2000.

[Kin13]   J. King. Fast vertex guarding for polygons with and without holes. *Comput. Geom.*, 46:219–231, 2013.

[Kle94]   J. Kleinberg. On-line search in a simple polygon. In *Proc. 5th ACM-SIAM Sympos. Discrete Algorithms*, pages 8–15, 1994.

[Kli80]   G.T. Klincsek. Minimal triangulations of polygonal domains. *Discrete Math.*, 9:121–123, 1980.

[KLU+17]   V. Keikha, M. Löffler, J. Urhausen, I. van der Hoog. Maximum-area triangle in a convex polygon, revisited. Preprint, `arXiv:1705.11035`, 2017.

[KS02]   J.M. Keil and J. Snoeyink. On the time bound for convex decomposition of simple polygons. *Int. J. Comput. Geom. Appl.*, 12,181–192, 2002.

[KS15a]   K. Klein and S. Suri. Capture bounds for visibility-based pursuit evasion. *Comput. Geom.*, 48:205–220, 2015.

[KS15b]   K. Klein and S. Suri. Pursuit evasion on polyhedral surfaces. *Algorithmica*, 73:730–747, 2015.

[Lac15]   M. Lackenby. A polynomial upper bound on Reidemeister moves. *Ann. Math. (2)*, 182:491–564, 2015.

[Lac16]   M. Lackenby. The efficient certification of knottedness and Thurston norm. Preprint, `arXiv:1604.00290`, 2016.

[Lit86]   J.E. Littlewood. *Littlewood's Miscellany*. Cambridge University Press, 1986.

[LS87]   D. Leven and M. Sharir. Planning a purely translational motion for a convex object in two-dimensional space using generalized Voronoi diagrams. *Discrete Comput. Geom.*, 2:9–31, 1987.

[LTL89]   W.T. Liou, J.J.M. Tan, and R.C.T. Lee. Minimum partitioning simple rectilinear polygons in $O(n \log \log n)$ time. In *Proc. 5th Sympos. Comput. Geom.*, pages 344–353, ACM Press, 1989.

[Mas78]   W.J. Masek. Some NP-complete set covering problems. Manuscript, MIT, Cambridge, 1978.

[Mat14]   B. Matschke. A survey on the square peg problem. *Notices Amer. Math. Soc.*, 61:346–352, 2014.

[Mou92]   D.M. Mount. Intersection detection and separators for simple polygons. In *Proc. 8th Sympos. Comput. Geom.*, pages 303–311, ACM Pres, 1992.

[MS90]   E.A. Melissaratos and D.L. Souvaine. On solving geometric optimization problems using shortest paths. In *Proc. 6th Sympos. Comput. Geom.*, pages 350–359, ACM Press, 1990.

[MS95a]   J.S.B. Mitchell, and S. Suri. Geometric algorithms. In M.O. Ball, T.L. Magnati, C.L. Monma, and G.L. Nemhauser, editors, *Handbook of Operations Research/Management Science*, pages 425–479, Elsevier, Amsterdam, 1995.

[MS95b]   J.S.B. Mitchell and S. Suri. Separation and approximation of polyhedral surfaces. *Comput. Geom.*, 5:95–114, 1995.

[MSD00]   A. Maheshwari, J.-R. Sack, and H.N. Djidjev. Link distance problems. In J.-R. Sack and J. Urrutia, editors, *Handbook of Computational Geometry*, pages 519–558, Elsevier, Amsterdam, 2000.

[MSW96]   D.M. Mount, R. Silverman, and A.Y. Wu. On the area of overlap of translated polygons. *Comp. Vis. Image Understanding* 64:5361, 1996.

[OAMB86]   J. O'Rourke, A. Aggarwal, S. Maddila, and M. Baldwin. An optimal algorithm for finding minimal enclosing triangles. *J. Algorithms*, 7:258–269, 1986.

[O'R87]   J. O'Rourke. *Art Gallery Theorems and Algorithms. Internat. Ser. Monogr. Comput. Sci.*, Oxford University Press, New York, 1987.

[O'R98]   J. O'Rourke. *Computational Geometry in C*, second edition. Cambridge University Press, 1998.

[OS83]   J. O'Rourke and K.J. Supowit. Some NP-hard polygon decomposition problems. *IEEE Trans. Inform. Theory*, IT-30:181–190, 1983.

[OT04]   J. O'Rourke and G. Tewari. The structure of optimal partitions of orthogonal polygons into fat rectangles. *Comput. Geom.*, 28(1):49–71, 2004.

[Pak10]   I. Pak. *Lectures on Discrete and Polyhedral Geometry*, Manuscript, `http://www.math.ucla.edu/~pak/book.htm`, 2010.

[Pal00]   L. Palios. A new competitive strategy for reaching the kernel of an unknown polygon. *Proc. 7th Scandinavian Workshop Algorithm Theory*, vol. 1851 of *LNCS*, pages 367–382, Springer, Berlin, 2000.

[Ror14]   D. Rorabaugh. A bound on a convexity measure for point sets. Preprint, `arXiv: 1409.4344`, 2014.

[RR94]   R. Ronfard and J. Rossignac. Triangulating multiply-connected polygons: A simple yet efficient algorithm. *Comput. Graph. Forum*, 13:C281–292, 1994.

[Sei91]   R. Seidel. A simple and fast incremental randomized algorithm for computing trapezoidal decompositions and for triangulating polygons. *Comput. Geom.*, 1:51–64, 1991.

[She92]   T.C. Shermer. A linear algorithm for bisecting a polygon. *Inform. Process. Lett.*, 41:135–140, 1992.

[SSW13]   M. Sharir, A. Sheffer, and E. Welzl. Counting plane graphs: perfect matchings, spanning cycles, and Kasteleyn's technique. *J. Combin. Theory Ser. A*, 120:777–794.

[ST94]   M. Sharir and S. Toledo. External polygon containment problems. *Comput. Geom.*, 4:99–118, 1994.

[Toe11]   O. Toeplitz. Über einige Aufgaben der Analysis situs. *Ver. Schweiz. Nat. Gesel. Solothurn*, 4:197, 1911.

[TTW11]   C.D. Tóth, G.T. Toussaint, and A. Winslow. Open guard edges and edge guards in simple polygons. In *Proc. XIV Spanish Meeting Comput. Geom.*, vol. 7579 of *LNCS*, pages 54–64, Springer, Berlin, 2011.

31 SHORTEST PATHS AND NETWORKS

Joseph S.B. Mitchell

INTRODUCTION

Computing an optimal path in a geometric domain is a fundamental problem in computational geometry, with applications in robotics, geographic information systems (GIS), wire routing, etc.

A taxonomy of shortest-path problems arises from several parameters that define the problem:

1. Objective function: the length of the path may be measured according to the Euclidean metric, an L_p metric, the number of links, a combination of criteria, etc.

2. Constraints on the path: the path may have to get from s to t while visiting a specified set of points or regions along the way.

3. Input geometry: the map of the geometric domain also specifies constraints on the path, requiring it to avoid various types of obstacles.

4. Type of moving object: the object to be moved along the path may be a single point or may be a robot of some specified geometry.

5. Dimension of the problem: often the problem is in 2 or 3 dimensions, but higher dimensions arise in some applications.

6. Single shot vs. repetitive mode queries.

7. Static vs. dynamic environments: in some cases, obstacles may be inserted or deleted or may be moving in time.

8. Exact vs. approximate algorithms.

9. Known vs. unknown map: the on-line version of the problem requires that the moving robot sense and discover the shape of the environment along its way. The map may also be known with some degree of uncertainty, leading to stochastic models of path planning.

We survey various forms of the problem, primarily in two and three dimensions, for motion of a single point, since most results have focused on these cases. We discuss shortest paths in a simple polygon (Section 31.1), shortest paths among obstacles (Section 31.2), and other metrics for length (Section 31.3). We also survey other related geometric network optimization problems (Section 31.4). Higher dimensions are discussed in Section 31.5.

GLOSSARY

Polygonal s-t path: A path from point s to point t consisting of a finite number of line segments (***edges***, or ***links***) joining a sequence of points (***vertices***).

Length of a path: A nonnegative number associated with a path, measuring its total cost according to some prescribed metric. Unless otherwise specified, the length will be the Euclidean length of the path.

Shortest/optimal/geodesic path: A path of minimum length among all paths that are feasible (satisfying all imposed constraints).

Shortest-path distance: The metric induced by a shortest-path problem. The shortest-path distance between s and t is the length of a shortest s-t path; in many geometric contexts, it is also referred to as ***geodesic distance***.

Locally shortest/optimal path: A path that cannot be improved by making a small change to it that preserves its ***combinatorial structure*** (e.g., the ordered sequence of triangles visited, for some triangulation of a polygonal domain P); also known as a ***taut-string*** path in the case of a shortest obstacle-avoiding path.

Simple polygon P of n vertices: A closed, simply-connected region whose boundary is a union of n (straight) line segments (edges), whose endpoints are the vertices of P.

Polygonal domain of n vertices and h holes: A closed, multiply-connected region whose boundary is a union of n line segments, forming $h+1$ closed (polygonal) cycles. A simple polygon is a polygonal domain with $h = 0$.

Triangulation of a simple polygon P: A decomposition of P into triangles such that any two triangles intersect in either a common vertex, a common edge, or not at all. A triangulation of P can be computed in $O(n)$ time. See Section 29.2.

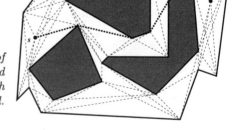

FIGURE 31.0.1

The visibility graph VG(P). Edges of VG(P) are of two types: (1) the heavy dark boundary edges of P, and (2) the edges that intersect the interior of P, shown with thin dashed segments. A shortest s-t path is highlighted.

Obstacle: A region of space whose interior is forbidden to paths. The complement of the set of obstacles is the ***free space***. If the free space is a polygonal domain P, the obstacles are the $h+1$ connected components (h ***holes***, plus the ***face at infinity***) of the complement of P.

Visibility graph VG(P):   A graph whose nodes are the vertices of P and whose edges join pairs of nodes for which the corresponding segment lies inside P. See Chapter 33. An example is shown in Figure 31.0.1.

Single-source query: A query that specifies a goal point t, and requests the length of a shortest path from a *fixed* source point s to t. The query may also require that a shortest s-t path be reported; in general, this can be done in additional time $O(k)$, where k is the number of edges in the output path. (Throughout this survey, when we report query times we omit the "$+O(k)$" that generally allows one to report a path after spending the initial query time to determine the length of a path.)

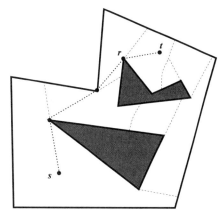

FIGURE 31.0.2
A shortest path map with respect to source point s within a polygonal domain. The dotted path indicates the short-est s-t path, which reaches t via the root r of its cell.

Shortest path map, **SPM(s):** A decomposition of free space into regions (***cells***) according to the "combinatorial structure" of shortest paths from a fixed source point s to points in the regions. Specifically, for shortest paths in a polygonal domain, SPM(s) is a decomposition of P into cells such that for all points t interior to a cell, the sequence of obstacle vertices along a shortest s-t path is fixed. In particular, the *last* obstacle vertex along a shortest s-t path is the ***root*** of the cell containing t. Each cell is ***star-shaped*** with respect to its root, which lies on the boundary of the cell. See Figure 31.0.2, where the root of the cell containing t is labeled r. If SPM(s) is preprocessed for point location (see Chapter 38), then single-source queries can be answered efficiently by locating the query point t within the decomposition.

Two-point query: A query that specifies two points, s and t, and requests the length of a shortest path between them. It may also request that a path be reported.

Geodesic Voronoi diagram **(VD):** A Voronoi diagram for a set of ***sites***, in which the underlying metric is the geodesic distance. See Chapters 27 and 29.

Geodesic center of P: A point within P that minimizes the maximum of the shortest-path lengths to any other point in P.

Geodesic diameter of P: The length of a longest shortest path between a pair of points $s, t \in P$; s and t are vertices for any longest s-t shortest path.

31.1 PATHS IN A SIMPLE POLYGON

The most basic geometric shortest-path problem is to find a shortest path inside a ***simple*** polygon P (having no holes), connecting two points, s and t. The complement of P serves as an obstacle through which the path is not allowed to travel. In this case, there is a unique taut-string path from s to t, since there is only one way to "thread" a string through a simply-connected region.

Algorithms for computing a shortest s-t path begin with a triangulation of P ($O(n)$ time; Section 29.2), whose dual graph is a tree. The **sleeve** is comprised of the triangles that correspond to the (unique) path in the dual that joins the triangle containing s to that containing t. By considering the effect of adding the triangles in order along the sleeve, it is not hard to obtain an $O(n)$ time algorithm for collapsing the sleeve into a shortest path. At a generic step of the algorithm, the sleeve has been collapsed to a structure called a **funnel** (with *base* ab and *root* r) consisting of the shortest path from s to a vertex r, and two (concave) shortest paths joining r to the endpoints of the segment ab that bounds the triangle abc processed next (see Figure 31.1.1). In adding triangle abc, we "split" the funnel in two according to the taut-string path from r to c, which will, in general, include a segment uc joining c to some (vertex) point of tangency u, along one of the two concave chains of the funnel. After the split, we keep that funnel (with base ac or bc) that contains the s-t taut-string path. The work needed to search for u can easily be charged off to those vertices that are discarded from further consideration. Thus, a shortest s-t path is found in time $O(n)$, which is worst-case optimal. See [GH89, GHL+87, LP84] for further details about computing shortest paths in simple polygons.

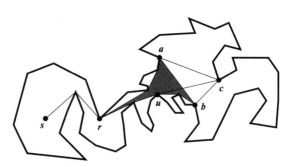

FIGURE 31.1.1
Splitting a funnel.

SHORTEST PATH MAPS

The shortest path map SPM(s) for a simple polygon has a particularly simple structure, since the boundaries between cells in the map are (line segment) chords of P obtained by extending appropriate edges of the visibility graph VG(P). It can be computed in time $O(n)$ by using somewhat more sophisticated data structures to do funnel splitting efficiently; in this case, we cannot discard one side of each split funnel. Single-source queries can be answered in $O(\log n)$ time, after storing the SPM(s) in an appropriate $O(n)$-size point location data structure (see Chapter 38). SPM(s) includes a tree of shortest paths from s to every vertex of P. For further details and proofs involving shortest path maps, see [GH89, GHL+87, LP84, Mit91].

TWO-POINT QUERIES

A simple polygon can be preprocessed in time $O(n)$, into a data structure of size $O(n)$, to support shortest-path queries between any two points $s, t \in P$. In time

$O(\log n)$ the length of the shortest path can be reported, and in additional time $O(k)$, the shortest path can be reported, where k is the number of vertices in the output path [GH89, Her91].

DYNAMIC VERSION

In the dynamic version of the problem, one allows the polygon P to change with addition and deletion of edges and vertices. If the changes are always made in such a way that the set of all edges yields a **connected planar subdivision** of the plane into simple polygons (i.e., no "islands" are created), then one can maintain a data structure of size $O(n)$ that supports two-point query time of $O(\log^2 n)$ (plus $O(k)$ if the path is to be reported), and update time of $O(\log^2 n)$ for each addition/deletion of an edge/vertex [GT97].

TABLE 31.1.1   Shortest paths and geodesic distance in simple polygons.

PROBLEM VERSION	COMPLEXITY	NOTES	SOURCE
Shortest s-t path	$O(n)$	space $O(n)$	[LP84]
	$O(\frac{n^2}{m} + n \log m \log^4(\frac{n}{m}))$ exp.	space m	[Har15]
	$O(n^2/m)$ expected	space $m = O(\frac{n}{\log^2 n})$	[Har15]
Single-source query	$O(\log n)$ query	builds SPM(s)	[GHL$^+$87]
	$O(n)$ preproc/space		
Two-point query	$O(\log n)$ query		[GH89]
	$O(n)$ preproc/space		
Two-polygon query	$O(\log k + \log n)$ query	between convex k-gons	[CT97]
	$O(n)$ space	in simple n-gon	
Dynamic two-point query	$O(\log^2 n)$ update/query		[GT97]
	$O(n)$ space		
Dynamic two-polygon	$O(\log k + \log^2 n)$ query	between convex k-gons	[CT97]
	$O(\log^2 n)$ update	in simple n-gon	
	$O(n)$ space		
Parallel algorithm	$O(\log n)$ time	in triangulated polygon	[Her95]
(CREW PRAM)	$O(n/\log n)$ processors	also builds SPM(s)	
Geodesic VD	$O((n+k)\log(n+k))$	k point sites	[PL98]
All nearest neighbors	$O(n)$	for set of vertices	[HS97]
Geodesic farthest-site VD	$O((n+k)\log(n+k))$ time	k point sites	[AFW93]
	$O(n+k)$ space		
Geodesic farthest-site VD	$O((n+k)\log\log n)$ time	k sites on ∂P	[OBA16]
All farthest neighbors	$O(n)$	for set of vertices	[HS97]
Geodesic diameter	$O(n)$		[HS97]
Geodesic center	$O(n)$		[ABB$^+$16]

OTHER RESULTS

Several other problems studied with respect to geodesic distances induced by a simple polygon are summarized in Table 31.1.1. See also Table 29.4.1.

Shortest paths within simple polygons yield a wealth of structural information about the polygon. In particular, they have been used to give an output-sensitive algorithm for constructing the visibility graph of a simple polygon ([Her89]) and can be used for constructing a *geodesic triangulation* of a simple polygon, which allows for efficient ray-shooting (see [CEG+94]). They also form a crucial step in solving *link distance* problems, as we will discuss later. Simplification of a simple polygon to preserve geodesic distances leads to algorithms whose running times can be written in terms of the number of reflex vertices (having internal angle greater than π), instead of the total number, n, of vertices; see [AHK+14]. An output-sensitive algorithm for computing geodesic disks within a simple polygon is given in [BKL11], along with a clustering algorithm in the geodesic metric within a simple polygon.

OPEN PROBLEMS

1. Can one devise a simple $O(n)$ time algorithm for computing the shortest path between two points in a simple polygon, *without* resorting to a (complicated) linear-time triangulation algorithm?

2. What are the best possible query/update times possible for the dynamic versions of the shortest path problem?

3. Can the geodesic Voronoi diagram for k sites within P be computed in time $O(n + k \log k)$?

31.2 PATHS IN A POLYGONAL DOMAIN

While in a simple polygon there is a unique taut-string path between two points, in a general polygonal domain P, there can be an exponential number of taut-string simple paths between two points.

The homotopy type of a path can be expressed as a sequence (with repetitions) of triangles visited, for some triangulation of P. For any given homotopy type, expressed with N triangles, a shortest path of that type can be computed in $O(N)$ time [HS94]. Efficient algorithms for computing a set of homotopic shortest paths among obstacles, for many pairs of start and goal points, are known [Bes03b, EKL06]. One can also efficiently test, in time $O(n \log n)$, if two simple paths are of the same homotopy type in a polygonal domain; here, n is the total number of vertices of the input paths and the polygonal domain [CLMS02]. For more details on shortest curves of a specified topology, see Chapter 23.

SEARCHING THE VISIBILITY GRAPH

Without loss of generality, we can assume that s and t are vertices of P (since we can make "point" holes in P at s and t). It is easy to show that any locally optimal s-t path must lie on the visibility graph VG(P) (Figure 31.0.1). We can construct VG(P) in output-sensitive time $O(E_{\text{VG}} + n \log n)$, where E_{VG} denotes the number of edges of VG(P) [GM91], even if we allow only $O(n)$ working space [PV96a]. In fact, a recent result [CW15b] shows that, after triangulation of P, VG(P) can be computed in time $O(E_{\mathrm{VG}} + n + h \log h)$. Given the graph VG($P$), whose edges are weighted by their Euclidean lengths, we can use Dijkstra's algorithm to construct a tree of shortest paths from s to all vertices of P, in time $O(E_{\mathrm{VG}} + n \log n)$ [FT87]. Thus, Euclidean shortest paths among obstacles in the plane can be computed in time $O(E_{\mathrm{VG}} + n \log n)$. This bound is worst-case quadratic in n, since $E_{\mathrm{VG}} \le \binom{n}{2}$; note too that domains exist with $E_{\mathrm{VG}} = \Omega(n^2)$.

Given the tree of shortest paths from s, we can compute SPM(s) in time $O(n \log n)$, by computing an additively weight Voronoi diagram (see Chapter 27) of the vertices, with each vertex weighted by its distance from s.

CONTINUOUS DIJKSTRA METHOD

Instead of searching the visibility graph (which may have quadratic size), an alternative paradigm for shortest-path problems is to construct the (linear-size) shortest path map directly. The ***continuous Dijkstra*** method was developed for this purpose.

Building on the success of the method in solving (in nearly linear time) the shortest-path problem for the L_1 metric, Mitchell [Nit96] developed a version of the continuous Dijkstra method applicable to the Euclidean shortest-path problem, obtaining the first subquadratic ($O(n^{1.5+\varepsilon})$) time bound. Subsequently, this result was improved by Hershberger and Suri [HS99], who achieve a nearly optimal algorithm based also on the continuous Dijkstra method. They give an $O(n \log n)$ time and $O(n \log n)$ space algorithm, coming close to the lower bounds of $\Omega(n + h \log h)$ time and $O(n)$ space.

The continuous Dijkstra paradigm involves simulating the effect of a wavefront propagating out from the source point, s. The ***wavefront*** at distance δ from s is the set of all points of P that are at geodesic distance δ from s. It consists of a set of curve pieces, called ***wavelets***, which are arcs of circles centered at obstacle vertices that have already been reached. At certain critical "events," the structure of the wavefront changes due to one of the following possibilities:

(1) a wavelet disappears (due to the closure of a cell of the SPM);

(2) a wavelet collides with an obstacle vertex;

(3) a wavelet collides with another wavelet; or

(4) a wavelet collides with an obstacle edge at a point interior to that edge.

It is not difficult to see from the fact that SPM(s) has linear size, that the total number of such events is $O(n)$. The challenge in applying this propagation scheme is devising an efficient method to know *what* events are going to occur and in being able to *process* each event as it occurs (updating the combinatorial structure of the wavefront).

One approach, used in [Nit96], is to track a "pseudo-wavefront," which is allowed to run over itself, and to "clip" only when a wavelet collides with a vertex that has already been labeled due to an earlier event. Detection of when a wavelet collides with a vertex is accomplished with range-searching techniques. An alternative approach, used in [HS99], simplifies the problem by first decomposing the domain P using a *conforming subdivision*, which allows one to propagate an approximate wavefront on a cell-by-cell basis. A key property of a conforming subdivision is that any edge of length L of the subdivision has only a constant number of (constant-sized) cells within geodesic distance L.

APPROXIMATION ALGORITHMS

One can compute approximate Euclidean shortest paths using standard methods of discretizing the set of directions. Clarkson [Cla87] gives an algorithm that uses $O((n \log n)/\varepsilon)$ time to build a data structure of size $O(n/\varepsilon)$, after which a $(1 + \varepsilon)$-approximate shortest path query can be answered in time $O(n \log n + n/\varepsilon)$. (These bounds rely also on an observation in [Che95].) Using a related approach, based on approximating Euclidean distance with fixed orientation distances, Mitchell [Mit92] computes a $(1 + \varepsilon)$-approximate shortest path in time $O((n \log n)/\sqrt{\varepsilon})$ using $O(n/\sqrt{\varepsilon})$ space. Chen, Das, and Smid [CDM01] have shown an $\Omega(n \log n)$ lower bound, in the algebraic computation tree model, on the time required to compute any approximation to the shortest path.

TWO-POINT QUERIES

Two-point queries in a polygonal domain are much more challenging than in the case of simple polygons, where optimal algorithms are known. One natural approach (observed by Chen et al. [CDK01]) is to store the shortest path map, SPM(v), rooted at each vertex v; this requires $O(n^2)$ space. Then, for a query pair (s, t), we compute the set of k_s vertices visible to s and k_t vertices visible to t, in time $O(\min\{k_s, k_t\} \log n)$, using the visibility complex of Pocchiola and Vegter [PV96b]. Then, assuming that $k_s \leq k_t$, we simply locate t in each of the k_s SPM's rooted at the vertices visible from s. This permits two-point queries to be answered in time $O(\min\{k_s, k_t\} \log n)$, which is worst-case $\Omega(n \log n)$, making it no better than computing a shortest path from scratch, in the worst case.

Methods for exact two-point queries that are efficient in the worst case utilize an *equivalence decomposition* of the domain P, for which all points z within a cell of the decomposition have topologically equivalent shortest path maps. Given query points s and t, one locates s within the decomposition, and then uses the resulting SPM, along with a parametric point location data structure, to locate t within the SPM with respect to s. The complexity of the decomposition can be quite high; there can be $\Omega(n^4)$ topologically distinct shortest path maps with respect to points within P. Chiang and Mitchell [CM99] have utilized this approach to obtain various tradeoffs between space and query time; see Table 31.2.1. Unfortunately, the space bounds are all impractically high. Quadratic space is possible with a query time of $O(h \log n)$ [GMS08]. If the query points are restricted to lie on the boundary, ∂P, of the domain, then $O(\log n)$ query can be achieved with preprocessing time/space of $\tilde{O}(n^5)$ [BO12], where $\tilde{O}(\cdot)$ indicates that polylogarithmic factors are ignored.

More efficient methods allow one to approximately answer two-point queries. As observed in [Che95], the method of Clarkson [Cla87] can be used to construct a data structure of size $O(n^2 + n/\varepsilon)$ in $O(n^2 \log n + (n/\varepsilon) \log n)$ time, so that two-point $(1 + \varepsilon)$-optimal queries can be answered in time $O((\log n)/\varepsilon)$, for any fixed $\varepsilon > 0$. Chen [Che95] was the first to obtain nearly *linear*-space data structures for approximate shortest path queries; these were obtained, though, at the cost of a higher approximation factor. He obtains a $(6 + \varepsilon)$-approximation, using $O(n^{3/2}/\log^{1/2} n)$

time to build a data structure of size $O(n \log n)$, after which queries can be answered in time $O(\log n)$. Arikati et al. [ACC$^+$96] give a spectrum of results based on planar t-spanners (see Section 32.3), with tradeoffs among the approximation factor and the preprocessing time, storage space, and query time. One such result gives a $(3\sqrt{2}+\varepsilon)$-approximation in query time $O(\log n)$, after using $O(n^{3/2}/\log^{1/2} n)$ time to build a data structure of size $O(n \log n)$.

In the special case that the polygonal domain is "t-rounded," meaning that the shortest path distance between any two vertices is at most some constant t times the Euclidean distance between them, Gudmundsson et al. [GLNS08] show that in query time $O(\log n)$, one can give a $(1 + \varepsilon)$-approximate answer to a two-point shortest path query while using only $O(n \log n)$ space and preprocessing time. Their result utilizes approximate distance oracles in t-spanner graphs, giving $O(1)$-time approximate distance queries between pairs of vertices; see Section 32.3.

OTHER RESULTS

The geodesic Voronoi diagram of k sites inside P can be constructed in time $O((n+k)\log(n+k))$, using the continuous Dijkstra method, simply starting with multiple source points. While the geodesic center/diameter problem has been solved in linear time for the case of simple polygons, in polygonal domains, the problem becomes much harder and the time bounds are much worse. For computing the geodesic diameter, [BKO13] show time bounds of $O(n^{7.73})$ or $O(n^7(\log n+h))$; one complicating fact is that the endpoints of a shortest path achieving the geodesic diameter can be interior to the domain. The geodesic center can be computed in time $O(n^{12+\varepsilon})$ [BKO15], which has been recently improved to $O(n^{11} \log n)$ [Wan16].

In the case of a planar domain with h *curved* obstacles, specified as a set of *splinegons* (polygons in which edges are replaced by convex curved arcs), having a total complexity of n, recent results have generalized the methods that were developed for polygonal domains. In particular, [HSY13] employ the continuous Dijkstra paradigm to obtain shortest paths in time $O(n \log n)$, under certain assumptions on the curved arcs, and $(1 + \varepsilon)$-approximate shortest paths (and a shortest path map) in time $O(n \log n + n \log(1/\varepsilon))$, under mild assumptions on the curved arcs. Further, [CW15a] provide an algorithm with running time $O(n + k + h \log^{1+\varepsilon} h)$, where $k = O(h^2)$ is the number of free common tangents among curved obstacles (related to the size of the relevant visibility graph); while the running time is worst-case quadratic in the number of obstacles, it is *linear* in n. Shortest paths for a point moving among curved obstacles arises in optimal path planning for a robot (e.g., a circular disk) among polygonal obstacles.

In [AEK$^+$16], the notion of a shortest path map is generalized to consider geodesic length queries from a source point s to a query line segment within P, or to a query visibility polygon; this allows one to rapidly compute the geodesic distance from s to a point that sees the query point t. Another generalization of the notion of a shortest path map is the kth *shortest path map*, k-SPM, in which the domain is decomposed according to the combinatorial type of the kth shortest homotopically distinct (different "threading") path from source s to destination t. The combinatorial complexity of the k-SPM is $\Theta(k^2h+kn)$, and it can be computed in time $O((k^3h + k^2n)\log(kn))$ [EHP$^+$15].

Table 31.2.1 summarizes various results.

TABLE 31.2.1   Shortest paths among planar obstacles, in a polygonal domain.

PROBLEM	COMPLEXITY	NOTES	SOURCE
Shortest s-t path	$O(n \log n)$	$O(n \log n)$ space	[HS99]
	$O(n + h^2 \log n)$	$O(n)$ space	[KMM97]
	$O(n^{1.5+\varepsilon})$	$O(n)$ space	[Nit96]
Approx shortest s-t path	$O((n \log n)/\sqrt{\varepsilon})$	$O(n/\sqrt{\varepsilon})$ space	[Mit92]
SPM(s)/geodesic VD	$O(n \log n)$	$O(n \log n)$ space	[HS99]
	$O(n^{1.5+\varepsilon})$	$O(n)$ space	[Nit96]
Two-point query	$O(\log n)$ query	exact	[CM99]
	$O(n^{11})$ preproc/space		
Two-point query	$O(\log^2 n)$ query	exact	[CM99]
	$O(n^{10} \log n)$ preproc/space		
Two-point query	$O(n^{1-\delta} \log n)$ query	exact	[CM99]
	$O(n^{5+10\delta+\varepsilon})$ preproc/space	$0 < \delta \leq 1$	
Two-point query	$O(\log n + h)$ query	exact	[CM99]
	$O(n^5)$ preproc/space		
Two-point query	$O(h \log n)$ query	exact	[CM99]
	$O(n + h^5)$ preproc/space		
Two-point query	$O(h \log n)$ query	exact	[GMS08]
	$O(n^2)$ space		
Two-point query	$O(\log n)$ query	exact	[BO12]
	$\tilde{O}(n^5)$ space	queries on ∂P	
Approx two-point query	$O(\log n + \rho)$ query	$(1 + \varepsilon)$-approx	[Che13]
	$O(n^2/\rho)$ space	any integer ρ	
	$O(n^2/\rho)$ preproc	$1 \leq \rho \leq \sqrt{n}$	
Approx two-point query	$O(n)$ query	$(1 + \varepsilon)$-approx	[Che13]
	$O(n)$ space		
	$O(n \log n)$ preproc		
Approx two-point query	$O(\log n)$ query	$(2 + \varepsilon)$-approx	[Che13]
	$O(n^{3/2})$ space		
	$O(n^{3/2})$ preproc		
Approx two-point query	$O(\log n)$ query	$(3 + \varepsilon)$-approx	[Che13]
	$O(n \log n)$ space	q is cover number	
	$O(n \log n + q^{3/2}/\sqrt{\log q})$ preproc	$1 \leq q \leq n$	
Approx two-point query	$O(\log n)$ query	$(1 + \varepsilon)$-approx	[GLNS08]
	$O(n \log n)$ space	t-rounded domain	[GLNS08]
	$O(n \log n)$ preproc		
Geodesic diameter	$O(n^{7.73})$ or $O(n^7(\log n + h))$		[BKO13]
Geodesic center	$O(n^{12+\varepsilon})$, $O(n^{11} \log n)$		[BKO15, Wan16]

OPEN PROBLEMS

1. Can the Euclidean shortest-path problem be solved in $O(n + h \log h)$ time and $O(n)$ space? (The L_1 shortest path problem can be solved in time $O(n + h \log h)$ in a triangulated domain [CW13].)

2. How efficiently, and using what size data structure, can one preprocess a polygonal domain for exact two-point queries? Can one obtain sublinear queries using a reasonable amount of space (say, subquadratic)?

3. How efficiently can one compute a geodesic center/diameter for a polygonal domain? Current polynomial-time bounds are likely far from optimal.

31.3 OTHER METRICS FOR LENGTH

In the problems considered so far, the Euclidean metric has been used to measure the length of a path. We consider now several other possible objective functions for measuring path length. Tables 31.3.1 and 31.3.2 summarize results.

GLOSSARY

L_p **metric:** The L_p distance between $q = (q_x, q_y)$ and $r = (r_x, r_y)$ is given by $d_p(q, r) = [|q_x - r_x|^p + |q_y - r_y|^p]^{1/p}$. The L_p length of a polygonal path is the sum of the L_p lengths of each edge of the path. Special cases of the L_p metric include the L_1 metric (**Manhattan metric**) and the L_∞ metric ($d_\infty(q, r) = \max\{|q_x - r_x|, |q_y - r_y|\}$).

Rectilinear path: A polygonal path with each edge parallel to a coordinate axis; also known as an **isothetic** path.

C-oriented path: A polygonal path with each edge parallel to one of a finite set C of $c = |C|$ **fixed orientations**.

Link distance: The minimum number of edges in a polygonal path from s to t within a polygonal domain P. If the paths are restricted to be rectilinear or C-oriented, then we obtain the **rectilinear link distance** or **C-oriented link distance**.

Min-link s-t path: A polygonal path from s to t that achieves the link distance.

Weighted region problem: Given a piecewise-constant function $f : \mathbb{R}^2 \to \mathbb{R}$ that is defined by assigning a nonnegative **weight** to each face of a given triangulation in the plane. The **weighted length** of an s-t path π is the path integral, $\int_\pi f(x, y)d\sigma$, of the weight function along π. The **weighted region metric** associated with f defines the distance $d_f(s, t)$ to be the infimum over all s-t paths π of the weighted length of π. The **weighted region problem** (WRP) asks for an s-t path of minimum weighted length.

Anisotropic path problem: Compute a minimum-cost path, where the cost of motion is *direction-dependent*. Additionally, the cost of motion may depend on a weight function f, as in the WRP.

Sailor's problem: Compute a minimum-cost path, where the cost of motion is *direction-dependent*, and there is a cost L per turn (in a polygonal path).

Bounded curvature shortest-path problem: Compute a shortest obstacle-avoiding smooth (C^1) path joining point s, with prescribed velocity orientation, to point t, with prescribed velocity orientation, such that at each point of the path the radius of curvature is at least 1.

Maximum concealment path: A path within polygonal domain P that minimizes the length during which the robot is exposed to a given set of "enemy" observers. This problem is a special case of the weighted region problem, in which weights are 0 (for travel in concealed free space), 1 (for travel in exposed free space), or ∞ (for travel through obstacles).

Total turn for an s-t path: The sum of the absolute values of all turn angles for a polygonal s-t path.

Minimum-time path problem: Find a path to minimize the total time required
to move from an initial position, at an initial velocity, to a goal position and
velocity, subject to bounds on the allowed acceleration and velocity along the
path. This problem is also known as the **kinodynamic motion planning
problem.**

LINK DISTANCE

In the min-link path problem, our goal is to minimize the number of links (and hence
the number of turns) in a path connecting s and t. In many problems, the link
distance provides a more natural measure of path complexity than the Euclidean
length, as well as having applications to curve simplification.

In a simple polygon P, a min-link path can be computed in time $O(n)$; see
also [LSD00] for a survey on link distance. In fact, in time $O(n)$ a **window partition** of P with respect to a point s can be computed, after which a min-link
path from s to t can be reported in time proportional to the link distance. The
algorithm, due to Suri [Sur90], computes the partition via "staged illumination,"
essentially a form of the continuous Dijkstra method under the link distance metric.

In a polygonal domain with holes, min-link paths can also be computed using
a staged illumination method, but the algorithm is not simple: it relies on efficient
methods for computing a single face in an arrangement of line segments (see Chapter 28). A min-link s-t path can be computed in time $O(E_{VG}\alpha^2(n)\log n)$, where
$\alpha(n)$ is the inverse Ackermann function; see Section 28.10. Computing link distance
in a polygonal domain in significantly subquadratic time may not be possible; deciding if the link distance between two points is 3 is 3SUM-hard [MPS14]. This
implies that obtaining a $4/3 - \varepsilon$ factor approximation is also 3SUM-hard; in fact,
obtaining an $O(1)$ additive approximation or a factor $(2 - \varepsilon)$ approximation is also
3SUM-hard, even in rectilinear domains [MPS14]. If we consider C-oriented and
rectilinear link distance, in which edges of the polygonal path must be from among
a given set of C directions (axis-parallel in the rectilinear case), then significantly
better time/space bounds are possible, and some of these apply also to combined
metrics, in which there is a cost for length as well as links.

Refer to Table 31.3.1 for many related results on link distance, including rectilinear link distance, and on two-point queries.

L_1 METRIC

Instead of measuring path length according to the L_2 (Euclidean) metric, consider
the problem of computing shortest paths in a polygonal domain P that are short
according to the L_1 metric.

A method based on visibility graph principles allows one to construct a sparse
graph (with $O(n\log n)$ nodes and edges) that is **path-preserving** in that it is
guaranteed to contain a shortest path between any two vertices. Applying Dijkstra's
algorithm then gives an $O(n\log^{1.5} n)$ time ($O(n\log n)$ space) algorithm for L_1-
shortest paths.

A method based on the continuous Dijkstra paradigm allows the SPM(s) to
be constructed in time $O(n\log n)$, using $O(n)$ space [Mit92], and, most recently in
time $O(n + h\log h)$ for a triangulated domain with h holes [CW13]. The special

TABLE 31.3.1   Link distance shortest-path problems.

PROBLEM	COMPLEXITY	NOTES	SOURCE
Min-link path	$O(E_{\mathrm{VG}}\alpha^2(n)\log n)$	polygonal domain	[MRW92]
Min-link path	3SUM-hard	polygonal domain	[MPS14]
Min-link path	$O(\sqrt{h})$-approx	polygonal domain, h holes	[MPS14]
Min-link path	$O(n)$	simple polygon	[Sur86, Sur90]
Rectilinear link path	$O(n)$	rectilinear simple polygon	[Ber91, HS94]
Rectilinear link path	$O(n + h\log h)$ time	rectilinear domain	[MPSW15]
	$O(n)$ space, $O(\log n)$ query	h holes, triangulated	
C-oriented link path	$O(C^2 n\log n)$ time	polygonal domain	[MPS14]
	$O(Cn)$ space, $O(C\log n)$ query		
C-oriented link path	$O(Cn\log n)$ time	polygonal domain	[MPS14]
	$O(n)$ space, $O(\log n)$ query	2-approx	[MPS14]
Two-point link query	$O(\log n)$ query	simple polygon	[AMS95]
	$O(n^3)$ space, preproc		
Two-point rectilinear	$O(\log n)$ query	rectilinear simple polygon	[Sch96]
link query	$O(n\log n)$ preproc, $O(n)$ space	also is L_1-opt	
Shortest k-link path	$O(n^3 k^3 \log(Nk/\varepsilon^{1/k}))$	simple polygon	[MPA92]

property of the L_1 metric that is exploited in this algorithm is the fact that the wavefront in this case is piecewise-linear, with wavelets that are line segments of slope ± 1, so that the first vertex hit by a wavelet can be determined by rectangular range searching techniques (see Chapter 40).

Methods for finding L_1-shortest paths generalize to the case of C-oriented paths, in which $c = |C|$ fixed directions are given. Shortest C-oriented paths can be computed in time $O(cn\log n)$. Since the Euclidean metric is approximated to within accuracy $O(1/c^2)$ if we use c equally-spaced orientations, this results in an algorithm that computes, in time $O((n/\sqrt{\varepsilon})\log n)$, a path guaranteed to have length within a factor $(1+\varepsilon)$ of the Euclidean shortest path length.

WEIGHTED REGION METRIC

The weighted region problem (WRP) seeks an optimal s-t path according to the weighted region metric d_f induced by a given weight function f, often specified by a piecewise-constant (or piecewise linear) function on a given triangulated domain in two (or more) dimensions. This problem is a natural generalization of the shortest-path problem in a polygonal domain: consider a weight function that assigns weight 1 to P and weight ∞ (or a sufficiently large constant) to the obstacles (the complement of P). The WRP models the minimum-time path problem for a point robot moving in a terrain of varied types (e.g., grassland, brushland, black-top, bodies of water, etc.), where each type of terrain has an assigned weight equal to the reciprocal of the maximum speed of traversal for the robot.

A standard formulation of the WRP assumes a piecewise-constant weight function f, specified by a triangulation in the plane having n vertices, with each face assigned an integer weight $\alpha \in \{0, 1, \ldots, W, +\infty\}$. (We can allow each edge of the triangulation to have a weight that is possibly distinct from that of the triangular facets on either side of it; in this way, linear features such as roads can be mod-

TABLE 31.3.2 Shortest paths in other metrics.

PROBLEM	COMPLEXITY	NOTES	SOURCE
L_1-shortest path, SPM(s)	$O(n \log n)$	polygonal domain	[Mit92, Mit89]
L_1 geodesic diameter	$O(n)$	simple polygon	[BKOW15]
L_1 geodesic center	$O(n)$	simple polygon	[BKOW15]
L_1 geodesic diameter	$O(n^2 + h^4)$	polygon with h holes	[BAE+17]
L_1 geodesic center	$O((n^4 + n^2 h^4)\alpha(n))$	polygon with h holes	[BAE+17]
L_1 two-point query	$O(\log^2 n)$ query $O(n^2 \log n)$ space $O(n^2 \log^2 n)$ preproc	polygonal domain	[CKT00]
L_1 two-point query	$O(\log n + k)$ query $O(n + h^{2+\varepsilon})$ space, preproc	polygonal domain	[CIW14]
L_1 two-point query	$O(\log n)$ query $O(n^2)$ space, preproc	rectangle obstacles	[AC91, AC93] [EM94]
L_1 two-point query	$O(\sqrt{n})$ query $O(n^{1.5})$ space, preproc	rectangle obstacles	[EM94]
L_1 two-point query	$O(\log n)$ query $O(n \log n)$ space $O(n \log^2 n)$ preproc	3-approx rectangle obstacles	[CK96]
C-oriented shortest path	$O(cn \log n)$		[Mit92]
two-point query	$O(c^2 \log^2 n)$ query	$O(c^2 n^2 \log^2 n)$ preproc	[CDK01]
Weighted region problem	$O(ES)$, or $O(n^8 L)$ $L = O(\log \frac{nNW}{\varepsilon})$	$(1+\varepsilon)$-approx	[MP91]
Weighted region problem	$O(\frac{kn + k^4 \log(k/\varepsilon)}{\varepsilon} \log^2 \frac{Wn}{\varepsilon})$ parameter $3 \le k \le n$	$(1+\varepsilon)$-approx weights $[1, W] \cup \{\infty\}$	[CJV15] [CJV15]
Weighted region problem	$O(\frac{n}{\sqrt{\varepsilon}} \log \frac{n}{\varepsilon} \log \frac{1}{\varepsilon})$	$(1+\varepsilon)$-approx geometric parameters	[AMS05]
Weighted region problem	$O((W \log W)\frac{n^3}{\varepsilon} \log \frac{Wn}{\varepsilon})$	$(1+\varepsilon)$-approx indep of geometry also anisotropic cost	[CNVW08]
Weighted region problem	$O(n^2)$	weights 0, 1, ∞	[GMMN90]
L_1 weighted region prob	$O(n \log^{3/2} n)$ preproc $O(\log n)$ query $O(n \log n)$ space	rectilinear regions single-source queries	[CKT00]
L_1 WRP, two-point query	$O(\log^2 n)$ query $O(n^2 \log^2 n)$ space, preproc	rectilinear regions	[CKT00]
L_1 WRP, two-point query	$O(\log n)$ query $O(n^{2+\varepsilon})$ space, preproc	rectilinear regions	[CIW14]
Bounded curvature path	$O(n^4 \log n)$	moderate obstacles	[BL96]
Bounded curvature path	$O(n^2 \log n)$	within convex polygon	[ABL+02]
Anisotropic path problem	$O(\frac{\rho^2 n^3}{\varepsilon^2} (\log \frac{\rho n}{\varepsilon})^2)$ $O(\frac{\rho^2 n^3}{\varepsilon^2} \log \frac{\rho n}{\varepsilon})$ space	parameter $\rho \ge 1$ $O(\log \frac{\rho n}{\varepsilon})$ query	[CNVW10]
Sailor's problem ($L = 0$)	$O(n^2)$	polygonal domain	[Sel95]
Sailor's problem ($L > 0$)	poly(n, ε)	ε-approx	[Sel95]
Max concealment	$O(v^2(v + n)^2)$	simple polygon	[GMMN90]
v viewpoints	$O(v^4 n^4)$	polygonal domain	[GMMN90]
Min total turn	$O(E_{\text{VG}} \log n)$	polygonal domain	[AMP91]

eled.) The local optimality condition, which follows from basic calculus, is that an optimal path must be polygonal (for piecewise-constant f), bending according to *Snell's Law of Refraction* when crossing a region boundary, and, if it utilizes a portion of an edge of the triangulation, it must turn to enter/leave the edge at the "critical angle" of refraction determined by the weight of the edge and the weight of the adjacent face.

Exact solution of the general WRP in the plane seems to be very difficult for algebraic reasons; the problem cannot be solved in the "algebraic computational model over the rational numbers" [DCG$^+$14]. Algorithms are, therefore, focused on approximation and generally fall into two categories: (1) those based on the continuous Dijkstra paradigm, propagating "intervals of optimality," which partition edges according to the combinatorial type of paths that optimally reach the edge from either side, while utilizing the local optimality condition during a breadth-first propagation; and, (2) those based on placing discrete sample points along edges or interior to faces, and searching for a shortest path in a corresponding network of edges interconnecting the sample points.

An algorithm of type (1) can be viewed as "exact" in the sense that it would give exactly optimal paths if the underlying predicates could be performed exactly. The predicates require determining a refraction path through a specified edge sequence in order to reach a specified vertex, or to find a "bisector" point b along an edge where the refraction paths to b through two specified edge sequences have the same lengths. The first provable result for the WRP was of type (1) [MP91]; it computes a $(1+\varepsilon)$-approximate optimal path, for any fixed $\varepsilon > 0$, in time $O(E \cdot S)$, where E is the number of "events" in the continuous Dijkstra algorithm, and S is the complexity of performing a numerical search to approximately solve the refraction-path predicates. It is shown that $O(n^4)$ is an upper bound on E, and that this bound is tight in the worst case, since there are instances in which the total number of intervals of optimality is $\Omega(n^4)$. The numerical search can be accomplished in time $S = O(k^2 \log(nNW/\varepsilon))$, on k-edge sequences, and it is shown that the maximum length k of an edge sequence is $O(n^2)$; thus, the overall time bound is at most $O(n^8 \log(nNW/\varepsilon))$ [MP91]. (A new variant utilizing a "discretized wavefront" approach, saving a factor of $O(n^3)$, has recently been announced [IK15].)

Algorithms of type (2) carefully place Steiner points on the edges (or, possibly interior to faces) of the input subdivision. Using a logarithmic discretization (as in [Pap85]), with care in how Steiner points are placed near vertices, provable approximation guarantees are obtained. A time bound with near-linear dependence on n, specifically $O((n/\sqrt{\varepsilon}) \log(n/\varepsilon) \log(1/\varepsilon))$, is possible [AMS05]; however, this bound has a hidden constant (in the big-O) that depends on the geometry (smallest angles) of the triangulation in such a way that a single tiny angle can cause the bounds to go to infinity. Using a different discretization method, [CNVW08] give an algorithm with running time $O(((W \log W)/\varepsilon)n^3 \log(Wn/\varepsilon))$, independent of the angles of the triangulation; in fact, this algorithm solves also the *anisotropic* generalization, in which a (possibly asymmetric) convex distance function, specified for each face of the triangulation, is utilized. Single-source approximate optimal path queries can be answered efficiently, even in anisotropic weighted regions, using a type of shortest path map data structure [CNVW10]. In [CJV15], new bounds that depend on k, the smallest integer so that the sum of the k smallest angles in the triangular faces is at least π; specifically, they obtain a $(1 + \varepsilon)$-approximation in time $O((kn + k^4 \log(k/\varepsilon))(1/\varepsilon) \log^2(Wn/\varepsilon))$. If the triangulation is "nice," one expects k to be a small constant, and the running time is near-linear in n. It should be noted that in algorithms of type (2) the dependence on $1/\varepsilon$ is polynomial (versus logarithmic in algorithms of type (1)), and the methods are not "exact" in that no matter how accurately predicates are evaluated, the algorithms do not guarantee to find the correct combinatorial type of an optimal path.

There are special cases of the weighted region problem that admit faster and simpler algorithms. For example, if the weighted subdivision is rectilinear, and path length is measured according to weighted L_1 length, then efficient algorithms for single-source and two-point queries can be based on searching a path-preserving graph [CKT00]. Similarly, if the region weights are restricted to $\{0, 1, \infty\}$ (while edges may have arbitrary nonnegative weights), then an $O(n^2)$ time algorithm can be based on a path-preserving graph similar to a visibility graph [GMMN90]. This also leads to an efficient method for performing **lexicographic** optimization, in which one prioritizes various types of regions according to which is most important for path length minimization.

The anisotropic path problem includes a generalization to the case in which each face of a given polygonal subdivision may have a different cost function, and the cost may depend on the direction of movement. In the model of Cheng et al. [CNVW10], distance in face f is measured according to a (possibly asymmetric) convex distance function whose unit disk D_f is contained within a (concentric) Euclidean unit disk and contains a (concentric) Euclidean disk of radius $1/\rho$, for a real parameter $\rho \geq 1$ that quantifies the degree of directionality of asymmetry in the cost function. Their algorithm uses time $O(\frac{\rho^2 n^3}{\varepsilon^2}(\log \frac{\rho n}{\varepsilon})^2)$ to compute a data structure of size $O(\frac{\rho^2 n^3}{\varepsilon^2} \log \frac{\rho n}{\varepsilon})$ that enables $(1+\varepsilon)$-approximate optimal path queries in time $O(\log \frac{\rho n}{\varepsilon})$.

The weighted region model applies also to the problem of computing "high-quality" paths among obstacles. In particular, it is natural to consider the cost of motion that is very close to an obstacle to be more costly than motion that has high clearance from obstacles. Letting the cost of motion at clearance δ from the nearest obstacle be weighted by $1/\delta$, and the cost of a path to be the weighted length (path integral of the cost function), one can obtain a fully polynomial time approximation scheme to compute a path with cost at most $(1 + \varepsilon)$ times optimal in time $O((n^2/\varepsilon^2) \log(n/\varepsilon))$ in a planar polygonal environment with n vertices [AFS16].

MINIMUM-TIME PATHS

The *kinodynamic motion planning problem* (also known as the *minimum-time path problem*) is a nonholonomic motion planning problem in which the objective is to compute a *trajectory* (a time-parameterized path, $(x(t), y(t))$) within a domain P that minimizes the total time necessary to move from an initial configuration (position and initial velocity) to a goal configuration (position and velocity), subject to bounds on the allowed acceleration and velocity along the path. (Algorithmic motion planning is discussed in detail in Chapter 50.) The minimum-time path problem is a difficult optimal control problem; optimal paths will be complicated curves given by solutions to differential equations.

The bounds on acceleration and velocity are most often given by upper bounds on the L_∞ norm (the "decoupled case") or the L_2 norm (the "coupled case").

If there is an upper bound on the L_∞ norm of the velocity and acceleration vectors, one can obtain an *exact*, exponential-time, polynomial-space algorithm, based on characterizing a set of "canonical solutions" (related to "bang-bang" controls) that are guaranteed to include an optimal solution path. This leads to an expression in the first-order theory of the reals, which can be solved exactly; see Chapter 37. However, it remains an open question whether or not a polynomial-time algorithm exists.

Donald et al. [DXCR93, DX95, RW00] developed approximation methods, including a polynomial-time algorithm that produces a trajectory requiring time at most $(1 + \varepsilon)$ times optimal, for the decoupled case. Their approach is to discretize (uniformly) the four-dimensional phase space that represents position and velocity, with special care to ensure that the size of the grid is bounded by a polynomial in $1/\varepsilon$ and n. Approximation algorithms for the coupled case are also known [DX95, RT94].

Optimal paths for a "car-like" robot (a "Dubins car") leads to the closely related shortest-path problem, the **bounded curvature shortest-path problem**, in which we require that no point of the path have a radius of curvature less than 1. For this problem, $(1+\varepsilon)$-approximation algorithms are known, with polynomial $(O(\frac{n^2}{\varepsilon^2} \log n))$ running time [AW01]. The problem is known to be NP-hard in a polygonal domain [KKP11, RW98]; further, deciding if there exists a simple curvature-constrained path is NP-hard [KKP11]. For the special case in which the obstacles are "moderate" (have differentiable boundary curves, with radius of curvature at least 1), both an approximation algorithm and an exact $O(n^4 \log n)$ algorithm have been found [BL96]. Within a convex polygon, one can determine if a curvature constrained path exists and, if so, compute one in time $O(n^2 \log n)$ [ABL$^+$02]. See also [KO13] for an analysis of how bounded curvature impacts path length.

MULTIPLE CRITERION OPTIMAL PATHS

The standard shortest-path problem asks for paths that minimize some *one* objective (length) function. Frequently, however, an application requires us to find paths to minimize *two or more* objectives; the resulting problem is a **bicriterion** (or **multi-criterion**) shortest-path problem. A path is called **efficient** or **Pareto optimal** if no other path has a better value for one criterion without having a worse value for the other criterion.

Multi-criterion optimization problems tend to be hard. Even the bicriterion path problem in a graph is NP-hard: Does there exist a path from s to t whose length is less than L and whose weight is less than W? Pseudo-polynomial-time algorithms are known, and many heuristics have been devised.

In geometric problems, various optimality criteria are of interest, including any pair from the following list: Euclidean (L_2) length, rectilinear (L_1) length, other L_p metrics, link distance, total turn, and so on. NP-hardness is known for several versions [AMP91]. One problem of particular interest is to compute a Euclidean shortest path within a polygonal domain, constrained to have at most k links. No exact solution is currently known for this problem. Part of the difficulty is that a minimum-link path will not, in general, lie on the visibility graph (or on any simple discrete graph). Furthermore, the computation of the turn points of such an optimal path appears to require the solution to high-degree polynomials. A $(1 + \varepsilon)$-approximation to the shortest k-link path in a simple polygon P can be found in time $O(n^3 k^3 \log (Nk/\varepsilon^{1/k}))$, where N is the largest integer coordinate of any vertex of P [MPA92]. In a *simple* polygon, one can always find an s-t path that simultaneously is within a factor 2 of optimal in link distance and within a factor $\sqrt{2}$ of optimal in Euclidean length; a corresponding result is not possible for polygons with holes. However, in $O(kE_{VG}^2)$ time, one can compute a path in a polygonal domain having at most $2k$ links and length at most that of a shortest k-link path.

In a rectilinear polygonal domain, efficient algorithms are known for the bicriterion path problem that combines *rectilinear* link distance and L_1 length [LYW96]. For example, efficient algorithms are known in two or more dimensions for computing optimal paths according to a *combined metric*, defined to be a linear combination of rectilinear link distance and L_1 path length [BKNO92]. (Note that this is not the same as computing the Pareto-optimal solutions.) Chen et al. [CDK01] give efficient algorithms for computing a shortest k-link rectilinear path, a minimum-link shortest rectilinear path, or any combined objective that uses a monotonic function of rectilinear link length and L_1 length in a rectilinear polygonal domain. Single-source queries can be answered in time $O(\log n)$, after $O(n \log^{3/2} n)$ preprocessing time to construct a data structure of size $O(n \log n)$; two-point queries can be answered in time $O(\log^2 n)$, using $O(n^2 \log^2 n)$ preprocessing time and space [CDK01].

OPEN PROBLEMS

1. Can one approximate link distance in a polygonal domain with a factor better than $O(\sqrt{h})$ in significantly subquadratic time? (Obtaining approximation factor $(2 - \varepsilon)$ is 3SUM-hard [MPS14].)

2. What is the smallest size data structure for a simple polygon P that allows logarithmic-time two-point link distance queries?

3. For a polygonal domain (with holes), what is the complexity of computing a shortest k-link path between two given points?

4. What is the complexity of the ladder problem for a polygonal domain, in which the cost of motion is the total work involved in translation/rotation?

5. Is it NP-hard to minimize the d_1-distance of a ladder endpoint?

6. What is the complexity of the bounded curvature shortest-path problem in a simple polygon?

31.4 GEOMETRIC NETWORK OPTIMIZATION

All of the problems considered so far involved computing a shortest path from one point to another (or from one point to all other points). We consider now some other network optimization problems, in which the objective is to compute a shortest path, cycle, tree, or other graphs, subject to various constraints. A summary of results is given in Table 31.4.1.

GLOSSARY

Minimum spanning tree (MST) of S: A tree of minimum total length whose nodes are a given set S of n points, and whose edges are line segments joining pairs of points.

Minimum Steiner spanning tree (Steiner tree) of S: A tree of minimum total length whose nodes are a superset of a given set S of n points, and whose edges are line segments joining pairs of points. Those nodes that are not points of S are called **Steiner points**.

Minimum Steiner forest of n point pairs: A forest of minimum total length such that each of the given point pairs lies within the same tree of the forest. The forest is allowed to utilize (Steiner) points that are not among the input points.

k-Minimum spanning tree (k-MST):   A minimum-length tree that spans some subset of $k \leq n$ points of S.

Traveling salesman problem (TSP):   Find a shortest cycle that visits every point of a set S of n points.

MAX TSP: Find a *longest* cycle that visits every point of a set S of n points.

Minimum latency tour problem: Find a tour on S that minimizes the sum of the "latencies," where the latency of $p \in S$ is the length of the tour from the given depot to p. Also known as the **deliveryman problem**, the **school-bus driver problem**, or the **traveling repairman problem**.

k-Traveling repairman problem: Find k tours covering S for k repairmen, minimizing the total latency. The repairmen may originate at a single common depot or at multiple depots.

Min/max-area TSP: Find a cycle on a given set S of points such that the cycle defines a simple polygon of minimum/maximum area.

TSP with neighborhoods: Find a shortest cycle that visits at least one point in each of a set of neighborhoods (e.g., polygons), $\{P_1, P_2, \ldots, P_k\}$.

Touring polygons problem: Find a shortest path/cycle that visits **in order** at least one point of each polygon in a sequence $(P_1, P_2, \ldots, P_k)$.

Watchman route (path) problem: Find a shortest cycle (path) within a polygonal domain P such that every point of P is visible from some point of the cycle.

Lawnmowing problem: Find a shortest cycle (path) for the center of a disk (a "lawnmower" or "cutter") such that every point of a given (possibly disconnected) region is covered by the disk at some position along the cycle (path).

Milling problem: Similar to the lawnmowing problem, but with the constraint that the cutter must at all times remain inside the given region (the "pocket" to be milled). When milling a polygonal region with a circular cutter, the portion that must be covered is the union of all disks within the polygonal region; the cutter cannot reach into convex corners of the polygon.

Zookeeper's problem: Find a shortest cycle in a simple polygon P (the **zoo**) through a given vertex v such that the cycle visits every one of a set of k disjoint convex polygons (**cages**), each sharing an edge with P.

Aquarium-keeper's problem: Find a shortest cycle in a simple polygon P (the **aquarium**) such that the cycle touches every edge of P.

Safari route problem: Find a shortest tour visiting a set of convex polygonal cages attached to the inside wall of a simple polygon P.

Relative convex hull of point set S within simple polygon P: The shortest cycle within P that surrounds S. The relative convex hull is necessarily a simple polygon, with vertices among the points of S and the vertices of P.

Monotone path problem: Find a shortest monotone path (if any) from s to t in a polygonal domain P. A polygonal path is **monotone** if there exists a direction vector d such that every directed edge of the path has a nonnegative inner product with d.

Doubling dimension (ddim):   A metric space $\mathcal{X}$ is said to have *doubling constant* c_d if any ball of radius r can be covered by c_d balls of radius $r/2$; the logarithm of c_d is the *doubling dimension (ddim)* of $\mathcal{X}$. Euclidean d-space, $\mathbb{R}^d$, has ddim $O(d)$.

MINIMUM SPANNING TREES

The (Euclidean) minimum spanning tree problem can be solved to optimality in the plane in time $O(n \log n)$ by appealing to the fact that the MST is a subgraph of the Delaunay triangulation; see Chapters 27 and 29. Efficient approximations in $\mathbb{R}^d$ are based on spanners (Section 32.3).

The Steiner tree and k-MST problems, however, are NP-hard; both have polynomial-time approximation schemes [Aro98, Mit99]. (In comparison, in graphs a 2-approximation is known for k-MST, as well as k-TSP [Gar05].) A PTAS for Steiner tree that takes time $O(n \log n)$ in any fixed dimension has been devised based on the concept of *banyans*, a generalization of the notion of t-spanners (Section 32.3), in combination with the PTAS techniques developed for the TSP and related problems [Aro98, Mit99, RS98]. A "t-banyan" approximates to within factor t the interconnection cost (allowing Steiner points) for subsets of sites of *any* cardinality (not just 2 sites, as in the case of t-spanners); [RS98] show that for any fixed $\varepsilon > 0$ and $d \geq 1$, there exists a $(1 + \varepsilon)$-banyan having $O(n)$ vertices and $O(n)$ edges, computable in $O(n \log n)$ time.

TRAVELING SALESMAN PROBLEM

The traveling salesman problem is a classical problem in combinatorial optimization, and has been studied extensively in its geometric forms. The problem is NP-hard, but has a simple 2-approximation algorithm based on "doubling" the minimum spanning tree. The Christofides heuristic augments a minimum spanning tree with a minimum-length matching on the odd-degree nodes of the tree, thereby obtaining an Eulerian graph from which a tour can be extracted; this yields a 1.5-approximation algorithm. (For the s-t path TSP, of computing a shortest Hamiltonian path between specified endpoints s and t, there is a $(1 + \sqrt{5})/2$-approximation in metric spaces [AKS15].) For the *graphic TSP*, in which distances are given by shortest path lengths in an unweighted graph, there is a $(13/9)$-approximation [Muc14].

Geometry helps in obtaining improved approximations: There are polynomial-time approximation schemes for geometric versions of the TSP, allowing one, for

TABLE 31.4.1   Other optimal path/cycle/network problems.

PROBLEM	COMPLEXITY	NOTES	SOURCE
Min spanning tree	$O(n \log n)$	exact, in $\mathbb{R}^2$	[PS85]
(MST) in $\mathbb{R}^d$	$O(n \log n)$	$(1+\varepsilon)$-approx, fixed d	[CK95]
Steiner tree in $\mathbb{R}^d$	$O(n \log n)$	$(1+\varepsilon)$-approx, fixed d	[RS98]
Steiner forest in $\mathbb{R}^2$	$O(n \log^c n)$	$(1+\varepsilon)$-approx	[BKM15]
k-MST in $\mathbb{R}^d$	$O(n \log n)$	$(1+\varepsilon)$-approx, fixed d	[RS98]
Min bicon. subgraph	$(1+\varepsilon)$-approx	$O(n \log n)$	[CL00]
Traveling salesman prob	$O(n \log n)$	$(1+\varepsilon)$-approx, fixed d	[Aro98, Mit99, RS98]
(TSP) in $\mathbb{R}^d$	$O(n)$	$(1+\varepsilon)$-approx, fixed d	[BG13]
	(randomized)	real RAM, atomic floor	[BG13]
TSP in low-dimensional	$2^{(k/\varepsilon)^{O(k^2)}} n+$	$(1+\varepsilon)$-approx	[BGK12, Got15]
metric space	$+(k/\varepsilon)^{O(k)} n \log^2 n$	ddim k, randomized	
MAX TSP	NP-hard in $\mathbb{R}^3$	$(1+\varepsilon)$-approx	[BFJ03]
	$O(n)$	L_1, L_∞ in $\mathbb{R}^2$	
	$O(n^{f-2} \log n)$	f-facet polyhedral norm	
Min-area TSP	NP-complete		[Fek00]
Max-area TSP	NP-complete	$(1/2)$-approx	[Fek00]
TSP w/neighborhoods	NP-hard	$O(\log n)$-approx	[MM95, GL99]
	no $O(1)$-approx	disconnected regions, $\mathbb{R}^2$	[SS06]
	APX-hard	connected regions, $\mathbb{R}^2$	[SS06]
	$O(1)$-approx	disjoint regions, $\mathbb{R}^2$	[Mit10]
	$(1+\varepsilon)$-approx	disjoint fat regions, $\mathbb{R}^2$	[Mit07]
	$(1+\varepsilon)$-approx	fat, weakly disjoint, ddim k	[CJ16]
Touring polygons prob	NP-hard	$(1+\varepsilon)$-approx	[DELM03, AMZ14]
	$O(nk^2 \log n)$	convex polygons	[DELM03]
	$O(nk \log(n/k))$	disjoint convex polygons	[DELM03]
Minimum latency prob	3.59-approx	metric space	[CGRT03, GK97]
	$(1+\varepsilon)$-approx	polytime in $\mathbb{R}^2$	[Sit14]
k-Traveling repairman	8.497-approx	single depot	[CGRT03, FHR07]
k-Traveling repairman	$(12+\varepsilon)$-approx	multidepot	[CK04]
Watchman route	$O(n^4 \log n)$	simple polygon	[DELM03]
(fixed source)	$O(n^3 \log n)$	simple polygon	[DELM03]
	$O(n)$	rectilinear simple polygon	[CN91]
	NP-hard	polygonal domain	[CN88]
	$O(\log^2 n)$-approx	polygonal domain	[Mit13]
Min-link watchman	NP-hard	$O(\log n)$-approx	[AMP03]
	NP-hard	simple polygon	[AL93]
	$O(1)$-approx	simple polygon	[AL95]
Lawnmowing problem	NP-hard	$O(1)$-approx, PTAS	[AFM00, FMS12]
Milling problem	$O(1)$-approx, PTAS	simple polygon	[AFM00, FMS12]
	NP-hard, $O(1)$-approx	polygonal domain	[AFM00, AFI+09]
Simple Hamilton path	$O(n^2 m^2)$	m points in simple n-gon	[CCS00]
	NP-complete	polygonal domain	[CCS00]
Aquarium-keeper's prob	$O(n)$	simple polygon	[CEE+91]
Zookeeper's problem	$O(n \log n)$	simple polygon	[Bes03a]
Relative convex hull	$\Theta(n + k \log kn)$	k points in simple n-gon	[GH89]
Monotone path prob	$O(n^3 \log n)$		[ACM89]

any fixed $\varepsilon > 0$, to get within a factor $(1+\varepsilon)$ of optimality [Aro98, Mit99]. Using the key idea of using t-spanners, the running time was improved to $O(n \log n)$ in any fixed dimension [RS98]. In fact, in the real RAM model (with atomic floor or mod function), a randomized linear-time PTAS is known in fixed dimension [BG13]. More generally, the TSP in metric spaces of bounded doubling dimension (ddim) also has a PTAS; specifically, for ddim k there is a PTAS running in time $2^{(k/\varepsilon)^{O(k^2)}} n + (k/\varepsilon)^{O(k)} n \log^2 n$, based on computing light spanners [BGK12, Got15]. The geometric TSP has also been studied with the objective of minimizing the sum of the direction changes at the points along the tour; this angular metric TSP is known to be NP-hard and to have a polynomial-time $O(\log n)$-approximation [ACK+00].

The **TSP-*with-neighborhoods*** (TSPN) problem arises when we require the tour/path to visit a set of regions, rather than a set of points. An $O(\log n)$-approximation algorithm is known for general connected regions in the plane [MM95, GL99]. It is NP-hard to approximate TSPN to within factor $2 - \varepsilon$ for connected regions that are allowed to overlap in the plane [SS06]. TSPN is APX-hard even for regions that are (intersecting) line segments, all of about the same length [EFS06]. Constant-factor approximation algorithms are known for some special cases [AH94, BGK+05, DM03], as well as for the general case of disjoint connected regions in the plane [Mit10]. Polynomial-time approximation schemes are known for disjoint (or "weakly disjoint") fat regions in the plane [Mit07], and, more generally, in metric spaces of bounded doubling dimension [CJ16]. For general disconnected regions in the plane, no constant factor approximation is possible (unless P=NP) [SS06]. For regions consisting each of k discrete points, a $(3/2)k$-approximation holds (even in general metric spaces) [Sla97]; for instances in the Euclidean plane, there is a lower bound of $\Omega(\sqrt{k})$, for $k > 4$, on the approximation factor [SS06], and the problem is APX-hard for $k = 2$ [DO08]. Results in higher dimensions include $O(1)$-approximation for neighborhoods that are planes or unit disks and $O(\log^3 n)$-approximation for lines [DT16]. It is NP-hard to approximate within any constant factor, even for connected, disjoint regions in 3-space [SS06].

LAWNMOWER AND WATCHMAN ROUTE PROBLEM

The lawnmowing problem is a TSP variant that seeks an optimal path for a lawnmower, modeled as, say, a circular cutter that must sweep out a region that covers a given domain of "grass." The milling problem requires that the cutter remains within the given domain. These problems are NP-hard in general, but constant-factor approximation algorithms are known [AFM00, AFI+09], and some variants have a PTAS [FMS12].

The watchman route problem seeks a shortest-path/tour so that every point of a domain is seen from some point along the path/tour; i.e., the path/tour must visit the visibility region associated with each point of the domain. In the case of a simple polygonal domain, the watchman route problem has an $O(n^4 \log n)$ time algorithm to compute an exact solution and $O(n^3 \log n)$ is possible if we are given a point through which the tour must pass [DELM03]. In the case of a polygonal domain with holes, the problem is easily seen to be NP-hard (from Euclidean TSP); an $O(\log^2 n)$-approximation algorithm is given by [Mit13], as well as a lower bound of $\Omega(\log n)$ on the approximation factor (assuming P$\neq$NP).

OPEN PROBLEMS

1. Is the MAX TSP NP-hard in the Euclidean plane? What if the tour is required to be noncrossing?

2. Is there a PTAS for the minimum latency problem and for the k-traveling repairman problem for points in any fixed dimension? (In $\mathbb{R}^2$, the minimum latency problem has a PTAS [Sit14].)

3. Can one obtain a PTAS for the TSP with neighborhoods (TSPN) problem in the plane if the regions are disjoint and each is a connected set? (Without disjointness, the problem is APX-hard even for regions that are line segments all of very nearly the same length [EFS06].) For (disconnected) regions each consisting of a discrete set of k points in the plane, what approximation factor (as a function of k) can be achieved? (The known $(3/2)k$-approximation does not exploit geometry [Sla97]; the lower bound $\Omega(\sqrt{k})$ applies to instances in the plane [SS06].) Is there a polynomial-time exact algorithm for TSPN in 3-space for neighborhoods that are planes?

4. Is the milling problem in simple polygons NP-hard?

5. Can the (Euclidean) watchman route in a simple polygon be computed in (near) linear time? (The fixed-source version is currently solved in $O(n^3 \log n)$ time [DELM03] in general, but in time $O(n)$ in rectilinear polygons [CN91].)

31.5 HIGHER DIMENSIONS

GLOSSARY

Polyhedral domain: A set $P \subset \mathbb{R}^3$ whose interior is connected and whose boundary consists of a union of a finite number of triangles. (The definition is readily extended to d dimensions, where the boundary must consist of a union of $(d-1)$-simplices.) The complement of P consists of connected (polyhedral) components, which are the **obstacles**.

Orthohedral domain: A polyhedral domain having each boundary facet orthogonal to one of the coordinate axes.

Polyhedral surface: A connected union of triangles, with any two triangles intersecting in a common edge, a common vertex, or not at all, and such that every point in the relative interior of the surface has a neighborhood homeomorphic to a disk.

Polyhedral terrain surface: A polyhedral surface given by an altitude function of position (x, y); a vertical line meets the surface in at most one point.

Edge sequence: The ordered list of obstacle edges that are intersected by a path.

COMPLEXITY

In three or more dimensions, most shortest-path problems become very difficult. In particular, there are two sources of complexity, even in the most basic Euclidean shortest-path problem in a polyhedral domain P.

One difficulty arises from algebraic considerations. In general, the shortest path in a polyhedral domain need not lie on any kind of discrete graph. Shortest paths in a polyhedral domain will be polygonal, with bend points that generally lie *interior* to obstacle edges, obeying a simple "unfolding" property: The path must enter and leave at the same angle to the edge. It follows that any locally optimal subpath joining two consecutive obstacle vertices can be "unfolded" at each edge along its edge sequence, thereby obtaining a straight segment. Given an edge sequence, this local optimality property uniquely identifies a shortest path through that edge sequence. However, to compare the lengths of two paths, each one shortest with respect to two (different) edge sequences, requires exponentially many bits, since the algebraic numbers that describe the optimal path lengths may have exponential degree.

A second difficulty arises from combinatorial considerations. The number of combinatorially distinct (i.e., having distinct edge sequences) shortest paths between two points may be exponential. This fact leads to a proof of the NP-hardness of the shortest-path problem [CR87]. In fact, the problem is NP-hard even for the case of obstacles that are disjoint axis-aligned boxes, even if the obstacles are all "stacked" axis-aligned rectangles (i.e., horizontal rectangles, orthogonal to the z-axis, with edges parallel to the x- and y-axes), and even if the rectangles are quadrants, each of which is unbounded to the northeast or to the southwest [MS04].

Thus, it is natural to consider approximation algorithms for the general case, or to consider special cases for which polynomial bounds are achievable.

SPECIAL CASES

If the polyhedral domain P has only a small number k of convex obstacles, a shortest path can be found in $n^{O(k)}$ time [Sha87]. If the obstacles are known to be vertical "buildings" (prisms) having only k different heights, then shortest paths can be found in time $O(n^{6k-1})$ [GHT89], but it is not known if this version of the problem is NP-hard if k is allowed to be large.

If we require paths to stay on a polyhedral surface (i.e., the domain P is essentially 2D), then the unfolding property of optimal paths can be exploited to yield polynomial-time algorithms. The continuous Dijkstra paradigm leads to an algorithm requiring $O(n^2 \log n)$ time to construct a shortest path map (or a geodesic Voronoi diagram), where n is the number of vertices of the surface [MMP87]. The worst-case running time has been improved to $O(n^2)$ by Chen and Han [CH96]. For the case of shortest paths on a convex polyhedral surface (or avoiding a single convex polytope obstacle in 3-space), Schreiber and Sharir [SS08] have given an optimal time $O(n \log n)$ algorithm based on the continuous Dijkstra paradigm; Schreiber [Sch07] has extended the methods to yield an $O(n \log n)$ time algorithm on "realistic" polyhedral surfaces. Kapoor [Kap99, O'R99] has announced an $O(n \log^2 n)$ time algorithm for general polyhedral surfaces, also based on the continuous Dijkstra paradigm. Since shortest paths on polyhedral surfaces are critical in many applications in computer graphics, practical experimental studies of shortest path algorithms have been conducted; see, e.g., [SSK$^+$05]. There has been considerable study of shortest paths and cycles on surfaces of complex topology; see Chapter 23.

Several facts are known about the set of edge sequences corresponding to shortest paths on the surface of a *convex* polytope P in $\mathbb{R}^3$. In particular, the worst-case number of distinct edge sequences that correspond to a shortest path between some pair of points is $\Theta(n^4)$, and the exact set of such sequences can be computed in time $O(n^6 \beta(n) \log n)$, where $\beta(n) = o(\log^* n)$ [AAOS97]. (A simpler $O(n^6)$ algorithm can compute a small superset of the sequences.) The number of **maximal** edge sequences for shortest paths is $\Theta(n^3)$. Some of these results depend on a careful study of the **star unfolding** with respect to a point p on the boundary, ∂P, of P. The star unfolding is the (nonoverlapping) cell complex obtained by subtracting from ∂P the shortest paths from p to the vertices of P, and then flattening the resulting boundary.

Results on exact algorithms for special cases are summarized in Table 31.5.1.

TABLE 31.5.1   Shortest paths in 3-space, d-space: exact algorithms.

OBSTACLES/DOMAIN	COMPLEXITY	NOTES	SOURCE
Polyhedral domain	NP-hard	convex obstacles	[CR87]
Axis-parallel stacked rect.	NP-hard	L_2 metric	[MS04]
One convex 3D polytope	$O(n \log n)$ time	$O(n \log n)$ space	[SS08]
k convex polytopes	$n^{O(k)}$	fixed k	[Sha87]
Vertical buildings	$O(n^{6k-1})$	k different heights	[GHT89]
Axis-parallel boxes	$O(n^2 \log^3 n)$	L_1 metric	[CKV87]
Axis-parallel	$O(n^2 \log n)$	L_1 metric	[CY95]
(disjoint)		path monotonicity, $\mathbb{R}^d$	[CY96]
Axis-parallel boxes, $\mathbb{R}^d$	$O(n^d \log n)$ preproc	combined L_1, link dist	[BKNO92]
	$O(\log^{d-1} n)$ query	single-source queries	
	$O((n \log n)^{d-1})$ space		
Above a terrain	$O(n^3 \log n)$ time	L_1 metric	[MS04]
Polyhedral surface	$O(n^2)$ time	builds SPM(s)	[CH96, MMP87]
		geodesic Voronoi	
Polyhedral surface	$O(n^2 \log^4 n)$ time	L_1, L_∞ metric	[CJ14b]
Two-point query	$O((\sqrt{n}/m^{1/4}) \log n)$ query	convex polytope	[AAOS97]
	$O(n^6 m^{1+\delta})$ space, preproc	$1 \le m \le n^2, \delta > 0$	
Geodesic diameter	$O(n^8 \log n)$	convex polytope	[AAOS97]

APPROXIMATION ALGORITHMS

Papadimitriou [Pap85] was the first to study the general problem from the point of view of approximations, giving a fully polynomial approximation scheme that produces a path guaranteed to be no longer than $(1+\varepsilon)$ times the length of a shortest path, in time $O(n^3(L + \log(n/\varepsilon))^2/\varepsilon)$, where L is the number of bits necessary to represent the value of an integer coordinate of a vertex of P. An alternate bound of Clarkson [Cla87] improves the running time in the case that $n\varepsilon^3$ is large. Choi, Sellen, and Yap [CSY95, CSY97] introduce the notion of "precision-sensitivity," writing the complexity in terms of a parameter, δ, that measures the implicit precision of the input instance, while drawing attention to the distinction between bit complexity and algebraic complexity.

Har-Peled [Har99b] shows how to compute an ***approximate shortest path map*** in polyhedral domains, computing, for fixed source s and $0 < \varepsilon < 1$, a subdivision of size $O(n^2/\varepsilon^{4+\delta})$ in time roughly $O(n^4/\varepsilon^6)$, so that for any point $t \in \mathbb{R}^3$ a $(1+\varepsilon)$-approximation of the length of a shortest s-t path can be reported in time $O(\log(n/\varepsilon))$.

Considerable effort has been devoted to approximation algorithms for shortest paths on polyhedral surfaces. Given a convex polytope obstacle, Agarwal et al. [AHPSV97] show how to surround the polytope with a constant-size ($O(\varepsilon^{-3/2})$, now improved to $O(\varepsilon^{-5/4})$ [CLM05]) convex polytope having the property that shortest paths are approximately preserved (within factor $(1+\varepsilon)$) on the outer polytope. This results in an approximation algorithm of time complexity $O(n \log(1/\varepsilon) + f(\varepsilon^{-5/4}))$, where $f(m)$ denotes the time complexity of solving exactly a shortest-path problem on an m-vertex convex surface (e.g., $f(m) = O(m^2)$ using [CH96]). Har-Peled [Har99a] gives an $O(n)$-time algorithm to preprocess a convex polytope so that a two-point query can be answered in time $O((\log n)/\varepsilon^{3/2} + 1/\varepsilon^3)$, yielding the $(1+\varepsilon)$-approximate shortest path distance, as well as a path having $O(1/\varepsilon^{3/2})$ segments that avoids the interior of the input polytope.

Varadarajan and Agarwal [VA99] obtained the first subquadratic-time algorithms for approximating shortest paths on general (nonconvex) polyhedral surfaces, computing a $(7 + \varepsilon)$-approximation in $O(n^{5/3} \log^{5/3} n)$ time, or a $(15 + \varepsilon)$-approximation in $O(n^{8/5} \log^{8/5} n)$ time. Their method is based on a partitioning of the surface into $O(n/r)$ patches, each having at most r faces, using a planar separator theorem. (The parameter r is chosen to be $n^{1/3} \log^{1/3} n$ or $n^{2/5} \log^{2/5} n$.) Then, on the boundary of each patch, a carefully selected set of points ("portals") is chosen, and these are interconnected with a graph that approximates shortest paths within each patch.

For a polyhedral terrain surface, one often wants to model the cost of motion that is anisotropic, depending on the steepness (gradient) of the ascent/descent. Cheng and Jin [CJ14b] give a $(1 + \varepsilon)$-approximation, taking time $O(\frac{1}{\sqrt{\varepsilon}} n^2 \log n + n^2 \log^4 n)$, in this model in which the cost of movement is a linear combination of path length and total ascent, and there are constraints on the steepness of the path. For the problem of computing a shortest path on a polyhedral terrain surface, subject to the constraint that the path be *descending* in altitude along the path, Cheng and Jin [CJ14a] have given a $(1 + \varepsilon)$-approximation algorithm with running time $O(n^4 \log(n/\varepsilon))$. Exact solution of the shortest descending path problem on a polyhedral terrain seems to be quite challenging; see, e.g., [AL09, AL11, RDN07].

Practical approximation algorithms are based on searching a discrete graph (an "edge subdivision graph," or a "pathnet")[LMS01, MM97] by placing Steiner points judiciously on the edges (or, possibly interior to faces) of the input surface. This approach applies also to the case of weighted surfaces and weighted convex decompositions of $\mathbb{R}^3$; see the earlier discussion of the weighted region problem. One can obtain provable results on the approximation factor; see Table 31.5.2. It is worth noting, however, that these complexity bounds are under the assumption that certain geometric parameters are "constants"; these parameters may be unbounded in terms of ε and the combinatorial input size n.

TABLE 31.5.2   Shortest paths in 3-space: approximation algorithms.

OBSTACLES/DOMAIN	COMPLEXITY	NOTES	SOURCE
Polyhedral domain	$O(n^4(L+\log(\frac{n}{\varepsilon}))^2/\varepsilon^2)$	$(1+\varepsilon)$-approx	[Pap85]
	$O(n^2\text{polylog }n/\varepsilon^4)$	$(1+\varepsilon)$-approx	[Cla87]
Polyhedral domain	$O(\frac{n^2}{\varepsilon^3}\log\frac{1}{\varepsilon}\log n)$	$(1+\varepsilon)$-approx	[AMS00]
		geometric parameters	
Polyhedral domain	NP-hard in 3-space	PTAS	[KLPS16]
Orthohedral poly. dom.	$O(\frac{n^2}{\varepsilon^3}\log\frac{1}{\varepsilon}\log n)$	$(1+\varepsilon)$-approx	[AMS00]
Orthohedral poly. dom.	$O(n^2\log^2 n)$	3D, rectilinear link dist.	[PS11]
Poly. terrain surface	$O(\frac{1}{\sqrt{\varepsilon}}n^2\log n + n^2\log^4 n)$	$(1+\varepsilon)$-approx	[CJ14b]
	cost(length,ascent)	gradient constraints	
Weighted poly. dom.	$O(K\frac{n}{\varepsilon^3}\log\frac{1}{\varepsilon}(\frac{1}{\sqrt{\varepsilon}}+\log n))$	$(1+\varepsilon)$-approx	[ADMS13]
	n tetrahedra	geometric parameter K	
Weighted poly. dom.	$O(2^{2^{O(\kappa)}}\frac{n}{\varepsilon^7}\log^2\frac{W}{\varepsilon}\log^2\frac{NW}{\varepsilon})$	$(1+\varepsilon)$-approx	[CCJV15]
	n tetrahedra	weights $\{1,\dots,W\}$	
	coordinates $\{1,\dots,N\}$	κ skinny tetra per comp	
One convex obstacle	$O(\varepsilon^{-5/4}\sqrt{n})$ expected	$(1+\varepsilon)$-approx	[CLM05]
k convex polytopes	$O(n)$	$2k$-approx	[HS98]
Convex poly. surface	$O(n\log\frac{1}{\varepsilon}+\frac{1}{\varepsilon^3})$	$(1+\varepsilon)$-approx	[AHPSV97]
Convex poly. surface	$O(\frac{1}{\varepsilon^{1.5}}\log n+\frac{1}{\varepsilon^3})$ query	$(1+\varepsilon)$-approx	[Har99a]
	$O(n)$ preproc (pp.)	two-point query	
Convex poly. surface	$O(\log\frac{n}{\varepsilon})$ query	single-source queries	[Har99b]
	$O(\frac{n}{\varepsilon^3}\log\frac{1}{\varepsilon}+\frac{n}{\varepsilon^{1.5}}\log\frac{1}{\varepsilon}\log n)$ pp.	$O(\frac{n}{\varepsilon}\log\frac{1}{\varepsilon})$ size SPM	
Nonconv. poly. surface	$O(\log\frac{n}{\varepsilon})$ query	single-source queries	[Har99b]
	$O(n^2\log n+\frac{n}{\varepsilon}\log\frac{1}{\varepsilon}\log\frac{n}{\varepsilon})$ pp.	$O(\frac{n}{\varepsilon}\log\frac{1}{\varepsilon})$ size SPM	
Convex poly. surface	$O(n+\frac{1}{\varepsilon^6})$	$(1-\varepsilon)$-approx diameter	[Har99a]
Nonconv. poly. surface	$O(n^{5/3}\log^{5/3}n)$	$(7+\varepsilon)$-approx	[VA99]
	$O(n^{8/5}\log^{8/5}n)$	$(15+\varepsilon)$-approx	[VA99]
Nonconv. poly. surface	$O(\frac{n}{\varepsilon}\log\frac{1}{\varepsilon}\log n)$	$(1+\varepsilon)$-approx	[AMS00]
		geometric parameters	
Vertical buildings	$O(n^2)$	1.1-approx	[GHT89]

OTHER METRICS

Link distance in a polyhedral domain in $\mathbb{R}^d$ can be approximated (within factor 2) in polynomial time by searching a weak visibility graph whose nodes correspond to simplices in a simplicial decomposition of the domain. Computing the exact link distance is NP-hard in $\mathbb{R}^3$, even on terrains; however, a PTAS is known [KLPS16]. If the domain is orthohedral, rectilinear link distance can be computed efficiently, in time $O(n^2\log^2 n)$ [PS11].

For the case of orthohedral domains and rectilinear (L_1) shortest paths, the shortest-path problem in $\mathbb{R}^d$ becomes relatively easy to solve in polynomial time, since the grid graph induced by the facets of the domain serves as a path-preserving graph that we can search for an optimal path. In $\mathbb{R}^3$, we can do better than to use the $O(n^3)$ grid graph induced by $O(n)$ facets; an $O(n^2\log^2 n)$ size subgraph suffices, which allows a shortest path to be found using Dijkstra's algorithm in time $O(n^2\log^3 n)$ [CKV87]. More generally, in $\mathbb{R}^d$ one can compute a data structure of size $O((n\log n)^{d-1})$, in $O(n^d\log n)$ preprocessing time, that supports fixed-source link distance queries in $O(\log^{d-1} n)$ time [BKNO92]. In fact, this last result can be extended, within the same complexities, to the case of a combined metric, in which path cost is measured as a linear combination of L_1 length and rectilinear link distance.

For the special case of disjoint rectilinear box obstacles and rectilinear (L_1) shortest paths, a structural result may help in devising very efficient algorithms: There always exists a coordinate direction such that *every* shortest path from s to t is monotone in this direction [CY96]. In fact, this result has led to an $O(n^2 \log n)$ algorithm for the case $d = 3$.

OPEN PROBLEMS

1. Can one compute shortest paths on a polyhedral surface in $\mathbb{R}^3$ in $O(n \log n)$ time using $O(n)$ space?

2. Can one compute a shortest path map for a polyhedral domain in output-sensitive time?

3. Is it 3SUM-hard to compute minimum-link rectilinear paths in 3D, or can one obtain a subquadratic-time algorithm? A nearly-quadratic-time algorithm is known [PS11].

4. What is the complexity of the shortest-path problem in 3-space among disjoint unit disk obstacles or disjoint axis-aligned unit cubes?

5. Can two-point queries be solved efficiently for Euclidean (or L_1) shortest paths among obstacles in 3-space?

31.6 SOURCES AND RELATED MATERIAL

SURVEYS

Some other related surveys offer additional material and references:

[BMSW11]: A survey of shortest paths on 3D surfaces. [Har11]: A book on geometric approximation algorithms.

[Mit00]: Another survey on geometric shortest paths and network optimization.

[Mit15]: A survey on approximation schemes for geometric network optimization.

RELATED CHAPTERS

Chapter 10: Geometric graph theory
Chapter 23: Computational topology of graphs on surfaces
Chapter 29: Triangulations and mesh generation
Chapter 30: Polygons

REFERENCES

[AAOS97]   P.K. Agarwal, B. Aronov, J. O'Rourke, and C.A. Schevon. Star unfolding of a poly-
            tope with applications. *SIAM J. Comput.*, 26:1689–1713, 1997.

[ABB⁺16]   H.-K. Ahn, L. Barba, P. Bose, J.-L. De Carufel, M. Korman, and E. Oh. A linear-time
            algorithm for the geodesic center of a simple. *Discrete Comput. Geom.*, 56:836–859,
            2016.

[ABL⁺02]   P.K. Agarwal, T. Biedl, S. Lazard, S. Robbins, S. Suri, and S. Whitesides. Curvature-
            constrained shortest paths in a convex polygon. *SIAM J. Comput.*, 31:1814–1851,
            2002.

[AC91]   M.J. Atallah and D.Z. Chen. Parallel rectilinear shortest paths with rectangular
            obstacles. *Comput. Geom.*, 1:79–113, 1991.

[AC93]   M.J. Atallah and D.Z. Chen. On parallel rectilinear obstacle-avoiding paths. *Comput.
            Geom.*, 3:307–313, 1993.

[ACC⁺96]   S.R. Arikati, D.Z. Chen, L.P. Chew, G. Das, M. Smid, and C.D. Zaroliagis. Planar
            spanners and approximate shortest path queries among obstacles in the plane. In
            Proc. 4th Eur. Sympos. Algorithms, vol. 1136 of *LNCS*, pages 514–528, Springer,
            Berlin, 1996.

[ACM89]   E.M. Arkin, R. Connelly, and J.S.B. Mitchell. On monotone paths among obstacles,
            with applications to planning assemblies. In *Proc. 5th Sympos. Comput. Geom.*, pages
            334–343, ACM Press, 1989.

[ACK⁺00]   A. Aggarwal, D. Coppersmith, S. Khanna, R. Motwani, and B. Schieber. The angular-
            metric traveling salesman problem. *SIAM J. Comput.*, 29:697–711, 2000.

[ADMS13]   L. Aleksandrov, H. Djidjev, A. Maheshwari, and J.-R. Sack. An approximation al-
            gorithm for computing shortest paths in weighted 3-d domains. *Discrete Comput.
            Geom.*, 50:124–184, 2013.

[AEK⁺16]   E.M. Arkin, A. Efrat, C. Knauer, J.S.B. Mitchell, V. Polishchuk, G. Rote, L. Schlipf,
            and T. Talvitie. Shortest path to a segment and quickest visibility queries. *J. Comput.
            Geom.*, 7:77–100, 2016.

[AFI⁺09]   E.M. Arkin, S.P. Fekete, K. Islam, H. Meijer, J.S.B. Mitchell, Y. Núñez-Rodríguez,
            V. Polishchuk, D. Rappaport, and H. Xiao. Not being (super)thin or solid is hard:
            A study of grid Hamiltonicity. *Comput. Geom.*, 42:582–605, 2009.

[AFM00]   E.M. Arkin, S.P. Fekete, and J.S.B. Mitchell. Approximation algorithms for lawn
            mowing and milling. *Comput. Geom.*, 17:25–50, 2000.

[AFW93]   B. Aronov, S.J. Fortune, and G. Wilfong. Furthest-site geodesic Voronoi diagram.
            Discrete Comput. Geom., 9:217–255, 1993.

[AFS16]   P.K. Agarwal, K. Fox, and O. Salzman. An efficient algorithm for computing high-
            quality paths amid polygonal obstacles. In *Proc. 27th ACM-SIAM Sympos. Discrete
            Algorithms*, pages 1179–1192, 2016.

[AH94]   E.M. Arkin and R. Hassin. Approximation algorithms for the geometric covering
            salesman problem. *Discrete Appl. Math.*, 55:197–218, 1994.

[AHK+14]   O. Aichholzer, T. Hackl, M. Korman, A. Pilz, and B. Vogtenhuber. Geodesic-preserving polygon simplification. *Internat. J. Comput. Geom. Appl.*, 24:307–323, 2014.

[AHPSV97]   P.K. Agarwal, S. Har-Peled, M. Sharir, and K.R. Varadarajan. Approximate shortest paths on a convex polytope in three dimensions. *J. ACM*, 44:567–584, 1997.

[AKS15]   H.-C. An, R. Kleinberg, and D.B. Shmoys. Improving Christofides' algorithm for the *s-t* path TSP. *J. ACM*, 62:34, 2015.

[AL93]   M.H. Alsuwaiyel and D.T. Lee. Minimal link visibility paths inside a simple polygon. *Comput. Geom.*, 3:1–25, 1993.

[AL95]   M.H. Alsuwaiyel and D.T. Lee. Finding an approximate minimum-link visibility path inside a simple polygon. *Inform. Process. Lett.*, 55:75–79, 1995.

[AL09]   M. Ahmed and A. Lubiw. Shortest descending paths through given faces. *Comput. Geom.*, 42:464–470, 2009.

[AL11]   M. Ahmed and A. Lubiw. Shortest descending paths: Towards an exact algorithm. *Internat. J. Comput. Geom. Appl.*, 21:431–466, 2011.

[AMP03]   E.M. Arkin, J.S.B. Mitchell, and C.D. Piatko. Minimum-link watchman tours. *Inform. Process. Lett.*, 86:203–207, 2003.

[AMP91]   E.M. Arkin, J.S.B. Mitchell, and C.D. Piatko. Bicriteria shortest path problems in the plane. In *Proc. 3rd Canad. Conf. Comput. Geom.*, pages 153–156, 1991.

[AMS95]   E.M. Arkin, J.S.B. Mitchell, and S. Suri. Logarithmic-time link path queries in a simple polygon. *Internat. J. Comput. Geom. Appl.*, 5:369–395, 1995.

[AMS00]   L. Aleksandrov, A. Maheshwari, and J.-R. Sack. Approximation algorithms for geometric shortest path problems. In *Proc. 32nd Sympos. Theory Comput.*, pages 286–295, ACM Press, 2000.

[AMS05]   L. Aleksandrov, A. Maheshwari, and J.-R. Sack. Determining approximate shortest paths on weighted polyhedral surfaces. *J. ACM*, 52:25–53, 2005.

[AMZ14]   A. Ahadi, A. Mozafari, and A. Zarei. Touring a sequence of disjoint polygons: Complexity and extension. *Theoret. Comput. Sci.*, 556:45–54, 2014.

[Aro98]   S. Arora. Polynomial time approximation schemes for Euclidean traveling salesman and other geometric problems. *J. ACM*, 45:753–782, 1998.

[AW01]   P.K. Agarwal and H. Wang. Approximation algorithms for curvature-constrained shortest paths. *SIAM J. Comput.*, 30:1739–1772, 2001.

[BAE+17]   S.W. Bae, M. Korman, J.S.B. Mitchell, Y. Okamoto, V. Polishchuk, and H. Wang. Computing the L_1 geodesic diameter and center of a polygonal domain. *Discrete Comput. Geom.*, 57:674–701, 2017.

[Ber91]   M. de Berg. On rectilinear link distance. *Comput. Geom.*, 1:13–34, 1991.

[Bes03a]   S. Bespamyatnikh. An $O(n \log n)$ algorithm for the zoo-keeper's problem. *Comput. Geom.*, 24:63–74, 2002.

[Bes03b]   S. Bespamyatnikh. Computing homotopic shortest paths in the plane. In *Proc. 14th ACM-SIAM Sympos. Discrete Algorithms*, pages 609–617, 2003.

[BFJ03]   A. Barvinok, S.P. Fekete, D.S. Johnson, A. Tamir, G.J. Woeginger, and R. Woodroofe. The geometric maximum traveling salesman problem. *J. ACM*, 50:641–664, 2003.

[BG13]   Y. Bartal and L.-A. Gottlieb. A linear time approximation scheme for Euclidean TSP. In *Proc. 54th Sympso. Found. Comp. Sci.*, pages 698–706, 2013.

[BGK12]   Y. Bartal, L. Gottlieb, and R. Krauthgamer. The traveling salesman problem: low-dimensionality implies a polynomial time approximation scheme. In *Proc. 44th ACM Sympos. Theory Comput.*, pages 663–672, 2012.

[BGK⁺05]   M. de Berg, J. Gudmundsson, M.J. Katz, C. Levcopoulos, M.H. Overmars, and A.F. van der Stappen. TSP with neighborhoods of varying size. *J. Algorithms*, 57:22—36, 2005.

[BKL11]   M.G. Borgelt, M. van Kreveld, and J. Luo. Geodesic disks and clustering in a simple polygon. *Internat. J. Comput. Geom. Appl.*, 21:595–608, 2011.

[BKM15]   G. Borradaile, P.N. Klein, and C. Mathieu. A polynomial-time approximation scheme for euclidean Steiner forest. *ACM Trans. Algorithms*, 11:19, 2015.

[BKNO92]   M. de Berg, M. van Kreveld, B.J. Nilsson, and M.H. Overmars. Shortest path queries in rectilinear worlds. *Internat. J. Comput. Geom. Appl.*, 2:287–309, 1992.

[BKO13]   S.W. Bae, M. Korman, and Y. Okamoto. The geodesic diameter of polygonal domains. *Discrete Comput. Geom.*, 50:306–329, 2013.

[BKO15]   S.W. Bae, M. Korman, and Y. Okamoto. Computing the geodesic centers of a polygonal domain. *Comput. Geom.*, to appear, 2015.

[BKOW15]   S.W. Bae, M. Korman, Y. Okamoto, and H. Wang. Computing the L_1 geodesic diameter and center of a simple polygon in linear time. *Comput. Geom.*, 48:495–505, 2015.

[BL96]   J.-D. Boissonnat and S. Lazard. A polynomial-time algorithm for computing a shortest path of bounded curvature amidst moderate obstacles. In *Proc. 12th Sympos. Comput. Geom.*, pages 242–251, ACM Press, 1996.

[BMSW11]   P. Bose, A. Maheshwari, C. Shu, and S. Wuhrer. A survey of geodesic paths on 3d surfaces. *Comput. Geom.*, 44:486–498, 2011.

[BO12]   S.W. Bae and Y. Okamoto. Querying two boundary points for shortest paths in a polygonal domain. *Comput. Geom.*, 45:284–293, 2012.

[CCS00]   Q. Cheng, M. Chrobak, and G. Sundaram. Computing simple paths among obstacles. *Comput. Geom.*, 16:223–233, 2000.

[CCJV15]   S.-W. Cheng, M.-K. Chiu, J. Jin, and A. Vigneron. Navigating weighted regions with scattered skinny tetrahedra. In *Proc. 26th Int. Sympos. Algorithms and Computation*, vol. 9472 of *LNCS*, pages 35–45, Springer, Berlin, 2015.

[CDK01]   D.Z. Chen, O. Daescu, and K.S. Klenk. On geometric path query problems. *Internat. J. Comput. Geom. Appl.*, 11:617–645, 2001.

[CDM01]   D.Z. Chen, G. Das, and M. Smid. Lower bounds for computing geometric spanners and approximate shortest paths. *Discrete Appl. Math.*, 110:151–167, 2001.

[CEE⁺91]   J. Czyzowicz, P. Egyed, H. Everett, D. Rappaport, T. Shermer, D. Souvaine, G. Toussaint, and J. Urrutia. The aquarium keeper's problem. In *Proc. 2nd ACM-SIAM Sympos. Discrete Algorithms*, pages 459–464, 1991.

[CEG⁺94]   B. Chazelle, H. Edelsbrunner, M. Grigni, L.J. Guibas, J. Hershberger, M. Sharir, and J. Snoeyink. Ray shooting in polygons using geodesic triangulations. *Algorithmica*, 12:54–68, 1994.

[CGRT03]   K. Chaudhuri, B. Godfrey, S. Rao, and K. Talwar. Paths, trees, and minimum latency tours. In *Proc. 44th IEEE Sympos. Found. Comp. Sci.*, pages 36–45, 2003.

[CH96]   J. Chen and Y. Han. Shortest paths on a polyhedron. *Internat. J. Comput. Geom. Appl.*, 6:127–144, 1996.

[Che95]   D.Z. Chen. On the all-pairs Euclidean short path problem. In *Proc. 6th ACM-SIAM Sympos. Discrete Algorithms*, pages 292–301, 1995.

[Che13]   D.Z. Chen. Efficient algorithms for geometric shortest path query problems. In P.M. Pardalos, D.-Z. Du, and R.L. Graham, editors, *Handbook of Combinatorial Optimization*, pages 1125–1154. Springer, New York, 2013.

[CIW14]   D.Z. Chen, R. Inkulu, and H. Wang. Two-point L_1 shortest path queries in the plane. In *Proc. 30th Sympos. Comput. Geom.*, page 406, ACM Press, 2014.

[CJ14a]   S.-W. Cheng and J. Jin. Approximate shortest descending paths. *SIAM J. Comput.*, 43:410–428, 2014.

[CJ14b]   S.-W. Cheng and J. Jin. Shortest paths on polyhedral surfaces and terrains. In *Proc. 46th ACM Sympos. Theory Comput.*, pages 373–382, 2014.

[CJ16]   T.H. Chan and S.H. Jiang. Reducing curse of dimensionality: Improved PTAS for TSP (with neighborhoods) in doubling metrics. In *Proc. 27th ACM-SIAM Sympos. Discrete Algorithms*, pages 754–765, 2016.

[CJV15]   S.-W. Cheng, J. Jin, and A. Vigneron. Triangulation refinement and approximate shortest paths in weighted regions. In *Proc. 26th ACM-SIAM Sympos. Discrete Algorithms*, pages 1626–1640, 2015.

[CK95]   P.B. Callahan and S.R. Kosaraju. A decomposition of multidimensional point sets with applications to k-nearest-neighbors and n-body potential fields. *J. ACM*, 42:67–90, 1995.

[CK96]   D.Z. Chen and K.S. Klenk. Rectilinear short path queries among rectangular obstacles. *Inform. Process. Lett.*, 57:313–319, 1996.

[CK04]   C. Chekuri and A. Kumar. Maximum coverage problem with group budget constraints and applications. In *Approximation, Randomization, and Combinatorial Optimization. Algorithms and Techniques*, vol. 3122 of *LNCS*, pages 72–83, Springer, Berlin, 2004.

[CKT00]   D.Z. Chen, K.S. Klenk, and H.-Y.T. Tu. Shortest path queries among weighted obstacles in the rectilinear plane. *SIAM J. Comput.*, 29:1223–1246, 2000.

[CKV87]   K.L. Clarkson, S. Kapoor, and P.M. Vaidya. Rectilinear shortest paths through polygonal obstacles in $O(n(\log n)^2)$ time. In *Proc. 3rd Sympos. Comput. Geom.*, pages 251–257, ACM Press, 1987.

[CL00]   A. Czumaj and A. Lingas. Fast approximation schemes for Euclidean multi-connectivity problems. In *Proc. 27th Int. Conf. on Automata, Languages and Prog.*, vol. 1853 of *LNCS*, pages 856–868, Springer, Berlin, 2000.

[Cla87]   K.L. Clarkson. Approximation algorithms for shortest path motion planning. In *Proc. 19th Sympos. Theory Comput.*, pages 56–65, ACM Press, 1987.

[CLM05]   B. Chazelle, D. Liu, and A. Magen. Sublinear geometric algorithms. *SIAM J. Comput.*, 35:627–646, 2005.

[CLMS02]   S. Cabello, Y. Liu, A. Mantler, and J. Snoeyink. Testing homotopy for paths in the plane. In *Proc. 18th Sympos. Comput. Geom.*, pages 160–169, ACM Press, 2002.

[CM99]   Y.-J. Chiang and J.S.B. Mitchell. Two-point Euclidean shortest path queries in the plane. In *Proc. 10th ACM-SIAM Sympos. Discrete Algorithms*, pages 215–224, 1999.

[CN88]   W.-P. Chin and S. Ntafos. Optimum watchman routes. *Inform. Process. Lett.*, 28:39–44, 1988.

[CN91]   W.-P. Chin and S. Ntafos. Shortest watchman routes in simple polygons. *Discrete Comput. Geom.*, 6:9–31, 1991.

[CNVW08]   S.-W. Cheng, H.-S. Na, A. Vigneron, and Y. Wang. Approximate shortest paths in anisotropic regions. *SIAM J. Comput.*, 38:802–824, 2008.

[CNVW10]   S.-W. Cheng, H.-S. Na, A. Vigneron, and Y. Wang. Querying approximate shortest paths in anisotropic regions. *SIAM J. Comput.*, 39:1888–1918, 2010.

[CR87]   J. Canny and J.H. Reif. New lower bound techniques for robot motion planning problems. In *Proc. 28th IEEE Sympos. Found. Comput. Sci.*, pages 49–60, 1987.

[CSY95]   J. Choi, J. Sellen, and C.K. Yap. Precision-sensitive Euclidean shortest path in 3-space. In *Proc. 11th Sympos. Comput. Geom.*, pages 350–359, ACM Press, 1995.

[CSY97]   J. Choi, J. Sellen, and C.K. Yap. Approximate Euclidean shortest paths in 3-space. *Internat. J. Comput. Geom. Appl.*, 7:271–295, 1997.

[CT97]   Y.-J. Chiang and R. Tamassia. Optimal shortest path and minimum-link path queries between two convex polygons inside a simple polygonal obstacle. *Internat. J. Comput. Geom. Appl.*, 7:85–121, 1997.

[CW13]   D.Z. Chen and H. Wang. L_1 shortest paths among polygonal obstacles in the plane. In *Proc. 30th Sympos. Theoret. Aspects Comp. Sci.*, pages 293–304, Schloss Dagstuhl, 2013.

[CW15a]   D.Z. Chen and H. Wang. Computing shortest paths among curved obstacles in the plane. *ACM Trans. Algorithms*, 11:26, 2015.

[CW15b]   D.Z. Chen and H. Wang. A new algorithm for computing visibility graphs of polygonal obstacles in the plane. *J. Comput. Geom.*, 6:316–345, 2015.

[CY95]   J. Choi and C.K. Yap. Rectilinear geodesics in 3-space. In *Proc. 11th Sympos. Comput. Geom.*, pages 380–389, ACM Press, 1995.

[CY96]   J. Choi and C.K. Yap. Monotonicity of rectilinear geodesics in d-space. In *Proc. 12th Sympos. Comput. Geom.*, pages 339–348, ACM Press, 1996.

[DCG⁺14]   J.-L. De Carufel, C. Grimm, A. Maheshwari, M. Owen, and M. Smid. A note on the unsolvability of the weighted region shortest path problem. *Comput. Geom.*, 47:724–727, 2014.

[DELM03]   M. Dror, A. Efrat, A. Lubiw, and J.S.B. Mitchell. Touring a sequence of polygons. In *Proc. 35th ACM Sympos. Theory Comput.*, pages 473–482, 2003.

[DM03]   A. Dumitrescu and J.S.B. Mitchell. Approximation algorithms for TSP with neighborhoods in the plane. *J. Algorithms*, 48:135–159, 2003.

[DO08]   M. Dror and J.B. Orlin. Combinatorial optimization with explicit delineation of the ground set by a collection of subsets. *SIAM J. Discrete Math.*, 21:1019–1034, 2008.

[DT16]   A. Dumitrescu and C.D. Tóth. The traveling salesman problem for lines, balls, and planes. *ACM Trans. Algorithms*, 12:43, 2016.

[DX95]   B.R. Donald and P.G. Xavier. Provably good approximation algorithms for optimal kinodynamic planning for Cartesian robots and open chain manipulators. *Algorithmica*, 14:480–530, 1995.

[DXCR93]   B. Donald, P. Xavier, J. Canny, and J. Reif. Kinodynamic motion planning. *J. ACM*, 40:1048–1066, 1993.

[EFS06]   K. Elbassioni, A.V. Fishkin, and R. Sitters. On approximating the TSP with intersecting neighborhoods. In *Proc. Int. Sympos. Algorithms Comput.*, vol. 4288 of *LNCS*, pages 213–222, Springer, Berlin, 2006.

[EHP⁺15]   S. Eriksson-Bique, J. Hershberger, V. Polishchuk, B. Speckmann, S. Suri, T. Talvitie, K. Verbeek, and H. Yıldız. Geometric k shortest paths. In *Proc. 26th ACM-SIAM Sympos. Discrete Algorithms*, pages 1616–1625, 2015.

[EKL06]   A. Efrat, S.G. Kobourov, and A. Lubiw. Computing homotopic shortest paths efficiently. *Comput. Geom.*, 35:162–172, 2006.

[EM94]   H. ElGindy and P. Mitra. Orthogonal shortest route queries among axis parallel rectangular obstacles. *Internat. J. Comput. Geom. Appl.*, 4:3–24, 1994.

[Fek00]   S.P. Fekete. On simple polygonalizations with optimal area. *Discrete Comput. Geom.*, 23:73–110, 2000.

[FHR07]   J. Fakcharoenphol, C. Harrelson, and S. Rao. The k-traveling repairmen problem. *ACM Trans. Algorithms*, 3:40, 2007.

[FMS12]   S.P. Fekete, J.S.B. Mitchell, and C. Schmidt. Minimum covering with travel cost. *J. Comb. Optim.*, 24:32–51, 2012.

[FT87]   M.L. Fredman and R.E. Tarjan. Fibonacci heaps and their uses in improved network optimization algorithms. *J. ACM*, 34:596–615, 1987.

[Gar05]   N. Garg. Saving an epsilon: a 2-approximation for the k-mst problem in graphs. In *Proc. 37th ACM Sympos. Theory Comput.*, pages 396–402, 2005.

[GH89]   L.J. Guibas and J. Hershberger. Optimal shortest path queries in a simple polygon. *J. Comput. Syst. Sci.*, 39:126–152, 1989.

[GHL⁺87]   L.J. Guibas, J. Hershberger, D. Leven, M. Sharir, and R.E. Tarjan. Linear-time algorithms for visibility and shortest path problems inside triangulated simple polygons. *Algorithmica*, 2:209–233, 1987.

[GHT89]   L. Gewali, S. Ntafos, and I.G. Tollis. Path planning in the presence of vertical obstacles. Technical report, Computer Science, University of Texas at Dallas, 1989.

[GK97]   M.X. Goemans and J. Kleinberg. An improved approximation ratio for the minimum latency problem. *Math. Prog.*, 82:111–124, 1999.

[GL99]   J. Gudmundsson and C. Levcopoulos. A fast approximation algorithm for TSP with neighborhoods. *Nord. J. Comput.*, 6:469, 1999.

[GLNS08]   J. Gudmundsson, C. Levcopoulos, G. Narasimhan, and M. Smid. Approximate distance oracles for geometric spanners. *ACM Trans. Algorithms*, 4:10, 2008.

[GM91]   S.K. Ghosh and D.M. Mount. An output-sensitive algorithm for computing visibility graphs. *SIAM J. Comput.*, 20:888–910, 1991.

[GMMN90]   L. Gewali, A. Meng, J.S.B. Mitchell, and S. Ntafos. Path planning in 0/1/∞ weighted regions with applications. *ORSA J. Comput.*, 2:253–272, 1990.

[GMS08]   H. Guo, A. Maheshwari, and J.-R. Sack. Shortest path queries in polygonal domains. In *Proc. 4th Int. Conf. Algorithmic Aspects Inform. Management*, vol. 5034 of *LNCS*, pages 200–211, Springer, Berlin, 2008.

[Got15]   L.-A. Gottlieb. A light metric spanner. In *Proc. 56th Sympos. Found. Comp. Sci.*, pages 759–772, 2015.

[GT97]   M.T. Goodrich and R. Tamassia. Dynamic ray shooting and shortest paths in planar subdivisions via balanced geodesic triangulations. *J. Algorithms*, 23:51–73, 1997.

[Har99a]   S. Har-Peled. Approximate shortest paths and geodesic diameters on convex polytopes in three dimensions. *Discrete Comput. Geom.*, 21:216–231, 1999.

[Har99b]   S. Har-Peled. Constructing approximate shortest path maps in three dimensions. *SIAM J. Comput.*, 28:1182–1197, 1999.

[Har11]   S. Har-Peled. *Geometric Approximation Algorithms*. Vol. 173 of *Math. Surveys Monogr.*, AMS, Providence, 2011.

[Har15]   S. Har-Peled. Shortest path in a polygon using sublinear space. *J. Comput. Geom.*, 7:19–45, 2015.

[Her89]   J. Hershberger. An optimal visibility graph algorithm for triangulated simple polygons. *Algorithmica*, 4:141–155, 1989.

[Her91]   J. Hershberger. A new data structure for shortest path queries in a simple polygon. *Inform. Process. Lett.*, 38:231–235, 1991.

[Her95]   J. Hershberger. Optimal parallel algorithms for triangulated simple polygons. *Internat. J. Comput. Geom. Appl.*, 5:145–170, 1995.

[HS94]   J. Hershberger and J. Snoeyink. Computing minimum length paths of a given homotopy class. *Comput. Geom.*, 4:63–98, 1994.

[HS97]   J. Hershberger and S. Suri. Matrix searching with the shortest path metric. *SIAM J. Comput.*, 26:1612–1634, 1997.

[HS98]   J. Hershberger and S. Suri. Practical methods for approximating shortest paths on a convex polytope in $\mathbb{R}^3$. *Comput. Geom.*, 10:31–46, 1998.

[HS99]   J. Hershberger and S. Suri. An optimal algorithm for Euclidean shortest paths in the plane. *SIAM J. Comput.*, 28:2215–2256, 1999.

[HSY13]   J. Hershberger, S. Suri, and H. Yıldız. A near-optimal algorithm for shortest paths among curved obstacles in the plane. In *Proc. 29th Sympos. Comput. Geom.*, pages 359–368, ACM Press, 2013.

[IK15]   R. Inkulu and S. Kapoor. A polynomial time algorithm for finding an approximate shortest path amid weighted regions. Preprint, `arXiv:1501.00340`, 2015.

[Kap99]   S. Kapoor. Efficient computation of geodesic shortest paths. In *Proc. 31st ACM Sympos. Theory Comput.*, pages 770–779, 1999.

[KMM97]   S. Kapoor, S.N. Maheshwari, and J.S.B. Mitchell. An efficient algorithm for Euclidean shortest paths among polygonal obstacles in the plane. *Discrete Comput. Geom.*, 18:377–383, 1997.

[KKP11]   D.G. Kirkpatrick, I. Kostitsyna, and V. Polishchuk. Hardness results for two-dimensional curvature-constrained motion planning. In *Proc. 23rd Canad. Conf. Comput. Geom.*, 2011.

[KLPS16]   I. Kostitsyna, M. Löffler, V. Polishchuk, and F. Staals. On the complexity of minimum-link path problems. In *Proc. 32nd Sympos. Comput. Geom.*, vol. 51 of *LIPIcs*, article 49, Schloss Dagstuhl, 2016.

[KO13]   H.-S. Kim and O. Cheong. The cost of bounded curvature. *Comput. Geom.*, 46:648–672, 2013.

[LMS01]   M. Lanthier, A. Maheshwari, and J.-R. Sack. Approximating shortest paths on weighted polyhedral surfaces. *Algorithmica*, 30:527–562, 2001.

[LP84]   D.T. Lee and F.P. Preparata. Euclidean shortest paths in the presence of rectilinear barriers. *Networks*, 14:393–410, 1984.

[LSD00]   A. Maheshwari, J.-R. Sack, and H.N. Djidjev. Link distance problems. In J.-R. Sack and J. Urrutia, editors, *Handbook of Computational Geometry*, pages 519–558, Elsevier, Amsterdam, 2000.

[LYW96]   D.T. Lee, C.D. Yang, and C.K. Wong. Rectilinear paths among rectilinear obstacles. *Discrete Appl. Math.*, 70:185–215, 1996.

[Mit13]   J.S.B. Mitchell. Approximating watchman routes. In *Proc. 24th ACM-SIAM Sympos. Discrete Algorithms*, pages 844–855, 2013.

[Mit89]   J.S.B. Mitchell. An optimal algorithm for shortest rectilinear paths among obstacles. In *Proc. 1st Canad. Conf. Comput. Geom.*, page 22, 1989.

[Mit91]   J.S.B. Mitchell. A new algorithm for shortest paths among obstacles in the plane. *Ann. Math. Artif. Intell.*, 3:83–106, 1991.

[Mit92]   J.S.B. Mitchell. L_1 shortest paths among polygonal obstacles in the plane. *Algorithmica*, 8:55–88, 1992.

[Nit96]   J.S.B. Mitchell. Shortest paths among obstacles in the plane. *Internat. J. Comput. Geom. Appl.*, 6:309–332, 1996.

[Mit99]   J.S.B. Mitchell. Guillotine subdivisions approximate polygonal subdivisions: A simple polynomial-time approximation scheme for geometric TSP, k-MST, and related problems. *SIAM J. Comput.*, 28:1298–1309, 1999.

[Mit00]   J.S.B. Mitchell. Geometric shortest paths and network optimization. In J.-R. Sack and J. Urrutia, editors, *Handbook of Computational Geometry*, pages 633–701, Elsevier, Amsterdam, 2000.

[Mit07]   J.S.B. Mitchell. A PTAS for TSP with neighborhoods among fat regions in the plane. In *Proc. 18th ACM-SIAM Sympos. Discrete Algorithms*, pages 11–18, 2007.

[Mit10]   J.S.B. Mitchell. A constant-factor approximation algorithm for TSP with pairwise-disjoint connected neighborhoods in the plane. In *Proc. 26th Sympos. Comput. Geom.*, pages 183–191, ACM Press, 2010.

[Mit15]   J.S.B. Mitchell. Approximation schemes for geometric network optimization problems. M.-Y. Kao, editor, *Encyclopedia of Algorithms*, Springer, Berlin, 2015.

[MM95]   C. Mata and J.S.B. Mitchell. Approximation algorithms for geometric tour and network design problems. In *Proc. 11th Sympos. Comput. Geom.*, pages 360–369, ACM Press, 1995.

[MM97]   C. Mata and J.S.B. Mitchell. A new algorithm for computing shortest paths in weighted planar subdivisions. In *Proc. 13th Sympos. Comput. Geom.*, pages 264–273, ACM Press, 1997.

[MMP87]   J.S.B. Mitchell, D.M. Mount, and C.H. Papadimitriou. The discrete geodesic problem. *SIAM J. Comput.*, 16:647–668, 1987.

[MP91]   J.S.B. Mitchell and C.H. Papadimitriou. The weighted region problem: finding shortest paths through a weighted planar subdivision. *J. ACM*, 38:18–73, 1991.

[MPA92]   J.S.B. Mitchell, C. Piatko, and E.M. Arkin. Computing a shortest k-link path in a polygon. In *Proc. 33rd IEEE Sympos. Found. Comput. Sci.*, pages 573–582, 1992.

[MPS14]   J.S.B. Mitchell, V. Polishchuk, and M. Sysikaski. Minimum-link paths revisited. *Comput. Geom.*, 47:651–667, 2014.

[MPSW15]   J.S.B. Mitchell, V. Polishchuk, M. Sysikaski, and H. Wang. An optimal algorithm for minimum-link rectilinear paths in triangulated rectilinear domains. In *Proc. 42nd Int. Conf. Automata, Languages, Progr.*, Part I, vol. 9134 of *LNCS*, pages 947–959, Springer, Berlin, 2015.

[MRW92]   J.S.B. Mitchell, G. Rote, and G. Woeginger. Minimum-link paths among obstacles in the plane. *Algorithmica*, 8:431–459, 1992.

[MS04]   J.S.B. Mitchell and M. Sharir. New results on shortest paths in three dimensions. In *Proc. 20th Sympos. Comput. Geom.*, pages 124–133, ACM Press, 2004.

[Muc14]   M. Mucha. $\frac{13}{9}$-approximation for graphic TSP. *Theory Comput. Syst.*, 55:640–657, 2014.

[OBA16]   E. Oh, L. Barba, and H.-K. Ahn. The farthest-point geodesic Voronoi diagram of points on the boundary of a simple polygon. In *Proc. 32nd Sympos. Comput. Geom.*, vol. 51 of *LIPIcs*, article 56, Schloss Dagstuhl, 2016.

[O'R99]   J. O'Rourke. Computational geometry column 35. *Internat. J. Comput. Geom. Appl.*, 9:513–516, 1999.

[Pap85]   C.H. Papadimitriou. An algorithm for shortest-path motion in three dimensions. *Inform. Process. Lett.*, 20:259–263, 1985.

[PL98]   E. Papadopoulou and D.T. Lee. A new approach for the geodesic Voronoi diagram of points in a simple polygon and other restricted polygonal domains. *Algorithmica*, 20:319–352, 1998.

[PV96a]   M. Pocchiola and G. Vegter. Topologically sweeping visibility complexes via pseudo-triangulations. *Discrete Comput. Geom.*, 16:419–453, 1996.

[PV96b]   M. Pocchiola and G. Vegter. The visibility complex. *Internat. J. Comput. Geom. Appl.*, 6:279–308, 1996.

[PS85]   F.P. Preparata and M.I. Shamos. *Computational Geometry: An Introduction.* Springer-Verlag, New York, 1985.

[PS11]   V. Polishchuk and M. Sysikaski. Faster algorithms for minimum-link paths with restricted orientations. In *Proc. 12th Sympos. Algorithms Data Structures*, vol. 6844 of *LNCS*, pages 655–666, Springer, Berlin, 2011.

[RS98]   S.B. Rao and W.D. Smith. Approximating geometrical graphs via "spanners" and "banyans." In *Proc. 30th ACM Sympos. Theory Comput.*, pages 540–550, 1998.

[RDN07]   S. Roy, S. Das, and S.C. Nandy. Shortest monotone descent path problem in polyhedral terrain. *Comput. Geom.*, 37:115–133, 2007.

[RT94]   J.H. Reif and S.R. Tate. Approximate kinodynamic planning using L_2-norm dynamic bounds. *Comput. Math. Appl.*, 27:29–44, 1994.

[RW98]   J. Reif and H. Wang. The complexity of the two dimensional curvature-constrained shortest-path problem. In *Proc. 3rd Workshop Algorithmic Found. Robot.*, pages 49–57, A.K. Peters, Natick, 1998.

[RW00]   J. Reif and H. Wang. Non-uniform discretization for kinodynamic motion planning and its applications. *SIAM J. Comput.*, 30:161–190, 2000.

[Sch96]   S. Schuierer. An optimal data structure for shortest rectilinear path queries in a simple rectilinear polygon. *Internat. J. Comput. Geom. Appl.*, 6:205–226, 1996.

[Sch07]   Y. Schreiber. Shortest paths on realistic polyhedra. In *Proc. 23rd Sympos. Comput. Geom.*, pages 74–83, ACM Press, 2007.

[Sel95]   J. Sellen. Direction weighted shortest path planning. In *Proc. IEEE Internat. Conf. Robot. Autom.*, pages 1970–1975, 1995.

[Sha87]   M. Sharir. On shortest paths amidst convex polyhedra. *SIAM J. Comput.*, 16:561–572, 1987.

[Sit14]   R. Sitters. Polynomial time approximation schemes for the traveling repairman and other minimum latency problems. In *Proc. 25th ACM-SIAM Sympos. Discrete Algorithms*, pages 604–616, 2014.

[Sla97]   P. Slavík. The errand scheduling problem. Technical report 97-2, Department of Computer Science, SUNY, Buffalo, 1997.

[SS06]   S. Safra and O. Schwartz. On the complexity of approximating TSP with neighborhoods and related problems. *Comput. Complexity*, 14:281–307, 2006.

[SS08]   Y. Schreiber and M. Sharir. An optimal-time algorithm for shortest paths on a convex polytope in three dimensions. *Discrete Comput. Geom.*, 1:500–579, 2008.

[SSK+05]   V. Surazhsky, T. Surazhsky, D. Kirsanov, S.J. Gortler, and H. Hoppe. Fast exact and approximate geodesics on meshes. *ACM Trans. Graph.*, 24:553–560, 2005.

[Sur86]   S. Suri. A linear time algorithm for minimum link paths inside a simple polygon. *Comput. Vision Graph. Image Process.*, 35:99–110, 1986.

[Sur90]   S. Suri. On some link distance problems in a simple polygon. *IEEE Trans. Robot. Autom.*, 6:108–113, 1990.

[VA99]   K.R. Varadarajan and P.K. Agarwal. Approximating shortest paths on a nonconvex polyhedron. *SIAM J. Comput.*, 30:1321–1340, 1999.

[Wan16]   H. Wang. On the geodesic centers of polygonal domains. In *Proc. 24th Eur. Sympos. Algorithms*, vol. 57 of *LIPIcs*, article 77, Schloss Dagstuhl, 2016.

32 PROXIMITY ALGORITHMS

Joseph S. B. Mitchell and Wolfgang Mulzer

INTRODUCTION

The notion of distance is fundamental to many aspects of computational geometry. A classic approach to characterize the distance properties of planar (and high-dimensional) point sets that has been studied since the early 1980s uses *proximity graphs* (Section 32.1). Proximity graphs are geometric graphs in which two vertices p, q are connected by an edge (p, q) if and only if a certain *exclusion region* for p, q contains no points from the vertex set. Depending on the specific exclusion region, many variants of proximity graphs can be defined, such as relative neighborhood graphs, Delaunay triangulations, β-skeletons, empty-strip graphs, etc. Since proximity graphs encode interesting information on the intrinsic structure of the point set, they have found many applications. From an algorithmic point of view, it is extremely useful to have a compact representation of the distance structure of a point set. The *well-separated pair decomposition* (WSPD) offers one way to achieve this (Section 32.2). WSPDs have numerous algorithmic applications, and the notion generalizes to certain non-Euclidean metrics. Furthermore, several variants of the WSPD have been developed to address its shortcomings, e.g., *semi-separated pair decompositions* and (α, β)-*pair decompositions*. *Geometric spanners* provide another means to approximate the complete Euclidean metric (Section 32.3). Here, the distance function is approximated by the shortest path distance in a sparse geometric graph. There are four basic constructions for geometric spanners: the greedy spanner, the Yao graph, the Θ-graph and the WSPD-spanner. To optimize various parameters, many variants have been defined, and the notion can be generalized beyond the Euclidean setting. Finally, we discuss work on making proximity structures *dynamic*, allowing for insertions and deletions of points (Section 32.4). The fundamental problem here is the *dynamic nearest neighbor problem*, which serves as a starting point for other structures. Additionally, there are several results on making geometric spanners dynamic.

32.1 PROXIMITY GRAPHS

GLOSSARY

Geometric graph: A graph $G = (V, E)$ together with an embedding in $\mathbb{R}^d$ that maps V to points and E to straight line segments that do not pass through nonincident vertices. (See Chapter 10.)

Planar straight-line graph (PSLG): A geometric graph $G = (V, E)$ embedded in $\mathbb{R}^2$ with noncrossing edges.

$\delta(p, q)$:   The distance between two points p and q.

Diameter of a point set V: The maximum distance $\delta(p, q)$ between two points $p, q \in V$. A pair that achieves the diameter is called a *diametric pair*.

Closest pair of a point set V: A pair $\{p, q\}$ of two distinct points in V with minimum distance $\delta(p, q)$. The distance $\delta(p, q)$ is called the *closest pair distance*.

Spread $\Phi(V)$ of a point set $V \subset \mathbb{R}^d$:   The ratio between the diameter and the closest pair distance of V: $\Phi(V) = \max_{s, t \in V} \delta(s, t) / \min_{s \neq t \in V} \delta(s, t)$.

L_p-metric: Let $x = (x_1, \ldots, x_d)$ and $y = (y_1, \ldots, y_d)$ be two d-dimensional points. For $p \in [1, \infty]$, we set $\delta_p(x, y) = (\sum_{i=1}^{d} |x_i - y_i|^p)^{1/p}$.

In particular, we have $\delta_1(x, y) = \sum_{i=1}^{d} |x_i - y_i|$ and $\delta_\infty(x, y) = \max_{i=1}^{d} |x_i - y_i|$.

Ball $B(x, r)$:   Let x be a point and $r \geq 0$. Then, we define the open ball $B(x, r) = \{y \mid \delta(x, y) < r\}$.

Nearest-neighbor graph $\mathrm{NNG}(V)$:   The directed graph with vertex set V and an edge (p, q) if and only if $B(p, \delta(p, q)) \cap V = \emptyset$.

Lune $L(p, q)$:   For two points p and q, we set

$$L(p, q) = B(p, \delta(p, q)) \cap B(q, \delta(p, q)).$$

Some authors prefer the term *lens* instead of lune.

Relative neighborhood graph $\mathrm{RNG}(V)$:   The graph with vertex set V and an edge (p, q) if and only if $L(p, q) \cap V = \emptyset$. Thus, the edge is present if and only if

$$\delta(p, q) = \min_{v \in V} \max\{\delta(p, v), \delta(q, v)\}.$$

Gabriel graph $\mathrm{GG}(V)$:   The graph with vertex set V and an edge (p, q) if and only if

$$B\left(\frac{p + q}{2}, \frac{\delta(p, q)}{2}\right) \cap V = \emptyset.$$

β-lune $L_\beta(p, q)$:   Let p and q be two points. For $\beta = 0$, $L_\beta(p, q)$ is the open line segment pq.

For $\beta \in (0, 1)$, $L_\beta(p, q)$ is the intersection of the two open disks of radius $\delta(p, q)/(2\beta)$ having bounding circles passing through both p and q.

For $\beta \geq 1$, we set

$$L_\beta(p, q) = B\left(p\left(1 - \frac{\beta}{2}\right) + q\frac{\beta}{2}, \frac{\beta}{2}\delta(p, q)\right) \cap B\left(q\left(1 - \frac{\beta}{2}\right) + p\frac{\beta}{2}, \frac{\beta}{2}\delta(p, q)\right).$$

Lune-based β-skeleton $G_\beta^l(V)$:   Let $\beta \geq 0$. We define $G_\beta^l(V)$ as the graph with vertex set V and an edge (p, q) if and only if $L_\beta(p, q) \cap V = \emptyset$.

Circle-based β-skeleton $G_\beta^c(V)$:   For $\beta = 0$, we define $G_\beta^c(V)$ as the graph with vertex set V and an edge (p, q) if and only if the open line segment pq contains no other points from V.

For $\beta \in (0, 1)$, we define $G_\beta^c(V)$ as the graph with vertex set V and an edge (p, q) if and only if the intersection of the two open disks with radius $\delta(p, q)/(2\beta)$ passing through p and q does not contain any other points from V.

For $\beta \geq 1$, we define $G_\beta^c(V)$ as the graph with vertex set V and an edge (p, q) if and only if the union of the two open disks with radius $\beta\delta(p, q)/2$ passing through p and q does not contain any other points from V.

L_p-**Delaunay triangulation** $D_p(V)$**:**   The graph with vertex set V that is the straight-line dual of the Voronoi diagram of V with respect to the L_p-norm.

Empty strip graph: The graph with vertex set V and an edge (p, q) if and only if the open infinite strip bounded by the two lines through p and through q orthogonal to the line segment pq contains no points from V.

Sphere of influence graph SIG(V)**:**   Let C_p be a circle centered at p with radius equal to the distance to a nearest neighbor of p. Then, SIG(V) is the graph with vertex set V and an edge (p, q) if and only if C_p and C_q intersect in at least two points.

Minimum-weight triangulation MWT(V)**:**   A geometric triangulation (i.e., an edge maximal planar straight-line graph) with vertex set V and the minimum total edge length.

BASIC STRUCTURES

Let V be a finite set in the Euclidean plane. The *nearest neighbor graph* NNG(V) connects each point in V to its nearest neighbor. It is usually defined as a directed graph, but some authors treat it as undirected. In general, NNG(V) is not connected, but each point in V has at least one incident edge.

The *relative neighborhood graph* RNG(V) connects two points p and q if and only if the *lune* $L(p, q)$ is empty of points from V. It was defined by Toussaint [Tou80]. The *Gabriel graph* GG(V) was first introduced by Gabriel and Sokal [GS69]. It is defined similarly as RNG(V), but two points p and q are connected by an edge if and only if their *diameter sphere* (i.e., the sphere with diameter pq) is empty. The RNG and the GG are always connected. (The RNG can be disconnected if one defines the exclusion region as a *closed* set.)

The β-*skeletons* are a continuous generalization of the Gabriel graph and the relative neighborhood graph [KR85]. They come in two variants, *circle-based* and *lune-based*, depending on the region that needs to be empty for an edge to be present. Both circle- and lune-based β-skeletons depend on a parameter $\beta \geq 0$. In circle-based β-skeletons, the union of two open generalized diameter circles needs to be empty of other points from V. In lune-based β-skeletons, an open β-lune needs to be empty; see Figure 32.1.1. The lune-based β-skeleton can be defined for any L_p-metric. Unless stated otherwise, we refer to the Euclidean case. For $\beta \in [0, 1]$, the circle-based and the lune-based β-skeleton coincide. For $\beta = 0$, the β-skeleton is the complete graph, provided that no three points of V lie on a line. For $\beta = 1$, we have $G_1^c(V) = G_1^l(V) = GG(V)$. For $\beta > 1$, the circle-based β-skeleton is a subgraph of the (Euclidean) lune-based β-skeleton. For $\beta = 2$, the lune-based β-skeleton coincides with the relative-neighborhood graph. For $\beta = \infty$, the circle-based β-skeleton becomes the empty graph and the lune-based β-skeleton becomes the *empty-strip graph*. For $0 \leq \beta_1 \leq \beta_2 \leq \infty$, we have $G_{\beta_2}^l(V) \subseteq G_{\beta_1}^l(V)$ and $G_{\beta_2}^c(V) \subseteq G_{\beta_1}^c(V)$.

These graph definitions capture the internal structure of a point set and are motivated by various applications, such as computer vision, texture discrimination, geographic analysis, pattern analysis, cluster analysis, and others. The following theorem states some relationships between proximity graphs. A version of this theorem was first established by Toussaint [Tou80]; see also [KR85, O'R82, MS80].

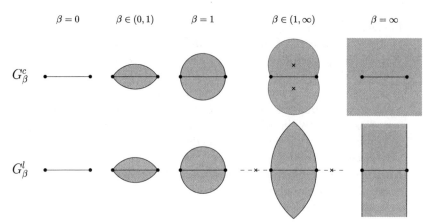

$\beta = 0$ $\beta \in (0,1)$ $\beta = 1$ $\beta \in (1,\infty)$ $\beta = \infty$

G_β^c

G_β^l

FIGURE 32.1.1

The exclusion regions for the circle-based and Euclidean lune-based β-skeleton for various values of β (cf. [Vel91]). For $\beta = 1$, the exclusion region coincides with the exclusion region for the Gabriel graph. For $\beta = 2$, the exclusion region for the lune-based β-skeleton is the exclusion region of the relative neighborhood graph. For $\beta = \infty$, the exclusion region of the lune-based β-skeleton is the exclusion region of the empty strip graph and the exclusion region of the circle-based β-skeleton is the whole plane.

THEOREM 32.1.1   *Hierarchy Theorem*

In any L_p metric, $p \in (1,\infty)$, for any finite point set V and for any $1 \leq \beta \leq 2$, we have

$$\text{NNG} \subseteq \text{MST}_p \subseteq \text{RNG} \subseteq G_\beta^l \subseteq \text{GG} \subseteq \text{DT}_p,$$

where MST_p is a minimum spanning tree of V in the L_p-norm and DT_p is the Delaunay triangulation of V.

O'Rourke showed that for $p = 1$ and $p = \infty$, the inclusion $\text{RNG} \subseteq \text{DT}_p$ does not necessarily hold; however, with a slightly different definition of Delaunay triangulation (in terms of empty open balls, instead of being the dual of the Voronoi diagram), the inclusion can be rescued [O'R82]. The MST is always connected, so by the Hierarchy Theorem, this also holds for the RNG, the GG and the lune-based β-skeleton with $\beta \in [0,2]$. In general, the circle-based β-skeleton is not connected for $\beta > 1$.

Clearly, neighborhood graphs on n vertices can have at most $\binom{n}{2}$ edges. In many cases, this is also attained, for example, for the L_1 and L_∞ metric [Kat88], for Gabriel graphs in three and more dimensions [CEG+94], and for RNGs in four dimensions and higher. In the plane, the Euclidean RNG has at most $3n - 8$ edges [BDH+12]; an earlier upper bound of $3n - 10$ [Urq83] turned out to be incorrect [BDH+12]. The planar Gabriel graph has at most $3n - 8$ edges [MS80]. For $1 < p < \infty$, the fact that the RNG is contained in the Delaunay triangulation yields an upper bound of $3n - 6$ edges [JT92]. In three dimensions, Euclidean RNGs have size at most $O(n^{4/3})$ [AM92]. No matching lower bound is known. See Table 32.1.1. Bose et al. [BDH+12] give bounds on a large number of parameters in various proximity graphs.

TABLE 32.1.1   Size of RNGs and Gabriel graphs.

DIM	METRIC	SIZE	REFERENCE	COMMENT
2	L_2	$\leq 3n - 8$	[BDH$^+$12]	
2	L_2	$\leq 3n - 8$	[MS80]	Gabriel Graphs
2	$L_p, 1 < p < \infty$	$\in [n-1, 3n-6]$	[JT92]	
≥ 2	L_1, L_∞	$\Theta(n^2)$	[Kat88]	
3	L_2	$O(n^{4/3})$	[AM92]	
≥ 3	L_2	$\Theta(n^2)$	[CEG$^+$94]	Gabriel Graphs
≥ 4	L_2	$\Theta(n^2)$	[JT92]	

TABLE 32.1.2   RNG algorithms.

DIM	METRIC	COMPLEXITY	REFERENCE	COMMENT
2	L_2	$O(n \log n)$	[Sup83]	arbitrary points
2	$L_p, 1 < p < \infty$	$O(n^2)$	[Tou80, Kat88]	arbitrary points
2	L_1, L_∞	$O(n \log n)$	[Lee85]	general position
2	L_1, L_∞	$O(n \log n + m)$	[Kat88]	m output size arbitrary points
3	L_2	$O(n^{3/2+\varepsilon})$	[AM92]	general position
3	L_2	$O(n^{7/4+\varepsilon})$	[AM92]	arbitrary points
3	$L_p, 1 < p < \infty$	$O(n^2)$	[Sup83, JK87, KN87]	general position
3	L_1, L_∞	$O(n \log^2 n)$	[Smi89]	general position
d	L_2	$O(n^{2(1-1/(d+1))+\varepsilon})$	[AM92]	general position
d	L_1, L_∞	$O(n \log^{d-1} n)$	[Smi89]	general position

ALGORITHMS

It is an interesting algorithmic problem to construct proximity graphs efficiently. Using the definition, $O(n^3)$ time complexity is trivial. In the case of general L_p-metrics, for $1 < p < \infty$, the fact that the Delaunay triangulation is a superset of the RNG leads to an $O(n^2)$ time algorithm in the plane. A faster algorithm for the Euclidean case was given by Supowit, who showed that in this case the RNG can be computed in $O(n \log n)$ time [Sup83]. For the L_1 and L_∞ metric, Lee described an $O(n \log n)$ time algorithm for planar point sets in general position, improving a previous $O(n^2 \log n)$ algorithm by O'Rourke [O'R82]. Katajainen [Kat88] gives an output-sensitive algorithm for the L_1 and L_∞ that achieves $O(n \log n + m)$ time, where m is the size of the resulting RNG. In three dimensions, Agarwal and Matoušek obtain $O(n^{3/2+\varepsilon})$ time for computing the Euclidean RNG of points in general position, and $O(n^{7/4+\varepsilon})$ for arbitrary points [AM92]. Their approach generalizes to higher dimensions, yielding time $O(n^{2(1-1/(d+1))+\varepsilon})$ for d-dimensional point sets in general position. For the L_p-norm, $1 < p < \infty$, several algorithms with running time $O(n^2)$ are available for three-dimensional point sets in general position [Sup83, JK87, KN87]. Finally, in the L_1 and L_∞-norm, the d-dimensional RNG for points in general position can be found in time $O(n \log^{d-1} n)$ [Smi89]. See Table 32.1.2.

There are also many results that describe algorithmic relationships between different proximity structures. Since planar graphs are closed under the minors relation, one can use the Borůvka-Sollin algorithm to find an MST of a planar point set V in $O(n)$ time, once $\mathrm{DT}(V)$ is available [CT76]. This works in any L_p-norm, $1 < p < \infty$. Similarly, given $\mathrm{DT}(V)$, we can compute the lune-based β-skeleton for V with $\beta \in [1, 2]$ in $O(n)$ additional time [Lin94]. This result holds in any L_p-metric, for $1 < p < \infty$ [Lin94]. By setting $\beta = 1$ and $\beta = 2$, this result applies in particular to $\mathrm{GG}(V)$ and $\mathrm{RNG}(V)$. A linear time algorithm to construct $\mathrm{GG}(V)$ from $\mathrm{DT}(V)$ was also described by Matula and Sokal [MS80]. The circle-based β-skeleton for a planar point set V with $\beta \geq 1$ can be found in $O(n)$ additional time given $\mathrm{DT}(V)$ [KR85, Vel91]. The fastest algorithm for computing a β-skeleton for a planar point set with $\beta \in [0, 1]$ takes $O(n^2)$ time [HLM03]. In some cases, the β-skeleton for $\beta \in [0, 1]$ can have $\Theta(n^2)$ edges. Given the Euclidean MST (or any connected subgraph of $\mathrm{DT}(V)$), we can compute $\mathrm{DT}(V)$ in $O(n)$ additional time [CW98, KL96]. There exists a general reduction from computing Delaunay triangulations to computing NNGs: Suppose we are given an algorithm that computes the NNG of any planar m-point set in $T(m)$ time, where $m \mapsto T(m)/m$ is monotonically increasing. Then, we can compute the Delaunay triangulation of a planar n-point set in $O(T(n))$ expected time [BM11]. This reduction is useful in settings where faster NNG algorithms are available, e.g., in transdichotomous models that allow manipulations at the bit-level [BM11].

APPLICATIONS AND VARIANTS

An important connection between the circle-based β-skeleton and the minimum-weight triangulation (MWT) was discovered by Keil: for $\beta = \sqrt{2} \approx 1.41421$, we have $G_\beta^c(V) \subseteq \mathrm{MWT}(V)$ [Kei94]. Cheng and Xu later improved this to $\beta = \sqrt{1 + \sqrt{4/27}} \approx 1.17682$ [CX01]. For $\beta = \sqrt{5/4 + \sqrt{1/108}} \approx 1.16027$, the circle-based β-skeleton need not be a subgraph of the MWT [WY01]. Even though it is NP-hard to compute the minimum weight triangulation [MR08], the β-skeleton provides a good heuristic for well-behaved point sets.

Just as Delaunay triangulations/Voronoi diagrams have been generalized to kth-order diagrams (see Chapter 27), the relative neighborhood graph and the Gabriel graph have kth-order generalizations, k-RNG and k-GG, in which the exclusion region may contain up to k points of V. The k-GG has $O(k(n - k))$ edges and can be constructed in time $O(k^2 n \log n)$ [SC90]; the k-GG is $(k + 1)$-connected [BCH+13], and the 10-GG is Hamiltonian (while the 1-GG is not necessarily Hamiltonian) [KSVC15]. The 17-RNG contains the Euclidean bottleneck matching, which leads to an efficient (roughly $O(n^{1.5})$) algorithm for computing a bottleneck matching [CTL92].

There are many ways to generalize the proximity graphs described in this section. The *sphere-of-influence* (SIG) graph was defined by Toussaint as a graph-theoretical "primal sketch" [Tou88]. In the SIG, two points are connected by an edge if and only if their nearest-neighbor circles intersect. In the plane, the SIG has at most $15n = O(n)$ edges [Sos99], and it can be computed in $\Theta(n \log n)$ time [AH85]. Veltkamp defines a family of γ-*neighborhood graphs* [Vel91]. Veltkamp's graphs are parameterized by two parameters γ_0 and γ_1, and they provide a common generalization for the Delaunay triangulation, the convex hull, the Gabriel graph, and the circle-based β-skeleton [Vel91]. *Empty-ellipse graphs* [DEG08] are a more recent variant of proximity graphs. They were defined by Devillers, Erickson and Goaoc to study the local behavior of Delaunay triangulations on surfaces in three-dimensional space. Here, two points are connected if and only if they lie on an axis-aligned ellipse with no other points from V in its interior. Devillers et al. show that the empty ellipse graph for a point set with stretch Φ has $O(n\Phi)$ edges [DEG08]. Cardinal, Collette and Langerman reverse the viewpoint of previous work on proximity graphs [CCL09]: in an *empty region graph*, two points form an edge if and only if some neighborhood around them—derived from a *template region*—is empty. Instead of analyzing the properties of certain given proximity graphs, Cardinal et al. start with certain desirable graph properties, such as connectivity, planarity, bipartiteness, or cycle-freeness, and they characterize maximal and minimal template regions that ensure these properties.

The field of *proximity drawings* studies which graphs can be represented as proximity graphs. For example, a tree can be represented as an RNG if and only if it has maximum vertex degree at most 5 [BLL96]. There is also a characterization of the trees representable as Gabriel graphs [BLL96]. Refer to [DBLL94, Lio13] for many more results. Chapter 54 discusses applications of proximity graphs in pattern recognition.

OPEN PROBLEMS

1. The complexity of Euclidean RNGs in $\mathbb{R}^3$ has still not been settled. Agarwal and Matoušek showed an upper bound of $O(n^{4/3})$, where n is the number of points [AM92]. No super-linear lower bound is known.

2. What is the complexity of the SIG? The best upper bound for n vertices is $15n$, but no lower bound exceeding $9n$ is known [Sos99]. The SIG has a linear number of edges in any fixed dimension [GPS94], and bounds on the expected number of edges are known [Dwy95]. However, there are no tight results.

32.2 QUADTREES AND WSPDS

GLOSSARY

Quadtree T associated with a set $S \subset \mathbb{R}^d$: A tree in which each inner node has exactly 2^d children, with each node ν having an associated subset $S(\nu) \subseteq S$ and an axis-parallel bounding hypercube $R(\nu)$ for $S(\nu)$, such that (i) $|S(\nu)| = 1$ if ν is a leaf; and (ii) for each internal node ν, the hypercubes for the 2^d children of ν constitute a partition of $R(\nu)$ into 2^d congruent hypercubes.

Compressed quadtree T associated with a set $S \subset \mathbb{R}^d$: A tree in which each inner node has exactly one or 2^d children, with each node ν having an associated subset $S(\nu) \subseteq S$ and an axis-parallel bounding hypercube $R(\nu)$ for $S(\nu)$, such that (i) $|S(\nu)| = 1$ if ν is a leaf; (ii) if ν has 2^d children, the hypercubes for the 2^d children partition $R(\nu)$ into 2^d congruent hypercubes; and (iii) if ν has one child ν', then $R(\nu') \subset R(\nu)$ and $R(\nu')$ is smaller than $R(\nu)$ by at least a constant factor. Usually, a compressed quadtree is obtained from a quadtree by contracting long paths in which each node has only one non-empty child square.

Fair-split tree T associated with a set $S \subset \mathbb{R}^d$:   A binary tree, where each node ν has an associated subset $S(\nu) \subseteq S$ and the axis-parallel bounding box $R(\nu)$ of $S(\nu)$, such that (i) $|S(\nu)| = 1$ if ν is a leaf; and (ii) for each internal node ν, let ν_1 and ν_2 be the two children of ν. Then, there exists a hyperplane h_ν orthogonal to the longest edge, ξ, of $R(\nu)$ separating $S(\nu_1)$ and $S(\nu_2)$ such that h_ν is at distance at least $|\xi|/3$ from each of the sides of $R(\nu)$ parallel to it.

s-well-separated pair: Let $s \geq 1$ be a fixed *separation constant*. Two nonempty point sets X and Y constitute an s-well-separated pair if and only if there are two radius-r enclosing balls, $B_X \supset X$ and $B_Y \supset Y$, such that the distance between B_X and B_Y is at least sr.

Well-separated pair decomposition (WSPD) of a set $S \subset \mathbb{R}^d$ of points for a fixed separation constant $s \geq 1$:   A set, $\{\{A_1, B_1\}, \{A_2, B_2\}, \ldots, \{A_m, B_m\}\}$, of pairs of nonempty subsets of S such that (i) $A_i \cap B_i = \emptyset$, for $i = 1, 2, \ldots, m$; (ii) each pair of distinct elements $\{a, b\} \subseteq S$ has a unique pair $\{A_i, B_i\}$ with $a \in A_i$, $b \in B_i$; and (iii) A_i and B_i are s-well-separated. The *size* of the WSPD is m.

Doubling dimension of a metric space (S, δ):   The doubling parameter $\lambda \in \mathbb{N}$ is the smallest integer such that for every $r \geq 0$ and every $p \in S$, the ball $B(p, r)$ can be covered by at most λ balls of radius $r/2$. The *doubling dimension* of (S, δ) is $\log \lambda$. A family of metric spaces has *bounded doubling dimension c* if the doubling dimension of all spaces in the family is at most c.

(Unit) disk graph: The graph with vertex set V where each $p \in V$ has an *associated radius* $r_p > 0$. There is an edge (p, q) if and only if $\delta(p, q) \leq r_p + r_q$, i.e., if the closed balls with radius r_p around p and with radius r_q around q intersect. The graph is called a *unit* disk graph if $r_p = 1/2$, for all $p \in V$.

s-semi-separated pair: Let $s > 1$ be a fixed *separation constant*. Two nonempty point sets X and Y constitute an s-semi-separated pair if and only if there are two enclosing balls, $B_X \supset X$ and $B_Y \supset Y$ with radius r_A and r_B, respectively such that the distance between B_X and B_Y is at least $s \min\{r_A, r_B\}$.

Semi-separated pair decomposition (SSPD) of a set $S \subset \mathbb{R}^d$ of points for a fixed separation constant $s > 1$:   A set, $\{\{A_1, B_1\}, \{A_2, B_2\}, \ldots, \{A_m, B_m\}\}$, of pairs of nonempty subsets of S such that (i) $A_i \cap B_i = \emptyset$, for $i = 1, 2, \ldots, m$; (ii) each pair of distinct elements $\{a, b\} \subseteq S$ has a unique pair $\{A_i, B_i\}$ with $a \in A_i$, $b \in B_i$; and (iii) A_i and B_i are s-semi-separated. The *size* of the SSPD is m.

QUADTREES

For every set S of n points, there is a quadtree with $O(n \log \Phi(S))$ nodes and depth $O(\log \Phi(S))$, and it can be computed in the same time [BCKO08, Sam90]. In general, the size and depth of a (regular) quadtree can be unbounded in n. To address this issue, one can define *compressed quadtrees*. The precise definition of a compressed quadtree varies in the literature [HP11, LM12], but the essential idea is to take a (regular) quadtree and to contract long paths in which each node has

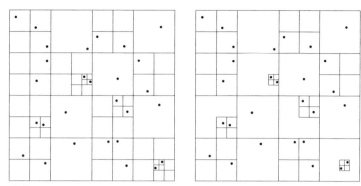

FIGURE 32.2.1
A regular and a compressed quadtree for a planar point set.

only a single non-empty child into single edges; see Figure 32.2.1. For every set S of n points, there is a compressed quadtree with $O(n)$ nodes. The depth may also be $\Theta(n)$. A compressed quadtree can be computed in $O(n \log n)$ time [HP11, LM12]. In fact, given the Delaunay triangulation $\mathrm{DT}(S)$ of a planar point set S, we can find a compressed quadtree for S in $O(n)$ additional time [KL98]. Conversely, given a suitable compressed quadtree for a planar point set S, we can find $\mathrm{DT}(S)$ in $O(n)$ additional time [BM11, LM12].

COMPUTATIONAL MODELS

When dealing with compressed quadtrees, it is important to keep the computational model in mind. Algorithms that use compressed quadtrees often rely on the real RAM model of computation and require the floor function $x \mapsto \lfloor x \rfloor, x \in \mathbb{R}$ [PS85]. In fact, Har-Peled pointed out that if we want the squares of a compressed quadtree to be aligned to a grid, some kind of non-standard operation is inevitable [HP11]. Nonetheless, the floor function provides unexpected computational power. It can be used to circumvent established lower bounds in the algebraic decision tree model [Ben83]. For example, using the floor function, we can use Rabin's algorithm to find the closest pair of a set of n points in $O(n)$ expected time, despite an $\Omega(n \log n)$ lower bound for algebraic decision trees [Ben83]. Not only that, the floor function lets us solve PSPACE-complete problems in polynomial time [Sch79]. Despite these issues, algorithms that use the floor function are often simple, efficient and practical. There is also a way to define compressed quadtrees in a way that is compatible with the algebraic decision model, but this usually comes at the cost of increased algorithmic complexity [BLMM11, LM12]. When comparing results that involve quadtrees, we should be aware of the details of the underlying computational model.

WELL-SEPARATED PAIR DECOMPOSITION

Callahan and Kosaraju [CK95] defined the notion of a well-separated pair decomposition (WSPD) for a point set S. They also showed the remarkable theorem that a WSPD of size $O(n)$ can be constructed in time $O(n)$, given a fair split tree of an

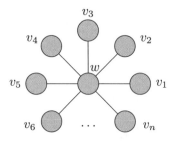

FIGURE 32.2.2

Let $s > 1$. Any WSPD for the shortest path metric of the star graph with separation parameter s has size $\Omega(n^2)$. For every pair of distinct vertices v_i, v_j, there must be a distinct s-well-separated pair $\{A, B\}$ with $v_i \in A, v_j \in B$: if there were another v_k with, say, $v_k \in A$, then A would have diameter 2 and distance at most 2 from B, making $\{A, B\}$ not s-well-separated.

input set S of n points in $\mathbb{R}^d$, for any fixed dimension d and separation constant $s \geq 1$. (More precisely, the size of the WSPD is $O(s^d n)$.) A fair split tree can be constructed using quadtree methods in time $O(n \log n)$ for any fixed dimension. Alternatively, the WSPD can be computed from a (compressed) quadtree in $O(n)$ additional time [BM11, Cha08, HP11, LM12]. In the algebraic decision tree model, it takes $\Omega(n \log n)$ steps to compute the WSPD.

Well-separated pair decompositions have countless applications in proximity problems [Smi07]. For example, let S be a set of n points. Given a WSPD for S of size m with separation parameter $s > 2$, we can find a closest pair in S in $O(m)$ additional time, since the closest pair occurs as a well-separated pair in the decomposition [CK95]. In fact, given a fair-split tree or a compressed quadtree that represents a WSPD for S with separation parameter $s > 2$, we can compute $NNG(S)$ in $O(n)$ additional time [CK95]. This fact, together with the connection between fast algorithms for $NNG(S)$ and fast algorithms for $DT(S)$, can be used to obtain improved running times for computing Delaunay triangulations in various models of computation, such as the word RAM [BM11]. WSPDs are also extremely useful in the context of approximation algorithms. For example, they can be used to approximate the diameter and the minimum-spanning tree of high-dimensional point sets. As we will see in the next section, they also play an important role in spanner construction. The survey of Smid contains further applications [Smi07].

Not every metric space admits a WSPD of subquadratic size. For example, in the shortest path metric of the unweighted star graph, every WSPD with separation parameter $s > 1$ must have $\Omega(n^2)$ pairs; see Figure 32.2.2. Notwithstanding, there are large families of finite metrics that have WSPDs with linear or near-linear size.

A family of metric spaces has *bounded doubling dimension c* if the doubling dimension of all spaces in the family is at most c. This notion was defined by Gupta, Krauthgamer and Lee [GKL03], following earlier work by Assouad [Ass83]. The family of finite subsets of $\mathbb{R}^d$ has bounded doubling dimension $\Theta(d)$. Talwar gave an algorithm to compute a WSPD with separation parameter $s \geq 1$ of size $O(s^{\log \lambda} n \log \Phi(S))$, where $\Phi(S)$ is the spread of S [Tal04]. This was improved by Har-Peled and Mendel, who showed how to compute a WSPD of size $O(s^{\log \lambda} n)$ in time $O(\lambda n \log n + s^{\log \lambda} n)$, asymptotically matching the bounds for the Euclidean case [HP11, HPM06].

Another interesting metric is given by *unit disk graphs*. In a unit disk graph, there is an edge (p, q) between two vertices if and only $\delta(p, q) \leq 1$. Even though the shortest path metric in unit disk graphs (with Euclidean edge lengths) does not have bounded doubling dimension, Gao and Zhang showed that it admits WSPDs of near-linear size [GZ05]. More specifically, they showed that the planar unit disk graphs with n vertices have WSPDs of size $O(s^4 n \log n)$ that can be computed in the same time. In dimension $d \geq 3$, there is a WSPD of size $O(n^{2-2/d})$ that can be found in time $O(n^{4/3} \log^{O(1)} n)$ for $d = 3$ and in time $O(n^{2-2/d})$ for $d \geq 4$. The bound on the size is tight for $d \geq 4$. The result of Gao and Zhang can be extended to general disk graphs where the ratio between the largest and the smallest radius is bounded [GZ05, Wil16]. For an unbounded radius ratio, however, no WSPD of subquadratic size is possible in general. Generalizations to the shortest path metric in *unweighted* disk graphs (also known as *hop distance*) are also possible, albeit with somewhat weaker results [GZ05].

For Euclidean point sets, there is always a WSPD with a linear number of pairs, but the total number of points in the sets of the pairs may be quadratic. In many applications, this is not an issue, because the sets can be represented implicitly and the algorithms do not need to inspect all sets in the WSPD. However, if this becomes necessary, a quadratic running time becomes hard to avoid. To address this, Varadarajan [Var98] introduced the notion of *s-semi-separated pair decomposition* (s-SSPD). In d dimensions, an s-SSPD with $O(s^d n)$ pairs whose sets contain $O(s^d n \log n)$ points in total can be computed in $O(s^d n \log n)$ time [ABFG09, AH12]. SSPDs have numerous applications, e.g., in computing the min-cost perfect matching of a planar point set [Var98] or in constructing spanners with certain properties [ABFG09, AH12, ACFS13]. Abu-Affash et al. [AACKS14] introduce the (α, β)-*pair decomposition*, another variant of the WSPD. They provide several applications, including its application to the Euclidean bottleneck Steiner path problem.

In general, the WSPD is the method of choice when we need to represent approximately the pairwise distances in a point set. The SSPD offers weaker guarantees, but it can be useful if we need to inspect all sets of the decomposition explicitly. The (α, β)-pair decomposition does not give a general approximation of distances, but it works only at a fixed scale. It is simpler to work with than the other two decompositions and it can provide stronger guarantees for certain problems.

OPEN PROBLEMS

1. What is the right bound of the size for WSPDs for unit disk graphs in the plane? Gao and Zhang's result gives an upper bound of $O(n \log n)$ [GZ05], but no super-linear lower bound is known.

2. Disk graphs with an unbounded radius ratio generally do not admit a WSPD of subquadratic size. Is there another way to represent the pairwise distances in these graphs compactly?

32.3 GEOMETRIC SPANNERS

GLOSSARY

Euclidean graph: A geometric graph with Euclidean lengths associated with the edges.

Complete Euclidean graph $\mathcal{E}_d$: A d-dimensional Euclidean graph (V, E) whose edge set E joins each pair of points of $V \subset \mathbb{R}^d$.

Yao-graph Y_k: For integer $k \geq 2$, a geometric graph in which each $v \in V$ is joined by an edge to the closest point $u \in V \cap C_i$, where, in dimension $d = 2$, each C_i is one of $k = 2\pi/\theta$ equal-sized *sectors* (cones) with apex v and angle θ.

Theta-graph Θ_k: For integer $k \geq 2$, a geometric graph, similar to the Yao-graph, in which each $v \in V$ is joined by an edge to the "closest" point $u \in V \cap C_i$, where "closest" is based on the projections of the points $V \cap C_i$ onto a ray with apex v within C_i, a sector of angle $2\pi/k$; typically, the ray is the bisector of C_i.

t-Spanner: A subgraph $G' = (V, E')$ of a graph $G = (V, E)$ such that for any $u, v \in V$ the distance $\delta_{G'}(u, v)$ within G' is at most t times the distance $\delta_G(u, v)$ within G. We focus on **Euclidean t-spanners** for which the underlying graph G is the complete Euclidean graph in $\mathbb{R}^d$.

Plane t-spanner: A Euclidean t-spanner that is a PSLG in $\mathbb{R}^2$. (Also known as a *planar t-spanner*.)

Stretch factor, t^*, of a Euclidean graph $G = (V, E)$:

$$t^* = \max_{u,v \in V, u \neq v} \left\{ \frac{\delta_G(u, v)}{\delta_2(u, v)} \right\}$$

where $\delta_2(u, v)$ is the Euclidean distance between u and v. Thus, t^* is the smallest value of t for which G is a Euclidean t-spanner. The stretch factor is also known as the **spanning ratio** or the **dilation** of G.

Size of a Euclidean graph $G = (V, E)$: The number of edges, $|E|$.

Weight of a Euclidean graph $G = (V, E)$: The sum of the Euclidean lengths of all edges $e \in E$.

Degree of a graph $G = (V, E)$: The maximum number of edges incident on a common vertex $v \in V$.

k-vertex fault-tolerant t-spanner: A t-spanner with the property that the removal of any subset of at most k nodes, along with the incident edges, results in a subgraph that remains a t-spanner on the remaining set of points.

t-SPANNERS

A natural greedy algorithm, similar to Kruskal's minimum spanning tree algorithm, can be used to construct t-spanners:

Given an input geometric graph $G = (V, E)$ and a real number $t > 1$. Initialize edge set $E' \leftarrow \emptyset$. For each edge $(u, v) \in E$, considered in nondecreasing order of length $\delta_2(u, v)$, if $\delta_{G'}(u, v) > t \cdot \delta_2(u, v)$, then $E' \leftarrow E' \cup \{(u, v)\}$. Output the graph $G' = (V, E')$.

The greedy algorithm results in a t-spanner of size $O(n)$, weight $O(\log n) \cdot |\text{MST}|$, and degree $O(1)$, for any fixed dimension d and spanning ratio $t > 1$ [ADD+93, CDNS95]. It can be applied also to general (nongeometric) graphs with weighted edges. The greedy spanner can be computed in $O(n^2 \log n)$ time with $O(n^2)$ space [BCF+10] and in $O(n^2 \log^2 n)$ time with $O(n)$ space [ABBB15].

The Yao-graphs Y_k and theta-graphs Θ_k explicitly take advantage of geometry, and each yields a t-spanner with spanning ratio $t = 1 + O(1/k)$ arbitrarily close to 1, for sufficiently large k [ADD+93, Cla87, Kei92, RS91, Yao82]. The Yao-graphs Y_k and theta-graphs Θ_k are connected for $k \geq 2$; see [ABB+14].

TABLE 32.3.1   Spanning ratios of Yao-graphs and Θ-graphs in $\mathbb{R}^2$.

GRAPH	SPANNING RATIO	REFERENCE
Y_2, Y_3	unbounded	[EM09]
Y_4	$\leq 8(29 + 23\sqrt{2}) \approx 663$	[BDD+12]
Y_5	$\in [2.87, 2 + \sqrt{3} \approx 3.74]$	[BBD+15]
Y_6	$\in [2, 5.8]$	[BBD+15]
$Y_k, k \geq 7$	lower/upper bounds depending on $k \pmod 4$	[BBD+15]
Θ_2, Θ_3	unbounded	[EM09]
Θ_4	[7,237]	[BBC+13]
Θ_5	[3.79, 9.96]	[BMRV15]
$\Theta_k, k \geq 6, k \equiv 2 \pmod 4$	$1 + 2\sin(\pi/k)$ (tight)	[BDCM+16, BGHI10]
$\Theta_k, k \geq 7$	lower/upper bounds depending on $k \pmod 4$	[BDCM+16]

The known bounds on spanning ratios of Yao-graphs and Θ-graphs are shown in Table 32.3.1; see also the detailed tables in [BDCM+16] for Θ-graphs and in [BBD+15] for Yao-graphs. Note that Y_5 has spanning ratio $3.74 < 3.79$, making it the only known case ($k = 5$) in which there is a strict separation between the spanning ratio of a theta-graph Θ_k and that of a Yao-graph Y_k; for other values of $k \geq 4$, it is not known which graph (Y_k or Θ_k) has a smaller spanning ratio. The spanning ratio for Θ-graphs does not necessarily decrease with an increase in k; for $k = 6$, the spanning ratio is 2 (and this is tight), while for $k = 8$, it is known that the spanning ratio is at least 2.17 [BDCM+16].

By selecting a representative edge from each pair in a WSPD, one obtains a t-spanner of size $O(n)$ with a spanning ratio that can be made arbitrarily close to 1, depending on the separation constant s.

One can in fact obtain t-spanners for n points in $\mathbb{R}^d$ that are simultaneously good with respect to size, weight, and degree—size $O(n)$, weight $O(|\text{MST}|)$, and bounded degree (independent of the dimension d). Gudmundsson et al. [GLN02] show that such spanners can be computed in time $O(n \log n)$, improving the previous bound of $O(n \log^2 n)$ [DN97] and re-establishing the time bound claimed in Arya et al. [ADM+95] (which was found to be flawed). $\Omega(n \log n)$ time is required for constructing *any* t-spanner for n points in $\mathbb{R}^d$ in the algebraic decision tree model [CDS01].

It was shown by Levcopoulos, Narasimhan, and Smid [LNS02] that k-vertex fault-tolerant spanners of size $O(k^2 n)$ can be constructed in time $O(n \log n + k^2 n)$; alternatively, spanners of size $O(kn \log n)$ can be constructed in time $O(kn \log n)$. Lukovszki [Luk99] and Czumaj and Zhao [CZ04] showed how to obtain even smaller, degree-bounded low-weight k-vertex fault-tolerant spanners; degree $O(k)$ and weight $O(k^2 |\text{MST}|)$ can be obtained, and these bounds are asymptotically optimal.

The spanning ratio of a given graph $G = (V, E)$ can be computed exactly in worst-case time $O(n^2 \log n + n|E|)$ using an all-pairs shortest path computation. Given a Euclidean graph with n vertices and m edges, its spanning ratio (stretch factor) can be $(1+\varepsilon)$-approximated in time $O(m + n \log n)$ [GLNS08]. Narasimhan and Smid [NS02] have studied the *bottleneck stretch factor problem*, in which the goal is to be able to compute quickly, for any given $b > 0$, an approximate stretch factor of the *bottleneck graph* $G_b = (V, E_b)$ whose edge set E_b consists of those edges of the complete graph whose length is at most b. We say that t is a (c_1, c_2)-*approximate stretch factor* of a graph if the true stretch factor, t^*, satisfies $t/c_1 \le t^* \le c_2 t$. A data structure of size $O(\log n)$ can be constructed that supports $O(\log \log n)$-time queries, for any $b > 0$, yielding a (c_1, c_2)-approximate stretch factor of G_b. The construction of the data structure, which is based on a WSPD, is done using a randomized algorithm with expected running time that is slightly subquadratic.

Spanners can be computed for geodesic distances in a polygonal domain P: a $(1+\varepsilon)$-spanner of the visibility graph $\text{VG}(P)$ can be computed in time $O(n \log n)$, for any $\varepsilon > 0$ [ACC$^+$96]. Geometric spanners can be used to obtain very efficient approximate two-point shortest path distance queries: for any constant $t > 1$, a t-spanner G for n points in $\mathbb{R}^d$ with m edges can be processed in time $O(m \log n)$, building a structure of size $O(n \log n)$, to support $(1+\varepsilon)$-approximate shortest path (in G) distance queries in $O(1)$ time between any two vertices of G. (A path can be reported in additional time proportional to the number of its edges.) Then, if the visibility graph $\text{VG}(P)$ is a t-spanner of the vertices of P, for some constant t, one obtains $O(1)$-time (resp., $O(\log n)$-time) $(1+\varepsilon)$-approximate shortest path distance queries between any two vertices (resp., points) of P. The assumption on $\text{VG}(P)$ holds if P has the "t-rounded" property for some t: the shortest path distance between any pair of vertices is at most t times the Euclidean distance between them; such is the case if the obstacles are *fat*, as shown by Chew et al. [CDKK02].

PLANE t-SPANNERS

For finite point sets in the plane it is natural to consider constructing *plane t-spanner* networks, whose edges do not cross. One cannot hope, in general, to obtain plane t-spanners with t arbitrarily close to 1: four points at the corners of a square have no plane t-spanner with $t < \sqrt{2}$.

The first result on plane t-spanners is due to Chew [Che86], who showed that the Delaunay triangulation in the L_1 metric is a $\sqrt{10}$-spanner for the complete Euclidean graph. (It is a $\sqrt{5}$-spanner for the complete graph whose edge lengths are measured in the L_1 metric.) Chew [Che89] improved this result, showing that the Delaunay triangulation in the convex distance function based on an equilateral triangle (also known as the triangular-distance Delaunay or TD-Delaunay graph) is a plane graph with spanning ratio at most 2; this bound is now known to be tight [BGHI10]. This had been the best known spanning ratio for a plane t-spanner until the work of Xia [Xia13], who showed that the Euclidean Delaunay triangulation has a spanning ratio less than 1.998. The lower bound of $\sqrt{2}$ on the spanning ratio of a plane t-spanner, given by the four corners of a square, has been improved by Mulzer [Mul04] to 1.41611 (by considering vertices of a regular 21-gon), and then by Dumitrescu and Ghosh [DG16] to 1.4308 (by considering vertices of a regular 23-gon). For points in convex position a spanning ratio of 1.88 can always be achieved [ABB+16].

The spanning ratio, τ_{Del}, of the Euclidean (L_2) Delaunay triangulation is not less than $\pi/2$, as shown by the example of placing points around a circle. This lower bound has been improved recently to $1.5932 > \pi/2$ [XZ11]. Dobkin, Friedman, and Supowit [DFS90] were able to show that $\tau_{\text{Del}} \le \phi\pi$, where $\phi = (1 + \sqrt{5})/2$ is the golden ratio. This upper bound was improved by Keil and Gutwin [KG92] to $\frac{4\pi}{3\sqrt{3}} \approx 2.42$. The current best known upper bound on τ_{Del} is 1.998, as shown by Xia [Xia13]. For the Delaunay triangulation in the L_1 metric, the original bound of $\sqrt{10}$ [Che86] on the spanning ratio has been improved to $\sqrt{4 + 2\sqrt{2}}$, and this is tight [BGHP15]. More generally, the Delaunay triangulation defined with respect to any convex distance function is a t-spanner [BCCS10].

Minimum spanning trees do not have a bounded spanning ratio. It is also known that lune-based β-skeletons, for any $\beta > 0$, can have an unbounded spanning ratio; see [Epp00]. Since for $\beta \ge 1$, the lune-based β-skeleton is a subgraph of the Gabriel graph ($\beta = 1$), it is a plane graph. In particular, the Gabriel graph ($\beta = 1$) and the relative neighborhood graph ($\beta = 2$) are not t-spanners for any constant t. Growth rates, as a function of n, for the spanning ratios of Gabriel graphs and β-skeletons for other values of β are given by [BDEK06].

The minimum weight triangulation and the greedy triangulation (see Chapter 31) are t-spanners for constant t. This follows from a more general result of Das and Joseph [DJ89], who show that a PSLG is a t-spanner if it has the "*diamond property*" and the "*good polygon property*." This result is similar to the empty region graphs of Cardinal et al. discussed in Section 32.1, where certain graph properties were obtained by requiring that an edge is present if and only if certain template regions were empty [CCL09]. The difference is that the diamond propery by Das and Joseph is only a necessary condition. A *fat* triangulation of S, for which the aspect ratio (ratio of the length of the longest side to the corresponding height) of every triangle is at most α, is known to be a 2α-spanner [KG01].

All of the plane spanners mentioned above have potentially unbounded degree. Bounded degree plane spanners are important in wireless network applications, especially in routing. One needs degree at least 3 to achieve a bounded spanning ratio (a Hamiltonian path has an unbounded spanning ratio). While Das and Heffernan [DH96] showed that t-spanners exist of maximum degree 3, their spanner is not necessarily plane. The best upper bounds on spanning ratio bounds currently known for plane spanners of degree at most δ are 20 for degree $\delta = 4$ [KPT16], 6 for $\delta = 6$ [BGHP10], 2.91 for $\delta = 14$ [KP08], and ≈ 4.414 for $\delta = 8$ [BHS16]; see [KPT16] for more details and a comprehensive table. While for points in convex position, degree-3 plane spanners are known [BBC+17, KPT16], it is not clear if there exist plane t-spanners of degree 3 for general point sets in the plane.

One can compute plane t-spanners of low weight. In linear time, for any $r > 0$, a plane t-spanner, with $t = (1+1/r)\tau_{\text{Del}}$, of weight at most $(2r+1)|\text{MST}|$ can be computed from a Delaunay triangulation, where τ_{Del} is the spanning ratio of the

Delaunay triangulation [LL92]. From any t-spanner, [GLN02] show that one can compute a subgraph of it that is low weight ($O(|\text{MST}|)$) and a t'-spanner (for a larger constant factor t'); thus, in order to find spanners that are of both bounded degree and of low weight, it suffices to focus on bounding the degree. One can compute in time $O(n \log n)$ a plane t-spanner that is simultaneously low weight ($O(|\text{MST}|)$) and low degree (degree at most k), with $t = (1+2\pi(k\cos(\pi/k))^{-1})\cdot\tau_{\text{Del}}$ for any integer $k \geq 14$ [KPX10]; in particular, for degree 14, the stretch factor is at most ≈ 2.918.

Planar t-spanners are also known for geodesic distances. A *conforming triangulation* for a polygonal domain P having triangles of aspect ratio at most α is a 2α-spanner for geodesic distances between vertices of P [KG01]. (A triangulation is *conforming* for P if each vertex of P is a vertex of the triangulation and each edge of P is the union of some edges of the triangulation.) The *constrained Delaunay triangulation* of P is a $\phi\pi$-spanner [KG01].

NON-EUCLIDEAN METRICS

The WSPD-construction of Har-Peled and Mendel implies that for every $\varepsilon > 0$, a space of bounded doubling dimension d has a spanner with $n\varepsilon^{-O(d)}$ edges and stretch factor $1 + \varepsilon$ that can be found in $2^{O(d)}n \log n + n\varepsilon^{-O(d)}$ time [HPM06]. Independently, Chan et al. [CGMZ16] obtained a similar result. They showed the existence of a spanner with stretch factor $1 + \varepsilon$ in which every vertex has degree at most $\varepsilon^{-O(d)}$. Subsequently, several improved constructions of spanners for bounded doubling metrics were described [CLNS15]. The weight was also considered in this context. Smid showed that the greedy $(1 + \varepsilon)$-spanner in spaces of bounded doubling dimension has $O(n)$ edges and weight $O(\log n|\text{MST}|)$ [Smi09]. Gottlieb provided an intricate construction of $(1 + \varepsilon)$-spanners with weight $O(|\text{MST}|)$ and $O(n)$ edges [Got15]. Filtser and Solomon proved that this is also achieved by the greedy spanner: for any $\varepsilon > 0$, the greedy $(1 + \varepsilon)$-spanner in a space of doubling dimension d has weight $(d/\varepsilon)^{O(d)}|\text{MST}|$ and $n(1/\varepsilon)^{O(d)}$ edges [FS16]. Moreover, an approximate version of the greedy spanner shows that for any $\varepsilon > 0$, one can construct in time $\varepsilon^{-O(d)}n \log n$ a $(1 + \varepsilon)$-spanner with weight $(d/\varepsilon)^{O(d)}|\text{MST}|$ and degree $\varepsilon^{-O(d)}$ [FS16]. This matches the best result for the Euclidean case [GLN02].

Sparse spanners also exist for disk graphs. Fürer and Kasiviswanathan [FK12] used a modification of the Yao graph to show that for fixed $\varepsilon > 0$, every disk graph has a spanner with $O(n)$ edges and stretch factor $1 + \varepsilon$. They also described an algorithm to find such spanners in time $O(n^{4/3+\tau} \log^{2/3} \Psi)$, where $\tau > 0$ can be made arbitrarily small and Ψ is the *radius ratio* between the largest and the smallest radius of a vertex in V. Kaplan et al. extended this result to *transmission graphs*, a directed version of disk graphs in which each vertex $p \in V$ has an associated radius $r_p > 0$ and we have a directed edge (p, q) if and only if $\delta(p, q) \leq r_p$, i.e., if q lies in the closed radius-r_p disk around p [KMRS15]. For any fixed $\varepsilon > 0$, every transmission graph has a spanner with stretch factor $1 + \varepsilon$ and $O(n)$ edges. This spanner can be found in $O(n(\log n + \log \Psi))$ time, where Ψ is again the radius ratio [KMRS15, Sei16]. Alternatively, the spanner can be found in $O(n \log^5 n)$ time, independent of Ψ. The results of Kaplan et al. can also be applied to general disk graphs. Here, the time to construct the spanner described by Fürer and Kasiviswanathan [FK12] can be improved to $O(n2^{\alpha(n)} \log^{10} n)$ expected time, where $\alpha(n)$ is the inverse Ackermann function [KMR+17, Sei16].

OPEN PROBLEMS

1. What is the best possible spanning ratio for a plane t-spanner? It is known to be between 1.4308 and 1.998. For points in convex position an upper bound of 1.88 is known. The upper bound of 1.998 comes from the Euclidean Delaunay graph; is there a plane t-spanner with a spanning ratio better than the Delaunay?

2. Determine tight bounds for the spanning ratio of theta-graphs Θ_k for $k \geq 7$.

3. What exactly is the spanning ratio of the Euclidean Delaunay triangulation? It is known to be between 1.5932 and 1.998.

4. What are the best possible spanning ratios for bounded degree plane spanners, for various degree bounds? Are there plane spanners of bounded spanning ratio having degree at most 3?

5. For a given set of points in the plane, can one compute in polynomial time a plane graph having the minimum possible spanning ratio?

6. How efficiently can Yao and theta-graphs be constructed in higher dimensions?

7. Is it possible to compute the greedy spanner in subquadratic time?

32.4 DYNAMIC PROXIMITY ALGORITHMS

GLOSSARY

Additively weighted Euclidean distance: Let $S \subset \mathbb{R}$ be a set of sites, such that each $s \in S$ has an associated weight $w_s \in \mathbb{R}$. The additively weighted Euclidean distance $\delta : \mathbb{R} \times S \to \mathbb{R}$ is defined as $\delta(p, s) = w_s + \delta_2(p, s)$.

Dynamic nearest neighbor: The problem of maintaining a set S of sites under the following operations: (i) insert a new site into S; (ii) delete a site from S; and (iii) given a query point q, find the site in S that minimizes the distance to q.

Dynamic bichromatic closest pair: The problem of maintaining two sets R and B of red and blue points such that points can be inserted into and deleted from R and B and such that we always have available a pair $(r, b) \in R \times B$ that minimizes the distance $\delta(r, b)$ among all pairs in $R \times B$.

In dynamic proximity algorithms, we would like to maintain some proximity structure for a point set that changes through insertions and deletions. The quintessential problem is the *dynamic nearest neighbor* problem: maintain a point set S under the following operations: (i) insert a new point into S; (ii) delete a point from S; (iii) given a query point q, determine a point $p \in S$ with $\delta(p, q) = \min_{r \in S} \delta(r, q)$.

For the Euclidean case, the first solution to this problem is due to Agarwal and Matoušek [AM95], who obtained amortized update time $O(n^\varepsilon)$, for every $\varepsilon > 0$, and worst-case query time $O(\log n)$. This was dramatically improved more than ten years later by Chan [Cha10]. His data structure achieves worst-case query time $O(\log^2 n)$ with amortized insertion time $O(\log^3 n)$ and amortized deletion time $O(\log^6 n)$ (Chan's original data structure was randomized, but a recent result of Chan and Tsakalidis yields a deterministic structure [CT16]). Kaplan et al. provide a variant of Chan's data structure that improves the amortized deletion time to $O(\log^5 n)$ [KMR$^+$17]. Similar results hold for more general metrics. Extending the results by Agarwal and Matoušek [AM95] for the Euclidean case, Agarwal, Efrat, and Sharir describe a dynamic nearest neighbor structure with amortized update time $O(n^\varepsilon)$, for any fixed $\varepsilon > 0$, and worst-case query time $O(\log n)$ [AES99]. The result by Agarwal, Efrat, and Sharir holds for a wide range of distance functions, including L_p-metrics and the additively weighted Euclidean distance. This result was improved by Kaplan et al. [KMR$^+$17]. Kaplan et al. built on Chan's data structure [Cha10] to construct a dynamic nearest neighbor structure for general metrics with worst-case query time $O(\log^2 n)$, amortized expected insertion time $O(\log^5 n \lambda_{s+2}(\log n))$ and amortized expected deletion time $O(\log^9 n \lambda_{s+2}(\log n))$. Here, s is a constant that depends on the metric under consideration, and $\lambda_t(\cdot)$ is the function that bounds the maximum length of a Davenport-Schinzel sequence of order t [SA95]. See Section 28.10 for more details on Davenport-Schinzel sequences.

Eppstein describes several reductions that provide fast dynamic algorithms for proximity problems once an efficient dynamic nearest neighbor structure is at hand [Epp95]. In particular, he showed that if there is a dynamic nearest neighbor structure whose queries all run in time $T(n)$, where $T(n)$ is monotonically increasing and $T(3n) = O(T(n))$, then the *dynamic bichromatic nearest neighbor problem* can be solved with amortized insertion time $O(T(n) \log n)$ and amortized deletion time $O(T(n) \log^2 n)$ [Epp95]. This implies that the MST of a planar point set can be maintained in $O(T(n) \log^4 n)$ amortized time per update (this result is not stated in the original paper, since it also needs a fast data structure for maintaining an MST in a general dynamic graph [HdLT01] that was not available when Eppstein wrote his paper).

Dynamic algorithms for geometric spanners have also been considered. The first result in this direction is due to Arya et al. [AMS99] who show how to construct a data structure of size $O(n \log^d n)$ that maintains a d-dimensional Euclidean spanner in $O(\log^d n \log \log n)$ expected amortized time per insertion and deletion in a model of random updates. A dynamic spanner by Gao et al. [GGN06] can handle arbitrary update sequences, with the performance bounds depending on the spread of the point set. The first dynamic spanner whose performance depends only on the number of points is due to Roditty [Rod12]. This result was improved several times [GR08a, GR08b]. An optimal construction was eventually obtained by Gottlieb and Roditty [GR08b]. Their spanner has stretch factor $1 + \varepsilon$, for any $\varepsilon > 0$, constant degree $\varepsilon^{-O(d)}$, and update time $\varepsilon^{-O(d)} \log n$. It also works in general metric spaces of constant doubling dimension d.

OPEN PROBLEMS

1. Can the Euclidean dynamic nearest neighbor problem be solved with amortized update time $O(\log n)$?

2. Can Eppstein's reduction from the bichromatic nearest neighbor problem to the dynamic nearest neighbor problem be improved?

32.5 SOURCES AND RELATED MATERIAL

SURVEYS

Several other surveys offer a wealth of additional material and references:

[BE97]: A survey of approximation algorithms for geometric optimization problems.

[BS13]: A survey on plane geometric spanners, with 22 highlighted open problems.

[Epp00]: A survey of results on spanning trees and t-spanners.

[GK07]: A survey on geometric spanners and related problems.

[HP11]: A book on geometric approximation algorithms.

[JT92]: A survey on relative neighborhood graphs.

[Lio13]: A survey on proximity drawings.

[NS07]: A book on geometric spanners.

[Sam90]: A book on quadtrees and related structures.

[Smi07]: A survey on well-separated pair decompositions.

[Tou14]: A recent survey on sphere-of-influence graphs.

RELATED CHAPTERS

Chapter 27: Voronoi diagrams and Delaunay triangulations
Chapter 31: Shortest paths and networks
Chapter 38: Point location
Chapter 39: Collision and proximity queries
Chapter 40: Range searching
Chapter 54: Pattern recognition

REFERENCES

[AACKS14]   A.K. Abu-Affash, P. Carmi, M.J. Katz, and M. Segal. The Euclidean bottleneck Steiner path problem and other applications of (α, β)-pair decomposition. *Discrete Comput. Geom.*, 51:1–23, 2014.

[ABB+14]   O. Aichholzer, S.W. Bae, L. Barba, P. Bose, M. Korman, A. van Renssen, P. Taslakian, and S. Verdonschot. Theta-3 is connected. *Comput. Geom.*, 47:910–917, 2014.

[ABB+16]   M. Amani, A. Biniaz, P. Bose, J.-D. De Carufel, A. Maheshwari, and M. Smid. A plane 1.88-spanner for points in convex position. *J. Comput. Geom.*, 7:520–539, 2016.

[ABBB15]   S.P.A. Alewijnse, Q.W. Bouts, A.P. ten Brink, and K. Buchin. Computing the greedy spanner in linear space. *Algorithmica*, 73:589–606, 2015.

[ABFG09]   M.A. Abam, M. de Berg, M. Farshi, and J. Gudmundsson. Region-fault tolerant geometric spanners. *Discrete Comput. Geom.*, 41:556–582, 2009.

[ACC+96]   S.R. Arikati, D.Z. Chen, L.P. Chew, G. Das, M. Smid, and C.D. Zaroliagis. Planar spanners and approximate shortest path queries among obstacles in the plane. In *Proc. 4th Eur. Sympos. Algorithms*, vol. 1136 of *LNCS*, pages 514–528, Springer, Berlin, 1996.

[ACFS13]   M.A. Abam, P. Carmi, M. Farshi, and M. Smid. On the power of the semi-separated pair decomposition. *Comput. Geom.*, 46:631–639, 2013.

[ADD+93]   I. Althöfer, G. Das, D.P. Dobkin, D. Joseph, and J. Soares. On sparse spanners of weighted graphs. *Discrete Comput. Geom.*, 9:81–100, 1993.

[ADM+95]   S. Arya, G. Das, D.M. Mount, J.S. Salowe, and M. Smid. Euclidean spanners: short, thin, and lanky. In *Proc. 27th ACM Sympos. Theory Comput.*, pages 489–498, 1995.

[AES99]   P.K. Agarwal, A. Efrat, and M. Sharir. Vertical decomposition of shallow levels in 3-dimensional arrangements and its applications. *SIAM J. Comput.*, 29:912–953, 1999.

[AH85]   D. Avis and J. Horton. Remarks on the sphere of influence graph. *Ann. New York Acad. Sci.*, 440:323-327, 1985.

[AH12]   M.A. Abam and S. Har-Peled. New constructions of SSPDs and their applications. *Comput. Geom.*, 45:200–214, 2012.

[AM92]   P.K. Agarwal and J. Matoušek. Relative neighborhood graphs in three dimensions. *Comput. Geom.*, 2:1–14, 1992.

[AM95]   P.K. Agarwal and J. Matoušek. Dynamic half-space range reporting and its applications. *Algorithmica*, 13:325–345, 1995.

[AMS99]   S. Arya, D.M. Mount, and M. Smid. Dynamic algorithms for geometric spanners of small diameter: Randomized solutions. *Comput. Geom.*, 13:91–107, 1999.

[Ass83]   P. Assouad. Plongements lipschitziens dans $\mathbb{R}^n$. *Bull. Soc. Math. France*, 111:429–448, 1983.

[BBC+13]   L. Barba, P. Bose, J.-D. de Carufel, A. van Renssen, and S. Verdonschot. On the stretch factor of the Theta-4 graph. In *Proc. 13th Int. Sympos. Algorithms Data Structures*, vol. 8037 of *LNCS*, pages 109–120, Springer, Berlin, 2013.

[BBC+17]   A. Biniaz, P. Bose, J.-D. de Carufel, C. Gavoille, A. Maheshwari, and M. Smid. Towards plane spanners of degree 3. *J. Comput. Geom.*, 8:11–31, 2017.

[BBD+15]   L. Barba, P. Bose, M. Damian, R. Fagerberg, W.L. Keng, J. O'Rourke, A. van Renssen, P. Taslakian, S. Verdonschot, and G. Xia. New and improved spanning ratios for Yao graphs. *J. Comput. Geom.*, 6:19–53, 2015.

[BCCS10]   P. Bose, P. Carmi, S. Collette, and M. Smid. On the stretch factor of convex Delaunay graphs. *J. Comput. Geom.*, 1:41–56, 2010.

[BCF+10]   P. Bose, P. Carmi, M. Farshi, A. Maheshwari, and M. Smid. Computing the greedy spanner in near-quadratic time. *Algorithmica*, 58:711–729, 2010.

[BCH+13]   P. Bose, S. Collette, F. Hurtado, M. Korman, S. Langerman, V. Sacristán, and M. Saumell. Some properties of k-Delaunay and k-Gabriel graphs. *Comput. Geom.*, 46:131–139, 2013.

[BCKO08]   M. de Berg, O. Cheong, M. van Kreveld, and M. Overmars. *Computational Geometry. Algorithms and Applications*, third edition. Springer-Verlag, Berlin, 2008.

[BDCM+16]   P. Bose, J.L. de Carufel, P. Morin, A. van Renssen, and S. Verdonschot. Towards tight bounds on Theta-graphs: More is not always better. *Theoret. Comput. Sci.*, 616:70–93, 2016.

[BDD+12]   P. Bose, M. Damian, K. Douïeb, J. O'Rourke, B. Seamone, M. Smid, and S. Wuhrer. $\pi/2$-angle Yao graphs are spanners. *Internat. J. Comput. Geom. Appl.*, 22:61–82, 2012.

[BDEK06]   P. Bose, L. Devroye, W. Evans, and D.G. Kirkpatrick. On the spanning ratio of Gabriel graphs and beta-skeletons. *SIAM J. Discrete Math.*, 20:412–427, 2006.

[BDH+12]   P. Bose, V. Dujmović, F. Hurtado, J. Iacono, S. Langerman, H. Meijer, V. Sacristán, M. Saumell, and D.R. Wood. Proximity graphs: E, δ, Δ, χ and ω. *Internat. J. Comput. Geom. Appl.*, 22:439–470, 2012.

[BE97]   M. Bern and D. Eppstein. Approximation algorithms for geometric problems. In D. S. Hochbaum, editor, *Approximation Algorithms for NP-hard problems*, pages 296–345, PWS Publishing Company, 1997.

[Ben83]   M. Ben-Or. Lower bounds for algebraic computation trees. In *Proc. 15th ACM Sympos. Theory Comput.*, pages 80–86, 1983.

[BGHI10]   N. Bonichon, C. Gavoille, N. Hanusse, and D. Ilcinkas. Connections between Theta-graphs, Delaunay triangulations, and orthogonal surfaces. In *Proc. 36th Workshop Graph Theoretic Concepts in Comp. Sci.*, vol. 6410 of *LNCS*, pages 266–278, Springer, Berlin, 2010.

[BGHP10]   N. Bonichon, C. Gavoille, N. Hanusse, and L. Perkovic. Plane spanners of maximum degree six. In *Proc. 37th Internat. Colloq. Automata Lang. Program.*, vol. 6198 of *LNCS*, pages 19–30, Springer, Berlin, 2010.

[BGHP15]   N. Bonichon, C. Gavoille, N. Hanusse, and L. Perkovic. Tight stretch factors for L_1- and L_∞-Delaunay triangulations. *Comput. Geom.*, 48:237–250, 2015.

[BHS16]   P. Bose, D. Hill, and M. Smid. Improved spanning ratio for low degree plane spanners. In *Proc. 12th Latin American Symp. Theoret. Informatics*, vol. 9644 of *LNCS*, pages 249–262, Springer, Berlin, 2016.

[BLL96]   P. Bose, W. Lenhart, and G. Liotta. Characterizing proximity trees. *Algorithmica*, 16:83–110, 1996.

[BLMM11]   K. Buchin, M. Löffler, P. Morin, and W. Mulzer. Preprocessing imprecise points for Delaunay triangulation: Simplified and extended. *Algorithmica*, 61:674–693, 2011.

[BM11]   K. Buchin and W. Mulzer. Delaunay triangulations in $O(\text{sort}(n))$ time and more. *J. ACM*, 58:6, 2011.

[BMRV15]   P. Bose, P. Morin, A. van Renssen, and S. Verdonschot. The θ_5-graph is a spanner. *Comput. Geom.*, 48:108–119, 2015.

[BS13]   P. Bose and M. Smid. On plane geometric spanners: A survey and open problems. *Comput. Geom.*, 46:818–830, 2013.

[CCL09]   J. Cardinal, S. Collette, and S. Langerman. Empty region graphs. *Comput. Geom.*, 42:183–195, 2009.

[CDKK02]   L.P. Chew, H. David, M.J. Katz, and K. Kedem. Walking around fat obstacles. *Inform. Process. Lett.*, 83:135–140, 2002.

[CDNS95]   B. Chandra, G. Das, G. Narasimhan, and J. Soares. New sparseness results on graph spanners. *Internat. J. Comput. Geom. Appl.*, 5:125–144, 1995.

[CDS01]   D.Z. Chen, G. Das, and M. Smid. Lower bounds for computing geometric spanners and approximate shortest paths. *Discrete Appl. Math.*, 110:151–167, 2001.

[CEG+94]   B. Chazelle, H. Edelsbrunner, L.J. Guibas, J. Hershberger, R. Seidel, and M. Sharir. Selecting heavily covered points. *SIAM J. Comput.*, 23:1138–1151, 1994.

[CGMZ16]   T.-H.H. Chan, A. Gupta, B. M. Maggs, and S. Zhou. On hierarchical routing in doubling metrics. *ACM Trans Algorithms*, 12:55:1–55:22, 2016.

[Cha08]   T.M. Chan. Well-separated pair decomposition in linear time? *Inform. Process. Lett.*, 107:138–141, 2008.

[Cha10]   T.M. Chan. A dynamic data structure for 3-D convex hulls and 2-D nearest neighbor queries. *J. ACM*, 57:16, 2010.

[Che86]   L.P. Chew. There is a planar graph almost as good as the complete graph. In *Proc. 2nd Sympos. Comput. Geom.*, pages 169–177, ACM Press, 1986.

[Che89]   L.P. Chew. There are planar graphs almost as good as the complete graph. *J. Comput. System Sci.*, 39:205–219, 1989.

[CK95]   P.B. Callahan and S.R. Kosaraju. A decomposition of multidimensional point sets with applications to k-nearest-neighbors and n-body potential fields. *J. ACM*, 42:67–90, 1995.

[Cla87]   K.L. Clarkson. Approximation algorithms for shortest path motion planning (extended abstract). In *Proc. 19th ACM Sympos. Theory Comput.*, pages 56–65, 1987.

[CLNS15]   T.H. Chan, M. Li, L. Ning, and S. Solomon. New doubling spanners: Better and simpler. *SIAM J. Comput.*, 44:37–53, 2015.

[CT76]   D. Cheriton and R.E. Tarjan. Finding minimum spanning trees. *SIAM J. Comput.*, 5:724–742, 1976.

[CT16]   T.M. Chan and K. Tsakalidis. Optimal deterministic algorithms for 2-d and 3-d shallow cuttings. *Discrete Comput. Geom.*, 56:866–881, 2016.

[CTL92]   M.-S. Chang, C.Y. Tang, and R.C.T. Lee. Solving the Euclidean bottleneck matching problem by k-relative neighborhood graphs. *Algorithmica*, 8:177–194, 1992.

[CW98]   F.Y.L. Chin and C.A. Wang. Finding the constrained Delaunay triangulation and constrained Voronoi diagram of a simple polygon in linear time. *SIAM J. Comput.*, 28:471–486, 1998.

[CX01]   S.-W Cheng and Y.-F. Xu. On β-skeleton as a subgraph of the minimum weight triangulation. *Theoret. Comput. Sci.*, 262:459–471, 2001.

[CZ04]   A. Czumaj and H. Zhao. Fault-tolerant geometric spanners. *Discrete Comput. Geom.*, 32:207–230, 2004.

[DBLL94]   G. Di Battista, W. Lenhart, and G. Liotta. Proximity drawability: a survey. In *Proc. DIMACS Int. Workshop on Graph Drawing*, vol. 894 of *LNCS*, pages 328–339, Springer, Berlin, 1994.

[DEG08]   O. Devillers, J. Erickson, and X. Goaoc. Empty-ellipse graphs. In *Proc. 19th ACM-SIAM Sympos. Discrete Algorithms*, pages 1249–1257, 2008.

[DFS90]   D.P. Dobkin, S.J. Friedman, and K.J. Supowit. Delaunay graphs are almost as good as complete graphs. *Discrete Comput. Geom.*, 5:399–407, 1990.

[DG16]   A. Dumitrescu and A. Ghosh. Lower bounds on the dilation of plane spanners. *Internat. J. Comput. Geom. Appl.*, 26:89–110, 2016.

[DH96]   G. Das and P.J. Heffernan. Constructing degree-3 spanners with other sparseness properties. *Int. J. Found. Comput. Sci.*, 7:121–136, 1996.

[DJ89]   G. Das and D. Joseph. Which triangulations approximate the complete graph? In *Proc. Int. Sympos. Optimal Algorithms*, pages 168–192, 1989.

[DN97]   G. Das and G. Narasimhan. A fast algorithm for constructing sparse Euclidean spanners. *Internat. J. Comput. Geom. Appl.*, 7:297–315, 1997.

[Dwy95]   R.A. Dwyer. The expected size of the sphere-of-influence graph. *Comput. Geom.*, 5:155–164, 1995.

[EM09]   N.M. El Molla. Yao spanners for wireless ad hoc networks. Master's thesis, Villanova University, 2009.

[Epp95]   D. Eppstein. Dynamic Euclidean minimum spanning trees and extrema of binary functions. *Discrete Comput. Geom.*, 13:111–122, 1995.

[Epp00]   D. Eppstein. Spanning trees and spanners. In J.-R. Sack and J. Urrutia, editors, *Handbook of Computational Geometry*, pages 425–461, Elsevier, Amsterdam, 2000.

[FK12]   M. Fürer and S.P. Kasiviswanathan. Spanners for geometric intersection graphs with applications. *J. Comput. Geom.*, 3:31–64, 2012.

[FS16]   A. Filtser and S. Solomon. The greedy spanner is existentially optimal. In *Proc. ACM Sympos. Principles Dist. Comput.)*, pages 9–17, 2016.

[GGN06]   J. Gao, L.J. Guibas, and A.T. Nguyen. Deformable spanners and applications. *Comput. Geom.*, 35:2–19, 2006.

[GK07]   J. Gudmundsson and C. Knauer. Dilation and detours in geometric networks. In T. Gonzalez, editor, *Handbook of Approximation Algorithms and Metaheuristics*, chap. 52, CRC Press, Boca Raton, 2007.

[GKL03]   A. Gupta, R. Krauthgamer, and J.R. Lee. Bounded geometries, fractals, and low-distortion embeddings. In *Proc. 44th IEEE Sympos. Found. Comput. Sci.*, pages 534–543, 2003.

[GLN02]   J. Gudmundsson, C. Levcopoulos, and G. Narasimhan. Fast greedy algorithms for constructing sparse geometric spanners. *SIAM J. Comput.*, 31:1479–1500, 2002.

[GLNS08]   J. Gudmundsson, C. Levcopoulos, G. Narasimhan, and M. Smid. Approximate distance oracles for geometric graphs. *ACM Trans. Alg.*, 4:10, 2008.

[Got15]   L.-A. Gottlieb. A light metric spanner. In *Proc. 56th IEEE Sympos. Found. Comput. Sci.*, pages 759–772, 2015.

[GPS94]   L.J. Guibas, J. Pach, and M. Sharir. Sphere-of-influence graphs in higher dimensions. In K. Böröcky and G. Fejes Tóth, editors, *Intuitive Geometry*, vol. 63 of *Coll. Math. Soc. J. Bolyai*, pages 131–137, Budapest, 1994.

[GR08a]   L.-A. Gottlieb and L. Roditty. Improved algorithms for fully dynamic geometric spanners and geometric routing. In *Proc. 19th ACM-SIAM Sympos. Discrete Algorithms*, pages 591–600, 2008.

[GR08b]   L.-A. Gottlieb and L. Roditty. An optimal dynamic spanner for doubling metric spaces. In *Proc. 16th Eur. Sympos. Algorithms*, vol. 5193 of *LNCS*, pages 478–489, Springer, Berlin, 2008.

[GS69]   K.R. Gabriel and R.R. Sokal. A new statistical approach to geographic variation analysis. *Systematic Biology*, 18:259–278, 1969.

[GZ05]   J. Gao and L. Zhang. Well-separated pair decomposition for the unit-disk graph metric and its applications. *SIAM J. Comput.*, 35:151–169, 2005.

[HdLT01]   J. Holm, K. de Lichtenberg, and M. Thorup. Poly-logarithmic deterministic fully-dynamic algorithms for connectivity, minimum spanning tree, 2-edge, and biconnectivity. *J. ACM*, 48:723–760, 2001.

[HLM03]   F. Hurtado, G. Liotta, and H. Meijer. Optimal and suboptimal robust algorithms for proximity graphs. *Comput. Geom.*, 25:35–49, 2003.

[HP11]   S. Har-Peled. *Geometric Approximation Algorithms*, vol. 173 of *Math. Surveys Monogr.*, AMS, Providence, 2011.

[HPM06]   S. Har-Peled and M. Mendel. Fast construction of nets in low-dimensional metrics and their applications. *SIAM J. Comput.*, 35:1148–1184, 2006.

[JK87]   J.W. Jaromczyk and M. Kowaluk. A note on relative neighborhood graphs. In *Proc. 3rd Sympos. Comput. Geom.*, pages 233–241, ACM Press, 1987.

[JT92]   J.W. Jaromczyk and G.T. Toussaint. Relative neighborhood graphs and their relatives. *Proc of IEEE*, 80:1502–1517, 1992.

[Kat88]   J. Katajainen. The region approach for computing relative neighbourhood graphs in the L_p metric. *Computing*, 40:147–161, 1988.

[Kei92]   J.M. Keil. Approximating the complete Euclidean graph. *Discrete Comput. Geom.*, 7:13–28, 1992.

[Kei94]   J.M. Keil. Computing a subgraph of the minimum weight triangulation. *Comput. Geom.*, 4:18–26, 1994.

[KG92]   J.M. Keil and C.A. Gutwin. Classes of graphs which approximate the complete Euclidean graph. *Discrete Comput. Geom.*, 7:13–28, 1992.

[KG01]   M.I. Karavelas and L.J. Guibas. Static and kinetic geometric spanners with applications. In *Proc. 12th ACM-SIAM Sympos. Discrete Algorithms*, pages 168–176, 2001.

[KL96]   R. Klein and A. Lingas. A linear-time randomized algorithm for the bounded Voronoi diagram of a simple polygon. *Internat. J. Comput. Geom. Appl.*, 6:263–278, 1996.

[KL98]   D. Krznaric and C. Levcopoulos. Computing a threaded quadtree from the Delaunay triangulation in linear time. *Nord. J. Comput.*, 5:1–18, 1998.

[KMR+17]   H. Kaplan, W. Mulzer, L. Roditty, P. Seiferth, and M. Sharir. Dynamic planar Voronoi diagrams for general distance functions and their algorithmic applications. In *Proc. 28th ACM-SIAM Sympos. Discrete Algorithms*, pages 2495–2504, 2017.

[KMRS15]   H. Kaplan, W. Mulzer, L. Roditty, and P. Seiferth. Spanners and reachability oracles for directed transmission graphs. In *Proc. 31st Sympos. Comput. Geom.*, vol. 34 of *LIPIcs*, pages 156–170, Schloss Dagstuhl, 2015.

[KN87]   J. Katajainen and O. Nevalainen. An almost naive algorithm for finding relative neighbourhood graphs in L_p metrics. *ITA*, 21:199–215, 1987.

[KP08]   I.A. Kanj and L. Perkovic. On geometric spanners of Euclidean and unit disk graphs. In *Proc. 25th Sympos. Theoret. Aspects Comput. Sci.*, vol. 1 of *LIPIcs* pages 409–420, Schloss Dagstuhl, 2008.

[KPT16]   I.A. Kanj, L. Perkovic, and D. Türkoglu. Degree four plane spanners: Simpler and better. In *Proc. 32nd Int. Sympos. Comput. Geom.*, vol. 51 of *LIPIcs*, article 45, Schloss Dagstuhl, 2016.

[KPX10]   I.A. Kanj, L. Perkovic, and G. Xia. On spanners and lightweight spanners of geometric graphs. *SIAM J. Comput.*, 39:2132–2161, 2010.

[KR85]   D.G. Kirkpatrick and J.D. Radke. A framework for computational morphology. In G.T. Toussaint, editor, *Computational Geometry*, vol. 2 of *Machine Intelligence and Pattern Recognition*, pages 217–248. North-Holland, Amsterdam, 1985.

[KSVC15]   T. Kaiser, M. Saumell, and N. van Cleemput. 10-Gabriel graphs are Hamiltonian. *Inform. Process. Lett.*, 115:877–881, 2015.

[Lee85]   D.T. Lee. Relative neighborhood graphs in the L_1-metric. *Pattern Recogn.*, 18:327–332, 1985.

[Lin94]   A. Lingas. A linear-time construction of the relative neighborhood graph from the Delaunay triangulation. *Comput. Geom.*, 4:199–208, 1994.

[Lio13]   G. Liotta. Proximity drawings. In R. Tamassia, editor, *Handbook on Graph Drawing and Visualization*, pages 115–154, CRC Press, Boca Raton, 2013.

[LL92]   C. Levcopoulos and A. Lingas. There are planar graphs almost as good as the complete graphs and almost as cheap as minimum spanning trees. *Algorithmica*, 8:251–256, 1992.

[LM12]   M. Löffler and W. Mulzer. Triangulating the square and squaring the triangle: Quadtrees and Delaunay triangulations are equivalent. *SIAM J. Comput.*, 41:941–974, 2012.

[LNS02]   C. Levcopoulos, G. Narasimhan, and M. Smid. Improved algorithms for constructing fault-tolerant spanners. *Algorithmica*, 32:144–156, 2002.

[Luk99]   T. Lukovszki. New results of fault tolerant geometric spanners. In *Proc. 6th Int. Workshop Algorithms Data Structures*, vol. 1663 of *LNCS*, pages 193–204, Springer, Berlin, 1999.

[MR08]   W. Mulzer and G. Rote. Minimum-weight triangulation is NP-hard. *J. ACM*, 55:11, 2008.

[MS80]   D.W. Matula and R.R. Sokal. Properties of Gabriel graphs relevant to geographic variation research and the clustering of points in the plane. *Geogr. Anal.*, 12:205–222, 1980.

[Mul04]   W. Mulzer. Minimum dilation triangulations for the regular n-gon. Master's thesis, Freie Universität Berlin, 2004.

[NS02]   G. Narasimhan and M. Smid. Approximation algorithms for the bottleneck stretch factor problem. *Nord. J. Comput.*, 9:13–31, 2002.

[NS07]   G. Narasimhan and M. Smid. *Geometric Spanner Networks*. Cambridge University Press, 2007.

[O'R82]   J. O'Rourke. Computing the relative neighborhood graph in the L_1 and L_∞ metrics. *Pattern Recogn.*, 15:189–192, 1982.

[PS85]   F.P. Preparata and M.I. Shamos. *Computational Geometry. An Introduction.* Springer-Verlag, New York, 1985.

[Rod12]   L. Roditty. Fully dynamic geometric spanners. *Algorithmica*, 62:1073–1087, 2012.

[RS91]   J. Ruppert and R. Seidel. Approximating the d-dimensional Euclidean graph. In *Proc. 3rd Canad. Conf. Comput. Geom.*, pages 207–210, 1991.

[SA95]   M. Sharir and P.K. Agarwal. *Davenport-Schinzel Sequences and Their Geometric Applications*. Cambridge University Press, 1995.

[Sam90]   H. Samet. *The Design and Analysis of Spatial Data Structures*. Addison-Wesley, Reading, 1990.

[SC90]   T.-H. Su and R.-C. Chang. The k-Gabriel graphs and their applications. In *SIGAL Int. Sympos. Algorithms*, pages 66–75. Springer-Verlag, Berlin, 1990.

[Sch79]   A. Schönhage. On the power of random access machines. In *Proc. 6th Int. Colloq. Automata Lang. Program.*, vol. 71 of *LNCS*, pages 520–529, Springer, Berlin, 1979.

[Sei16]   P. Seiferth. *Disk Intersection Graphs: Models, Data Structures, and Algorithms.* PhD thesis, Freie Universität Berlin, 2016.

[Smi89]   W.D. Smith. *Studies in Computational Geometry Motivated by Mesh Generation.* PhD thesis, Princeton University, 1989.

[Smi07]   M. Smid. The well-separated pair decomposition and its applications. In T. Gonzalez, editor, *Handbook of Approximation Algorithms and Metaheuristics*, chap. 53, CRC Press, Boca Raton, 2007.

[Smi09]   M. Smid. The weak gap property in metric spaces of bounded doubling dimension. In *Efficient Algorithms, Essays Dedicated to Kurt Mehlhorn on the Occasion of His 60th Birthday*, pages 275–289, Springer, Berlin, 2009.

[Sos99]   M.A. Soss. On the size of the Euclidean sphere of influence graph. In *Proc. 11th Canad. Conf. Comput. Geom.*, pages 43–46, 1999.

[Sup83]   K.J. Supowit. The relative neighborhood graph, with an application to minimum spanning trees. *J. ACM*, 30:428–448, 1983.

[Tal04]   K. Talwar. Bypassing the embedding: algorithms for low dimensional metrics. In *Proc. 36th ACM Sympos. Theory Comput*, pages 281–290, 2004.

[Tou80]   G.T. Toussaint. The relative neighbourhood graph of a finite planar set. *Pattern Recogn.*, 12:261–268, 1980.

[Tou88]   G.T. Toussaint. A graph-theoretical primal sketch. In G.T. Toussaint, editor, *Computational Morphology*, pages 229–260, North-Holland, Amsterdam, 1988.

[Tou14]   G.T. Toussaint. The sphere of influence graph: Theory and applications. *Information Technology & Computer Science*, 14:2091–1610, 2014.

[Urq83]   R.B. Urquhart. Some properties of the planar Euclidean relative neighbourhood graph. *Pattern Recogn. Lett.*, 1:317–322, 1983.

[Var98]   K.R. Varadarajan. A divide-and-conquer algorithm for min-cost perfect matching in the plane. In *Proc. 39th IEEE Sympos. Found. Comput. Sci.*, pages 320–331, 1998.

[Vel91]   R.C. Veltkamp. The γ-neighborhood graph. *Comput. Geom.*, 1:227–246, 1991.

[Wil16]   M. Willert. Routing schemes for disk graphs and polygons. Master's thesis, Freie Universität Berlin, 2016.

[WY01]   C.A. Wang and B. Yang. A lower bound for β-skeleton belonging to minimum weight triangulations. *Comput. Geom.*, 19:35–46, 2001.

[Xia13]   G. Xia. The stretch factor of the Delaunay triangulation is less than 1.998. *SIAM J. Comput.*, 42:1620–1659, 2013.

[XZ11]   G. Xia and L. Zhang. Toward the tight bound of the stretch factor of Delaunay triangulations. In *Proc. 23rd Canad. Conf. Comput. Geom.*, 2011.

[Yao82]   A.C.-C. Yao. On constructing minimum spanning trees in k-dimensional spaces and related problems. *SIAM J. Comput.*, 11:721–736, 1982.

33 VISIBILITY

Joseph O'Rourke

INTRODUCTION

In a geometric context, two objects are "visible" to each other if there is a line segment connecting them that does not cross any obstacles. Over 500 papers have been published on aspects of visibility in computational geometry in the last 40 years. The research can be broadly classified as primarily focused on combinatorial issues, or primarily focused on algorithms. We partition the combinatorial work into "art gallery theorems" (Section 33.1) and illumination of convex sets (33.2), and research on visibility graphs (33.3) and the algorithmic work into that concerned with polygons (33.4), more general planar environments (33.5) paths (33.6), and mirror reflections (33.7). All of this work concerns visibility in two dimensions. Investigations in three dimensions, both combinatorial and algorithmic, are discussed in Section 33.8, and the final section (33.9) touches on visibility in $\mathbb{R}^d$.

33.1 ART GALLERY THEOREMS

A typical "art gallery theorem" provides combinatorial bounds on the number of guards needed to visually cover a polygonal region P (the art gallery) defined by n vertices. Equivalently, one can imagine light bulbs instead of guards and require full direct-light illumination.

GLOSSARY

Guard: A point, a line segment, or a line—a source of visibility or illumination.

Vertex guard: A guard at a polygon vertex.

Point guard: A guard at an arbitrary point.

Interior visibility: A guard $x \in P$ can see a point $y \in P$ if the segment xy is nowhere exterior to P: $xy \subseteq P$.

Exterior visibility: A guard x can see a point y outside of P if the segment xy is nowhere interior to P; xy may intersect ∂P, the boundary of P.

Star polygon: A polygon visible from a single interior point.

Diagonal: A segment inside a polygon whose endpoints are vertices, and which otherwise does not touch ∂P.

Floodlight: A light that illuminates from the apex of a cone with aperture α.

Vertex floodlight: One whose apex is at a vertex (at most one per vertex).

MAIN RESULTS

The most general combinatorial results obtained to date are summarized in Table 33.1.1. Tight bounds, and ranges between lower and upper bounds are listed for the minimum number of guards sufficient for all polygons with n vertices (and possibly h holes).

TABLE 33.1.1   Number of guards needed.

PROBLEM NAME	POLYGONS	INT/EXT	GUARD	NUMBER
Art gallery theorem	simple	interior	vertex	$\lfloor n/3 \rfloor$
Fortress problem	simple	exterior	point	$\lceil n/3 \rceil$
Prison yard problem	simple	int & ext	vertex	$\lceil n/2 \rceil$
Prison yard problem	orthogonal	int & ext	vertex	$[\lceil 5n/16 \rceil, \lfloor 5n/12 \rfloor + 1]$
Orthogonal polygons	simple orthogonal	interior	vertex	$\lfloor n/4 \rfloor$
Orthogonal with holes	orthogonal with h holes	interior	vertex	$[\lfloor 2n/7 \rfloor, \lfloor (17n-8)/52 \rfloor]$
Orthogonal with holes	orthogonal with h holes	interior	vertex	$[\lfloor (n+h)/4 \rfloor, \lfloor (n+2h)/4 \rfloor]$
Polygons with holes	polygons with h holes	interior	point	$\lfloor (n+h)/3 \rfloor$

Of special note is the difficult **orthogonal prison yard problem:** How many vertex guards are needed to cover both the interior and the exterior of an orthogonal polygon? See Figure 33.1.1. An upper bound of $\lfloor 5n/12 \rfloor + 2$ was obtained by [HK96] via the following graph-coloring theorem: Every plane, bipartite, 2-connected graph has an **even triangulation** (all nodes have even degree) and therefore the resulting graph is 3-colorable. This bound was subsequently improved to $\lfloor 5n/12 \rfloor + 1$ in [MP12].

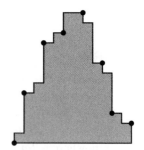

FIGURE 33.1.1
A pyramid polygon with $n = 24$ vertices whose interior and exterior are covered by 8 guards. Repeating the pattern establishes a lower bound of $5n/16 + c$ on the orthogonal prison yard problem [HK93].

COVERS AND PARTITIONS

Each art gallery theorem above implies a cover result, a cover by star polygons. Many of the theorem proofs rely on particular partitions. For example, the orthogonal polygon result depends on the theorem that every orthogonal polygon may be partitioned via diagonals into convex quadrilaterals.

Most cover problems are NP-hard, and finding a minimum guard set for a simple polygon is NP-complete. Approximation algorithms have only achieved $O(\log n)$ times the fewest guards [Gho10]. See Section 30.2 for more on covers and partitions.

EDGE GUARDS

A variation permits guards (***mobile guards***) to patrol segments, diagonals, or edges; equivalent is illumination by line segment/diagonal/edge light sources (fluorescent light bulbs). Here there are fewer results; see Table 33.1.2. Toussaint conjectures that the last line of this table should be $\lfloor n/4 \rfloor$ for sufficiently large n.

TABLE 33.1.2   Edge guards.

POLYGONS	GUARD	BOUNDS	SOURCE
Polygon	diagonal	$\lfloor n/4 \rfloor$	[O'R83]
Orthogonal polygons	segment	$\lfloor (3n+4)/16 \rfloor$	[Agg84, O'R87]
Orthogonal polygons with h holes	segment	$\lfloor (3n+4h+4)/16 \rfloor$	[GHKS96]
Polygon ($n > 11$)	closed edge	$[\lfloor n/4 \rfloor, \lfloor 3n/10 \rfloor + 1]$	[She94]
Polygon	open edge	$[\lfloor n/3 \rfloor, \lfloor n/2 \rfloor]$	[TTW12]

33.1.1 1.5D TERRAIN GUARDING

A 1.5D terrain is an x-monotone chain of edges, guarded by points on the chain. Guarding such a terrain has application to placing communication devices to cover the terrain. Although known to be NP-hard [KK11], it required further work to find a polynomial discretization, thereby establishing its NP-completeness [FHKS16].

TABLE 33.1.3   Floodlights.

APEX	ALPHA	BOUNDS	SOURCE
Any point	$[180°, 360°]$	$\lfloor n/3 \rfloor$	[Tót00]
Any point	$[90°, 180°]$	$2\lfloor n/3 \rfloor$	[Tót00]
Any point	$[45°, 60°)$	$[n-2, n-1]$	[Tót03d]
Vertex	$< 180°$	not always possible	[ECOUX95]
Vertex	$180°$	$[9n/14 - c, \lfloor 2n/3 \rfloor - 1]$	[ST05]

33.1.2 FLOODLIGHT ILLUMINATION

Urrutia introduced a class of questions involving guards with restricted vision, or, equivalently, illumination by floodlights: How many floodlights, each with aperture α, and with their apexes at distinct nonexterior points, are sufficient to cover any polygon of n vertices? One surprise is that $\lfloor n/3 \rfloor$ half-guards/π-floodlights suffice, although not when restricted to vertices. A second surprise is that, for any $\alpha < \pi$, there is a polygon that cannot be illuminated by an α floodlight at every vertex. See Table 33.1.3. A third surprise is that the best result on vertex π-floodlights employs pointed pseudotriangulations (cf. Chapter 5) in an essential way.

33.2 ILLUMINATION OF PLANAR CONVEX SETS

A natural extension of exterior visibility is illumination of the plane in the presence of obstacles. Here it is natural to use "illumination" in the same sense as "visibility." Under this model, results depend on whether light sources are permitted to lie on obstacle boundaries: $\lfloor 2n/3 \rfloor$ lights are necessary and sufficient (for $n > 5$) if they may [O'R87], and $\lfloor 2(n+1)/3 \rfloor$ if they may not [Tót02]. More work has been done on illuminating the boundary of the obstacles, under a stronger notion of illumination, corresponding to "clear visibility."

GLOSSARY

Illuminate: x illuminates y if xy does not include a point strictly interior to an obstacle, and does not cross a segment obstacle.

Cross: xy crosses segment s if they have exactly one point p in common, and p is in the relative interior of both xy and s.

Clearly illuminate: x clearly illuminates y if the open segment (x, y) does not include any point of an obstacle.

Compact: Closed and bounded in $\mathbb{R}^d$.

Homothetic: Obtained by dilation and translation.

Isothetic: Sides parallel to orthogonal coordinate axes.

MAIN RESULTS

A third, even stronger notion of illumination is considered in Section 33.9 below. The main question that has been investigated is: How many point lights strictly exterior to a collection of n pairwise disjoint compact, convex objects in the plane are needed to clearly illuminate every object boundary point? Answers for a variety of restricted sets are shown in Table 33.2.1.

TABLE 33.2.1   Illuminating convex sets in plane.

FAMILY	BOUNDS	SOURCE
Convex sets	$4n - 7$	[Fej77]
Circular disks	$2n - 2$	[Fej77]
Isothetic rectangles	$[n - 1, n + 1]$	[Urr00]
Homothetic triangles	$[n, n + 1]$	[CRCU93]
Triangles	$[n, \lfloor (5n + 1)/4 \rfloor]$	[Tót03b]
Segments (one side)	$[4n/9 - 2, \lfloor (n + 1)/2 \rfloor]$	[CRC$^+$95, Tót03c]
Segments (both sides)	$\lfloor 4(n + 1)/5 \rfloor$	[Tót01]

The most interesting open problem here is to close the gap for triangles. Urrutia conjectures [Urr00] that $n + c$ lights suffice for some constant c.

33.3 VISIBILITY GRAPHS

Whereas art gallery theorems seek to encapsulate an environment's visibility into one function of n, the study of visibility graphs endeavors to uncover the more fine-grained structure of visibility. The original impetus for their investigation came from pattern recognition, and its connection to shape continues to be one of its primary sources of motivation; see Chapter 54. Another application is graphics (Chapter 52): illumination and radiosity depend on 3D visibility relations (Section 33.8.)

GLOSSARY

Visibility graph: A graph with a node for each object, and arcs between objects that can see one another.

Vertex visibility graph: The objects are the vertices of a simple polygon.

Endpoint visibility graph: The objects are the endpoints of line segments in the plane. See Figure 33.3.1b.

Segment visibility graph: The objects are whole line segments in the plane, either open or closed.

Object visibility: Two objects A and B are visible to one another if there are points $x \in A$ and $y \in B$ such that x sees y.

Point visibility: Two points x and y can see one another if the segment xy is not "obstructed," where the meaning of "obstruction" depends on the problem.

ϵ-visibility: Lines of sight are finite-width beams of visibility.

Hamiltonian: A graph is Hamiltonian if there is a simple cycle that includes every node.

OBSTRUCTIONS TO VISIBILITY

For polygon vertices, x sees y if xy is nowhere exterior to the polygon, just as in art gallery visibility; this implies that polygon edges are part of the visibility graph. For segment endpoints, x sees y if the closed segment xy intersects the union of all the segments either in just the two endpoints, or in the entire closed segment. This disallows grazing contact with a segment, but includes the segments themselves in the graph.

GOALS

Four goals can be discerned in research on visibility graphs:

1. Characterization: asks for a precise delimiting of the class of graphs realizable by a certain class of geometric objects.

2. Recognition: asks for an algorithm to recognize when a graph is a visibility graph.

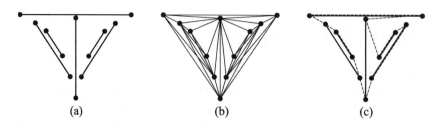

FIGURE 33.3.1
(a) *A set of 6 pairwise disjoint line segments.* (b) *Their endpoint visibility graph G.* (c) *A Hamiltonian cycle in G.*

3. Reconstruction: asks for an algorithm that will take a visibility graph as input, and output a geometric realization.

4. Counting: concerned with the number of visibility graphs under various restrictions [HN01].

POINT VISIBILITY GRAPHS

Given a set P of n points in the plane, visibility between $x, y \in P$ may be blocked by a third point in P. The recognition of point visibility graph is NP-hard [Roy16], in fact it is complete for the existential theory of the reals [CH17]. However, for planar graphs, there is complete characterization, and an $O(n)$-time recognition algorithm [GR15]. Pfender [Pfe08] constructed point visibility graphs of clique number 6 and arbitrary high chromatic number.

For example, it was established in [PPVW12] that every visibility graph with minimum degree δ has vertex connectivity of at least $\delta/2 + 1$, and if the number of collinear points is no more than 4, then G has connectivity of at least $2\delta/3 + 1$. This later quantity is conjectured to hold without the collinearity restriction. Related Ramsey-type problems and results are surveyed in [PW10].

VERTEX VISIBILITY GRAPHS

A complete characterization of vertex visibility graphs of polygons has remained elusive, but progress has been made by:

1. Restricting the class of polygons: polynomial-time recognition and reconstruction algorithms for orthogonal staircase polygons have been obtained. See Figure 33.3.2.

2. Restricting the class of graphs: every 3-connected vertex visibility graph has a 3-clique ordering, i.e., an ordering of the vertices so that each vertex is part of a triangle composed of preceding vertices.

3. Adding information: assuming knowledge of the boundary Hamiltonian circuit, four necessary conditions have been established by Ghosh and others [Gho97], and conjectured to be sufficient.

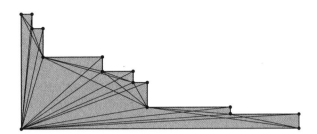

FIGURE 33.3.2
A staircase polygon and its vertex visibility graph.

ENDPOINT VISIBILITY GRAPHS

A set of n pairwise disjoint line segments forms a noncrossing perfect matching on the $2n$ endpoints in the plane. For segment endpoint visibility graphs, there have been three foci:

1. Are the graphs Hamiltonian? See Figure 33.3.1c. Posed by Mirzaian, this was settled in the affirmative [HT03]: YES, there is always a Hamiltonian polygon (i.e., a noncrossing circuit) for pairwise disjoint line segments, not all lying on a line.

2. In the quest for generating a *random* noncrossing perfect matching, Aichholzer et al. [ABD+09] conjecture that any two such matchings are connected by sequence of noncrossing perfect matchings in which consecutive matching are *compatible* (the union of the two matchings is also noncrossing). Every matching on $4n$ vertices is known to have a compatible matching [IST13].

3. Size questions: there must be at least $5n - 4$ edges [SE87], and at least $6n - 6$ when no segment is a "chord" splitting the convex hull [GOH+02]; the smallest clique cover has size $\Omega(n^2/\log^2 n)$ [AAAS94].

SEGMENT VISIBILITY GRAPHS

Whole segment visibility graphs have been investigated most thoroughly under the restriction that the segments are all (say) vertical and visibility is horizontal. Such segments are often called **bars**. The visibility is usually required to be ϵ-visibility. Endpoints on the same horizontal line often play an important role here, as does the distinction between closed segments and intervals (which may or may not include their endpoints). There are several characterizations:

1. G is representable by segments, with no two endpoints on the same horizontal line, iff there is a planar embedding of G such that, for every interior k-face F, the induced subgraph of F has exactly $2k - 3$ edges.

2. G is representable by segments, with endpoints on the same horizontal permitted, iff there is a planar embedding of G with all cutpoints on the exterior face.

3. Every 3-connected planar graph is representable by intervals.

OBSTACLE NUMBER

An interesting variant of visibility graphs has drawn considerable attention. Given a graph G, an *obstacle representation* of G is a mapping of its nodes to the plane such that edge (x, y) is in G if and only if the segment xy does not intersect any "obstacle." An obstacle is any connected subset of $\mathbb{R}^2$. The obstacle number of G is the minimum number of obstacles in an obstacle representation of G. At least one obstacle is needed to represent any graph other than the complete graph. There are graphs with obstacle number $\Omega(n/(\log \log n)^2)$ [DM13]. No upper bound better than $O(n^2)$ is known.

When the obstacles are points and G is the empty graph on n vertices, this quantity is known as the *blocking number* $b(n)$; see [Mat09, PW10]. It is conjectured that $\lim_{n \to \infty} b(n)/n = \infty$, but the best bound is $b(n) \geq (25/8 - o(1))n$ [DPT09].

INVISIBILITY GRAPHS

For a set $X \subseteq \mathbb{R}^d$, its *invisibility graph* $\mathcal{I}(\mathcal{X})$ has a vertex for each point in X, and an edge between two vertices u and v if the segment uv is not completely contained in X. The chromatic number $\chi(X)$ and clique number $\omega(X)$ of $\mathcal{I}(\mathcal{X})$ have been studied, primarily in the context of the covering number, the fewest convex sets whose union is X. It is clear that $\omega(X) \leq \chi(X)$, and it was conjectured in [MV99] that for planar sets X, there is no upper bound on χ as a function of ω. This conjecture was settled positively in [CKM+10].

OTHER VISIBILITY GRAPHS

The notion of a visibility graph can be extended to objects such as disjoint disks: each disk is a node, with an arc if there is a segment connecting them that avoids touching any other disk. Rappaport proved that the visibility graph of disjoint congruent disks is Hamiltonian [Rap03]. *Rectangle visibility graphs*, which restrict visibility to vertical or horizontal lines of sight between disjoint rectangles, have been studied for their role in graph drawing (Chapter 55). A typical result is that any graph with a maximum vertex degree 4 can be realized as a rectangle visibility graph [BDHS97].

OPEN PROBLEMS

Ghosh and Goswami list 44 open problems on visibility graphs in their survey [GG13]. Below we list just three.

1. Given a visibility graph G and a Hamiltonian circuit C, construct in polynomial time a simple polygon such that its vertex visibility graph is G, with C corresponding to the polygon's boundary.

2. Given a visibility graph G of a simple polygon P, find the Hamiltonian cycle that corresponds to the boundary of P.

3. Develop an algorithm to recognize whether a polygon vertex visibility graph is planar. Necessary and sufficient conditions are known [LC94].

33.4 ALGORITHMS FOR VISIBILITY IN A POLYGON

Designing algorithms to compute aspects of visibility in a polygon P was a major focus of the computational geometry community in the 1980s. For most of the basic problems, optimal algorithms were found, several depending on Chazelle's linear-time triangulation algorithm [Cha91]. See [Gho07] for a book-length survey.

GLOSSARY

Throughout, P is a simple polygon.

Kernel: The set of points in P that can see all of P. See Figure 33.4.4.

Point visibility polygon: The region visible from a point in P.

Segment visibility polygon: The region visible from a segment in P.

MAIN RESULTS

The main algorithms are listed in Table 33.4.1. We discuss two of these algorithms below to illustrate their flavor.

TABLE 33.4.1   Polygon visibility algorithms.

ALGORITHM TO COMPUTE	TIME COMPLEXITY	SOURCE
Kernel	$O(n)$	[LP79]
Point visibility polygon	$O(n)$	[JS87]
Segment visibility polygon	$O(n)$	[GHL$^+$87]
Shortest illuminating segment	$O(n)$	[DN94]
Vertex visibility graph	$O(E)$	[Her89]

VISIBILITY POLYGON ALGORITHM

Let $x \in P$ be the visibility source. Lee's linear-time algorithm [JS87] processes the vertices of P in a single counterclockwise boundary traversal. At each step, a vertex is either pushed on or popped off a stack, or a *wait* event is processed. The latter occurs when the boundary at that point is invisible from x. At any stage, the stack represents the visible portion of the boundary processed so far.

Although this algorithm is elementary in its tools, it has proved delicate to implement correctly.

VISIBILITY GRAPH ALGORITHM

In contrast, Hershberger's vertex visibility algorithm [Her89] uses sophisticated tools to achieve output-size sensitive time complexity $O(E)$, where E is the number of edges of the graph. His algorithm exploits the intimate connection between shortest paths and visibility in polygons. It first computes the *shortest path map* (Chapter 31) in $O(n)$ time for a vertex, and then systematically transforms this into the map of an adjacent vertex in time proportional to the number of changes. Repeating this achieves $O(E)$ time overall.

Most of the above algorithms have been parallelized; see, for example, [GSG92].

33.5 ALGORITHMS FOR VISIBILITY AMONG OBSTACLES

The shortest path between two points in an environment of polygonal obstacles follows lines of sight between obstacle vertices. This has provided an impetus for developing efficient algorithms for constructing visibility regions and graphs in such settings. The obstacles most studied are noncrossing line segments, which can be joined end-to-end to form polygonal obstacles. Many of the questions mentioned in the previous section can be revisited for this environment.

The major results are shown in Table 33.5.1; the first three are described in [O'R87]; the fourth is discussed below.

TABLE 33.5.1   Algorithms for visibility among obstacles.

ALGORITHM TO COMPUTE	TIME COMPLEXITY
Point visibility region	$O(n \log n)$
Segment visibility region	$\Theta(n^4)$
Endpoint visibility graph	$O(n^2)$
Endpoint visibility graph	$O(n \log n + E)$

ENDPOINT VISIBILITY GRAPH

The largest effort has concentrated on constructing the endpoint visibility graph. Worst-case optimal algorithms were first discovered by constructing the line arrangement dual to the endpoints in $O(n^2)$ time. Since many visibility graphs have less than a quadratic number of edges, an output-size sensitive algorithm was a significant improvement: $O(n \log n + E)$ where E is the number of edges of the graph [GM91]. This was further improved to $O(n + h \log h + E)$ for h polygonal obstacles with a total of n vertices [DW15].

A polygon with n vertices and h holes can be preprocessed into a data structure with space complexity $O(\min(|E|, nh) + n)$ in time $O(E + n \log n)$, and can report the visibility polygon $V(q)$ of a query point q in time $O(|V(q)| \log n + h)$ [IK09].

33.6 VISIBILITY PATHS

A fruitful idea was introduced to visibility research in the mid-1980s: the notion of "link distance" between two points, which represents the smallest number of mutually visible relay stations needed to communicate from one point to another; see Section 31.3. A related notion called "watchman tours" was introduced a bit later, mixing shortest paths and visibility problems, and employing many of the concepts developed for link-path problems (Section 31.4).

33.7 MIRROR REFLECTIONS

GLOSSARY

Light ray reflection: A light ray reflects from an interior point of a mirror with reflected angle equal to incident angle; a ray that hits a mirror endpoint is absorbed.

Mirror polygon: A polygon all of whose edges are mirrors reflecting light rays.

Periodic light ray: A ray that reflects from a collection of mirrors and, after a finite number of reflections, rejoins its path (and thenceforth repeats that path).

Trapped light ray: One that reflects forever, and so never "reaches" infinity.

Klee asked whether every polygonal room whose walls are mirrors (a mirror polygon) is illuminable from every interior point [Kle69, KW91]. Tokarsky answered NO by constructing rooms that leave one point dark when the light source is located at a particular spot [Tok95]. Complementing Tokarsky's result, it is now known that if P is a *rational polygon* (angles rational multiples of π), for every x there are at most finitely many dark points y in P [LMW16]. However, a second question of Klee remains open: Is every mirror polygon illuminable from *some* interior point?

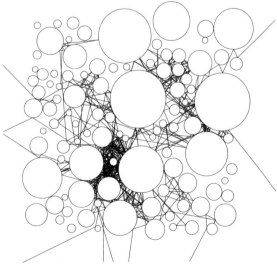

FIGURE 33.7.1
100 *mirror disks fail to trap* 10 *rays from a point source (near the center)* [OP01].

The behavior of light reflecting in a polygon is complex. Aronov et al. [ADD+98] proved that after k reflections, the boundary of the illuminated region has combinatorial complexity $O(n^{2k})$, with a matching lower bound for any fixed k. Even determining whether every triangle supports a periodic ray is unresolved; see [HH00].

Pach asked whether a finite set of disjoint circular mirrors can trap all the rays from a point light source [Ste96]. See Fig. 33.7.1. This and many other related questions [OP01] remain open.

33.8 VISIBILITY IN THREE DIMENSIONS

Research on visibility in three dimensions (3D) has concentrated on three topics: hidden surface removal, polyhedral terrains, and various 3D visibility graphs.

33.8.1 HIDDEN SURFACE REMOVAL

"Hidden surface removal" is one of the key problems in computer graphics (Chapter 52), and has been the focus of intense research for two decades. The typical problem instance is a collection of (planar) polygons in space, from which the view from $z = \infty$ must be constructed. Traditionally, hidden-surface algorithms have been classified as either *image-space* algorithms, exploiting the ultimate need to compute visible colors for image pixels, and *object-space* algorithms, which perform exact computations on object polygons. We only discuss the latter.

The complexity of the output scene can be quadratic in the number of input vertices n. A worst-case optimal $\Theta(n^2)$ algorithm can be achieved by projecting the lines containing each polygon edge to a plane and constructing the resulting arrangement of lines [Dév86, McK87]. More recent work has focused on obtaining output-size sensitive algorithms, whose time complexity depends on the number of vertices k in the output scene (the complexity of the **visibility map**), which is often less than quadratic in n. See Table 33.8.1 for selected results. In the table, k is the complexity of the **visibility map**, the "wire-frame" projection of the scene. A notable example is based on careful construction of "visibility maps," which leads, e.g., to a complexity of $O((n+k)\log^2 n)$ for performing hidden surface removal on nonintersecting spheres, where k is the complexity of the output map.

TABLE 33.8.1   Hidden-surface algorithm complexities.

ENVIRONMENT	COMPLEXITY	SOURCE
Isothetic rectangles	$O((n+k)\log n)$	[BO92]
Polyhedral terrain	$O((n+k)\log n \log\log n)$	[RS88]
Nonintersecting polyhedra	$O(n\sqrt{k}\log n)$	[SO92]
	$O(n^{1+\epsilon}\sqrt{k})$	[BHO$^+$94]
	$O(n^{2/3+\epsilon}k^{2/3} + n^{1+\epsilon})$	[AM93]
Arbitrary intersecting spheres	$O(n^{2+\epsilon})$	[AS00]
Nonintersecting spheres	$O(k + n^{3/2}\log n)$	[SO92]
Restricted-intersecting spheres	$O((n+k)\log^2 n)$	[KOS92]

33.8.2 BINARY SPACE PARTITION TREES

Binary Space Partition (BSP) trees are a popular method of implementing the basic *painter's algorithm*, which displays objects back-to-front to obtain proper occlusion of front-most surfaces. A **BSP** partitions $\mathbb{R}^d$ into empty, open convex sets by hyperplanes in a recursive fashion. A BSP for a set S of n line segments in $\mathbb{R}^2$ is a partition such that all the open regions corresponding to leaf nodes of the tree are empty of points from S: all the segments in S lie along the boundaries of the regions. An example is shown in Fig. 33.8.1. In general, a BSP for S will "cut up" the segments in S, in the sense that a particular $s \in S$ will not lie in the boundary of a single leaf region. In the figure, partitions 1 and 2 both cut segments, but partition 3 does not.

An attractive feature of BSPs is that an implementation to construct them is easy: In $\mathbb{R}^3$, select a polygon, partition all objects by the plane containing it, and recurse. Bounding the size (number of leaves) of BSP trees has been a challenge. The long-standing conjecture that $O(n)$ size in $\mathbb{R}^2$ is achievable was shown to be false. See Table 33.8.2 for selected results.

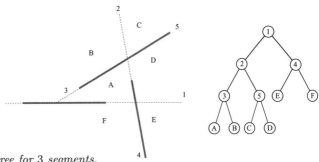

FIGURE 33.8.1
A binary space partition tree for 3 segments.

TABLE 33.8.2　BSP complexities.

DIM	CLASS	BOUND	SOURCE
2	segments	$O(n \log n)$	[PY90]
2	isothetic	$\Theta(n)$	[PY92]
2	fat	$\Theta(n)$	[BGO97]
2	segments	$\Theta(n \log n / \log \log n)$	[Tót03a, Tót11]
3	polyhedra	$O(n^2)$	[PY90]
3	polyhedra	$\Omega(n^2)$	[Cha84]
3	isothetic	$\Theta(n^{3/2})$	[PY92]
3	fat orthog. rects.	$O(n \log^8 n)$	[Tót08]

33.8.3 POLYHEDRAL TERRAINS

Polyhedral terrains are an important special class of 3D surfaces, arising in a variety of applications, most notably geographic information systems (Chapter 59).

GLOSSARY

Polyhedral terrain: A polyhedral surface that intersects every vertical line in at most a single point.

Perspective view: A view from a point.

Orthographic view: A view from infinity (parallel lines of sight).

Ray-shooting query: A query asking which terrain face is first hit by a ray shooting in a given direction from a given point. (See Chapter 41.)

$\alpha(n)$: The inverse Ackermann function (nearly a constant). See Section 28.10.

COMBINATORIAL BOUNDS

Several almost-tight bounds on the maximum number of combinatorially different views of a terrain have been obtained, as listed in Table 33.8.3.

TABLE 33.8.3   Bounds for polyhedral terrains.

VIEW TYPE	BOUND	SOURCE
Along vertical	$O(n^2 2^{\alpha(n)})$	[CS89]
Orthographic	$O(n^{5+\epsilon})$	[AS94]
Perspective	$O(n^{8+\epsilon})$	[AS94]

Bose et al. established that $\lfloor n/2 \rfloor$ vertex guards are sometimes necessary and always sufficient to guard a polyhedral terrain of n vertices [BSTZ97, BKL96].

ALGORITHMS

Algorithms seek to exploit the terrain constraints to improve on the same computations for general polyhedra:

1. To compute the orthographic view from above the terrain:
   time $O((k+n)\log n \log \log n)$, where k is the output size [RS88].

2. To preprocess for $O(\log n)$ ray-shooting queries for rays with origin on a vertical line [BDEG94].

33.8.4 3D VISIBILITY GRAPHS

GLOSSARY

Aspect graph: A graph with a node for each combinatorially distinct view of a collection of polyhedra, with two nodes connected by an arc if the views can be reached directly from one another by a continuous movement of the viewpoint.

Isothetic: Edges parallel to Cartesian coordinate axes.

Box visibility graph: A graph realizable by disjoint isothetic boxes in 3D with orthogonal visibility.

K_n: The complete graph on n nodes.

There have been three primary motivations for studying visibility graphs of objects in three dimensions.

1. Computer graphics: Useful for accelerating interactive "walkthroughs" of complex polyhedral scenes [TS91], and for radiosity computations [TH93]. See Chapter 52.

2. Computer vision: "Aspect graphs" are used to aid image recognition. The maximum number of nodes in an aspect graph for a polyhedron of n vertices depends on both convexity and the type of view. See Table 33.8.4. Note that the nonconvex bounds are significantly larger than those for terrains.

TABLE 33.8.4   Combinatorial complexity of visibility graphs.

CONVEXITY	ORTHOGRAPHIC	PERSPECTIVE	SOURCE
Convex polyhedron	$\Theta(n^2)$	$\Theta(n^3)$	[PD90]
Nonconvex polyhedron	$\Theta(n^6)$	$\Theta(n^9)$	[GCS91]

3. Combinatorics: It has been shown that K_{22} is realizable by disjoint isothetic rectangles in "$2\frac{1}{2}$D" with vertical visibility (all rectangles are parallel to the xy-plane), but that K_{56} (and therefore all larger complete graphs) cannot be so represented [BEF+93]. It is known that K_{42} is a box visibility graph [BJMO94] but that K_{184} is not [FM99].

33.9 PENETRATING ILLUMINATION OF CONVEX BODIES

A rich vein of problems was initiated by Hadwiger, Levi, Gohberg, and Markus; see [MS99] for the complex history. The problems employ a different notion of exterior illumination, which could be called *penetrating illumination* (or perhaps "stabbing"), and focuses on a single convex body in $\mathbb{R}^d$.

GLOSSARY

Penetrating illumination: An exterior point x penetratingly illuminates a point y on the boundary ∂K of an object K if the ray from x through y has a non-empty intersection with the interior int K of K.

Direction illumination: A point $y \in \partial K$ is illuminated from direction $\mathbf{v}$ if the ray from the exterior through y with direction $\mathbf{v}$ has a non-empty intersection with int K.

Affine symmetry: An object in $\mathbb{R}^3$ has affine symmetry if it unchanged after reflection through a point, reflection in a plane, or rotation about a line by angle $2\pi/n$, $n = 2, 3, \ldots$.

The central problem may be stated: What is the minimum number of exterior points sufficient to penetratingly illuminate any compact, convex body K in $\mathbb{R}^d$? The problem is only completely solved in 2D: 4 lights are needed for a parallelogram, and 3 for all other convex bodies. In 3D it is known that 8 lights are needed for a parallelepiped (Fig. 33.9.1), and conjectured that 7 suffice for all other convex bodies. Bezdek proved that 8 lights suffice for any 3-polytope with an affine symmetry [Bez93]. Lassak proved that no more than 20 lights are needed for any compact, convex body in 3D [Bol81].

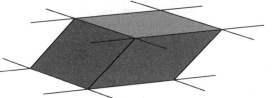

FIGURE 33.9.1

A parallelepiped requires $2^3 = 8$ lights for penetrating illumination of its boundary.

One reason for the interest in this problem is its connection to other problems, particularly covering problems. Define:

$I_0(K)$: the minimum number of points sufficient to penetratingly illuminate K.

$I_\infty(K)$: the minimum number of directions sufficient to direction-illuminate K.

$H(K)$: the minimum number of smaller homothetic copies of K that cover K.

$i(K)$: the minimum number of copies of int K that cover K.

Remarkably,

$$I_0(K) = I_\infty(K) = H(K) = i(K) \,,$$

as established by Boltjanski, Hadwiger, and Soltan; see again [MS99]. Several have conjectured that these quantities are $\leq 2^d$ for compact, convex bodies in $\mathbb{R}^d$, with equality only for the d-parallelotope. The conjecture has been established only for special classes of bodies in 3 and higher dimensions, e.g., [Bol01, Bez11].

33.10 SOURCES AND RELATED MATERIAL

SURVEYS

All results not given an explicit reference above may be traced in these surveys.

[O'R87]: A monograph devoted to art gallery theorems and visibility algorithms.

[She92]: A survey of art gallery theorems and visibility graphs, updating [O'R87].

[O'R92]: A short update to [She92].

[Urr00]: Art gallery results, updating [She92].

[GG13]: Survey of open problems on visibility graphs.

[O'R93]: Survey of visibility graph results.

[Gho07]: Survey of visibility algorithms in $\mathbb{R}^2$.

[MSD00]: Survey of link-distance algorithms.

[Mur99]: A Ph.D. thesis on hidden-surface removal algorithms.

[Tót05]: Survey of binary space partitions.

[MS99, Bez06, BK16]: Surveys of illumination of convex bodies.

RELATED CHAPTERS

Chapter 29: Triangulations and mesh generation
Chapter 30: Polygons
Chapter 31: Shortest paths and networks
Chapter 41: Ray shooting and lines in space
Chapter 42: Geometric intersection
Chapter 52: Computer graphics
Chapter 54: Pattern recognition
Chapter 59: Geographic information systems

REFERENCES

[AAAS94]   P.K. Agarwal, N. Alon, B. Aronov, and S. Suri. Can visibility graphs be represented compactly? *Discrete Comput. Geom.*, 12:347–365, 1994.

[ABD⁺09]   O. Aichholzer, S. Bereg, A. Dumitrescu, A. García, C. Huemer, F. Hurtado, M. Kano, A. Márquez, D. Rappaport, S. Smorodinsky, D.L. Souvaine, J. Urrutia, and D.R. Wood. Compatible geometric matchings. *Comput. Geom.* 42:617–626, 2009.

[ADD⁺98]   B. Aronov, A.R. Davis, T.K. Dey, S.P. Pal, and D.C. Prasad. Visibility with multiple reflections. *Discrete Comput. Geom.*, 20:61–78, 1998.

[Agg84]   A. Aggarwal. *The Art Gallery Problem: Its Variations, Applications, and Algorithmic Aspects.* Ph.D. thesis, Dept. of Comput. Sci., Johns Hopkins Univ., Baltimore, 1984.

[AM93]   P.K. Agarwal and J. Matoušek. Ray shooting and parametric search. *SIAM J. Comput.*, 22:794–806, 1993.

[AS94]   P.K. Agarwal and M. Sharir. On the number of views of polyhedral terrains. *Discrete Comput. Geom.*, 12:177–182, 1994.

[AS00]   P.K. Agarwal and M. Sharir. Pipes, cigars, and kreplach: The union of Minkowski sums in three dimensions. *Discrete Comput. Geom.*, 24:645–685, 2000.

[BDEG94]   M. Bern, D.P. Dobkin, D. Eppstein, and R. Grossman. Visibility with a moving point of view. *Algorithmica*, 11:360–378, 1994.

[BDHS97]   P. Bose, A.M. Dean, J.P. Hutchinson, and T.C. Shermer. On rectangle visibility graphs. *Proc. Graph Drawing*, vol. 1190 of *LNCS*, pages 25–35, Springer, Berlin, 1997.

[BEF⁺93]   P. Bose, H. Everett, S.P. Fekete, M.E. Houle, A. Lubiw, H. Meijer, K. Romanik, T.C. Shermer, S. Whitesides, and C. Zelle. On a visibility representation for graphs in three dimensions. *J. Graph Algorithms Appl.*, 2:1–16, 1998.

[Bez93]   K. Bezdek. Hadwiger-Levi's covering problem revisited. In J. Pach, editor, *New Trends in Discrete and Computational Geometry*, vol. 10 of *Algorithms Combin.*, pages 199–233, Springer-Verlag, Berlin, 1993.

[Bez06]   K. Bezdek. The illumination conjecture and its extensions. *Period. Math. Hungar.*, 53:59–69, 2006.

[Bez11]   K. Bezdek. The illumination conjecture for spindle convex bodies. *Proc. Steklov Inst. Math.*, 275:169–176, 2011.

[BGO97]   M. de Berg, M. de Groot, and M.H. Overmars. New results on binary space partitions in the plane. *Comput. Geom.*, 8:317–333, 1997.

[BHO⁺94]   M. de Berg, D. Halperin, M.H. Overmars, J. Snoeyink, and M. van Kreveld. Efficient ray shooting and hidden surface removal. *Algorithmica*, 12:30–53, 1994.

[BJMO94]   P. Bose, A. Josefczyk, J. Miller, and J. O'Rourke. K_{42} is a box visibility graph. In *Snapshots in Comput. Geom.*, pages 88–91, Univ. Saskatchewan, 1994.

[BK16]   K. Bezdek and M.A. Khan. The geometry of homothetic covering and illumination. Preprint, arXiv:1602.06040v2, 2016.

[BKL96]   P. Bose, D.G. Kirkpatrick. and Z. Li. Efficient algorithms for guarding or illuminating the surface of a polyhedral terrain. *Proc. 8th Canad. Conf. Comput. Geom.*, pages 217–222, 1996.

[BO92]   M. de Berg and M.H. Overmars. Hidden surface removal for *c*-oriented polyhedra. *Comput. Geom.*, 1:247–268, 1992.

[Bol81]   V. Boltjansky. Combinatorial geometry. *Algebra Topol. Geom.*, 19:209–274, 1981. In Russian. Cited in [MS99].

[Bol01]   V. Boltjansky. Solution of the illumination problem for bodies with md $M = 2$. *Discrete Comput. Geom.*, 26:527–541, 2001.

[BSTZ97]   P. Bose, T.C. Shermer, G.T. Toussaint, B. Zhu. Guarding polyhedral terrains. *Comput. Geom.*, 7:173–185, 1997.

[CH17]   J. Cardinal and U. Hoffmann. Recognition and complexity of point visibility graphs. *Discrete Comput. Geom.*, 57:164–178, 2017.

[Cha84]   B. Chazelle. Convex partitions of polyhedra: A lower bound and worst-case optimal algorithm. *SIAM J. Comput.*, 13:488–507, 1984.

[Cha91]   B. Chazelle. Triangulating a simple polygon in linear time. *Discrete Comput. Geom.*, 6:485–524, 1991.

[CKM⁺10]   J. Cibulka, J. Kynčl, V. Mészáros, R. Stolař, and P. Valtr. On three parameters of invisibility graphs. In *Internat. Comput. Combinatorics Conf.*, pages 192–198. Springer, 2010.

[CRCU93]   J. Czyzowicz, E. Rivera-Campo, and J. Urrutia. Illuminating rectangles and triangles in the plane. *J. Combin. Theory Ser. B*, 57:1–17, 1993.

[CRC⁺95]   J. Czyzowicz, E. Rivera-Campo, J. Urrutia, and J. Zaks. On illuminating line segments in the plane. *Discrete Math.*, 137:147–153, 1995.

[CS89]   R. Cole and M. Sharir. Visibility problems for polyhedral terrains. *J. Symbolic Comput.*, 7:11–30, 1989.

[DW15]   D.Z. Chen and H. Wang. A new algorithm for computing visibility graphs of polygonal obstacles in the plane. *J. Comput. Geom.*, 6:316-345, 2015.

[Dév86]   F. Dévai. Quadratic bounds for hidden line elimination. In *Proc. 2nd Sympos. Comput. Geom.*, pages 269–275, ACM Press, 1986.

[DM13]   V. Dujmovic and P. Morin. On obstacle numbers. *Electron. J. Combin*, 22:P3, 2013.

[DN94]   G. Das and G. Narasimhan. Optimal linear-time algorithm for the shortest illuminating line segment. In *Proc. 10th Sympos. Comput. Geom.*, pages 259–266, ACM Press, 1994.

[DPT09]   A. Dumitrescu, J. Pach, and G. Tóth. A note on blocking visibility between points. *Geombinatorics*, 19:67–73, 2009.

[ECOUX95]   V. Estivill-Castro, J. O'Rourke, J. Urrutia, and D. Xu. Illumination of polygons with vertex floodlights. *Inform. Process. Lett.*, 56:9–13, 1995.

[Fej77]   L. Fejes Tóth. Illumination of convex discs. *Acta Math. Acad. Sci. Hungar.*, 29:355–360, 1977.

[FHKS16]   S. Friedrichs, M. Hemmer, J. King, and C. Schmidt. The continuous 1.5D terrain guarding problem: Discretization, optimal solutions, and PTAS. *J. Comput. Geom.*, 7:256–284, 2016.

[FM99]   S.P. Fekete and H. Meijer. Rectangle and box visibility graphs in 3d. *Internat. J. Comput. Geom. Appl.*, 9:1–27, 1999.

[GCS91]   Z. Gigus, J.F. Canny, and R. Seidel. Efficiently computing and representing aspect graphs of polyhedral objects. *IEEE Trans. Pattern Anal. Mach. Intell.*, 13:542–551, 1991.

[GG13]   S.K. Ghosh and P.P. Goswami. Unsolved problems in visibility graphs of points, segments, and polygons. *ACM Comput. Surv.*, 46:22, 2013.

[GHKS96]   E. Győri, F. Hoffmann, K. Kriegel, and T.C. Shermer. Generalized guarding and partitioning for rectilinear polygons. *Comput. Geom.*, 6:21–44, 1996.

[GHL+87]   L.J. Guibas, J. Hershberger, D. Leven, M. Sharir, and R.E. Tarjan. Linear-time algorithms for visibility and shortest path problems inside triangulated simple polygons. *Algorithmica*, 2:209–233, 1987.

[Gho97]   S.K. Ghosh. On recognizing and characterizing visibility graphs of simple polygons. *Discrete Comput. Geom.*, 17:143–162, 1997.

[Gho07]   S.K. Ghosh. *Visibility Algorithms in the Plane.* Cambridge Univ. Press, 2007.

[Gho10]   S.K. Ghosh. Approximation algorithms for art gallery problems in polygons. *Discrete Appl. Math.*, 158:718–722, 2010.

[GM91]   S.K. Ghosh and D.M. Mount. An output-sensitive algorithm for computing visibility graphs. *SIAM J. Comput.*, 20:888–910, 1991.

[GOH+02]   A. García-Olaverri, F. Hurtado, M. Noy and J. Tejel. On the minimum size of visibility graphs. *Inform. Proc. Lett.*, 81:223–230, 2002.

[GR15]   S.K. Ghosh and B. Roy. Some results on point visibility graphs. *Theor. Comput. Sci.*, 575:17–32, 2015.

[GSG92]   M.T. Goodrich, S. Shauck, and S. Guha. Parallel methods for visibility and shortest path problems in simple polygons. *Algorithmica*, 8:461–486, 1992.

[Her89]   J. Hershberger. An optimal visibility graph algorithm for triangulated simple polygons. *Algorithmica*, 4:141–155, 1989.

[HH00]   L. Halbelsen and N. Hungerbühler. On periodic billiard trajectories in obtuse triangles. *SIAM Rev.*, 42:657–670, 2000.

[HK93]   F. Hoffmann and K. Kriegel. A graph coloring result and its consequences for some guarding problems. In *Proc. 4th Internat. Sympos. Algorithms Comput.*, vol. 762 of *LNCS*, pages 78–87, Springer-Verlag, Berlin, 1993.

[HK96]   F. Hoffmann and K. Kriegel. A graph coloring result and its consequences for some guarding problems. *SIAM J. Discrete Math.*, 9:210–224, 1996.

[HN01]   F. Hurtado and M. Noy. On the number of visibility graphs of simple polygons. *Discrete Math.*, 232:139–144, 2001.

[HT03]   M. Hoffmann and C.D. Tóth. Segment endpoint visibility graphs are Hamiltonian. *Comput. Geom.*, 26:47–68, 2003.

[IK09]   R. Inkulu and S. Kapoor. Visibility queries in a polygonal region. *Comput. Geom.*, 42:852–864, 2009.

[IST13]   M. Ishaque, D.L. Souvaine, and C.D. Tóth. Disjoint compatible geometric matchings. *Discrete Comput. Geom.*, 49:89–131, 2013.

[JS87]   B. Joe and R.B. Simpson. Correction to Lee's visibility polygon algorithm. *BIT*, 27:458–473, 1987.

[KK11]   J. King and E. Krohn. Terrain guarding is NP-hard. *SIAM J. Comput.*, 40:1316–1339, 2011.

[Kle69]   V. Klee. Is every polygonal region illuminable from some point? *Amer. Math. Monthly*, 76:180, 1969.

[KOS92]   M.J. Katz, M.H. Overmars, and M. Sharir. Efficient hidden surface removal for objects with small union size. *Comput. Geom.*, 2:223–234, 1992.

[KW91]   V. Klee and S. Wagon. *Old and New Unsolved Problems in Plane Geometry*. Math. Assoc. Amer., 1991.

[LC94]   S.-Y. Lin and C. Chen. Planar visibility graphs. In *Proc. 6th Canad. Conf. Comput. Geom.*, pages 30–35, 1994.

[LMW16]   S. Lelievre, T. Monteil, and B. Weiss. Everything is illuminated. *Geom. Topol.* 20:1737–62, 2016.

[LP79]   D.T. Lee and F.P. Preparata. An optimal algorithm for finding the kernel of a polygon. *J. Assoc. Comput. Mach.*, 26:415–421, 1979.

[Mat09]   J. Matoušek. Blocking visibility for points in general position. *Discrete Comput. Geom.*, 42:219–223, 2009.

[MV99]   J. Matoušek and P. Valtr. On visibility and covering by convex sets. *Israel J. Math.*, 113:341–379, 1999.

[McK87]   M. McKenna. Worst-case optimal hidden-surface removal. *ACM Trans. Graph.*, 6:19–28, 1987.

[MP12]   T.S. Michael and V. Pinciu. Guarding orthogonal prison yards: An upper bound. *Congr. Numerantium*, 211:57–64, 2012.

[MS99]   H. Martini and V. Soltan. Combinatorial problems on the illumination of convex bodies. *Aequationes Math.*, 57:121–152, 1999.

[MSD00]   A. Maheshwari, J.-R. Sack, and H.N. Djidjev. Link distance problems. In J.-R. Sack and J. Urrutia, editors, *Handbook of Computational Geometry*, pages 519–558, Elsevier, Amsterdam, 2000.

[Mur99]   T.M. Murali. *Efficient Hidden-Surface Removal in Theory and in Practice*. Ph.D. thesis, Brown University, Providence, 1999.

[O'R83]   J. O'Rourke. Galleries need fewer mobile guards: A variation on Chvátal's theorem. *Geom. Dedicata*, 14:273–283, 1983.

[O'R87]   J. O'Rourke. *Art Gallery Theorems and Algorithms. Internat. Series Monographs Computer Science*. Oxford University Press, New York, 1987.

[O'R92]   J. O'Rourke. Computational geometry column 15. *Internat. J. Comput. Geom. Appl.*, 2:215–217, 1992. Also in *SIGACT News*, 23:2, 1992.

[O'R93]   J. O'Rourke. Computational geometry column 18. *Internat. J. Comput. Geom. Appl.*, 3:107–113, 1993. Also in *SIGACT News*, 24:1:20–25, 1993.

[OP01]   J. O'Rourke and O. Petrovici. Narrowing light rays with mirrors. In *Proc. 13th Canad. Conf. Comput. Geom.*, pages 137–140, 2001.

[PD90]   H. Plantinga and C.R. Dyer. Visibility, occlusion, and the aspect graph. *Internat. J. Comput. Vision*, 5:137–160, 1990.

[Pfe08]   F. Pfender. Visibility graphs of point sets in the plane. *Discrete Comput. Geom.*, 39:455–459, 2008.

[PPVW12]   M.S. Payne, A. Pór, P. Valtr, and D.R. Wood. On the connectivity of visibility graphs. *Discrete Comput. Geom.*, 48:669–681, 2012.

[PW10]   A. Pór and D.R. Wood. On visibility and blockers. *J. Comput. Geom.*, 1:29–40, 2010.

[PY90]   M.S. Paterson and F.F. Yao. Efficient binary space partitions for hidden-surface removal and solid modeling. *Discrete Comput. Geom.*, 5:485–503, 1990.

[PY92]   M.S. Paterson and F.F. Yao. Optimal binary space partitions for orthogonal objects. *J. Algorithms*, 13:99–113, 1992.

[Rap03]   D. Rappaport. The visibility graph of congruent discs is Hamiltonian. *Internat. J. Comput. Geom. Appl.*, 25:257–265, 2003.

[Roy16]   B. Roy. Point visibility graph recognition is NP-hard. *Internat. J. Comput. Geom. Appl.*, 26:1–32 , 2016.

[RS88]   J.H. Reif and S. Sen. An efficient output-sensitive hidden-surface removal algorithms and its parallelization. In *Proc. 4th Sympos. Comput. Geom.*, pages 193–200, ACM Press, 1988.

[SE87]   X. Shen and H. Edelsbrunner. A tight lower bound on the size of visibility graphs. *Inform. Process. Lett.*, 26:61–64, 1987.

[She94]   T.C. Shermer. A tight bound on the combinatorial edge guarding problem. In *Snapshots in Comput. Geom.*, pages 191–223, Univ. Saskatchewan, 1994.

[She92]   T.C. Shermer. Recent results in art galleries. *Proc. IEEE*, 80:1384–1399, 1992.

[SO92]   M. Sharir and M.H. Overmars. A simple output-sensitive algorithm for hidden-surface removal. *ACM Trans. Graph.*, 11:1–11, 1992.

[ST05]   B. Speckmann and C.D. Tóth. Allocating vertex π-guards in simple polygons via pseudo-triangulations. *Discrete Comput. Geom.*, 33:345–364, 2005.

[Ste96]   I. Stewart. Mathematical recreations. *Sci. Amer.*, 275:100–103, 1996. Includes light in circular forest problem due to J. Pach.

[TH93]   S. Teller and P. Hanrahan. Global visibility algorithms for illumination computations. In *Proc. ACM Conf. SIGGRAPH 93*, pages 239–246, 1993.

[Tok95]   G.W. Tokarsky. Polygonal rooms not illuminable from every point. *Amer. Math. Monthly*, 102:867–879, 1995.

[Tót00]   C.D. Tóth. Art gallery problem with guards whose range of vision is 180°. *Comput. Geom.*, 17:121–134, 2000.

[Tót01]   C.D. Tóth. Illuminating both sides of line segments. In J. Akiyama, M. Kano, and M. Urabe, editors, *Discrete and Computational Geometry*, vol. 2098 of *LNCS*, pages 370–380, Springer, Berlin, 2001.

[Tót02]   C.D. Tóth. Illumination in the presence of opaque line segments in the plane. *Comput. Geom.*, 21:193–204, 2002.

[Tót03a]   C.D. Tóth. A note on binary plane partitions. *Discrete Comput. Geom.*, 30:3–16, 2003.

[Tót03b]   C.D. Tóth. Guarding disjoint triangles and claws in the plane. *Comput. Geom.*, 25:51–65, 2003.

[Tót03c]   C.D. Tóth. Illuminating disjoint line segments in the plane. *Discrete Comput. Geom.*, 30:489–505, 2003.

[Tót03d]   C.D. Tóth. Illumination of polygons by 45°-floodlights. *Discrete Math.*, 265:251–260, 2003.

[Tót05]   C.D. Tóth. Binary space partitions: Recent developments. In J.E. Goodman, J. Pach, and E. Welz, editors, *Combinatorial and Computational Geometry*, vol. 52 of MSRI Publications, Cambridge University Press, pages 529–556, 2005.

[Tót08]   C.D. Tóth. Binary space partitions for axis-aligned fat rectangles. *SIAM J. Comput.*, 38:429–447, 2008.

[Tót11]   C.D. Tóth. Binary plane partitions for disjoint line segments. *Discrete Comput. Geom.*, 45:617–646, 2011.

[TS91]   S.J. Teller and C.H. Séquin. Visibility preprocessing for interactive walkthroughs. *Comput. Graph.*, 25:61–69, 1991.

[TTW12]   C.D. Tóth, G. Toussaint, and A. Winslow. Open guard edges and edge guards in simple polygons. In A. Marquez, P. Ramos, and J. Urrutia, editors, *Computational Geometry*, vol. 7579 of *LNCS*, pages 54–64, Springer, Berlin, 2012.

[Urr00]   J. Urrutia. Art gallery and illumination problems. In J.-R. Sack and J. Urrutia, editors, *Handbook of Computational Geometry*, pages 973–1027, Elsevier, Amsterdam, 2000.

34 GEOMETRIC RECONSTRUCTION PROBLEMS
Yann Disser and Steven S. Skiena

INTRODUCTION

Many problems from mathematics and engineering can be described in terms of reconstruction from geometric information. ***Reconstruction*** is the algorithmic problem of combining the results of measurements of some aspect of a physical or mathematical object to obtain desired information about the object.

In this chapter, we consider three different classes of geometric reconstruction problems. In Section 34.1, we examine static reconstruction problems, where we are given a geometric structure derived from an original structure, and seek to invert this transformation. In Section 34.2, we consider interactive reconstruction problems, where we are permitted to repeatedly "probe" an unknown object at arbitrary places and seek to reconstruct the object using the fewest such probes. Finally, in Section 34.3, we turn to exploration and mapping of unknown environments, where we take the perspective of a mobile agent that locally observes its surroundings and aims to infer information about the global layout of the environment.

Our focus in this chapter is on *exact, theoretical* reconstruction problems from a perspective of computational geometry. In contrast, a significant body of theoretical work in computational geometry is concerned with the *approximate* reconstruction of shapes and surfaces (see Chapter 35). Practical reconstruction problems beyond the scope of this chapter arise in many fields, with examples including computer vision, computer-aided tomography, and the reconstruction of 3D objects from 2D images. In these settings, quality criteria for good solutions are not always mathematically well-defined and may rely on aesthetics, practical applicability, or consistency with reference data.

34.1 STATIC RECONSTRUCTION PROBLEMS

Here we consider inverse problems of the following type. Let A be a geometric structure, and T a transformation such that $T(A) \to B$, where B is some different geometric structure. Now, given T and B, construct a structure A' such that $T(A') \to B$. If T is one-to-one, then $A = A'$. If not, we may be interested in finding or counting all solutions.

GLOSSARY

Gabriel graph: A graph whose vertices are points in $\mathbb{R}^2$, with an edge (x, y) if points x and y define the diameter of an empty circle.

Relative neighborhood graph: A graph whose vertices are points in $\mathbb{R}^2$, with an edge (x, y) if there exists no point z such that z is closer to x than y is and z is closer to y than x is. See Section 32.1.

Interpoint distance: Distance between a pair of points in $\mathbb{R}^2$. The distance is ***labeled*** if the identities of the two points defining the distance are associated with the distance, and ***unlabeled*** otherwise.

Stabbing Information: For every vertex v of a polygon the (at most) two edges that are first intersected by the rays wv and uv, where w, u are the neighbors of v along the boundary.

Line cross-sections: A set of lines $\mathcal{L}$ together with the line segments that constitute the intersection of $\mathcal{L}$ with a polygon P.

Visibility polygon: The subset of points in a polygon P visible to a fixed point $x \in P$, i.e., all points $y \in P$ for which the line segment xy is contained in P.

Direction edge/face count: A vector d together with a number k of edges/faces visible in an orthogonal projection in direction d.

Cross-ratio in a triangulation: The ratio $\frac{bd}{ce}$, where a, b, c and a, d, e are the lengths of the edges (in ccw order) of two touching triangles.

Vertex visibility graph: The graph with a node for every vertex of a polygon, and with edges between pairs of vertices that mutually see each other, i.e., whose straight-line connection lies inside the polygon. See Section 33.3.

Point visibility graph: The graph for a set of points with an edge between two points that mutually see each other, i.e., whose straight-line connection does not contain other points. See Section 33.3.

Angle measurement: The ordered list of angles in counter-clockwise order between the edges of the visibility graph at a vertex.

Distance measurement: The ordered list of distances in counter-clockwise order to the vertices visible from a given vertex.

Corner: A vertex of a polygon with an interior angle different from π.

Complex moment of order k***:***   The complex value $\iint_B z^k \, dx \, dy$ for a region $B \subset \mathbb{C}$ and $z = x + iy$.

Extended Gaussian image: A transform that maps each face of a convex polyhedron to a vector normal to the face whose length is proportional to the area of the face.

X-ray projection: The length of the intersection of a line with a convex body.

Determination: A class of sets is determined by n directions if there are n fixed directions such that all sets can be reconstructed from X-ray projections along these directions.

Verification: A class of sets is verified by n directions if, for each particular set, there are n X-ray projections that distinguish this set from any other.

MAIN RESULTS

An example of an important class of reconstruction problems is visibility graph reconstruction, i.e., given a graph G, construct a polygon P whose visibility graph is G (see Section 33.3). Results for this and other static reconstruction problems are summarized in Table 34.1.1. We characterize each problem by its input and the inverted structure we wish to reconstruct. We also specify whether the corresponding transformation is one-to-one, i.e., the result of the reconstruction is unique.

TABLE 34.1.1   Static reconstruction problems.

INPUT	INVERTED STRUCTURE	RESULT	UNIQ	SOURCE
MST with degree ≤ 5	point embedding in $\mathbb{R}^2$	always realizable	no	[MS91]
MST with degree 6	point embedding in $\mathbb{R}^2$	NP-hard	no	[EW96]
MST with degree ≥ 7	point embedding in $\mathbb{R}^2$	never realizable	–	[MS91]
Gabriel graph	point embedding in $\mathbb{R}^2$	partial charact.	no	[MS80]
rel. neighborhood graph	point embedding in $\mathbb{R}^2$	partial charact.	no	[LS93]
Delaunay triangulation	point embedding in $\mathbb{R}^2$	partial charact.	no	[Dil90][Sug94]
Voronoi diagram	point embedding in $\mathbb{R}^2$	partial charact.	no	[AB85]
point visibility graph	point embedding in $\mathbb{R}^2$	$\exists\mathbb{R}$-complete	no	[GR15][CH17]
labeled interpoint distances	points realizing these in $\mathbb{R}^d$	NP-hard	no	[Sax79]
all unlab. interpoint dists	points realizing these in $\mathbb{R}^1$	$O(2^n n \log n)$	no	[LSS03]
all unlab. interpoint dists	points realizing these in $\mathbb{R}^d$	NP-hard	no	[LSS03]
vertex visibility graph	polygon realizing it	$\in$PSPACE	no	[Eve90]
distance visibility graph	polygon realizing it	$O(n^2)$	yes	[CL92]
endpoints $V \subset \mathbb{R}^2$	orthogonal line segments	$O(n \log n)$	no	[RW93]
endpoints $V \subset \mathbb{R}^2$	disjoint orth. line segments	NP-hard	no	[RW93]
corners $V \subset \mathbb{R}^2$	orthogonal polygon	$O(n \log n)$	yes	[O'R88]
corners $V \subset \mathbb{R}^2$, 3 slopes	polygon with these slopes	NP-hard	no	[FW90]
vertices $V \subset \mathbb{R}^2$	orthogonal polygon	NP-hard	no	[Rap89]
angle measurements	compatible polygon	$O(n^2)$	yes	[DMW11][CW12]
stabbing information	compatible orthogonal poly	$O(n \log n)$	no	[JW02]
set of s line cross-sections	all compatible polygons	$O(s \log s)$ /poly	no	[SBG06]
cross-ratios, bound. angles	compatible polygon	uniqueness	yes	[Sno99]
set of visibility polygons	compatible polygon	NP-hard	no	[BDS11]
direction edge counts	convex polygon	charact., algo.	no	[BHL11]
direction face counts	convex 3D polyhedron	NP-hard	no	[BHL11]
ext. Gaussian image	convex 3D polyhedron	$O(n \log n)$ /iter.	yes	[Lit85]
4 complex moments	triangle in $\mathbb{C}$	uniqueness	yes	[Dav77]
$2n$ complex moments	vertices of polygon in $\mathbb{C}$	algorithm	yes	[MVK+95]
X-ray projections	convex body if unique	algorithm	yes	[GK07]

Another class of problems concerns proximity drawability. Given a graph G, we seek a set of points corresponding to vertices of G such that two points are "sufficiently" close if and only if there is an edge in G for the corresponding vertices. Examples of proximity drawability problems include finding points to realize graphs as minimum spanning trees (MST), Delaunay triangulations (Chapter 29), Gabriel graphs, and relative neighborhood graphs (RNGs) (Chapter 32). Although many of the results are quite technical, Liotta [Lio13] provides an excellent survey of results on these and other classes of proximity drawings; see also Chapter 55.

To provide some intuition about the minimum spanning tree results, observe that low degree graphs are easily embedded as point sets. If the maximum degree is 2, i.e., the graph is a simple path, then any straight line embedding will work. One can show that any two line segments vu, vw corresponding to adjacent edges of the tree need to form an angle not smaller than $\pi/3$, since the segment uw cannot be shorter than vu or vw. This implies that degrees larger than 6 cannot be realized, and forces the neighbors of degree 6 vertices to be spaced at equal angles of $\pi/3$, a very restrictive condition leading to the hardness result.

Other typical reconstruction problems are concerned with constructing polygons that are compatible with given geometrical parameters. See Figure 34.1.1 for three such examples (taken from [CD+13b]). In the first example, the angles between lines-of-sight are known at each vertex, and it turns out that this information uniquely determines the polygon (up to scaling and rotation) [DMW11]. In the second example, a polygon has to be constructed from its intersection with a set of lines. In this case, the polygon is not uniquely determined, but all compatible polygons can be enumerated efficiently [SBG06]. The third example shows the "orthogonal connect-the-dots" problem, where an orthogonal polygon has to be recovered from the coordinates of its vertices. This is uniquely and efficiently possible if vertices of degree π are forbidden [O'R88], and otherwise it is NP-hard to find any compatible polygon [Rap89].

FIGURE 34.1.1

From left to right: Reconstruction from angle measurements, line cross-sections, and vertices.

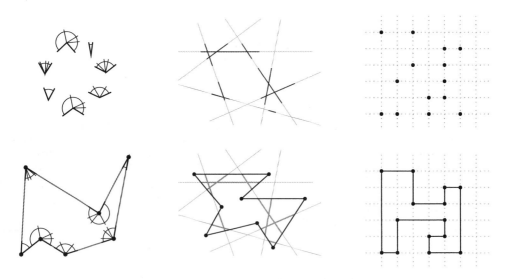

Another important set of problems concern reconstructing objects from a fixed set of X-ray projections, conventionally called Hammer's X-ray problem [Ham63]. Different problems arise depending upon whether the X-rays originate from a point or line source, and whether we seek to verify or determine the object. A selection of results on parallel X-rays (line sources) are listed in Table 34.1.2. For example, parallel X-rays in certain sets of four directions suffice to determine any convex body if the directions are not a subset of the edges of an affinely regular polygon.

If the directions do form such a subset, then there exist noncongruent polygons that are not distinguished by any number n of parallel X-rays in these directions. Nevertheless, any pair of nonparallel directions suffice to determine "most" (in the sense of Baire category) convex sets.

There is also a collection of results on *point source X-rays*. For example, convex sets in $\mathbb{R}^2$ are determined by directed X-rays from three noncollinear point sources. The substantial literature on such X-ray problems is very well covered by Gardner's monograph [Gar06], from which two of the open problems listed below are drawn.

The related field of *discrete tomography* is inspired by the use of electron microscopy to reconstruct the positions of atoms in crystal structures. A typical problem is placing integers in a matrix so as to realize a given set of row and column sums. The problem becomes more complex when the reconstructed body must satisfy connectivity constraints or simultaneously satisfy row/column sums of multiple colors. Collections of survey articles on discrete tomography include Herman and Kuba [HK99, HK07].

TABLE 34.1.2   Selected results on Hammer's X-ray problem.

DIM	PROBLEM	SETS	RESULT	SOURCE
2	verify	convex polygons	2 parallel X-rays do not suffice	[Gar83]
	verify	convex set	3 parallel X-rays suffice	[Gie62]
	determine	convex set	4 parallel X-rays suffice	[GM80][GG97]
	determine	convex set	n arb. paral. X-rays do not suffice	[Gie62]
	determine	star-shaped poly.	finite num. paral. X-rays insufficient	[Gar92]
	determine	convex body	3 point X-rays suffice	[Vol86]
3	determine	convex body	4 parallel, coplanar X-rays suffice	[Gar06]
	determine	convex body	4 arb., paral. X-rays do not suffice	[Gar06]
d	determine	convex body	2 parallel X-rays "usually" suffice	[VZ89]
	verify	compact sets	no finite number of directions suffice	[Gar92]

OPEN PROBLEMS

1. Give an efficient algorithm to reconstruct a set of n points on the line from the set of $\binom{n}{2}$ unlabeled interpoint distances it defines: see [LSS03]. Note that the problem indeed remains open as of this writing, despite published comments to the contrary: see [DGN05].

2. Is a polygon uniquely determined by its distance measurements?

3. Give an algorithm to determine whether a graph is the visibility graph of a simple polygon [GG13, Problem 29].

4. Characterize the convex sets in $\mathbb{R}^2$ that can be determined by two parallel X-rays [Gar06, Problem 1.1].

5. Are convex bodies in $\mathbb{R}^3$ determined by parallel X-rays in some set of five directions [Gar06, Problem 2.2]?

34.2 INTERACTIVE RECONSTRUCTION PROBLEMS

In *static* reconstruction problems all available data about the structure that has to be reconstructed is revealed in a one-shot fashion. In contrast, *interactive* reconstruction allows to request data in multiple rounds, and allows each request to depend on the data gathered so far. This process is generally modeled via **geometric probing**, which defines access to the unknown geometric structure via a mathematical or physical measuring device, a **probe**. A variety of problems from robotics, medical instrumentation, mathematical optimization, integral and computational geometry, graph theory, and other areas fit into this paradigm.

The model of geometric probing was introduced by Cole and Yap [CY87] and inspired by work in robotics and tactile sensing. A substantial body of work has followed, which is extensively surveyed in [Ski92]. A collection of open problems in probing appears in [Ski89a]. More recent probing models include proximity probes [ABG15], wedge probes [BCSS15], and distance probes [AM15].

GLOSSARY

Determination: The algorithmic problem of computing how many probes of a certain type are necessary to completely determine or reconstruct an object drawn from a particular class of objects.

Verification: The algorithmic problem of, given a supposed description of an object, computing how many probes of a certain type are necessary to test whether the description is valid.

Model-based: A problem where any object is constrained to be one of a known, finite set of m possible objects.

Point probe: An oracle that tests whether a given point is within the object.

Finger probe: An oracle that returns the first point of intersection between a directed line and the object.

Hyperplane probe: An oracle that returns the first time when a hyperplane moving parallel to itself intersects the object.

X-ray probe: An oracle that measures the length of the intersection between a line and the object.

Silhouette probe: An oracle that returns a $(d-1)$-dimensional projection (in a given direction) of the d-dimensional object.

Halfspace probe: An oracle that measures the area or volume of the intersection between a halfspace and the object.

Cut-set probe: An oracle that, for a specified graph and partition of the vertices, returns the size of the cut-set determined by the partition.

Proximity probe: An oracle that returns the nearest point of the object to a specified origin point.

Wedge probe: An oracle that, for a specified origin point and translation direction, returns the first contact points between the object and a moving wedge with angle ω.

Distance probe: An oracle that returns the distance between two named points.

Fourier probe: An oracle that for a given vector $\xi \in \mathbb{R}^2$ returns $\int_D e^{-i\langle \xi, x \rangle} \, \mathrm{d}x$ with respect to a region $D \subset \mathbb{R}^2$.

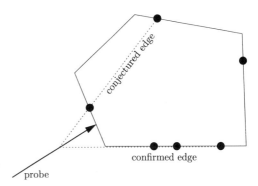

FIGURE 34.2.1
Determining the next edge of P using finger probes.

MAIN RESULTS

For a particular probing model, the determination problem asks how many probes are sufficient to completely reconstruct an object from a given class. For example, Cole and Yap's strategy for reconstructing a convex polygon P from finger probes is based on the observation that three collinear contact points must define an edge. The strategy, illustrated in Figure 34.2.1, repeatedly aims a probe at the intersection point between a confirmed edge (defined by three collinear points) and a conjectured edge (defined by two contact points). If this intersection point is indeed a contact point, another vertex is determined due to convexity; if not, the existence of another edge can be inferred. Since we avoid probing the interior of any edge that has been determined, roughly $3n$ probes suffice in total, since not more than one edge can be hit four times. Table 34.2.1 summarizes probing results for a wide variety of models. In the table, f_i denotes the number of i-dimensional faces of P.

Cole and Yap's finger probing model is not powerful enough to determine nonconvex objects. There are three major reasons for this. A tiny crack in an edge can go forever undetected, since no finite strategy can explore the entire surface of the polygon. Second, it is easy to construct nonconvex polygons whose features cannot be entirely contacted with straight-line probes originating from infinity. Finally, for nonconvex polygons there exists no constant k such that k collinear probes determine an edge. To generalize the class of objects, enhanced finger probes have been considered. One such probe [ABY90] returns surface normals as well as contact points, eliminating the second problem. When restricted to polygons with no two edges defined by the same supporting line, the first and third problems are eliminated as well.

In the verification problem, we are given a description of a putative object, and charged with using a small number of probes to prove that the description is correct. Verification is clearly no harder than determination, since we are free to ignore the description in planning the probes, and could simply compare the determined object to its description. Sometimes significantly fewer probes suffice for verification. For example, we can verify a putative convex polygon with $2n$

TABLE 34.2.1   Upper and lower bounds for determination for various probing models.

PROBE	OBJECT	LOWER	UPPER	SOURCE
finger	convex polygon	$3n$	$3n$	[CY87]
finger (n known)	convex polygon	$2n+1$	$3n-1$	[CY87]
finger	convex polyhedron in $\mathbb{R}^d$	$df_0 + f_{d-1}$	$f_0 + (d+2)f_{d-1}$	[LB88][DEY90]
finger (model based)	convex polygon	$n-1$	$n+4$	[JS92]
$k = 2$ or 3 fingers	convex polygon	$2n-k$	$2n$	[LB92]
4 or 5 fingers	convex polygon	$(4n-5)/3$	$\lfloor (4n+2)/3 \rfloor$	[LB92]
$k \geq 6$ fingers	convex polygon	n	$n+1$	[LB92]
enh. fingers	nondegenerate polygon	$3n-3$	$3n-3$	[ABY90]
Line	convex polygon	$3n+1$	$3n+1$	[Li88]
Line (model based)	convex polygon	$2n-3$	$2n+4$	[JS92]
Silhouette	convex polygon	$3n-2$	$3n-2$	[Li88]
Silhouette	convex polyhedron in $\mathbb{R}^3$	$f_2/2$	$5f_0 + f_2$	[DEY90]
X-ray	convex polygon	$3n-3$	$5n+19$	[ES88]
Parallel X-ray	convex polygon	3	3	[ES88]
Parallel X-ray	nondegenerate polygon	$\lfloor \log n \rfloor - 2$	$2n+2$	[MS96]
Halfplane	convex polygon	$2n$	$7n+7$	[Ski91]
Proximity	convex polygon	$2n$	$3.5n+5$	[ABG15]
Wedge ($\omega \leq \pi/2$)	convex polygon	$2n+2$	$2n+5$	[BCSS15]
Fourier	nondegenerate polygon		$3n$	[WP16]
Cut-set	embedded graph	$\binom{n}{2}$	$\binom{n}{2}$	[Ski89b]
Cut-set	unembedded graph	$\Omega(n^2/\log n)$	$O(n^2/\log n)$	[Ski89b]
Distance (2 rounds)	points in $\mathbb{R}^1$	$9n/8$	$9n/7 + O(1)$	[AM15]

probes by sending one finger probe to contact each vertex and the interior of each edge. This gives three contact points on each edge, which, by convexity, suffices to verify the polygon. Table 34.2.2 summarizes results in verification.

Of course, there are other classes of problems that do not fit so easily into the confines of these tables. Verification is closely related to approximate geometric testing; see [ABM+97, Rom95]. An interesting application of probing to nonconvex polygons is presented in [HP99]. See [Ric97, Ski92] for discussions of probing with uncertainty and tactile sensing in robotics.

TABLE 34.2.2   Upper and lower bounds for verification for various probing models.

PROBE	OBJECT	LOWER	UPPER	SOURCE
Finger	convex polygon	$2n$	$2n$	[CY87]
Finger (n known)	convex polygon	$3\lceil n/2 \rceil$	$3\lceil n/2 \rceil$	[Ski88]
Line	convex polygon	$2n$	$2n$	[DEY90]
X-ray	convex polygon	$3n/2$	$3n/2 + 6$	[ES88]
Halfplane	convex polygon	$2n/3$	$n+1$	[Ski91]

OPEN PROBLEMS

1. Tighten the gap between the lower and upper bounds for determination for finger probes in higher dimensions [DEY90].

2. Tighten the bounds for determination of convex n-gons with X-ray probes. Does a finite number (i.e., $f(n)$) of parallel X-ray probes suffice to verify or determine simple n-gons? Since each parallel X-ray probe provides a representation of the complete polygon, there is hope to detect arbitrarily small cracks in a finite number of probes; see [MS96].

3. Consider generalizations of halfplane probes to higher dimensions. How many probes are necessary to determine convex (or nonconvex) polyhedra?

4. Silhouette probes return the shadow cast by a polytope in a specified direction. These dualize to ***cross-section probes*** that return a slice of the polytope. Tighten the current bounds [DEY90] on determination with silhouettes in $\mathbb{R}^3$.

34.3 GEOMETRIC EXPLORATION AND MAPPING

In ***geometric mapping*** we face the problem of reconstructing a surrounding geometric structure using local perception. We take the perspective of an agent ***exploring*** an initially unknown environment, while trying to piece together information gathered through its ***sensors*** in order to (partially) infer the global structure, i.e., a ***map***. Settings vary in the types of environments that are considered, the movement and sensor capabilities of the agent, its initial knowledge of the environment, and the type of map that needs to be inferred. Similarly to interactive reconstruction, the movements of the agent may depend on its past observations, and we can view the setting as geometric probing with restricted transitions between consecutive probes.

Importantly, we generally require the map to be ***uniquely*** reconstructed, and thus the first question when studying a specific setting is whether mapping is feasible, i.e., whether the movement and sensor capabilities suffice to uniquely infer the map at some point (irrespective of running time). If this is the case, we are interested in mapping strategies that minimize the required movement of the agent (irrespective of computing time).

GLOSSARY

Exploration: The problem of navigating and covering an initially unknown environment using local sensing.

Mapping: The exploration problem with the additional objective of (uniquely) reconstructing a representation (map) of the environment.

Graph exploration: The problem of visiting all vertices of an initially unknown graph with an agent moving between vertices along edges of the graph. The edges of the graph are labeled with locally unique labels, and, in each step, the agent chooses a label of an outgoing edge and moves to its other end.

Anonymous graph: A graph with vertices that cannot be distinguished (unless their degrees differ). In contrast, a ***labeled*** graph has unique node identifiers.

Combinatorial visibility vector (cvv): A vector $c \in \{0,1\}^{d-1}$ for a vertex v of degree d of a polygon, with $c_i = 1$ exactly if the i-th and $(i+1)$-st vertex visible from v (in ccw order) are neighbors along the boundary of the polygon.

Cvv sensor: Provides the combinatorial visibility vector at the current vertex.

Look-back sensor: Provides the label of the edge leading back to the previous location of the agent.

Pebble: A device that can be dropped at a vertex of an anonymous graph to make the vertex distinguishable, and can be picked up and reused later.

Angle sensor: Provides the angle measurement (see Section 34.1) at the current vertex.

Angle type sensor: Provides a bit $t \in \{0,1\}$ for each pair of vertices u, w visible from the current vertex v, with $t = 1$ exactly if the angle between the segments vu and vw is larger than π.

Direction sensor: Provides the angle between some globally fixed line and the line segments connecting the current vertex to each visible vertex (in ccw order).

Distance sensor: Provides the lengths of the line segments connecting the current vertex to each visible vertex (in ccw order). A ***continuous*** distance sensor provides the distance to the boundary of the environment in each direction, i.e., it provides the visibility polygon (see Section 34.1) of the current location.

Contact sensor: Provides a bit $c \in \{0,1\}$, with $c = 1$ exactly if the agent's location corresponds to a point on the boundary of the environment.

Cut: The maximal extension vx of a boundary edge uv of a polygon P, such that v is a reflex vertex of P, and vx is collinear to uv and lies inside P.

Cut diagram: A graph associated with a polygon, with a node for each point where (two or more) cuts and/or boundary edges of the polygon intersect (in particular for each vertex of the polygon), and an edge between two points that are neighbors along a cut or a boundary edge.

FIGURE 34.3.1

From left to right: angle, angle type, distance, and direction sensor.

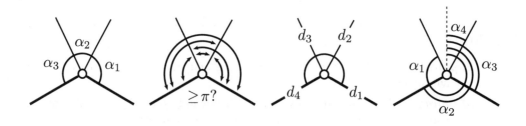

MAIN RESULTS

Research on geometric exploration and mapping has mainly considered polygonal environments, either with a focus on feasibility (weak sensors) or efficiency (strong sensors). With regards to feasibility, a key question is how minimalistic an agent model may be to still allow inferring a meaningful map of the environment. Suri et al. [SVW08] introduced such a model, where an agent moves from vertex to vertex along lines-of-sight in a simple polygon, and only observes the incident lines-of-sight in counter-clockwise (ccw) order when at a vertex. Obviously, such a minimalistic agent cannot hope to reconstruct the full geometry of the environment. Instead, the goal in this model is to infer the visibility graph that has an edge for each line-of-sight (see Section 34.1). Note that the visibility graph is a reasonable topological map, because, for example, it contains all shortest vertex-to-vertex paths in the polygon (see Chapter 31). Suri et al. [SVW08] showed that, if the agent is additionally equipped with a pebble, it can always reconstruct the visibility graph. On the other hand, Brunner et al. [BMS$^+$08] showed that without pebbles the problem is infeasible, and not even the total number n of vertices can be inferred. It remains open, whether knowledge of n alone already allows mapping. Results for various extensions of the basic model are in Table 34.3.1.

TABLE 34.3.1   Summary of results on visibility graph mapping.

SENSOR	INFO	FEASIBLE	RUNTIME	SOURCE
cvv, look-back	–	no		[BMS$^+$08]
pebble	–	yes	poly	[SVW08]
angle	–	yes	poly	[DMW11]
look-back	n	yes	poly	[CD$^+$13a]
angle type	n	yes	exp	[CDD$^+$15]
directions	n	yes	exp	[DGM$^+$14]
distance	n	open	exp	
none	n	open	exp	

TABLE 34.3.2   Summary of results on mapping rooms with obstacles.

ROOM	OBSTACLES	COMP. RATIO	SOURCE
orthogonal polygon	none	≤ 2	[DKP98]
orthogonal polygon	none	$\geq 5/4$	[Kle94]
polygon	none	≤ 26.5	[HIKK01]
orthogonal polygon	orthogonal	$O(n)$	[DKP98]
rectangle	rectangular	$\Omega(\sqrt{n})$	[AKS02]

Another simplistic model was studied by Katsev et al. [KYT$^+$11]. In their model, the agent can only move along the boundary and across cuts of the polygon, and the objective is to reconstruct the cut diagram of the environment. They show that this is possible if the agent can distinguish convex from reflex vertices and distinguish the two cut edges at a reflex vertex in ccw order.

A much more powerful model was studied by Deng et al. [DKP98]. Here the agent has a global sense of direction, can move freely in the interior of a polygonal environment (the "room") with polygonal obstacles, and has a continuous distance sensor that provides the exact geometry of the visible portion of the environment from the current location. Results in this model concern the competitive ratio between the length of the exploration path and an offline optimum path (of minimum length) ensuring that all interior points of the environment are visible at some point (see Table 34.3.2). Note the difference to the search problem where an object needs to be located in the environment and the offline optimum only needs to establish visibility to the corresponding location.

The general problem of mapping unknown discrete environments can be formulated in terms of graph exploration (see Table 34.3.3). In this abstract setting, the agent moves between vertices of an initially unknown, directed (strongly connected) graph, with the goal of inferring the graph up to isomorphism. For this purpose, we assume the outgoing edges at a vertex to have locally unique labels that the agent sees and uses to specify its moves. Note that, in this model, there is no immediate way to distinguish vertices with the same degrees, and, in particular, a single agent cannot hope to distinguish two 3-regular graphs, even if it knows the number of vertices. Bender and Slonim [BS94] showed that mapping is feasible for two agents in polynomial time, and Bender et al. [BFR$^+$02] showed that $\Theta(\log \log n)$ pebbles are necessary and sufficient for a single agent to achieve polynomial time, i.e., "a friend is only worth $\Theta(\log \log n)$ pebbles." The main result of Bender et al. [BFR$^+$02] is that a single pebble suffices if (a bound on) n is known.

TABLE 34.3.3 Summary of results on graph exploration and mapping.

GRAPH	#AGENTS	EXTRAS	RESULT	SOURCE
anonymous digraph	1	n known	infeasible	
anonymous digraph	2	randomized	$O(n^5\Delta^2)$ algorithm	[BS94]
anonymous digraph	1	1 pebble, n known	$O(n^8\Delta^2)$ algorithm	[BFR$^+$02]
anonymous digraph	1	$O(\log \log n)$ pebbles	poly time algorithm	[BFR$^+$02]
anonymous digraph	1	$o(\log \log n)$ pebbles	exp time needed	[BFR$^+$02]
labeled graph	const		comp. ratio: $O(1)$	DFS
labeled tree	$k < \sqrt{n}$		CR: $\Omega(\log k / \log \log k)$	[DLS07]
labeled graph	$k = \sqrt{n}$	randomized	CR: $\Omega(\sqrt{\log k} / \log \log k)$	[OS12]
labeled tree	k		comp. ratio: $O(k / \log k)$	[FGKP06]
labeled graph	$n^{2+\varepsilon}$		comp. ratio: $O(1)$	[DDK$^+$15]
labeled graph	$\exp(n)$		comp. ratio: 1	BFS

In case the vertices of the graph are distinguishable and edges are undirected, a single agent can map any graph simply using depth-first search until every edge was visited. This strategy visits every edge at most twice, and thus trivially yields a competitive ratio of 2, compared with an offline optimal traversal that visits all edges. On the other hand, a team of exponentially many agents can execute a breadth-first search style strategy by splitting all agents at a vertex evenly among all unexplored neighbors in each step. Obviously, this strategy needs an optimal number of steps. In general, a team of k agents needs at least $O(D + n/k)$ steps, where D is the maximum shortest path distance from the starting location to an un-explored vertex. Dereniowski et al. [DDK$^+$15] showed that a constant competitive ratio can already be obtained with (roughly) quadratic team size $k = Dn^{1+\varepsilon}$. The asymptotically best-possible competitive ratios for smaller, super-constant team sizes remain open. The best known lower bound on the competitive ratio of deterministic algorithms of $\Omega(\log k/\log\log k)$ for the domain $k < \sqrt{n}$ (with $n/k > D$) is due to Dynia et al. [DLS07]. This bound holds already on trees. Fraigniaud et al. [FGKP06] gave an algorithm for trees that achieves a ratio of $O(k/\log k)$.

OPEN PROBLEMS

1. Can a visibility graph be mapped by an agent without additional sensors, i.e., by observing only degrees, if the number n of vertices is known? Note that knowledge of (some bound on) n is necessary [BMS$^+$08].

2. Can a visibility graph be mapped with an agent using a distance sensor?

3. Close the gaps for the mapping of rooms with/without obstacles.

4. What is the best-possible competitive ratio for mapping labeled graphs with k agents in the domain $k \in \omega(1) \cap o(n^{2+\varepsilon})$?

34.4 SOURCES AND RELATED MATERIAL

SURVEYS

[Lio13]: Survey on embedding proximity graphs (Table 34.1.1).

[Gar06]: Survey of Hammer's X-ray problem and related work in geometric tomography (Table 34.1.2).

[HK99, HK07]: Surveys on discrete tomography.

[Rom95]: Survey on geometric testing.

[Ski92]: Survey on geometric probing (Table 34.2.1).

[CD$^+$13b]: Survey on mapping polygons (Table 34.3.1).

RELATED CHAPTERS

Chapter 31: Shortest paths and networks
Chapter 32: Proximity algorithms
Chapter 33: Visibility
Chapter 35: Curve and surface reconstruction
Chapter 50: Algorithmic motion planning
Chapter 51: Robotics
Chapter 55: Graph drawing
Chapter 61: Rigidity and scene analysis

REFERENCES

[AB85]   P.F. Ash and E.D. Bolker. Recognizing Dirichlet tessellations. *Geom. Dedicata*, 19:175–206, 1985.

[ABG15]   A. Adler, F. Banahi, and K. Goldberg. Efficient proximity probing algorithms for metrology. *IEEE Trans. Autom. Sci. Eng.*, 12:84–95, 2015.

[ABM+97]   E.M. Arkin, P. Belleville, J.S.B. Mitchell, D.M. Mount, K. Romanik, S. Salzberg, and D.L. Souvaine. Testing simple polygons. *Comput. Geom.*, 8:97–114, 1997.

[ABY90]   P.D. Alevizos, J.-D. Boissonnat, and M. Yvinec. Non-convex contour reconstruction. *J. Symbolic Comput.*, 10:225–252, 1990.

[AKS02]   S. Albers, K. Kursawe, and S. Schuierer. Exploring unknown environments with obstacles. *Algorithmica*, 32:123–143, 2002.

[AM15]   M. Alam and A. Mukhopadhyay. Three paths to point placement. In *Proc. 1st Conf. Algorithms Discrete Appl. Math.*, vol. 8959 of *LNCS*, pages. 33–44, Springer, Berlin, 2015.

[BCSS15]   P. Bose, J.-L. De Carufel, A. Shaikhet, and M. Smid. Probing convex polygons with a wedge. *Comput. Geom.*, 58:34–59, 2016.

[BDS11]   T. Biedl, S. Durocher, and J. Snoeyink. Reconstructing polygons from scanner data. *Theoret. Comput. Sci.*, 414:4161–4172, 2011.

[BFR+02]   M.A. Bender, A. Fernández, D. Ron, A. Sahai, and S. Vadhan. The power of a pebble: Exploring and mapping directed graphs. *Information and Computation*, 176:1–21, 2002.

[BHL11]   T. Biedl, M. Hasan, and A. López-Ortiz. Reconstructing convex polygons and convex polyhedra from edge and face counts in orthogonal projections. *Internat. J. Comput. Geom. Appl.*, 21:215–239, 2011.

[BMS+08]   J. Brunner, M. Mihalák, S. Suri, E. Vicari, and P. Widmayer. Simple robots in polygonal environments: A hierarchy. In *Proc. 4th Workshop on Algorithmic Aspects of Wireless Sensor Networks*, vol. 5389 of *LNCS*, pages 111–124, Springer, Berlin, 2008.

[BS94]   M.A. Bender and D.K. Slonim. The power of team exploration: Two robots can learn unlabeled directed graphs. In *Proc. 35th IEEE Sympos. Found. Comp. Sci.*, pp. 75–85, 1994.

[CD+13a]   J. Chalopin, S. Das, Y. Disser, M. Mihalák, and P. Widmayer. Mapping simple polygons: How robots benefit from looking back. *Algorithmica*, 65:43–59, 2013.

[CD+13b]   J. Chalopin, S. Das, Y. Disser, M. Mihalák, and P. Widmayer. Simple agents learn to find their way: An introduction on mapping polygons. *Discrete Appl. Math.*, 161:1287–1307, 2013.

[CDD+15]   J. Chalopin, S. Das, Y. Disser, M. Mihalák, and P. Widmayer. Mapping simple polygons: The power of telling convex from reflex. *ACM Trans. Algorithms*, 11:33–49, 2015.

[CH17]   J. Cardinal and U. Hoffmann. Recognition and complexity of point visibility graphs. *Discrete Comput. Geom.*, 57:164–178, 2017.

[CL92]   C.R. Coullard and A. Lubiw. Distance visibility graphs. *Internat. J. Comput. Geom. Appl.*, 2:349–362, 1992.

[CW12]   D.Z. Chen and H. Wang. An improved algorithm for reconstructing a simple polygon from its visibility angles. *Comput. Geom.*, 45:254–257, 2012.

[CY87]   R. Cole and C.K. Yap. Shape from probing. *J. Algorithms*, 8:19–38, 1987.

[Dav77]   P.J. Davis. Plane regions determined by complex moments. *J. Approximation Theory*, 19:148–153, 1977.

[DDK⁺15] D. Dereniowski, Y. Disser, A. Kosowski, D. Pajak, and P. Uznański. Fast collaborative graph exploration. *Information and Computation*, 243:37–49, 2015.

[DEY90]   D. Dobkin, H. Edelsbrunner, and C.K. Yap. Probing convex polytopes. In I.J. Cox and G.T. Wilfong, editors, *Autonomous Robot Vehicles*, pages 328–341, Springer, Berlin, 1990.

[DGM⁺14] Y. Disser, S.K. Ghosh, M. Mihalák, P. Widmayer. Mapping a polygon with holes using a compass. *Theoret. Comput. Sci.*, 553:106–113, 2014.

[DGN05]   A. Daurat, Y. Gerard, and M. Nivat. Some necessary clarifications about the chords' problem and the Partial Digest Problem. *Theoret. Comput. Sci.*, 347:432–436, 2005.

[Dil90]   M.B. Dillencourt. Realizability of Delaunay triangulations. *Inform. Process. Lett.*, 33:424–432, 1990.

[DKP98]   X. Deng, T. Kameda, and C. Papadimitriou. How to learn an unknown environment I: The rectilinear case. *J. ACM*, 45:215–245, 1998.

[DLS07]   M. Dynia, J. Łopuszański, and C. Schindelhauer. Why robots need maps. In *Proc. 14th Coll. Struct. Inform. Comm. Complexity (SIROCCO)*, vol. 4474 of *LNCS*, pages 41–50, Springer, Berlin, 2007.

[DMW11]   Y. Disser, M. Mihalák, and P. Widmayer. A polygon is determined by its angles. *Comput. Geom.*, 44:418–426, 2011.

[ES88]   H. Edelsbrunner and S.S. Skiena. Probing convex polygons with X-rays. *SIAM J. Comput.*, 17:870–882, 1988.

[Eve90]   H. Everett. *Visibility Graph Recognition*. PhD Thesis, University of Toronto, 1990.

[EW96]   P. Eades and S. Whitesides. The realization problem for Euclidean minimum spanning trees is NP-hard. *Algorithmica*, 16:60–82, 1996.

[FGKP06] P. Fraigniaud, L. Gąsieniec, D.R. Kowalski, and A. Pelc. Collective tree exploration. *Networks*, 48:166–177, 2006.

[FW90]   F. Formann and G.J. Woeginger. On the reconstruction of simple polygons. *Bull. Eur. Assoc. Theor. Comput. Sci. ETACS*, 40:225–230, 1990.

[Gar83]   R.J. Gardner. Symmetrals and X-rays of planar convex bodies. *Arch. Math. (Basel)*, 41:183–189, 1983.

[Gar92]   R.J. Gardner. X-rays of polygons. *Discrete Comput. Geom.*, 7:281–293, 1992.

[Gar06]   R.J. Gardner. *Geometric Tomography*, second edition. Cambridge Univ. Press, 2006.

[GG97]   R.J. Gardner and P. Gritzmann. Discrete tomography: Determination of finite sets by X-rays. *Trans. Amer. Math. Soc.*, 349:2271–2295, 1997.

[GG13]   S.K. Ghosh and P.P. Goswami. Unsolved problems in visibility graphs of points, segments, and polygons. *ACM Comput. Surv.*, 46:1–29, 2013.

[Gie62]   O. Giering. Bestimmung von Eibereichen und Eikörpern durch Steiner-Symmetrisierungen. In *Sitzungsberichte der Bayerischen Akademie der Wissenschaften, Math.-Nat. Kl.*, pages 225–253, 1962.

[GK07]   R.J. Gardner and M. Kinderlen. A solution to Hammer's X-ray reconstruction problem. *Adv. Math.*, 214:323–343, 2007.

[GM80]   R.J. Gardner and P. McMullen. On Hammer's X-ray problem. *J. London Math Soc.*, 21:171–175, 1980.

[GR15]   S.K. Ghosh and B. Roy. Some results on point visibility graphs. *Theoret. Comput. Sci.*, 575:17–32, 2015.

[Ham63]   P.C. Hammer. Problem 2. *Proc. Symposia in Pure Mathematics*, 7:498–499, AMS, Providence, 1963.

[HIKK01] F. Hoffmann, C. Icking, R. Klein, and K. Kriegel. The polygon exploration problem. *SIAM J. Comput.*, 31:577–600, 2001.

[HK99]   G.T. Herman and A. Kuba. *Discrete Tomography: Foundations, Algorithms, and Applications*. Springer, Berlin, 1999.

[HK07]   G.T. Herman and A. Kuba, editors. *Advances in Discrete Tomography and its Applications*. Birkhäuser, Basel, 2007.

[HP99]   K. Hunter and T. Pavlidis. Non-interactive geometric probing: Reconstructing non-convex polygons. *Comput. Geom.* 14:221–240, 1999.

[JS92]   E. Joseph and S.S. Skiena. Model-based probing strategies for convex polygons. *Comput. Geom.*, 2:209–221, 1992.

[JW02]   L. Jackson and S.K. Wismath. Orthogonal polygon reconstruction from stabbing information. *Comput. Geom.*, 23:69–83, 2002.

[Kle94]   J.M. Kleinberg. On-line search in a simple polygon. In *Proc. 5th ACM-SIAM Sympos. Discrete Algorithms*, pages 8–15, 1994.

[KYT+11] M. Katsev, A. Yershova, B. Tovar, R. Ghrist, and S.M. LaValle. Mapping and pursuit-evasion strategies for a simple wall-following robot. *IEEE Trans. Robotics*, 27:113–128, 2011.

[LB88]   M. Lindenbaum and A. Bruckstein. Reconstructing convex sets from support hyperplane measurements. Tech. Report 673, Dept. Electrical Engineering, Technion, 1988.

[LB92]   M. Lindenbaum and A. Bruckstein. Parallel strategies for geometric probing. *J. Algorithms*, 13:320–349, 1992.

[Li88]   R. Li Shuo-Yen. Reconstruction of polygons from projections. *Inform. Process. Lett.*, 28:235–240, 1988.

[Lio13]   G. Liotta. Proximity drawings. In R. Tamassia, editor, *Handbook of Graph Drawing*, pages. 115–154, CRC Press, Boca Raton, 2013.

[Lit85]   J.J. Little. Extended Gaussian images, mixed volumes, shape reconstruction. In *Proc. 1st Sympos. Comput. Geom.*, pages 15–23, ACM Press, 1985.

[LS93]   A. Lubiw and N. Sleumer. Maximal outerplanar graphs are relative neighbourhood graphs. In *Proc. 5th Canadian Conf. Comput. Geom.*, pages 198–203, 1993.

[LSS03]   P. Lemke, S.S. Skiena, and W.D. Smith. Reconstructing sets from interpoint distances. In *Discrete and Computational Geometry—The Goodman-Pollack Festschrift (B. Aronov et al., editors)*, pages 597–691, Springer, Heidelberg, 2003.

[MS80]   D.W. Matula and R.R. Sokal. Properties of Gabriel graphs relevant to geographic variation research and the clustering of points in the plane. *Geographical Analysis*, 12:205–222, 1980.

[MS91]   C.L. Monma, S. Suri. Transitions in geometric minimum spanning trees. *Discrete Comput. Geom.*, 8:265–293, 1992.

[MS96]   H. Meijer and S.S. Skiena. Reconstructing polygons from X-rays. *Geom. Dedicata*, 61:191–204, 1996.

[MVK+95]   P. Milanfar, G.C. Verghese, W.C. Karl, and A.S. Willsky. Reconstructing polygons from moments with connections to array processing. *IEEE Trans. Signal Processing*, 43:432–443, 1995.

[O'R88]   J. O'Rourke. Uniqueness of orthogonal connect-the-dots. *Mach. Intell. Pattern Recogn.*, 6:97–104, 1988.

[OS12]   C. Ortolf and C. Schindelhauer. Online multi-robot exploration of grid graphs with rectangular obstacles. In *Proc, 24th ACM Sympos. Parall. Algorithms Architectures*, pages 27–36, 2012.

[Rap89]   D. Rappaport. Computing simple circuits from a set of line segments is NP-complete. *SIAM J. Comput.*, 18:1128–1139, 1989.

[Ric97]   T. Richardson. Approximation of planar convex sets from hyperplane probes. *Discrete Comput. Geom.*, 18:151–177, 1997.

[Rom95]   K. Romanik. Geometric probing and testing—A survey. DIMACS Tech. Report 95-42, Rutgers University, 1995.

[RW93]   F. Rendel and G. Woeginger. Reconstructing sets of orthogonal line segments in the plane. *Discrete Math.*, 119:167–174, 1993.

[Sax79]   J.B. Saxe. Embeddability of weighted graphs in *k*-space is strongly NP-hard. In *Proc. 17th Allerton Conference on Communication, Control, and Computing*, pages 480–489, 1979.

[SBG06]   A. Sidlesky, G. Barequet, and C. Gotsman. Polygon reconstruction from line cross-sections. In *Proc. 18th Canadian Conf. Comput. Geom.*, pp. 81–84, 2006.

[Ski88]   S.S. Skiena. *Geometric Probing*. PhD thesis, University of Illinois, Department of Computer Science, 1988.

[Ski89a]   S.S. Skiena. Problems in geometric probing. *Algorithmica*, 4:599–605, 1989.

[Ski89b]   S.S. Skiena. Reconstructing graphs from cut-set sizes. *Inform. Process. Lett.*, 32:123–127, 1989.

[Ski91]   S.S. Skiena. Probing convex polygons with half-planes. *J. Algorithms*, 12:359–374, 1991.

[Ski92]   S.S. Skiena. Interactive reconstruction via geometric probing. *Proc. IEEE*, 80:1364–1383, 1992.

[Sno99]   J. Snoeyink. Cross-ratios and angles determine a polygon. *Discrete Comput. Geom.*, 22:619–631, 1999.

[Sug94]   K. Sugihara. Simpler proof of a realizability theorem on Delaunay triangulations. *Inform. Process. Lett.*, 50:173–176, 1994.

[SVW08]   S. Suri, E. Vicari, and P. Widmayer. Simple robots with minimal sensing: from local visibility to global geometry. *I. J. Robotics Res.*, 27:1055–1067, 2008.

[Vol86]   A. Volčič. A three-point solution to Hammer's X-ray problem. *J. London Math. Soc.*, 34:349–359, 1986.

[VZ89]   A. Volčič and T. Zamfirescu. Ghosts are scarce. *J. London Math. Soc.*, 40:171–178, 1989.

[WP16]   M. Wischerhoff and G. Plonka. Reconstruction of polygonal shapes from sparse Fourier samples. *J. Comput. Appl. Math.*, 297:117–131, 2016.

35 CURVE AND SURFACE RECONSTRUCTION
Tamal K. Dey

INTRODUCTION

The problem of reconstructing a shape from its sample appears in many scientific and engineering applications. Because of the variety in shapes and applications, many algorithms have been proposed over the last three decades, some of which exploit application-specific information and some of which are more general. We focus on techniques that apply to the general setting and have geometric and topological guarantees on the quality of reconstruction.

GLOSSARY

Simplex: A k-simplex in $\mathbb{R}^m$, $0 \leq k \leq m$, is the convex hull of $k + 1$ affinely independent points in $\mathbb{R}^m$ where $0 \leq k \leq m$. The 0-, 1-, 2-, and 3-simplices are also called *vertices*, *edges*, *triangles*, and *tetrahedra* respectively.

Simplicial complex: A simplicial complex $\mathcal{K}$ is a collection of simplices with the conditions that, (i) all sub-simplices spanned by the vertices of a simplex in $\mathcal{K}$ are also in $\mathcal{K}$, and (ii) if $\sigma_1, \sigma_2 \in \mathcal{K}$ intersect, then $\sigma_1 \cap \sigma_2$ is a sub-simplex of both. The underlying space $|\mathcal{K}|$ of $\mathcal{K}$ is the set of all points in its simplices. (Cf. Chapter 15.)

Distance: Given two subsets $X, Y \subseteq \mathbb{R}^m$, the Euclidean distance between them is given by $d(X, Y) = \inf_{x \in X, y \in Y} \|x - y\|_2$. Additionally, $d(x, y)$ denotes the Euclidean distance between two points x and y in $\mathbb{R}^m$.

k-manifold: A k-manifold is a topological space where each point has a neighborhood homeomorphic to $\mathbb{R}^k$ or the halfspace $\mathbb{H}^k$. The points with $\mathbb{H}^k$ neighborhood constitute the boundary of the manifold.

Voronoi diagram: Given a point set $P \in \mathbb{R}^m$, a Voronoi cell V_p for each point $p \in P$ is defined as

$$V_p = \{x \in \mathbb{R}^m \mid d(x, p) \leq d(x, q), \forall q \in P\}.$$

The Voronoi diagram Vor P of P is the collection of all such Voronoi cells and their faces.

Delaunay triangulation: The Delaunay triangulation of a point set $P \in \mathbb{R}^m$ is a simplicial complex Del P such that a simplex with vertices $\{p_0, .., p_k\}$ is in Del P if and only if $\bigcap_{i=0,k} V_{p_i} \neq \emptyset$. (Cf. Chapter 27.)

Shape: A shape Σ is a subset of an Euclidean space.

Sample: A sample P of a shape Σ is a finite set of points from Σ.

Medial axis: The medial axis of a shape $\Sigma \in \mathbb{R}^m$ is the closure of the set of points in $\mathbb{R}^m$ that have more than one closest point in Σ. See Figure 35.1.1(a) for an illustration.

Local feature size: The local feature size for a shape $\Sigma \subseteq \mathbb{R}^m$ is a continuous function $f : \Sigma \to \mathbb{R}$ where $f(x)$ is the distance of $x \in \Sigma$ to the medial axis of Σ. See Figure 35.1.1(a).

ϵ-sample: A sample P of a shape Σ is an ϵ-sample if for each $x \in \Sigma$ there is a sample point $p \in P$ so that $d(p, x) \leq \epsilon f(x)$.

ϵ-local sample: A sample P of a shape Σ is ϵ-local if it is an ϵ-sample and for every pair of points $p, q \in P \times P$, $d(p, q) \geq \frac{\epsilon}{c}$ for some fixed constant $c \geq 1$.

ϵ-uniform sample: A sample P of a shape Σ is ϵ-uniform if for each $x \in \Sigma$ there is a sample point $p \in P$ so that $d(p, x) \leq \epsilon f_{min}$ where $f_{min} = \min\{f(x), x \in \Sigma\}$ and $\epsilon > 0$ is a constant.

35.1 CURVE RECONSTRUCTION

In its simplest form the reconstruction problem appeared in applications such as pattern recognition (Chapter 54), computer vision, and cluster analysis, where a curve in two dimensions is to be approximated from a set of sample points. In the 1980s several geometric graph constructions over a set of points in plane were discovered which reveal a pattern among the points. The influence graph of Toussaint [AH85], the β-skeleton of Kirkpatrick and Radke [KR85], the α-shapes of Edelsbrunner, Kirkpatrick, and Seidel [EKS83] are such graph constructions. Since then several algorithms have been proposed that reconstruct a curve from its sample with guarantees under some sampling assumption.

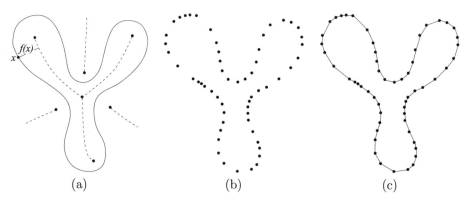

FIGURE 35.1.1
A smooth curve (solid), its medial axis (dashed) (a), sample (b), reconstruction (c).

GLOSSARY

Curve: A curve C in plane is the image of a function $p : [0, 1] \to \mathbb{R}^2$ where $p(t) = (x(t), y(t))$ for $t \in [0, 1]$ and $p[t] \neq p[t']$ for any $t \neq t'$ except possibly $t, t' \in \{0, 1\}$. It is *smooth* if p is differentiable and the derivative $\frac{d}{dt}p(t) = (\frac{dx(t)}{dt}, \frac{dy(t)}{dt})$ does not vanish.

Boundary: A curve C is said to have no boundary if $p[0] = p[1]$; otherwise, it is a curve with boundary.

Reconstruction: The reconstruction of C from its sample P is a geometric graph $G = (P, E)$ where an edge pq belongs to E if and only if p and q are adjacent sample points on C. See Figure 35.1.1.

Semiregular curve: One for which the left tangent and right tangent exist at each point of the curve, though they may be different.

UNIFORM SAMPLE

α-***shapes***: Edelsbrunner, Kirkpatrick, and Seidel [EKS83] introduced the concept of α-shape of a finite point set $P \subset \mathbb{R}^2$. It is the underlying space of a simplicial complex called the α-***complex***. The α-complex of P is defined by all simplices with vertices in P that have an empty circumscribing disk of radius α. Bernardini and Bajaj [BB97] show that the α-shapes reconstruct curves from ϵ-uniform samples if ϵ is sufficiently small and α is chosen appropriately.

r-***regular shapes***: Attali considered r-regular shapes that are constructed using certain morphological operations with r as a parameter [Att98]. It turns out that these shapes are characterized by requiring that any circle passing through the points on the boundary has radius greater than r. A sample P from the boundary curve $C \subset \mathbb{R}^2$ of such a shape is called γ-dense if each point $x \in C$ has a sample point within γr distance. Let η_{pq} be the sum of the angles opposite to pq in the two incident Delaunay triangles at a Delaunay edge $pq \in \mathrm{Del}\, P$. The main result in [Att98] is that if $\gamma < \sin \frac{\pi}{8}$, Delaunay edges with $\eta_{pq} < \pi$ reconstruct C.

EMST: Figueiredo and Gomes [FG95] show that the Euclidean minimum spanning tree (EMST) reconstructs curves with boundaries when the sample is sufficiently dense. The sampling density condition that is used to prove this result is equivalent to that of ϵ-uniform sampling for an appropriate $\epsilon > 0$. Of course, EMST cannot reconstruct curves without boundaries and/or multiple components.

NONUNIFORM SAMPLE

Crust: Amenta, Bern, and Eppstein [ABE98] proposed the first algorithm called Crust to reconstruct a curve with guarantee from a sample that is not necessarily uniform. The algorithm operates in two phases. The first phase computes the Voronoi diagram of the sample points in P. Let V be the set of Voronoi vertices in this diagram. The second phase computes the Delaunay triangulation of the larger set $P \cup V$. The Delaunay edges that connect only sample points in this triangulation constitute the crust; see Figure 35.1.2.

   The theoretical guarantee of the Crust algorithm is based on the notion of dense sampling that respects features of the sampled curve. The important concepts of local feature size and ϵ-sample were introduced by Amenta, Bern, and Eppstein [ABE98]. They prove:

THEOREM 35.1.1

For $\epsilon < \frac{1}{5}$, given an ϵ-sample P of a smooth curve $C \subset \mathbb{R}^2$ without boundary, the Crust reconstructs C from P.

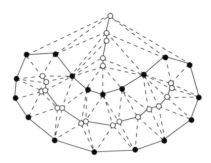

FIGURE 35.1.2
Crust edges (solid) among the Delaunay triangulation of a sample and their Voronoi vertices.

The two Voronoi diagram computations of the Crust are reduced to one by Gold and Snoeyink [GS01].

Nearest neighbor: After the introduction of the Crust, Dey and Kumar [DK99] proposed a curve reconstruction algorithm called NN-Crust based on nearest neighbors. They showed that all nearest neighbor edges that connect a point to its Euclidean nearest neighbor must be in the reconstruction if the input is $\frac{1}{3}$-sample. However, not all edges of the reconstruction are necessarily nearest neighbor edges. The remaining edges are characterized as follows. Let p be a sample point with only one nearest neighbor edge pq incident to it. Consider the halfplane with pq being an outward normal to its bounding line through p, and let r be the nearest to p among all sample points lying in this halfplane. Call pr the half-neighbor edge of p. Dey and Kumar show that all half-neighbor edges must also be in the reconstruction for a $\frac{1}{3}$-sample.

The algorithm first computes all nearest neighbor edges and then computes the half-neighbor edges to complete the reconstruction. Since all edges in the reconstruction must be a subset of Delaunay edges if the sample is sufficiently dense, all nearest neighbor and half-neighbor edges can be computed from the Delaunay triangulation. Thus, as Crust, this algorithm runs in $O(n \log n)$ time for a sample of n points.

THEOREM 35.1.2

For $\epsilon \leq \frac{1}{3}$, given an ϵ-sample P of a smooth curve $C \subset \mathbb{R}^2$ without boundary, NN-Crust reconstructs C from P.

NONSMOOTHNESS, BOUNDARIES

The crust and nearest neighbor algorithms assume that the sampled curve is smooth and has no boundary. Nonsmoothness and boundaries make reconstruction harder.

Traveling Salesman Path: Giesen [Gie00] considered a fairly large class of nonsmooth curves and showed that the Traveling Salesman Path (or Tour) reconstructs them from sufficiently dense samples. A semiregular curve C is ***benign*** if the angle between the two tangents at each point is less than π. Giesen proved the following:

THEOREM 35.1.3

For a benign curve $C \subset \mathbb{R}^2$, there exists an $\epsilon > 0$ so that if P is an ϵ-uniform sample of C, then C is reconstructed by the Traveling Salesman Path (or Tour) in case C has boundary (or no boundary).

The uniform sampling condition for the Traveling Salesman approach was later removed by Althaus and Mehlhorn [AM02], who also gave a polynomial-time algorithm to compute the Traveling Salesman Path (or Tour) in this special case of curve reconstruction. Obviously, the Traveling Salesman approach cannot handle curves with multiple components. Also, the sample points representing the boundary need to be known a priori to choose between path or tour.

Conservative Crust: In order to allow boundaries in curve reconstruction, it is essential that the sample points representing boundaries are detected. Dey, Mehlhorn, and Ramos presented such an algorithm, called the *conservative crust* [DMR00].

Any algorithm for handling curves with boundaries faces a dilemma when an input point set samples a curve without boundary densely and simultaneously samples another curve with boundary densely. This dilemma is resolved in conservative crust by a justification on the output. For any input point set P, the graph output by the algorithm is guaranteed to be the reconstruction of a smooth curve $C \subset \mathbb{R}^2$ possibly with boundary for which the input point set is a dense sample. The main idea of the algorithm is that an edge pq is chosen in the output only if there is a large enough ball centering the midpoint of pq which is empty of all Voronoi vertices in the Voronoi diagram of P. The rationale behind this choice is that these edges are small enough with respect to local feature size of C since the Voronoi vertices approximate its medial axis.

With a certain sampling condition tailored to handle nonsmooth curves, Funke and Ramos used conservative crust to reconstruct nonsmooth curves that may have boundaries [FR01].

SUMMARIZED RESULTS

The strengths and deficiencies of the discussed algorithms are summarized in Table 35.1.1.

TABLE 35.1.1   Curve reconstruction algorithms.

ALGORITHM	SAMPLE	SMOOTHNESS	BOUNDARY	COMPONENTS
α-shape	uniform	required	none	multiple
r-regular shape	uniform	required	none	multiple
EMST	uniform	required	exactly two	single
Crust	non-uniform	required	none	multiple
Nearest neighbor	non-uniform	required	none	multiple
Traveling Salesman	non-uniform	not required	must be known	single
Conservative crust	non-uniform	required	any number	multiple

OPEN PROBLEM

All algorithms described above assume that the sampled curve does not cross itself. It is open to devise an algorithm that can reconstruct such curves under some reasonable sampling condition.

35.2 SURFACE RECONSTRUCTION

A number of surface reconstruction algorithms have been designed in different application fields. The problem appeared in medical imaging where a set of cross sections obtained via CAT scan or MRI needs to be joined with a surface. The points on the boundary of the cross sections are already joined by a polygonal curve and the output surface needs to join these curves in consecutive cross sections. A dynamic programming based solution for two such consecutive curves was first proposed by Fuchs, Kedem, and Uselton [FKU77]. A negative result by Gitlin, O'Rourke, and Subramanian [GOS96] shows that, in general, two polygonal curves cannot be joined by a nonself-intersecting surface with only those vertices; even deciding its feasibility is NP-hard. Several solutions with the addition of Steiner points have been proposed to overcome the problem, see [MSS92, BG93]. The most general version of the surface reconstruction problem does not assume any information about the input points other than their 3D coordinates, and requires a piecewise linear approximation of the sampled surface; see Figure 35.2.1. In the context of computer graphics and vision, this problem has been investigated intensely with emphasis on the practical effectiveness of the algorithms [BMR+99, Boi84, CL96, GCA13, GKS00, HDD+92]. In computational geometry, several algorithms have been designed based on Voronoi/Delaunay diagrams that have guarantees on geometric proximity (Hausdorff closeness) and topological equivalence (homeomorphism/isotopy). We focus mainly on them.

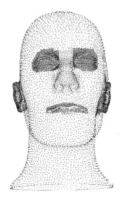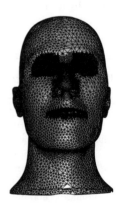

FIGURE 35.2.1
A point sample and the reconstructed surface.

GLOSSARY

Surface: A surface $S \subset \mathbb{R}^3$ is a 2-manifold embedded in $\mathbb{R}^3$. Thus each point $p \in S$ has a neighborhood homeomorphic to $\mathbb{R}^2$ or halfplane $\mathbb{H}^2$. The points with neighborhoods homeomorphic to $\mathbb{H}^2$ constitute the boundary ∂S of S.

Smooth Surface: A surface $S \subset \mathbb{R}^3$ is smooth if for each point $p \in S$ there is a neighborhood $W \subseteq \mathbb{R}^3$ and a map $\pi : U \to W \cap S$ of an open set $U \subset \mathbb{R}^2$ onto $W \cap S$ so that

(i) π is differentiable,

(ii) π is a homeomorphism,

(iii) for each $q \in U$ the differential $d\pi_q$ is one-to-one.

A surface $S \subset \mathbb{R}^3$ with boundary ∂S is smooth if its interior $S \setminus \partial S$ is a smooth surface and ∂S is a smooth curve.

Smooth Closed Surface: We call a smooth surface $S \subset \mathbb{R}^3$ closed if it is compact and has no boundary.

Restricted Voronoi: Given a subspace $\mathbb{N} \subseteq \mathbb{R}^3$ and a point set $P \subseteq \mathbb{R}^3$, the restricted Voronoi diagram of P w.r.t $\mathbb{N}$ is $\mathrm{Vor}\, P|_{\mathbb{N}} = \{F \cap \mathbb{N} \mid F \in \mathrm{Vor}\, P\}$.

Restricted Delaunay: The dual of $\mathrm{Vor}\, P|_{\mathbb{N}}$ is called the restricted Delaunay triangulation $\mathrm{Del}\, P|_{\mathbb{N}}$ defined as

$$\mathrm{Del}\, P|_{\mathbb{N}} = \{\sigma \mid \sigma = \mathrm{Conv}\,\{p_0, ..., p_k\} \in \mathrm{Del}\, P \text{ where } (\bigcap_{i=0,k} V_{p_i}) \cap \mathbb{N} \neq \emptyset\}.$$

Watertight surface: A 2-complex $\mathcal{K}$ embedded in $\mathbb{R}^3$ is called watertight if the underlying space $|\mathcal{K}|$ of $\mathcal{K}$ is the boundary of the closure of some 3-manifold in $\mathbb{R}^3$.

Hausdorff ϵ-close: A subset $X \subset \mathbb{R}^3$ is Hausdorff ϵ-close to a surface $S \subset \mathbb{R}^3$ if every point $y \in S$ has a point $x \in X$ with $d(x,y) \leq \epsilon f(y)$, and similarly every point $x \in X$ has a point $y \in S$ with $d(x,y) \leq \epsilon f(y)$.

Homeomorphism: Two topological spaces (e.g., surfaces) are homeomorphic if there is a continuous bijective map between them with continuous inverse.

Steiner points: The points used by an algorithm that are not part of the finite input point set are called Steiner points.

α-SHAPES

Generalization of α-shapes to 3D by Edelsbrunner and Mücke [EM94] can be used for surface reconstruction in case the sample is more or less uniform. An alternate definition of α-shapes in terms of the restricted Delaunay triangulation is more appropriate for surface reconstruction. Let $\mathbb{N}$ denote the space of all points covered by open balls of radius α around each sample point $p \in P$. The α-shape for P is the underlying space of the α-complex which is the restricted Delaunay triangulation $\mathrm{Del}\, P|_{\mathbb{N}}$; see Figure 35.2.2 below for an illustration in 2D. It is shown that the α-shape is always homotopy equivalent to $\mathbb{N}$. If P is a sample of a surface $S \subset \mathbb{R}^3$, the space $\mathbb{N}$ becomes homotopy equivalent to S if α is chosen appropriately and P

is sufficiently dense [EM94]. Therefore, by transitivity of homotopy equivalence, the α-shape is homotopy equivalent to S if α is appropriate and the sample P is sufficiently dense.

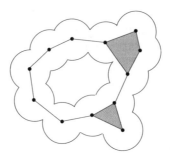

FIGURE 35.2.2
Alpha shape of a set of points in $\mathbb{R}^2$.

The major drawback of α-shapes is that they require a nearly uniform sample and an appropriate parameter α for reconstruction.

CRUST

The Crust algorithm for curve reconstruction was generalized for surface reconstruction by Amenta and Bern [AB99]. In case of curves in 2D, Voronoi vertices for a dense sample lie close to the medial axis. That is why a second Voronoi diagram with the input sample points together with the Voronoi vertices is used to separate the Delaunay edges that reconstruct the curve. Unfortunately, Voronoi vertices in 3D can lie arbitrarily close to the sampled surface. One can place four arbitrarily close points on a smooth surface which lie near the diametric plane of the sphere defined by them. This sphere can be made empty of any other input point and thus its center as a Voronoi vertex lies close to the surface. With this important observation Amenta and Bern forsake the idea of putting all Voronoi vertices in the second phase of crust and instead identify a subset of Voronoi vertices called *poles* that lie far away from the surface, and in fact close to the medial axis.

Let P be an ϵ-sample of a smooth closed surface $S \subset \mathbb{R}^3$. Let V_p be a Voronoi cell in the Voronoi diagram Vor P. The farthest Voronoi vertex of V_p from p is called the positive pole of p. Call the vector from p to the positive pole the *pole vector* for p; this vector approximates the surface normal $\mathbf{n}_p$ at p. The Voronoi vertex of V_p that lies farthest from p in the opposite direction of the pole vector is called its negative pole. The opposite direction is specified by the condition that the vector from p to the negative pole must make an angle more than $\frac{\pi}{2}$ with the pole vector. Figure 35.2.3(a) illustrates these definitions. If V_p is unbounded, the positive pole is taken at infinity and the direction of the pole vector is taken as the average of all directions of the unbounded Voronoi edges in V_p.

The Crust algorithm in 3D proceeds as follows. First, it computes Vor P and then identifies the set of poles, say L. The Delaunay triangulation of the point set $P \cup L$ is computed and the set of Delaunay triangles, T, is filtered that have all three vertices only from P. This set of triangles almost approximates S but may not form a surface. Nevertheless, the set T includes all restricted Delaunay triangles in $\text{Del}\,P|_S$. According to a result by Edelsbrunner and Shah [ES97], the underlying space $|\text{Del}\,P|_S|$ of $\text{Del}\,P|_S$ is homeomorphic to S if each Voronoi face satisfies a topological condition called the "closed ball property." Amenta and Bern show that if P is an ϵ-sample for $\epsilon \leq 0.06$, each Voronoi face in $\text{Vor}\,P$ satisfies this property. This means that, if the triangles in $\text{Del}\,P|_S$ could be extracted from T, we would have a surface homeomorphic to S. Unfortunately, it is impossible to detect the restricted Delaunay triangles of $\text{Del}\,P|_S$ since S is unknown. However, the fact that T contains them is used in a manifold extraction step that computes a manifold out of T after a normal filtering step. This piecewise linear manifold surface is output by Crust. The Crust guarantees that the output surface lies very close to S.

THEOREM 35.2.1

For $\epsilon \leq 0.06$, given an ϵ-sample P of a smooth closed surface $S \subset \mathbb{R}^3$, the Crust algorithm produces a 2-complex that is Hausdorff ϵ-close to S.

Actually, the output of Crust is also homeomorphic to the sampled surface under the stated condition of the theorem above, a fact which was proved later in the context of the Cocone algorithm discussed next.

COCONE

The Cocone algorithm was developed by Amenta, Choi, Dey, and Leekha [ACDL02]. It simplified the reconstruction by Crust and enhanced its proof of correctness.

A ***cocone*** C_p for a sample point p is defined as the complement of the double cone with p as apex and the pole vector as axis and an opening angle of $\frac{3\pi}{4}$; see Figure 35.2.3(b). Because the pole vector at p approximates the surface normal $\mathbf{n}_p$, the cocone C_p (clipped within V_p) approximates a thin neighborhood around the tangent plane at p. For each point p, the algorithm then determines all Voronoi edges in V_p that are intersected by the cocone C_p. The dual Delaunay triangles of these Voronoi edges constitute the set of candidate triangles T.

It can be shown that the circumscribing circles of all candidate triangles are small [ACDL02]. Specifically, if $pqr \in T$ has circumradius r, then

(i) $r = O(\epsilon)f(x)$ where $f(x) = \min\{f(p), f(q), f(r)\}$.

It turns out that any triangle with such small circumradius must lie almost parallel to the surface, i.e., if $\mathbf{n}_{pqr}$ is the normal to a candidate triangle pqr, then

(ii) $\angle(\mathbf{n}_{pqr}, \mathbf{n}_x) = O(\epsilon)$ up to orientation where $x \in \{p, q, r\}$.

Also, it is proved that

(iii) T includes all restricted Delaunay triangles in $\text{Del}\,P|_S$.

These three properties of the candidate triangles ensure that a *manifold extraction* step, as in the Crust algorithm, extracts a piecewise-linear surface which is homeomorphic to the original surface S.

Cocone uses a single Voronoi diagram as opposed to two in the Crust algorithm and also eliminates the normal filtering step. It provides the following guarantees.

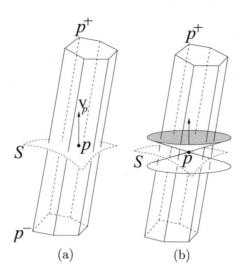

(a)   (b)

FIGURE 35.2.3

A long thin Voronoi cell V_p, the positive pole p^+, the pole vector $\mathbf{v}_p$ and the negative pole p^- (a), the cocone (b).

THEOREM 35.2.2

For $\epsilon \leq 0.06$, given a sample P of a smooth closed surface $S \subset \mathbb{R}^3$, the Cocone algorithm computes a Delaunay subcomplex $N \subseteq \operatorname{Del} P$ where $|N|$ is Hausdorff ϵ-close and is homeomorphic to S.

Actually, the homeomorphism property can be strengthened to *isotopy*, a stronger topological equivalence condition. Because of the Voronoi diagram computation, the Cocone runs in $O(n^2)$ time and space. Funke and Ramos [FR02] improved its complexity to $O(n \log n)$ though the resulting algorithm seems impractical. Cheng, Jin, and Lau [CJL17] simplified this approach making it more practical.

NATURAL NEIGHBOR

Boissonnat and Cazals [BC02] revisited the approach of Hoppe et al. [HDD$^+$92] by approximating the sampled surface as the zero set of a signed distance function. They used natural neighbors and an ϵ-sampling condition to provide output guarantees.

Given an input point set $P \subset \mathbb{R}^3$, the **natural neighbors** $N_{x,P}$ of a point $x \in \mathbb{R}^3$ are the Delaunay neighbors of x in $\operatorname{Del}(P \cup x)$. Letting $V(x)$ denote the Voronoi cell of x in $\operatorname{Vor}(P \cup x)$, this means

$$N_{x,P} = \{p \in P \mid V(x) \cap V_p \neq \emptyset\}.$$

Let $A(x,p)$ denote the volume stolen by x from V_p, i.e.,

$$A(x,p) = \operatorname{Vol}(V(x) \cap V_p).$$

The natural coordinate associated with a point p is a continuous function $\lambda_p : \mathbb{R}^3 \to \mathbb{R}$ where

$$\lambda_p(x) = \frac{A(x,p)}{\Sigma_{q \in P} A(x,q)}.$$

Some of the interesting properties of λ_p are that it is continuously differentiable except at p, and any point $x \in \mathbb{R}^3$ is a convex combination of its natural neighbors: $\Sigma_{p \in N_{x,P}} \lambda_p(x)p = x$. Boissonnat and Cazals assume that each point p is equipped with a unit normal $\mathbf{n}_p$ which can either be computed via pole vectors, or is part of the input. A distance function $h_p : \mathbb{R}^3 \to \mathbb{R}$ for each point p is defined as $h_p(x) = (p - x) \cdot \mathbf{n}_p$. A global distance function $h : \mathbb{R}^3 \to \mathbb{R}$ is defined by interpolating these local distance functions with natural coordinates. Specifically,

$$h(x) = \Sigma_{p \in P} \lambda_p^{1+\delta}(x)h_p(x).$$

The δ term in the exponent is added to make h continuously differentiable. By definition, $h(x)$ locally approximates the signed distance from the tangent plane at each point $p \in P$ and, in particular, $h(p) = 0$.

Since h is continuously differentiable, $\widehat{S} = h^{-1}(0)$ is a smooth surface unless 0 is a critical value. A discrete approximation of $\widehat{S}$ can be computed from the Delaunay triangulation of P as follows. All Voronoi edges that intersect $\widehat{S}$ are computed via the sign of h at their two endpoints. The dual Delaunay triangles of these Voronoi edges constitute a piecewise linear approximation of $\widehat{S}$. If the input sample P is an ϵ-sample for sufficiently small ϵ, then a theorem similar to that for Cocone holds.

MORSE FLOW

Morse theory is concerned with the study of critical points of real-valued functions on manifolds. Although the original theory was developed for smooth manifolds, various extensions have been made to incorporate more general settings. The theory builds upon a notion of gradient of the Morse function involved. The critical points where the gradient vanishes are mainly of three types, local minima, local maxima, and saddles. The gradient vector field usually defines a flow that can be thought of as a mechanism for moving points along the steepest ascent. Some surface reconstruction algorithms build upon this concept of **Morse flow**.

Given a point sample P of a smooth closed surface $S \subseteq \mathbb{R}^3$, consider the distance function $d : \mathbb{R}^3 \to \mathbb{R}$ where $d(x) := d(x, P)$ is the distance of x to the nearest sample point in P. This function is not differentiable everywhere. Still, one can define a flow for d. For every point $x \in \mathbb{R}^3$, let $A(x) \subseteq P$ be the set of sample points closest to x. The driver $r(x)$ for x is the closest point in the convex hull of $A(x)$. In the case where $r(x) = x$, we say x is critical and regular otherwise. The normalized gradient of d at a regular point x is defined as the unit vector in the direction $x - r(x)$. Notice that the gradient vanishes at a critical point x where $x = r(x)$. The flow induced by this vector field is a map $\phi : \mathbb{R}^3 \times \mathbb{R}^+ \to \mathbb{R}^3$ such that the right derivative of $\phi(x, t)$ at every point x with respect to time t equals the gradient vector. This flow defines flow curves (integral lines) along which points move toward the steepest ascent of d and arrive at critical points in the limit. See Figure 35.2.4 for an illustration in $\mathbb{R}^2$.

It follows from the definition that the critical points of d occur at the points where a Delaunay simplex in $\mathrm{Del}\,P$ intersects its dual Voronoi face in $\mathrm{Vor}\,P$. Given a critical point c of d, the set of all points that flow into c constitute the *stable manifold* of c. The set of all stable manifolds partitions $\mathbb{R}^3$ into cells that form a cell complex together. Giesen and John [GJ08] named it as **flow complex** and studied its properties. The dimension of each stable manifold is the index of its associated critical point. Index-0 critical points are minima which are the Delaunay vertices, or equivalently the points in P. Index-1 critical points, also called the 1-saddles, are the intersections of Delaunay edges and their dual Voronoi facets. Index-2 critical points, also called the 2-saddles, are the intersections of Delaunay triangles and their dual Voronoi edges. Index-3 critical points are maxima which are a subset of Voronoi vertices.

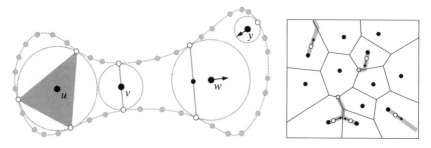

FIGURE 35.2.4

On left: Point sample from a curve, sets $A(x)$ are shown with hollow circles, driver for w is the smaller black circle, driver of y is the hollow circle, u and v are critical; On right: some flow curves for a point set in $\mathbb{R}^2$.

A surface reconstruction algorithm based on the flow complex was proposed by Dey, Giesen, Ramos, and Sadri [DGRS08]. They observe that the critical points of d separate into two groups, one near the surface S, called the *surface critical points*, and the other near the medial axis M, called the *medial axis critical points*. Let $S_\eta = \{x \in \mathbb{R}^3 \mid d(x, S) \leq \eta\}$ and $M_\eta = \{x \in \mathbb{R}^3 \mid d(x, M) = \eta\}$ denote the η-offset of S and M respectively. For any point $c \in \mathbb{R}^3$, let $\tilde{c}$ be its orthogonal projection on S, and $\check{c}$ be the point where the ray $\tilde{c}c$ intersects M first time. Let $\rho(\tilde{c}) = d(\tilde{c}, \check{c})$. An important result proved in [DGRS08] is:

THEOREM 35.2.3

For $\epsilon < \frac{1}{3}$, let P be an ϵ-sample of a smooth closed surface $S \subset \mathbb{R}^3$. Let c be any critical point of the distance function d. Then, either $c \in S_{\epsilon^2 f(\tilde{c})}$, or in $M_{2\epsilon\rho(\tilde{c})}$.

The algorithm in [DGRS08] first separates the medial axis critical points from the surface ones using an angle criterion. The union of the stable manifolds for the medial axis critical points separates further into two connected components, one for the outer medial axis critical points and the other for the inner medial axis critical points. These two connected components can be computed using a union-find data structure. The boundary of any one of these connected components is output as the reconstructed surface.

The main disadvantage of the flow complex based surface reconstruction is that the stable manifolds constituting this complex are not necessarily subcomplexes of the Delaunay complex. Consequently, its construction is more complicated. A Morse theory based reconstruction that sidesteps this difficulty is the Wrap algorithm of Edelsbrunner [Ede03]. A different distance function is used in Wrap. A Delaunay circumball $B(c, r)$ that circumscribes a Delaunay simplex can be treated as a weighted point $\hat{c} = (c, r)$. For any point $x \in \mathbb{R}^3$, one can define the weighted distance which is also called the *power distance* as $\pi(x, \hat{c}) = d(x, c)^2 - r^2$. For a point set P, let C denote the centers of the Delaunay balls for simplices in Del P and $\widehat{C}$ denote the corresponding weighted points. These Delaunay balls also include those of the infinite tetrahedra formed by the convex hull triangles and a point at infinity. Their centers are at infinity and their radii are infinite.

Define a distance function $g : \mathbb{R}^3 \to \mathbb{R}$ as $g(x) = \min_{\hat{c} \in \widehat{C}} \pi(x, \hat{c})$. Consider the Voronoi diagram of $\widehat{C}$ with the power distance metric. This diagram, also called the power diagram of $\widehat{C}$, denoted as $\mathrm{Pow}\,\widehat{C}$, coincides with the Delaunay triangulation $\mathrm{Del}\,P$ extended with the infinite tetrahedra. A point $x \in \mathbb{R}^3$ contained in a Delaunay tetrahedron $\sigma \in \mathrm{Del}\,P$ has distance $g(x) = \pi(x, \hat{c})$ where c is the Delaunay ball of σ. Analogous to d, one can define a flow for the distance function g whose critical points coincide with those of d. Edelsbrunner defined a *flow relation* among the Delaunay simplices using the flow for g. Let σ be a proper face of two simplices τ and ζ. We say $\tau < \sigma < \zeta$ if there is a point x in the interior of σ such that the flow curve through x proceeds from the interior of τ into the interior of ζ. The Wrap algorithm, starting from the infinite tetrahedra, collapses simplices according to the flow relation. It finds a simplex σ with a coface ζ where $\sigma < \zeta$ and ζ is the only coface adjacent to σ. The collapse modifies the current complex $\mathcal{K}$ to $\mathcal{K} \setminus \{\sigma, \zeta\}$, which is known to maintain a homotopy equivalence between the two complexes. The algorithm stops when it can no longer find a simplex to collapse. The output is a subcomplex of $\mathrm{Del}\,P$ and is necessarily homotopy equivalent to a 3-ball. The algorithm can be modified to create an output complex of higher genus by starting the collapse from other source tetrahedra. The boundary of this complex can be taken as the output approximating S. There is no guarantee for topological equivalence between the output complex and the surface S for the original Wrap algorithm. Ramos and Sadri [RS07] proposed a version of the Wrap algorithm that ensures this topological equivalence under a dense ϵ-sampling assumption.

WATERTIGHT SURFACES

Most of the surface reconstruction algorithms face a difficulty while dealing with undersampled surfaces and noise. While some heuristics such as in [DG03] can detect undersampling, it leaves holes in the surface near the vicinity of undersampling. Although this may be desirable for reconstructing surfaces with boundaries, many applications such as CAD designs require that the output surface be *watertight*, i.e., a surface that bounds a solid.

The natural neighbor algorithm of [BC02] can be adapted to guarantee a watertight surface. Recall that this algorithm approximates a surface $\widehat{S}$ implicitly defined by the zero set of a smooth map $h : \mathbb{R}^3 \to \mathbb{R}$. This surface is a smooth 2-manifold without boundary in $\mathbb{R}^3$. However, if the input sample P is not dense for this surface, the reconstructed output may not be watertight. Boissonnat and Cazals suggest to sample more points on $\widehat{S}$ to obtain a dense sample for $\widehat{S}$ and then reconstruct it from the new sample.

Amenta, Choi, and Kolluri [ACK01] use the crust approach to design the Power Crust algorithm to produce watertight surfaces. This algorithm first distinguishes the inner poles that lie inside the solid bounded by the sampled surface S from the outer poles that lie outside. A consistent orientation of the pole vectors is used to decide between inner and outer poles. To prevent outer poles at infinity, eight corners of a large box containing the sample are added. Let $\mathbb{B}_O$ and $\mathbb{B}_I$ denote the Delaunay balls centered at the outer and inner poles respectively. The union of Delaunay balls in $\mathbb{B}_I$ approximate the solid bounded by S. The union of Delaunay balls in $\mathbb{B}_O$ do not approximate the entire exterior of S although one of its boundary components approximates S. The implication is that the cells in the power diagram $\text{Pow}\,(\mathbb{B}_O \cup \mathbb{B}_I)$ can be partitioned into two sets, with the boundary between approximating S. The facets in the power diagram $\text{Pow}\,(\mathbb{B}_O \cup \mathbb{B}_I)$ that separate cells generated by inner and outer poles form this boundary which is output by Power Crust.

Dey and Goswami [DG03] announced a watertight surface reconstructor called Tight Cocone. This algorithm first computes the surface with Cocone, which may leave some holes in the surface due to anomalies in sampling. A subsequent sculpting [Boi84] in the Delaunay triangulation of the input points recover triangles that fill the holes. Unlike Power Crust, Tight Cocone does not add Steiner points.

BOUNDARY

The algorithms described so far work on the assumption that the sampled surface has no boundary. The reconstruction of smooth surfaces with boundary is more difficult because the algorithm has to reconstruct the boundary from the sample as well. The algorithms such as Crust and Cocone cannot be extended to surfaces with boundary because they employ a manifold extraction step which iteratively prunes triangles with edges adjacent to a single triangle. Surfaces with non-empty boundaries necessarily contain such triangles in their reconstruction, and thus cannot withstand such a pruning step. Another difficulty arises on the theoretical front because the restricted Delaunay triangulation $\text{Del}\,P|_S$ of a sample P for a surface S with boundary may not be homeomorphic to S no matter how dense P is [DLRW09]. This property is a crucial ingredient for proving the correctness of the Crust and Cocone algorithms. Specifically, the manifold extraction step draws upon this property. Dey, Li, Ramos, and Wenger [DLRW09] sidestepped this difficulty by replacing the prune-and-walk step with a peeling step.

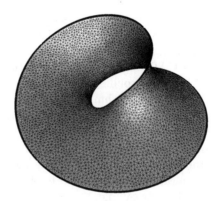

FIGURE 35.2.5
Reconstruction of a non-orientable surface with boundary (Möbius band) with Peel algorithm.

The Peel algorithm in [DLRW09] works as follows. It takes a point sample P and a positive real α as input. First, it computes the α-complex of P with the input parameter $\alpha > 0$. If P is dense, the Delaunay tetrahedra retained in this complex is proven to be "flat" meaning that they all have two non-adjacent edges, called flat edges, that subtend an internal angle close to π. A flat tetrahedron t is peelable if one of its flat edges, say e, is not adjacent to any tetrahedron other than t. A peeling of t means that the two triangles adjacent to e are removed along with e and t from the current complex. The algorithm successively finds peelable tetrahedra and peels them. It stops when there is no peelable tetrahedron and outputs the current complex. It is shown that the algorithm can always find a peelable tetrahedron as long as there is any tetrahedron at all. This means that the output is a set of triangles that are not adjacent to any tetrahedra. Dey et al. prove that the underlying space of this set of triangles is isotopic to S if P is sufficiently dense. Figure 35.2.5 shows an output surface computed by Peel.

The correctness of the Peel algorithm depends on the assumption that the input P is a globally uniform, that is, ϵ-uniform sample of S. A precise statement of the guarantee is given in the following theorem.

THEOREM 35.2.4

Let S be a smooth closed surface with boundary. For a sufficiently small $\epsilon > 0$ and $6\epsilon < \alpha \le 6\epsilon + O(\epsilon)$, if P is an ϵ-uniform sample of S, $Peel(P, \alpha)$ produces a Delaunay sub-complex isotopic and Hausdorff ϵ-close to S.

NOISE

The input points are assumed to lie on the surface for all of the surface reconstruction algorithms discussed so far. In reality, the points can be a little off from the sampled surface. The noise introduced by these perturbations is often referred to as the Hausdorff noise because of the assumption that the Hausdorff distance between the surface S and its sample P is small. In the context of surface reconstruction, Dey and Goswami [DG06] first modeled this type of noise and presented an algorithm with provable guarantees.

A point set $P \subset \mathbb{R}^3$ is called an ϵ-noisy sample of a smooth closed surface $S \subset \mathbb{R}^3$ if conditions 1 and 2 below are satisfied. It is called an (ϵ, κ)-sample if additionally condition 3 is satisfied.

1. The orthogonal projection $\tilde{P}$ of P on S is an ϵ-sample of S.

2. $d(p, \tilde{p}) \le \epsilon^2 f(\tilde{p})$.

3. For every point $p \in P$, its distance to its κ-th nearest point is at least $\epsilon f(\tilde{p})$.

The first two conditions say that the point sample is dense and sufficiently close to S. The third condition imposes some kind of relaxed version of local uniformity by considering the κ-th nearest neighbor instead of the nearest neighbor. Dey and Goswami [DG06] observe that, under the (ϵ, κ)-sampling condition, some of the Delaunay balls of tetrahedra in $\mathrm{Del}\,P$ remain almost as large as the Delaunay balls centering the poles in the noise-free case. The small Delaunay balls cluster near the sample points. See Figure 35.2.6. With this observation, they propose to separate out the 'big' Delaunay balls from the small ones by thresholding. Let $B(c, r)$ be a Delaunay ball of a tetrahedron $pqrs \in \mathrm{Del}\,P$ and let ℓ be the smallest among the distances of p, q, r, and s to their respective κ-th nearest neighbors. If $r > k\ell$ for a suitably chosen fixed constant k, the ball $B(c, r)$ is marked as big. After marking all such big Delaunay balls, the algorithm starts from any of the infinite tetrahedra and continues collecting any big Delaunay ball that has a positive power distance (intersects deeply) to any of the balls collected so far. As in the power crust algorithm, this process separates the inner big Delaunay balls from the outer ones. It is proved that, if the thresholds are chosen right and an appropriate sampling condition holds, the boundary ∂D of the union D of outer (or inner) big Delaunay balls is homeomorphic to S. To make the output a Delaunay subcomplex, the authors [DG06] suggest to collect the points $P' \subseteq P$ contained in the boundary of the big outer Delaunay balls and compute the restricted Delaunay triangulation of P' with respect to a smooth *skin surface* [Ede99] approximating ∂D. An even easier option which seems to work in practice is to take the restricted Delaunay triangulation $\mathrm{Del}\, P'|_{\partial D}$ which coincides with the boundary of the union of tetrahedra circumscribed by the big outer Delaunay balls.

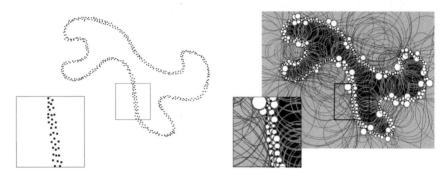

FIGURE 35.2.6
On left: Noisy point sample from a curve; on right: big and small Delaunay balls.

Mederos, Amenta, Vehlo, and Figueiredo [MAVF05] adapted the idea in [DG06] to the framework of power crust. They do not require the third uniformity condition, but instead rely on an input parameter c that needs to be chosen appropriately. Their algorithm starts by identifying the polar balls. A subset of the polar balls whose radii are larger than the chosen parameter c is selected. These "big" polar balls are further marked as inner and outer using the technique described before. Let $\mathbb{B}_O$ and $\mathbb{B}_I$ denote the set of these inner and outer polar balls respectively. Considering the polar balls as weighted points, the algorithm computes the power crust, that is, the facets in the power diagram of $\mathrm{Pow}\,(\mathbb{B}_O \cup \mathbb{B}_I)$ that separate a Voronoi cell corresponding to a ball in $\mathbb{B}_O$ from a cell corresponding to a ball in $\mathbb{B}_I$.

THEOREM 35.2.5

There exist $c > 0$, $\epsilon > 0$ so that if $P \subset \mathbb{R}^3$ is an ϵ-noisy sample of a smooth closed surface $S \subset \mathbb{R}^3$, the above algorithm with parameter c returns the power crust in $\mathrm{Pow}\,(\mathbb{B}_O \cup \mathbb{B}_I)$ which is homeomorphic to S.

SUMMARIZED RESULTS

The properties of the above discussed surface reconstruction algorithms are summarized in Table 35.2.1.

TABLE 35.2.1   Surface reconstruction algorithms.

ALGORITHM	SAMPLE	PROPERTIES	SOURCE
α-shape	uniform	α to be determined.	[EM94]
Crust	non-uniform	Theoretical guarantees from Voronoi structures, two Voronoi computations.	[AB99]
Cocone	non-uniform	Simplifies crust, single Voronoi computation with topological guarantee, detects undersampling.	[ACDL02] [DG03]
Natural Neighbor	non-uniform	Theoretical guarantees using Voronoi diagram and implicit functions.	[BC02]
Morse Flow	non-uniform	Draws upon the gradient flow of distance functions, flow complex introduces Steiner points, Wrap computes Delaunay subcomplex.	[DGRS08] [Ede03]
Power Crust	non-uniform	Watertight surface using power diagrams, introduces Steiner points.	[ACK01]
Tight Cocone	non-uniform	Watertight surface using Delaunay triangulation.	[DG03]
Peel	uniform	Theoretical guarantees for surfaces with boundaries.	[DLRW09]
Noise	uniform	Filters big Delaunay balls, assumes Hausdorff noise.	[DG06] [MAVF05]

OPEN PROBLEMS

All guarantees given by various surface reconstruction algorithms depend on the notion of dense sampling. Watertight surface algorithms can guarantee a surface without holes, but no theoretical guarantees exist under any type of undersampling. Also, non-smoothness and noise still need more investigation.

1. Design an algorithm that reconstructs nonsmooth surfaces under reasonable sampling conditions.

2. Design a surface reconstruction algorithm that handles noise more general than Hausdorff.

35.3 SHAPE RECONSTRUCTION

All algorithms discussed above are designed for reconstructing a shape of specific dimension from the samples. Thus, the curve reconstruction algorithms cannot handle samples from surfaces and the surface reconstruction algorithms cannot handle samples from curves. Therefore, if a sample is derived from shapes of mixed dimensions, i.e., both curves and surfaces in $\mathbb{R}^3$, none of the curve and surface reconstruction algorithms are adequate. General shape reconstruction algorithms should be able to handle any shape embedded in Euclidean spaces. However, this goal may be too ambitious, as it is not clear what would be a reasonable definition of dense samples for general shapes that are nonsmooth or nonmanifold. The

ϵ-sampling condition would require infinite sampling in these cases. We therefore distinguish two cases: (i) smooth manifold reconstruction for which a computable sampling criterion can be defined, (ii) shape reconstruction for which it is currently unclear how a computable sampling condition could be defined to guarantee reconstruction. This leads to a different definition for the general shape reconstruction problem in the glossary below.

GLOSSARY

Shape reconstruction: Given a set of points $P \subseteq \mathbb{R}^m$, compute a shape that best approximates P.

Manifold reconstruction: Compute a piecewise-linear approximation to a manifold M, given a sample P of M.

MAIN RESULTS

Shape reconstruction: Not many algorithms are known to reconstruct shapes. The definition of α-shapes is general enough to be applicable to shape reconstruction. In Figure 35.2.2, the α-shape reconstructs a shape in $\mathbb{R}^2$ which is not a manifold. Similarly, it can reconstruct curves, surfaces and solids and their combinations in three dimensions. Melkemi [Mel97] proposed $\mathcal{A}$-shapes that can reconstruct shapes in $\mathbb{R}^2$. Its class of shapes includes α-shapes. Given a set of points P in $\mathbb{R}^2$, a member in this class of shapes is identified with another finite set $\mathcal{A} \subseteq \mathbb{R}^2$. The $\mathcal{A}$-shape of S is generated by edges that connect points $p, q \in P$ if there is a circle passing through p, q and a point in $\mathcal{A}$, and all other points in $P \cup \mathcal{A}$ lie outside the circle. The α-shape is a special case of $\mathcal{A}$-shapes where $\mathcal{A}$ is the set of all points on Voronoi edges that span empty circles with points in P. The crust is also a special case of $\mathcal{A}$-shape where $\mathcal{A}$ is the set of Voronoi vertices.

For points sampled from compact subsets of $\mathbb{R}^m$, Chazal, Cohen-Steiner, and Lieutier [CCL09] developed a sampling theory that guarantees a weaker topological equivalence. The output is guaranteed to be homotopy equivalent (instead of homeomorphic) to the sampled space when the input satisfies an appropriate sampling condition.

Manifold reconstruction: When the sample P is derived from a smooth manifold M embedded in some Euclidean space $\mathbb{R}^m$, the curse of dimensionality makes reconstruction harder. Furthermore, unlike in two and three dimensions, the restricted Delaunay triangulation $\mathrm{Del}\,P|_M$ may not have underlying space homeomorphic to M no matter how dense P is. As observed by Cheng, Dey, and Ramos [CDR05], the normal spaces of restricted Delaunay simplices in dimensions four and above can be arbitrarily oriented instead of aligning with those of M. This leads to the topological discrepancy between $\mathrm{Del}\,P|_M$ and M as observed by Boissonnat, Guibas, and Oudot [BGO09].

To overcome the difficulty with the topological discrepancy, Cheng, Dey, and Ramos [CDR05] proposed an algorithm that utilizes the restricted Delaunay triangulation of a weighted version $\widehat{P}$ of P. Using the concept of weight pumping in sliver exudations [CDE+00], they prove that $\mathrm{Del}\,\widehat{P}|_M$ becomes homeomorphic to M under appropriate weight assignments if P is sufficiently dense. Since it is possible for P to be dense for a curve and a surface simultaneously already in three dimensions [DGGZ03, BGO09], one needs some uniformity condition on P to disambiguate the multiple possibilities. Cheng et al. assumed P to be locally uniform, or ϵ-local. The algorithm in [CDR05] assigns some appropriate weights to P and then computes a Delaunay sub-complex of Del $\hat{P}$ restricted to a space consisting of the union of cocone-like spaces around each point in P. This restricted complex is shown to be homeomorphic and Hausdorff close to M. Similar result was later obtained by Boissonnat, Guibas, and Oudot [BGO09] using the witness complex [CdS04] and weights.

The above algorithms compute complexes of dimensions equal to that of the embedding dimension $\mathbb{R}^m$. Consequently, they have complexity exponential in m. For example, it is known that Delaunay triangulation of n points in $\mathbb{R}^m$ has size $\Theta(n^{\lceil \frac{m}{2} \rceil})$. Usually, the intrinsic dimension k of the k-manifold M is small compared to m. Therefore, manifold reconstruction algorithms that replace the dependence on m with that on k have better time and size complexity. Boissonnat and Ghosh [BG14] achieve this goal. They compute the restricted Delaunay complex with respect to the approximate tangent spaces of M at each sample point. Each such individual tangent complex may not agree globally to form a triangulation of M. Nonetheless, Boissonnat and Ghosh [BG14] show that under appropriate weight assignment and sampling density, the collection of all tangent complexes over all sample points become consistent to form a reconstruction of M. The running time of this algorithm is exponential in k, but linear in m.

All known algorithms for reconstructing manifolds in high dimensions use data structures that are computationally intensive. A practical algorithm with geometric and topological guarantee is still elusive.

OPEN PROBLEMS

1. Design an algorithm that reconstructs k-manifolds in $\mathbb{R}^m$, $m \geq 4$, with guarantees and is practical.

2. Reconstruct shapes with guarantees.

35.4 SOURCES AND RELATED MATERIAL

BOOKS/SURVEYS

[Dey07]: Curve and surface reconstruction: Algorithms with mathematical analysis.

[Ede98]: Shape reconstruction with Delaunay complex.

[OR00]: Computational geometry column 38 (recent results on curve reconstruction).

[MSS92]: Surfaces from contours.

[MM98]: Interpolation and approximation of surfaces from 3D scattered data points.

RELATED CHAPTERS

REFERENCES

[AM02]   E. Althaus and K. Mehlhorn. Traveling salesman-based curve reconstruction in poly-nomial time. *SIAM J. Comput.*, 31: 27–66, 2002.

[AB99]   N. Amenta and M. Bern. Surface reconstruction by Voronoi filtering. *Discrete Comput. Geom.*, 22:481–504, 1999.

[ACDL02]   N. Amenta, S. Choi, T.K. Dey, and N. Leekha. A simple algorithm for homeomorphic surface reconstruction. *Internat. J. Comput. Geom. Appl.*, 12:125–141, 2002.

[ABE98]   N. Amenta, M. Bern, and D. Eppstein. The crust and the β-skeleton: Combinatorial curve reconstruction. *Graphical Models and Image Processing*, 60:125–135, 1998.

[ACK01]   N. Amenta, S. Choi, and R.K. Kolluri. The power crust, union of balls, and the medial axis transform. *Comput. Geom.*, 19:127–153, 2001.

[Att98]   D. Attali. r-regular shape reconstruction from unorganized points. *Comput. Geom.*, 10:239–247, 1998.

[AH85]   D. Avis and J. Horton. Remarks on the sphere of influence graph. *Ann. New York Acad. Sci.*, 440:323–327, 1985.

[BB97]   F. Bernardini and C.L. Bajaj. Sampling and reconstructing manifolds using α-shapes. In *Proc. 9th Canad. Conf. Comput. Geom.*, pages 193–198, 1997.

[BMR+99]   F. Bernardini, J. Mittleman, H. Rushmeier, C. Silva, and G. Taubin. The ball-pivoting algorithm for surface reconstruction. *IEEE Trans. Visual. Comput. Graphics*, 5:349–359, 1999.

[Boi84]   J.-D. Boissonnat. Geometric structures for three-dimensional shape representation. *ACM Trans. Graphics*, 3:266–286, 1984.

[BC02]   J.-D. Boissonnat and F. Cazals. Smooth surface reconstruction via natural neighbor interpolation of distance functions. *Comput. Geom.*, 22:185–203, 2002.

[BG93]   J.-D. Boissonnat and B. Geiger. Three-dimensional reconstruction of complex shapes based on the Delaunay triangulation. In *Proc. Biomed. Image Process. Biomed. Visualization*, vol. 1905 of SPIE, pages 964–975, 1993.

[BG14]   J.-D. Boissonnat and A. Ghosh. Manifold reconstruction using tangential Delaunay complexes. *Discrete Comput. Geom.*, 51:221–267, 2014.

[BGO09]   J.-D. Boissonnat, L.J. Guibas, and S.Y. Oudot. Manifold reconstruction in arbitrary dimensions using witness complexes. *Discrete Comput. Geom.*, 42:37–70, 2009.

[CdS04]   G. Carlsson and V. de Silva. Topological estimation using witness complexes. In *Proc. Eurographics Conf. Point-Based Graphics*, pages 157–166, ACM Press, 2004.

[CCL09]   F. Chazal, D. Cohen-Steiner, and A. Lieutier. A sampling theory for compact sets in Euclidean space. *Discrete Comput. Geom.*, 41:461–479, 2009.

[CDE⁺00]   S.-W. Cheng, T.K. Dey, H. Edelsbrunner, M.A. Facello, and S.-H. Teng. Sliver exudation. *J. ACM*, 47:883–904, 2000.

[CDR05]   S.-W. Cheng, T.K. Dey, and E.A. Ramos. Manifold reconstruction from point samples. In *Proc. 16th ACM-SIAM Sympos. Discrete Algorithms*, pages 1018–1027, 2005.

[CJL17]   S.-W. Cheng, J. Jin, and M.-K. Lau. A fast and simple surface reconstruction algorithm. *ACM Trans. Algorithms*, 13:27, 2017.

[CL96]   B. Curless and M. Levoy. A volumetric method for building complex models from range images. In *Proc. 23rd Conf. Comput. Graphics Interactive Tech. (SIGGRAPH)*, pages 306–312, ACM Press, 1996.

[Dey07]   T.K. Dey. *Curve and Surface Reconstruction: Algorithms with Mathematical Analysis*. Cambridge University Press, 2007.

[DG03]   T.K. Dey and J. Giesen. Detecting undersampling in surface reconstruction. In B. Aronov, S. Basu, J. Pach, and M. Sharir, editors, *Discrete and Computational Geometry: The Goodman-Pollack Festschrift*, vol. 25 of *Algorithms Combin.*, pages 329–345, Springer, Berlin, 2003.

[DGGZ03]   T.K. Dey, J. Giesen, S. Goswami, and W. Zhao. Shape dimension and approximation from samples. *Discrete Comput. Geom.*, 29:419–434, 2003.

[DGRS08]   T.K. Dey, J. Giesen, E.A. Ramos, and B. Sadri. Critical points of the distance to an epsilon-sampling on a surface and flow-complex-based surface reconstruction. *Internat. J. Comput. Geom. Appl.*, 18:29–61 , 2008.

[DG03]   T.K. Dey and S. Goswami. Tight cocone: A water-tight surface reconstructor. *J. Comput. Inf. Sci. Eng*, 3:302–307, 2003.

[DG06]   T.K. Dey and S. Goswami. Provable surface reconstruction from noisy samples. *Comput. Geom.*, 35:124–141, 2006.

[DK99]   T.K. Dey and P. Kumar. A simple provable curve reconstruction algorithm. In *Proc. 10th ACM-SIAM Sympos. Discrete Algorithms*, pages 893–894, 1999.

[DMR00]   T.K. Dey, K. Mehlhorn, and E.A. Ramos. Curve reconstruction: Connecting dots with good reason. *Comput. Geom.*, 15:229–244, 2000.

[DLRW09]   T.K. Dey, K. Li, E.A. Ramos, and R. Wenger. Isotopic reconstruction of surfaces with boundaries. *Computer Graphics Forum*, 28:1371–1382, 2009.

[Ede98]   H. Edelsbrunner. Shape reconstruction with Delaunay complex. In *Proc. 3rd Latin American Sympos. Theoret. Inf.*, vol. 1380 of *LNCS*, pages 119–132, Springer, Berlin, 1998.

[Ede99]   H. Edelsbrunner. Deformable smooth surface design. *Discrete Comput. Geom.*, 21:87–115, 1999.

[Ede03]   H. Edelsbrunner. Surface reconstruction by wrapping finite point sets in space. In B. Aronov, S. Basu, J. Pach, and M. Sharir, editors, *Discrete and Computational Geometry: The Goodman-Pollack Festschrift*, vol. 25 of *Algorithms Comb.*, pages 379–404, Springer, Berlin, 2003.

[EKS83]   H. Edelsbrunner, D.G. Kirkpatrick, and R. Seidel. On the shape of a set of points in the plane. *IEEE Trans. Inf. Theory*, 29:551–559, 1983.

[EM94]   H. Edelsbrunner and E.P. Mücke. Three-dimensional alpha shapes. *ACM Trans. Graphics*, 13:43–72, 1994.

[ES97]   H. Edelsbrunner and N.R. Shah. Triangulating topological spaces. *Internat. J. Comput. Geom. Appl.*, 7:365–378, 1997.

[FG95]   L.H. de Figueiredo and J. de Miranda Gomes. Computational morphology of curves. *Visual Computer*, 11:105–112, 1995.

[FKU77]   H. Fuchs, Z.M. Kedem, and S.P. Uselton. Optimal surface reconstruction from planar contours. *Commun. ACM*, 20:693–702, 1977.

[FR01]   S. Funke and E.A. Ramos. Reconstructing a collection of curves with corners and endpoints. In *Proc. 12th ACM-SIAM Sympos. Discrete Algorithms*, pages 344–353, 2001.

[FR02]   S. Funke and E.A. Ramos. Smooth-surface reconstruction in near-linear time. In *13th ACM-SIAM Sympos. Discrete Algorithms*, pages 781–790, 2002.

[Gie00]   J. Giesen. Curve reconstruction, the traveling salesman problem and Menger's theorem on length. *Discrete Comput. Geom.*, 24:577–603, 2000.

[GJ08]   J. Giesen and M. John. The flow complex: A data structure for geometric modeling. *Comput. Geom.*, 39:178–190, 2008.

[GCA13]   S. Giraudot, D. Cohen-Steiner, and P. Alliez. Noise-adaptive shape reconstruction from raw point sets. *Comput. Graphics Forum*, 32:229–238, 2013.

[GOS96]   C. Gitlin, J. O'Rourke, and V. Subramanian. On reconstruction of polyhedra from slices. *Internat. J. Comput. Geom. Appl.*, 6:103–112, 1996.

[GS01]   C.M. Gold and J. Snoeyink. Crust and anti-crust: A one-step boundary and skeleton extraction algorithm. *Algorithmica*, 30:144–163, 2001.

[GKS00]   M. Gopi, S. Krishnan, and C.T. Silva. Surface reconstruction based on lower dimensional localized Delaunay triangulation. *Computer Graphics Forum*, 19:467–478, 2000.

[HDD$^+$92] H. Hoppe, T.D. DeRose, T. Duchamp, J. McDonald, and W. Stützle. Surface reconstruction from unorganized points. In *Proc. 19th Conf. Comput. Graphics Interactive Tech. (SIGGRAPH)*, pages 71–78, ACM Press, 1992.

[KR85]   D.G. Kirkpatrick and J.D. Radke. A framework for computational morphology. In G.T. Toussaint, editor, *Computational Geometry*, vol. 2 of *Mach. Intell. Pattern Recog.*, pages 217–248, North-Holland, Amsterdam, 1985.

[MAVF05]   B. Mederos, N. Amenta, L. Velho, and H. de Figueiredo. Surface reconstruction from noisy point clouds. In *Proc. 3rd Eurographics Sympos. Geom. Processing*, pages 53–62, 2005.

[Mel97]   M. Melkemi. $\mathcal{A}$-shapes of a finite point set. Correspondence in *Proc. 13th Sympos. Comput. Geom.*, pages 367–369, ACM Press, 1997.

[MM98]   R. Mencl and H. Müller. Interpolation and approximation of surfaces from three-dimensional scattered data points. In *State of the Art Reports (Eurographics)*, pages 51–67, 1998.

[MSS92]   D. Meyers, S. Skinner, and K. Sloan. Surfaces from contours. *ACM Trans. Graphics*, 11:228–258, 1992.

[OR00]   J. O'Rourke. Computational geometry column 38. *Internat. J. Comput. Geom. Appl.*, 10:221–223, 2000. Also in *SIGACT News*, 31:28–30, 2000.

[RS07]   E.A. Ramos and B. Sadri. Geometric and topological guarantees for the WRAP reconstruction algorithm. In *Proc. 18th ACM-SIAM Sympos. Discrete Algorithms*, pages 1086–1095, 2007.

36 COMPUTATIONAL CONVEXITY

Peter Gritzmann and Victor Klee

INTRODUCTION

The subject of Computational Convexity draws its methods from discrete mathematics and convex geometry, and many of its problems from operations research, computer science, data analysis, physics, material science, and other applied areas. In essence, it is the study of the computational and algorithmic aspects of high-dimensional convex sets (especially polytopes), with a view to applying the knowledge gained to convex bodies that arise in other mathematical disciplines or in the mathematical modeling of problems from outside mathematics.

The name *Computational Convexity* is of more recent origin, having first appeared in print in 1989. However, results that retrospectively belong to this area go back a long way. In particular, many of the basic ideas of *Linear Programming* have an essentially geometric character and fit very well into the conception of Computational Convexity. The same is true of the subject of *Polyhedral Combinatorics* and of the *Algorithmic Theory of Polytopes and Convex Bodies*.

The emphasis in Computational Convexity is on problems whose underlying structure is the convex geometry of normed vector spaces of finite but generally *not* restricted dimension, rather than of fixed dimension. This leads to closer connections with the optimization problems that arise in a wide variety of disciplines. Further, in the study of Computational Convexity, the underlying model of computation is mainly the binary (Turing machine) model that is common in studies of computational complexity. This requirement is imposed by prospective applications, particularly in mathematical programming. For the study of algorithmic aspects of convex bodies that are not polytopes, the binary model is often augmented by additional devices called "oracles." Some cases of interest involve other models of computation, but the present discussion focuses on aspects of computational convexity for which binary models seem most natural. Many of the results stated in this chapter are qualitative, in the sense that they classify certain problems as being solvable in polynomial time, or show that certain problems are NP-hard or harder. Typically, the tasks remain to find optimal exact algorithms for the problems that are polynomially solvable, and to find useful approximation algorithms or heuristics for those that are NP-hard. In many cases, the known algorithms, even when they run in polynomial time, appear to be far from optimal from the viewpoint of practical application. Hence, the qualitative complexity results should in many cases be regarded as a guide to future efforts but not as final words on the problems with which they deal.

Some of the important areas of computational convexity, such as linear and convex programming, packing and covering, and geometric reconstructions, are covered in other chapters of this Handbook. Hence, after some remarks on presentations of polytopes in Section 36.1, the present discussion concentrates on the following areas that are not covered elsewhere in the Handbook: 36.2, Algorithmic Theory of Convex Bodies; 36.3, Volume Computations; 36.4, Mixed Volumes; 36.5, Containment Problems; 36.6, Radii; 36.7, Constrained Clustering. There are various other classes of problems in computational convexity that will not be covered e.g., projections of polytopes [Fil90, BGK96], sections of polytopes [Fil92], Minkowski addition of polytopes [GS93], geometric tomography [Gar95, GG94, GG97, GGH17] or the Minkowski reconstruction of polytopes [GH99].

Because of the diversity of topics covered in this chapter, each section has a separate bibliography.

FURTHER READING

[GJ79]   M.R. Garey and D.S. Johnson. *Computers and Intractability. A Guide to the Theory of NP-Completeness.* Freeman, San Francisco, 1979.

[GK93a]   P. Gritzmann and V. Klee. Mathematical programming and convex geometry. In P.M. Gruber and J.M. Wills, editors, *Handbook of Convex Geometry*, Volume A, pages 627–674, North-Holland, Amsterdam, 1993.

[GK94a]   P. Gritzmann and V. Klee. On the complexity of some basic problems in computational convexity: I. Containment problems. *Discrete Math.*, 136:129–174, 1994. Reprinted in W. Deuber, H.-J. Prömel, and B. Voigt, editors, *Trends in Discrete Mathematics*, pages 129–174, North-Holland, Amsterdam, 1994.

[GK94b]   P. Gritzmann and V. Klee. On the complexity of some basic problems in computational convexity: II. Volume and mixed volumes. In T. Bisztriczky, P. McMullen, R. Schneider, and A. Ivić Weiss, editors, *Polytopes: Abstract, Convex and Computational,* vol. 440 of *NATO ASI Series*, pages 373–466, Kluwer, Dordrecht, 1994.

[Gru07]   P.M. Gruber. *Convex and Discrete Geometry.* Springer-Verlag, Berlin, 2007.

RELATED CHAPTERS

REFERENCES

[BGK96]   T. Burger, P. Gritzmann, and V. Klee. Polytope projection and projection polytopes. *Amer. Math. Monthly*, 103:742–755, 1996.

[Fil90]   P. Filliman. Exterior algebra and projections of polytopes. *Discrete Comput. Geom.*, 5:305–322, 1990.

[Fil92]   P. Filliman. Volumes of duals and sections of polytopes. *Mathematika*, 39:67–80, 1992.

[GGH17]   U. Grimm, P. Gritzmann, and C. Huck. Discrete tomography of model sets: Reconstruction and uniqueness. In M. Baake and U. Grimm, editors, *Aperiodic Order, Vol. 2: Chrystallography and Almost Periodicity*, pages 39–71, Cambridge Univ. Press, 2017.

[Gar95]   R.J. Gardner. *Geometric Tomography.* Cambridge University Press, New York, 1995; 2nd edition, 2006.

[GG94]   R.J. Gardner and P. Gritzmann. Successive determination and verification of polytopes by their X-rays. *J. London Math. Soc.*, 50:375–391, 1994.

[GG97]   R.J. Gardner and P. Gritzmann. Discrete tomography: Determination of finite sets by X-rays. *Trans. Amer. Math. Soc.*, 349:2271–2295, 1997.

[GH99]   P. Gritzmann and A. Hufnagel. On the algorithmic complexity of Minkowski's reconstruction theorem. *J. London Math. Soc.* (2), 59:1081–1100, 1999.

[GS93]   P. Gritzmann and B. Sturmfels. Minkowski addition of polytopes: computational complexity and applications to Gröbner bases. *SIAM J. Discrete Math.*, 6:246–269, 1993.

36.1 PRESENTATIONS OF POLYTOPES

A convex polytope $P \subset \mathbb{R}^n$ can be represented in terms of its vertices or in terms of its facet inequalities. From a theoretical viewpoint, the two possibilities are equivalent. However, as the dimension increases, the number of vertices can grow exponentially in terms of the number of facets, and vice versa, so that different presentations may lead to different classifications concerning polynomial-time computability or NP-hardness. (See Sections 15.1 and 26.3 of this Handbook.)

For algorithmic purposes it is usually not the polytope P as a *geometric* object that is relevant, but rather its *algebraic presentation*. The discussion here is based mainly on the *binary* or *Turing machine* model of computation, in which the *size of the input* is defined as the length of the binary encoding needed to present the input data to a Turing machine and the *time-complexity* of an algorithm is also defined in terms of the operations of a Turing machine. Hence the algebraic presentation of the objects at hand must be finite.

Among important special classes of polytopes, the zonotopes are particularly interesting because they can be so compactly presented.

GLOSSARY

Convex body in $\mathbb{R}^n$: A compact convex subset of $\mathbb{R}^n$.

$\mathcal{K}^n$: The family of all convex bodies in $\mathbb{R}^n$.

Proper convex body in $\mathbb{R}^n$: A convex body in $\mathbb{R}^n$ with nonempty interior.

Polytope: A convex body that has only finitely many extreme points.

$\mathcal{P}^n$: The family of all convex polytopes in $\mathbb{R}^n$.

n-polytope: Polytope of dimension n.

Face of a polytope P: P itself, the empty set, or the intersection of P with some supporting hyperplane; $f_i(P)$ is the number of i-dimensional faces of P.

Facet of an n-polytope P: Face of dimension $n-1$.

Simple n-polytope: Each vertex is incident to precisely n edges or, equivalently, to precisely n facets.

Simplicial polytope: A polytope in which each facet is a simplex.

$\mathcal{V}$-***presentation*** of a polytope P: A string $(n, m; v_1, \ldots, v_m)$, where $n, m \in \mathbb{N}$ and $v_1, \ldots, v_m \in \mathbb{R}^n$ such that $P = \operatorname{conv}\{v_1, \ldots, v_m\}$.

$\mathcal{H}$-presentation of a polytope P: A string $(n, m; A, b)$, where $n, m \in \mathbb{N}$, A is a real $m \times n$ matrix, and $b \in \mathbb{R}^m$ such that $P = \{x \in \mathbb{R}^n \mid Ax \leq b\}$.

irredundant $\mathcal{V}$- or $\mathcal{H}$-presentation of a polytope P: A $\mathcal{V}$- or $\mathcal{H}$-presentation $(n, m; v_1, \ldots, v_m)$ or $(n, m; A, b)$ of P with the property that none of the points $v_1, \ldots, v_m$ or none of the inequalities $Ax \leq b$ can be omitted without altering P, respectively.

$\mathcal{V}$-polytope P: A string $(n, m; v_1, \ldots, v_m)$, where $n, m \in \mathbb{N}$ and $v_1, \ldots, v_m \in \mathbb{Q}^n$. P is usually identified with the geometric object $\mathrm{conv}\{v_1, \ldots, v_m\}$.

$\mathcal{H}$-polytope P: A string $(n, m; A, b)$, where $n, m \in \mathbb{N}$, A is a rational $m \times n$ matrix, $b \in \mathbb{Q}^m$, and the set $\{x \in \mathbb{R}^n \mid Ax \leq b\}$ is bounded. P is usually identified with this set.

Size of a $\mathcal{V}$- or an $\mathcal{H}$-polytope P: Number of binary digits needed to encode the string $(n, m; v_1, \ldots, v_m)$ or $(n, m; A, b)$, respectively.

Zonotope: The vector sum (Minkowski sum) of a finite number of line segments; equivalently, a polytope of which each face has a center of symmetry.

$\mathcal{S}$-presentation of a zonotope Z in $\mathbb{R}^n$:   A string $(n, m; c; z_1, \ldots, z_m)$, where $n, m \in \mathbb{N}$ and $c, z_1, \ldots, z_m \in \mathbb{R}^n$, such that $Z = c + \sum_{i=1}^{m} [-1, 1] z_i$.

Parallelotope in $\mathbb{R}^n$:   A zonotope $Z = c + \sum_{i=1}^{m} [-1, 1] z_i$, with $z_1, \ldots, z_m$ linearly independent.

$\mathcal{S}$-zonotope Z in $\mathbb{R}^n$:   A string $(n, m; c; z_1, \ldots, z_m)$, where $n, m \in \mathbb{N}$ and $c, z_1, \ldots, z_m \in \mathbb{Q}^n$. Z is usually identified with the geometric object $c + \sum_{i=1}^{m} [-1, 1] z_i$.

36.1.1 CONVERSION OF ONE PRESENTATION INTO THE OTHER

Note, first, that from a given $\mathcal{V}$- or $\mathcal{H}$-representation of a polytope P, an irredundant $\mathcal{V}$- or $\mathcal{H}$-representation of P can be computed in polynomial time by means of linear programming, respectively; see also Section 26.2.

The following results indicate the difficulties that may be expected in converting the $\mathcal{H}$-presentation of a polytope into a $\mathcal{V}$-presentation or vice versa.

For $\mathcal{H}$-presented n-polytopes with m facets, the *maximum* possible number of vertices is

$$\mu(m, n) = \binom{m - \lfloor (n+1)/2 \rfloor}{m - n} + \binom{m - \lfloor (n+2)/2 \rfloor}{m - n},$$

and this is also the maximum possible number of facets for a $\mathcal{V}$-presented n-polytope with m vertices. The first maximum is attained within the family of simple n-polytopes, the second within the family of simplicial n-polytopes.

When n is fixed, the number of vertices is bounded by a polynomial in the number of facets, and vice versa, and it is possible to pass from either sort of presentation to the other in polynomial time. However, the degree of the polynomial goes to infinity with n. A consequence of this is that when the dimension n is permitted to vary in a problem concerning polytopes, the manner of presentation is often influential in determining whether the problem can be solved in polynomial time or is NP-hard. For the case of variable dimension, it is #P-hard even to determine the number of facets of a given $\mathcal{V}$-polytope, or to determine the number of vertices of a given $\mathcal{H}$-polytope, [Lin86].

For *simple* $\mathcal{H}$-presented n-polytopes with m facets, the *minimum* possible number of vertices is $(m - n)(n - 1) + 2$. The large gap between this number and the above sum of binomial coefficients makes it clear that, from a practical standpoint, the worst-case behavior of any conversion algorithm should be measured in terms of *both* input size and output size. In this respect, the following problem seems fundamental.

OPEN PROBLEM 36.1.1

What is the computational complexity of POLYTOPE VERIFICATION: *Given an* $\mathcal{H}$-*polytope* P *and a* $\mathcal{V}$-*polytope* Q *in* $\mathbb{R}^n$, *decide whether* $P = Q$.

The maximum number of j-dimensional faces of an n-dimensional zonotope formed as the sum of m segments is

$$2 \binom{m}{j} \sum_{k=0}^{n-1-j} \binom{m-1-j}{k},$$

and hence, the number of vertices or of facets (or of faces of any dimension) of an $\mathcal{S}$-zonotope is not bounded by any polynomial in the size of the $\mathcal{S}$-presentation.

In combinatorial optimization one is particularly interested in "perfect formulations" of 0-1-polytopes in $\mathbb{R}^n$ associated with the underlying problems. A well-studied example is that of the traveling salesman polytopes, the convex hull of the incidence vectors of Hamiltonian cycles of the complete graph on the given number of cities. Since formulations in the "natural space" of the application often have an exponential number of inequalities, one tries to find small *extended formulations*, i.e., formulations with a polynomial number of inequalities, after allowing a polynomial number of extra variables; see e.g., [CCZ13]. In effect, one is asking for a polytope in higher (but not too high) dimension whose projection on $\mathbb{R}^n$ coincides with the original polytope. For some "oracular" results (in the spirit of the next section) see [BV08].

We end this section by mentioning two other ways of presenting polytopes.

A general result of Bröcker and Scheiderer (see [BCR98]) on semi-algebraic sets implies that for each n-polytope P in $\mathbb{R}^n$ (no matter how complicated its facial structure may be), there exists a system of $n(n+1)/2$ polynomial inequalities that has P as its solution-set, and that n polynomial inequalities suffice to describe the interior of P. More recently, Bröcker showed (see [AH11]) that for polytopes n polynomial inequalities actually always suffice. [AH11] give a fully constructive proof that any *simple* n-polytope can be described by n polynomial inequalities.

For a polytope P in $\mathbb{R}^n$ whose interior is known to contain the origin, [GKW95] shows that the entire face-lattice of P can be reconstructed with the aid of at most

$$f_0(P) + (n-1)f_{n-1}^2(P) + (5n-4)f_{n-1}(P)$$

queries to the *ray-oracle* of P. In each such query, one specifies a ray issuing from the origin and the oracle is required to tell where the ray hits the boundary of P. Related results were obtained in [DEY90].

For more on oracles, see Section 36.2 of this Handbook.

FURTHER READING

[BL93]   M.M. Bayer and C.W. Lee. Combinatorial aspects of convex polytopes. In P.M. Gruber and J.M. Wills, editors, *Handbook of Convex Geometry*, Volume A, pages 251–305, North-Holland, Amsterdam, 1993.

[BCR98]   J. Bochnak, M. Coste, and M.-F. Roy. *Real Algebraic Geometry.* Springer-Verlag, Berlin, 1998.

[Brø83]   A. Brøndsted. *An Introduction to Convex Polytopes.* Springer-Verlag, New York, 1983.

[GK94a]   P. Gritzmann and V. Klee. On the complexity of some basic problems in computational convexity: I. Containment problems. *Discrete Math.*, 136:129–174, 1994. Reprinted in W. Deuber, H.-J. Prömel, and B. Voigt, editors, *Trends in Discrete Mathematics*, pages 129–174, North-Holland, Amsterdam, 1994.

[Grü67]   B. Grünbaum. *Convex Polytopes.* Wiley-Interscience, London, 1967. (Second edition prepared in collaboration with V. Kaibel, V. Klee, and G. Ziegler, Springer-Verlag, New York, 2003.)

[KK95]   V. Klee and P. Kleinschmidt. Convex polytopes and related complexes. In R.L. Graham, M. Grötschel, and L. Lovász, editors, *Handbook of Combinatorics*, Volume I., pages 875–917. North-Holland, Amsterdam, 1995.

[MS71]   P. McMullen and G.C. Shephard. *Convex Polytopes and the Upper Bound Conjecture.* Cambridge University Press, 1971.

[Zie95]   G.M. Ziegler. *Lectures on Polytopes.* Volume 152 of *Graduate Texts in Math.*, Springer-Verlag, New York, 1995.

RELATED CHAPTERS

REFERENCES

[AH11]   G. Averkov and M. Henk. Representing simple d-dimensional polytopes by d polynomials. *Math. Prog.*, 126:203–230, 2011.

[BV08]   A. Barvinok and E. Veomett. The computational complexity of convex bodies. In *Surveys on Discrete and Computational Geometry: Twenty Years Later*, vol. 453 of *Contemp. Math.*, pages 117–137, ACM, Providence, 2008.

[CCZ13]   M. Conforti, G. Cornuéjols, and G. Zambelli. Extended formulations in combinatorial optimization. *Ann. Oper. Res.*, 204:97–143, 2013.

[DEY90]   D. Dobkin, H. Edelsbrunner, and C.K. Yap. Probing convex polytopes. In I.J. Cox and G.T. Wilfong, editors, *Autonomous Robot Vehicles*, pages 328–341. Springer, New York, 1990.

[GKW95]   P. Gritzmann, V. Klee, and J. Westwater. Polytope containment and determination by linear probes. *Proc. London Math. Soc.*, 70:691–720, 1995.

[Lin86]   N. Linial. Hard enumeration problems in geometry and combinatorics. *SIAM J. Alg. Disc. Meth.*, 7:311–335, 1986.

36.2 ALGORITHMIC THEORY OF CONVEX BODIES

Polytopes may be $\mathcal{V}$-presented or $\mathcal{H}$-presented. However, a different approach is required to deal with convex bodies K that are not polytopes, since an enumeration of all the extreme points of K or of its polar is not possible. A convenient way to deal with the general situation is to assume that the convex body in question is given by an algorithm (called an *oracle*) that answers certain sorts of questions about the body. A small amount of a priori information about the body may be known, but aside from this, all information about the specific convex body must be obtained from the oracle, which functions as a "black box." In other words, while it is assumed that the oracle's answers are always correct, nothing is assumed about the manner in which it produces those answers. The algorithmic theory of convex bodies was developed in [GLS88] with a view to proper (i.e., n-dimensional) convex bodies in $\mathbb{R}^n$. For many purposes, provisions can be made to deal meaningfully with improper bodies as well, but that aspect is largely ignored in what follows.

GLOSSARY

Outer parallel body of a convex body K: $K(\epsilon) = K + \epsilon B^n$, where B^n is the Euclidean unit ball in $\mathbb{R}^n$.

Inner parallel body of a convex body K: $K(-\epsilon) = K \setminus \big((\mathbb{R}^n \setminus K) + \epsilon B^n \big)$.

Weak membership problem for a convex body K in $\mathbb{R}^n$:   Given $y \in \mathbb{Q}^n$, and a rational number $\epsilon > 0$, conclude with one of the following: *report that $y \in K(\epsilon)$*; or *report that $y \notin K(-\epsilon)$*.

Weak separation problem for a convex body K in $\mathbb{R}^n$:   Given a vector $y \in \mathbb{Q}^n$, and a rational number $\epsilon > 0$, conclude with one of the following: *report that $y \in K(\epsilon)$; or find a vector $z \in \mathbb{Q}^n$ such that $\| z \|_{(\infty)} = 1$ and $z^T x < z^T y + \epsilon$ for every $x \in K(-\epsilon)$*.

Weak (linear) optimization problem for a convex body K in $\mathbb{R}^n$:   Given a vector $c \in \mathbb{Q}^n$ and a rational number $\epsilon > 0$, conclude with one of the following: *find a vector $y \in \mathbb{Q}^n \cap K(\epsilon)$ such that $c^T x \le c^T y + \epsilon$ for every $x \in K(-\epsilon)$; or report that $K(-\epsilon) = \emptyset$*.

Circumscribed convex body K: A positive rational number R is given explicitly such that $K \subset RB^n$.

Well-bounded convex body K: Positive rational numbers r, R are given explicitly such that $K \subset RB^n$ and K contains a ball of radius r.

Centered well-bounded convex body K: Positive rational numbers r, R and a vector $b \in \mathbb{Q}^n$ are given explicitly such that $b + rB^n \subset K$ and $K \subset RB^n$.

Weak membership oracle for a convex body K: Algorithm that solves the weak membership problem for K.

Weak separation oracle for K: Algorithm that solves the weak separation problem for K.

Weak (linear) optimization oracle for K: Algorithm that solves the weak (linear) optimization problem for K.

The three problems above are very closely related in the sense that when the classes of proper convex bodies are appropriately restricted to those that are circumscribed, well-bounded, or centered, and when input sizes are properly defined, an algorithm that solves any one of the problems in polynomial time can be used as a subroutine to solve the others in polynomial time also. The definition of input size involves the size of ϵ, the dimension of K, the given a priori information (size(r), size(R), and/or size(b)), and the input required by the oracle. The following theorem of [GLS88] contains a list of the precise relationships among the three basic oracles for proper convex bodies. The notation "$(\mathcal{A}; prop) \rightarrow_\pi \mathcal{B}$" indicates the existence of an (oracle-) polynomial-time algorithm that solves problem $\mathcal{B}$ for every proper convex body that is given by the oracle $\mathcal{A}$ and has all the properties specified in $prop$. ($prop= \emptyset$ means that the statement holds for general proper convex bodies.)

(WEAK MEMBERSHIP; centered, well-bounded) $\rightarrow_\pi$ WEAK SEPARATION;
(WEAK MEMBERSHIP; centered, well-bounded) $\rightarrow_\pi$ WEAK OPTIMIZATION;
(WEAK SEPARATION; $\emptyset$) $\rightarrow_\pi$ WEAK MEMBERSHIP;
(WEAK SEPARATION; circumscribed) $\rightarrow_\pi$ WEAK OPTIMIZATION;
(WEAK OPTIMIZATION; $\emptyset$) $\rightarrow_\pi$ WEAK MEMBERSHIP;
(WEAK OPTIMIZATION; $\emptyset$) $\rightarrow_\pi$ WEAK SEPARATION.

It should be emphasized that there are polynomial-time algorithms that, accepting as input a set P that is a proper $\mathcal{V}$-polytope, a proper $\mathcal{H}$-polytope, or a proper $\mathcal{S}$-zonotope, produce membership, separation, and optimization oracles for P, and also compute a lower bound on the inradius of P, an upper bound on its circumradius, and a "center" b_P for P. This implies that if an algorithm performs certain tasks for convex bodies given by some of the above (appropriately specified) oracles, then the same algorithm can also serve as a basis for procedures that perform these tasks for $\mathcal{V}$- or $\mathcal{H}$-polytopes and for $\mathcal{S}$-zonotopes. Hence the oracular framework, in addition to being applicable to convex bodies that are not polytopes, serves also to modularize the approach to algorithmic aspects of polytopes. On the other hand, there are lower bounds on the performance of approximate algorithms for the oracle model that do not carry over to the case of $\mathcal{V}$- or $\mathcal{H}$-polytopes or $\mathcal{S}$-zonotopes [BF87, BGK+01]. However, if in polyhedral combinatorics certain tasks are known to be NP-hard then the above $\rightarrow_\pi$ implications can be used to show that certain other tasks are also hard. For instance, if the membership or the separation problem for the traveling salesman polytopes could be solved in polynomial time then optimization would also be tractable. Since the traveling salesman problem is known to be NP-hard, so are the membership or the separation problem for the traveling salesman polytopes. See [BV08] for approximations of convex bodies by sets for which a polynomial-time membership oracle is available.

FURTHER READING

[GLS88]   M. Grötschel, L. Lovász, and A. Schrijver. *Geometric Algorithms and Combinatorial Optimization*. Springer-Verlag, Berlin, 1988; 2nd edition, 1993.

[Sch95]   A. Schriver. Polyhedral combinatorics. In R.L. Graham, M. Grötschel, and L. Lovász, editors, *Handbook of Combinatorics*, Volume II, pages 1649–1704, North-Holland, Amsterdam, and MIT Press, Cambridge, 1995.

[Sch03]   A. Schriver. *Combinatorial Optimization: Polyhedra and Efficiency.* Vol. 24 of Algorithms Combin., Springer-Verlag, Berlin, 2003.

RELATED CHAPTERS

Chapter 7: Lattice points and lattice polytopes

REFERENCES

[BF87]   I. Bárány and Z. Füredi. Computing the volume is difficult. *Discrete Comput. Geom.*, 2:319–326, 1987.

[BV08]   A. Barvinok and E. Veomett. The computational complexity of convex bodies. In *Surveys on Discrete and Computational Geometry: Twenty Years Later*, vol. 453 of *Contemp. Math.*, pages 117–137, ACM, Providence, 2008.

[BGK+01]   A. Brieden, P. Gritzmann, R. Kannan, V. Klee, L. Lovász, and M. Simonovits. Deterministic and randomized polynomial-time approximation of radii. *Mathematika*, 48:63–105, 2001.

36.3 VOLUME COMPUTATIONS

It may be fair to say that the modern study of volume computations began with Kepler [Kep15] who derived the first *cubature formula* for measuring the capacities of wine barrels, and that it was the task of volume computation that motivated the general field of integration. The problem of computing or approximating volumes of convex bodies is certainly one of the basic problems in mathematics.

GLOSSARY

In the following, G is a subgroup of the group of all affine automorphisms of $\mathbb{R}^n$.

Dissection of an n-polytope P into n-polytopes $P_1, \ldots, P_k$: $P = P_1 \cup \cdots \cup P_k$, where the polytopes P_i have pairwise disjoint interiors.

Polytopes $P, Q \subset \mathbb{R}^n$ are **G-equidissectable:**   For some k there exist dissections $P_1, \ldots, P_k$ of P and $Q_1, \ldots, Q_k$ of Q, and elements $g_1, \ldots, g_k$ of G, such that $P_i = g_i(Q_i)$ for all i.

Polytopes $P, Q \subset \mathbb{R}^n$ are **G-equicomplementable:**   There are polytopes P_1, P_2 and Q_1, Q_2 such that P_2 is dissected into P and P_1, Q_2 is dissected into Q and Q_1, P_1 and Q_1 are G-equidissectable, and P_2 and Q_2 are G-equidissectable.

Decomposition of a set S: $S = S_1 \cup \cdots \cup S_k$, where the sets S_i are pairwise disjoint.

Sets S, T are **G-equidecomposable:**   For some k there are decompositions $S_1, \ldots, S_k$ of S and $T_1, \ldots, T_k$ of T, and elements $g_1, \ldots, g_k$ of G, such that $S_i = g_i(T_i)$ for all i.

Valuation on a family $\mathcal{S}$ of subsets of $\mathbb{R}^n$: A functional $\varphi : \mathcal{S} \to \mathbb{R}$ with the property that $\varphi(S_1) + \varphi(S_2) = \varphi(S_1 \cup S_2) + \varphi(S_1 \cap S_2)$ whenever the sets $S_1, S_2, S_1 \cup S_2, S_1 \cap S_2 \in \mathcal{S}$.

G-invariant valuation φ: $\varphi(S) = \varphi(g(S))$ for all $S \in \mathcal{S}$ and $g \in G$.

Simple valuation φ: $\varphi(S) = 0$ whenever $S \in \mathcal{S}$ and S is contained in a hyperplane.

Monotone valuation φ: $\varphi(S_1) \leq \varphi(S_2)$ whenever $S_1, S_2 \in \mathcal{S}$ with $S_1 \subset S_2$.

Class $\mathcal{P}$ of $\mathcal{H}$-polytopes is **near-simplicial:** There is a nonnegative integer σ such that $\mathcal{P} = \bigcup_{n \in \mathbb{N}} \mathcal{P}_{\mathcal{H}}(n, \sigma)$, where $\mathcal{P}_{\mathcal{H}}(n, \sigma)$ is the family of all n-dimensional $\mathcal{H}$-polytopes P in $\mathbb{R}^n$ such that each facet of P has at most $n + 1 + \sigma$ vertices.

Class $\mathcal{P}$ of $\mathcal{V}$-polytopes is **near-simple:** There is a nonnegative integer τ such that $\mathcal{P} = \bigcup_{n \in \mathbb{N}} \mathcal{P}_{\mathcal{V}}(n, \tau)$, where $\mathcal{P}_{\mathcal{V}}(n, \tau)$ is the family of all n-dimensional $\mathcal{V}$-polytopes P in $\mathbb{R}^n$ such that each vertex of P is incident to at most $n + \tau$ edges.

Class $\mathcal{P}$ of $\mathcal{V}$-polytopes is **near-parallelotopal:** There is a nonnegative integer ζ such that $\mathcal{Z} = \bigcup_{n \in \mathbb{N}} \mathcal{Z}_{\mathcal{S}}(n, \zeta)$, where $\mathcal{Z}_{\mathcal{S}}(n, \zeta)$ is the family of all $\mathcal{S}$-zonotopes in $\mathbb{R}^n$ that are represented as the sum of at most $n + \zeta$ segments.

V: The functional that associates with a convex body K its volume.

$\mathcal{H}$-VOLUME: For a given $\mathcal{H}$-polytope P and a nonnegative rational ν, decide whether $V(P) \leq \nu$.

$\mathcal{V}$-VOLUME, $\mathcal{S}$-VOLUME: Similarly for $\mathcal{V}$-polytopes and $\mathcal{S}$-zonotopes.

λ-APPROXIMATION for some functional ρ: Given a positive integer n and a well-bounded convex body K given by a weak separation oracle, determine a nonnegative rational μ such that

$$\rho(K) \leq (1 + \lambda)\mu \quad \text{and} \quad \mu \leq (1 + \lambda)\rho(K).$$

EXPECTED VOLUME COMPUTATION: Given a positive integer n, a centered well-bounded convex body K in $\mathbb{R}^n$ given by a weak membership oracle, and positive rationals β and ϵ, determine a positive rational random variable μ such that

$$\text{prob}\left\{ \left| \frac{\mu}{V(K)} - 1 \right| \leq \epsilon \right\} \geq 1 - \beta.$$

36.3.1 CLASSICAL BACKGROUND, CHARACTERIZATIONS

The results in this subsection connect the subject matter of volume computation with related "classical" problems. In the following, G is a group of affine automorphisms of $\mathbb{R}^n$, as above, and D is the group of isometries.

(i) Two polytopes are G-equidissectable if and only if they are G-equicomplementable.

(ii) Two polytopes P and Q are G-equidissectable if and only if $\varphi(P) = \varphi(Q)$ for all G-invariant simple valuations on $\mathcal{P}^n$.

(iii) Two plane polygons are of equal area if and only if they are D-equidissectable.

(iv) If one agrees that an a-by-b rectangle should have area ab, and also agrees that the area function should be a D-invariant simple valuation, it then follows from the preceding result that the area of any plane polygon P can be determined (at least in theory) by finding a rectangle R to which P is equidissectable. This provides a satisfyingly geometric theory of area that does not require any limiting considerations. The third problem of Hilbert [Hil00] asked, in effect, whether such a result extends to 3-polytopes. A negative answer was supplied by [Deh00], who showed that a regular tetrahedron and a cube of the same volume are not D-equidissectable.

(v) If P and Q are n-polytopes in $\mathbb{R}^n$, then for P and Q to be equidissectable under the group of all isometries of $\mathbb{R}^n$, it is necessary that $f^*(P) = f^*(Q)$ for each additive real function f such that $f(\pi) = 0$, where $f^*(P)$ is the so-called *Dehn invariant* of P associated with f. The condition is also sufficient for equidissectability when $n \leq 4$, but the matter of sufficiency is unsettled for $n \geq 5$.

(vi) Two plane polygons are of equal area if and only if they are D-equidecomposable.

(vii) In [Lac90], it was proved that any two plane polygons of equal area are equidecomposable under the group of translations. That paper also settled Tarski's old problem of "*squaring the circle*" by showing that a square and a circular disk of equal area are equidecomposable; there too, translations suffice. On the other hand, a disk and a square cannot be *scissors congruent*; i.e., there is no equidissection (with respect to rigid motions) into pieces that, roughly speaking, could be cut out with a pair of scissors.

(viii) If X and Y are bounded subsets of $\mathbb{R}^n$ (with $n \geq 3$), and each set has nonempty interior, then X and Y are D-equidecomposable. This is the famous *Banach-Tarski paradox*.

(ix) Under the group of all volume-preserving affinities of $\mathbb{R}^n$, two n-polytopes are equidissectable if and only if they are of equal volume.

(x) If φ is a translation-invariant, nonnegative, simple valuation on $\mathcal{P}^n$ (resp. $\mathcal{K}^n$), then there exists a nonnegative real α such that $\varphi = \alpha V$.

(xi) A translation-invariant valuation on $\mathcal{P}^n$ that is homogeneous of degree n is a constant multiple of the volume.

(xii) A continuous, rigid-motion-invariant, simple valuation on $\mathcal{K}^n$ is a constant multiple of the volume.

(xiii) A nonnegative simple valuation on $\mathcal{P}^n$ (resp. $\mathcal{K}^n$) that is invariant under all volume-preserving linear maps of $\mathbb{R}^n$ is a constant multiple of the volume.

36.3.2 SOME VOLUME FORMULAS

Since simplex volumes can be computed so easily, the most natural approach to the problem of computing the volume of a polytope P is to produce a *triangulation* of P (see Chapter 16). Then compute the volumes of the individual simplices and add them up to find the volume of P. (This uses the fact that the volume is a simple valuation.) As a consequence, one sees that when the dimension n is fixed, the volume of $\mathcal{V}$-polytopes and of $\mathcal{H}$-polytopes can be computed in polynomial time.

Another equally natural method is to dissect P into pyramids with common apex over its facets. Since the volume of such a pyramid is just $1/n$ times the product of its height and the $(n-1)$-volume of its base, the volume can be computed recursively.

Another approach that has become a standard tool for many algorithmic questions in geometry is the *sweep-plane* technique. The general idea is to "sweep" a hyperplane through a polytope P, keeping track of the changes that occur when the hyperplane sweeps through a vertex. As applied to volume computation, this leads to the volume formula given below that does not explicitly involve triangulations, [BN83, Law91].

Suppose that $(n, m; A, b)$ is an irredundant $\mathcal{H}$-presentation of a simple polytope P. Let $b = (\beta_1, \ldots, \beta_m)^T$ and denote the row-vectors of A by $a_1^T, \ldots, a_m^T$. Let $M = \{1, \ldots, m\}$ and for each nonempty subset I of M, let A_I denote the submatrix of A formed by rows with indices in I and let b_I denote the corresponding right-hand side. Let $\mathcal{F}_0(P)$ denote the set of all vertices of the polytope $P = \{x \in \mathbb{R}^n \mid Ax \le b\}$. For each $v \in \mathcal{F}_0(P)$, there is a set $I = I_v \subset M$ of cardinality n such that $A_I v = b_I$ and $A_{M \setminus I} v \le b_{M \setminus I}$. Since P is assumed to be simple and its $\mathcal{H}$-presentation to be irredundant, the set I_v is unique.

Let $c \in \mathbb{R}^n$ be such that $\langle c, v_1 \rangle \ne \langle c, v_2 \rangle$ for any pair of vertices v_1, v_2 that form an edge of P. Then it turns out that

$$V(P) = \frac{1}{n!} \sum_{v \in \mathcal{F}_0(P)} \frac{\langle c, v \rangle^n}{\prod_{i=1}^n e_i^T A_{I_v}^{-1} c |\det(A_{I_v})|}.$$

The ingredients of this volume formula are those that are computed in the (dual) simplex algorithm. More precisely, $\langle c, v \rangle$ is just the value of the objective function at the current basic feasible solution v, $\det(A_{I_v})$ is the determinant of the current basis, and $A_{I_v}^{-1} c$ is the vector of reduced costs, i.e., the (generally infeasible) dual point that belongs to v.

For practical computations, this volume formula has to be combined with some vertex enumeration technique. Its closeness to the simplex algorithm suggests the use of a *reverse search* method [AF92], which is based on the simplex method with Bland's pivoting rule.

As it stands, the volume formula does not involve triangulation. However, when interpreted in a polar setting, it is seen to involve the faces of the simplicial polytope P° that is the polar of P. Accordingly, generalization to nonsimple polytopes involves polar triangulation. In fact, for general polytopes P, one may apply a "lexicographic rule" for moving from one basis to another, but this amounts to a particular triangulation of P°.

Another possibility for computing the volume of a polytope P is to study the *exponential integral* $\int_P e^{\langle c, x \rangle} dx$, where c is an arbitrary vector of $\mathbb{R}^n$; see [Bar93]. (Note that for $c = 0$, this integral just gives the volume of P.) Exponential integrals satisfy certain relations that make it possible to compute the integrals efficiently in some important cases. In particular, exponential sums can be used to obtain the tractability result for near-simple $\mathcal{V}$-polytopes stated in the next subsection.

36.3.3 TRACTABILITY RESULTS

The volume of a polytope P can be computed in polynomial time in the following cases:

(i) when the dimension is fixed and P is a $\mathcal{V}$-polytope, an $\mathcal{H}$-polytope, or an $\mathcal{S}$-zonotope;

(ii) when the dimension is part of the input and P is a near-simple $\mathcal{V}$-polytope, a near-simplicial $\mathcal{H}$-polytope, or a near-parallelotopal $\mathcal{S}$-zonotope.

36.3.4 INTRACTABILITY RESULTS

(i) Since the output can have super polynomial size [Law91] there is no polynomial-space algorithm for exact computation of the volume of $\mathcal{H}$-polytopes.

(ii) $\mathcal{H}$-VOLUME is #P-hard even for the intersections of the unit cube with one rational halfspace.

(iii) $\mathcal{H}$-VOLUME is #P-hard in the strong sense. (This follows from the result of [BW92] that the problem of computing the number of linear extensions of a given partially ordered set $\mathcal{O} = (\{1, \ldots, n\}, \prec)$ is #P-complete, in conjunction with the fact that this number is equal to $n! V(P_{\mathcal{O}})$, where the set $P_{\mathcal{O}} = \{x = (\xi_1, \ldots, \xi_n)^T \in [0, 1]^n \mid \xi_i \leq \xi_j \iff i \prec j\}$ is the *order polytope* of $\mathcal{O}$ [Sta86].)

(iv) The problem of computing the volume of the convex hull of the regular $\mathcal{V}$-cross-polytope and an additional integer vector is #P-hard.

(v) $\mathcal{S}$-VOLUME is #P-hard.

36.3.5 DETERMINISTIC APPROXIMATION

(i) There exists an oracle-polynomial-time algorithm that, for any convex body K of $\mathbb{R}^n$ given by a weak optimization oracle, and for each $\epsilon > 0$, finds rationals μ_1 and μ_2 such that

$$\mu_1 \leq V(K) \leq \mu_2 \quad \text{and} \quad \mu_2 \leq n!(1 + \epsilon)^n \mu_1.$$

(ii) Suppose that

$$\lambda(n) < \left(\frac{n}{\log n}\right)^{n/2} - 1 \qquad \text{for all } n \in \mathbb{N}.$$

Then there exists no deterministic oracle-polynomial-time algorithm for λ-APPROXIMATION of the volume [BF87].

[DV13] give better approximations at higher computational cost. More precisely, it is shown that there is a deterministic algorithm, that accepts as input a well-bounded centrally symmetric convex body K given by a weak membership oracle and an $\epsilon \in (0, 1]$, and computes a $(1 + \epsilon)^n$-approximation of $V(K)$ in time $O(1/\epsilon)^{O(n)}$ and polynomial space. In view of the results of [BF87], this is optimal up to the constant in the exponent.

36.3.6 RANDOMIZED ALGORITHMS

[DFK89] proved that there is a randomized algorithm for EXPECTED VOLUME COMPUTATION that runs in time that is oracle-polynomial in n, $1/\epsilon$, and $\log(1/\beta)$.

The first step is a rounding procedure, using an algorithmic version of John's theorem; see Section 36.5.4. For the second step, one may therefore assume that $B^n \subset K \subset (n+1)\sqrt{n}B^n$. Now, let

$$k = \left\lceil \frac{3}{2}(n+1)\log(n+1) \right\rceil, \quad \text{and} \quad K_i = K \cap \left(1 + \frac{1}{n}\right)^i B^n \quad \text{for } i = 0, \ldots, k.$$

Then it suffices to estimate each ratio $V(K_i)/V(K_{i-1})$ up to a relative error of order $\epsilon/(n\log n)$ with error probability of order $\beta/(n\log n)$.

The main step of the algorithm of [DFK89] is based on a method for sampling nearly uniformly from within certain convex bodies K_i. It superimposes a chessboard grid of small cubes (say of edge length δ) on K_i, and performs a random walk over the set $\mathcal{C}_i$ of cubes in this grid that intersect a suitable parallel body $K_i + \alpha B^n$, where α is small. This walk is performed by moving through a facet with probability $1/f_{n-1}(C_n) = (2n)^{-1}$ if this move ends up in a cube of $\mathcal{C}_i$, and staying at the current cube if the move would lead outside of $\mathcal{C}_i$. The random walk gives a *Markov chain* that is irreducible (since the moves are connected), aperiodic, and hence ergodic. But this implies that there is a unique stationary distribution, the limit distribution of the chain, which is easily seen to be a *uniform distribution*. Thus after a sufficiently large (but polynomially bounded) number of steps, the current cube in the random walk can be used to sample nearly uniformly from $\mathcal{C}_i$. Having obtained such a uniformly sampled cube, one determines whether it belongs to $\mathcal{C}_{i-1}$ or to $\mathcal{C}_i \setminus \mathcal{C}_{i-1}$.

Now note that if ν_i is the number of cubes in $\mathcal{C}_i$, then the number $\mu_i = \nu_i/\nu_{i-1}$ is an estimate for the volume ratio $V(K_i)/V(K_{i-1})$. It is this number μ_i that can now be "randomly approximated" using the approximation constructed above of a uniform sampling over $\mathcal{C}_i$. In fact, a cube C that is reached after sufficiently many steps in the random walk will lie in $\mathcal{C}_{i-1}$ with probability approximately $1/\mu_i$; hence this probability can be approximated closely by repeated sampling.

This algorithm has been improved significantly by various authors. [LV06] achieved a bound where (except for logarithmic factors) n enters only to the fourth power—this is denoted by writing $O^*(n^4)$—which is currently the best running time in general. Recently, [CV16b] gave an $O^*(n^3)$ algorithm for convex bodies K containing B^n and being "mostly contained" in $O^*(\sqrt{n})B^n$, i.e., the expected value of $\|X\|^2$ for a uniform random point X of K is $O^*(n)$. Currently such a "well-rounding" can, however, only be achieved in time $O^*(n^4)$, [LV06].

The remarkable improvements over the original $O(n^{23})$ bound for the running time of [DFK89] rely on better initial rounding of the convex body and on improved sampling methods. In particular, the chain of bodies K_i (or equivalently their characteristic functions) were replaced by more general distributions f_i starting with one that is highly concentrated around a point close to an incenter of K (playing the role of K_0) and ending with a near uniform distribution. (In analogy to simulated annealing this process is called "cooling.") [CV16b] use Gaussian functions of the type

$$f_i(x) = \begin{cases} e^{-\frac{\|x\|^2}{2\sigma_i^2}} & \text{for } x \in K; \\ 0 & \text{otherwise,} \end{cases}$$

where the parameters σ_i are suitably adapted. The random walk then picks a random point q from a ball of suitable radius centered at the previous point x_j which is accepted as the next point x_{j+1} with probability $\min\{1, f_i(q)/f_i(x_j)\}$. (This is referred to as Gaussian sampling using the ball walk with Metropolis filter.) The ratio of the integrals of f_{i+1} and f_i is finally estimated by

$$\frac{1}{k} \sum_{j=1}^{k} \frac{f_{i+1}(x_j)}{f_i(x_j)}.$$

With this kind of running time, randomized volume computations are getting close to being practical; see [CV16a] for some corresponding study.

FURTHER READING

[Bol78]   V.G. Boltyanskii. *Hilbert's Third Problem* (Transl. by R. Silverman). Winston, Washington, 1978.

[GW89]   R.J. Gardner and S. Wagon. At long last, the circle has been squared. *Notices Amer. Math. Soc.*, 36:1338–1343, 1989.

[GK94b]   P. Gritzmann and V. Klee. On the complexity of some basic problems in computational convexity: II. Volume and mixed volumes. In T. Bisztriczky, P. McMullen, R. Schneider, and A. Ivić Weiss, editors, *Polytopes: Abstract, Convex and Computational*, vol. 440 of *NATO ASI Series*, pages 373–466, Kluwer, Dordrecht, 1994.

[Had57]   H. Hadwiger. *Vorlesungen über Inhalt, Oberfläche und Isoperimetrie*. Springer-Verlag, Berlin, 1957.

[McM93]   P. McMullen. Valuations and dissections. In P.M. Gruber and J.M. Wills, editors, *Handbook of Convex Geometry*, Volume B, pages 933–988, North-Holland. Amsterdam, 1993.

[MS83]   P. McMullen and R. Schneider. Valuations on convex bodies. In P.M. Gruber and J.M. Wills, editors, *Convexity and Its Applications*, pages 170–247. Birkhäuser, Basel, 1983.

[Sah79]   C.-H. Sah. *Hilbert's Third Problem: Scissors Congruence*. Pitman, San Francisco, 1979.

[Sim03]   M. Simonovits. How to compute the volume in high dimension? *Math. Prog.*, 97:237–374, 2003.

[Vem07]   S. Vempala. Geometric random walks: A survey. In *Combinatorial and Computational Geometry*, vol. 52 of *MSRI Publ.*, pages 577–616. Cambridge University Press, 2007.

[Wag85]   S. Wagon. *The Banach-Tarski Paradox*. Cambridge University Press, 1985.

RELATED CHAPTERS

REFERENCES

[AF92]   D. Avis and K. Fukuda. A pivoting algorithm for convex hulls and vertex enumeration of arrangements of polyhedra. *Discrete Comput. Geom.*, 8:295–313, 1992.

[BF87]   I. Bárány and Z. Füredi. Computing the volume is difficult. *Discrete Comput. Geom.*, 2:319–326, 1987.

[Bar93]   A. Barvinok. Computing the volume, counting integral points, and exponential sums. *Discrete Comput. Geom.*, 10:123–141, 1993.

[BN83]   H. Bieri and W. Nef. A sweep-plane algorithm for computing the volume of polyhedra represented in Boolean form. *Linear Algebra Appl.*, 52/53:69–97, 1983.

[BW92]   G. Brightwell and P. Winkler. Counting linear extensions. *Order*, 8:225–242, 1992.

[CV16a]   B. Cousins and S. Vempala. A practical volume algorithm. *Math. Prog. Comp.*, 8:133–160, 2016.

[CV16b]   B. Cousins and S. Vempala. Gaussian cooling and $O^*(n^3)$ algorithms for volume and Gaussian volume. Preprint, `arXiv:1409.6011v3`, version 3, 2016.

[DV13]   D. Dadush and S. Vempala. Near-optimal deterministic algorithms for volume computation via M-ellipsoids. *Proc. Natl. Acad. Sci. USA.*, 110: 19237–19245, 2013.

[Deh00]   M. Dehn. Über raumgleiche Polyeder. *Nachr. Akad. Wiss. Göttingen Math.-Phys. Kl.*, 345–354, 1900.

[DFK89]   M.E. Dyer, A.M. Frieze, and R. Kannan. A random polynomial time algorithm for approximating the volumes of convex bodies. *J. ACM*, 38:1–17, 1989.

[Hil00]   D. Hilbert. Mathematische Probleme. *Nachr. Königl. Ges. Wiss. Göttingen Math.-Phys. Kl.*, 253–297, 1900; *Bull. Amer. Math. Soc.*, 8:437–479, 1902.

[Kep15]   J. Kepler. *Nova Stereometria doliorum vinariorum.* 1615. See M. Caspar, editor, *Johannes Kepler Gesammelte Werke*, Beck, München, 1940.

[Lac90]   M. Laczkovich. Equidecomposability and discrepancy: A solution of Tarski's circle-squaring problem. *J. Reine Angew. Math.*, 404:77–117, 1990.

[Law91]   J. Lawrence. Polytope volume computation. *Math. Comp.*, 57:259–271, 1991.

[LV06]   L. Lovász and S. Vempala. Simulated annealing in convex bodies and an $O^*(n^4)$ volume algorithm. *J. Comput. Syst. Sci.*, 72:392–417, 2006.

[Sta86]   R.P. Stanley. Two order polytopes. *Discrete Comput. Geom.*, 1:9–23, 1986.

36.4 MIXED VOLUMES

The study of mixed volumes, the *Brunn-Minkowski theory*, forms the backbone of classical convexity theory. It is also useful for applications in other areas, including combinatorics and algebraic geometry. A relationship to solving systems of polynomial equations is described at the end of this section.

GLOSSARY

Mixed volume: Let $K_1, \ldots, K_s$ be convex bodies in $\mathbb{R}^n$, and let $\xi_1, \ldots, \xi_s$ be non-negative reals. Then the function $V\left(\sum_{i=1}^{s} \xi_i K_i\right)$ is a homogeneous polynomial of degree n in the variables $\xi_1, \ldots, \xi_s$, and can be written in the form

$$V\left(\sum_{i=1}^{s} \xi_i K_i\right) = \sum_{i_1=1}^{s} \sum_{i_2=1}^{s} \cdots \sum_{i_n=1}^{s} \xi_{i_1} \xi_{i_2} \cdots \xi_{i_n} V(K_{i_1}, K_{i_2}, \ldots, K_{i_n}),$$

where the coefficients $V(K_{i_1}, K_{i_2}, \ldots, K_{i_n})$ are invariant under permutations of their argument. The coefficient $V(K_{i_1}, K_{i_2}, \ldots, K_{i_n})$ is called the mixed volume of the convex bodies $K_{i_1}, K_{i_2}, \ldots, K_{i_n}$.

36.4.1 MAIN RESULTS

Mixed volumes are nonnegative, monotone, multilinear, and continuous valuations. They generalize the ordinary volume in that $V(K) = V(\overbrace{K, \ldots, K}^{n})$. If A is an affine transformation, then $V(A(K_1), \ldots, A(K_n)) = |\det(A)| V(K_1, \ldots, K_n)$.

Among the most famous inequalities in convexity theory is the ***Aleksandrov-Fenchel inequality***,

$$V(K_1, K_2, K_3, \ldots, K_n)^2 \geq V(K_1, K_1, K_3, \ldots, K_n) \, V(K_2, K_2, K_3, \ldots, K_n),$$

and its consequence, the ***Brunn-Minkowski theorem***, which asserts that for each $\lambda \in [0, 1]$,

$$V^{\frac{1}{n}}((1-\lambda)K_0 + \lambda K_1) \geq (1-\lambda)V^{\frac{1}{n}}(K_0) + \lambda V^{\frac{1}{n}}(K_1).$$

OPEN PROBLEM 36.4.1

Provide a useful geometric characterization of the sequences $(K_1, \ldots, K_n)$ for which equality holds in the Aleksandrov-Fenchel inequality.

36.4.2 TRACTABILITY RESULTS

When n is fixed, there is a polynomial-time algorithm whereby, given s ($\mathcal{V}$- or $\mathcal{H}$-) polytopes $P_1, \ldots, P_s$ in $\mathbb{R}^n$, all the mixed volumes $V(P_{i_1}, \ldots, P_{i_n})$ can be computed.

When the dimension is part of the input, it follows at least that mixed volume computation is not harder than volume computation. In fact, computation (for $\mathcal{V}$-polytopes or $\mathcal{S}$-zonotopes) or approximation (for $\mathcal{H}$-polytopes) of any single mixed volume is #P-easy.

There is a polynomial-time algorithm for approximating the mixed volume of n convex bodies up to a simply exponential error, [Gur09].

36.4.3 INTRACTABILITY RESULTS

Since mixed volumes generalize the ordinary volume, it is clear that mixed volume computation cannot be easier, in general, than volume computation. In addition, there are hardness results for mixed volumes that do not trivially depend on the hardness of volume computations. One such result is described next.

As the term is used here, a *box* is a rectangular parallelotope with axis-aligned edges. Since the vector sum of boxes $V(Z_1, \ldots, Z_n)$ is again a box, the volume of the sum is easy to compute. Nevertheless, computation of the mixed volume $V(Z_1, \ldots, Z_n)$ is hard; see [DGH98]. This is in interesting contrast to the fact that the volume of a sum of segments (a zonotope) is hard to compute even though each of the mixed volumes can be computed in polynomial time.

36.4.4 RANDOMIZED ALGORITHMS

Since the mixed volumes of convex bodies $K_1, \ldots, K_s$ are coefficients of the polynomial $\varphi(\xi_1, \ldots, \xi_s) = V(\sum_{i=1}^{s} \xi_i K_i)$, it seems natural to estimate these coefficients by combining an interpolation method with a randomized volume algorithm. However, there are significant obstacles to this approach, even for the case of two bodies. First, for a general polynomial φ there is *no* way of obtaining *relative* estimates of its coefficients from *relative* estimates of the values of φ. This can be overcome in the case of two bodies by using the special structure of the polynomial $p(x) = V(K_1 + xK_2)$. However, even then the absolute values of the entries of the "inversion" that is used to express the coefficients of the polynomial in terms of its approximate values are not bounded by a polynomial, while the randomized volume approximation algorithm is polynomial only in $\frac{1}{\tau}$ but not in $\text{size}(\tau)$.

Suppose that $\psi : \mathbb{N} \to \mathbb{N}$ is nondecreasing with

$$\psi(n) \le n \qquad \text{and} \qquad \psi(n) \log \psi(n) = o(\log n).$$

Then there is a polynomial-time algorithm for the problem whose instance consists of $n, s \in \mathbb{N}$, $m_1, \ldots, m_s \in \mathbb{N}$ with $m_1 + m_2 + \cdots + m_s = n$ and $m_1 \ge n - \psi(n)$, of well-bounded convex bodies $K_1, \ldots, K_s$ of $\mathbb{R}^n$ given by a weak membership oracle, and of positive rational numbers ϵ and β, and whose output is a random variable $\hat{V}_{m_1, \ldots, m_s} \in \mathbb{Q}$ such that

$$\text{prob} \left\{ \frac{|\hat{V}_{m_1, \ldots, m_s} - V_{m_1, \ldots, m_s}|}{V_{m_1, \ldots, m_s}} \ge \epsilon \right\} \le \beta,$$

where

$$V_{m_1, \ldots, m_s} = V(\overbrace{K_1, \ldots, K_1}^{m_1}, \ldots, \overbrace{K_s, \ldots, K_s}^{m_s}).$$

Note that the hypotheses above require that m_1 is close to n, and hence that the remaining m_i's are relatively small. A special feature of an interpolation method as used for the proof of this result is that in order to compute *a specific* coefficient of the polynomial under consideration, it computes essentially *all previous* coefficients. Since there can be a polynomial-time algorithm for computing *all such* mixed volumes only if $\psi(n) \le \log n$, the above result is essentially best-possible for any interpolation method.

OPEN PROBLEM 36.4.2   [DGH98]

Is there a polynomial-time randomized algorithm that, for any $n, s \in \mathbb{N}$, $m_1, \ldots, m_s \in \mathbb{N}$ with $m_1 + m_2 + \cdots + m_s = n$, well-bounded convex bodies $K_1, \ldots, K_s$ in $\mathbb{R}^n$ given *by a weak membership oracle, and positive rationals ϵ and β, computes a random variable $\hat{V}_{m_1,\ldots,m_s} \in \mathbb{Q}$ such that* $\mathrm{prob}\{|\hat{V}_{m_1,\ldots,m_s} - V_{m_1,\ldots,m_s}|/V_{m_1,\ldots,m_s} \geq \epsilon\} \leq \beta$?

Even the case $s = n$, $m_1 = \cdots = m_s = 1$ is open in general. See, however, [Bar97] for some partial results and [Mal16] for performance bounds in terms of geometric invariants.

AN APPLICATION

Let $S_1, S_2, \ldots, S_n$ be subsets of $\mathbb{Z}^n$, and consider a system $F = (f_1, \ldots, f_n)$ of Laurent polynomials in n variables, such that the exponents of the monomials in f_i are in S_i for all $i = 1, \ldots, n$. For $i = 1, \ldots, n$, let

$$f_i(x) = \sum_{q \in S_i} c_q^{(i)} x^q,$$

where $f_i \in \mathbb{C}[x_1, x_1^{-1}, \ldots, x_n, x_n^{-1}]$, and x^q is an abbreviation for the monomial $x_1^{q_1} \cdots x_n^{q_n}$; $x = (x_1, \ldots, x_n)$ is the vector of indeterminates and $q = (q_1, \ldots, q_n)$ the vector of exponents. Further, let $\mathbb{C}^* = \mathbb{C} \setminus \{0\}$.

Now, if the coefficients $c_q^{(i)}$ ($q \in S_i$) are chosen "generically," then the number $L(F)$ of distinct common roots of the system F in $(\mathbb{C}^*)^n$ depends only on the **Newton polytopes** $P_i = \mathrm{conv}(S_i)$ of the polynomials. More precisely,

$$L(F) = n! \cdot V(P_1, P_2, \ldots, P_n).$$

In general, $L(F) \leq n! \cdot V(P_1, P_2, \ldots, P_n)$. These connections can be utilized to develop a numerical continuation method for computing the isolated solutions of sparse polynomial systems; see [CLO98, DE05, Stu02].

FURTHER READING

[BZ88]   Y.D. Burago and V.A. Zalgaller. *Geometric Inequalities*. Springer-Verlag, Berlin, 1988.

[CLO98]   D. Cox, J. Little, and D. O'Shea. Solving polynomial equations. In *Using Algebraic Geometry*, vol. 185 of *Grad. Texts Math.*, pages 24–70, Springer, New York, 1998.

[DE05]   A. Dickenstein and I.Z. Emiris. *Solving Polynomial Equations*. Springer-Verlag, Berlin, 2005.

[GK94b]   P. Gritzmann and V. Klee. On the complexity of some basic problems in computational convexity: II. Volume and mixed volumes. In T. Bisztriczky, P. McMullen, R. Schneider, and A. Ivić Weiss, editors, *Polytopes: Abstract, Convex and Computational*, vol. 440 of *NATO ASI Series*, pages 373–466. Kluwer, Dordrecht, 1994.

[San93]   J.R. Sangwine-Yager. Mixed volumes. In P.M. Gruber and J.M. Wills, editors, *Handbook of Convex Geometry*, Volume A, pages 43–72, North-Holland, Amsterdam, 1993.

[Sch93]   R. Schneider. *Convex Bodies: The Brunn-Minkowski Theory*. Vol. 44 of *Encyclopedia Math. Appl.*, Cambridge University Press, 1993; 2nd expanded edition, 2013.

[Stu02]   B. Sturmfels. *Solving Systems of Polynomial Equations*. Vol. 97 of *CBMS Regional Conf. Ser. in Math.*, AMS, Providence, 2002.

RELATED CHAPTERS

Chapter 15: Basic properties of convex polytopes
Chapter 44: Randomization and derandomization

REFERENCES

[Bar97]   A. Barvinok. Computing mixed discriminants, mixed volumes and permanents. *Discrete Comput. Geom.*, 18:205–237, 1997.

[DGH98]   M.E. Dyer, P. Gritzmann, and A. Hufnagel. On the complexity of computing mixed volumes. *SIAM J. Comput.*, 27:356–400, 1998.

[Gur09]   L. Gurvits. A polynomial-time algorithm to approximate the mixed volume within a simply exponential factor. *Discrete Comput. Geom.*, 41:533–555, 2009.

[Mal16]   G. Malajovich. Computing mixed volume and all mixed cells in quermassintegral time. *Found. Comput. Math.*, in press, 2016.

36.5 CONTAINMENT PROBLEMS

Typically, containment problems involve two fixed sequences, Γ and Ω, that are given as follows: for each $n \in \mathbb{N}$, let $\mathcal{C}_n$ denote a family of closed convex subsets of $\mathbb{R}^n$, and let $\omega_n : \mathcal{C}_n \longrightarrow \mathbb{R}$ be a functional that is nonnegative and is monotone with respect to inclusion. Then $\Gamma = (\mathcal{C}_n)_{n\in\mathbb{N}}$ and $\Omega = (\omega_n)_{n\in\mathbb{N}}$.

GLOSSARY

(Γ, Ω)-INBODY: Accepts as input a positive integer n, a body K in $\mathbb{R}^n$ that is given by an oracle or is an $\mathcal{H}$-polytope, a $\mathcal{V}$-polytope, or an $\mathcal{S}$-zonotope, and a positive rational λ. It answers the question of whether there is a $C \in \mathcal{C}_n$ such that $C \subset K$ and $\omega_n(C) \geq \lambda$.

(Γ, Ω)-CIRCUMBODY is defined similarly for $C \supset K$.

j-simplex S **bound to a polytope** P: Each vertex of S is a vertex of P.

Largest j-simplex in a given polytope: One of maximum j-measure.

36.5.1 THE GENERAL CONTAINMENT PROBLEM

The general containment problem deals with the question of computing, approximating, or measuring extremal bodies of a given class that are contained in or contain a given convex body. The broad survey [GK94a] can (yet being somewhat older) still be used as a starting point for getting acquainted with the subject and its many applications. Here we want to minimize the overlap with other chapters, and restrict the exposition to some selected examples. In particular we do not focus on *coresets* (see AHV07), as they are covered in Chapter 48.

The emphasis here will be on containment under homothety and affinity. For some results on containment under similarity see [GK94a, Sec. 7]; see also [Fir15] for numerical computations in the case of $\mathcal{V}$-polytopes in $\mathcal{H}$-polytopes.

36.5.2 OPTIMAL CONTAINMENT UNDER HOMOTHETY

The results on (Γ, Ω)-INBODY and (Γ, Ω)-CIRCUMBODY are summarized below for the case in which each C_n is a fixed polytope,

$$\mathcal{C}_n = \{g(C_n) \mid g \text{ is a homothety}\},$$

and

$$\omega_n(g(C_n)) = \rho, \quad \text{when } g(C_n) = a + \rho C_n \text{ for some } a \in \mathbb{R}^n \text{ and } \rho \geq 0.$$

As an abbreviation, these specific problems are denoted by $\mathcal{E}^{\text{Hom}}$-INBODY and $\mathcal{E}^{\text{Hom}}$-CIRCUMBODY, respectively, where $\mathcal{E} = (C_n)_{n \in \mathbb{N}}$ and a subscript ($\mathcal{V}$ or $\mathcal{H}$) is used to indicate the manner in which each C_n is presented.

There are polynomial-time algorithms for the following problems:

$\mathcal{E}_{\mathcal{V}}^{\text{Hom}}$-INBODY for $\mathcal{V}$-polytopes P; $\mathcal{E}_{\mathcal{V}}^{\text{Hom}}$-CIRCUMBODY for $\mathcal{V}$-polytopes P;

$\mathcal{E}_{\mathcal{V}}^{\text{Hom}}$-INBODY for $\mathcal{H}$-polytopes P; $\mathcal{E}_{\mathcal{H}}^{\text{Hom}}$-CIRCUMBODY for $\mathcal{V}$-polytopes P;

$\mathcal{E}_{\mathcal{H}}^{\text{Hom}}$-INBODY for $\mathcal{H}$-polytopes P; $\mathcal{E}_{\mathcal{H}}^{\text{Hom}}$-CIRCUMBODY for $\mathcal{H}$-polytopes P.

These positive results are best possible in the sense that the cases not listed above contain instances of NP-hard problems. In fact, the problem $\mathcal{E}_{\mathcal{H}}^{\text{Hom}}$-INBODY is coNP-complete even when C_n is the standard unit $\mathcal{H}$-cube while P is restricted to the class of all affinely regular $\mathcal{V}$-cross-polytopes centered at the origin. The problem $\mathcal{E}_{\mathcal{V}}^{\text{Hom}}$-CIRCUMBODY is coNP-complete even when C_n is the standard $\mathcal{V}$-cross-polytope while P is restricted to the class of all $\mathcal{H}$-parallelotopes centered at the origin.

There are some results for bodies that are more general than polytopes. Suppose that for each $n \in \mathbb{N}$, C_n is a centrally symmetric body in $\mathbb{R}^n$, and that there exists a number μ_n whose size is bounded by a polynomial in n and an n-dimensional $\mathcal{S}$-parallelotope Z that is strictly inscribed in $\mu_n C_n$ (i.e., the intersection of Z with the boundary of $\mu_n C_n$ consists of the vertex set of Z), the size of the presentation being bounded by a polynomial in n. Then with $\mathcal{E} = (C_n)_{n \in \mathbb{N}}$, (an appropriate variant of) the problem $\mathcal{E}^{\text{Hom}}$-CIRCUMBODY is NP-hard for the classes of all centrally symmetric $(n-1)$-dimensional $\mathcal{H}$-polytopes in $\mathbb{R}^n$. With the aid of polarity, similar results for $\mathcal{E}^{\text{Hom}}$-INBODY can be obtained. A particularly important special case is that of *norm maximization*, i.e., maximizing the Euclidean (or some other norm) over a polytope.

Besides the obvious examples of unit balls of norms and polytopes, containment problems have also been studied for *spectrahedra* which arise in convex algebraic geometry [HN12] and generalize the class of polyhedra. [KTT13] extend known complexity results to spectrahedra. For instance, they show that deciding whether a given $\mathcal{V}$-polytope is contained in a given spectrahedron can be decided in polynomial time, while deciding whether a spectrahedron is contained in a $\mathcal{V}$-polytope is co-NP-hard. As spectrahedra arise as feasible regions of semidefinite programs they also give semidefinite conditions to certify containment.

36.5.3 OPTIMAL CONTAINMENT UNDER AFFINITY: SIMPLICES

This section focuses on the problem of finding a largest j-dimensional simplex in a given n-dimensional polytope, where *largest* means of maximum j-measure.

When an n-polytope P has m vertices, it contains at most $\binom{m}{j+1}$ bound j-simplices. There is always a largest j-simplex that is bound, and hence there is a finite algorithm for finding a largest j-simplex contained in P.

Each largest j-simplex in P contains at least two vertices of P. However, there are polytopes P of arbitrarily large dimension, with an arbitrarily large number of vertices, such that some of the largest n-simplices in P have only two vertices in the vertex-set of P. Hence for $j \geq 2$ it is not clear whether there is a finite algorithm for producing a useful presentation of *all* the largest j-simplices in a given n-polytope.

The problem of finding a largest j-simplex in a $\mathcal{V}$- or $\mathcal{H}$-polytope can be solved in polynomial time when the dimension n of the polytope is fixed. Further, for fixed j, the volumes of all bound j-simplices in a given $\mathcal{V}$-polytope can be computed in polynomial time (even for varying n).

Suppose that the functions $\psi : \mathbb{N} \to \mathbb{N}$ and $\gamma : \mathbb{N} \to \mathbb{N}$ are both of order $\Omega(n^{1/k})$ for some $k \in \mathbb{N}$, and that $1 \leq \gamma(n) \leq n$ for each $n \in \mathbb{N}$. Then the following problem is NP-complete: Given $n, \lambda \in \mathbb{N}$, and the vertex set V of an n-dimensional $\mathcal{V}$-polytope $P \subset \mathbb{R}^n$ with $|V| \leq n + \psi(n)$, and given $j = \gamma(n)$, decide whether P contains a j-simplex S such that $(j!)^2 \mathrm{vol}(S)^2 \geq \lambda$. Note that the conditions for γ are satisfied when $\gamma(n) = \max\{1, n - \mu\}$ for a nonnegative integer constant μ, and also when $\gamma(n) = \max\{1, \lfloor \mu n \rfloor\}$ for a fixed rational μ with $0 < \mu \leq 1$.

A similar hardness result holds for $\mathcal{H}$-polytopes. There the question is the same, but the growth condition on the function γ is that $1 \leq \gamma(n) \leq n$ and that there exists a function $f : \mathbb{N} \to \mathbb{N}$, bounded by a polynomial in n, such that for each $n \in \mathbb{N}$, $f(n) - \gamma(f(n)) = n$. Note that such an f exists when the function γ is constant, and also when $\gamma(n) = \lfloor \mu n \rfloor$ for fixed rational μ with $0 < \mu < 1$.

Under the assumption that the function $\gamma : \mathbb{N} \to \mathbb{N}$ is such that $\gamma(n) = \Omega(n^{1/k})$ for some fixed $k > 0$, [Pac02] gives a unifying approach for proving the NP-hardness of the problems, for which an instance consists of $n \in \mathbb{N}$, an $\mathcal{H}$-polytope or $\mathcal{V}$-polytope P in $\mathbb{R}^n$, and a rational $\lambda > 0$, and the question is whether there exists an $\gamma(n)$-simplex $S \subset P$ with $\mathrm{V}^2(S) \geq \lambda$.

The following conjecture is, however, still open.

CONJECTURE 36.5.1   [GKL95]

For each function $\gamma : \mathbb{N} \to \mathbb{N}$ with $1 \leq \gamma(n) \leq n$, the problem of finding a largest j-simplex in a given n-dimensional $\mathcal{H}$-parallelotope P is NP-hard.

The "dual" problem of finding smallest simplices containing a given polytope P seems even harder, since the relationship between a smallest such simplex and the faces of P is much weaker. However, [Pac02] gives the following hardness results for *j-simplicial cylinders* C which are cylinders of the form $C = S + L$, where S is a j-simplex with $0 \in \mathrm{aff}(S)$ and $L = \mathrm{aff}(S)^\perp$. Let again the function $\gamma : \mathbb{N} \to \mathbb{N}$ be such that $\gamma(n) = \Omega(n^{1/k})$ for some fixed $k > 0$. Then it is NP-hard to decide whether for given $n \in \mathbb{N}$, positive rational λ and an $\mathcal{H}$-polytope or $\mathcal{V}$-polytope P in $\mathbb{R}^n$ there exists an $\gamma(n)$-simplicial cylinder C with $P \subset C$ and $\mathrm{V}^2(S) \leq \lambda$. Note that the condition on γ particularly includes the case $j = n$, i.e., that of an ordinary n-dimensional simplex.

These results have been complemented in various ways. [Kou06] shows that the decision problem related to finding a largest j-simplex in a given $\mathcal{V}$-polytope is $W[1]$-complete with respect to the parameter j. See e.g., [FG06] for background information on parametrized complexity. Also deterministic approximation and nonapproximability results have been given [BGK00a, Pac04, Kou06, DEFM15, Nik15] showing in particular that the problem of finding a largest j-simplex in a given $\mathcal{V}$-polytope can on the one hand be approximated in polynomial time up to a factor of $e^{j/2+o(j)}$; yet, on the other hand, there is a constant $\mu > 1$ such that it is NP-hard to approximate within a factor of μ^j.

APPLICATIONS

Applications of this problem and its relatives include that of finding submatrices of maximum determinant, and, in particular, the Hadamard determinant problem, of finding optimal weighing designs, and bounding the growth of pivots in Gaussian elimination with complete pivoting; see [GK94a], [Nik15].

36.5.4 OPTIMAL CONTAINMENT UNDER AFFINITY: ELLIPSOIDS

For an arbitrary proper body K in $\mathbb{R}^n$, there is a unique ellipsoid E_0 of maximum volume contained in K, and it is concentric with the unique ellipsoid E of minimum volume containing K. If a is the common center, then $K \subset a + n(E_0 - a)$, where the factor n can be replaced by $\sqrt{n}$ when K is centrally symmetric. E is called the **Löwner-John ellipsoid** of K, and it plays an important role in the algorithmic theory of convex bodies.

Algorithmic approximations of the Löwner-John ellipsoid can be obtained by use of the ellipsoid method [GLS88]: There exists an oracle-polynomial-time algorithm that, for any well-bounded body K of $\mathbb{R}^n$ given by a weak separation oracle, finds a point a and a linear transformation A such that

$$a + A(B^n) \subset K \subset a + (n+1)\sqrt{n}A(B^n).$$

Further, the dilatation factor $(n+1)\sqrt{n}$ can be replaced by $\sqrt{n(n+1)}$ when K is symmetric, by $(n+1)$ when K is an $\mathcal{H}$-polytope, and by $\sqrt{n+1}$ when K is a symmetric $\mathcal{H}$-polytope.

[TKE88] and [KT93] give polynomial-time algorithms for approximating the ellipsoid of maximum volume E_0 that is contained in a given $\mathcal{H}$-polytope. For each rational $\gamma < 1$, there exists a polynomial-time algorithm that, given $n, m \in \mathbb{N}$ and $a_1, \ldots, a_m \in \mathbb{Q}^n$, computes an ellipsoid $E = a + A(B^n)$ such that

$$E \subset P = \{x \in \mathbb{R}^n \mid \langle a_i x \rangle \leq 1, \text{ for } i = 1, \ldots, m\} \quad \text{and} \quad \frac{V(E)}{V(E_0)} \geq \gamma.$$

The running time of the algorithm is

$$O\left(m^{3.5} \log\big(mR/(r\log(1/\gamma))\big) \log\big(nR/(r\log(1/\gamma))\big)\right),$$

where the numbers r and R are, respectively, a lower bound on the inradius of P and an upper bound on its circumradius.

It is not known whether a similar result holds for $\mathcal{V}$-polytopes.

As shown in [TKE88], an approximation of E_0 of the kind given above leads to the following inclusion:

$$a + A(B^n) \subset K \subset a + \frac{n(1 + 3\sqrt{1 - \gamma})}{\gamma} A(B^n).$$

Other important ellipsoids related to convex bodies K are the *M-ellipsoids*; see e.g., [Pis89]. Intuitively an M-ellipsoid E is an ellipsoid with small covering number with respect to K. More precisely, for two sets A, B let $N(A; B)$ denote the number of translates of B needed to cover A. Then every convex body K in $\mathbb{R}^n$ admits an ellipsoid E for which $N(K; E)N(E; K)$ is bounded by $2^{O(n)}$, [Mil86]; this is best possible up to the constant in the exponent. [DV13] give a deterministic algorithm for computing an M-ellipsoid for a well-bounded convex body K in $\mathbb{R}^n$ given by a weak membership oracle in time $2^{O(n)}$. This is best possible up to the constant in the exponent.

FURTHER READING

[AHV07]   P.K. Agarwal, S. Har-Peled, and K.R. Varadarajan. Geometric approximation via coresets. *Combinatorial and Computational Geometry*, vol. 52 of *MSRI Publ.*, pages 1–30, Cambridge University Press, 2007.

[GK94a]   P. Gritzmann and V. Klee. On the complexity of some basic problems in computational convexity: I. Containment problems. *Discrete Math.*, 136:129–174, 1994. Reprinted in W. Deuber, H.-J. Prömel, and B. Voigt, editors, *Trends in Discrete Mathematics*, pages 129–174, North-Holland, Amsterdam, 1994.

[GLS88]   M. Grötschel, L. Lovász, and A. Schrijver. *Geometric Algorithms and Combinatorial Optimization*. Springer-Verlag, Berlin, 1988; second edition, 1993.

[FG06]   J. Flum and M. Grohe. *Parameterized Complexity Theory*. Springer, Berlin, 2006.

[Pis89]   G. Pisier. *The Volume of Convex Bodies and Banach Space Geometry*. Cambridge University Press, 1989.

RELATED CHAPTERS

Chapter 48: Coresets and sketches

REFERENCES

[BGK00a] A. Brieden, P. Gritzmann, and V. Klee. Oracle-polynomial-time approximation of largest simplices in convex bodies. *Discrete Math.*, 221:79–92, 2000.

[DV13]   D. Dadush and S. Vempala. Near-optimal deterministic algorithms for volume computation via M-ellipsoids. *Proc. Natl. Acad. Sci. USA.*, 110: 19237–19245, 2013.

[DEFM15] M. Di Summa, F. Eisenbrand, Y. Faenza, and C. Moldenhauer. On largest volume simplices and sub-determinants. In *Proc. 26th ACM-SIAM Sympos. Discrete Algorithms*, pages 315–323, 2015.

[Fir15]   M. Firsching. Computing maximal copies of polyhedra contained in a polyhedron. *Exp. Math.*, 24:98–105, 2015.

[GKL95]   P. Gritzmann, V. Klee, and D.G. Larman. Largest j-simplices in n-polytopes. *Discrete Comput. Geom.*, 13:477–515, 1995.

[HN12]   J.W. Helton and J. Nie. Semidefinite representation of convex sets and convex hulls. In J.B. Lasserre and M.F. Anjos, editors, *Handbook of Semidefinite, Conic and Polynomial Programming*, pages 77112, Springer, New York, 2012.

[KTT13]   K. Kellner, T. Theobald, and C. Trabandt. Containment problems for polytopes and spectrahedra. *SIAM J. Optim.*, 23:1000–1020, 2013.

[KT93]   L.G. Khachiyan and M.J. Todd. On the complexity of approximating the maximal inscribed ellipsoid in a polytope. *Math. Prog.*, 61:137–160, 1993.

[Kou06]   I. Koutis. Parameterized complexity and improved inapproximability for computing the largest j-simplex in a V-polytope. *Inform. Process. Lett.*, 100:8–13, 2006.

[Mil86]   V.D. Milman. Inegalites de Brunn-Minkowski inverse et applications at la theorie locales des espaces normes. *C. R. Acad. Sci. Paris*, 302:25–28, 1986.

[Nik15]   A. Nikolov. Randomized rounding for the largest simplex problem. In *Proc. 47th ACM Sympos. Theory Comput.*, pages 861–870, 2015.

[Pac02]   A. Packer. NP-hardness of largest contained and smallest containing simplices for V- and H-polytopes. *Discrete Comput. Geom.*, 28:349–377, 2002.

[Pac04]   A. Packer. Polynomial-time approximation of largest simplices in V-polytopes. *Discrete Appl. Math.*, 134:3213–237, 2004.

[TKE88]   S.P. Tarasov, L.G. Khachiyan, and I.I. Erlich. The method of inscribed ellipsoids. *Soviet Math. Dokl.*, 37:226–230, 1988.

36.6 RADII

The diameter, width, circumradius, and inradius of a convex body are classical functionals that play an important role in convexity theory and in many applications. For other applications, generalizations have been introduced. Here we focus on the case that the underlying space is a ***Minkowski space*** (i.e., a finite-dimensional normed space) $\mathbb{M} = (\mathbb{R}^n, \| \ \|)$. Let B denote its unit ball, j a positive integer, and K a convex body. For some generalizations to the case of non symmetric "unit balls" see [BK13, BK14].

GLOSSARY

Outer j-radius $R_j(K)$ **of** K: Infimum of the positive numbers ρ such that the space contains an $(n-j)$-flat F for which $K \subset F + \rho B$.

j-ball of radius ρ: Set of the form $(q + \rho B) \cap F = \{x \in F \mid \|x - q\| \leq \rho\}$ for some j-flat F in $\mathbb{R}^n$ and point $q \in F$.

Inner j-radius $r_j(K)$ **of** K: Maximum of the radii of the j-balls contained in K.

Diameter **of** K: $2r_1(K)$.

Width **of** K: $2R_1(K)$.

Inradius **of** K: $r_n(K)$.

Circumradius **of** K: $R_n(K)$.

Note that for a convex body K that is symmetric about the origin $r_1(K)$ coincides with the norm-maximum $\max_{x \in K} \|x\|$ over K.

For the case of variable dimension (i.e., the dimension is part of the input), Tables 36.6.1, 36.6.2, and 36.6.3 provide a rapid indication of the main complexity results for the most important radii: r_1, R_1, r_n, and R_n; and for the three most important ℓ_p spaces: $\mathbb{R}_2^n$, $\mathbb{R}_1^n$, and $\mathbb{R}_\infty^n$. The designations P, NPC, and NPH indicate respectively polynomial-time computability, NP-completeness, and NP-hardness. The tables provide only a rough indication of results. They are imprecise in the following respects: (i) the diameter and width are actually equal to $2r_1$ and $2R_1$ respectively; (ii) the results for $\mathbb{R}_2^n$ involve the square of the radius rather than the radius itself; (iii) some of the P entries are based on polynomial-time approximability rather than polynomial-time computability; (iv) the designations NPC and NPH do not refer to computability per se, but to the appropriately related decision problems involving the establishment of lower or upper bounds for the radii in question.

TABLE 36.6.1   Complexity of radii in $\mathbb{R}_2^n$.

Polytope functional		$\mathcal{H}$-polytopes		$\mathcal{V}$-polytopes	
		general	symmetric	general	symmetric
Diameter	r_1^2	NPC	NPC	P	P
Inradius	r_n^2	P	P	NPH	NPC
Width	R_1^2	NPC	P	NPC	NPC
Circumradius	R_n^2	NPC	NPC	P	P

TABLE 36.6.2   Complexity of radii in $\mathbb{R}_1^n$.

Polytope functional		$\mathcal{H}$-polytopes		$\mathcal{V}$-polytopes	
		general	symmetric	general	symmetric
Diameter	r_1	NPC	NPC	P	P
Inradius	r_n	P	P	P	P
Width	R_1	P	P	P	P
Circumradius	R_n	NPC	NPC	P	P

TABLE 36.6.3   Complexity of radii in $\mathbb{R}_\infty^n$.

Polytope functional		$\mathcal{H}$-polytopes		$\mathcal{V}$-polytopes	
		general	symmetric	general	symmetric
Diameter	r_1	P	P	P	P
Inradius	r_n	P	P	NPC	NPC
Width	R_1	NPC	P	NPC	NPC
Circumradius	R_n	P	P	P	P

For inapproximability results in the Turing machine model see [BGK00b] and [Bri02]; for sharp bounds on the approximation error of polynomial-time algorithms in the oracle model see [BGK$^+$01]. In view of the results in Section 36.3 on volume computation where there is a sharp contrast between the performance of deterministic and randomized algorithms it may be worth noting that, generally, for radii *randomization does not help!* This means that the same limitations on the error of polynomial-time approximations that apply for deterministic algorithms also apply for randomized algorithms, [BGK$^+$01].

Parametrized complexity (see e.g., [FG06]) has been used in [KKW15] to analyze more sharply how the hardness of norm maximization and radius computation depends on the dimension n. In particular, it is show that for $p = 1$ the problem of maximizing the p-th power of the ℓ_p-norm over $\mathcal{H}$-polytopes is fixed parameter tractable but that for each $p \in \mathbb{N} \setminus \{1\}$ norm maximization is $W[1]$-hard.

APPLICATIONS

Applications of radii include conditioning in global optimization, sensitivity analysis of linear programs, orthogonal minimax regression, computer graphics and computer vision, chromosome classification, set separation, and design of membranes and sieves; see [GK93b].

FURTHER READING

[BGK$^+$01]   A. Brieden, P. Gritzmann, R. Kannan, V. Klee, L. Lovász, and M. Simonovits. Deterministic and randomized polynomial-time approximation of radii. *Mathematika*, 48:63–105, 2001.

[GK94a]   P. Gritzmann and V. Klee. On the complexity of some basic problems in computational convexity: I. Containment problems. *Discrete Math.*, 136:129–174, 1994. Reprinted in W. Deuber, H.-J. Prömel, and B. Voigt, editors, *Trends in Discrete Mathematics*, pages 129–174, North-Holland, Amsterdam, 1994.

[FG06]   J. Flum and M. Grohe. *Parameterized Complexity Theory*. Springer-Verlag, Berlin, 2006.

REFERENCES

[BK13]   R. Brandenberg and S. König. No dimension-independent core-set for containment under homothetics. *Discrete Comput. Geom.*, 49:3–21, 2013.

[BK14]   R. Brandenberg and S. König. Sharpening geometric inequalities using computable symmetry measures. *Mathematika*, 61:559–580, 2014.

[Bri02]   A. Brieden. On geometric optimization problems likely not contained in APX. *Discrete Comput. Geom.*, 28:201–209, 2002.

[BGK00b]   A. Brieden, P. Gritzmann and V. Klee. Inapproximability of some geometric and quadratic optimization problem. In: P.M. Pardalos, editor, *Approximation and Complexity in Numerical Optimization: Continuous and Discrete Problems*, pages 96–115, Kluwer, Dordrecht, 2000.

[GK93b]   P. Gritzmann and V. Klee. Computational complexity of inner and outer j-radii of polytopes in finite dimensional normed spaces. *Math. Prog.*, 59:163–213, 1993.

[KKW15]   C. Knauer, S. König, and D. Werner. Fixed parameter complexity and approximability of norm maximization. *Discrete Comput. Geom.*, 53:276–295, 2015.

36.7 CONSTRAINED CLUSTERING

Clustering has long been known as an instrumental part of data analytics. In increasingly many applications additional constraints are imposed on the clusters, for instance, bounding the cluster sizes. As has been observed in numerous fields, good clusterings are closely related to geometric diagrams, i.e., various generalizations of Voronoi diagrams. Applications of constrained clustering include the representation of polycrystals (grain maps) in material science, [ABG+15], farmland consolidation, [BBG14], facility and robot network design [Cor10], and electoral district design, [BGK17].

As Voronoi diagrams are covered in great detail in Chapter 27, we will concentrate in the following on some geometric aspects of the relation between diagrams and constrained clusterings.

GLOSSARY

Instance of a constrained clustering problem: $(k, m, n, X, \omega, \kappa^-, \kappa^+)$ (*weakly balanced*) or $(k, m, n, X, \omega, \kappa)$ (*strongly balanced*), where $k, m, n \in \mathbb{N}$, $X = \{x_1, \ldots, x_m\} \subset \mathbb{R}^n$, $\omega : X \to (0, \infty)$, $\kappa, \kappa^-, \kappa^+ : \{1, \ldots, k\} \to (0, \infty)$, such that $\kappa^- \leq \kappa^+$ and $\sum_{i=1}^k \kappa^-(i) \leq \sum_{j=1}^m \omega(x_j) \leq \sum_{i=1}^k \kappa^+(i)$ in the weakly balanced case and $\sum_{j=1}^m \omega(x_j) = \sum_{i=1}^k \kappa(i)$ in the strongly balanced case.

Of course, n is again the dimension of space, m is the number of points of the given set X in $\mathbb{R}^n$, and $\omega(x)$ specifies the weight of each point $x \in X$. The set X has to be split (in a fractional or integer fashion that will be specified explicitly next) into clusters $C_1, \ldots, C_k$ whose total weights lie in the given intervals $[\kappa^-(i), \kappa^+(i)]$ in the weakly balanced case or actually coincide with the prescibed weight $\kappa(i)$ in the strongly balanced case, respectively.

Balanced clustering $\mathcal{C}$ for an instance $(k, m, n, X, \omega, \kappa^-, \kappa^+)$, or $(k, m, n, X, \omega, \kappa)$: $\mathcal{C} = \{C_1, \ldots, C_k\}$ with $C_i = (\xi_{i,1}, \ldots, \xi_{i,m}) \in [0,1]^m$ for $i \in \{1, \ldots, k\}$, such that $\sum_{i=1}^k \xi_{i,j} = 1$ for $j \in \{1, \ldots, m\}$ and $\kappa^-(i) \leq \sum_{j=1}^m \omega(x_j)\xi_{i,j} \leq \kappa^+(i)$ for $i \in \{1, \ldots, k\}$ in the weakly balanced case and $\sum_{j=1}^m \omega(x_j)\xi_{i,j} = \kappa(i)$ in the strongly balanced case. C_i is the ith **cluster**. Note that $\xi_{i,j}$ is the fraction of x_j assigned to the cluster C_i.

Integer clustering $\mathcal{C}$: $C_i \in \{0,1\}^m$ for $i \in \{1, \ldots, k\}$.

Constrained clustering problem: Given an instance of a constrained clustering problem, find a balanced (integer) clustering $\mathcal{C}$ which optimizes some given objective function (examples of which will be given later).

Gravity vector of a given clustering $\mathcal{C}$: $\mathbf{g}(\mathcal{C}) = (g(C_1)^T, \ldots, g(C_k)^T)^T$, where $g(C_i) = \left(\sum_{j=1}^m \xi_{i,j}\omega(x_j)x_j\right) / \left(\sum_{j=1}^m \xi_{i,j}\omega(x_j)\right)$ is the **center of gravity** of C_i for $i \in \{1, \ldots, k\}$.

Gravity body for a given instance $(k, m, n, X, \omega, \kappa^-, \kappa^+)$, or $(k, m, n, X, \omega, \kappa)$: $Q = \mathrm{conv}\{\mathbf{g}(\mathcal{C}) \mid \mathcal{C} \text{ is a balanced clustering}\}$.

Support $\operatorname{supp}(C_i)$ ***of a cluster*** C_i***:*** $\operatorname{supp}(C_i) = \{x_j \in X \mid \xi_{i,j} > 0\}$.

Support multi-graph $G(\mathcal{C})$ of a clustering $\mathcal{C}$: vertices: $C_1, \ldots, C_k$; edges: $\{C_i, C_l\}$ for every j for which $x_j \in \operatorname{supp}(C_i) \cap \operatorname{supp}(C_l)$; label of an edge $\{C_i, C_l\}$: x_j. A cycle in $G(\mathcal{C})$ is ***colored*** if not all of its labels coincide. $G(\mathcal{C})$ is ***c-cycle-free*** if it does not contain any colored cycle.

Given $\mathcal{F} = \{\varphi_1, \ldots, \varphi_k\}$ with functions $\varphi_i : \mathbb{R}^n \to \mathbb{R}$ for $i \in \{1, \ldots, k\}$; $\mathcal{F}$-***diagram*** $\mathcal{P}$***:*** $\mathcal{P} = \{P_1, \ldots, P_k\}$ with $P_i = \{x \in \mathbb{R}^n \mid \varphi_i(x) \leq \varphi_l(x) \; \forall l \in \{1, \ldots, k\}\}$ for all i.

An $\mathcal{F}$-diagram $\mathcal{P}$ is ***feasible*** for a clustering $\mathcal{C}$: $\operatorname{supp}(C_i) \subset P_i$ for all i.

An $\mathcal{F}$-diagram $\mathcal{P}$ ***supports*** a clustering $\mathcal{C}$: $\operatorname{supp}(C_i) = P_i \cap X$ for all i.

An $\mathcal{F}$-diagram $\mathcal{P}$ is ***strongly feasible*** for a clustering $\mathcal{C}$: $\mathcal{P}$ supports $\mathcal{C}$ and $G(\mathcal{C})$ is c-cycle-free.

$(\mathcal{D}, h, \mathcal{S}, \mathcal{M})$-***diagram***: $\mathcal{F}$-diagram for the functions $\varphi_i(x) := h(d_i(s_i, x)) - \mu_i$, where $\mathcal{D} = (d_1, \ldots, d_k)$ is a k-tuple of metrics (or more general distance measures) in $\mathbb{R}^n$, $h : [0, \infty) \to [0, \infty)$ is monotonically increasing, $\mathcal{S} = \{s_1, \ldots, s_k\} \subset \mathbb{R}^n$, and $\mathcal{M} = (\mu_1, \ldots, \mu_k)^T \in \mathbb{R}^k$. The vectors s_i are called ***sites***. If the metrics d_i are all identical, the resulting diagram is ***isotropic***, otherwise it is ***anisotropic***.

Centroidal: A $(\mathcal{D}, h, \mathcal{S}, \mathcal{M})$-diagram that supports a balanced clustering $\mathcal{C}$ is ***centroidal*** if the sites s_i coincide with the centers of gravity $g(C_i)$ of the clusters.

IMPORTANT SPECIAL CASES

Additively weighted Voronoi diagram: $d_1, \ldots, d_k = \|\,.\,\|_{(2)}$, $h = \text{id}$, i.e., $\varphi_i(x) = \|x - s_i\|_{(2)} - \mu_i$ for all i.

Power diagram: $d_1, \ldots, d_k = \|\,.\,\|_{(2)}$, $h = (.)^2$, i.e., $\varphi_i(x) = \|x - s_i\|_{(2)}^2 - \mu_i$ for all i.

Anisotropic additively weighted Voronoi diagram (with ellipsoidal norms): For $i = 1, \ldots, k$ each d_i is induced by an ellipsoidal norm $\|\,.\,\|_{M_i}$, i.e., $\|x\|_{M_i} = \sqrt{x^T M_i x}$ for a symmetric positive definite matrix M_i, $h = \text{id}$, i.e., $\varphi_i(x) = \|x - s_i\|_{M_i} - \mu_i$.

Anisotropic power diagram (with ellipsoidal norms): For $i = 1, \ldots, k$ each d_i is induced by an ellipsoidal norm $\|\,.\,\|_{M_i}$, $h = (.)^2$, i.e., $\varphi_i(x) = \|x - s_i\|_{M_i}^2 - \mu_i$.

Anisotropic additively weighted Voronoi diagrams have been used by [LS03] and [CG11] for mesh generation. Anisotropic power diagrams were used in [ABG+15] for the reconstruction of polycrystals from information about grain volumes, centers and moments. In [CCD14] districts are designed so as to balance the workload of service vehicles. The effect of different diagrams for electoral district design is studied in [BGK17].

BASIC FACTS

In general, an $\mathcal{F}$-diagram $\mathcal{P}$ does not constitute a dissection of $\mathbb{R}^n$. If, however, $\mathcal{D}$ is a family of metrics induced by strictly convex norms, $h : \mathbb{R} \to \mathbb{R}$ is injective, and

$\mathcal{S} = \{s_1, \ldots, s_k\} \subset \mathbb{R}^n$ is such that $s_i \neq s_l$ for $i \neq l$, then the $(\mathcal{D}, h, \mathcal{S}, \mathcal{M})$-diagram $\mathcal{P} = \{P_1, \ldots, P_k\}$ has the property that $\mathrm{int}(P_i) \cap \mathrm{int}(P_l) = \emptyset$ whenever $i \neq l$.

Given an instance of a constrained clustering problem and any choice of metrics $\mathcal{D}$, functions h and sites $\mathcal{S}$, and let $\mathcal{C}^*$ be a minimizer of

$$\sum_{i=1}^{k} \sum_{j=1}^{m} \xi_{i,j} \cdot \omega(x_j) \cdot \varphi_i(x_j).$$

Then there exists a choice of the additive parameter tuple $\mathcal{M}$, such that the corresponding $(\mathcal{D}, h, \mathcal{S}, \mathcal{M})$-diagram supports $\mathcal{C}^*$; see [BGK17].

Thus, $\mathcal{D}, h, \mathcal{S}$ can be regarded as *structural parameters* while $\mathcal{M}$ is the *feasibility parameter*. Typically, $\mathcal{D}$ and h are defined by requirements on the clusters for a given specific application. Optimization over $\mathcal{S}$ can be done with respect to different criteria. Natural choices involve the total variances or the intercluster distance. For any choice of structural parameters, the feasibility parameter $\mathcal{M}$ is then provided by the dual variables of a certain linear program; see [BG12], [CCD16], [BGK17].

Unless the weights are all the same, this approach does not automatically yield integral assignments in general but may require subsequent rounding. However, the number of fractionally assigned points can be controlled to be at most $k - 1$.

POWER DIAGRAMS AND GRAVITY BODIES

Power diagrams constitute cell-decompositions of space into polyhedra. As it turns out all relevant power diagrams in $\mathbb{R}^n$ can be encoded in a certain convex body in $\mathbb{R}^{nk}$. More precisely, we are dealing with the power cells

$$P_i = \{x \in \mathbb{R}^n \mid \|x - s_i\|_{(2)}^2 - \mu_i \leq \|x - s_l\|_{(2)}^2 - \mu_l \; \forall l \in \{1, \ldots, k\}\}.$$

For given sites s_i, the inequalities for x are in fact linear. Hence P_i is a polyhedron and thus, particularly, convex. The relation of power diagrams to *least-square clustering*, i.e., clusterings minimizing the objective function

$$\sum_{i=1}^{k} \sum_{j=1}^{m} \xi_{i,j} \cdot \omega(x_j) \cdot \|x_j - s_i\|_{(2)}^2,$$

has been studied in [BHR92, AHA98, BG12]; see also [BGK17].

In the strongly balanced case, the gravity body Q is a polytope. Its vertices are precisely the gravity vectors of all strongly balanced clusterings that admit a strongly feasible power diagram. In the weakly balanced case, Q does in general have more than finitely many extreme points. However, each extreme point is still the gravity vector of a clustering that admits a strongly feasible power diagram, [BG12].

CENTROIDAL POWER DIAGRAMS AND CLUSTERING BODIES

A quite natural (and in spite of the NP-hardness of the problem practically efficient) approach was introduced in [BG04] for the problem of consolidation of farmland. It models optimal balanced clustering as a convex maximization problem that involves two norms, a norm $\| \cdot \|$ on $\mathbb{R}^n$ and a second norm $\| \cdot \|_\diamond$ on $\mathbb{R}^{k(k-1)/2}$. $\| \cdot \|_\diamond$

is required to be *monotone*, i.e., $\|x\|_\diamond \leq \|y\|_\diamond$ whenever $x, y \in \mathbb{R}^{k(k-1)/2}$ with $0 \leq x \leq y$. A balanced clustering is desired that maximizes

$$\left\| \left(\|c_1 - c_2\|, \|c_1 - c_3\|, \ldots, \|c_{k-1} - c_k\| \right)^T \right\|_\diamond$$

where c_i is a suitable approximation of the center $g(C_i)$ which actually coincides with $g(C_i)$ in the strongly balanced case. In this model, intuitively, a feasible clustering is optimal, if the corresponding "inexact" centers of gravity c_i are pushed apart as far as possible.

This model leads to the study of *clustering bodies*

$$C = \left\{ \mathbf{z} = (z_1^T, \ldots, z_k^T)^T \in \mathbb{R}^{kn} \mid \left\| \left(\|z_1 - z_2\|, \ldots, \|z_{k-1} - z_k\| \right)^T \right\|_\diamond \leq 1 \right\}$$

in $\mathbb{R}^{kn}$. Note that these sets live in $\mathbb{R}^{kn}$ rather than in the typically much higher dimensional space $\mathbb{R}^{km}$ of the optimization problem. Further, note that C has a non-trivial lineality space $\mathrm{ls}(C)$ since a translation applied to all component vectors leaves it invariant. Of course, C and $C \cap (\mathrm{ls}(C))^\perp$ can be regarded as the unit ball of a seminorm or norm, respectively. Hence we are, in effect dealing with the problem of (semi-) norm maximization over polytopes (as in Sections 36.5 and 36.6). In particular, in the strongly balanced case, centroidal power diagrams correspond to the local maxima of the ellipsoidal function $\psi : \mathbb{R}^{kn} \to [0, \infty)$ defined by $\psi(\mathbf{z}) = \sum_{i=1}^k \kappa_i \|z_i\|_{(2)}^2$ for $\mathbf{z} = (z_1^T, \ldots, z_k^T)^T \in \mathbb{R}^{kn}$; see [BG12] for additional results.

As it turns out, clustering bodies provide a rich class of sets which include polytopal and smooth bodies but also "mixtures." For some choices of norms one can find permutahedral substructures. [BG10] gives tight bounds for the approximability of such clustering bodies by polyhedra with only polynomially many facets. The proposed algorithm then solves a linear program in $\mathbb{R}^{km}$ for each facet of such an approximating polyhedron. In spite of the NP-hardness of the general balanced clustering problem one obtains good approximate solutions very efficiently, [BG04]; see also [BBG14].

FURTHER READING

[AKL13]   F. Aurenhammer, R. Klein, and D.T. Lee. *Voronoi Diagrams and Delaunay Triangulations*. World Scientific, Singapore, 2013.

[HG12]   F.K. Hwang and U.G. Rothblum. *Partitions: Optimality and Clustering - Vol. 1: Single-Parameter*. World Scientific, Singapore, 2012.

[HGC13]   F.K. Hwang, U.G. Rothblum and H.B. Chen. *Partitions: Optimality and Clustering - Vol. 2: Multi-Parameter*. World Scientific, Singapore, 2013.

RELATED CHAPTERS

Chapter 27: Voronoi diagrams and Delaunay triangulations

REFERENCES

[ABG+15]   A. Alpers, A. Brieden, P. Gritzmann, A. Lyckegaard, and H.F. Poulsen. Generalized balanced power diagrams for 3D representations of polycrystals. *Phil. Mag.*, 95:1016–1028, 2015.

[AHA98]   F. Aurenhammer, F. Hoffmann, and B. Aronov. Minkowski-type theorems and least-squares clustering. *Algorithmica*, 20:61–76, 1998.

[BHR92]   E.R. Barnes, A.J. Hoffman, and U.G. Rothblum. Optimal partitions having disjoint convex and conic hulls. *Math. Prog.*, 54:69–86, 1992.

[BBG14]   S. Borgwardt, A. Brieden, and P. Gritzmann. Geometric clustering for the consolidation of farmland and woodland. *Math. Intelligencer*, 36:37–44, 2014.

[BG04]   A. Brieden and P. Gritzmann. A quadratic optimization model for the consolidation of farmland by means of lend-lease agreements. In D. Ahr, R. Fahrion, M. Oswald, and G. Reinelt, editors, *Operations Research Proceedings 2003*, pages 324–331, Springer, Berlin, 2004.

[BG10]   A. Brieden and P. Gritzmann. On clustering bodies: Geometry and polyhedral approximation. *Discrete Comput. Geom.*, 44:508–534, 2010.

[BG12]   A. Brieden and P. Gritzmann. On optimal weighted balanced clusterings: Gravity bodies and power diagrams. *SIAM J. Discrete Math.*, 26:415–434, 2012.

[BGK17]   A. Brieden, P. Gritzmann, and F. Klemm. Constrained clustering via diagrams: A unified theory and its application to electoral district design. *European J. Oper. Res.*, 263:18–34, 2017.

[CG11]   G.D. Canas and S.J. Gortler. Orphan-free anisotropic Voronoi diagrams. *Discrete Comput. Geom.*, 46:526–541, 2011.

[CCD14]   J.G. Carlsson, E. Carlsson, and R. Devulapalli. Balancing workloads of service vehicles over a geographic territory. In *Proc. IEEE/RSJ Int. Conf. Intell. Robots and Systems*, 209–216, 2014.

[CCD16]   J.G. Carlsson, E. Carlsson, and R. Devulapalli. Shadow prices in territory division. *Networks and Spatial Economics*, 16:893-931, 2016.

[Cor10]   J. Cortées. Coverage optimization and spatial load balancing by robotic sensor networks. *IEEE Trans. Automatic Control*, 55:749–754, 2010.

[LS03]   F. Labelle and J.R. Shewchuk. Anisotropic Voronoi diagrams and guaranteed-quality anisotropic mesh generation. In *Proc. 19th Sympos. Comput. Geom.*, pages 191–200, ACM Press, 2003.

37 COMPUTATIONAL AND QUANTITATIVE REAL ALGEBRAIC GEOMETRY

Saugata Basu and Bhubaneswar Mishra

INTRODUCTION

Computational and quantitative real algebraic geometry studies various algorithmic and quantitative questions dealing with the *real solutions* of a system of equalities, inequalities, and inequations of polynomials over the real numbers. This emerging field is largely motivated by the power and elegance with which it solves a broad and general class of problems arising in robotics, vision, computer-aided design, geometric theorem proving, etc.

The algorithmic problems that arise in this context are formulated as decision problems for the *first-order theory of reals* and the related problems of *quantifier elimination* (Section 37.1), as well as problems of computing topological invariants of semi-algebraic sets. The associated geometric structures are then examined via an exploration of the *semialgebraic sets* (Section 37.2). Algorithmic problems for semialgebraic sets are considered next. In particular, Section 37.2 discusses real algebraic numbers and their representation, relying on such classical theorems as Sturm's theorem and Thom's Lemma (Section 37.3). This discussion is followed by a description of semialgebraic sets using the concept of *cylindrical algebraic decomposition* (CAD) in both one and higher dimensions (Sections 37.4 and 37.5). This concept leads to brief descriptions of two algorithmic approaches for the decision and quantifier elimination problems (Section 37.6): namely, Collins's algorithm based on CAD, and some more recent approaches based on critical points techniques and on reducing the multivariate problem to easier univariate problems. These new approaches rely on the work of several groups of researchers: Grigor'ev and Vorobjov [Gri88, GV88], Canny [Can87, Can90], Heintz et al. [HRS90], Renegar [Ren91, Ren92a, Ren92b, Ren92c], and Basu et al. [BPR96]. In Section 37.7 we describe certain mathematical results on bounding the topological complexity of semi-algebraic sets, and in Section 37.8 we discuss some algorithms for computing topological invariants of semi-algebraic sets. In Section 37.9 we describe some quantitative results from metric semi-algebraic geometry. These have proved useful in applications in computer science. In Section 37.10 we discuss the connection between quantitative bounds on the topology of semi-algebraic sets, and the polynomial partitioning method that have gained prominence recently in discrete and computational geometry. Finally, we give a few representative applications of computational semi-algebraic geometry in Section 37.11.

37.1 FIRST-ORDER THEORY OF REALS

The *decision problem* for the first-order theory of reals is to determine if a *Tarski sentence* in the first-order theory of reals is true or false. The *quantifier elimination problem* is to determine if there is a logically equivalent quantifier-free formula for an arbitrary Tarski formula in the first-order theory of reals. As a result of Tarski's work, we have the following theorem.

THEOREM 37.1.1 [Tar51]

- Let Ψ be a Tarski sentence. There is an effective decision procedure for Ψ.

- Let Ψ be a Tarski formula. There is a quantifier-free formula ϕ logically equivalent to Ψ. If Ψ involves only polynomials with rational coefficients, then so does the sentence ϕ.

Tarski formulas are formulas in a first-order language (defined by Tarski in 1930 [Tar51]) constructed from equalities, inequalities, and inequations of polynomials over the reals. Such formulas may be constructed by introducing logical connectives and universal and existential quantifiers to the atomic formulas. *Tarski sentences* are Tarski formulas in which all variables are bound by quantification.

GLOSSARY

Term: A constant, variable, or term combining two terms by an arithmetic operator: $\{+, -, \cdot\}$. A constant is a real number. A variable assumes a real number as its value. A term contains finitely many such algebraic variables: $x_1, x_2, \ldots, x_k$.

Atomic formula: A formula comparing two terms by a binary relational operator: $\{=, \neq, >, <, \geq, \leq\}$.

Quantifier-free formula: An atomic formula, a negation of a quantifier-free formula given by the unary Boolean connective $\{\neg\}$, or a formula combining two quantifier-free formulas by a binary Boolean connective: $\{\Rightarrow, \wedge, \vee\}$. *Example*: The formula $(x^2 - 2 = 0) \wedge (x > 0)$ defines the (real algebraic) number $+\sqrt{2}$.

Tarski formula: If $\phi(y_1, \ldots, y_r)$ is a quantifier-free formula, then it is also a Tarski formula. All the variables y_i are *free* in ϕ. Let $\Phi(y_1, \ldots, y_r)$ and $\Psi(z_1, \ldots, z_s)$ be two Tarski formulas (with free variables y_i and z_i, respectively); then a formula combining Φ and Ψ by a Boolean connective is a Tarski formula with free variables $\{y_i\} \cup \{z_i\}$. Lastly, if $\mathcal{Q}$ stands for a quantifier (either universal $\forall$ or existential $\exists$) and if $\Phi(y_1, \ldots, y_r, x)$ is a Tarski formula (with free variables x and $y_1, \ldots, y_r$), then

$$\left(\mathcal{Q}\, x\right) \left[\Phi(y_1, \ldots, y_r, x)\right]$$

is a Tarski formula with only the y's as free variables. The variable x is *bound* in $(\mathcal{Q}\, x)[\Phi]$.

Tarski sentence: A Tarski formula with no free variable. *Example*: $(\exists\, x)\, (\forall\, y)\, [y^2 - x < 0]$. This Tarski sentence is false.

Prenex Tarski formula: A Tarski formula of the form

$$\left(\mathcal{Q}\, x_1\right)\left(\mathcal{Q}\, x_2\right)\cdots\left(\mathcal{Q}\, x_k\right)\left[\phi(y_1, y_2, \ldots, y_r, x_1, \ldots, x_k)\right],$$

where ϕ is quantifier-free. The string of quantifiers $(\mathcal{Q}\, x_1)(\mathcal{Q}\, x_2)\cdots(\mathcal{Q}\, x_k)$ is called the **prefix** and ϕ is called the **matrix**.

Prenex form of a Tarski formula, Ψ: A prenex Tarski formula logically equivalent to Ψ. For every Tarski formula, one can find its prenex form using a simple procedure that works in four steps: (1) eliminate redundant quantifiers; (2) rename variables so that the same variable does not occur as free and bound; (3) move negations inward; and finally, (4) push quantifiers to the left.

THE DECISION PROBLEM

The general **decision problem** for the first-order theory of reals is to determine if a given Tarski sentence is true or false. A particularly interesting special case of the problem is when all the quantifiers are existential. We refer to the decision problem in this case as the **existential problem** for the first-order theory of reals.

The general decision problem was shown to be decidable by Tarski [Tar51]. However, the complexity of Tarski's original algorithm could only be given by a very rapidly-growing function of the input size (e.g., a function that could not be expressed as a bounded tower of exponents of the input size). The first algorithm with substantial improvement over Tarski's algorithm was due to Collins [Col75]; it has a doubly-exponential time complexity in the number of variables appearing in the sentence. Further improvements have been made by a number of researchers (Grigor'ev-Vorobjov [Gri88, GV88], Canny [Can88, Can93], Heintz et al. [HRS89, HRS90], Renegar [Ren92a, Ren92b, Ren92c]) and most recently by Basu et al. [BPR96, Bas99a].

In the following, we assume that our Tarski sentence is presented in its prenex form:

$$\left(\mathcal{Q}_1\mathbf{x}^{[1]}\right)\left(\mathcal{Q}_2\mathbf{x}^{[2]}\right)\cdots\left(\mathcal{Q}_\omega\mathbf{x}^{[\omega]}\right)\left[\psi(\mathbf{x}^{[1]}, \ldots, \mathbf{x}^{[\omega]})\right],$$

where the $\mathcal{Q}_i$'s form a sequence of alternating quantifiers (i.e., $\forall$ or $\exists$, with every pair of consecutive quantifiers distinct), with $\mathbf{x}^{[i]}$ a partition of the variables

$$\bigcup_{i=0}^{\omega}\mathbf{x}^{[i]} = \{x_1, x_2, \ldots, x_k\} \triangleq \mathbf{x}, \quad \text{and} \quad |\mathbf{x}^{[i]}| = k_i,$$

and where ψ is a quantifier-free formula with atomic predicates consisting of polynomial equalities and inequalities of the form

$$g_i\left(\mathbf{x}^{[1]}, \ldots, \mathbf{x}^{[\omega]}\right) \gtreqless 0, \quad i = 1, \ldots, s.$$

Here, g_i is a multivariate polynomial (over $\mathbb{R}$ or $\mathbb{Q}$, as the case may be) of total degree bounded by d. There are a total of s such polynomials. The special case $\omega = 1$ reduces the problem to that of the existential problem for the first-order theory of reals.

If the polynomials of the basic equalities, inequalities, inequations, etc., are over the rationals, then we assume that their coefficients can be stored with at most L bits. Thus the arithmetic complexity can be described in terms of k, k_i, ω, s, and d, and the bit complexity will involve L as well.

TABLE 37.1.1   Selected time complexity results.

GENERAL OR EXISTENTIAL	TIME COMPLEXITY	SOURCE
General	$L^3(sd)2^{O(\Sigma k_i)}$	[Col75]
Existential	$L^{O(1)}(sd)^{O(k^2)}$	[GV92]
General	$L^{O(1)}(sd)^{(O(\Sigma k_i))^{4\omega-2}}$	[Gri88]
Existential	$L^{1+o(1)}(s)^{(k+1)}(d)^{O(k^2)}$	[Can88, Can93]
General	$(L\log L\log\log L)(sd)^{(2^{O(\omega)})\Pi k_i}$	[Ren92a, Ren92b, Ren92c]
Existential	$(L\log L\log\log L)s\,(s/k)^k\,(d)^{O(k)}$	[BPR96]
General	$(L\log L\log\log L)(s)^{\Pi(k_i+1)}(d)^{\Pi O(k_i)}$	[BPR96]

Table 37.1.1 highlights a representative set of known bit-complexity results for the decision problem.

QUANTIFIER ELIMINATION PROBLEM

Formally, given a Tarski formula of the form,

$$\Psi(\mathbf{x}^{[0]}) = (\mathcal{Q}_1\mathbf{x}^{[1]})\,(\mathcal{Q}_2\mathbf{x}^{[2]})\,\cdots\,(\mathcal{Q}_\omega\mathbf{x}^{[\omega]})\,[\psi(\mathbf{x}^{[0]},\mathbf{x}^{[1]},\ldots,\mathbf{x}^{[\omega]})],$$

where ψ is a quantifier-free formula, the **quantifier elimination problem** is to construct another quantifier-free formula, $\phi(\mathbf{x}^{[0]})$, such that $\phi(\mathbf{x}^{[0]})$ holds if and only if $\Psi(\mathbf{x}^{[0]})$ holds. Such a quantifier-free formula takes the form

$$\phi(\mathbf{x}^{[0]}) \equiv \bigvee_{i=1}^{I}\bigwedge_{j=1}^{J_i}\left(f_{i,j}(\mathbf{x}^{[0]}) \gtreqless 0\right),$$

where $f_{i,j} \in \mathbb{R}[\mathbf{x}^{[0]}]$ is a multivariate polynomial with real coefficients.

The current best bounds were given by Basu et. al. [BPR96] and are summarized as follows:

$$I \leq (s)^{\Pi(k_i+1)}(d)^{\Pi\,O(k_i)}$$
$$J_i \leq (s)^{\Pi_{i>0}(k_i+1)}(d)^{\Pi_{i>0}\,O(k_i)}.$$

The total degrees of the polynomials $f_{i,j}(\mathbf{x}^{[0]})$ are bounded by

$$(d)^{\Pi_{i>0}\,O(k_i)}.$$

Nonetheless, comparing the above bounds to the bounds obtained in *semilinear geometry*, it appears that the "combinatorial part" of the complexity of both the formula and the computation could be improved to $(s)^{\Pi_{i>0}(k_i+1)}$. Basu [Bas99a] proved the following bound (with improved combinatorial part) for the size of the equivalent quantifier-free formula

$$I, J_i \leq (s)^{\Pi_{i>0}(k_i+1)}(d)^{k_0'\,\Pi_{i>0}\,O(k_i)},$$

where $k'_0 = \min(k_0 + 1, \tau \prod_{i>0}(k_i + 1))$ and τ is a bound on the number of free-variables occurring in any polynomial in the original Tarski formula. The total degrees of the polynomials $f_{i,j}(\mathbf{x}^{[0]})$ are still bounded by

$$(d)^{\prod_{i>0} O(k_i)}.$$

Furthermore, the algorithmic complexity of Basu's procedure involves only $(s)^{\prod_{i>0}(k_i+1)}(d)^{k'_0 \prod_{i>0} O(k_i)}$ arithmetic operations.

Lower bound results for the quantifier elimination problem can be found in Davenport and Heintz [DH88]. They showed that for every k, there exists a Tarski formula Ψ_k with k quantifiers, of length $O(k)$, and of constant degree, such that any quantifier-free formula ψ_k logically equivalent to Ψ_k must involve polynomials of

$$\text{degree} = 2^{2^{\Omega(k)}} \quad \text{and} \quad \text{length} = 2^{2^{\Omega(k)}}.$$

Note that in the simplest possible case (i.e., $d = 2$ and $k_i = 2$), upper and lower bounds are doubly-exponential and match well. This result, however, does not imply a similar lower bound for the decision problems.

From the point of view of hardness, it is easily seen that the problem of deciding a sentence in the existential theory of the reals (with integer coefficients) is **NP**-hard in the Turing model of computation, and it was proved by Blum, Shub and Smale [BSS89], that deciding whether a real polynomial of degree 4 has a real zero in $\mathbb{R}^n$ is **NP$_\mathbb{R}$**-complete in the Blum-Shub-Smale model of computations [BCSS98].

37.2 SEMIALGEBRAIC SETS

Every quantifier-free formula composed of polynomial inequalities and Boolean connectives defines a semialgebraic set. Thus, these semialgebraic sets play an important role in real algebraic geometry.

GLOSSARY

Semialgebraic set: A subset $S \subseteq \mathbb{R}^k$ defined by a set-theoretic expression involving a system of polynomial inequalities

$$S = \bigcup_{i=1}^{I} \bigcap_{j=1}^{J_i} \Big\{ (\xi_1, \ldots, \xi_k) \in \mathbb{R}^k \mid \mathrm{sgn}(f_{i,j}(\xi_1, \ldots, \xi_k)) = s_{i,j} \Big\},$$

where the $f_{i,j}$'s are multivariate polynomials over $\mathbb{R}$ and the $s_{i,j}$'s are corresponding sets of signs in $\{-1, 0, +1\}$.

Real algebraic set, real algebraic variety: A subset $Z \subseteq \mathbb{R}^k$ defined by a system of algebraic equations.

$$Z = \Big\{ (\xi_1, \ldots, \xi_k) \in \mathbb{R}^n \mid f_1(\xi_1, \ldots, \xi_k) = \cdots = f_m(\xi_1, \ldots, \xi_k) = 0 \Big\},$$

where the f_i's are multivariate polynomials over $\mathbb{R}$. For any finite set of polynomials $\mathcal{P} \subset \mathbb{R}[x_1, \ldots, x_k]$ we denote by $\mathrm{Zer}(\mathcal{P}, \mathbb{R}^k)$ the associated real algebraic set in $\mathbb{R}^k$. Real algebraic sets are also referred to as real algebraic varieties, or just as varieties if the ambient space is clear from context.

Semialgebraic map: A map $\theta : S \to T$, from a semialgebraic set $S \subseteq \mathbb{R}^k$ to a semialgebraic set $T \subseteq \mathbb{R}^\ell$, such that its graph $\{(x, \theta(x)) \in \mathbb{R}^{k+\ell} : s \in S\}$ is a semialgebraic set in $\mathbb{R}^{k+\ell}$. Note that projection, being linear, is a semialgebraic map.

Dimension of a semi-algebraic set: For any semi-algebraic set S, the dimension of S is the maximum number i, such that there exists a semi-algebraic injective map from $(0,1)^i$ to S.

Real dimension of a real algebraic variety: The real dimension of a real algebraic variety $V \subset \mathbb{R}^k$ is the dimension of V as a semi-algebraic set and is at most k. Note also that the real dimension of V is at most the *complex dimension* of the Zariski closure of V in $\mathbb{C}^k$, i.e., the smallest complex sub-variety of $\mathbb{C}^k$ which contains V. We refer the reader to [BCR98] for intricacies of dimension theory of real and complex varieties.

TARSKI-SEIDENBERG THEOREM

Equivalently, semialgebraic sets can be defined as

$$S = \left\{ (\xi_1, \ldots, \xi_k) \in \mathbb{R}^k \quad | \quad \psi(\xi_1, \ldots, \xi_k) = \mathsf{True} \right\},$$

where $\psi(x_1, \ldots, x_k)$ is a quantifier-free formula involving n algebraic variables. As a direct corollary of Tarski's theorem on quantifier elimination, we see that extensions of Tarski formulas are also semialgebraic sets.

While real algebraic sets are quite interesting and would be natural objects of study in this context, *they are not closed under projection onto a subspace*. Hence they tend to be unwieldy. However, *semialgebraic sets are closed under projection*. This follows from a more general result: the famous **Tarski-Seidenberg theorem** which is an immediate consequence of quantifier elimination, since images are described by formulas involving only existential quantifiers.

THEOREM 37.2.1 *(Tarski-Seidenberg Theorem)* [Tar51]

Let S be a semialgebraic set in $\mathbb{R}^k$, and let $\theta : \mathbb{R}^k \to \mathbb{R}^\ell$ be a semialgebraic map. Then $\theta(S)$ is semialgebraic in $\mathbb{R}^\ell$.

In fact, semialgebraic sets can be defined simply as the smallest class of subsets of $\mathbb{R}^k, k \geq 0$, containing real algebraic sets and closed under projection.

GLOSSARY

Connected component of a semialgebraic set: A maximal connected subset of a semialgebraic set. Semialgebraic sets have a finite number of connected components and these are also semialgebraic.

Semialgebraic decomposition of a semialgebraic set S: A finite collection $\mathcal{K}$ of disjoint connected semialgebraic subsets of S whose union is S. The collection of connected components of a semialgebraic set forms a semialgebraic decomposition. Thus, every semialgebraic set admits a semialgebraic decomposition.

Set of sample points for S: A finite number of points meeting every nonempty connected component of S.

Sign assignment: A vector of sign values of a set of polynomials at a point p. More formally, let $\mathcal{P} \subset \mathbb{R}[X_1, \ldots, X_k]$.

Any point $p = (\xi_1, \ldots, \xi_k) \in \mathbb{R}^k$ has a ***sign assignment*** with respect to $\mathcal{P}$ as follows:

$$\mathrm{sgn}_{\mathcal{P}}(p) = \Big(\mathrm{sgn}(f(\xi_1, \ldots, \xi_k)) \mid f \in \mathcal{P}\Big).$$

A sign assignment induces an equivalence relation: Given two points $p, q \in \mathbb{R}^k$, we say

$$p \sim_{\mathcal{P}} q, \quad \text{if and only if} \quad \mathrm{sgn}_{\mathcal{P}}(p) = \mathrm{sgn}_{\mathcal{P}}(q).$$

Sign class of $\mathcal{P}$***:*** An equivalence class in the partition of $\mathbb{R}^k$ defined by the equivalence relation $\sim_{\mathcal{P}}$.

Semialgebraic decomposition for $\mathcal{P}$***:*** A finite collection of disjoint connected semialgebraic subsets $\{C_i\}$ such that each C_i is contained in some semialgebraic sign class of $\mathcal{P}$. That is, the sign of each $f \in \mathcal{P}$ is ***invariant*** in each C_i. The collection of connected components of the sign-invariant sets for $\mathcal{P}$ forms a semialgebraic decomposition for $\mathcal{P}$.

Cell decomposition for $\mathcal{P}$***:*** A semialgebraic decomposition for $\mathcal{P}$ into finitely many disjoint semialgebraic subsets $\{C_i\}$ called ***cells***, such that each cell C_i is homeomorphic to $\mathbb{R}^{\delta(i)}$, $0 \leq \delta(i) \leq k$. $\delta(i)$ is called the ***dimension of the cell*** C_i, and C_i is called a ***$\delta(i)$-cell***.

Cellular decomposition for $\mathcal{P}$***:*** A cell decomposition for $\mathcal{P}$ such that the closure $\overline{C_i}$ of each cell C_i is a union of cells C_j: $\overline{C_i} = \cup_j C_j$.

Realization of a first-order formula: For any finite family of polynomials $\mathcal{P} \subset \mathbb{R}[x_1, \ldots, x_k]$, we call an element $\sigma \in \{0, 1, -1\}^{\mathcal{P}}$, a *sign condition* on $\mathcal{P}$. For any semi-algebraic set $Z \subset \mathbb{R}^k$, and a sign condition $\sigma \in \{0, 1, -1\}^{\mathcal{P}}$, we denote by $\mathrm{Reali}(\sigma, Z)$ the semi-algebraic set defined by

$$\{\mathbf{x} \in Z \mid \mathrm{sign}(P(\mathbf{x})) = \sigma(P), P \in \mathcal{P}\},$$

and call it the *realization* of σ on Z.

$\mathcal{P}$- and $\mathcal{P}$-closed semi-algebraic sets: More generally, we call any Boolean formula Φ with atoms, $P\{=, >, <\}0, P \in \mathcal{P}$, a *$\mathcal{P}$-formula*. We call the realization of Φ, namely the semi-algebraic set

$$\mathrm{Reali}\left(\Phi, \mathbb{R}^k\right) \quad = \quad \{\mathbf{x} \in \mathbb{R}^k \mid \Phi(\mathbf{x})\}$$

a *$\mathcal{P}$-semi-algebraic set*. Finally, we call a Boolean formula without negations, and with atoms $P\{\geq, \leq\}0$, $P \in \mathcal{P}$, a *$\mathcal{P}$-closed formula*, and we call the realization, $\mathrm{Reali}\left(\Phi, \mathbb{R}^k\right)$, a *$\mathcal{P}$-closed semi-algebraic set*.

Betti numbers of semi-algebraic sets: For any semi-algebraic set S we denote by $b_i(S)$, the rank of the i-th homology group, $\mathrm{H}_i(S, \mathbb{Z})$, of S, and by $b(S) = \sum_i b_i(S)$. In particular, $b_0(S)$ equals the number of connected components of S.

Generalized Euler-Poincaré characteristic of semi-algebraic sets: The *Euler-Poincaré characteristic*, $\chi(S)$, of a closed and bounded semi-algebraic set $S \subset \mathbb{R}^k$ is defined as $\chi(S) = \sum_i (-1)^i b_i(S)$. The Euler-Poincaré characteristic defined above for closed and bounded semi-algebraic set can be extended additively to all semi-algebraic sets, and this additive invariant is then known as the generalized Euler-Poincaré characteristic.

37.3   REAL ALGEBRAIC NUMBERS

Real algebraic numbers are real roots of rational univariate polynomials and provide finitary representation for some of the basic objects (e.g., sample points). Furthermore, we note that (1) real algebraic numbers have effective finitary representation, (2) field operations and polynomial evaluation on real algebraic numbers are efficiently (polynomially) computable, and (3) conversions among various representations of real algebraic numbers are efficiently (polynomially) computable. The key machinery used in describing and manipulating real algebraic numbers relies upon techniques based on the Sturm-Sylvester theorem, Thom's lemma, resultant construction, and various bounds for real root separation.

GLOSSARY

Real algebraic number: A real root α of a univariate polynomial $p(t) \in \mathbb{Z}[t]$ with integer coefficients.

Polynomial for α: A univariate polynomial p such that α is a real root of p.

Minimal polynomial of α: A univariate polynomial p of minimal degree defining α as above.

Degree of a nonzero real algebraic number: The degree of its minimal polynomial. By convention, the degree of the 0 polynomial is $-\infty$.

Real closed field: An ordered field in which every positive element is a square, and every odd degree polynomial has a root.

OPERATIONS ON REAL ALGEBRAIC NUMBERS

Note that if α and β are real algebraic numbers, then so are $-\alpha$, α^{-1} (assuming $\alpha \neq 0$), $\alpha + \beta$, and $\alpha \cdot \beta$. These facts can be constructively proved using the algebraic properties of a resultant construction.

THEOREM 37.3.1

The real algebraic numbers form a field.

A real algebraic number α can be represented by a polynomial for α and a component that identifies the root. There are essentially three types of information that may be used for this identification: *order* (where we assume the real roots are indexed from left to right), *sign* (by a vector of signs), or *interval* (an interval that contains exactly one root).

A classical technique due to Sturm and Sylvester shows how to compute the number of real roots of a univariate polynomial $p(t)$ in an interval $[a, b]$. One important use of this classical theorem is to compute a sequence of relatively small (nonoverlapping) intervals that isolate the real roots of p.

GLOSSARY

Sturm sequence of a pair of polynomials $p(t)$ and $q(t) \in \mathbb{R}[t]$:

$$\overline{\text{STURM}}(p, q) = \Big\langle \hat{r}_0(t), \hat{r}_1(t), \dots, \hat{r}_s(t) \Big\rangle,$$

where
$$
\begin{aligned}
\hat{r}_0(t) &= p(t) \\
\hat{r}_1(t) &= q(t) \\
&\vdots \\
\hat{r}_{i-1}(t) &= \hat{q}_i(t)\, \hat{r}_i(t) - \hat{r}_{i+1}(t), \quad \deg(\hat{r}_{i+1}) < \deg(\hat{r}_i) \\
&\vdots \\
\hat{r}_{s-1}(t) &= \hat{q}_s(t)\, \hat{r}_s(t).
\end{aligned}
$$

Number of variations in sign of a finite sequence $\bar{c}$ of real numbers: Number of times the entries change sign when scanned sequentially from left to right; denoted $\text{Var}(\bar{c})$.

For a vector of polynomials $\overline{P} = \langle p_1(t), \dots, p_m(t) \rangle$ and a real number a:

$$\text{Var}_a(\overline{P}) = \text{Var}(\overline{P}(a)) = \text{Var}(\langle p_1(a), \dots, p_m(a) \rangle).$$

Formal derivative: $p'(t) = D(p(t))$, where $D \colon \mathbb{R}[t] \to \mathbb{R}[t]$ is the (formal) derivative map, taking t^n to nt^{n-1} and $a \in \mathbb{R}$ (a constant) to 0.

STURM-SYLVESTER THEOREM

THEOREM 37.3.2 *Sturm-Sylvester Theorem* [Stu35, Syl53]

Let $p(t)$ and $q(t) \in \mathbb{R}[t]$ be two real univariate polynomials. Then, for any interval $[a, b] \subseteq \mathbb{R} \cup \{\pm\infty\}$ (where $a < b$):

$$\text{Var}\Big[\overline{P}\Big]_a^b = c_p\Big[q > 0\Big]_a^b - c_p\Big[q < 0\Big]_a^b,$$

where

$$\overline{P} \triangleq \overline{\text{STURM}}(p, p'q),$$

$$\text{Var}\Big[\overline{P}\Big]_a^b \triangleq \text{Var}_a(\overline{P}) - \text{Var}_b(\overline{P}),$$

and $c_p[\mathcal{P}]_a^b$ counts the number of distinct real roots (without counting multiplicity) of p in the interval (a, b) at which the predicate $\mathcal{P}$ holds.

Note that if we take $S_p \triangleq \overline{\text{STURM}}(p, p')$ (i.e., $q = 1$) then

$$
\begin{aligned}
\text{Var}\Big[S_p\Big]_a^b &= c_p\Big[\text{True}\Big]_a^b - c_p\Big[\text{False}\Big]_a^b \\
&= \# \text{ of distinct real roots of } p \text{ in } (a, b).
\end{aligned}
$$

COROLLARY 37.3.3

Let $p(t)$ and $q(t)$ be two polynomials with coefficients in a real closed field K. For any interval $[a, b]$ as before, we have

$$
\begin{bmatrix} 1 & 1 & 1 \\ 0 & 1 & -1 \\ 0 & 1 & 1 \end{bmatrix}
\begin{bmatrix} c_p\big[q = 0\big]_a^b \\ c_p\big[q > 0\big]_a^b \\ c_p\big[q < 0\big]_a^b \end{bmatrix}
=
\begin{bmatrix} \mathrm{Var}\big[\overline{\mathrm{STURM}}(p, p')\big]_a^b \\ \mathrm{Var}\big[\overline{\mathrm{STURM}}(p, p'q)\big]_a^b \\ \mathrm{Var}\big[\overline{\mathrm{STURM}}(p, p'q^2)\big]_a^b \end{bmatrix}.
$$

These identities as well as some related algorithmic results (the so-called BKR-algorithm) are based on results of Ben-Or et al. [BOKR86] and their extensions by others. Using this identity, it is a fairly simple matter to decide the sign conditions of a single univariate polynomial q at the roots of a univariate polynomial p. It is possible to generalize this idea to decide the sign conditions of a sequence of univariate polynomials $q_0(t)$, $q_1(t)$, ..., $q_n(t)$ at the roots of a single polynomial $p(t)$ and hence give an efficient (both sequential and parallel) algorithm for the decision problem for Tarski sentences involving univariate polynomials. Further applications in the context of general decision problems are described below.

GLOSSARY

Fourier sequence of a real univariate polynomial $p(t)$ of degree n:

$$
\overline{\mathrm{FOURIER}}(p) = \big\langle p^{(0)}(t) = p(t),\ p^{(1)}(t) = p'(t),\ \ldots, p^{(n)}(t) \big\rangle,
$$

where $p^{(i)}$ is the ith derivative of p with respect to t.

Sign-invariant region of $\mathbb{R}$ determined by a sign sequence $\bar{s}$ with respect to $\overline{\mathrm{FOURIER}}(p)$: The region $R(\bar{s})$ with the property that $\xi \in R(\bar{s})$ if and only if $\mathrm{sgn}(p^{(i)}(\xi)) = s_i$.

THOM'S LEMMA

LEMMA 37.3.4   Thom's Lemma   [Tho65]

Every nonempty sign-invariant region $R(\bar{s})$ (determined by a sign sequence $\bar{s}$ with respect to $\overline{\mathrm{FOURIER}}(p)$) must be connected, i.e., consists of a single interval.

Let $\mathrm{sgn}_\xi(\overline{\mathrm{FOURIER}}(p))$ be the sign sequence obtained by evaluating the polynomials of $\overline{\mathrm{FOURIER}}(p)$ at ξ. Then as an immediate corollary of Thom's lemma, we have:

COROLLARY 37.3.5

Let ξ and ζ be two real roots of a real univariate polynomial $p(t)$ of positive degree $n > 0$. Then $\xi = \zeta$, if

$$
\mathrm{sgn}_\xi(\overline{\mathrm{FOURIER}}(p')) = \mathrm{sgn}_\zeta(\overline{\mathrm{FOURIER}}(p')).
$$

REPRESENTATION OF REAL ALGEBRAIC NUMBERS

Let $p(t)$ be a univariate polynomial of degree d with integer coefficients. Assume that the distinct real roots of $p(t)$ have been enumerated as follows:

$$\alpha_1 < \alpha_2 < \cdots < \alpha_{j-1} < \alpha_j = \alpha < \alpha_{j+1} < \cdots < \alpha_l,$$

where $l \leq d = \deg(p)$. Then we can represent any of its roots uniquely and in a finitary manner.

GLOSSARY

Order representation of an algebraic number: A pair consisting of its polynomial p and its index j in the monotone sequence enumerating the real roots of p: $\langle\alpha\rangle_o = \langle p, j\rangle$. *Example:* $\langle\sqrt{2}+\sqrt{3}\rangle_o = \langle x^4 - 10x^2 + 1, 4\rangle$.

Sign representation of an algebraic number: A pair consisting of its polynomial p and a sign sequence $\bar{s}$ representing the signs of its Fourier sequence evaluated at the root: $\langle\alpha\rangle_s = \langle p, \bar{s} = \text{sgn}_\alpha(\overline{\text{FOURIER}}(p'))\rangle$. *Example:* $\langle\sqrt{2}+\sqrt{3}\rangle_s = \langle x^4 - 10x^2 + 1, (+1, +1, +1)\rangle$. The validity of this representation follows easily from Thom's Lemma.

Interval representation of an algebraic number: A triple consisting of its polynomial p and the two endpoints of an isolating interval, (l, r) $(l, r \in \mathbb{Q}, l < r)$, containing only α: $\langle\alpha\rangle_i = \langle p, l, r\rangle$. *Example:* $\langle\sqrt{2}+\sqrt{3}\rangle_i = \langle x^4 - 10x^2 + 1, 3, 7/2\rangle$.

37.4 UNIVARIATE DECOMPOSITION

In the one-dimensional case, a semialgebraic set is the union of finitely many intervals whose endpoints are real algebraic numbers. For instance, given a set of univariate defining polynomials:

$$\mathcal{F} = \left\{ f_i(x) \in \mathbb{Q}[x] \mid i = 1, \ldots, m \right\},$$

we may enumerate all the real roots of the f_i's (i.e., the real roots of the single polynomial $F = \prod f_i$) as

$$-\infty < \xi_1 < \xi_2 < \cdots < \xi_{i-1} < \xi_i < \xi_{i+1} < \cdots < \xi_s < +\infty,$$

and consider the following finite set $\mathcal{K}$ of elementary intervals defined by these roots:

$$[-\infty, \xi_1), \quad [\xi_1, \xi_1], \quad (\xi_1, \xi_2), \quad \ldots,$$
$$(\xi_{i-1}, \xi_i), \quad [\xi_i, \xi_i], \quad (\xi_i, \xi_{i+1}), \quad \ldots, \quad [\xi_s, \xi_s], \quad (\xi_s, +\infty].$$

Note that $\mathcal{K}$ is, in fact, a cellular decomposition for $\mathcal{F}$. Any semialgebraic set S defined by $\mathcal{F}$ is simply the union of a subset of elementary intervals in $\mathcal{K}$. Furthermore, for each interval $C \in \mathcal{K}$, we can compute a sample point α_C as follows:

$$\alpha_C = \begin{cases} \xi_1 - 1, & \text{if } C = [-\infty, \xi_1); \\ \xi_i, & \text{if } C = [\xi_i, \xi_i]; \\ (\xi_i + \xi_{i+1})/2, & \text{if } C = (\xi_i, \xi_{i+1}); \\ \xi_s + 1, & \text{if } C = (\xi_s, +\infty]. \end{cases}$$

Now, given a first-order formula involving a single variable, its validity can be checked by evaluating the associated univariate polynomials at the sample points. Using the algorithms for representing and manipulating real algebraic numbers, we see that the bit complexity of the decision algorithm is bounded by $(Lmd)^{O(1)}$. The resulting cellular decomposition has no more than $2md + 1$ cells.

Using variants of the theorem due to Ben-Or et al. [BOKR86], Thom's lemma, and some results on parallel computations in linear algebra, one can show that this univariate decision problem is "well-parallelizable," i.e., the problem is solvable by uniform circuits of bounded depth and polynomially many "gates" (simple processors).

37.5 MULTIVARIATE DECOMPOSITION

A straightforward generalization of the standard univariate decomposition to higher dimensions is provided by Collins's cylindrical algebraic decomposition [Col75]. In order to represent a semialgebraic set $S \subseteq \mathbb{R}^k$, we may assume recursively that we can construct a cell decomposition of its projection $\pi(S) \subseteq \mathbb{R}^{k-1}$ (also a semialgebraic set), and then decompose S as a union of the **sectors** and **sections** in the cylinders above each cell of the projection, $\pi(S)$.

This procedure also leads to a cell decomposition of S. One can further assign an algebraic sample point in each cell of S recursively in a straightforward manner.

If $\mathcal{F}$ is a set of polynomials defining the semialgebraic set $S \subseteq \mathbb{R}^k$, then at no additional cost, we may in fact compute a cell decomposition for $\mathcal{P}$ using the procedure described above. Such a decomposition leads to a *cylindrical algebraic decomposition* for $\mathcal{P}$.

GLOSSARY

Cylindrical algebraic decomposition (CAD): A recursively defined cell decomposition of $\mathbb{R}^k$ for $\mathcal{P}$. The decomposition is a cellular decomposition if the set of defining polynomials $\mathcal{P}$ satisfies certain nondegeneracy conditions.

In the recursive definition, the cells of k-dimensional CAD are constructed from a $(k-1)$-dimensional CAD: Every $(k-1)$-dimensional CAD cell C' has the property that the distinct real roots of $\mathcal{F}$ over C' vary continuously as a function of the points of C'.

Moreover, the following quantities remain invariant over a $(k-1)$-dimensional cell: (1) the total number of complex roots of each polynomial of $\mathcal{F}$; (2) the number of distinct complex roots of each polynomial of $\mathcal{F}$; and (3) the total number of common complex roots of every distinct pair of polynomials of $\mathcal{F}$.

These conditions can be expressed by a set $\Phi(\mathcal{F})$ of at most $O(sd)^2$ polynomials in $k-1$ variables, obtained by considering *principal subresultant coefficients* (PSC's). Thus, they correspond roughly to *resultants* and *discriminants*, and ensure that the polynomials of $\mathcal{P}$ do not intersect or "fold" in a cylinder over a $(k-1)$-dimensional cell. The polynomials in $\Phi(\mathcal{P})$ are each of degree no more than d^2.

More formally, a $\mathcal{P}$-sign-invariant cylindrical algebraic decomposition of $\mathbb{R}^k$ is:

- BASE CASE: $k = 1$. A univariate cellular decomposition of $\mathbb{R}^1$ as in the previous section.

- INDUCTIVE CASE: $k > 1$. Let $\mathcal{K}'$ be a $\Phi(\mathcal{P})$-sign-invariant CAD of $\mathbb{R}^{k-1}$. For each cell $C' \in \mathcal{K}'$, define an ***auxiliary polynomial*** $g_{C'}(x_1, \ldots, x_{k-1}, x_k)$ as the product of those polynomials of $\mathcal{P}$ that do not vanish over the $(k-1)$-dimensional cell, C'. The real roots of the auxiliary polynomial g'_C over C' give rise to a finite number (perhaps zero) of semialgebraic continuous functions, which partition the cylinder $C' \times (\mathbb{R} \cup \{\pm\infty\})$ into finitely many $\mathcal{P}$-sign-invariant "slices." The auxiliary polynomials are of degree no larger than sd.

  Assume that the polynomial $g_{C'}(p', x_k)$ has l distinct real roots for each $p' \in C'$: $r_1(p'), r_2(p'), \ldots, r_l(p')$, each r_i being a continuous function of p'. The following sectors and sections are cylindrical over C' (see Figure 37.5.1.):

$$
\begin{aligned}
C_0^* &= \left\{ \langle p', x_k \rangle \mid p' \in C' \wedge x_k \in [-\infty, r_1(p')) \right\}, \\
C_1 &= \left\{ \langle p', x_k \rangle \mid p' \in C' \wedge x_k \in [r_1(p'), r_1(p')] \right\}, \\
C_1^* &= \left\{ \langle p', x_k \rangle \mid p' \in C' \wedge x_k \in (r_1(p'), r_2(p')) \right\}, \\
&\ \vdots \\
C_l^* &= \left\{ \langle p', x_k \rangle \mid p' \in C' \wedge x_k \in (r_l(p'), +\infty] \right\}.
\end{aligned}
$$

The k-dimensional CAD is thus the union of all the sections and sectors computed over the cells of the $(k-1)$-dimensional CAD.

A straightforward recursive algorithm to compute a CAD follows from the above description.

CYLINDRICAL ALGEBRAIC DECOMPOSITION

If we assume that the dimension k is a fixed constant, then the preceding cylindrical algebraic decomposition algorithm is polynomial in $s = |\mathcal{P}|$ and $d = \deg(\mathcal{P})$. However, the algorithm can be easily seen to be doubly-exponential in k as the number of polynomials produced at the lowest dimension is $(sd)^{2^{O(k)}}$, each of degree no larger than $d^{2^{O(k)}}$. The number of cells produced by the algorithm is also *doubly-exponential*. This bound can be seen to be tight by a result due to Davenport and Heintz [DH88], and is related to their lower bound for the quantifier elimination problem (Section 37.1).

CONSTRUCTING SAMPLE POINTS

Cylindrical algebraic decomposition provides a sample point in every sign-invariant connected component for $\mathcal{P}$. However, the total number of sample points generated is doubly-exponential, while the number of connected components of all sign conditions is only singly-exponential. In order to avoid this high complexity (both algebraic and combinatorial) of a CAD, many recent techniques for constructing sample points use a single projection to a line instead of a sequence of cascading pro-

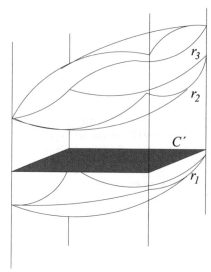

FIGURE 37.5.1
Sections and sectors "slicing" the cylinder over a lower dimensional cell.

jections. For instance, if one chooses a height function carefully then one can easily enumerate its critical points and then associate at least two such critical points to every connected component of the semialgebraic set. From these critical points, it will be possible to create at least one sample point per connected component. Using Bézout's bound, it is seen that only a singly-exponential number of sample points is created, thus improving the complexity of the underlying algorithms.

However, in order to arrive at the preceding conclusion using critical points, one requires certain genericity conditions that can be achieved by symbolically deforming the underlying semialgebraic sets. These infinitesimal deformations can be handled by extending the underlying field to a field of *Puiseux series*. Many of the significant complexity improvements based on these techniques have been due to a careful choice of the symbolic perturbation schemes which results in keeping the number of perturbation variables small.

Currently, the best algorithm for computing a finite set of points of bounded size that intersects *every connected component* of each nonempty sign condition is due to Basu et al. [BPR98] and has an arithmetic time-complexity of $s(s/k)^k d^{O(k)}$, where s is the number of polynomials, d the maximum degree, and k the number of variables.

37.6 ALGORITHMIC APPROACHES

COLLINS' APPROACH

The decision problem for the first-order theory of reals can be solved easily using a cylindrical algebraic decomposition. First consider the existential problem for a sentence with only existential quantifiers,

$$(\exists\, \mathbf{x}^{[0]})\, [\psi(\mathbf{x}^{[0]})].$$

This sentence is true if and only if there is a $q \in C$, a sample point in the cell C,

$$q \;=\; \alpha^{[0]} = (\alpha_1, \ldots, \alpha_k) \;\in\; \mathbb{R}^k,$$

such that $\psi(\alpha^{[0]})$ is true. Thus we see that the decision problem for the purely existential sentence can be solved by simply evaluating the matrix ψ over the finitely many sample points in the associated CAD. This also implies that the existential quantifiers could be replaced by finitely many disjunctions ranging over all the sample points. Note that the same arguments hold for any semialgebraic decomposition with at least one sample point per sign-invariant connected component.

In the general case, one can describe the decision procedure by means of a search process that proceeds *only on* the coordinates of the sample points in the cylindrical algebraic decomposition. This follows because a sample point in a cell acts as a representative for any point in the cell as far as the sign conditions are concerned.

Consider a Tarski sentence

$$(\mathcal{Q}_1 \mathbf{x}^{[1]})\, (\mathcal{Q}_2 \mathbf{x}^{[2]}) \;\cdots\; (\mathcal{Q}_\omega \mathbf{x}^{[\omega]})\, [\psi(\mathbf{x}^{[1]}, \ldots, \mathbf{x}^{[\omega]})],$$

with $\mathcal{F}$ the set of polynomials appearing in the matrix ψ. Let $\mathcal{K}$ be a cylindrical algebraic decomposition of $\mathbb{R}^k$ for $\mathcal{F}$. Since the cylindrical algebraic decomposition produces a sequence of decompositions:

$$\mathcal{K}_1 \text{ of } \mathbb{R}^1,\ \ \mathcal{K}_2 \text{ of } \mathbb{R}^2,\ \ \ldots,\ \ \mathcal{K}_k \text{ of } \mathbb{R}^k,$$

such that each cell $C_{i-1,j}$ of $\mathcal{K}_i$ is cylindrical over some cell C_{i-1} of $\mathcal{K}_{i-1}$, the search progresses by first finding cells C_1 of $\mathcal{K}_1$ such that

$$(\mathcal{Q}_2 x_2) \;\cdots\; (\mathcal{Q}_k x_k)\, [\psi(\alpha_{C_1}, x_2, \ldots, x_k)] = \text{True}.$$

For each C_1, the search continues over cells C_{12} of $\mathcal{K}_2$ cylindrical over C_1 such that

$$(\mathcal{Q}_3 x_3) \;\cdots\; (\mathcal{Q}_k x_k)\, [\psi(\alpha_{C_1}, \alpha_{C_{12}}, x_3, \ldots, x_k)] = \text{True},$$

etc. Finally, at the bottom level the truth properties of the matrix ψ are determined by evaluating at all the coordinates of the sample points.

This produces a tree structure, where each node at the $(i{-}1)$th level corresponds to a cell $C_{i-1} \in \mathcal{K}_{i-1}$ and its children correspond to the cells $C_{i-1,j} \in \mathcal{K}_i$ that are cylindrical over C_{i-1}. The leaves of the tree correspond to the cells of the final decomposition $\mathcal{K} = \mathcal{K}_k$. Because we only have finitely many sample points, the universal quantifiers can be replaced by finitely many conjunctions and the existential quantifiers by disjunctions. Thus, we label every node at the $(i{-}1)$th level "AND" (respectively, "OR") if $\mathcal{Q}_i$ is a universal quantifier $\forall$ (respectively, $\exists$) to produce a so-called AND-OR tree. The truth of the Tarski sentence is thus determined by simply evaluating this AND-OR tree.

A quantifier elimination algorithm can be devised by a similar reasoning and a slight modification of the CAD algorithm described above.

NEW APPROACHES USING CRITICAL POINTS

In order to avoid the cascading projections inherent in Collins's algorithm, the new approaches employ a single projection to a one-dimensional set by using critical points in the manner described above. As before, we start with a sentence with only existential quantifiers,

$$(\exists \, \mathbf{x}^{[0]}) \, [\psi(\mathbf{x}^{[0]})].$$

Let $\mathcal{P} = \{f_1, \ldots, f_s\}$ be the set of polynomials appearing in the matrix ψ.

Under certain genericity conditions, it is possible to produce a set of sample points such that every sign-invariant connected component of the decomposition induced by $\mathcal{F}$ contains at least one such point. Furthermore, these sample points are described by a set of univariate polynomial sequences, where each sequence is of the form

$$p(t), q_0(t), q_1(t), \ldots, q_k(t),$$

and encodes a sample point $\langle \frac{q_1(\alpha)}{q_0(\alpha)}, \ldots, \frac{q_k(\alpha)}{q_0(\alpha)} \rangle$. Here α is a root of p. Now the decision problem for the existential theory can be solved by deciding the sign conditions of the sequence of univariate polynomials

$$f_1(q_1/q_0, \ldots, q_k/q_0), \ \ldots, \ f_s(q_1/q_0, \ldots, q_k/q_0),$$

at the roots of the univariate polynomial $p(t)$. Note that we have now reduced a multivariate problem to a univariate problem and can solve this by the BKR approach.

In order to keep the complexity reasonably small, one needs to ensure that the number of such sequences is small and that these polynomials are of low degree. Assuming that the polynomials in $\mathcal{F}$ are in general position, one can achieve this complexity and compute the polynomials p and q_i (for example, by the u-resultant method in Renegar's algorithm [Ren92a, Ren92b, Ren92c]).

If the genericity conditions are violated, one needs to symbolically deform the polynomials and carry out the computations on these polynomials with additional perturbation parameters. The Basu-Pollack-Roy (BPR) algorithm [BPR98, BPR96] differs from Renegar's algorithm primarily in the manner in which these perturbations are made so that their effect on the algorithmic complexity is controlled.

Next consider an existential Tarski formula of the form

$$(\exists \, \mathbf{x}^{[0]}) \, [\psi(\mathbf{y}, \mathbf{x}^{[0]})],$$

where $\mathbf{y}$ represents the free variables. If we carry out the same computation as before over the ambient field $\mathbb{R}(\mathbf{y})$, we get a set of *parameterized* univariate polynomial sequences, each of the form

$$p(\mathbf{y}, t), q_0(\mathbf{y}, t), q_1(\mathbf{y}, t), \ldots, q_k(\mathbf{y}, t).$$

For a fixed value of $\mathbf{y}$, say $\bar{y}$, the polynomials

$$p(\bar{y}, t), q_0(\bar{y}, t), q_1(\bar{y}, t), \ldots, q_k(\bar{y}, t)$$

can then be used as before to decide the truth or falsity of the sentence

$$(\exists \, \mathbf{x}^{[0]}) \, [\psi(\bar{y}, \mathbf{x}^{[0]})].$$

Also, one may observe that the *parameter space* **y** can be partitioned into semialgebraic sets so that all the necessary information can be obtained by computing at sample values $\bar{y}$.

This process can be extended to ω blocks of quantifiers, by replacing each block of variables by a finite number of cases, each involving only one new variable; the last step uses a CAD method for these ω-many variables.

37.7 TOPOLOGICAL COMPLEXITY OF SEMI-ALGEBRAIC SETS

One useful feature of semi-algebraic sets is that they admit uniform bounds on their topological complexity (measured by their Betti numbers) in terms of the number and degrees of polynomials appearing in the first-order formulas defining them. The uniformity refers to the fact that these bounds are independent of the coefficients of the defining polynomials (unlike for example the bounds in Theorems 37.9.1, 37.9.2, 37.9.3, and 37.9.4).

The first results on bounding the Betti numbers of real varieties were proved by Oleĭnik and Petrovskiĭ [PO49], Thom [Tho65] and Milnor [Mil64]. Using a Morse-theoretic argument and Bézout's theorem they proved (using $b(S)$ to denote the sum of the Betti numbers of a semi-algebraic set S):

THEOREM 37.7.1   [PO49, Tho65, Mil64]

Let $\mathcal{Q} \subset \mathbb{R}[x_1, \ldots, x_k]$ with $\deg(Q) \leq d, Q \in \mathcal{Q}$. Then,

$$b(\mathrm{Zer}(\mathcal{Q}, \mathbb{R}^k)) \leq d(2d-1)^{k-1}. \tag{37.7.1}$$

More generally, if $S \subset \mathbb{R}^k$ is defined by $P_1 \geq 0, \ldots, P_s \geq 0$, with

$$P_i \in \mathbb{R}[x_1, \ldots, x_k], \deg(P_i) \leq d, 1 \leq i \leq d,$$

then

$$b(S) \leq sd(2sd-1)^{k-1}. \tag{37.7.2}$$

Theorem 37.7.1 was later generalized to arbitrary semi-algebraic sets defined by quantifier-free formulas in two steps. In the first step, Theorem 37.7.1 was extended to a particular class of semi-algebraic sets, namely $\mathcal{P}$-closed semi-algebraic sets, where $\mathcal{P} \subset \mathbb{R}[x_1, \ldots, x_k]$ is a finite family of polynomials. The following theorem appears in [BPR05a] and sharpens an earlier result of [Bas99b].

THEOREM 37.7.2   [BPR05a]

If $S \subset \mathbb{R}^k$ is a $\mathcal{P}$-closed semi-algebraic set, then

$$b(S) \leq \sum_{i=0}^{k} \sum_{j=0}^{k-i} \binom{s+1}{j} 6^j d(2d-1)^{k-1}, \tag{37.7.3}$$

where $s = \mathrm{card}(\mathcal{P}) > 0$ and $d = \max_{P \in \mathcal{P}} \deg(P)$.

Using an additional ingredient (namely, a technique to replace an arbitrary semi-algebraic set by a locally closed one with a very controlled increase in the number of polynomials used to describe the given set), Gabrielov and Vorobjov [GV05] extended Theorem 37.7.2 to arbitrary $\mathcal{P}$-semi-algebraic sets with only a small increase in the bound. Their result in conjunction with Theorem 37.7.2 gives the following theorem.

THEOREM 37.7.3 [GV09]

If $S \subset \mathbb{R}^k$ is a $\mathcal{P}$-semi-algebraic set, then

$$b(S) \leq \sum_{i=0}^{k} \sum_{j=0}^{k-i} \binom{2ks+1}{j} 6^j d(2d-1)^{k-1}, \qquad (37.7.4)$$

where $s = \mathrm{card}(\mathcal{P})$ and $d = \max_{P \in \mathcal{P}} \deg(P)$.

The bounds mentioned above have many applications, for example, providing upper bounds on the number of distinct configurations of n points in $\mathbb{R}^k$ as well as on the number of combinatorial types of polytopes [GP86a, GP86b], in geometric transversal theory [GPW95], and for proving lower bounds for membership testing in certain algebraic models of computations [Yao97, MMP96, GV15].

37.8 COMPUTING TOPOLOGICAL INVARIANTS OF SEMI-ALGEBRAIC SETS

The problem of computing topological invariants of semi-algebraic sets is an important problem that has attracted much attention. Unlike the problem of quantifier-elimination or deciding Tarski sentences, an effective algorithm for deciding connectivity of semi-algebraic sets does not automatically follow from the Tarski-Seidenberg principle. However, one can decide questions about connectivity (as well as compute other topological invariants such as the Betti numbers) using effective triangulation of semi-algebraic sets via Cylindrical Algebraic Decomposition. For instance, in one of their seminal papers, Schwartz and Sharir [SS83] gave an algorithm for computing the homology groups of S using cylindrical algebraic decomposition. Their algorithm necessarily has doubly exponential complexity.

Most of the recent work in algorithmic semi-algebraic geometry has focused on obtaining *singly exponential time* algorithms, that is, algorithms with complexity of the order of $(sd)^{k^{O(1)}}$ rather than $(sd)^{2^k}$, where s is the number of polynomials in the input, d the maximum degree of the polynomials in the input and k the number of variables.

ROADMAPS OF SEMI-ALGEBRAIC SETS

Singly-exponential time algorithms for counting the number of connected components of semi-algebraic sets, or for computing a semi-algebraic path connecting two points in the same connected component, all use the notion of a "roadmap," first introduced by Canny [Can87], who also gave a singly exponential time algorithm to compute it. A roadmap of a semi-algebraic set S is a one-dimensional connected subset which is connected inside each connected component of the S. Once these roadmaps are available, they can be used to link up two points in the same connected component. The main geometric idea is to construct roadmaps starting from the critical sets of some projection function. The basic roadmap algorithm has been improved and extended by several researchers (Heintz et al. [HRS90], Gournay and Risler [GR93], Grigor'ev and Vorobjov [Gri88, GV88], and Canny [Can87, Can90]). Using the same ideas as above and some additional techniques for controlling the combinatorial complexity of the algorithm it is possible to extend the roadmap algorithm to the case of semi-algebraic sets. The following theorem appears in [BPR00, BPR16].

THEOREM 37.8.1 [BPR00, BPR16]

Let $Q \in \mathbb{R}[x_1, \ldots, x_k]$ with $\mathrm{Zer}(Q, \mathbb{R}^k)$ of dimension k' and let $\mathcal{P} \subset \mathbb{R}[x_1, \ldots, x_k]$ be a set of at most s polynomials for which the degrees of the polynomials in $\mathcal{P}$ and Q are bounded by d. Let S be a $\mathcal{P}$-semi-algebraic subset of $\mathrm{Zer}(Q, \mathbb{R}^k)$. There is an algorithm which computes a roadmap $\mathrm{RM}(S)$ for S with complexity $s^{k'+1}d^{O(k^2)}$.

Theorem 37.8.1 immediately implies that there is an algorithm whose output is exactly one point in every semi-algebraically connected component of S and whose complexity is bounded by $s^{k'+1}d^{O(k^2)}$. In particular, this algorithm counts the number of semi-algebraically connected components of S within the same time bound.

Schost and Safey el Din [SS10] have given a *probabilistic* algorithm for computing the roadmap of a smooth, bounded real algebraic hyper-surface in $\mathbb{R}^k$ defined by a polynomial of degree d, whose complexity is bounded by $d^{O(k^{3/2})}$. Complex algebraic techniques related to the geometry of polar varieties play an important role in this algorithm. More recently, a *deterministic* algorithm for computing roadmaps of *arbitrary* real algebraic sets with the same complexity bound, has also been obtained [BRS$^+$14].

The main new idea is to consider the critical points of projection maps onto a co-ordinate subspace of dimension bigger than 1 (in fact, of dimension $\sqrt{k}$). As a result the dimension in the recursive calls to the algorithm decreases by $\sqrt{k}$ at each step of the recursion (compared to the case of the ordinary roadmap algorithms where it decreases by 1 in each step). This strategy results in the improved complexity. One also needs to prove suitable generalizations of the results guaranteeing the connectivity of the roadmap (see [BPR16, Chapter 15]) in this more general situation.

The recursive schemes used in the algorithms in [SS10] and [BRS$^+$14] have a common defect in that they are unbalanced in the following sense. The dimension of the fibers in which recursive calls are made is equal to $k - \sqrt{k}$ (in the classical case this dimension is $k - 1$), which is much larger than the dimension of the "polar variety" which is $\sqrt{k}$ (this dimension is equal to 1 in the classical case). While being less unbalanced than in the classical case (which accounts for the improvement in the complexity), there is scope for further improvement if these two dimensions can be made roughly equal. There are certain formidable technical obstructions to be overcome to achieve this.

This challenge was tackled in [BR14] where an algorithm based on a balanced (divide-and-conquer) scheme is given for computing a roadmap of an algebraic set. The following theorem is proved.

THEOREM 37.8.2 [BR14]

There exists an algorithm that takes as input:

1. a polynomial $P \in \mathbb{R}[x_1, \ldots, x_k]$, with $\deg(P) \leq d$;

2. a finite set, A, of real univariate representations whose associated set of points, $\mathcal{A} = \{p_1, \ldots, p_m\}$, is contained in $V = \mathrm{Zer}(P, \mathbb{R}^k)$, and such that the degree of the real univariate representation representing p_i is bounded by D_i for $1 \leq i \leq m$;

and computes a roadmap of V containing $\mathcal{A}$. The complexity of the algorithm is bounded by

$$\left(1 + \sum_{i=1}^{m} D_i^{O(\log^2(k))} \right) (k^{\log(k)} d)^{O(k \log^2(k))}.$$

The size of the output is bounded by $(\mathrm{card}(\mathcal{A}) + 1)(k^{\log(k)} d)^{O(k \log(k))}$, while the degrees of the polynomials appearing in the descriptions of the curve segments and points in the output are bounded by

$$(\max_{1 \leq i \leq m} D_i)^{O(\log(k))} (k^{\log(k)} d)^{O(k \log(k))}.$$

A probabilistic algorithm based on a similar divide-and-conquer strategy which works for smooth, bounded algebraic sets, and with a slightly better complexity is given in [SS17]. As in [SS10], complex algebraic (as opposed to semi-algebraic) techniques play an important role in this algorithm.

COMPUTING BETTI NUMBERS OF SEMI-ALGEBRAIC SETS

Algorithms for computing roadmaps of semi-algebraic sets immediately yield a (singly exponential complexity) algorithm for computing the number of connected components (or in other words, the zero-th Betti number) of a semi-algebraic set S. Computing higher Betti numbers of semi-algebraic sets with singly exponential complexity is more difficult. The best known result in this direction is the following, which generalizes an earlier result [BPR08] giving a singly exponential algorithm for computing the first Betti number of a semi-algebraic set.

THEOREM 37.8.3   [Bas06]

For any given ℓ, there is an algorithm that takes as input a $\mathcal{P}$-formula describing a semi-algebraic set $S \subset \mathbb{R}^k$, and outputs $b_0(S), \ldots, b_\ell(S)$. The complexity of the algorithm is $(sd)^{k^{O(\ell)}}$, where $s = \mathrm{card}\ (\mathcal{P})$ and $d = \max_{P \in \mathcal{P}} \deg(P)$.

Note that the complexity is singly exponential in k for every fixed ℓ.

COMPUTING GENERALIZED EULER-POINCARÉ CHARACTERISTIC OF SEMI-ALGEBRAIC SETS

Another topological invariant of semi-algebraic sets for which a singly exponential algorithm is known is the generalized Euler-Poincaré characteristic. It was shown in [Bas99b] that the Euler-Poincaré characteristic of a given $\mathcal{P}$-closed semi-algebraic set can be computed in $(ksd)^{O(k)}$ time.

The following result (which should be viewed as a generalization of the univariate sign determination algorithm) appears in [BPR05b].

THEOREM 37.8.4   [BPR05b]

There exists an algorithm which given an algebraic set $Z = \mathrm{Zer}(Q, \mathbb{R}^k) \subset \mathbb{R}^k$ and a finite set of polynomials $\mathcal{P} = \{P_1, \ldots, P_s\} \subset \mathbb{R}[x_1, \ldots, x_k]$, computes the list of the generalized Euler-Poincaré characteristics of the realizations of all realizable sign conditions of $\mathcal{P}$ on Z.

If the degrees of the polynomials in $\mathcal{P} \cup \{Q\}$ are bounded by d, and the real dimension of $Z = \mathrm{Zer}(Q, \mathbb{R}^k)$ is k', then the complexity of the algorithm is

$$s^{k'+1} O(d)^k + s^{k'}((k' \log_2(s) + k \log_2(d))d)^{O(k)}.$$

37.9 QUANTITATIVE RESULTS IN METRIC SEMI-ALGEBRAIC GEOMETRY

There are certain metric results that follow from the algorithmic results mentioned above which are very useful in practice. These results have several applications. For example, they are needed for efficient stopping criteria in algorithms that use sub-division methods to compute certificates of positivity of a given polynomial over some compact semi-algebraic subset of $\mathbb{R}^k$ (for example, over the standard simplex) [BCR08].

The following theorem is an example of the metric upper bounds referred to above. It provides an explicit upper bound on the radius of a ball centered at the origin which is guaranteed to meet every semi-algebraically connected component of any given semi-algebraic set in terms of the number s, the maximum degree d, and a bound τ on the bit sizes of the coefficients of the defining polynomials.

To state the result precisely we first introduce the following notation. Given an integer n, we denote by $\mathrm{bit}(n)$ the number of bits of its absolute value in the binary representation. Note that

$$\mathrm{bit}(nm) \le \mathrm{bit}(n) + \mathrm{bit}(m), \tag{37.9.1}$$

$$\mathrm{bit}\left(\sum_{i=1}^{n} m_i\right) \le \mathrm{bit}(n) + \sup_{i=1}^{n}(\mathrm{bit}(m_i)). \tag{37.9.2}$$

THEOREM 37.9.1   [BR10]

Let $\mathcal{P} = \{P_1, \ldots, P_s\} \subset \mathbb{Z}[x_1, \ldots, x_k]$ and suppose that $P \in \mathcal{P}$ have degrees at most d, and the coefficients of $P \in \mathcal{P}$ have bit sizes at most τ. Then there exists a ball centered at the origin of radius

$$\left((2DN(2N-1)+1)2^{(2N-1)(\tau''+\mathrm{bit}(2N-1)+\mathrm{bit}(2DN+1))}\right)^{1/2}$$

where

$$
\begin{aligned}
d' &= \sup(2(d+1), 6), \\
D &= k(d'-2)+2, \\
N &= d'(d'-1)^{k-1}, \\
\tau'' &= N(\tau_2' + \mathrm{bit}(N) + 2\,\mathrm{bit}(2D+1)+1), \\
\tau_2' &= \tau_1' + 2(k-1)\mathrm{bit}(N) + (2k-1)\mathrm{bit}(k), \\
\tau_1' &= D(\tau_0' + 4\,\mathrm{bit}(2D+1) + \mathrm{bit}(N)) - 2\,\mathrm{bit}(2D+1) - \mathrm{bit}(N), \\
\tau_0' &= 2\tau + k\,\mathrm{bit}(d+1) + \mathrm{bit}(2d') + \mathrm{bit}(s)
\end{aligned}
$$

intersecting every semi-algebraically connected component of the realization of every realizable sign condition on $\mathcal{P}$.

Note that asymptotic bounds of the form $2^{\tau d^{O(k)}}$ for the same problem were known before [BPR16, GV88, Ren92a]. One point which needs some explanation is the fact that the number of polynomials, s, plays a role in the estimate in Theorem 37.9.1, while it does not appear in the formula $2^{\tau d^{O(k)}}$. The explanation for this situation is that the total number of polynomials of degree at most d in k variables with bit sizes bounded by τ is bounded by $(2^{\tau+1})^{\binom{d+k}{k}} = 2^{\tau d^{O(k)}}$.

A related result is the following bound on the minimum value attained by an integer polynomial restricted to a compact connected component of a basic closed semi-algebraic subset of $\mathbb{R}^k$ defined by polynomials with integer coefficients in terms of the degrees and the bit sizes of the coefficients of the polynomials involved.

THEOREM 37.9.2 [JPT13]

Let $\mathcal{P} = \{P_1, \ldots . P_s\} \subset \mathbb{Z}[x_1, \ldots, x_k]$ with $\deg(P) \leq d$ for all $P \in \mathcal{P}$, and let the coefficients of P have bit sizes bounded by τ. Let $Q \in \mathbb{Z}[x_1, \ldots, x_k]$, $\deg(Q) \leq d$, and let the bit sizes of the coefficients of Q be also bounded by τ. Let C be a compact connected component of the basic closed semi-algebraic set defined by $P_1 = \cdots = P_\ell = 0, P_{\ell+1} \geq 0, \ldots, P_s \geq 0$. Then the minimum value attained by Q over C is a real algebraic number of degree at most $2^{n-1}d^n$, and if it is not equal to 0, then its absolute value is bounded from below by

$$(2^{4-k/2}Hd^k)^{-k2^k d^k},$$

where $H = \max(2^\tau, 2k + 2s)$.

We discuss next a generalization of Theorem 37.9.1 which has proved to be useful in practice. Theorem 37.9.1 gives an upper bound on the radius of a ball meeting every connected component of the realizations of every realizable sign condition on a family of polynomials with integer coefficients. It is well known that the intersections of any semi-algebraic set $S \subset \mathbb{R}^k$ with open balls of large enough radius are semi-algebraically homeomorphic. This result is a consequence of the local conic structure of semi-algebraic sets [BCR98, Theorem 9.3.6].

Given a semi-algebraic set $S \subset \mathbb{R}^k$ defined by polynomials with coefficients in $\mathbb{Z}$, the following theorem gives a singly exponential upper bound on the radius of a ball having the following property. The *homotopy type* (not necessarily the homeomorphism type) of S is preserved upon intersection of S with all balls of larger radii.

THEOREM 37.9.3 [BV07]

Let $\mathcal{P} = \{P_1, \ldots, P_s\} \subset \mathbb{Z}[x_1, \ldots, x_k]$ and suppose that $P \in \mathcal{P}$ have degrees at most d, and the coefficients of $P \in \mathcal{P}$ have bit sizes at most τ. Let $S \subset \mathbb{R}^k$ be a $\mathcal{P}$-semi-algebraic set.

There exists a constant $c > 0$, such that for any $R_1 > R_2 > 2^{\tau d^{ck}}$ we have,

1. $S \cap \mathbf{B}_k(0, R_1) \simeq S \cap \mathbf{B}_k(0, R_2)$, and

2. $S \setminus \mathbf{B}_k(0, R_1) \simeq S \setminus \mathbf{B}_k(0, R_2)$

(where $\simeq$ denotes homotopy equivalence, and for $R > 0$, $\mathbf{B}_k(0, R)$ denotes the closed ball of radius R centered at the origin).

THEOREM 37.9.4 [BV07]

Let $S \subset \mathbb{R}^m$ be a $\mathcal{P}$-semi-algebraic set, with $\mathcal{P} \subset \mathbb{Z}[x_1, \ldots, x_k]$ and $\mathbf{0} \in S$. Let $\deg(P) < d$ for each $P \in \mathcal{P}$, and the bit sizes of the coefficients of $P \in \mathcal{P}$ be less than τ. Then, there exists a constant $c > 0$ such that for every $0 < r < 2^{-\tau d^{ck}}$ the set $S \cap \mathbf{B}_k(0, r)$ is contractible.

37.10 CONNECTION TO INCIDENCE GEOMETRY

Recently there has been a major impetus to take a new look at the problem of bounding the topological complexity of semi-algebraic sets. It has come from the injection of algebraic techniques to attack certain long-standing open questions in incidence geometry (see for example [KMSS12, AMS13, SSS13, Zah13, SS14, Gut15b, Gut15a, SSZ15, MP15, Zah15]). The source of this impetus is the pioneering work of Guth and Katz [GK15], who following prior ideas of Elekes and Sharir [ES11], made a breakthrough in a long-standing problem of Erdős on the number of distinct distances between n points in a plane. Their main tool was a certain kind of decomposition (called "polynomial partitioning") using the polynomial ham-sandwich theorem due to Stone and Tukey [ST42], which played a somewhat analogous role that cuttings or trapezoidal decomposition [Mat02] played in the more classical Clarkson-Shor (see [CEG+90]) type divide-and-conquer arguments for such problems. The polynomial partitioning result proved by Guth and Katz can be summarized as follows.

THEOREM 37.10.1 [GK15]

Let $\mathcal{S} \subset \mathbb{R}^k$ be a finite set of cardinality n, and $0 \leq r \leq n$. Then, there exists a polynomial $P \in \mathbb{R}[x_1, \ldots, x_k]$, with $\deg(P) \leq r^{1/k}$, having the property that for each connected component C of $\mathbb{R}^k \setminus \mathrm{Zer}(P, \mathbb{R}^k)$, $\mathrm{card}(C \cap \mathcal{S}) \leq O(n/r)$.

The "partitioning" refers to the partitioning of the complement of the hypersurface $\mathrm{Zer}(P, \mathbb{R}^k)$ (the zeros of P in $\mathbb{R}^k$) into its (open) connected components. The theorem ensures that each such connected component does not contain too many of the points. Moreover, using Theorem 37.7.1 one also has that the number of such connected components is bounded by $O(r)$. Notice that the theorem allows for the possibility that most (in fact all) the points in $\mathcal{S}$ are contained in the hypersurface $\mathrm{Zer}(P, \mathbb{R}^k)$.

The application of the partitioning theorem to concrete problems (say bounding incidences) leads to a dichotomy caused by the last observation. The open pieces of the decomposition are handled by induction, while a separate argument is needed for handling the co-dimension one piece, i.e., the hypersurface $\mathrm{Zer}(P, \mathbb{R}^k)$ itself. However, notice that P can have a degree not bounded by any constant when r is allowed to grow with n. The fact that this degree could depend on n (the number of points) and is not constant distinguishes the Guth-Katz technique from earlier decomposition based techniques (such as cuttings or trapezoidal decomposition [Mat02]). Because of this reason, when applying Guth-Katz technique to higher dimensional problems it is necessary to have more refined quantitative upper bounds in real algebraic geometry where the dependence on the degrees is more explicit. More precisely, one needs better bounds on the number of connected components of the realizations of different sign conditions of a finite family of poly-nomials, $\mathcal{P}$, restricted to a given variety V. A bound on this number that took into account the real dimension of the variety V was known before (due to Basu, Pollack and Roy [BPR05a]), but this bound did not make a distinction between the dependence on the degrees of the polynomials in $\mathcal{P}$ and that of the polynomials defining V. In the Guth-Katz framework, such a distinction becomes crucial, since both these degrees could be large but the degree of the polynomials defining V is asymptotically much smaller than those of $\mathcal{P}$.

In [BB12] a stronger bound was proved on the number of connected components of the realizable sign conditions of a family of polynomials restricted to a variety, having a much more refined dependence on the degrees.

THEOREM 37.10.2 [BB12]

Let $\mathcal{Q}, \mathcal{P} \subset \mathbb{R}[x_1, \ldots, x_k]$ *be finite subsets of polynomials such that* $\deg(Q) \leq d_0$ *for all* $Q \in \mathcal{Q}$, $\deg P = d_P$ *for all* $P \in \mathcal{P}$, *and the real dimension of* $\mathrm{Zer}(\mathcal{Q}, \mathbb{R}^k)$ *is* $k' \leq k$. *Suppose also that* $\mathrm{card}\,\mathcal{P} = s$, *and* $\deg(P) \leq d, P \in \mathcal{P}$.

Then, the number of connected components of the realizations of all realizable sign conditions of $\mathcal{P}$ *on* $\mathrm{Zer}(\mathcal{Q}, \mathbb{R}^k)$ *is bounded by* $(sd)^{k'} d_0^{k-k'} O(1)^k$.

This result opened the doors to using the polynomial partitioning theorem in higher dimensions, and was used to that effect in several papers, for example [Zah12, Zah13, KMSS12, MP15] (see also the survey by Tao [Tao14]). It is also an ingredient in the analysis of the data structure for range searching with semi-algebraic sets due to Agarwal, Matoušek and Sharir [AMS13].

In more sophisticated applications of the polynomial partitioning method a further improvement was needed, in which the bound on the number of connected components depends on a sequence of degrees of arbitrary length. Since dimensions of real varieties (in contrast to complex varieties) can be rather badly behaved (for example, the real variety defined by one non-constant polynomial can be empty), a bound depending on the degree sequence must also take into account the dimensions of the intermediate varieties not just the final variety. This strategy leads to the following theorem.

THEOREM 37.10.3 [BB16]

Let $Q_1, \ldots, Q_\ell \in \mathbb{R}[x_1, \ldots, x_k]$ *such that for each* $i, 1 \leq i \leq \ell$, $\deg(Q_i) \leq d_i$. *For* $1 \leq i \leq \ell$, *denote by* $\mathcal{Q}_i = \{Q_1, \ldots, Q_i\}$, $V_i = \mathrm{Zer}(\mathcal{Q}_i, \mathbb{R}^k)$, *and* $\dim_\mathbb{R}(V_i) \leq k_i$. *We set* $V_0 = \mathbb{R}^k$, *and adopt the convention that* $k_i = k$ *for* $i \leq 0$. *Suppose that* $2 \leq d_1 \leq d_2 \leq \frac{1}{k+1} d_3 \leq \frac{1}{(k+1)^2} d_4 \leq \cdots \leq \frac{1}{(k+1)^{\ell-2}} d_\ell$. *Then,*

$$b_0(V_\ell) \leq O(1)^\ell O(k)^{2k} \left(\prod_{1 \leq j < \ell} d_j^{k_{j-1} - k_j} \right) d_\ell^{k_{\ell-1}},$$

and in particular if $\ell \leq k$,

$$b_0(V_\ell) \leq O(k)^{2k} \left(\prod_{1 \leq j < \ell} d_j^{k_{j-1} - k_j} \right) d_\ell^{k_{\ell-1}}.$$

We remark that the (complex) dimension of complex varieties behaves better than the real dimension of real varieties. To see the contrast, consider the following example. Using the same notation as in Theorem 37.10.3, let for $1 \leq i \leq \ell$, $W_i \subset \mathbb{C}^k$ denote the complex variety defined by the set of polynomials $\mathcal{Q}_i$, and suppose that $W_i \neq \emptyset$. Then, for $1 \leq i < \ell$, $\dim(W_i) - 1 \leq \dim(W_{i+1}) \leq \dim(W_i)$, where $\dim(W_j)$ denotes the complex dimension of W_j, $1 \leq j \leq \ell$. While the second inequality is also true for the sequence, $(\dim(V_i))_{1 \leq i \leq \ell}$, of real dimensions of the sequence of real varieties V_i, the first inequality is definitely not true in general.

Theorem 37.10.3 enabled progress on several incidence questions (see [Zah15, BS16]).

Aside from its applications in polynomial partitioning, Theorem 37.10.3 also remedies a well-known anomaly, which is that the naive statement of Bézout inequality, that the number of isolated solutions in $\mathbb{C}^n$ of a system of n polynomial equations in n variables is bounded by the product of their degrees, is no longer true for isolated solutions in $\mathbb{R}^n$. This is illustrated by the following well-known example [Ful98]. Let $k = 3$, and let

$$
\begin{aligned}
Q_1 &= x_3, \\
Q_2 &= x_3, \\
Q_3 &= \sum_{i=1}^{2} \left(\prod_{j=1}^{d} (x_i - j)^2 \right).
\end{aligned}
$$

The real variety defined by $\{Q_1, Q_2, Q_3\}$ is 0-dimensional, and has d^2 isolated (in $\mathbb{R}^3$) points, whereas the degree sequence is $(d_1, d_2, d_3) = (1, 1, 2d)$, and thus the bound predicted by the naive Bézout inequality is equal to $2d$.

37.11 APPLICATIONS

Computational real algebraic geometry finds applications in robotics, vision, computer-aided design, geometric theorem proving, and other fields. Important problems in robotics include kinematic modeling, the inverse kinematic solution, computation of the workspace and workspace singularities, and the planning of an obstacle-avoiding motion of a robot in a cluttered environment—all arising from the algebro-geometric nature of robot kinematics. In solid modeling, graphics, and vision, almost all applications involve the description of surfaces, the generation of various auxiliary surfaces such as blending and smoothing surfaces, the classification of various algebraic surfaces, the algebraic or geometric invariants associated with a surface, the effect of various affine or projective transformations of a surface, the description of surface boundaries, and so on.

To give examples of the nature of the solutions demanded by various applications, we discuss a few representative problems from robotics, engineering, and computer science.

ROBOT MOTION PLANNING

Given the initial and desired configurations of a robot (composed of rigid subparts) and a set of obstacles, find a collision-free continuous motion of the robot from the initial configuration to the final configuration.

The algorithm proceeds in several steps. The first step translates the problem to **configuration space**, a parameter space modeled as a low-dimensional algebraic manifold (assuming that the obstacles and the robot subparts are bounded by piecewise algebraic surfaces). The second step computes the set of configurations that avoid collisions and produces a semialgebraic description of this so-called "free space" (subspaces of the configuration space). Since the initial and final configurations correspond to two points in the configuration space, we simply have to test whether they lie in the same connected component of the free space. If so, they can be connected by a piecewise algebraic path. Such a path gives rise to an obstacle-avoiding motion of the robot(s) and can be computed using either Collins's CAD [SS83], yielding an algorithm with doubly-exponential time complexity, or by computing a roadmap (Theorem 37.8.1) which yields a singly exponential algorithm.

OFFSET SURFACE CONSTRUCTION IN SOLID MODELING

*Given a polynomial $f(x, y, z)$, whose zeros define an algebraic surface in three-dimensional space, compute the envelope of a family of spheres of radius r whose centers lie on the surface f. Such a surface is called a (two-sided) **offset surface** of f.*

Let $p = \langle x, y, z \rangle$ be a point on the offset surface and $q = \langle u, v, w \rangle$ be a **footprint** of p on f; that is, q is the point at which a normal from p to f meets f. Let $\vec{t_1} = \langle t_{1,1}, t_{1,2}, t_{1,3} \rangle$ and $\vec{t_2} = \langle t_{2,1}, t_{2,2}, t_{2,3} \rangle$ be two linearly independent tangent vectors to f at the point q. Then, we see that the system of polynomial equations

$$
\begin{aligned}
(x - u)^2 + (y - v)^2 + (z - w)^2 - r^2 &= 0, \\
f(u, v, w) &= 0, \\
(x - u)t_{1,1} + (y - v)t_{1,2} + (z - w)t_{1,3} &= 0, \\
(x - u)t_{2,1} + (y - v)t_{2,2} + (z - w)t_{2,3} &= 0,
\end{aligned}
$$

describes a surface in the (x, y, z, u, v, w) six-dimensional space, which, when projected into the three-dimensional space with coordinates (x, y, z), gives the offset surface in an implicit form. The offset surface is computed by simply eliminating the variables u, v, w from the preceding set of equations.

This approach (the **envelope method**) of computing the offset surface has several problematic features: the method does not deal with self-intersection in a clean way and, sometimes, generates additional points not on the offset surface. For a discussion of these and several other related problems in solid modeling, see [Hof89] and Chapter 57 of this Handbook.

GEOMETRIC THEOREM PROVING

Given a geometric statement consisting of a finite set of hypotheses and a conclusion,

$$
\begin{aligned}
\text{Hypotheses} \quad &: \quad f_1(x_1, \ldots, x_k) = 0, \ldots, f_r(x_1, \ldots, x_k) = 0 \\
\text{Conclusion} \quad &: \quad g(x_1, \ldots, x_k) = 0
\end{aligned}
$$

decide whether the conclusion $g = 0$ is a consequence of the hypotheses $((f_1 = 0) \wedge \cdots \wedge (f_r = 0))$.

Thus we need to determine whether the following universally quantified first-order sentence holds:

$$\left(\forall \, x_1, \ldots, x_k\right) \left[\left((f_1 = 0) \wedge \cdots \wedge (f_r = 0)\right) \Rightarrow g = 0 \right].$$

One way to solve the problem is by first translating it into the form: decide if the following existentially quantified first-order sentence is unsatisfiable:

$$\left(\exists \, x_1, \ldots, x_k, z\right) \left[(f_1 = 0) \wedge \cdots \wedge (f_r = 0) \wedge (gz - 1) = 0 \right].$$

When the underlying domain is assumed to be the field of real numbers, then we may simply check whether the following multivariate polynomial (in $x_1, \ldots, x_k, z$) has no real root:

$$f_1^2 + \cdots + f_r^2 + (gz - 1)^2.$$

If, on the other hand, the underlying domain is assumed to be the field of complex numbers (an algebraically closed field), then other tools from computational algebra are used (e.g., techniques based on Hilbert's Nullstellensatz). In the general setting, some techniques based on Ritt-Wu characteristic sets have proven very powerful. See [Cho88].

For another approach to geometric theorem proving, see Section 57.4.

CONNECTION TO SEMIDEFINITE PROGRAMMING

Checking *global nonnegativity* of a function of several variables occupies a central role in many areas of applied mathematics, e.g., optimization problems with polynomial objectives and constraints, as in quadratic, linear and Boolean programming formulations. These problems have been shown to be NP-hard in the most general setting, but do admit good approximations involving polynomial-time computable relaxations (see Parillo [Par00]).

Provide checkable conditions or procedure for verifying the validity of the proposition

$$F(x_1, \ldots, x_k) \geq 0, \quad \forall x_1, \ldots, x_k,$$

where F is a multivariate polynomial in the ring of multivariate polynomials over the reals, $\mathbb{R}[x_1, \ldots, x_k]$.

An obvious necessary condition for F to be globally nonnegative is that it has even degree. On the other hand, a rather simple sufficient condition for a real-valued polynomial $F(x)$ to be globally nonnegative is the existence of a *sum-of-squares decomposition*:

$$F(x_1, \ldots, x_k) = \sum_i f_i^2(x_1, \ldots, x_k), \quad f_i(x_1, \ldots, x_k) \in \mathbb{R}[x_1, \ldots, x_k].$$

Thus one way to solve the global nonnegativity problem is by finding a sum-of-squares decomposition. Note that since there exist globally nonnegative polynomials not admitting a sum-of-squares decomposition (e.g., the Motzkin form $x^4y^2 + x^2y^4 + z^6 - 3x^2y^2z^2$), the procedure suggested below does not give a solution to the problem in all situations.

The procedure can be described as follows: express the given polynomial $F(x_1, \ldots, x_k)$ of degree $2d$ as a quadratic form in all the monomials of degree less than or equal to d:

$$F(x_1, \ldots, x_k) = z^T Q z, \quad z = [1, x_1, \ldots, x_k, x_1 x_2, \ldots x_k^d],$$

where Q is a constant matrix to be determined. If the above quadratic form can be solved for a positive semidefinite Q, then $F(x_1, \ldots, x_k)$ is globally nonnegative. Since the variables in z are not algebraically independent, the matrix Q is not unique, but lives in an affine subspace. Thus, we need to determine if the intersection of this affine subspace and the positive semidefinite matrix cone is nonempty. This problem can be solved by a **semidefinite programming** feasibility problem

$$\begin{aligned} \text{trace}(zz^T Q) &= F(x_1, \ldots, x_k), \\ Q &\succeq 0. \end{aligned}$$

The dimensions of the matrix inequality are $\binom{k+d}{d} \times \binom{k+d}{d}$ and are polynomial for a fixed number of variables (k) or fixed degree (d). Thus our question reduces to efficiently solvable semidefinite programming (SDP) problems.

37.12 SOURCES AND RELATED MATERIAL

SURVEYS

[Mis93]: A textbook for algorithmic algebra covering Gröbner bases, characteristic sets, resultants, and real algebra. Chapter 8 gives many details of the classical results in computational real algebra.

[BPR16] A textbook on algorithm in real algebraic geometry covering modern techniques and algorithms for solving the fundamental algorithmic problems in semi-algebraic geometry. Contains all relevant mathematical background including the theory of general real closed fields, real algebra, algebraic topology and Morse theory etc.

[Bas08] A survey of algorithms in semi-algebraic geometry and topology with particular emphasis on the connections to discrete and computational geometry.

[BHK+11] A collection of survey articles on modern developments in real algebraic geometry including modern developments in algorithmic real algebraic geometry.

[CJ98]: An anthology of key papers in computational real algebra and real algebraic geometry. Contains reprints of the following papers cited in this chapter: [BPR98, Col75, Ren91, Tar51].

[AB88]: A special issue of the *J. Symbolic Comput.* on computational real algebraic geometry. Contains several papers ([DH88, Gri88, GV88] cited here) addressing many key research problems in this area.

[BR90]: A very accessible and self-contained textbook on real algebra and real algebraic geometry.

[BCR98]: A self-contained textbook on real algebra and real algebraic geometry.

[HRR91]: A survey of many classical and recent results in computational real algebra.

[Cha94]: A survey of the connections among computational geometry, computational algebra, and computational real algebraic geometry.

[Tar51]: Primary reference for Tarski's classical result on the decidability of elementary algebra.

[Col75]: Collins's work improving the complexity of Tarski's solution for the decision problem [Tar51]. Also, introduces the concept of cylindrical algebraic decomposition (CAD).

[Ren91]: A survey of some recent results, improving the complexity of the decision problem and quantifier elimination problem for the first-order theory of reals. This is mostly a summary of the results first given in a sequence of papers by Renegar [Ren92a, Ren92b, Ren92c].

[Lat91]: A comprehensive textbook covering various aspects of robot motion planning problems and different solution techniques. Chapter 5 includes a description of the connection between the motion planning problem and computational real algebraic geometry.

[SS83]: A classic paper in robotics showing the connection between the robot motion planning problem and the connectivity of semialgebraic sets using CAD. Contains several improved algorithmic results in computational real algebra.

[Can87]: Gives a singly-exponential time algorithm for the robot motion planning problem and provides complexity improvement for many key problems in computational real algebra.

[Hof89]: A comprehensive textbook covering various computational algebraic techniques with applications to solid modeling. Contains a very readable description of Gröbner bases algorithms.

[Cho88]: A monograph on geometric theorem proving using Ritt-Wu characteristic sets. Includes computer-generated proofs of many classical geometric theorems.

RELATED CHAPTERS

Chapter 50: Algorithmic motion planning
Chapter 51: Robotics
Chapter 57: Solid modeling
Chapter 60: Geometric applications of the Grassmann-Cayley algebra

REFERENCES

[AB88]   D. Arnon and B. Buchberger, editors, *Algorithms in Real Algebraic Geometry.* Special Issue: *J. Symbolic Comput.*, 5(1-2), 1988.

[AMS13]   P.K. Agarwal, J. Matoušek, and M. Sharir. On range searching with semialgebraic sets. II. *SIAM J. Comput.*, 42:2039–2062, 2013.

[Bas99a]   S. Basu. New results on quantifier elimination over real closed fields and applications to constraint databases. *J. ACM*, 46:537–555, 1999.

[Bas99b]   S. Basu. On bounding the Betti numbers and computing the Euler characteristic of semi-algebraic sets. *Discrete Comput. Geom.*, 22:1–18, 1999.

[Bas06]   S. Basu. Computing the first few Betti numbers of semi-algebraic sets in single exponential time. *J. Symbolic Comput.*, 41:1125–1154, 2006.

[Bas08]   S. Basu. Algorithmic semi-algebraic geometry and topology—recent progress and open problems. In *Surveys on Discrete and Computational Geometry: Twenty Years Later*, vol. 453 of *Contemporary Math.*, pages 139–212, AMS, Providence, 2008.

[BB12]   S. Barone and S. Basu. Refined bounds on the number of connected components of sign conditions on a variety. *Discrete Comput. Geom.*, 47:577–597, 2012.

[BB16]   S. Barone and S. Basu. On a real analog of Bezout inequality and the number of connected components of sign conditions. *Proc. Lond. Math. Soc. (3)*, 112:115–145, 2016.

[BCR98]   J. Bochnak, M. Coste, and M.-F. Roy. *Real Algebraic Geometry*. Springer-Verlag, Berlin, 1998. (Also in French, *Géométrie Algébrique Réelle*. Springer-Verlag, Berlin, 1987.)

[BCR08]   F. Boudaoud, F. Caruso, and M.-F. Roy. Certificates of positivity in the Bernstein basis. *Discrete Comput. Geom.*, 39:639–655, 2008.

[BCSS98]   L. Blum, F. Cucker, M. Shub, and S. Smale. *Complexity and Real Computation*. Springer-Verlag, New York, 1998.

[BHK+11]   S. Basu, J. Huisman, K. Krzysztof, V. Powers, and J.-P. Rolin. Real algebraic geometry. In *Real Algberaic Geometry*, Rennes, France, 2011.

[BOKR86]   M. Ben-Or, D. Kozen, and J. Reif. The complexity of elementary algebra and geometry. *J. Comp. Systems Sci.*, 18:251–264, 1986.

[BPR96]   S. Basu, R. Pollack, and M.-F. Roy. On the combinatorial and algebraic complexity of quantifier elimination. *J. ACM*, 43:1002–1045, 1996.

[BPR98]   S. Basu, R. Pollack, and M.-F. Roy. A new algorithm to find a point in every cell defined by a family of polynomials. In *Quantifier Elimination and Cylindrical Algebraic Decomposition, Texts Monogr. Symbol. Comput.*, pages 341–350, Springer-Verlag, Vienna, 1998.

[BPR00]   S. Basu, R. Pollack, and M.-F. Roy. Computing roadmaps of semi-algebraic sets on a variety. *J. Amer. Math. Soc.*, 13:55–82, 2000.

[BPR05a]   S. Basu, R. Pollack, and M.-F. Roy. On the Betti numbers of sign conditions. *Proc. Amer. Math. Soc.*, 133:965–974, 2005.

[BPR05b]   S. Basu, R. Pollack, and M.-F. Roy. Computing the Euler-Poincaré characteristics of sign conditions. *Comput. Complexity*, 14:53–71, 2005.

[BPR08]   S. Basu, R. Pollack, and M.-F. Roy. Computing the first Betti number of a semi-algebraic set. *Found. Comput. Math.*, 8:97–136, 2008.

[BPR16]   S. Basu, R. Pollack, and M.-F. Roy. *Algorithms in Real Algebraic Geometry*. Second edition, Springer, Berlin, 2006. Revised version of the second edition online at http://perso.univ-rennes1.fr/marie-francoise.roy/, 2016.

[BR10]   S. Basu and M.-F. Roy. Bounding the radii of balls meeting every connected component of semi-algebraic sets. *J. Symbolic Comput.*, 45:1270–1279, 2010.

[BR90]   R. Benedetti and J.-J. Risler. *Real Algebraic and Semi-algebraic Sets*. Actualités Mathématiques, Hermann, Paris, 1990.

[BR14]   S. Basu and M.-F. Roy. Divide and conquer roadmap for algebraic sets. *Discrete Comput. Geom.*, 52:278–343, 2014.

[BRS+14]   S. Basu, M.-F. Roy, M. Safey El Din, and É. Schost. A baby step–giant step roadmap algorithm for general algebraic sets. *Found. Comput. Math.*, 14:1117–1172, 2014.

[BS16]   S. Basu and M. Sombra. Polynomial partitioning on varieties of codimension two and point-hypersurface incidences in four dimensions. *Discrete Comput. Geom.*, 55:158-184, 2016.

[BSS89]   L. Blum, M. Shub, and S. Smale. On a theory of computation and complexity over the real numbers: NP-completeness, recursive functions and universal machines. *Bull. Amer. Math. Soc. (N.S.)*, 21:1–46, 1989.

[BV07]   S. Basu and N. Vorobjov. On the number of homotopy types of fibres of a definable map. *J. Lond. Math. Soc. (2)*, 76:757–776, 2007.

[Can87]   J.F. Canny. *The Complexity of Robot Motion Planning*. MIT Press, Cambridge, 1988.

[Can88]   J. Canny. Some algebraic and geometric computations in PSPACE. In *Proc. 20th ACM Sympos. Theory Comput.*, pages 460–467, 1988.

[Can90]   J. Canny. Generalised characteristic polynomials. *J. Symbolic Comput.*, 9:241–250, 1990.

[Can93]   J. Canny. Improved algorithms for sign determination and existential quantifier elimination. *Comput. J.*, 36:409–418, 1993.

[CEG+90]   K.L. Clarkson, H. Edelsbrunner, L.J. Guibas, M. Sharir, and E. Welzl. Combinatorial complexity bounds for arrangements of curves and spheres. *Discrete Comput. Geom.*, 5:99–160, 1990.

[Cha94]   B. Chazelle. Computational geometry: A retrospective. In *Proc. 26th ACM Sympos. Theory Comput.*, pages 75–94, 1994.

[Cho88]   S.-C. Chou. *Mechanical Geometry Theorem Proving*. Vol. 41 of *Mathematics and its Applications*, Reidel Publishing Co., Dordrecht, 1988.

[CJ98]   B.F. Caviness and J.R. Johnson, editors. *Quantifier Elimination and Cylindrical Algebraic Decomposition. Texts Monogr. Symbol. Comput.*, Springer-Verlag, Vienna, 1998.

[Col75]   G.E. Collins. Quantifier elimination for real closed fields by cylindric algebraic decomposition. In *Proc. 2nd GI Conf. on Automata Theory and Formal Languages*, vol. 33 of *Lecture Notes Comp. Sci.*, pages 134–183, Springer, Berlin, 1975.

[DH88]   J.H. Davenport and J. Heintz. Real quantifier elimination is doubly exponential. *J. Symbolic Comp.*, 5:29–35, 1988.

[ES11]   G. Elekes and M. Sharir. Incidences in three dimensions and distinct distances in the plane. *Combin. Probab. Comput.*, 20:571–608, 2011.

[Ful98]   W. Fulton. *Intersection theory*. Volume 2 of *Ergebnisse der Mathematik und ihrer Grenzgebiete. 3. Folge. A Series of Modern Surveys in Mathematics [Results in Mathematics and Related Areas. 3rd Series. A Series of Modern Surveys in Mathematics]*, 2nd edition, Springer, Berlin, 1998.

[GK15]   L. Guth and N.H. Katz. On the Erdős distinct distances problem in the plane. *Ann. of Math. (2)*, 181:155–190, 2015.

[GP86a]   J.E. Goodman and R. Pollack. There are asymptotically far fewer polytopes than we thought. *Bull. Amer. Math. Soc. (N.S.)*, 14:127–129, 1986.

[GP86b]   J.E. Goodman and R. Pollack. Upper bounds for configurations and polytopes in $\mathbb{R}^d$. *Discrete Comput. Geom.*, 1:219–227, 1986.

[GPW95]   J.E. Goodman, R. Pollack, and R. Wenger. On the connected components of the space of line transversals to a family of convex sets. *Discrete Comput. Geom.*, 13:469–476, 1995.

[GR93]   L. Gournay and J.-J. Risler. Construction of roadmaps of semi-algebraic sets. *Appl. Algebra Eng. Commun. Comput.*, 4:239–252, 1993.

[Gri88]   D.Y. Grigoriev. Complexity of deciding Tarski algebra. *J. Symbolic Comput.*, 5:65–108, 1988.

[Gut15a]   L. Guth. Distinct distance estimates and low degree polynomial partitioning. *Discrete Comput. Geom.*, 53:428–444, 2015.

[Gut15b]   L. Guth. Polynomial partitioning for a set of varieties. *Math. Proc. Cambridge Philos. Soc.*, 159:459–469, 2015.

[GV88]   D.Y. Grigoriev and N.N. Vorobjov, Jr. Solving systems of polynomial inequalities in subexponential time. *J. Symbolic Comput.*, 5:37–64, 1988.

[GV92]   D.Y. Grigoriev and N.N. Vorobjov, Jr. Counting connected components of a semi-algebraic set in subexponential time. *Comput. Complexity*, 2:133–186, 1992.

[GV05]   A. Gabrielov and N. Vorobjov. Betti numbers of semialgebraic sets defined by quantifier-free formulae. *Discrete Comput. Geom.*, 33:395–401, 2005.

[GV09]   A. Gabrielov and N. Vorobjov. Approximation of definable sets by compact families, and upper bounds on homotopy and homology. *J. Lond. Math. Soc. (2)*, 80:35–54, 2009.

[GV15]   A. Gabrielov and N. Vorobjov. On topological lower bounds for algebraic computation trees. *Found. Comput. Math.*, to appear.

[Hof89]   C.M. Hoffmann. *Geometric and Solid Modeling*. Morgan Kaufmann, San Mateo, 1989.

[HRR91]   J. Heintz, T. Recio, and M.-F. Roy. Algorithms in real algebraic geometry and applications to computational geometry. In J.E. Goodman, R. Pollack, and W. Steiger, editors, *Discrete and Computational Geometry*, vol. 6 of *DIMACS Ser. Discrete Math. Theoret. Comput. Sci.*, pages 137–163, AMS, Providence, 1991.

[HRS89]   J. Heintz, M.-F. Roy, and P. Solernó. On the complexity of semi-algebraic sets. In *Proc. IFIP 11th World Computer Congr.*, pages 293–298, North-Holland, Amsterdam, 1989.

[HRS90]   J. Heintz, M.-F. Roy, and P. Solernó. Sur la complexité du principe de Tarski-Seidenberg. *Bull. Soc. Math. France*, 118:101–126, 1990.

[JPT13]   G. Jeronimo, D. Perrucci, and E. Tsigaridas. On the minimum of a polynomial function on a basic closed semialgebraic set and applications. *SIAM J. Optim.*, 23:241–255, 2013.

[KMSS12]   H. Kaplan, J. Matoušek, Z. Safernová, and M. Sharir. Unit distances in three dimensions. *Combin. Probab. Comput.*, 21:597–610, 2012.

[Lat91]   J.-C. Latombe. *Robot Motion Planning*. Vol. 124 of *Internat. Ser. Engrg. Comp. Sci.*. Kluwer, Dordrecht, 1991.

[Mat02]   J. Matoušek. *Lectures on Discrete Geometry*. Vol. 212 of *Graduate Texts in Mathematics*, Springer-Verlag, New York, 2002.

[Mil64]   J. Milnor. On the Betti numbers of real varieties. *Proc. Amer. Math. Soc.*, 15:275–280, 1964.

[Mis93]   B. Mishra. *Algorithmic Algebra. Texts and Monographs in Computer Science*, Springer-Verlag, New York, 1993.

[MMP96]   J.L. Montaña, J.E. Morais, and L.M. Pardo. Lower bounds for arithmetic networks. II. Sum of Betti numbers. *Appl. Algebra Engrg. Comm. Comput.*, 7:41–51, 1996.

[MP15]   J. Matoušek and Z. Patáková. Multilevel polynomial partitions and simplified range searching. *Discrete Comput. Geom.*, 54:22–41, 2015.

[Par00]   P.A. Parrilo. *Structured Semidefinite Programs and Semialgebraic Geometry Methods in Robustness and Optimization*, PhD Thesis, California Institute of Technology, Pasadena, 2000.

[PO49]   I.G. Petrovskiĭ and O.A. Oleĭnik. On the topology of real algebraic surfaces. *Izvestiya Akad. Nauk SSSR. Ser. Mat.*, 13:389–402, 1949.

[Ren91]   J. Renegar. Recent progress on the complexity of the decision problem for the reals. In J.E. Goodman, R. Pollack, and W. Steiger, editors, *Discrete and Computational Geometry*, vol. 6 of *DIMACS Ser. Discrete Math. Theoret. Comput. Sci.*, pages 287–308, AMS, Providence, 1991.

[Ren92a]   J. Renegar. On the computational complexity and geometry of the first-order theory of the reals. I. Introduction. Preliminaries. The geometry of semi-algebraic sets. The decision problem for the existential theory of the reals. *J. Symbolic Comput.*, 13:255–299, 1992.

[Ren92b]   J. Renegar. On the computational complexity and geometry of the first-order theory of the reals. II. The general decision problem. Preliminaries for quantifier elimination. *J. Symbolic Comput.*, 13:301–327, 1992.

[Ren92c]   J. Renegar. On the computational complexity and geometry of the first-order theory of the reals. III. Quantifier elimination. *J. Symbolic Comput.*, 13:329–352, 1992.

[SS83]   J.T. Schwartz and M. Sharir. On the piano movers' problem. II. General techniques for computing topological properties of real algebraic manifolds. *Adv. Appl. Math.*, 4:298–351, 1983.

[SS10]   M. Safey el Din and É. Schost. A baby steps/giant steps probabilistic algorithm for computing roadmaps in smooth bounded real hypersurface. *Discrete Comput. Geom.*, 45:181–220, 2010.

[SS17]   M. Safey El Din and É. Schost. A nearly optimal algorithm for deciding connectivity queries in smooth and bounded real algebraic sets. *J. ACM*, 63:48, 2017.

[SS14]   M. Sharir and N. Solomon. Incidences between points and lines in $\mathbb{R}^4$. In *Proc. 30th Sympos. Comput. Geom.*, pages 189–197, ACM, New York, 2014.

[SSS13]   M. Sharir, A. Sheffer, and J. Solymosi. Distinct distances on two lines. *J. Combin. Theory Ser. A*, 120:1732–1736, 2013.

[SSZ15]   M. Sharir, A. Sheffer, and J. Zahl. Improved bounds for incidences between points and circles. *Combin. Probab. Comput.*, 24:490–520, 2015.

[ST42]   A.H. Stone and J.W. Tukey. Generalized "sandwich" theorems. *Duke Math. J.*, 9:356–359, 1942.

[Stu35]   C. Sturm. Mémoire sur la Résolution des Équations Numériques. *Mém. Savants Etrangers*, 6:271–318, 1835.

[Syl53]   J.J. Sylvester. On a theory of the syzygetic relations of two rational integral functions, comprising an application to the theory of Sturm's functions, and that of the greatest algebraic common measure. *Philos. Trans. Roy. Soc. London*, 143:407–548, 1853.

[Tao14]   T. Tao. Algebraic combinatorial geometry: The polynomial method in arithmetic combinatorics, incidence combinatorics, and number theory. *EMS Surveys Math. Sci.*, 1:1–46, 2014.

[Tar51]   A. Tarski. *A Decision Method for Elementary Algebra and Geometry*, second edition. University of California Press, 1951.

[Tho65]   R. Thom. Sur l'homologie des variétés algébriques réelles. In *Differential and Combinatorial Topology (A Symposium in Honor of Marston Morse)*, pages 255–265, Princeton University Press, 1965.

[Yao97]   A.C.-C. Yao. Decision tree complexity and Betti numbers. *J. Comput. Syst. Sci.*, 55:36–43, 1997.

[Zah12]   J. Zahl. On the Wolff circular maximal function. *Illinois J. Math.*, 56:1281–1295, 2012.

[Zah13]   J. Zahl. An improved bound on the number of point-surface incidences in three dimensions. *Contrib. Discrete Math.*, 8:100–121, 2013.

[Zah15]   J. Zahl. A Szemerédi–Trotter type theorem in $\mathbb{R}^4$. *Discrete Comput. Geom.*, 54:513–572, 2015.

Part V

GEOMETRIC DATA STRUCTURES AND SEARCHING

38 POINT LOCATION

Jack Snoeyink

INTRODUCTION

"Where am I?" is a basic question for many computer applications that employ geometric structures (e.g., in computer graphics, geographic information systems, robotics, and databases). Given a set of disjoint geometric objects, the ***point-location problem*** asks for the object containing a query point that is specified by its coordinates. Instances of the problem vary in the dimension and type of objects and whether the set is static or dynamic. Classical solutions vary in preprocessing time, space used, and query time. Recent solutions also consider entropy of the query distribution, or exploit randomness, external memory, or capabilities of the word RAM machine model.

Point location has inspired several techniques for structuring geometric data, which we survey in this chapter. We begin with point location in one dimension (Section 38.1) or in one polygon (Section 38.2). In two dimensions, we look at how techniques of persistence, fractional cascading, trapezoid graphs, or hierarchical triangulations can lead to optimal comparison-based methods for point location in static subdivisions (Section 38.3), the current best methods for dynamic subdivisions (Section 38.4), and at methods not restricted to comparison-based models (Section 38.5). There are fewer results on point location in higher dimensions; these we mention in (Section 38.6).

The vision/robotics term ***localization*** refers to the opposite problem of determining (approximate) coordinates from the surrounding local geometry. This chapter deals exclusively with point location.

38.1 ONE-DIMENSIONAL POINT LOCATION

The simplest nontrivial instance of point location is list searching. The objects are points $x_1 \leq \cdots \leq x_n$ on the real line, presented in arbitrary order, and the intervals between them, (x_i, x_{i+1}) for $1 \leq i < n$. The answer to a query q is the name of the object containing q.

The list-searching problem already illustrates several aspects of general point location problems and several data structure innovations.

GLOSSARY

Decomposable problem: A problem whose answer can be obtained from the answers to the same problem on the sets of an arbitrary partition of the input [Ben79, BS80]. The one-dimensional point location as stated above—find the interval containing q—is not decomposable, since partitioning into subsets of points gives a very different set of intervals. "Find the lower endpoint of the containing interval" is decomposable, however; one can report the highest "lowest point" returned from all subsets of the partition.

Preprocessing/queries: If one assumes that many queries will search the same input, then resources can profitably be spent building data structures to facilitate the search. Three resources are commonly analyzed:

Query time: Computation time to answer a single query, given a point location data structure. Usually a worst-case upper bound, expressed as a function of the number of objects in the structure, n.

Preprocessing time: Time required to build a point location structure for n objects.

Space: Memory used by the point location structure for n objects.

Dynamic point location: Maintaining a location data structure as points are inserted and deleted. The one-dimensional point location structures can be made dynamic without changing their asymptotic performances.

Randomized point location: Data structures whose preprocessing algorithms may make random choices in an attempt to avoid poor performance caused by pathological input data. Preprocessing and query times are reported as expectations over these random choices. Randomized algorithms make no assumptions on the input or query distributions. They often use a sample to obtain information about the input distribution, and can achieve good expected performance with simple algorithms.

Entropy bounds: If the probability of a query falling in region i is p_i, then Shannon entropy $H = \sum_i -p_i \log_2(p_i)$ is a lower bound for expected query time, where the expectation is over the query probability distribution.

Static optimality: A (self-adjusting) search structure has static optimality if, for any (infinite) sequence of searches, its cumulative search time is asymptotically bounded by cumulative time of the best static structure for those searches.

Transdichotomous: Machine models, such as the word RAM, that are not restricted to comparisons, are called transdichotomous if they support bit operations or other computations that allow algorithms to break through information-theoretic lower bounds that apply to comparison-based models, such as decision trees.

LIST SEARCH AS ONE-DIMENSIONAL POINT LOCATION

Table 38.1.1 reports query time, preprocessing time, and space for several search methods. Linear search requires no additional data structure if the problem is decomposable. Binary search trees or randomized search trees [SA96, Pug90] require a total order and an ability to do comparisons. An adversary argument shows that these comparison-based query algorithms require $\Omega(\log n)$ comparisons. If however, searches will be near each other, or near the ends, a finger search tree can find an element d intervals away in $O(\log d)$ time. If the probability distribution for queries is known, then the lower bound on expected query time is H, and expected $H + 2$ can be achieved by weight-balanced trees [Meh77]. Even if the distribution is not known, splay trees achieve static optimality [ST85].

Note that these one-dimensional structures can be built dynamically by operations that take the same amortized time as performing a query. So in theory, we need not report the preprocessing time; that will change in higher dimensions.

If we step away from comparison-based models, a useful method in practice is to partition the input range into b equal-sized buckets, and to answer a query by searching the bucket containing the query. If the points are restricted to integers $[1, \ldots, U]$, then van Emde Boas [EBKZ77] has shown how hashing techniques can be applied in stratified search trees to answer a query in $O(\log \log U)$ time. Combining these with Fredman and Willard's fusion trees [FW93] can achieve $O(\sqrt{\log n})$-time queries without the restriction to integers [CP09].

TABLE 38.1.1   List search as one-dimensional point location.

TECHNIQUE	QUERY	PREPROC	SPACE
Linear search	$O(n)$	none	data only
Binary search	$O(\log n)$	$O(n \log n)$	$O(n)$
Randomized tree	exp. $O(\log n)$	exp. $O(n \log n)$	$O(n)$
Finger search	$O(\log d)$	$O(n \log n)$	$O(n)$
Weight-balance tree	exp. H+2	$O(n \log n)$	$O(n)$
Splay tree	$O(OPT)$ in limit	$O(n)$	$O(n)$
Bucketing	$O(n)$	$O(n + b)$	$O(n + b)$
van Emde Boas tree	$O(\log \log U)$	exp. $O(n)$	$O(n)$
word RAM	$O(\sqrt{\log n})$	$O(n \log \log n)$	$O(n)$

38.2 POINT-IN-POLYGON

The second simplest form of point location is to determine whether a query point q lies inside a given n-sided polygon P [Hai94]. Without preprocessing the polygon, one may use parity of the winding or crossing numbers: count intersections of a ray from q with the boundary of polygon P. Point q is inside P iff the number is odd. A query takes $O(n)$ time.

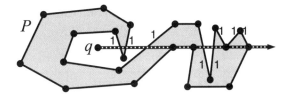

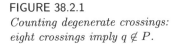

FIGURE 38.2.1
Counting degenerate crossings:
eight crossings imply $q \notin P$.

One must count carefully in degenerate cases when the ray passes through a vertex or edge of P. When the ray is horizontal, as in Figure 38.2.1, then edges of P can be considered to contain their lower but not their upper endpoints. Edges inside the ray can be ignored. This is consistent with the count obtained by perturbing the ray infinitesimally upward. Schirra [Sch08] experimentally observes which points are incorrectly classified using inexact floating point arithmetic in various algorithms. Stewart [Ste91] considered point-in-polygon algorithm design when vertex and edge positions may be imprecise.

To obtain sublinear query times, preprocess the polygon P using the more general techniques of the next sections.

38.3 PLANAR POINT LOCATION: STATIC

Theoretical research has produced a number of planar point location methods that are optimal for comparison-based models: $O(n \log n)$ time to preprocess a planar subdivision with n vertices for $O(\log n)$ time queries using $O(n)$ space. Preprocessing time reduces to linear if the input is given in an appropriate format, and some preprocessing schemes have been parallelized (see Chapter 46).

We focus on the data structuring techniques used to reach optimality: persistence, fractional cascading, trapezoid graphs, and hierarchical triangulations.

In a planar subdivision, point location can be made decomposable by storing with each edge the name of the face immediately above. If one knows for each subproblem the edge below a query, then one can determine the edge directly below and report the containing face, even for an arbitrary partition into subproblems.

GLOSSARY

Planar subdivision: A partitioning of a region of the plane into point *vertices,* line segment *edges,* and polygonal *faces.*

Size of a planar subdivision: The number of vertices, usually denoted by n. Euler's relation bounds the numbers of edges $e \leq 3n - 6$ and faces $f \leq 2n - 4$; often the constants are suppressed by saying that the number of vertices, edges, and faces are all $O(n)$.

Monotone subdivision: A planar subdivision whose faces are x-monotone polygons: i.e., the intersection of any face with any vertical line is connected.

Triangulation/trapezoidation: Planar subdivisions whose faces are triangles/ whose faces are trapezoids with parallel sides all in the same direction.

Dual graph: A planar subdivision can be viewed as a graph with vertices joined by edges. The dual graph has a node for each face and an arc joining two faces if they share a common edge.

SLABS AND PERSISTENCE

By drawing a vertical line through every vertex, as shown in Figure 38.3.1(a), we obtain vertical *slabs* in which point location is almost one-dimensional. Two binary searches suffice to answer a query: one on x-coordinates for the slab containing q, and one on edges that cross that slab. Query time is $O(\log n)$, but space may be quadratic if all edges are stored with the slabs that they cross.

The location structures for adjacent slabs are similar. We could sweep from left to right to construct balanced binary search trees on the edges for all slabs:

TABLE 38.3.1   A select few of the best static planar point location results known for subdivision with n edges. Expectations are over decisions made by the algorithm; averages are over a query distribution with entropy H. For distance sensitivity, scale the subdivision to have unit area and denote the distance from query q to the nearest boundary by Δ_q. The static optimality result is for regions of constant complexity.

TECHNIQUE	QUERY	PREPROC	SPACE
Slab + persistence [ST86]	$O(\log n)$	$O(n \log n)$	$O(n)$
Separating chain + fractional cascade [EGS86]	$O(\log n)$	$O(n \log n)$	$O(n)$
Randomized [HKH16]	$O(\log n)$	exp. $O(n \log n)$	$O(n)$
Weighted randomized [AMM07]	average $(5 \ln 2)H + O(1)$	exp. $O(n \log n)$	exp. $O(n)$
Optimal query [SA00]	$\log_2 n + \sqrt{\log_2 n} + \Theta(1)$	$O(2^{2\sqrt{\log n}})$	$O(2^{2\sqrt{\log n}})$
+ struct. sharing	$\log_2 n + \sqrt{\log_2 n} + O(\log_2^{1/4} n)$	exp. $O(n \log n)$	exp. $O(n)$
Optimal entropy [CDI+12]	avg. $H + O(H^{1/2} + 1)$	exp. $O(n \log n)$	$O(n)$
Distance sensitive [ABE+16]	$N \min\{\log n, -\log \Delta_q\}$	exp. $O(n \log n)$	$O(n)$
Static optimality [IM12]	$O(OPT)$ in limit	$O(n)$	$O(n)$

As we sweep over the right endpoint of an edge, we remove the corresponding tree node. As we sweep over the left endpoint of an edge, we add a node. This takes $O(n \log n)$ total time and a linear number of node updates. To store all slabs in linear space, Sarnak and Tarjan [ST86] add to this the idea of *persistence*.

Rather than modifying a node to update the tree, copy the $O(\log n)$ nodes on the path from the root to this node, then modify the copies. This ***node-copying persistence*** preserves the former tree and gives access to a new tree (through the new root) that represents the adjacent slab. The total space for n trees is $O(n \log n)$. Figure 38.3.1(a) provides an illustration. The initial tree contains 8 and 1. (Recall that edges are named by the face immediately above.) Then 2, 3, and 7 are added, 8 is copied during rebalancing, but node 1 is not changed. When 6 is added, 7 is copied in the rebalancing, but the two subtrees holding 1, 2, 3, and 8 are not changed.

Limited node copying reduces the space to linear. Give each node spare left and right pointers and left and right time-stamps. Build a balanced tree for the initial slab. When a pointer is to be modified, use a spare and time-stamp it, if there is a spare available. Future searches can use the time-stamp to determine whether to follow the pointer or the spare. Otherwise, copy the node and modify its ancestor to point to the copy. If the slab location structures are maintained with $O(1)$ rotations per update, then the amortized cost of copying is also $O(1)$ per update.

Preprocessing takes $O(n \log n)$ time to sort by x coordinates and build either persistent data structure. To compare constants with other methods, the data structure has about 12 entries per edge because of extra pointers and copying. Searches take about $4 \log_2 N$ comparisons, where N is the number of edges that can intersect a vertical line; this is because there are two comparisons per node and "$O(1)$ rotation" tree-balancing routines are balanced only to within a factor of two.

We will see two other slab-based data structures in Section 38.4 on dynamic point location: interval trees and segment trees recursively merge slabs, and save

FIGURE 38.3.1

Optimal static methods: (a) *Slab (persistent);* (b) *separating chain (fractional cascading).*

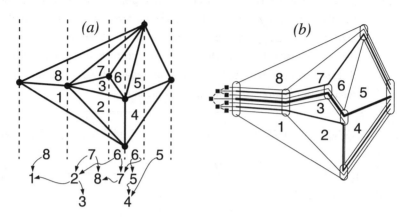

space by choosing where to store segments in the resulting slab tree. The other comparison-based point location schemes in this section do not represent slabs explicitly. Nevertheless, Chan and Pătraşcu [CP09] have convincingly argued that point location in a slab is a fundamental operation in computational geometry; by using a word RAM to perform point location in a slab faster than the comparison-based lower bounds (Section 38.5), they are able to speed up many classical geometric computations.

SEPARATING CHAINS AND FRACTIONAL CASCADING

If a subdivision is monotone, then its faces can be totally ordered consistent with aboveness; in other words, we can number faces $1, \ldots, f$ so that any vertical line encounters lower numbers below higher numbers. The **separating chain** between the faces $< k$ and those $\geq k$ is a monotone chain of edges [LP77]. Figure 38.3.1(b) shows all separating chains for a subdivision; the middle chain, $k = 5$, is shown darkest.

A balanced binary tree of separating chains can be used for point location: if query point q is above chain i and below chain $i+1$, then q is in face i. To preserve linear space we need to avoid the duplication of edges in chains that can be seen in Figure 38.3.1(b).

Note that the separating chains that contain an edge are defined by consecutive integers; we can store the first and last with each edge. Then form a binary tree in which each subtree stores the separating chains from some interval: at each node, store the edges of the median chain that have not been stored higher in the tree, and recursively store the intervals below and above the median in the left and right subtrees respectively. The root, for example, stores all edges of the middle chain. Since no edge is stored twice, this data structure takes $O(n)$ space.

As we search the tree for a query point q, we keep track of the edges found so far that are immediately above and below q. (Initially, no edges have been found.) Now, the root of the subtree to search is associated with a separating chain. If that chain does not contain one of the edges that we know is above or below q, then we search the x-coordinates of edges stored at the node and find the one on the vertical line through q. We then compare against the separating chain and recursively search the left or right subtree. Thus, this separating chain method [LP77] inspects $O(\log n)$ tree nodes at a cost of $O(\log n)$ each, giving $O(\log^2 n)$ query time.

To reduce the query time, we can use fractional cascading [CG86, EGS86] for efficient search in multiple lists. As we traverse our search tree, at every node we search a list by x-coordinates. We can make all searches after the first take constant time, if we increase the total size of these lists by 50%. Pass every fourth x-coordinate from a child list to its parent, and establish connections so that knowing one's position in the parent list gives one's position in the child to within four nodes.

Preprocessing takes $O(n)$ time on a monotone subdivision; arbitrary planar subdivisions can be made monotone by plane sweep in $O(n \log n)$ time. One can trade off space and query time in fractional cascading, but typical constants are 8 entries per edge for a query time of $4 \log_2 n$.

TRAPEZOID GRAPH METHODS

Preparata's [Pre81] trapezoid method is a simple, practical method that achieves $O(\log n)$ query time at the cost of $O(n \log n)$ space. Its underlying search structure, the **trapezoid graph**, is the basis for important variations: randomized point location in optimal expected time and space, a recursive application giving exact worst-case optimal query time, and, in the next section, average time point location achieving the entropy bound.

A trapezoid graph is a directed, acyclic graph (DAG) in which each non-leaf node ν is associated with a trapezoid τ_ν whose parallel sides are vertical and whose top and bottom are either a single subdivision edge or are at infinity. Node ν splits τ_ν either by a vertical line through a subdivision vertex (a vertical node) or by a subdivision edge (a horizontal node). The root is associated with a trapezoid that contains the entire subdivision; each leaf reports the region that contains its implied trapezoid.

Most planar point location structures can be represented as trapezoid graphs, including the slab and separating chain methods. Bucketing and some triangulation methods cannot, since they may make comparisons with coordinates or segments that are not in the input.

FIGURE 38.3.2

An example subdivision with its trapezoid graph, using circles for vertical splits at vertices, rectangles for horizontal splits at edges, and numbered leaves. Edges af, bf, and cf are cut and duplicated.

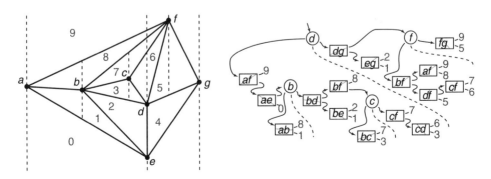

In Preparata's trapezoid method of point location, the trapezoid graph is a tree constructed top-down from the root. Figure 38.3.2 shows an example. If trapezoid τ_ν does not contain a subdivision vertex or intersect an edge, then node ν is a leaf. If every subdivision edge intersecting τ_ν has at least one endpoint inside τ_ν, then make ν a vertical node and split τ_ν by a vertical line through the median vertex. Otherwise, make ν a horizontal node and split τ_ν by the median of all edges cutting through τ_ν, and call ν a *horizontal split node*. This tree has depth (and query time) $3\log n$ [SA00]. Experiments [EKA84] suggest that this method performs well, although its worst-case size and preprocessing time are $O(n \log n)$.

In a delightful paper, Seidel and Adamy [SA00] give the exact number of comparisons for point location in a planar subdivision of n edges by establishing a tight bound of $\log_2 n + \sqrt{\log_2 n} + \Theta(1)$ on the worst-case height of a trapezoid graph. (The paper has an extra factor of $O(\log_2 \log_2 n)$ that was removed by Seidel and Kirkpatrick [unpublished].) The lower bound uses a stack of $n/2$ horizontal lines that are each cut into two along a diagonal.

The upper bound divides a trapezoid into $t = 2\sqrt{\log_2 n}$ slabs and uses horizontal splits to define trapezoids with point location subproblems to be solved recursively. Each subproblem with a location structure of depth d is given weight 2^d, and a weight balanced trapezoid tree is constructed to determine the relevant subproblem for a query. Query time in this trapezoid tree is optimal. Preprocessing time is determined by the number of tree nodes, which is $O(n2^t)$.

They also show that $\Omega(n \log n)$ space is required for a trapezoid tree, but that space can be reduced to linear by using cuttings to make the trapezoid graph into a DAG.

A space-efficient trapezoid graph can be most easily built as the history graph (a DAG) of the randomized incremental construction (RIC) of an arrangement of segments [Mul90, Sei91] (see Chapter 28 and Section 44.2). RIC gives an expected optimal point location scheme: $O(\log n)$ expected query time, $O(n \log n)$ expected preprocessing time, and $O(n)$ expected space, where the expectation is taken over random choices made by the construction algorithm. Hemmer et al. [HKH16] guarantee the query time and space bounds in the worst case: They develop efficient verification for space and query time of a structure by allowing the maximum path length, or depth of the DAG, to remain large as long as the longest query path remains logarithmic. By rerunning the randomized preprocessing if the space and query time bounds cannot be verified, the expectation remains only on the preprocessing time.

TRIANGULATIONS

Kirkpatrick [Kir83] developed the second optimal point-location method specifically for triangulations. This is not a restriction for subdivisions specified by vertex coordinates, since any planar subdivision can be triangulated, although it can be an added complication to do so. It can increase the required precision for subdivisions whose vertices are computed, such as Voronoi diagrams.

This scheme creates a hierarchy of subdivisions in which all faces, including the outer face, are triangles. Although point location based on hierarchical triangulations suffers from large constant factors, but the ideas are still of theoretical and practical importance. Hierarchical triangulations have become an important tool for algorithmic problems on convex polyhedra, terrain representation, and mesh

FIGURE 38.3.3

Hierarchical triangulation.

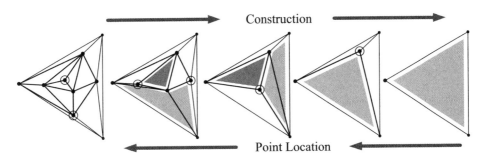

Construction

Point Location

simplification.

In every planar triangulation, one can find (in linear time) an independent set of low-degree vertices that consists of a constant fraction of all vertices. In Figure 38.3.3 these are circled and, in the next picture, are removed and the shaded hole is retriangulated if necessary. Repeating this "coarsening" operation a logarithmic number of times gives a constant-size triangulation.

To locate the triangle containing a query point q, start by finding the triangle in the coarsest triangulation, at right in Figure 38.3.3. Knowing the hole (shaded) that this triangle came from, one need only replace the missing vertex and check the incident triangles to locate q in the previous, finer triangulation.

Given a triangulation, preprocessing takes $O(n)$ time, but the hidden constants on time and space are large. For example, choosing the independent set by greedily taking vertices in order of increasing degree up to 10 guarantees 1/6th of the vertices [SK97], which leads to a data structure with $12n$ triangles in which a query could take $35 \log_2 n$ comparisons.

ENTROPY BOUNDS

The work of Malamotos with Arya, Mount, and co-authors initiated a fruitful exploration into how to modify the schemes above if we know something about the query distribution. By analogy to weights in a weighted binary search tree, suppose that we have a planar subdivision with regions of constant complexity (e.g., trapezoids or triangles) and that we know the probability p_i of a query falling in the ith region. The entropy is $H = \sum_i -p_i \log_2 p_i$. Arya et al. [AMM07] showed that a weighted randomized construction gives expected query times satisfying ***entropy bounds***. For a constant K, assign to a subdivision edge that is incident on regions with total probability P the weight $\lceil KPn \rceil$, and perform a randomized incremental construction. The use of integral weights ensures that ratios of weights are bounded by $O(n)$, which is important to achieve query time bounded by $O(H)$.

Entropy-preserving cuttings can be used to give a method whose query time of $H + O(1 + H^{1/2})$ approaches the optimal entropy bound [AMMW07], at the cost of increased space and programming complexity.

A subtlety related to decomposability has tripped up a few researchers: entropy is easy to work with only if the region descriptions have constant complexity, but triangulation or trapezoidation of complex regions can increase entropy. Collette et al. [CDI+12] work with connected planar subdivision G and define an entropy $\hat{H}(G, D)$ as the expected cost of any linear decision tree that solves point location for query distribution D. They show three things: how to create a Steiner triangulation that nearly minimizes entropy over all triangulations of G, that the minimum entropy over triangulations is a lower bound for $\hat{H}$ (meaning that the increased entropy may be necessary), and that the entropy-preserving cuttings give a query structure that matches the leading term of the lower bound.

Entropy-bounded query structures are also used in two interesting applications that do not assume that the query distribution is known in advance.

Aronov et al. [ABE+16] use them to give a **distance-sensitive** query algorithm that is faster for points far from the boundary. They show how to decompose a unit area subdivision into pieces of constant complexity such that any point q within distance Δ_q of the boundary is in a piece of area $\Omega(\Delta_p^2)$. (They construct a triangulation with this property for any convex polygon, and a decomposition into 7-gons for simple polygons.) Entropy bounds for a query in this decomposition give query time $O(\min\{\log n, -\log(\Delta_q)\})$.

Iacono and Mulzer [IM12] assume that regions have constant complexity and demonstrate how to prove **static optimality**: They show that in the limit they answer queries in asymptotically the same time as the best (static) decision tree by simply rebuilding, after every n^α queries, an entropy-bounded query structure for the n^β regions that have been most frequently accessed. Cheng and Lau [CL15] show that the analysis can extend to convex subdivisions, at an additional $O(\log \log n)$ time per query, by simply using balanced hierarchical triangulations of each convex region.

PLANAR SEPARATOR THEOREM

The first optimal point location scheme was based on Lipton and Tarjan's **planar separator theorem** [LT80] that every planar graph of n nodes has a set of $O(\sqrt{n})$ nodes that partition it into roughly equal pieces. Goodrich [Goo95] gave a linear-time construction of a family of planar separators in his parallel triangulation algorithm. The fact that embedded graphs have small separators continues to be important in theoretical work.

When applied to the dual graph of a planar subdivision, the nodes are a small set of faces that partition the remainder of the faces: simple methods taking up to quadratic space can be used to determine which set of the partition needs to be searched recursively. Bose et al. [BCH+12] combine separators with encodings of triangulations as permutations of points and bit-vector operations to build $o(n)$-size indices for point location in triangulations. (The bit-vectors are assumed to support rank and select operations, so their work implicitly assumes a RAM or cell-probe model of computation.) Their "succinct geometric indices" can be used to achieve the asymptotic bounds on the minimum number of comparisons, minimum entropy bounds, or $O(\log n)$-time query bounds for an implicit data structure that stores only a permutation of the input points.

38.4 PLANAR POINT LOCATION: DYNAMIC

In dynamic planar point location, the subdivision can be updated by adding or deleting vertices and edges. Unlike the static case, algorithms that match the performance of one-dimensional point location have not been found. (Except in special cases, like rectilinear subdivisions [GK09].) Like the static case, the search has produced interesting combinations of data structure ideas.

GLOSSARY

Updates: A dynamic planar subdivision is most commonly updated by inserting or deleting a vertex or edge. Update time usually refers to the worst-case time for a single insertion or deletion. Some methods support insertion or deletion of a chain of k vertices and edges faster than doing k individual updates.

Vertex expansion/contraction: Updating a planar subdivision by splitting a vertex into two vertices joined by an edge, or the inverse: contracting an edge and merging the two endpoints into one. This operation, supported by the "primal/dual spanning tree" (discussed below), is important for point location in three-dimensional subdivisions.

Amortized update time: When times are reported as amortized, then an individual operation may be expensive, but the total time for k operations, starting from an empty data structure, will take at most k times the amortized bound.

TABLE 38.4.1 Dynamic point location results.

TECHNIQUE	QUERY	UPDATE	SPACE	UPDATES SUPPORTED
Trapezoid method [CT92]	$O(\log n)$	$O(\log^2 n)$	$O(n \log n)$	ins/del vertex & edge
Separating chain [PT89]	$O(\log^2 n)$	$O(\log^2 n)$	$O(n)$	ins/del edge & edge
I/O-efficient [ABR12]	$O(\log_B^2 N)$	$O(\log_B N)$	$O(N/B)$	measures I/O blocks read
Pr/dual span tree [GT91]	$O(\log^2 n)$	$O(\log n)$	$O(n)$	ins/del edge & chain,
amortized	$O(\log n \log \log n)$	$O(1)$	$O(n)$	expand/contract vertex
Interval tree [CJ92]	$O(\log^2 n)$	$O(\log n)$	$O(n)$	ins/del edge & chain
with frac casc [BJM94]	$O(\log n \log \log n)$	$O(\log^2 n)$	$O(n)$	amort del, ins faster
Segment tree [CN15]	$O(\log n (\log \log n)^2)$	$O(\log n \log \log n)$	$O(n)$	many variants

Insertion/Deletion-only: When all updates are insertions or all are deletions, specialized structures can often be more efficient. Note that deletion-only structures for a decomposable problem support dynamic updates by this Bentley-Saxe transformation [BS80]: Maintain structures whose sizes are bounded by a geometric series, where only the smallest need support insertion. Whenever an update would make ith structure too large or small, rebuilding all structures through the $(i+1)$st. For k updates, an element participates in amortized $O(\log k)$ rebuilds.

Vertical ray shooting problem: Maintain a set of interior-disjoint segments in a structure that can report the segment directly below a query point q. Updates are insertion or deletion of segments. This decomposable problem does not require subdivisions to remain connected, but also does not maintain identity of faces.

I/O efficient algorithm: An algorithm whose asymptotic number of I/O operations is minimal. Model parameters are problem size N, disk block size B and memory size M, with typically $B \leq \sqrt{M}$. Sorting requires $O((N/B) \log_B N)$ time.

SEPARATING CHAIN AND TRAPEZOID GRAPH METHODS

The separating chain method of Section 38.3 was the first to be made fully dynamic [PT89]. Although both its asymptotics and its constant factors are larger than other methods, it has been made I/O-efficient [AV04]. This is an impressive theoretical accomplishment, but simpler algorithms that assume that the input is somewhat evenly distributed in the plane will be more practical.

Preparata's trapezoid graph method [Pre81] is one of the easiest to make dynamic. It preserves its optimal $O(\log n)$ query time, but also its suboptimal $O(n \log n)$ space. To support updates in $O(\log^2 n)$ time, Chiang and Tamassia [CT92, CPT96] store the binary tree on subdivision edges in a *link-cut* tree [ST83], which supports in $O(\log n)$ time the operation of linking two trees by adding an arc, and the inverse, cutting an arc to make two trees.

PRIMAL/DUAL SPANNING TREE

Goodrich and Tamassia [GT98] gave an elegant approach based on link-cut trees that takes linear space for the restricted case of dynamic point location in monotone subdivisions. A monotone subdivision has a ***monotone spanning tree*** in which all root-to-leaf paths are monotone. Each edge not in the tree closes a cycle and defines a monotone polygon.

In any planar graph whose faces are simple polygons, the duals of edges not in the spanning tree form a dual spanning tree of faces, as in Figure 38.4.1(b). Goodrich and Tamassia [GT98] use a centroid decomposition of the dual tree to guide comparisons with monotone polygons in the primal tree. The centroid edge, which breaks the dual tree into two nearly-equal pieces, is indicated in Figure 38.4.1(b). The primal edge creates the shaded monotone polygon; if the query is inside then we recursively explore the corresponding piece of the dual tree. Using link-cut trees, the centroid decomposition can be maintained in logarithmic time per update, giving a dynamic point-location structure with $O(\log^2 n)$ query time.

In the static setting, fractional cascading can turn this into an optimal point location method. Dynamic fractional cascading [MN90] can be used to reduce the dynamic query time and to obtain $O(1)$ amortized update time.

The dual nature of the structure supports insertion and deletion of dual edges, which correspond to expansion and contraction of vertices. These are needed to support static three-dimensional point location via persistence. Furthermore, a k-vertex monotone chain can be inserted/deleted in $O(\log n + k)$ time.

FIGURE 38.4.1
Dynamic methods: (a) *Priority search (interval tree);* (b) *primal/dual spanning tree.*

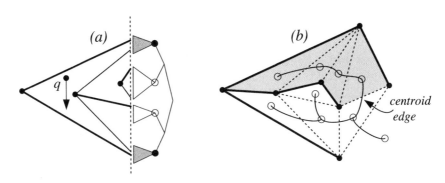

DYNAMIC INTERVAL OR SEGMENT TREES

The current best results solve the vertical ray shooting problem in a dynamic set of disjoint segments using interval or segment trees to store slabs. Let's consider the classic interval tree of Cheng and Janardan [CJ92], and the recent work of Chan and Nekrich [CN15], which presents many variants that reduce space and trade query and update times in a segment tree.

The key subproblem in both is vertical ray shooting for 1-sided segments: For a set of interior-disjoint segments S that intersect a common vertical line ℓ, report the segment of S directly below a query point q. Let's assume q is left of ℓ; we can make a separate structure for the right.

Cheng and Janardan [CJ92] solve this subproblem in linear space and $O(\log |S|)$ query and update time by a priority-tree search: they build a binary search tree on segments S in the vertical order along ℓ, and store in each subtree a pointer to the "priority segment" with endpoint farthest left of ℓ. At each level of this search tree, only two candidate subtrees may contain the segment below q—the ones whose priority segments are immediately above and below q. Figure 38.4.1(a) illustrates a case in which the search continues in the two shaded subtrees.

In their work, this subproblem arises naturally in a recursively defined *interval tree*: The root stores segments that cross a vertical line ℓ through the median endpoint. Segments to the left (right) are stored in an interval tree that is the left (right) child of the root, with the corresponding ray-shooting structure. Space is $O(n)$, since each segment is stored once.

To locate a query point q, visit the $O(\log n)$ nodes on the path to the slab containing q, and return the closest of the segments found by ray shooting for 1-sided segments in each node. Total query time is $O(\log^2 n)$. Constants are moderate, with only 4 or 5 entries per edge and 6 comparisons per search step. Updates to the priority search tree take $O(\log n)$ time with larger constants; they must maintain tree balance and segment priorities. To handle changes in the number of slabs, use a BB[α] or weight-balanced B-tree [AV03] and rebuild the affected priority search tree structures in linear time when nodes split. This makes the update cost amortized $O(\log n)$.

To speed up the query time, one would like work done in one 1-sided subproblem to make the rest easier. The fractional cascading idea of sharing samples of segments between subproblems requires that pairs of segments can be locally ordered, but segments in different nodes of an interval tree may share no x-coordinates. (If all segments share the same slope, they can be ordered by y-intercept to enable dynamic fractional cascading [MN90]; more on this below.)

Several researchers [ABG06, BJM94, Ble08, CN15, GK09] have used ideas from *segment trees*, which store segments in the balanced slab recursively as follows: Starting from the root, store any segment that crosses the slab for that node (the union of the leaf slabs in its subtree). Pass unstored segments to the children whose slabs they intersect; segments that straddle the median are sent to both children. A segment is stored in at most $2\log_2 n$ nodes, since to be stored, an endpoint must be in the parent's slab. Thus, we have a structure with $O(n\log n)$ space that can perform updates and queries in $O(\log^2 n)$ time apiece. Thus, there are extra logs on space, query, and update.

Baumgarten et al. [BJM94] observed that segments stored in nodes whose slabs contain query point q all intersect a vertical line through q, so dynamic fractional cascading [MN90] from the bottom can reduce query and insertion time to $O(\log n \log \log n)$. They create a linear space data structure with this query time by combining segment and interval trees, using fractional cascading on blocks of $O(\log^2 n)$ segments in each interval tree node. Deletion time remains $O(\log^2 n)$.

Chan and Nekrich [CN15] carefully combine many ideas that come closest to removing all three logs. First, they point out that a deletion-only structure for horizontal segments can reduce space for any dynamic point location structure, including the trapezoid graph. They maintain a dynamic structure with up to $n/\log n$ segments, then use the Bentley-Saxe transformation [BS80] to put the rest into $O(\log \log n)$ groups whose size limits double. For each group they 1) build a static point location structure on the segments, 2) rank each segment in a total order consistent with aboveness, and 3) maintain a deletion-only structure for horizontal segments made by replacing each segment's original y coordinates with its rank. A query q in each static structure either returns the segment below, or its rank r if it has been deleted. The horizontal structure queried with (q_x, r) then returns the candidate segment from that group.

To reduce the query time, they use dynamic fractional cascading like Baumgarten et al.: using ideas from the segment tree to pass samples to speed up searches in the interval tree. (They describe the details using random sampling and finger search trees to more easily consider dynamic updates for several variants.) They trade query time for update time by coloring each segment's $O(\log n)$ fragments with $O(\log \log n)$ colors and building separate fractional cascading for each color. The colors for a segment are determined by the levels crossed by subtree of slabs spanned by the segment in a manner like the tree interpretation of van Emde Boas queues, which allows an inserted or deleted segment to be found and updated in $O(\log n \log \log n)$ time in all its colored lists. Query time increases to $O(\log n (\log \log n)^2)$ because of the extra lists that must be searched. Their work suggests other variations that can trade query and update times, so one is $O(\log n)$ while the other is $O(\log^{1+\varepsilon} n)$, or can use word RAM tricks to shave a factor of $\log \log n$ from the query.

As mentioned above, the special case of horizontal segments is easier, as the y-order can be used in fractional cascading. Giyora and Kaplan [GK09] achieve a linear space structures with $O(\log^{1+\varepsilon} n)$ query and $O(\log n)$ update on a pointer machine and $O(\log n)$ query and update times on a word RAM.

OPEN PROBLEMS

1. Improve dynamic planar point location to simultaneously attain $O(n)$ space and $O(\log n)$ query and update time, or establish a lower bound.

2. Can persistent data structures be made dynamic? The fact that data are copied seems to work against maintaining a data structure under insertions and deletions.

3. Create a dynamic data structure for subdivisions that need not remain connected (may have holes) that can report in sublinear time whether two points are in the same face.

38.5 PLANAR POINT LOCATION: OTHER MODELS

Programming complexity and non-negligible asymptotic constants mean that optimal point location techniques are used less than might be expected. See [TV01] for a study of geometric algorithm engineering that uses point location schemes as its example.

PICK HARDWARE

Graphic workstations employ special "pick hardware" that draws objects on the screen and returns a list of objects that intersect a query pixel. The hardware imposes a minimum time of about 1/30th of a second on a pick operation, but hundreds of thousands of polygons may be considered in this time.

BUCKETING AND SPATIAL INDEX STRUCTURES

Because data in practical applications tend to be evenly distributed, bucketing techniques are far more effective [AEI+85, EKA84] than worst-case analysis would predict. For problems in two and three dimensions, a uniform grid will often trim data to a manageable size [MAFL16].

Adaptive data structures for more general spatial indexing, such as k-d trees, quadtrees, BANG files, R-trees, and their relatives [Sam90b], can be used as filters for point location—these techniques are common in databases and geographic information systems.

Various definitions for "fat regions" have used to explore theoretical bounds on schemes that use spatial indexing structures. To give one example, Löffler, Simons, and Strash [LSS13] use dynamic quadtrees to store a representative points near the middle of each region, ensuring that cells for large regions are large and that a query point will have to do efficient point-in-region tests for only a constant number of regions. Thus, for disjoint fat regions, they achieve $O(\log n)$-time insert, delete, and query operations. They also can perform $O(\log \log n)$-time "local updates," which replace a region by another of similar diameter and separation.

Chan and Pătraşcu [CP09] combine sampling and bucketing ideas in their ***transdichotomous*** structures for point location in a slab. Given a slab with an ordered list of m crossing segments whose left and right y-coordinates are $O(w)$-bit rationals that lie in intervals of length L and R, they select b evenly spaced segments from the list and partition the endpoints into h equal length buckets on left and right. Iterate to select segments separating buckets: find the highest segment from the first non-empty bucket on the left, round up the coordinates of both ends, discard all segments with an endpoint below, and repeat.

Knowing the location of query point q among the selected segments, two comparisons of q with original segments allows a recursive query in either a set of m/b segments, or segments with left interval length L/h, or segments with right interval length R/h. Shrinking the intervals is progress because small y-coordinate offsets can be packed into words for parallel evaluation in the real RAM model. Thus, they can build, in $o(n \log n)$ time, an $O(n)$-space structure to answer point location in a slab queries in $O(\log n / \log \log n)$ time. They combine this with the point location techniques of Section 38.3 to give point location within the same bounds. They get even better bounds for off-line point location [CP10], where the queries are known in advance, by packing query points into words. They use these techniques to give transdichotomous bounds for many computational geometry problems.

SUBDIVISION WALKING

Applications that store planar subdivisions with their adjacency relations, such as geographic information systems, can walk through the regions of the subdivision from a known position p to the query q.

To walk a subdivision with $O(n)$ edges, compute the intersections of $\overline{pq}$ with the current region and determine if q is inside. If not, let q' denote the intersection point closest to q. Advance to the region incident to q' that contains a point in the interior of $\overline{q'q}$ and repeat. In the worst case, this walk takes $O(n)$ time. The application literature typically claims $O(\sqrt{n})$ time, which is the average number of intersections with a line under the assumption that vertices and edges of the subdivision are evenly distributed. When combined with bucketing or hierarchical data structures (for example, maintaining a regular grid or quadtree with known positions and starting from the closest to answer a query), walking is an effective, practical location method.

For triangulations, the algorithm walking $\overline{pq}$ is easy to implement. Guibas and Stolfi's [GS85] incremental Delaunay triangulation uses an even simpler walk from edge to edge, but this depends on an acyclicity theorem (Sections 19.4 and 26.1) that does not hold for arbitrary triangulations. A robust walk should remember its starting point and handle vertices on the traversed segment as if they had been perturbed consistently. Broutin, Devillers, and Hemsley prove nice bounds for their "cone walk" in random Delaunay triangulations [BDH16]

There have been several analyses of Jump & Walk schemes in triangulations, both analytically and experimentally. Devroye et al. [DLM04] show expected query times of $O(n^{1/4})$ for a scheme that keeps $n^{1/4}$ points with known locations, and walks from the nearest to find a query. In their experiments, the combination of a 2-d search tree with walking performed the best. De Castro and Devillers [CD13] survey, compare, and tune many variations, including those that save space by building a hierarchy formed from small samples (a technique implemented in the

CGAL library [BDP⁺02, Dev02]) and those that are distribution sensitive by dynamically choosing points to keep. See also their Java Demo [DC11].

38.6 LOCATION IN HIGHER DIMENSIONS

In higher dimensions, known point location methods do not achieve both linear space and logarithmic query time. Linear space can be attained by relatively straightforward linear search, such as the point-in-polygon test.

Logarithmic time, or $O(d \log n)$ time, can be obtained by projection [DL76]: project the $(d-2)$-faces of a subdivision to an arrangement in $d-1$ dimensions and recursively build a point location structure for the arrangement in the projection. Knowing the cell in the projection gives a list of the possible faces that project to that cell, so an additional logarithmic search can return the answer. The worst-case space required is $O(n^{2^d})$.

Because point location is decomposable, batching can trade space for time: preprocessing n/k groups of k facets into structures with $S(k)$ space and $Q(k)$ time gives, in total, $O(nS(k)/k)$ space and $O(nQ(k)/k)$ query time.

Clever ways of batching can lead to better structures. Randomized methods can often reduce the dependence on dimension from doubly- to singly-exponential, since random samples can be good approximations to a set of geometric objects. They can also be used with objects that are implicitly defined.

We should mention that convex polyhedra can be preprocessed using the Dobkin-Kirkpatrick hierarchy (Section 38.3) so that the point-in-convex-polyhedron test does take $O(n)$ space and $O(\log n)$ query time.

THREE-DIMENSIONAL POINT LOCATION

Dynamic location structures can be used for static spatial point location in one higher dimension by employing persistence. If one swept a plane through a subdivision of three-space into polyhedra, one could see the intersection as a dynamic planar subdivision in which vertices (intersections of the sweep plane with edges) move along linear trajectories. Whenever the sweep plane passes through a vertex in space, vertices in the plane may join and split.

Goodrich and Tamassia's primal/dual method supports the necessary operations to maintain a point location structure for the sweeping plane. Using node-copying to make the structures persistent gives an $O(n \log n)$ space structure that can answer queries in $O(\log^2 n)$ time. Preprocessing takes $O(n \log n)$ time.

Devillers et al. [DPT02] tested several approaches to subdivision walking for Delaunay tetrahedralization, and established the practical effectiveness of the hierarchical Delaunay in three dimensions as well.

RECTILINEAR SUBDIVISIONS

Restricting attention to rectilinear (orthogonal) subdivisions permits better results via data structures for orthogonal range search. The ***skewer tree***, a multidimensional interval tree, gives static point location among n rectangular prisms with

$O(n)$ space and $O(\log^{d-1} n)$ query time after $O(n \log n)$ preprocessing [EHH86]. These can be made dynamic by using Giyora and Kaplan's [GK09] structure at the lowest level.

In dimensions two and three, stratified trees and perfect hashing [DKM+94] can be used to obtain $O((\log \log U)^{d-1})$ query time in a fixed universe $[1, \ldots, U]$, or $O(\log n)$ query time in general. Iacono and Langerman [IL00] use "justified hyperrectangles" to obtain $O(\log \log U)$ query times in every dimension d, but the space and preprocessing time, which are $O(fn \log \log U)$ and $O(fn \log U \log \log U)$, respectively, depend on a *fatness parameter* f that equals the average ratio of the dth power of smallest dimension to volume of all hyperrectangles in the subdivision.

POINT LOCATION AMONG ALGEBRAIC VARIETIES

Chazelle and Sharir [CS90] consider point location in a general setting, among n algebraic varieties of constant maximum degree b in d-dimensional Euclidean space. They augment Collins's cylindrical algebraic decomposition to obtain an $O(n^{2^{d-1}})$-space, $O(\log n)$-query time structure after $O(n^{2^{d+6}})$ preprocessing. Hidden constants depend on the degrees of projections and intersections, which can be b^{4^d}.

This method provides a general technique to obtain subquadratic solutions to optimization problems that minimize a function $\{F(a,b) \mid a \in A, b \in B\}$, where $F(a,b)$ has a constant-size algebraic description. For a fixed b, F is algebraic in a. Thus, small batches of points from B can be preprocessed in subquadratic time, and each a can be tested against each batch, again in subquadratic time.

OPEN PROBLEMS

1. Find an optimal method for static (or dynamic) point location in a three-dimensional subdivision with n vertices and $O(n)$ faces: $O(n)$ space and $O(\log n)$ query time.

RANDOMIZED POINT LOCATION

The techniques of Chapter 44 can lead to good point location methods when a random sample of a set of objects can be used to approximate the whole. Arrangements of hyperplanes in dimension d are a good example. A random sample of hyperplanes divides space into cells intersected by few hyperplanes; recursively sampling in each cell gives a point location structure for the arrangement. Table 38.6.1 lists the performance of some randomized point location methods for hyperplanes. Query time can be traded for space by choosing larger random samples.

The randomized incremental construction algorithms of Section 44.2 are simple because they naturally build randomized point location structures along with the objects that they aim to construct [Mul93, Sei93]. These have good "tail bounds" and work well as insertion-only location structures.

Randomized point location structures can be made fully dynamic by lazy deletion and randomized rebuild techniques [BDS95, MS91]; they maintain good ex-

TABLE 38.6.1   Randomized point location in arrangements.

TECHNIQUE	OBJECTS	QUERY	PREPROC	SPACE
Random sample [Cla87]	hyperplanes	$O(c^d \log n)$ exp	$O(n^{d+1+\epsilon})$ exp	$O(n^{d+\epsilon})$
Derandomized [CF94]	hyperplanes	$O(c^d \log n)$	$O(n^{2d+1})$	$O(n^d)$
Random sample [MS91]	dyn hpl $d \le 4$	$O(\log n)$ exp	$O(n^{d+\epsilon})$ exp	$O(n^{d+\epsilon})$
Epsilon nets [Mei93]	hyperplanes	$O(d^5 \log n)$ exp	$O(n^{d+1+\epsilon})$ exp	$O(n^{d+\epsilon})$

pected performance if random elements are chosen for insertion and deletion. That is, the sequence of insertions and deletions may be specified, but the elements are to be chosen independently of their roles in the data structure.

IMPLICIT POINT LOCATION

In some applications of point location, the objects are not given explicitly. A planar motion planning problem may ask whether a start and a goal point are in the same cell of an arrangement of constraint segments or curves, without having explicit representations of all cells.

Consider a simple example: an arrangement of n lines, which defines nearly n^2 bounded cells. Without storing all cells, we can determine whether two points p and q are in the same cell by preprocessing $\sqrt{n}$ subarrangements of $\sqrt{n}$ lines ($O(n\sqrt{n})$ cells in all) and making sure that p and q are together in each subarrangement. If the lines are put into batches by slope, then within the same asymptotic time, an algorithm can return the pair of lines defining the lowest vertex as a unique cell name.

Implicit location methods are often seen as special cases of range queries (Chapter 40) or vertical ray shooting [Aga91]. Table 38.6.2 lists results on implicit location among line segments, which depend upon tools discussed in Chapters 40, 44, and 47, specifically random sampling, ϵ-net theory, and spanning trees with low stabbing number.

TABLE 38.6.2   Implicit point location results for arrangements of n line segments.

TECHNIQUE	QUERY	PREPROC	SPACE
Span tree lsn [Aga92]	$O(\sqrt{n} \log^2 n)$	$O(n^{3/2} \log^\omega n)$	$O(n \log^2 n)$
Batch sp tree [AK94]	$O((n/\sqrt{s}) \log^2 (n/\sqrt{s}) + \log n)$	$O\left((sn(\log(n/\sqrt{s}) + 1)^{2/3}\right)$	$n\sqrt{\log n} \le s \le n^2$

38.7   SOURCES AND RELATED MATERIAL

SURVEYS

Graphic Gems IV has code for point in polygon algorithms. These recent papers have good overviews of the literature or present variations of ideas on their topics.

[Hai94, Wei94]: Point-in-polygon algorithms in *Graphics Gems IV*, with code.

[IM12]: Nice history of entropy bounds for point location.

[CP09]: Speeding up point location speeds up many geometric algorithms.

[CN15]: Variations for dynamic point location.

RELATED CHAPTERS

REFERENCES

[ABE+16]   B. Aronov, M. de Berg, D. Eppstein, M. Roeloffzen, and B. Speckmann. Distance-sensitive planar point location. *Comput. Geom.*, 54:17–31, 2016.

[ABG06]   L. Arge, G.S. Brodal, and L. Georgiadis. Improved dynamic planar point location. In *Proc. 47th IEEE Sympos. Found. Comp. Sci.*, pages 305–314, 2006.

[ABR12]   L. Arge, G.S. Brodal, and S.S. Rao. External memory planar point location with logarithmic updates. *Algorithmica*, 63:457–475, 2012.

[AEI+85]   Ta. Asano, M. Edahiro, H. Imai, M. Iri, and K. Murota. Practical use of bucketing techniques in computational geometry. In G.T. Toussaint, editor, *Computational Geometry*, pages 153–195, North-Holland, Amsterdam, 1985.

[Aga91]   P.K. Agarwal. Geometric partitioning and its applications. In J.E. Goodman, R. Pollack, and W. Steiger, editors, *Computational Geometry*, vol. 6 of *DIMACS Ser. Discrete Math. Theor. Comp. Sci.*, AMS, Providence, 1991.

[Aga92]   P.K. Agarwal. Ray shooting and other applications of spanning trees with low stabbing number. *SIAM J. Comput.*, 21:540–570, 1992.

[AK94]   P.K. Agarwal and M. van Kreveld. Implicit point location in arrangements of line segments, with an application to motion planning. *Internat. J. Comput. Geom. Appl.*, 4:369–383, 1994.

[AMM07]   S. Arya, T. Malamatos, and D.M. Mount. A simple entropy-based algorithm for planar point location. *ACM Trans. Algorithms*, 3:17, 2007.

[AMMW07]   S. Arya, T. Malamatos, D.M. Mount, and K.C. Wong. Optimal expected-case planar point location. *SIAM J. Comput.*, 37:584–610, 2007.

[AV03]   L. Arge and J.S. Vitter. Optimal external memory interval management. *SIAM Journal on Computing*, 32:1488–1508, 2003.

[AV04]   L. Arge and J. Vahrenhold. I/O-efficient dynamic planar point location. *Comput. Geom.*, 29:147–162, 2004.

[BCH+12]   P. Bose, E.Y. Chen, M. He, A. Maheshwari, and P. Morin. Succinct geometric indexes supporting point location queries. *ACM Trans. Algorithms*, 8:10, 2012.

[BDH16]   N. Broutin, O. Devillers, and R. Hemsley. Efficiently navigating a random Delaunay triangulation. *Random Structures Algorithms*, 49:95–136, 2016.

[BDP+02]   J.-D. Boissonnat, O. Devillers, S. Pion, M. Teillaud, and M. Yvinec. Triangulations in CGAL. *Comput. Geom.*, 22:5–19, 2002.

[BDS95]   M. de Berg, K. Dobrindt, and O. Schwarzkopf. On lazy randomized incremental construction. *Discrete Comput. Geom.*, 14:261–286, 1995.

[Ben79]   J.L. Bentley. Decomposable searching problems. *Inform. Process. Lett.*, 8:244–251, 1979.

[BJM94]   N. Baumgarten, H. Jung, and K. Mehlhorn. Dynamic point location in general subdivisions. *J. Algorithms*, 17:342–380, 1994.

[Ble08]   G.E. Blelloch. Space-efficient dynamic orthogonal point location, segment intersection, and range reporting. In *Proc. ACM-SIAM Sympos. Discrete Algorithms*, pages 894–903, 2008.

[BS80]   J.L. Bentley and J.B. Saxe. Decomposable searching problems I: Static-to-dynamic transformations. *J. Algorithms*, 1:301–358, 1980.

[CD13]   P.M.M. de Castro and O. Devillers. Practical distribution-sensitive point location in triangulations. *Comput. Aided Geom. Design*, 30:431–450, 2013.

[CDI+12]   S. Collette, V. Dujmović, J. Iacono, S. Langerman, and P. Morin. Entropy, triangulation, and point location in planar subdivisions. *ACM Trans. Algorithms*, 8:29, 2012.

[CF94]   B. Chazelle and J. Friedman. Point location among hyperplanes and unidirectional ray-shooting. *Comput. Geom.*, 4:53–62, 1994.

[CG86]   B. Chazelle and L.J. Guibas. Fractional cascading: I. A data structuring technique. *Algorithmica*, 1:133–162, 1986.

[CJ92]   S.-W. Cheng and R. Janardan. New results on dynamic planar point location. *SIAM J. Comput.*, 21:972–999, 1992.

[CL15]   S.-W. Cheng and M.-K. Lau. Adaptive point location in planar convex subdivisions. In *Proc. 26th Int. Sympos. Algorithms Comput.*, vol. 9472 of *LNCS*, pages 14–22, Springer, Berlin, 2015.

[Cla87]   K.L. Clarkson. New applications of random sampling in computational geometry. *Discrete Comput. Geom.*, 2:195–222, 1987.

[CN15]   T.M. Chan and Y. Nekrich. Towards an optimal method for dynamic planar point location. In *Proc 56th IEEE Sympos. Found. Comp. Sci.*, pages 390–409, 2015.

[CP09]   T.M. Chan and M. Pǎtraşcu. Transdichotomous results in computational geometry, I: Point location in sublogarithmic time. *SIAM J. Comput.*, 39:703–729, 2009.

[CP10]   T.M. Chan and M. Pǎtraşcu. Transdichotomous results in computational geometry, II: Offline search. Preprint, arXiv:1010.1948, 2010.

[CPT96]   Y.-J. Chiang, F.P. Preparata, and R. Tamassia. A unified approach to dynamic point location, ray shooting, and shortest paths in planar maps. *SIAM J. Comput.*, 25:207–233, 1996.

[CS90]   B. Chazelle and M. Sharir. An algorithm for generalized point location and its application. *J. Symbolic Comput.*, 10:281–309, 1990.

[CT92]   Y.-J. Chiang and R. Tamassia. Dynamization of the trapezoid method for planar point location in monotone subdivisions. *Internat. J. Comput. Geom. Appl.*, 2:311–333, 1992.

[DC11]   O. Devillers and P.M.M. de Castro. A pedagogic JavaScript program for point location strategies. In *Proc. 27th Sympos. Comput. Geom.*, pages 295–296, ACM Press, 2011.

[Dev02]   O. Devillers. The Delaunay hierarchy. *Int. J. Found. Comp. Sci.*, 13:163–180, 2002.

[DKM+94]   M. Dietzfelbinger, A. Karlin, K. Mehlhorn, F. Meyer auf der Heide, H. Rohnert, and R.E. Tarjan. Dynamic perfect hashing: upper and lower bounds. *SIAM J. Comput.*, 23:738–761, 1994.

[DL76]   D.P. Dobkin and R.J. Lipton. Multidimensional searching problems. *SIAM J. Comput.*, 5:181–186, 1976.

[DLM04]   L. Devroye, C. Lemaire, and J.-M. Moreau. Expected time analysis for Delaunay point location. *Comput. Geom.*, 29:61–89, 2004.

[DPT02]   O. Devillers, S. Pion, and M. Teillaud. Walking in a triangulation. *Int. J. Found. Comp. Sci.*, 13.:181–199, 2002.

[EBKZ77]   P. van Emde Boas, R. Kaas, and E. Zijlstra. Design and implementation of an efficient priority queue. *Math. Syst. Theory*, 10:99–127, 1977.

[EGS86]   H. Edelsbrunner, L.J. Guibas, and J. Stolfi. Optimal point location in a monotone subdivision. *SIAM J. Comput.*, 15:317–340, 1986.

[EHH86]   H. Edelsbrunner, G. Haring, and D. Hilbert. Rectangular point location in d dimensions with applications. *Comput. J.*, 29:76–82, 1986.

[EKA84]   M. Edahiro, I. Kokubo, and Ta. Asano. A new point-location algorithm and its practical efficiency: Comparison with existing algorithms. *ACM Trans. Graph.*, 3:86–109, 1984.

[FW93]   M.L. Fredman and D.E. Willard. Surpassing the information theoretic bound with fusion trees. *J. Comput. Syst. Sci.*, 47:424–436, 1993.

[Goo95]   M.T. Goodrich. Planar separators and parallel polygon triangulation. *J. Comput. Syst. Sci.*, 51:374–389, 1995.

[GK09]   Y. Giora and H. Kaplan. Optimal dynamic vertical ray shooting in rectilinear planar subdivisions. *ACM Trans. Algorithms*, 5:28:1–28:51, 2009.

[GS85]   L.J. Guibas and J. Stolfi. Primitives for the manipulation of general subdivisions and the computation of Voronoi diagrams. *ACM Trans. Graph.*, 4:74–123, 1985.

[GT91]   M.T. Goodrich and R. Tamassia. Dynamic trees and dynamic point location. In *Proc. 23rd ACM Sympos. Theory Comput.*, pages 523–533, 1991.

[GT98]   M.T. Goodrich and R. Tamassia. Dynamic trees and dynamic point location. *SIAM J. Comput.*, 28:612–636, 1998.

[Hai94]   E. Haines. Point in polygon strategies. In P. Heckbert, editor, *Graphics Gems IV*, pages 24–46, Academic Press, Boston, 1994.

[HKH16]   M. Hemmer, M. Kleinbort, and D. Halperin. Optimal randomized incremental construction for guaranteed logarithmic planar point location. *Comput. Geom.*, 58:110–123, 2016.

[IL00]   J. Iacono and S. Langerman. Dynamic point location in fat hyperrectangles with integer coordinates. In *Proc. 12th Canad. Conf. Comput. Geom.*, pages 181–186, 2000.

[IM12]   J. Iacono and W. Mulzer. A static optimality transformation with applications to planar point location. *Internat. J. Comput. Geom. Appl.*, 22:327–340, 2012.

[Kir83]   D.G. Kirkpatrick. Optimal search in planar subdivisions. *SIAM J. Comput.*, 12:28–35, 1983.

[LP77]   D.T. Lee and F.P. Preparata. Location of a point in a planar subdivision and its applications. *SIAM J. Comput.*, 6:594–606, 1977.

[LSS13]   M. Löffler, J.A. Simons, and D. Strash. Dynamic planar point location with sublogarithmic local updates. In *Proc. 13th Sympos. Algorithms Data Structures*, pages 499–511, vol. 8037 of *LNCS*, Springer, Berlin, 2013.

[LT80]   R.J. Lipton and R.E. Tarjan. Applications of a planar separator theorem. *SIAM J. Comput.*, 9:615–627, 1980.

[MAFL16]   S.V.G. Magalhães, M.V.A. Andrade, W.R. Franklin, and W. Li. PinMesh-fast and exact 3D point location queries using a uniform grid. *Computers and Graphics (Pergamon)*, 58:1–11, 2016.

[Meh77]   K. Mehlhorn. Best possible bounds on the weighted path length of optimum binary search trees. *SIAM J. Comput.*, 6:235–239, 1977.

[Mei93]   S. Meiser. Point location in arrangements of hyperplanes. *Inform. Comput.*, 106:286–303, 1993.

[MN90]   K. Mehlhorn and S. Näher. Dynamic fractional cascading. *Algorithmica*, 5:215–241, 1990.

[MS91]   K. Mulmuley and S. Sen. Dynamic point location in arrangements of hyperplanes. In *Proc. 7th Sympos. Comput. Geom.*, pages 132–141, ACM Press, 1991.

[Mul90]   K. Mulmuley. A fast planar partition algorithm, I. *J. Symbolic Comput.*, 10:253–280, 1990.

[Mul93]   K. Mulmuley. *Computational Geometry: An Introduction through Randomized Algorithms.* Prentice Hall, Englewood Cliffs, 1993.

[Pre81]   F.P. Preparata. A new approach to planar point location. *SIAM J. Comput.*, 10:473–482, 1981.

[PT89]   F.P. Preparata and R. Tamassia. Fully dynamic point location in a monotone subdivision. *SIAM J. Comput.*, 18:811–830, 1989.

[Pug90]   W. Pugh. Skip lists: a probabilistic alternative to balanced trees. *Commun. ACM*, 33:668–676, 1990.

[SA96]   R. Seidel and C.R. Aragon. Randomized search trees. *Algorithmica*, 16:464–497, 1996.

[SA00]   R. Seidel and U. Adamy. On the exact worst case query complexity of planar point location. *J. Algorithms*, 37:189–217, 2000.

[Sam90]   H. Samet. *The Design and Analysis of Spatial Data Structures.* Addison-Wesley, Reading, 1990.

[Sch08]   S. Schirra. How reliable are practical point-in-polygon strategies? In *Proc. 16th European Sympos. Algorithms*, vol. 5193 of *LNCS*, pages 744–755, Springer, Berlin, 2008.

[Sei91]   R. Seidel. A simple and fast incremental randomized algorithm for computing trapezoidal decompositions and for triangulating polygons. *Comput. Geom.*, 1:51–64, 1991.

[Sei93]   R. Seidel. Backwards analysis of randomized geometric algorithms. In J. Pach, editor, *New Trends in Discrete and Computational Geometry*, vol. 10 of *Algorithms and Combinatorics*, pages 37–68. Springer-Verlag, New York, 1993.

[SK97]   J. Snoeyink and M. van Kreveld. Linear-time reconstruction of Delaunay triangulations with applications. In *Proc. 5th European Sympos. Algorithms*, vol. 1284 of *LNCS*, pages 459–471, Springer, Berlin, 1997.

[ST83]   D.D. Sleator and R.E. Tarjan. A data structure for dynamic trees. *J. Comput. Syst. Sci.*, 26:362–381, 1983.

[ST85]   D.D. Sleator and R.E. Tarjan. Self-adjusting binary search trees. *J. ACM*, 32:652–686, 1985.

[ST86]   N. Sarnak and R.E. Tarjan. Planar point location using persistent search trees. *Commun. ACM*, 29:669–679, 1986.

[Ste91]   A.J. Stewart. Robust point location in approximate polygons. In *Proc. 3rd Canad. Conf. Comput. Geom.*, pages 179–182, 1991.

[TV01]   R. Tamassia and L. Vismara. A case study in algorithm engineering for geometric computing. *Internat. J. Comput. Geom. Appl.*, 11:15–70, 2001.

[Wei94]   K. Weiler. An incremental angle point in polygon test. In P. Heckbert, editor, *Graphics Gems IV*, pages 16–23, Academic Press, Boston, 1994.

39 COLLISION AND PROXIMITY QUERIES

Ming C. Lin, Dinesh Manocha, and Young J. Kim

INTRODUCTION

In a geometric context, a collision or proximity query reports information about the relative configuration or placement of two objects. Some of the common examples of such queries include checking whether two objects overlap in space, or whether their boundaries intersect, or computing the minimum Euclidean separation distance between their boundaries. Hundreds of papers have been published on different aspects of these queries in computational geometry and related areas such as robotics, computer graphics, virtual environments, and computer-aided design. These queries arise in different applications including robot motion planning, dynamic simulation, haptic rendering, virtual prototyping, interactive walkthroughs, computer gaming, and molecular modeling. For example, a large-scale virtual environment, e.g., a walkthrough, creates a model of the environment with virtual objects. Such an environment is used to give the user a sense of presence in a synthetic world and it should make the images of both the user and the surrounding objects feel solid. The objects should not pass through each other, and objects should move as expected when pushed, pulled, or grasped; see Fig. 39.0.1. Such actions require fast and accurate collision detection between the geometric representations of both real and virtual objects. Another example is rapid prototyping, where digital representations of mechanical parts, tools, and machines, need to be tested for interconnectivity, functionality, and reliability. In Fig. 39.0.2, the motion of the pistons within the combustion chamber wall is simulated to check for tolerances and verify the design.

This chapter provides an overview of different queries and the underlying algorithms. It includes algorithms for collision detection and distance queries among convex polytopes (Section 39.1), nonconvex polygonal models (Section 39.2), continuous collision detection (Section 39.3), penetration depth queries (Section 39.4), Hausdorff distance queries (Section 39.5), curved objects (Section 39.6), and large environments composed of multiple objects (Section 39.7). Finally, it briefly describes different software packages available to perform some of the queries (Section 39.8).

PROBLEM CLASSIFICATION

Collision Detection: Checks whether two objects overlap in space or their boundaries share at least one common point.

Separation Distance: Length of the shortest line segment joining two sets of points, A and B:

$$\text{dist}(A, B) = \min_{a \in A} \min_{b \in B} |a - b|.$$

FIGURE 39.0.1

A hand reaching toward a ring and a bunny at top. The corresponding image of the user in the real world is shown at bottom. Red finger tips indicate contacts between the user's hand and the virtual ring and bunny.

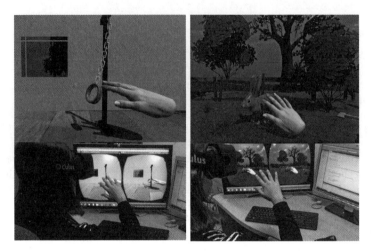

Hausdorff distance: Maximum deviation of one set from the other:

$$\text{haus}(A, B) = \max_{a \in A} \min_{b \in B} |a - b|.$$

Spanning Distance: Maximum distance between the points of two sets:

$$\text{span}(A, B) = \max_{a \in A} \max_{b \in B} |a - b|.$$

Penetration Depth: Minimum distance between two overlapping objects needed to translate one object to make it just touch the other:

$$\text{pd}(A, B) = \text{minimum } ||\mathbf{v}|| \text{ such that } \min_{a \in A} \min_{b \in B} |\mathbf{a} - \mathbf{b} + \mathbf{v}| {\geq} 0.$$

Generalized Penetration Depth: Minimal rigid motion under some distance metric d to make one object just touch the other:

$$\text{gpd}(A, B) = \text{minimum } ||\mathbf{M}||_d \text{ such that } \min_{a \in A} \min_{b \in B} |\mathbf{a} - \mathbf{M}\mathbf{b}| \geq 0.$$

There are two forms of collision detection query: **Boolean** and **enumerative**. The Boolean distance query computes whether the two sets have at least one point in common. The enumerative form yields some representation of the intersection set.

There are at least three forms of the distance queries: exact, approximate, and Boolean. The exact form asks for the exact distance between the objects. The approximate form yields an answer within some given error tolerance of the true measure—the tolerance could be specified as a relative or absolute error. The Boolean form reports whether the exact measure is greater or less than a given value. Furthermore, the norm by which distance is defined may be varied. The Euclidean norm is the most common, but in principle other norms are possible, such as the L_1 and L_∞ norms.

FIGURE 39.0.2

In this virtual prototyping application, the motion of the pistons is simulated to check for tolerances by performing distance queries.

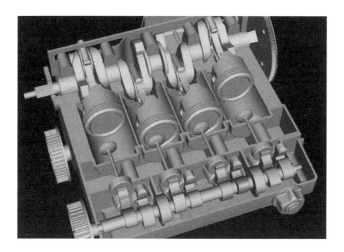

TABLE 39.0.1   Classification of Proximity Queries.

CRITERIA	TYPES
Report	Boolean, exact, approximate, enumerative
Measure	Separation, span, Hausdorff, penetration, collision
Multiplicity	2-body, n-body
Temporality	Static, dynamic
Representation	Polyhedra, convex objects, implicit, parametric, NURBS, quadrics, set-theoretic combinations
Dimension	2,3,d

Each of these queries can be augmented by adding the element of time. If the trajectories of two objects are known, then the next time can be determined at which a particular Boolean query (collision, separation distance, or penetration) will become TRUE or FALSE. In fact, this "time-to-next-event" query can have exact, approximate, and Boolean forms. These queries are called ***dynamic queries***, whereas the ones that do not use motion information are called ***static queries***. In a case where the motion of an object cannot be represented as a closed-form function of time, the underlying application often performs static queries at specific time steps in the application.

These measures, as defined above, apply only to pairs of sets. However, some applications work with many objects, and need to find the proximity information among all or a subset of the pairs. Thus, most of the query types listed above have associated N-body variants.

Finally, the primitives can be represented in different forms. They may be convex polytopes, general polygonal models, curved models represented using parametric or implicit surfaces, set-theoretic combination of objects, etc. Different set of algorithms are known for each representation. A classification of proximity queries based on these criteria is shown in Table 39.0.1.

39.1 CONVEX POLYTOPES

In this section, we give a brief survey of algorithms for collision detection and separation-distance computation between a pair of convex polytopes. A number of algorithms with good asymptotic performance have been proposed. The optimal runtime algorithm for Boolean collision queries takes $O(\log n)$ time, where n is the number of features [BL15]. It precomputes the bounded Dobkin-Kirkpatrick (BDK) hierarchy for each polytope in linear time and space and uses it to perform the query. In practice, three classes of algorithms are commonly used for convex polytopes: linear programming, Minkowski sums, and tracking closest features based on Voronoi diagrams.

LINEAR PROGRAMMING

The problem of checking whether two convex polytopes intersect or not can be posed as a linear programming (LP) problem. In particular, two convex polytopes do not overlap if and only if there exists a separation plane between them. The coefficients of the separation plane equation are treated as unknowns. Linear constraints result by requiring that all vertices of the first polytope lie in one halfspace of this plane and those of the other polytope lie in the other halfspace. The linear programming algorithms are used to check whether there is any feasible solution to the given set of constraints. Given the fixed dimension of the problem, some of the well-known linear programming algorithms (e.g., [Sei90]; cf. Chapter 49) can be used to perform the Boolean collision query in expected linear-time. By caching the last pair of witness points to compute the new separating planes, Chung and Wang [CW96] proposed an iterative method that can quickly update the separating axis or the separating vector in nearly "constant time" in dynamic applications with high motion coherence.

MINKOWSKI SUMS AND CONVEX OPTIMIZATION

Collision and distance queries can be performed based on the Minkowski sum of two objects. It has been shown [CC86] that the minimum separation distance between two objects is the same as the minimum distance from the origin of the **Minkowski sums** of A and $-B$ to the surface of the sums. The Minkowski sum is also referred to as the **translational C-space obstacle** (TCSO). While the Minkowski sum of two convex polytopes can have $O(n^2)$ features [DHKS93], a fast algorithm for separation-distance computation based on convex optimization that exhibits linear-time performance in practice has been proposed by Gilbert et al. [GJK88], also known as the GJK algorithm. It uses pairs of vertices from each object that define simplices within each polytope and a corresponding simplex in the TCSO. Initially the simplex is set randomly and the algorithm refines it using local optimization, until it computes the closest point on the TCSO from the origin of the Minkowski sums. The algorithm assumes that the origin is not inside the TCSO.

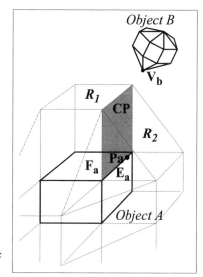

FIGURE 39.1.1
A walk across external Voronoi region of Object A. Vertex $\mathbf{V}_b$ *of Object B lies in the Voronoi region of* $\mathbf{E}_a$.

TRACKING CLOSEST FEATURES USING GEOMETRIC LOCALITY AND MOTION COHERENCE

Lin and Canny [LC91] proposed a distance-computation algorithm between non-overlapping convex polytopes. Often referred to as the LC algorithm, it tracks the closest features between the polytopes. This is the first approach that explicitly takes advantages of motion coherence and geometric locality. The features may correspond to a vertex, face, or an edge on each polytope. It precomputes the external Voronoi region for each polytope. At each time step, it starts with a pair of features and checks whether they are the closest features, based on whether they lie in each other's Voronoi region. If not, it performs a local walk on the boundary of each polytope until it finds the closest features. See Figure 39.1.1. In applications with high motion coherence, the local walk typically takes nearly "constant time" in practice. Typically the number of neighbors for each feature of a polytope is constant and the extent of "local walk" is proportional to the amount of the relative motion undergone by the polytopes.

Mirtich [Mir98] further optimized this algorithm by proposing a more robust variation that avoids some geometric degeneracies during the local walk, without sacrificing the accuracy or correctness of the original algorithm.

Guibas et al. [GHZ99] proposed an approach that exploits both coherence of motion using LC and hierarchical representations by Dobkin and Kirkpatrick [DK90] to reduce the runtime dependency on the amount of the local walks.

Ehmann and Lin [EL00] modified the LC algorithm and used an error-bounded level-of-detail (LOD) hierarchy to perform different types of proximity queries, using the progressive refinement framework (cf. Chapter 39). The implementation of this technique, "multi-level Voronoi Marching," outperforms the existing libraries for collision detection between convex polytopes. It also uses an initialization technique based on directional lookup using hashing, resembling that of [DZ93].

By taking a philosophy similar to that of LC, Cameron [Cam97] presented an extension to the basic GJK algorithm by exploiting motion coherence and geometric locality in terms of connectivity between neighboring features. The algorithm tracks the **witness points**, a pair of points from the two objects that realize the minimum separation distance between them. Rather than starting from a random simplex in the TCSO, the algorithm starts with the witness points from the previous iteration and performs hill climbing to compute a new set of witness points for the current configuration. The running time of this algorithm is a function of the number of refinement steps that the algorithm performs.

TABLE 39.1.1   Algorithms for convex polytopes.

METHOD	FEATURES
DK	$O(\log^2 n)$ query time, collision query only
LP	Linear running time, collision query
GJK	Linear-time behavior in practice, collision and separation-distance queries
LC	Expected constant-time in coherent environments, collision and separation-distance queries

39.2 GENERAL POLYGONAL MODELS

Algorithms for collision and separation-distance queries between general polygon models can be classified based on whether they assume closed polyhedral models, or are represented as a collection of polygons. The latter, also referred to as "polygon soups," makes no assumption related to the connectivity among different faces or whether they represent a closed set.

Some of the most common algorithms for collision detection and separation-distance computation use spatial partitioning or bounding volume hierarchies (BVHs). The spatial subdivisions are a recursive partitioning of the embedding space, whereas bounding volume hierarchies are based on a recursive partitioning of the primitives of an object. These algorithms are based on the divide-and-conquer paradigm. Examples of spatial partitioning hierarchies include k-D trees and octrees [Sam89], R-trees and their variants [HKM95], cone trees, BSPs [NAT90] and their extensions to multi-space partitions [BG91]. The BVHs use bounding volumes (BVs) to bound or contain sets of geometric primitives, such as triangles, polygons, curved surfaces, etc. In a BVH, BVs are stored at the internal nodes of a tree structure. The root BV contains all the primitives of a model, and children BVs each contain separate partitions of the primitives enclosed by the parent. Leaf node BVs typically contain one primitive. In some variations, one may place several primitives at a leaf node, or use several volumes to contain a single primitive. BVHs are used to perform collision and separation-distance queries. These include sphere-trees [Hub93, Qui94], ellipsoid-trees [LWH+07], AABB-trees [BKSS90, HKM95, PML97], OBB-trees [GLM96, BCG+96, Got00], spherical shell-trees [KPLM98, KGL+98], k-DOP-trees [HKM96, KHM+98], SSV-trees [LGLM99], multiresolution hierarchies [OL03], convex hull-trees [EL01], SCB-trees [LAM09], and k-IOS trees [ZK12], as shown in Table 39.2.1. Readers are referred to a book written by Ericson [Eri04] for more details on spatial partitioning and BVHs.

TABLE 39.2.1   Types of bounding volume hierarchies.

NAME	TYPE OF BOUNDING VOLUME
Sphere-tree	Sphere
Ellipsoid-tree	Ellipsoid
AABB-tree	Axis-aligned bounding box (AABB)
OBB-Tree	Oriented bounding box (OBB)
Spherical shell-tree	Spherical shell
k-DOP-tree	Discretely oriented polytope defined by k vectors (k-DOP)
SSV-Tree	Swept-sphere volume (SSV)
Convex hull-tree	Convex polytope
SCB-tree	Slab cut ball (SCB)
k-IOS-tree	Intersection of k-spheres (IOS)

COLLISION DETECTION

Collision queries are performed by traversing the BVHs. Two models are compared by recursively traversing their BVHs in tandem. Each recursive step tests whether BVs A and B, one from each hierarchy, overlap. If they do not, the recursion branch is terminated. But if A and B overlap, the enclosed primitives may overlap and the algorithm is applied recursively to their children. If A and B are both leaf nodes, the primitives within them are compared directly.

SEPARATION-DISTANCE COMPUTATION

The structure of the separation-distance query is very similar to the collision query. As the query proceeds, the smallest distance found by comparing primitives is maintained in a variable δ. At the start of the query, δ is initialized to ∞, or to the distance between an arbitrary pair of primitives. Each recursive call with BVs A and B must determine if some primitive within A and some primitive within B are closer than, and therefore will modify, δ. The call returns trivially if BVs A and B are farther than the current δ, as this precludes any primitives within them being closer than δ. Otherwise the algorithm is applied recursively to its children. For leaf nodes it computes the exact distance between the primitives, and if the new computed distance is less than δ, it updates δ.

To perform an approximate distance query, the distance between BVs A and B is used as a lower limit to the exact distances between their primitives. If this bound prevents δ from being reduced by more than the acceptable tolerance, that recursion branch is terminated.

QUERIES ON BOUNDING VOLUMES

Algorithms for collision detection and distance computation need to perform the underlying queries on the BVHs, including finding out whether two BVs overlap, or computing the separation distance between them. The performance of the overall proximity query algorithm is governed largely by the performance of the sub-algorithms used for proximity queries on a pair of BVs.

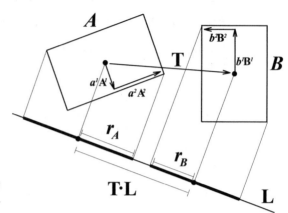

FIGURE 39.2.1
L *is a separating axis for OBBs A and B because projection onto* **L** *renders them disjoint intervals.*

A number of specialized and highly optimized algorithms have been proposed to perform these queries on different BVs. It is relatively simple to check whether two spheres overlap. Two AABBs can be checked for overlap by comparing their dimensions along the three axes. The separation distance between them can be computed based on the separation along each axis. The overlap test can be easily extended to k-DOPs, where their projections are checked along the k fixed axis [KHM$^+$98].

An efficient algorithm to test two OBBs for overlap based on the separating axis theorem (SAT) has been presented in [GLM96, Got00]. It computes the projection of each OBB along 15 axes in 3D. The 15 axes are computed from the face normals of the OBBs (6 face normals) and by taking the cross-products of the edges of the OBBs (9 cross-products). It is shown that two OBBs overlap if and only if their projection along each of these axes overlap. Furthermore, an efficient algorithm that performs overlap tests along each axis has been described. In practice, it can take anywhere from 80 to 240 arithmetic operations to check whether two OBBs overlap. The computation is robust and works well in practice [GLM96]. Figure 39.2.1 shows one of the separating axis tests for two rectangles in 2D.

Algorithms based on different ***swept-sphere volumes*** (SSVs) have been presented in [LGLM99]. Three types of SSVs are suggested: point swept-sphere (PSS), line swept-sphere (LSS), and a rectangular swept-sphere (RSS). Each BV is formulated by taking the Minkowski sum of the underlying primitive—a point, line, or a rectangle in 3D, respectively—with a sphere. Algorithms to perform collision or distance queries between these BVs can be formulated as computing the distance between the underlying primitives. Larsen et al. [LGLM99] have presented an efficient and robust algorithm to compute distance between two rectangles in 3D (as well rectangles degenerating to lines and points). Moreover, they used priority directed search and primitive caching to lower the number of bounding volume tests for separation-distance computations.

In terms of higher-order bounding volumes, fast overlap tests based on spherical shells have been presented in [KPLM98, KGL$^+$98]. Each spherical shell corresponds to a portion of the volume between two concentric spheres. The overlap test between two spherical shells takes into account their structure and reduces to checking whether there is a point contained in a circle that lies in the positive half-plane defined by two lines. The two lines and the circles belong to the same plane.

It is also possible to devise parallel algorithms to traverse the bounding volume hierarchy using multi-core CPUs or many-core GPUs. Lee and Kim [LK10] propose parallel proximity algorithms using heuristic task decomposition to perform both Euclidean distance calculation and collision detection between rigid models. Many parallel algorithms [LMM10, TMLT11] exploit thread and data parallelism on many-core GPUs to perform fast hierarchy construction, updates, and traversal using bounding volumes for rigid and deformable models.

PERFORMANCE OF BOUNDING VOLUME HIERARCHIES

The performance of BVHs on proximity queries is governed by a number of design parameters, including techniques to build the trees, the maximum number of children per node, and the choice of BV type. An additional design choice is the descent rule. This is the policy for generating recursive calls when a comparison of two BVs does not prune the recursion branch. For instance, if BVs A and B failed to prune, one may recursively compare A with each of the children of B, B with each of the children of A, or each of the children of A with each of the children of B. This choice does not affect the correctness of the algorithm, but may impact the performance. Some of the commonly used algorithms assume that the BVHs are binary trees and each primitive is a single triangle or a polygon. The cost of performing the proximity query is given as [GLM96, LGLM99]:

$$T = N_{bv} \times C_{bv} + N_p \times C_p,$$

where T is the total cost function for proximity queries, N_{bv} is the number of bounding volume pair operations, and C_{bv} is the total cost of a BV pair operation, including the cost of transforming each BV for use in a given configuration of the models, and other per BV-operation overhead. N_p is the number of primitive pairs tested for proximity, and C_p is the cost of testing a pair of primitives for proximity (e.g., overlaps or distance computation).

Typically, for tight-fitting bounding volumes, e.g., oriented bounding boxes (OBBs), N_{bv} and N_p are relatively small, whereas C_{bv} is relatively high. In contrast, C_{bv} is low while N_{bv} and N_p may be larger for simple BV types like spheres and axis-aligned bounding boxes (AABBs). Due to these opposing trends, no single BV yields optimum performance for proximity queries in all situations.

39.3 CONTINUOUS COLLISION DETECTION

Earlier collision detection algorithms only check for collisions at sample configurations. As a result, they may miss a collision that occurs between two successive configurations or time steps. This is sometimes known as the tunneling problem, because it typically occurs when a rapidly moving object passes undetected through a thin obstacle. Continuous collision detection (CCD) algorithms avoid the tunneling problem by interpolating a continuous motion trajectory between successive configurations and checking for collisions along that trajectory. If a collision occurs, the first time of contact (ToC) between the moving objects is reported.

RIGID MODELS

At a broad level, CCD algorithms can be classified into algebraic equation solvers, swept-volume formulations, adaptive bisection approaches, kinetic data structures (KDS), Minkowski sum formulations, and conservative advancement (CA).

Algebraic equation solvers [Can86, CWLK06, KR03, RKC00] compute the ToC by numerically finding roots of low-order polynomials, while swept-volume formulations [Cam90, AMBJ06] operate directly on 4D volumes swept out by object motion over time. *Kinetic data structures* [ABG+00, KGS97, KSS02] are based on the formal framework of KDS to keep track of collision events of models during their motion and exploits motion coherence and geometric locality. *Minkowski sum formulations* [Ber04] pose a CCD problem as ray shooting against the translational configuration space obstacle, which is equivalent to Minkowski sums between two translating models. *Adaptive bisection approaches* [SWF+93, HBZ90, RKC02, SSL02] find the ToC by adaptively subdividing the expected collision-time interval, and *conservative advancement* [Lin93, Mir96, ZKM08, TKM09, TKM10] conservatively advances the time step while avoiding collision until it reaches the ToC.

Most of these approaches are unable to perform fast CCD queries on general polygonal models, although some can handle polygon-soup models [RKC02, SSL02]. Redon et al. [RKC02] use a continuous version of the separating axis theorem to extend the static OBB-tree algorithm [GLM96] to CCD, and demonstrate real-time performance on polygonal models. But the algorithm becomes overly conservative when there is a large rotation between two configurations. For polyhedral models, there is a faster algorithm [ZKM08] using conservative advancement that finds the ToC by comparing an upper bound of motion trajectory against the velocity of the models. Tang et al. [TKM09] use controlled advancement to detect collision for polygon-soup model by adaptively adjusting the advancement step, and they extend this work by using different acceleration techniques. This approach is also applicable to articulated models [TMK14]. FCL [PCM12] also uses an implementation of [TKM09], which forms part of a generic collision and proximity detection library.

ARTICULATED MODELS

A conservative condition is used in [SSL02] to guarantee a collision-free motion between two configurations, but such a condition is likely to become overly conservative when an object slides over another object. Redon et al. [RKLM04] describe an extension of their previous algorithm [RKC02] to articulated models, but this algorithm is relatively slow for complicated objects. Zhang et al. have extended their approach [ZKM08] to articulated models [ZRLK07] by modeling each link as a polyhedron.

DEFORMABLE MODELS

CCD algorithms for deformable models compute all possible times of contact between pairs of overlapping primitive during the motion, including self-collisions between the primitives. In order to perform a continuous collision query between two triangle primitives, an approach based on algebraic equation solvers was first introduced by Moore and Wilhelms [MW88] using fifth-order algebraic equations. These equations were further reduced to cubic by taking into account co-planarity

FIGURE 39.3.1

In this simulation, a cloth falls into a funnel and passes through it under the pressure of a ball. This model has 47K vertices, 92K triangles, and a lot of self-collisions. The GPU-based CCD algorithm [TMLT11] takes 4.4ms and 10ms per frame to compute all the collisions.

constraints. In this case, detecting collisions between deforming triangles corresponds to performing pairwise six face-vertex and nine edge-edge elementary tests. Each elementary test reduces to solving a cubic algebraic equation [Pro97, BFA02]. Some filters can be used to accelerate these tests [TMT10a]. Another approach based on conservative advancement (CA) for deforming objects was presented by Tang et al. [TKM10], which was extended from [ZLK06, ZRLK07, TKM09]. The CA-based algorithms are able to avoid solving high-order algebraic equations. In order to avoid redundant elementary tests, the connectivity and adjacency information of a deformable mesh can be used [GKJ+05, TCYM08] or feature-based hierarchies [CTM08] can be also employed.

Most CCD algorithms for deformable objects try to reduce the number of self-collision queries [TCYM08] and redundant elementary tests [TMY+11]. The self-collisions can be classified into two types: self-collision between adjacent, and between non-adjacent triangles or the primitives. In practice, the cost of self-collisions can account for 50-90% of the total running time of the algorithm [GKJ+05]. The normal cone technique was proposed to cull non-self-colliding triangles [Pro97] for discrete collision detection and was extended to CCD by Tang et al. [TKM10]. The chromatic decomposition technique can also be used to accelerate self-collisions [GKJ+05]. Schvartzman et al. [SPO10] presented a star-contour test as a sufficient condition to determine whether the contour of a projected surface patch is collision-free or not. These approaches are designed to reduce the number of elementary tests. By exploiting the abundant parallelism some techniques have been proposed to utilize multi-core CPUs [KHY09] [TMT10b], GPUs [LGS+09, TMLT11], and their hybrid combinations [KHH+09].

RELIABLE METHODS

Most CCD algorithms are implemented using finite-precision arithmetic, potentially resulting in a false negative or a false positive. It is important to perform reliable CCD queries, as it is well known that even a single missed collision can affect the accuracy of the entire simulation system.

The most reliable algorithms are based on exact computations [BEB12] that can perform reliable queries with no false negatives or false positives. Recently, Tang et al. [TTWM14] presented an exact algorithm based on Bernstein sign classification. Both of these methods are based on an exact computation paradigm and use extended precision libraries to perform accurate CCD computations. In practice, these exact arithmetic operations can be expensive for real-time applications. Furthermore, it can be difficult to implement such exact arithmetic operations or libraries on GPUs or embedded processors. The second category of accurate solutions for CCD computations is based on performing floating-point error analysis and using appropriate error tolerances [Wan14]. This approach can be used on any processors that support IEEE floating-point arithmetic operations. The resulting CCD algorithm (SafeCCD) eliminates false negatives altogether but can still result in a high number of false positives. Wang et al. [WTTM15] describe a formulation based on Bernstein sign classification that takes advantage of the geometry properties of Bernstein basis and Bézier curves to perform Boolean collision queries. This approach eliminates all the false negatives and $90 \sim 95\%$ of the false positives.

39.4   PENETRATION-DEPTH COMPUTATION

In this section, we briefly review *penetration depth* (PD) computation algorithms between convex polytopes and general polyhedral models. The PD of two interpenetrating objects A and B is defined as the minimum translation distance that one object undergoes to make the interiors of A and B disjoint. It can be also defined in terms of the TCSO. When two objects are overlapping, the origin of the Minkowski sum of A and $-B$ is contained inside the TCSO. The penetration depth corresponds to the minimum distance from the origin to the surface of TCSO [Cam97]. PD computation is often used in motion planning [HKL+98], contact resolution for dynamic simulation [MZ90, ST96] and force computation in haptic rendering [KOLM02]. Fig. 39.4.1 shows a haptic rendering application of penetration-depth and separation-distance computation. For example, computation of dynamic response in penalty-based methods often needs to perform PD queries for imposing the non-penetration constraint for rigid body simulation. In addition, many applications (such as motion planning and dynamic simulation) require a continuous distance measure when two (nonconvex) objects collide for a well-posed computation.

Several algorithms for PD computation involve computing Minkowski sums and the closest point on its surface from the origin. The worst-case complexity of the overall PD algorithm is dominated by computing Minkowski sums, which can be $\Omega(n^2)$ for convex polytopes and $\Omega(n^6)$ for general (or nonconvex) polyhedral models [DHKS93]. Given the complexity of Minkowski sums, many approximation algorithms have been proposed in the literature for fast PD estimation.

CONVEX POLYTOPES

Dobkin et al. [DHKS93] proposed a hierarchical algorithm to compute the directional PD using Dobkin and Kirkpatrick polyhedral hierarchy. For any direction d,

FIGURE 39.4.1

Penetration depth is applied to virtual exploration of a digestive system using haptic interaction to feel and examine different parts of the model. The distance computation and penetration depth computation algorithms are used for disjoint (D) and penetrating (P) situations, respectively, to compute the forces at the contact areas.

 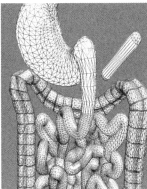 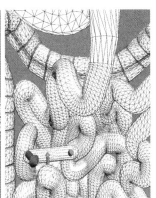

it computes the directional penetration depth in $O(\log n \log m)$ time for polytopes with m and n vertices. Agarwal et al. [AGHP+00] designed a randomized approach to compute the PD values [AGHP+00], achieving $O(m^{\frac{3}{4}+\epsilon}n^{\frac{3}{4}+\epsilon} + m^{1+\epsilon} + n^{1+\epsilon})$ expected time for any positive constant ϵ. Cameron [Cam97] presented an extension to the GJK algorithm [GJK88] to compute upper and lower bounds on the PD between convex polytopes. Bergen further elaborated this idea in an expanding polytope algorithm [Ber01]. The algorithm iteratively improves the result of the PD computation by expanding a polyhedral approximation of the Minkowski sums of two polytopes. Kim et al. [KLM02] presented an incremental algorithm that marches toward a "locally optimal" solution by walking on the surface of the Minkowski sum. The surface of the TCSO is implicitly computed by constructing a local Gauss map and performing a local walk on the polytopes.

POLYHEDRAL MODELS

Algorithms for penetration-depth estimation between general polygonal models are based on discretization of the object space containing the objects, or use of digital geometric algorithms that perform computations on a finite resolution grid. Fisher and Lin [FL01] presented a PD estimation algorithm based on the distance-field computation using the fast marching level-set method. It is applicable to all polyhedral objects as well as deformable models, and it can also check for self-penetration. Hoff et al. [HZLM01, HZLM02] proposed an approach based on performing discretized computations on graphics rasterization hardware. It uses multipass rendering techniques for different proximity queries between general rigid and deformable models, including penetration depth estimation. Kim et al. [KLM02] presented a fast approximation algorithm for general polyhedral models using a combination of object-space as well as discretized computations. Given the global nature of the PD problem, it decomposes the boundary of each polyhedron into convex pieces, computes the pairwise Minkowski sums of the resulting convex polytopes and uses graphics rasterization hardware to perform the closest-point query up to a given discretized resolution. The results obtained are refined using a local walking algorithm. To further speed up this computation and improve the estimate, the algorithm uses a hierarchical refinement technique that takes advantage of geometry culling, model simplification, accelerated ray-shooting, and local refinement with greedy walking. The overall approach combines discretized closest-point queries with geometry culling and refinement at each level of the hierarchy. Its accuracy can vary as a function of the discretization error. Recently, Je et al. [JTL$^+$12] propose a real-time algorithm that finds the PD between general polygonal models based on iterative and local optimization techniques. The iterative optimization consists of two projection steps, in- and out-projection. The in-projection and out-projection steps are formulated as a variant of the continuous collision detection algorithm, and a linear complementarity problem, respectively. A locally optimal solution is finally obtained using a type of Gauss-Seidel iterative algorithm. This algorithm can process complicated models consisting of tens of thousands triangles at interactive rates.

GENERALIZED PENERATION-DEPTH COMPUTATION

Conventional penetration-depth algorithms tend to overestimate the amount of penetration, as the resulting separating motion is often limited to translation only. Another set of algorithms [ZKVM07] use a different penetration measure, called generalized PD, which is defined as a minimal rigid motion used to separate overlapping objects; here, the minimality is based on some distance metric in configuration space such as D_g [ZKVM07], S and geodesic [NPR09], displacement [ZKM07b, ZKM07a], and object norm [ZKM07a]. However, the exact computation of generalized PD may require an arrangement of high dimensional contact surfaces, resulting in $O(n^{12})$ combinatorial complexity for a rigid model with n triangles in 3D [ZKVM07]. Thus, all existing approaches use an approximate method.

Very few algorithms exist to compute generalized PD. Zhang et al. [ZKVM07] first proposed the notion of generalized PD, and presented a method to compute only the lower and upper bounds of generalized PD for rigid models. Later, the same authors proposed a more efficient technique, using a contact-space sampling and displacement metric [ZKM07a]. Nawratil et al. [NPR09] presented methods based on kinematical geometry, using S and geodesic metrics. However, all of these methods are rather slow for interactive applications, and it is not clear whether they are applicable to articulated models. Recently, Tang and Kim [TK14] extended the idea of iterative, contact-space projection techniques [JTL$^+$12] to the generalized PD problem. This algorithm is capable of computing the generalized PD for both rigid and articulated models at interactive rates. Pan et al. [PZM13] presented an algorithm to approximate the boundary of C-obstacle space using active learning and use that approximation to compute the PD.

OTHER METRICS

Other metrics used to characterize the intersection between two objects include the **growth distance** defined by Gilbert and Ong [GO94]. This is a consistent distance measure regardless of whether the objects are disjoint or overlapping; it is differs from the PD between two interpenetrating convex objects.

39.5 HAUSDORFF DISTANCE

Since the seminal work by [Ata83], different algorithms for calculating Hausdorff distances have been proposed in the literature. In this section, we briefly survey work relevant to Hausdorff distance computation for polygonal models in $\mathbb{R}^2$ and in $\mathbb{R}^3$. We refer readers to [AG00, TLK09] for more extensive surveys of the field.

For $\mathbb{R}^2$, [Ata83] presented a linear-time algorithm for convex polygons. For simple, non-convex polygons with n and m vertices, [ABB95] presented an $O((n+m)\log(n+m))$-time algorithm based on an observation that the Hausdorff distance can only be realized at the points of intersection between Voronoi boundary surfaces and the polygon edges. In $\mathbb{R}^3$, for polygon-soup models with n triangles, [God98, ABG+03] presented deterministic and randomized algorithms that run respectively in $O(n^5)$ and $O(n^{3+\epsilon})$, where $\epsilon > 0$. [Lla05] proposed an algorithm using random covering, and also demonstrated implementation results for simple, convex ellipsoids. Since then, more practical algorithms have been put forward, but all these algorithms only approximate the Hausdorff distance due to the complexity of the exact computation. [GBK05] use polygon subdivision to approximate the solution within an error bound. [CRS98, ASCE02] sample a polygonal surface to approximate the distance, but no sampling analysis is provided. More recently, a real-time algorithm has been proposed to approximate the Hausdorff distance within a error bound between complex, polygonal models in $\mathbb{R}^3$ using bounding volume hierarchies [TLK09].

39.6 SPLINE AND ALGEBRAIC OBJECTS

Most of the algorithms highlighted above are limited to polygonal objects. In many applications of geometric and solid modeling, curved objects whose boundaries are described using rational splines or algebraic equations are used (cf. Chapter 56). Algorithms used to perform different proximity queries on these objects may be classified by subdivision methods, tracing methods, analytic methods and bounding volume hierarchy methods. See [Hof89, Man92, Far14] for surveys. Next, we briefly enumerate these methods.

SUBDIVISION METHODS

All subdivision methods for parametric surfaces work by recursively subdividing the domain of the two surface patches in tandem, and examining the spatial relationship between patches [LR80, SWF+93, MP09]. Depending on various criteria, the domains are further subdivided and recursively examined, or the given recursion branch is terminated. In all cases, whether it is the intersection curve or the distance function, the solution is known only to some finite precision.

TRACING METHODS

The tracing method begins with a given point known to be on the intersection curve [BFJP87, MC91, KM97]. Then the intersection curve is traced in sufficiently small steps until the edge of the patch is found, or until the curve loops back to itself. In practice, it is easy to check for intersections with a patch boundary, but difficult to know when the tracing point has returned to its starting position. Frequently this is posed as an initial-value differential equations problem [KPW92], or as solving a system of algebraic equations [MC91, KM97, LM97]. At the intersection point on the surfaces, the intersection curve must be mutually orthogonal to the normals of the surfaces. Consequently, the vector field that the tracing point must follow is given by the cross product of the normals.

ANALYTIC METHODS

Analytic methods usually involve implicitizing one of the parametric surfaces— obtaining an implicit representation of the model [SAG84, MC92]. The parametric surface is a mapping from (u, v)-space to (x, y, z)-space, and the implicit surface is a mapping from (x, y, z)-space to $\mathbb{R}$. Substituting the parametric functions $f_x(u, v), f_y(u, v), f_z(u, v)$ for x, y, z of the implicit function leads to a scalar function in u and v. The locus of roots of this scalar function map out curves in the (u, v) plane, which are the pre-images of the intersection curve [KPP90, MC91, KM97, Sar83]. Based on its representation as an algebraic plane curve, efficient algorithms have been proposed by a number of researchers [AB88, KM97, KCMh99]. Recently, a parallel technique utilizing the CPU and GPU multicore architecture has been introduced to find surface-surface intersections [PEK+11].

BOUNDING VOLUME HIERARCHY METHODS

For free-form surfaces, BVHs can be also constructed to accelerate distance calculation between surfaces. Kim et al. [KOY+11] proposed the Coons BVH of free-form surfaces by recursively computing the bilinear surface of the four corners of patches (i.e., Coons patch) comprising the underlying surface. The resulting Coons BVH can be used to accelerate collision detection and Euclidean distance calculation [KOY+11] as well as Hausdorff distance calculation [KOY+13] while saving the memory space of BVH considerably compared to conventional BVHs for a polyhedral surface.

39.7   LARGE ENVIRONMENTS

Large environments are composed of multiple moving objects. Different methods have been proposed to overcome the bottleneck of $O(n^2)$ pairwise tests in an environment composed of n objects. The problem of performing proximity queries in large environments is typically divided into two parts [Hub93, CLMP95]: the *broad phase*, in which we identify the pair of objects on which we need to perform different proximity queries, and the *narrow phase*, in which we perform the exact pairwise queries. An architecture for multi-body collision detection algorithm is shown in Figure 39.7.1. In this section, we present a brief overview of algorithms used in the broad phase.

The simplest algorithms for large environments are based on spatial subdi-

FIGURE 39.7.1

Typically, the object's motion is constrained by collisions with other objects in the simulated environment. Depending on the outcome of the proximity queries, the resulting simulation computes an appropriate response.

Architecture for Multi-body Collision Detection

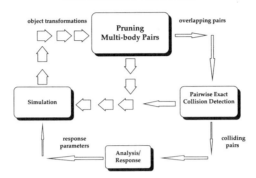

visions. The space is divided into cells of equal volume, and at each instance the objects are assigned to one or more cells. Collisions are checked between all object pairs belonging to each cell. In fact, Overmars presented an efficient algorithm based on a hash table to efficiently perform point location queries in fat subdivisions [Ove92] (see also Chapter 38). This approach works well for sparse environments in which the objects are uniformly distributed through the space. Another approach operates directly on 4D volumes swept out by object motion over time [Cam90]. Efficient algorithms for maintenance and self-collision tests for kinematic chains composed of multiple links have been presented in [LSHL02].

Several algorithms compute an axis-aligned bounding box (AABB) for each object, based on their extremal points along each direction. Given n bounding boxes, they check which boxes overlap in space. A number of efficient algorithms are known for the static version of this problem. In 2D, the problem reduces to checking 2D intervals for overlap using interval trees and can be performed in $O(n \log n + s)$ where s is the total number of intersecting rectangles [Ede83]. In 3D, algorithms of $O(n \log^2 n + s)$ complexity are known, where s is the number of overlapping pairwise bounding boxes [HSS83, HD82]. Algorithms for N-body proximity queries in dynamic environments are based on the **sweep and prune** approach [CLMP95]. This incrementally computes the AABBs for each object and checks them for overlap by computing the projection of the bounding boxes along each dimension and sorting the interval endpoints using insertion sort or bubble sort [SH76, Bar92, CLMP95]. In environments where the objects make relatively small movements between successive frames, the lists can be sorted in expected linear time, leading to expected-time $O(n+m)$, where m is the number of overlapping intervals along any dimension. A parallel version of sweep and prune has been implemented using graphics hardware by performing sweeps in a massively parallel fashion [LHLK10]. It is capable of finding all overlaps among one million AABB pairs at interactive rates. These algorithms are limited to environments where objects undergo rigid motion. Govindaraju et al. [GRLM03] have presented a general algorithm for large environments composed of rigid as well as nonrigid

FIGURE 39.7.2

GPU-based collision detection and simulation at interactive rates for massive bodies [LHLK10]. From left to right: N-body collision detection for 1M arbitrarily moving boxes, simulation of 0.3M particles, rigid-body dynamics for 16K torus models.

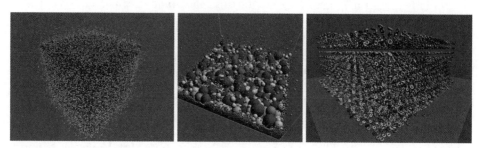

motion. This algorithm uses graphics hardware to prune the number of objects that are in close proximity and eventually checks for overlapping triangles between the objects. In practice, it works well in large environments composed of nonrigid and breakable objects. However, its accuracy is governed by the resolution of the rasterization hardware.

OUT-OF-CORE ALGORITHMS

In many applications, it may not be possible to load a massive geometric model composed of millions of primitives in the main memory for interactive proximity queries. In addition, algorithms based on spatial partitioning or bounding volume hierarchies also add additional memory overhead. Thus, it is important to develop proximity-query algorithms that use a relatively small or bounded memory footprint.

Wilson et al. [WLML99] presented an out-of-core algorithm to perform collision and separation-distance queries on large environments. It uses **overlap graphs** to exploit locality of computation. For a large model, the algorithm automatically encodes the proximity information between objects and represents it using an overlap graph. The overlap graph is computed off-line and preprocessed using graph partitioning, object decomposition, and refinement algorithms. At run time it traverses localized subgraphs and orders the computations to check the corresponding geometry for proximity tests, as well as pre-fetch geometry and associated hierarchical data structures. To perform interactive proximity queries in dynamic environments, the algorithm uses the BVHs, modifies the localized subgraph(s) on the fly, and takes advantage of spatial and temporal coherence.

39.8   PROXIMITY QUERY PACKAGES

Many systems and libraries have been developed for performing different proximity queries. These include:

PQP: PQP, a Proximity Query Package, supports collision detection, separation-distance computation or tolerance verification. It uses OBBTree for collision queries and a hierarchy of swept-sphere volumes to perform distance queries [LGLM99]. It assumes that each object is a collection of triangles and can handle polygon soup models. `http://gamma.cs.unc.edu/SSV`

SWIFT: SWIFT is a library for collision detection, distance computation, and contact determination between 3D polygonal objects undergoing rigid motion. It assumes that the input primitives are convex polytopes or a union of convex pieces. The underlying algorithm is based on a variation of LC [EL00]. The resulting system is faster, more robust, and more memory efficient than I-COLLIDE. `http://gamma.cs.unc.edu/SWIFT`

SWIFT++: SWIFT++ is a library for collision detection, approximate and exact distance computation, and contact determination between closed and bounded polyhedral models. It decomposes the boundary of each polyhedra into convex patches and precomputes a hierarchy of convex polytopes [EL01]. It uses the SWIFT library to perform the underlying computations between the bounding volumes. `http://gamma.cs.unc.edu/SWIFT++`

QuickCD: QuickCD is a general-purpose collision detection library, capable of performing exact collision detection on complex models. The input model is a collection of triangles, with assumptions on the structure or topologies of the model. It precomputes a hierarchy of k-DOPs for each object and uses them to perform fast collision queries [KHM+98]. `http://www.ams.sunysb.edu/~jklosow/quickcd/QuickCD.html`

OPCODE: OPCODE is a collision detection library between general polygonal models. It uses a hierarchy of AABBs. It is memory efficient in comparison to earlier collision packages. `http://www.codercorner.com/Opcode.htm`

DEEP: DEEP estimates the penetration depth and the associated penetration direction between two overlapping convex polytopes. It uses an incremental algorithm that computes a "locally optimal" solution by walking on the surface of the Minkowski sum of two polytopes [KLM02]. `http://gamma.cs.unc.edu/DEEP`

PIVOT: PIVOT computes generalized proximity information between arbitrary objects using graphics hardware. It uses multipass rendering techniques and accelerated distance computation, and provides an approximate solution for different proximity queries. These include collision detection, distance computation, local penetration depth, contact region and normals, etc. [HZLM01, HZLM02]. It involves no preprocessing and can handle deformable models. `http://gamma.cs.unc.edu/PIVOT`

FAST/C²A: These two packages perform continuous collision detection (CCD) between 2-manifold, rigid models and between non-manifold, rigid models, respectively, and report the first time of contact between them as well as their associated contact features. Both packages can perform CCD at interactive rates on a standard PC for complex models consisting of 10K $\sim$ 70K triangles [ZLK06, TKM09]. `http://graphics.ewha.ac.kr/FAST, ~/C2A`

CATCH: CATCH performs CCD between articulated models consisting of 2-manifold links and reports the first time of contact between them as well as their associated contact features. CATCH is also capable of reporting self-collisions [ZRLK07]. `http://graphics.ewha.ac.kr/CATCH`

SelfCCD: SelfCCD can be used for continuous collision detection between deformable models. It can also compute self-collisions and uses many acceleration techniques described in [CTM08, TCYM08, TMT10a]. It has been used for self-collisions in cloth and other deformable benchmarks. http://gamma.cs.unc.edu/SELFCCD

MCCD: MCCD is a parallel extension of SelfCCD that can exploit the multiple CPU cores to accelerate the collision computations. It has been used to obtain almost linear speedups on 8 or 16 cores for complex benchmarks [TMT10b]. http://gamma.cs.unc.edu/MCCD

BSC/TightCCD: These two packages are used to perform reliable continuous collision queries between triangulated models [TTWM14, WTTM15]. They are based on extended precision arithmetic or floating point error bounds. http://gamma.cs.unc.edu/BSC

FCL: FCL is a flexible collision library that can perform multiple queries including collision detection, separation distance, continuous collision detection and penetration depth [PCM12]. Furthermore, it can handle rigid, deformable, articulated and point-cloud models. It has also been integrated into ROS. http://gamma.cs.unc.edu/FCL

PolyDepth: PolyDepth estimates the penetration depth and the associated penetration direction between two overlapping polygonal models. PolyDepth calculates a "locally optimal" solution and is also capable of calculating multiple solutions for each overlapping region [JTL+12]. http://graphics.ewha.ac.kr/polydepth

gSaP: gSaP cull collisions between very large numbers of moving bodies using graphics processing units (GPUs). gSaP implemented entirely on GPUs using the CUDA framework can handle a million moving objects at interactive rates. [LHLK10]. http://graphics.ewha.ac.kr/gsap

Bullet: Bullet is an open source collision detection, rigid body and soft body dynamics library, primarily designed for use in computer games, visual effects and robotic simulation. It supports both discrete and continuous collision detection for convex and non-convex mesh models. http://bulletphysics.org

39.9 SOURCES AND RELATED MATERIAL

RELATED CHAPTERS

Chapter 27: Voronoi diagrams and Delaunay triangulations
Chapter 38: Point location
Chapter 42: Geometric intersection
Chapter 50: Algorithmic motion planning
Chapter 51: Robotics
Chapter 53: Modeling motion
Chapter 67: Software

REFERENCES

[AB88]   S.S. Abhyankar and C. Bajaj. Computations with algebraic curves. In *Proc. Int. Symp. Symbolic Algebraic Computation*, vol. 358 of *LNCS*, pages 279–284, Springer, Berlin, 1988.

[ABB95]   H. Alt, B. Behrends, and J. Blömer. Approximate matching of polygonal shapes. *Ann. Math. Artif. Intell.*, 13:251–266, 1995.

[ABG⁺00]   P.K. Agarwal, J. Basch, L.J. Guibas, J. Hershberger, and L. Zhang. Deformable free space tiling for kinetic collision detection. *I. J. Robot. Res.*, 21:179–198, 2002.

[ABG⁺03]   H. Alt, P. Braß, M. Godau, C. Knauer, and C. Wenk. Computing the Hausdorff distance of geometric patterns and shapes. *Discrete Comput. Geom.*, 25:65–76, 2003.

[AG00]   H. Alt and L.J. Guibas. Discrete geometric shapes: Matching, interpolation, and approximation. In J.-R. Sack and J. Urrutia, editors, *Handbook of Computational Geometry*, pages 121–153, Elsevier, Amsterdam, 2000.

[AGHP⁺00] P.K. Agarwal, L.J. Guibas, S. Har-Peled, A. Rabinovitch, and M. Sharir. Penetration depth of two convex polytopes in 3d. *Nordic J. Comput.*, 7:227–240, 2000.

[AMBJ06]   K. Abdel-Malekl, D. Blackmore, and K. Joy. Swept volumes: Foundations, perspectives, and applications. *Int. J. Shape Modeling*, 12:87-127, 2006.

[ASCE02]   N. Aspert, D. Santa-Cruz, and T. Ebrahimi. Mesh: Measuring errors between surfaces using the Hausdorff distance. In *Proc. IEEE Int. Conf. Multimedia Expo*, pages 705–708, 2002.

[Ata83]   M.J. Atallah. A linear time algorithm for the Hausdorff distance between convex polygons. *Inform. Process. Lett*, 17:207–209, 1983.

[Bar92]   D. Baraff. *Dynamic Simulation of Non-penetrating Rigid Bbody Simulation*. PhD thesis, Cornell University, 1992.

[BCG⁺96]   G. Barequet, B. Chazelle, L.J. Guibas, J.S.B. Mitchell, and A. Tal. Boxtree: A hierarchical representation of surfaces in 3D. *Comp. Graphics Forum*, 15:387–396, 1996.

[BEB12]   T. Brochu, E Edwards, and R. Bridson. Efficient geometrically exact continuous collision detection. *ACM Trans. Graph.*, 31:96, 2012.

[Ber01]   G. van den Bergen. Proximity queries and penetration depth computation on 3d game objects. *Game Developers Conference*, 2001.

[Ber04]   G. van den Bergen. Ray casting against general convex objects with application to continuous collision detection. Preprint, `http://dtecta.com/papers/jgt04raycast.pdf`, 2004.

[BFA02]   R. Bridson, R. Fedkiw, and J. Anderson. Robust treatment for collisions, contact and friction for cloth animation. *ACM Trans. Graph.*, 21:594–603, 2002.

[BFJP87]   R.E. Barnhill, G. Farin, M. Jordan, and B. Piper. Surface/surface intersection. *Comput. Aided Geom. Design*, 4:3–16, 1987.

[BKSS90]   N. Beckmann, H.-P. Kriegel, R. Schneider, and B. Seeger. The R^*-tree: An efficient and robust access method for points and rectangles. *Proc. SIGMOD Conf. Management of Data*, pages 322–331, 1990.

[BL15]   L. Barba and S. Langerman. Optimal detection of intersections between convex polyhedra. In *Proc. 26th ACM-SIAM Sympos. Discrete Algorithms*, pages 1641–1654, 2015.

[BG91]   W.J. Bouma and G. Vaněček. Collision detection and analysis in a physically based simulation. *Proc. Eurographics Workshop Animation and Simulation*, pages 191–203, 1991.

[Cam90]   S. Cameron. Collision detection by four-dimensional intersection testing. *IEEE Trans. Robot. Autom.*, 6:291–302, 1990.

[Cam97]   S. Cameron. Enhancing GJK: Computing minimum and penetration distance between convex polyhedra. In *Proc. IEEE Int. Conf. Robot. Autom.*, pages 3112–3117, 1997.

[Can86]   J. Canny. Collision detection for moving polyhedra. *IEEE Trans. Pattern Analysis Machine Intel.*, 8:200–209, 1986.

[CC86]   S. Cameron and R.K. Culley. Determining the minimum translational distance between two convex polyhedra. In *Proc. IEEE Int. Conf. Robot. Autom.*, pages 591–596, 1986.

[CLMP95]   J. Cohen, M. Lin, D. Manocha, and M. Ponamgi. I-COLLIDE: An interactive and exact collision detection system for large-scale environments. In *Proc. ACM Interactive 3D Graphics Conf.*, pages 189–196, 1995.

[CRS98]   P. Cignoni, C. Rocchini, and R. Scopigno. Metro: Measuring error on simplified surfaces. *Comput. Graphics Forum*, 17:167–174, 1998.

[CTM08]   S. Curtis, R. Tamstorf, and D. Manocha. Fast collision detection for deformable models using representative-triangles. *Proc. ACM Sympos. Interactive 3D Graphics Games*, pages 61–69, 2008.

[CW96]   K. Chung and W. Wang. Quick collision detection of polytopes in virtual environments. In *Proc. 21st ACM Sympos. Virtual Reality Software and Technology*, 1996.

[CWLK06]   Y.-K. Choi, W. Wang, Y. Liu, and M.-S. Kim. Continuous collision detection for elliptic disks. *IEEE Trans. Robot.*, 2006.

[DHKS93]   D. Dobkin, J. Hershberger, D. Kirkpatrick, and S. Suri. Computing the intersection-depth of polyhedra. *Algorithmica*, 9:518–533, 1993.

[DK90]   D.P. Dobkin and D.G. Kirkpatrick. Determining the separation of preprocessed polyhedra—a unified approach. In *Proc. 17th Int. Colloq. Automata Lang. Program.*, vol. 443 of *LNCS*, pages 400–413, Springer, Berlin, 1990.

[DZ93]   P. Dworkin and D. Zeltzer. A new model for efficient dynamics simulation. *Proc. Eurographics Workshop Animation and Simulation*, pages 175–184, 1993.

[Ede83]   H. Edelsbrunner. A new approach to rectangle intersections, Part I. *Int. J. Comput. Math.*, 13:209–219, 1983.

[EL00]   S. Ehmann and M.C. Lin. Accelerated proximity queries between convex polyhedra using multi-level Voronoi marching. *Proc. IEEE/RSJ Int. Conf. Intelligent Robots and Systems*, pages 2101–2106, 2000.

[EL01]   S. Ehmann and M.C. Lin. Accurate and fast proximity queries between polyhedra using convex surface decomposition. *Comput. Graphics Forum*, 20:500–510, 2001.

[Eri04]   C. Ericson. *Real-Time Collision Detection*. CRC Press, Boca Raton, 2004.

[Far14]   G. Farin. *Curves and Surfaces for Computer-Aided Geometric Design: a Practical Guide*. Elsevier, Amsterdam, 2014.

[FL01]   S. Fisher and M.C. Lin. Deformed distance fields for simulation of non-penetrating flexible bodies. *Proc. Eurographics Workshop Comp. Animation and Simulation*, pages 99–111, 2001.

[GBK05]   M. Guthe, P. Borodin, and R. Klein. Fast and accurate Hausdorff distance calculation between meshes. *J. WSCG* 13:41–48, 2005.

[GHZ99]   L.J. Guibas, D. Hsu, and L. Zhang. H-Walk: Hierarchical distance computation for moving convex bodies. In *Proc. 15th Sympos. Comput. Geom.*, pages 265–273, ACM Press, 1999.

[GJK88]   E.G. Gilbert, D.W. Johnson, and S.S. Keerthi. A fast procedure for computing the distance between objects in three-dimensional space. *IEEE J. Robot. Autom.*, RA-4:193–203, 1988.

[GKJ$^+$05]   N. Govindaraju, D. Knott, N. Jain, I. Kabal, R. Tamstorf, R. Gayle, M.C. Lin, and D. Manocha. Collision detection between deformable models using chromatic decomposition. *ACM Trans. Graph.*, 24:991–999, 2005.

[GLM96]   S. Gottschalk, M.C. Lin, and D. Manocha. OBB-Tree: A hierarchical structure for rapid interference detection. *Proc. 23rd Conf. Comput. Graphics Interactive Tech.*, pages 171–180, 1996.

[GO94]   E.G. Gilbert and C.J. Ong. New distances for the separation and penetration of objects. In *Proc. IEEE Int. Conf. Robot. Autom.*, pages 579–586, 1994.

[God98]   M. Godau. *On the Complexity of Measuring the Similarity between Geometric Objects in Higher Dimensions*. PhD thesis, Freien Universität, 1998.

[Got00]   S. Gottschalk. *Collision Queries Using Oriented Bounding Boxes*. PhD thesis, University of North Carolina, 2000.

[GRLM03]   N. Govindaraju, S. Redon, M.C. Lin, and D. Manocha. CULLIDE: Interactive collision detection between complex models in large environments using graphics hardware. *Proc. ACM SIGGRAPH/EUROGRAPHICS Workshop on Graphics Hardware*, pages 25–32, 2003.

[HBZ90]   B.V. Herzen, A.H. Barr, and H.R. Zatz. Geometric collisions for time-dependent parametric surfaces. *ACM SIGRAPH Computer Graphics*, 24:39–48, 1990.

[HD82]   H.W. Six and D. Wood. Counting and reporting intersections of *D*-ranges. *IEEE Trans. Computers*, C-31:181–187, 1982.

[HKL$^+$98]   D. Hsu, L.E. Kavraki, J. Latombe, R. Motwani, and S. Sorkin. On finding narrow passages with probabilistic roadmap planners. *Proc. 3rd Workshop Algorithmic Found. Robotics*, pages 25–32, A.K. Peters, Naticks, 1998.

[HKM95]   M. Held, J.T. Klosowski, and J.S.B. Mitchell. Evaluation of collision detection methods for virtual reality fly-throughs. In *Canadian Conf. Comput. Geom.*, pages 205–210, Quebec City, 1995.

[HKM96]   M. Held, J.T. Klosowski, and J.S.B. Mitchell. Real-time collision detection for motion simulation within complex environments. In *Proc. ACM SIGGRAPH Visual*, page 151, 1996.

[Hof89]   C.M. Hoffmann. *Geometric and Solid Modeling*. Morgan Kaufmann, San Mateo, 1989.

[HSS83]   J.E. Hopcroft, J.T. Schwartz, and M. Sharir. Efficient detection of intersections among spheres. *I. J. Robot. Res*, 2:77–80, 1983.

[Hub93]   P.M. Hubbard. Interactive collision detection. In *Proc. IEEE Sympos. Research Frontiers in Virtual Reality*, pages 24–31, 1993.

[HZLM01]   K.E. Hoff III, A. Zaferakis, M. Lin, and D. Manocha. Fast and simple 2D geometric proximity queries using graphics hardware. *Proc. ACM Sympos. Interactive 3D Graphics*, pages 145–148, 2001.

[HZLM02]   K. Hoff, A. Zaferakis, M. Lin, and D. Manocha. Fast 3d geometric proximity queries between rigid and deformable models using graphics hardware acceleration. Technical Report TR02-004, University of North Carolina, 2002.

[JTL+12]   C. Je, M. Tang, Y. Lee, M. Lee, and Y. Kim. PolyDepth: Real-time penetration depth computation using iterative contact-space projection. *ACM Trans. Graph.*, 31:5, 2012.

[KCMh99]   J. Keyser, T. Culver, D. Manocha, and S. Krishnan. MAPC: A library for efficient and exact manipulation of alge braic points and curves. In *Proc. 15th Sympos. Comput. Geom.*, pages 360–369, ACM Press, 1999.

[KGL+98]   S. Krishnan, M. Gopi, M.C. Lin, D. Manocha, and A. Pattekar. Rapid and accurate contact determination between spline models using shelltrees. *Comp. Graphcis Forum*, 17:C315–C326, 1998.

[KGS97]   D.-J. Kim, L.J. Guibas, and S.-Y. Shin. Fast collision detection among multiple moving spheres. In *Sympos. Comput. Geom.*, pages 373–375, ACM Press, 1997.

[KHH+09]   D. Kim, J.-P. Heo, J. Huh, J. Kim, and S.-E. Yoon. HPCCD: Hybrid parallel continuous collision detection using CPUs and GPUs. *Comp. Graphics Forum (Pacific Graphics)*, 28, 2009.

[KHM+98]   J.T. Klosowski, M. Held, J.S.B. Mitchell, H. Sowizral, and K. Zikan. Efficient collision detection using bounding volume hierarchies of k-dops. *IEEE Trans. Vis. Comp. Graph.*, 4:21–37, 1998.

[KHY09]   D. Kim, J.-P. Heo, and S.-E. Yoon. PCCD: parallel continuous collision detection. In *ACM SIGGRAPH Posters*, page 50, 2009.

[KLM02]   Y.J. Kim, M.C. Lin, and D. Manocha. DEEP: an incremental algorithm for penetration depth computation between convex polytopes. In *Proc. IEEE Conf. Robot. Autom.*, pages 921–926, 2002.

[KM97]   S. Krishnan and D. Manocha. An efficient surface intersection algorithm based on the lower dimensional formulation. *ACM Trans. Graph.*, 16:74–106, 1997.

[KOLM02]   Y.J. Kim, M.A. Otaduy, M.C. Lin, and D. Manocha. Six-degree-of-freedom haptic display using localized contact computations. *Proc. 10th Sympos. Haptic Interfaces for Virtual Environment and Teleoperator Systems.*, pages 209–216, 2002.

[KOY+11]   Y.J. Kim, Y.-T. Oh, S.-H. Yoon, M.-S. Kim, and G. Elber. Coons BHV for freeform geometric models. *ACM Trans. Graph.*, 30:169, 2011.

[KOY+13]   Y.J. Kim, Y.-T. Oh, S.-H. Yoon, M.-S. Kim, and G. Elber. Efficient Hausdorff distance computation for freeform geometric models in close proximity. *Computer-Aided Design*, 45:270–276, 2013.

[KPLM98]   S. Krishnan, A. Pattekar, M. Lin, and D. Manocha. Spherical shell: A higher order bounding volume for fast proximity queries. In *Proc. 3rd Int. Workshop Algorithmic Found. Robotics*, pages 177–190, A.K Peters, Natick, 1998.

[KPP90]   G.A. Kriezis, P.V. Prakash, and N.M. Patrikalakis. Method for intersecting algebraic surfaces with rational polynomial patches. *Computer-Aided Design*, 22:645–654, 1990.

[KPW90]   G.A. Kriezis, N.M. Patrikalakis, and F.E. Wolter. Topological and differential equation methods for surface intersections. *Computer-Aided Design*, 24:41–55, 1990.

[KR03]   B. Kim and J. Rossignac. Collision prediction for polyhedra under screw motions. In *ACM Conf. Solid Modeling Appl.*, pages 4–10, 2003.

[KSS02]   D. Kirkpatrick, J. Snoeyink, and B. Speckmann. Kinetic collision detection for simple polygons. *Internat. J. Comput. Geom. Appl.*, 12:3–27, 2002.

[LAM09]   T. Larsson and T. Akenine-Möller. Bounding volume hierarchies of slab cut balls. *Comp. Graphics Forum*, 28:2379–2395, 2009.

[LC91]   M.C. Lin and J.F. Canny. Efficient algorithms for incremental distance computation. In *Proc. IEEE Conf. Robot. Autom.*, pages 1008–1014, 1991.

[LGLM99]   E. Larsen, S. Gottschalk, M.C. Lin, and D. Manocha. Fast proximity queries with swept sphere volumes. Technical Report TR99-018, University of North Carolina, 1999.

[LGS⁺09]   C. Lauterbach, M. Garland, S. Sengupta, D. Luebke, and D. Manocha. Fast BVH construction on GPUs. In *Comp. Graphics Forum*, 28:375–384, 2009.

[LHLK10]   F. Liu, T. Harada, Y. Lee, and Y.J. Kim. Real-time collision culling of a million bodies on graphics processing units. *ACM Trans. Graph.*, 29:154:1–154:8, 2010.

[Lin93]   M.C. Lin. *Efficient Collision Detection for Animation and Robotics*. PhD thesis, University of California, Berkeley, 1993.

[LK10]   Y. Lee and Y.J. Kim. Simple and parallel proximity algorithms for general polygonal models. *Comp. Animation Virtual Worlds*, 21:365–374, 2010

[Lla05]   B. Llanas. Efficient computation of the Hausdorff distance between polytopes by exterior random covering. *Comput. Optim. Appl.*, 30:161–194, 2005.

[LM97]   M.C. Lin and D. Manocha. Efficient contact determination between geometric models. *Internat. J. Comput. Geom. Appl.*, 7:123–151, 1997.

[LMM10]   C. Lauterbach, Q. Mo, and D. Manocha. gProximity: hierarchical GPU-based operations for collision and distance queries. *Comp. Graphics Forum*, 29:419–428, 2010.

[LR80]   J.M. Lane and R.F. Riesenfeld. A theoretical development for the computer generation and display of piecewise polynomial surfaces. *IEEE Trans. Pattern Analysis Machine Intel.*, 2:150–159, 1980.

[LSHL02]   I. Lotan, F. Schwarzer, D. Halperin, and J.-C. Latombe. Efficient maintenance and self-collision testing for kinematic chains. In *Proc. 18th Sympos. Comput. Geom.*, pages 43–52, ACM Press, 2002.

[LWH⁺07]   S. Liu, C.C.L. Wang, K.-C. Hui, X. Jin, and H. Zhao. Ellipsoid-tree construction for solid objects. In *Proc. ACM Sympos. Solid and Physical Modeling*, pages 303–308, 2007.

[Man92]   D. Manocha. *Algebraic and Numeric Techniques for Modeling and Robotics*. PhD thesis, University of California, Berkeley, 1992.

[MC91]   D. Manocha and J.F. Canny. A new approach for surface intersection. *Internat. J. Comput. Geom. Appl.*, 1:491–516, 1991.

[MC92]   D. Manocha and J.F. Canny. Algorithms for implicitizing rational parametric surfaces. *Comp. Aided Geom. Design*, 9:25–50, 1992.

[Mir96]   B. Mirtich. *Impulse-based Dynamic Simulation of Rigid Body Systems*. PhD thesis, University of California, Berkeley, 1996.

[Mir98]   B. Mirtich. V-Clip: Fast and robust polyhedral collision detection. *ACM Trans. Graph.*, 17:177–208, 1998.

[MP09]   B. Mourrain and J.P. Pavone. Subdivision methods for solving polynomial equations. *Symbolic Comput.*, 44:292–306, 2009.

[MW88]   M. Moore and J. Wilhelms. Collision detection and response for computer animation. *ACM SIGGRAPH Comp. Graphics*, 22:289–298, 1988.

[MZ90]   M. McKenna and D. Zeltzer. Dynamic simulation of autonomous legged locomotion. *ACM SIGGRAPH Comp. Graphics*, 24:29–38, 1990.

[NAT90]   B. Naylor, J. Amanatides, and W. Thibault. Merging BSP trees yield polyhedral modeling results. In *ACM SIGGRAPH Comp. Graphics*, 24:115–124, 1990.

[NPR09]   G. Nawratil, H. Pottmann, and B. Ravani. Generalized penetration depth computation based on kinematical geometry. *Comput. Aided Geom. Des.*, 26:425–443, 2009.

[OL03]   M.A. Otaduy and M.C. Lin. Sensation preserving simplification for haptic rendering. *ACM Trans. Graph.*, 543–553, 2003.

[Ove92]   M.H. Overmars. Point location in fat subdivisions. *Inform. Process. Lett.*, 44:261–265, 1992.

[PCM12]   J. Pan, S. Chitta, and D. Manocha. Fcl: A general purpose library for collision and proximity queries. In *Proc. IEEE Int. Conf. Robot. Autom.*, pages 3859–3866, 2012.

[PEK+11]   C.-H. Park, G. Elber, K.-J. Kim, G.-Y. Kim, and J.-K. Seong. A hybrid parallel solver for systems of multivariate polynomials using CPUs and GPUs. *Computer-Aided Design*, 43:1360–1369, 2011.

[PML97]   M. Ponamgi, D. Manocha, and M.C. Lin. Incremental algorithms for collision detection between solid models. *IEEE Trans. Vis. Comp. Graph.*, 3:51–67, 1997.

[Pro97]   X. Provot. Collision and self-collision handling in cloth model dedicated to design garment. *Graphics Interface*, pages 177–189, 1997.

[PZM13]   J. Pan, X. Zhang, and D. Manocha. Efficient penetration computation using active learning. *ACM Trans. Graph.*, 32:1476–1484, 2013.

[Qui94]   S. Quinlan. Efficient distance computation between non-convex objects. In *Proc. IEEE Int. Conf. Robot. Autom.*, pages 3324–3329, 1994.

[RKC00]   S. Redon, A. Kheddar, and S. Coquillart. An algebraic solution to the problem of collision detection for rigid polyhedral objects. *Proc. IEEE Int. Conf. Robot. Autom.*, pages 3733–3738, 2000.

[RKC02]   S. Redon, A. Kheddar, and S. Coquillart. Fast continuous collision detection between rigid bodies. *Comp. Graphics Forum*, 21:279–287, 2002.

[RKLM04]   S. Redon, Y.J. Kim, M.C. Lin, and D. Manocha. Fast continuous collision detection for articulated models. In *Proc. ACM Sympos. Solid Modeling Appl.*, pages 145–156, 2004.

[SAG84]   T.W. Sederberg, D.C. Anderson, and R.N. Goldman. Implicit representation of parametric curves and surfaces. *Comp. Vis. Graphics Image Process.*, 28:72–84, 1984.

[Sam89]   H. Samet. *Spatial Data Structures: Quadtree, Octrees and Other Hierarchical Methods*. Addison Wesley, Reading, 1989.

[Sar83]   R.F. Sarraga. Algebraic methods for intersection. *Comp. Vis. Graphics Image Process.*, 22:222–238, 1983.

[SWF+93]   J.M. Snyder, A.R. Woodbury, K. Fleischer, B. Currin, and A.H. Barr. Interval methods for multi-point collisions between time dependent curved surfaces. In *Proc. 20th Conf. Comp. Graphics Interactive Techniques*, pages 321–334, 1993.

[Sei90]   R. Seidel. Linear programming and convex hulls made easy. In *Proc. 6th Sympos. Comput. Geom.*, pages 211–215, ACM Press, 1990.

[SH76]   M.I. Shamos and D. Hoey. Geometric intersection problems. *Proc. 17th IEEE Sympos. Found. Comp. Sci.*, pages 208–215, 1976.

[SPO10]   S.C. Schvartzman, Á.G. Pérez, and M.A. Otaduy. Star-contours for efficient hierarchical self-collision detection. *ACM Trans. Graph.*, 29:80, 2010.

[SSL02]   F. Schwarzer, M. Saha, and J.-C. Latombe. Exact collision checking of robot paths. In *Algorithmic Foundations of Robotics V*, vol. 7 of Springer Tracts Advanced Robotics, pages 25–41, 2002.

[ST96]   D.E. Stewart and J.C. Trinkle. An implicit time-stepping scheme for rigid body dynamics with inelastic collisions and coulomb friction. *Internat. J. Numer. Methods Eng.*, 39:2673–2691, 1996.

[SWF+93]   J.M. Snyder, A.R. Woodbury, K. Fleischer, B. Currin, and A.H. Barr. Interval methods for multi-point collisions between time-dependent curved surfaces. In *Proc. 20th Conf. Computer Graphics Interactive Techniques*, pages 321–334, ACM Press, 1993.

[TCYM08]   M. Tang, S. Curtis, S.-E. Yoon, and D. Manocha. Interactive continuous collision detection between deformable models using connectivity-based culling. In *Proc. ACM Sympos. Solid Physical Modeling*, pages 25–36, 2008.

[TK14]   M. Tang and Y.J. Kim. Interactive generalized penetration depth computation for rigid and articulated models using object norm. *ACM Trans. Graph.*, 33:1, 2014.

[TKM09]   M. Tang, Y.J. Kim, and D. Manocha. C^2A: Controlled conservative advancement for continuous collision detection of polygonal models. In *Proc. IEEE Int. Conf. Robot. Autom.*, pages 849–854, 2009.

[TKM10]   M. Tang, Y.J. Kim, and D. Manocha. Continuous collision detection for non-rigid contact computations using local advancement. In *Proc. IEEE Int. Conf. Robot. Autom.*, pages 4016–4021, 2010.

[TLK09]   M. Tang, M. Lee, and Y.J. Kim. Interactive Hausdorff distance computation for general polygonal models, *ACM Trans. Graph.*, 28:74, 2009.

[TMK14]   M. Tang, D. Manocha, and Y.J. Kim. Hierarchical and controlled advancement for continuous collision detection of rigid and articulated models. *IEEE Trans. Vis. Comp. Graph.*, 20:755–766, 2014.

[TMLT11]   M. Tang, D. Manocha, J. Lin, and R. Tong. Collision-streams: Fast GPU-based collision detection for deformable models. In *Proc. ACM SIGGRAPH Sympos. Interactive 3D Graphics and Games*, pages 63–70, 2011.

[TMT10a]   M. Tang, D. Manocha, and R. Tong. Fast continuous collision detection using deforming non-penetration filters. In *Proc. ACM SIGGRAPH Sympos. Interactive 3D Graphics and Games*, pages 7–13, 2010.

[TMT10b]   M. Tang, D. Manocha, and R Tong. MCCD: Multi-core collision detection between deformable models using front-based decomposition. *Graphical Models*, 72:7–23, 2010.

[TMY+11]   M. Tang, D. Manocha, S.-E. Yoon, P. Du, J.-P. Heo, and R. Tong. Volccd: Fast continuous collision culling between deforming volume meshes. *ACM Trans. Graph.*, 30:111, 2011.

[TTWM14]   M. Tang, R. Tong, Z. Wang, and D. Manocha. Fast and exact continuous collision detection with Bernstein sign classification. *ACM Trans. Graph.*, 33:186, 2014.

[Wan14]   H. Wang. Defending continuous collision detection against errors. *ACM Trans. Graph.*, 33:122, 2014.

[WLML99]   A. Wilson, E. Larsen, D. Manocha, and M.C. Lin. Partitioning and handling massive models for interactive collision detection. *Comp. Graphics Forum*, 18:319–329, 1999.

[WTTM15]   Z. Wang, M. Tang, R. Tong, and D. Manocha. TightCCD: Efficient and robust continuous collision detection using tight error bounds. In *Comp. Graphics Forum*, 34:289–298, 2015.

[ZK12]   X. Zhang and Y.J. Kim. *k*-IOS: Intersection of spheres for efficient proximity query. In *IEEE Int. Conf. Robot. Autom.*, pages 354–359, 2012.

[ZKM08]   L. Zhang, Y.J. Kim, and D. Manocha. A simple path non-existence algorithm using *c*-obstacle query. In *Algorithmic Foundation of Robotics VII*, vol. 47 of *Springer Tracts Advanced Robotics*, pages 269–284, 2008.

[ZKM07a]   L. Zhang, Y.J. Kim, and D. Manocha. A fast and practical algorithm for generalized penetration depth computation. In *Proc. Robotics: Science and Systems Conf.*, 2007.

[ZKM07b]   L. Zhang, Y.J. Kim, and D. Manocha. C-DIST: efficient distance computation for rigid and articulated models in configuration space. In *ACM Sympos. Solid Physical Modeling*, pages 159–169, 2007.

[ZKVM07]   L. Zhang, Y.J. Kim, G. Varadhan, and D. Manocha. Generalized penetration depth computation. *Computer-Aided Design*, 39:625–638, 2007.

[ZLK06]   X. Zhang, M. Lee, and Y.J. Kim. Interactive continuous collision detection for non-convex polyhedra. *The Visual Computer*, pages 749–760, 2006.

[ZRLK07]   X. Zhang, S. Redon, M. Lee, and Y.J. Kim. Continuous collision detection for articulated models using Taylor models and temporal culling. *ACM Trans. Graph.*, 26:15, 2007.

40 RANGE SEARCHING

Pankaj K. Agarwal

INTRODUCTION

A central problem in computational geometry, range searching arises in many applications, and a variety of geometric problems can be formulated as range-searching problems. A typical range-searching problem has the following form. Let S be a set of n points in $\mathbb{R}^d$, and let R be a family of subsets of $\mathbb{R}^d$; elements of R are called *ranges*. Typical examples of ranges include rectangles, halfspaces, simplices, and balls. *Preprocess S into a data structure so that for a query range $\gamma \in$ R, the points in $S \cap \gamma$ can be reported or counted efficiently.* A single query can be answered in linear time using linear space, by simply checking for each point of S whether it lies in the query range. Most of the applications, however, call for querying the same set S numerous times, in which case it is desirable to answer a query faster by preprocessing S into a data structure.

Range counting and range reporting are just two instances of range-searching queries. Other examples include *range-emptiness queries*: determine whether $S \cap \gamma = \emptyset$; and *range-min/max queries*: each point has a weight and one must return the point in the query range with the minimum/maximum weight. Many different types of range-searching queries can be encompassed in the following general formulation of the range-searching problem:

Let $(\mathbf{S}, +)$ be a commutative semigroup. Each point $p \in S$ is assigned a weight $w(p) \in \mathbf{S}$. For any subset $S' \subseteq S$, let $w(S') = \sum_{p \in S'} w(p)$, where addition is taken over the semigroup.[1] For a query range $\gamma \in$ R, compute $w(S \cap \gamma)$. For example, counting queries can be answered by choosing the semigroup to be $(\mathbb{Z}, +)$, where $+$ denotes standard integer addition, and setting $w(p) = 1$ for every $p \in S$; emptiness queries by choosing the semigroup to be $(\{0, 1\}, \vee)$ and setting $w(p) = 1$; reporting queries by choosing the semigroup to be $(2^S, \cup)$ and setting $w(p) = \{p\}$; and range-max queries by choosing the semigroup to be $(\mathbb{R}, \max)$ and choosing $w(p)$ to be the weight of p.

A more general (decomposable) *geometric-searching* problem can be defined as follows: Let S be a set of *objects* in $\mathbb{R}^d$ (e.g., points, hyperplanes, balls, or simplices), $(\mathbf{S}, +)$ a commutative semigroup, $w : S \to \mathbf{S}$ a weight function, R a set of ranges, and $\Diamond \subseteq S \times$ R a "spatial" relation between objects and ranges. For a query range $\gamma \in$ R, the goal is to compute $\sum_{p \Diamond \gamma} w(p)$. Range searching is a special case of this problem in which S is a set of points in $\mathbb{R}^d$ and $\Diamond = \in$. Another widely studied searching problem is ***intersection searching***, where $p \Diamond \gamma$ if p intersects γ. As we will see below, range-searching data structures are useful for many other geometric searching problems.

The performance of a data structure is measured by the time spent in answering a query, called the *query time* and denoted by $Q(n)$; by the *size* of the data structure, denoted by $S(n)$; and by the time spent in constructing in the data struc-

---

[1]Since $\mathbf{S}$ need not have an additive identity, we may need to assign a special value *nil* to the empty sum.

ture, called the *preprocessing time* and denoted by $P(n)$. Since the data structure is constructed only once, its query time and size are generally more important than its preprocessing time. If a data structure supports insertion and deletion operations, its *update time* is also relevant. The query time of a range-reporting query on any reasonable machine depends on the output size, so the query time for a range-reporting query consists of two parts—*search time*, which depends only on n and d, and *reporting time*, which depends on n, d, and the output size. Throughout this chapter, k will be used to denote the output size.

We assume d to be a small constant, and big-O and big-Ω notation hide constants depending on d. The dependence on d of the performance of almost all the data structures mentioned in this chapter is exponential, which makes them unsuitable in practice for large values of d.

The size of any range-searching data structure is at least linear, since it has to store each point (or its weight) at least once. Assuming the coordinates of input points to be real numbers, the query time in any reasonable model of computation such as pointer machines, RAM, or algebraic decision trees is $\Omega(\log n)$ even when $d = 1$ (faster query time is possible if the coordinates are integers, say, bounded by $n^{O(1)}$). Therefore, a natural question is whether a linear-size data structure with logarithmic query time exists for range searching. Although near-linear-size data structures are known for orthogonal range searching in any fixed dimension that can answer a query in polylogarithmic time, no similar bounds are known for range searching with more complex ranges such as simplices or disks. In such cases, one seeks a trade-off between the query time and the size of the data structure—how fast can a query be answered using $O(n\mathrm{polylog}(n))$ space, how much space is required to answer a query in $O(\mathrm{polylog}(n))$ time, and what kind of space/query-time trade-off can be achieved?

40.1   MODELS OF COMPUTATION

Most geometric algorithms and data structures are implicitly described in the familiar ***random access machine*** (RAM) model, or in the ***real RAM*** model. In the traditional RAM model, if the coordinates are integers in the range $[0{:}U]$,[2] for some $U \geq n$, then memory cells can contain arbitrary $\omega := O(\log U)$ bit long integers, called *words*, which can be added, multiplied, subtracted, divided (computing $\lfloor x/y \rfloor$), compared, and used as pointers to other memory cells in constant time. The real RAM model allows each memory cell to store arbitrary real numbers, and it supports constant-time arithmetic and relational operations between two real numbers, though conversions between integers and reals are not allowed. In the case of range searching over a semigroup other than integers, memory cells are allowed to contain arbitrary values from the semigroup, but only the semigroup-addition operations can be performed on them.

Many range-searching data structures are described in the more restrictive ***pointer-machine model***. The main difference between the RAM and the pointer-machine models is that on a pointer machine, a memory cell can be accessed only through a series of pointers, while in the RAM model, any memory cell can be accessed in constant time. In the basic pointer-machine model, a data structure

---

[2]For $b \geq a$, we use $[a{:}b]$ to denote the set of integers between a and b.

is a directed graph with out-degree 2; each node is associated with a label, which is an integer between 0 and n. Nonzero labels are indices of the points in S, and the nodes with label 0 store auxiliary information. The query algorithm traverses a portion of the graph and visits at least one node with label i for each point p_i in the query range. Chazelle [Cha88] defines several generalizations of the pointer-machine model that are more appropriate for answering counting and semigroup queries. In these models, nodes are labeled with arbitrary $O(\log n)$-bit integers, and the query algorithm is allowed to perform arithmetic operations on these integers.

The **cell probe model** is the most basic model for proving lower bounds on data structures [Yao81]. In this model, a data structure consists of a set of memory cells, each storing ω bits. Each cell is identified by an integer address, which fits in ω bits. The data structure answers a query by probing a number of cells from the data structure and returns the correct answer based on the contents of the probed cells. It handles an update operation by reading and updating (probing) a number of cells to reflect the changes. The cost of an operation is the number of cells probed by the data structure to perform that operation.

The best lower bound on the query time one can hope to prove in the cell probe model is $\Omega(\text{polylog}(n))$, which is far from the best-known upper bounds. Extensive work has been done on proving lower bounds in the **semigroup arithmetic model**, originally introduced by Fredman [Fre81a] and refined by Yao [Yao85]. In this model, a data structure can be regarded informally as a set of precomputed partial sums in the underlying semigroup. The size of the data structure is the number of sums stored, and the query time is the minimum number of semigroup operations required (on the precomputed sums) to compute the answer to a query. The query time ignores the cost of various auxiliary operations, including the cost of determining which of the precomputed sums should be added to answer a query.

The informal model we have just described is much too powerful. For example, the optimal data structure for range-counting queries in this semigroup model consists of the $n+1$ integers $0, 1, \ldots, n$. A counting query can be answered by simply returning the correct answer. Since no additions are required, a query can be answered in zero "time," using a linear-size data structure. The notion of **faithful** semigroup circumvents this problem: A commutative semigroup $(\mathbf{S}, +)$ is faithful if for each $n > 0$, for any sets of indices $I, J \subseteq [1:n]$ where $I \neq J$, and for every sequence of positive integers α_i, β_j ($i \in I, j \in J$), there are semigroup values $s_1, s_2, \ldots, s_n \in \mathbf{S}$ such that $\sum_{i \in I} \alpha_i s_i \neq \sum_{j \in J} \beta_j s_j$. For example, $(\mathbb{Z}, +)$, $(\mathbb{R}, \min)$, $(\mathbb{N}, \gcd)$, and $(\{0, 1\}, \vee)$ are faithful, but $(\{0, 1\}, + \bmod 2)$ is not faithful.

Let $S = \{p_1, p_2, \ldots, p_n\}$ be a set of objects, $\mathbf{S}$ a faithful semigroup, R a set of ranges, and $\Diamond$ a relation between objects and ranges. (Recall that in the standard range-searching problem, the objects in S are points, and $\Diamond$ is containment.) Let $x_1, x_2, \ldots, x_n$ be a set of n variables over $\mathbf{S}$, each corresponding to an object in S. A *generator* $g(x_1, \ldots, x_n)$ is a linear form $\sum_{i=1}^n \alpha_i x_i$, where α_i's are non-negative integers, not all zero. (In practice, the coefficients α_i are either 0 or 1.) A *storage scheme* for $(S, \mathbf{S}, \mathsf{R}, \Diamond)$ is a collection of generators $\{g_1, g_2, \ldots, g_s\}$ with the following property: For any query range $\gamma \in \mathsf{R}$, there is a set of indices $I_\gamma \subseteq [1:s]$ and a set of labeled nonnegative integers $\{\beta_i \mid i \in I_\gamma\}$ such that the linear forms $\sum_{p_i \Diamond \gamma} x_i$ and $\sum_{i \in I_\gamma} \beta_i g_i$ are identically equal. In other words, the equation

$$\sum_{p_i \Diamond \gamma} w(p_i) = \sum_{i \in I_\gamma} \beta_i g_i(w(p_1), w(p_2), \ldots, w(p_n))$$

holds for *any* weight function $w : S \to \mathbf{S}$. (Again, in practice, $\beta_i = 1$ for all $i \in I_\gamma$.) The size of the smallest such set I_γ is the query time for γ; the time to actually choose the indices I_γ is ignored. The space used by the storage scheme is measured by the number of generators. There is no notion of preprocessing time in this model.

A serious weakness of the semigroup model is that it does not allow subtractions even if the weights of points belong to a group (e.g., range counting). Therefore the **group model** has been proposed in which each point is assigned a weight from a commutative group and the goal is to compute the group sum of the weights of points lying in a query range. The data structure consists of a collection of group elements and auxiliary data, and it answers a query by adding and subtracting a subset of the precomputed group elements to yield the answer to the query. The query time is the number of group operations performed [Fre82, Cha98].

The lower-bound proofs in the semigroup model have a strong geometric flavor because subtractions are not allowed: the query algorithm can use a precomputed sum that involves the weight of a point p only if p lies in the query range. A typical proof basically reduces to arguing that not all query ranges can be "covered" with a small number of subsets of input objects [Cha01]. Unfortunately, no such property holds for the group model, which makes proving lower bounds in the group model much harder. Notwithstanding recent progress, the known lower bounds in the group model are much weaker than those under the semigroup model.

Almost all geometric range-searching data structures are constructed by subdividing space into several regions with nice properties and recursively constructing a data structure for each region. Queries are answered with such a data structure by performing a depth-first search through the resulting recursive space partition. The **partition-graph** model, introduced by Erickson [Eri96a, Eri96b], formalizes this divide-and-conquer approach. This model can be used to study the complexity of emptiness queries, which are trivial in semigroup and pointer-machine models.

We conclude this section by noting that most of the range-searching data structures discussed in this paper (halfspace range-reporting data structures being a notable exception) are based on the following general scheme. Given a point set S, the structure precomputes a family $\mathsf{F} = \mathsf{F}(S)$ of *canonical subsets* of S and stores the weight $w(C) = \sum_{p \in C} w(p)$ of each canonical subset $C \in \mathsf{F}$. For a query range γ, the query procedure determines a partition $\mathsf{C}_\gamma = \mathsf{C}(S, \gamma) \subseteq \mathsf{F}$ of $S \cap \gamma$ and adds the weights of the subsets in C_γ to compute $w(S \cap \gamma)$. We refer to such a data structure as a **decomposition scheme**.

There is a close connection between the decomposition schemes and the storage schemes of the semigroup model described earlier. Each canonical subset $C = \{p_i \mid i \in I\} \in \mathsf{F}$, where $I \subseteq [1{:}n]$, corresponds to the generator $\sum_{i \in I} x_i$. How exactly the weights of canonical subsets are stored and how C_γ is computed depends on the model of computation and on the specific range-searching problem. In the semigroup (or group) arithmetic model, the query time depends only on the number of canonical subsets in C_γ, regardless of how they are computed, so the weights of canonical subsets can be stored in an arbitrary manner. In more realistic models of computation, however, some additional structure must be imposed on the decomposition scheme in order to efficiently compute C_γ. In a *hierarchical* decomposition scheme, the canonical subsets and their weights are organized in a tree T. Each node v of T is associated with a canonical subset $C_v \in \mathsf{F}$, and if w is a child of v in T then $C_w \subseteq C_v$. Besides the weight of C_v, some auxiliary information is also stored at v, which is used to determine whether $C_v \in \mathsf{C}_\gamma$ for a query range γ. If the weight of each canonical subset can be stored in $O(1)$

memory cells and if one can determine in $O(1)$ time whether $C_w \in \mathsf{C}_\gamma$ where w is a descendent of a given node v, the hierarchical decomposition scheme is called *efficient*. The total size of an efficient decomposition scheme is simply $O(|\mathsf{F}|)$. For range-reporting queries, in which the "weight" of a canonical subset is the set itself, the size of the data structure is $O(\sum_{C \in \mathsf{F}} |C|)$, but it can be reduced to $O(|\mathsf{F}|)$ by storing the canonical subsets implicitly. Finally, let $r > 1$ be a parameter, and set $\mathsf{F}_i = \{C \in \mathsf{F} \mid r^{i-1} \leq |C| \leq r^i\}$. A hierarchical decomposition scheme is called *r-convergent* if there exist constants $\alpha \geq 1$ and $\beta > 0$ so that the degree of every node in T is $O(r^\alpha)$ and for all $i \geq 1$, $|\mathsf{F}_i| = O((n/r^i)^\alpha)$ and, for all query ranges γ, $|\mathsf{C}_\gamma \cap \mathsf{F}_i| = O((n/r^i)^\beta)$, i.e., the number of canonical subsets in the data structure and in any query output decreases exponentially with their size. We will see below in Section 40.5 that r-convergent hierarchical decomposition schemes can be cascaded together to construct multi-level structures that answer complex geometric queries.

To compute $\sum_{p_i \in \gamma} w(p_i)$ for a query range γ using a hierarchical decomposition scheme T, a query procedure performs a depth-first search on T, starting from its root. At each node v, using the auxiliary information stored at v, the procedure determines whether γ contains C_v, whether γ *crosses* C_v (i.e., γ intersects C_v but does not contain C_v), or whether γ is disjoint from C_v. If γ contains C_v, then C_v is added to C_γ (rather, the weight of C_v is added to a running counter). Otherwise, if γ intersects C_v, the query procedure identifies a subset of children of v, say $\{w_1, \ldots, w_a\}$, so that the canonical subsets $C_{w_i} \cap \gamma$, for $1 \leq i \leq a$, form a partition of $C_v \cap \gamma$. Then the procedure searches each w_i recursively. The total query time is $O(\log n + |\mathsf{C}_\gamma|)$ if the decomposition scheme is r-convergent for some constant $r > 1$ and constant time is spent at each node visited.

40.2 ORTHOGONAL RANGE SEARCHING

Query ranges in d-dimensional orthogonal range searching are axis-aligned rectangles of the form $\prod_{i=1}^d [a_i, b_i]$. Multi-key searching in database systems can be formulated as orthogonal range searching. For example, the points of S may correspond to employees of a company, each coordinate corresponding to a key such as age and salary. A query of the form—report all employees between the ages of 30 and 40 who earn between \$50,000 and \$75,000—can be formulated as an orthogonal range-reporting query with query rectangle being $[30, 40] \times [5 \times 10^4, 7.5 \times 10^4]$.

UPPER BOUNDS

Most orthogonal range-searching data structures with polylog(n) query time under the pointer-machine model are based on **range trees**, introduced by Bentley [Ben80]. For a set S of n points in $\mathbb{R}^2$, the range tree T of S is a minimum-height binary tree with n leaves whose ith leftmost leaf stores the point of S with the ith smallest x-coordinate. Each interior node v of T is associated with a canonical subset $C_v \subseteq S$ containing the points stored at leaves in the subtree rooted at v. Let a_v (resp. b_v) be the smallest (resp. largest) x-coordinate of any point in C_v. The interior node v stores the values a_v and b_v and the set C_v in an array sorted by the y-coordinates of its points. The size of T is $O(n \log n)$, and it can be constructed in time $O(n \log n)$. The range-reporting query for a rectangle $R = [a_1, b_1] \times [a_2, b_2]$ can be answered by traversing T in a top-down manner as follows. Suppose a node v is being visited by the query procedure. If v is a leaf and R contains the point of C_v, then report the point. If v is an interior node and the interval $[a_v, b_v]$ does not intersect $[a_1, b_1]$, there is nothing to do. If $[a_v, b_v] \subseteq [a_1, b_1]$, report all the points of C_v whose y-coordinates lie in the interval $[a_2, b_2]$, by performing a binary search. Otherwise, recursively visit both children of v. The query time of this procedure is $O(\log^2 n + k)$, which can be improved to $O(\log n + k)$ using *fractional cascading* [BCKO08]. A d-dimensional range tree can be extended to $\mathbb{R}^{d+1}$ by using a multi-level structure and by paying a $\log n$ factor in storage and query time.

Since the original range-tree paper by Bentley, several data structures with improved bounds have been proposed; see Table 40.2.1 for a summary of known results. If queries are "3-sided rectangles" in $\mathbb{R}^2$, say, of the form $[a_1, b_1] \times [a_2, \infty)$, then a *priority search tree* of size $O(n)$ can answer a range-reporting query in time $O(\log n + k)$ [McC85]. Chazelle [Cha86] showed that an orthogonal range reporting query in $\mathbb{R}^2$ can be answered in $O(\log n + k)$ time using $O(n\text{Lg}n)$ space, where $\text{Lg}n = \log_{\log n} n = \log n / \log \log n$, by constructing a range tree of $O(\log n)$ fanout and storing additional auxiliary structures at each node. For $d = 3$, Afshani et al. [AAL10] proposed a data structure of $O(n\text{Lg}^2 n)$ size with $O(\log n + k)$ query time. Afshani et al. [AAL09, AAL10] presented a recursive data structure for orthogonal range reporting, based on range tree, that can be extended from $\mathbb{R}^d$ to $\mathbb{R}^{d+1}$ by paying a cost of $\text{Lg}n$ in both space and query time. For $d = 4$, a query thus can be answered in $O(\log n\text{Lg}n + k)$ time using $O(n\text{Lg}^3 n)$ space. Afshani et al. [AAL12] presented a slightly different data structure of size $O(n\text{Lg}^4 n)$ that answers a 4D query in $O(\log n\sqrt{\text{Lg}n} + k)$ time. For $d \geq 4$, a query can be answered in $O(\log(n)\text{Lg}^{d-3}n + k)$ time using $O(n\text{Lg}^{d-1}n)$ space, or in $O(\log(n)\text{Lg}^{d-4+1/(d-2)}n + k)$ time using $O(n\text{Lg}^d n)$ space. The preprocessing time of these data structures is $O(n \log^2 n\text{Lg}^{d-3}n)$ and $O(n \log^3 n\text{Lg}^{d-3}n)$, respectively [AT14]. More refined bounds on the query time can be found in [AAL10].

There is extensive work on range-reporting data structures in the RAM model when $S \subset [0{:}U]^d$ for some integer $U \geq n$, assuming that each memory cell can store a word of length $\omega = O(\log U)$. For $d = 1$, a range-reporting query can be answered in $O(\log \omega + k)$ time using the van Emde Boas tree of linear size [EBKZ77]. Alstrup et al. [ABR01] proposed a linear-size data structure that can answer a range reporting query in $O(k + 1)$ time. In contrast, any data structure of size $n^{O(1)}$ for finding the predecessor in S of a query point has $\Omega(\log \omega)$ query time [Ajt88].

Using standard techniques, the universe $[0{:}U]^d$ can be reduced to $[0{:}n]^d$ by paying $O(n)$ space and $O(\log \log U)$ additional time in answering a query. So for $d \geq 2$, we assume that $U = n$ and that the coordinates of the query rectangle are also integers in the range $[0{:}n]$; we refer to this setting as the *rank space*.

For $d = 2$, the best known approach can answer a range-reporting query in $O((1 + k) \log \log n)$ time using $O(n \log \log n)$ space, in $O((1 + k) \log^\epsilon n)$ time using $O(n)$ space, or in $O(\log \log n + k)$ time using $O(n \log^\epsilon n)$ space [CLP11], where $\epsilon > 0$ is an arbitrarily small constant. For $d = 3$, a query can be answered in $O(\log \log n + k)$ time using $O(n \log^{1+\epsilon} n)$ space; if the query range is an octant, then the size of the data structure is only linear. These data structures can be extended to higher dimensions by paying a factor of $\log^{1+\epsilon} n$ in space and $\text{Lg}n$ in query time per dimension. Therefore a d-dimensional query can be answered in $O(\text{Lg}^{d-3}n \log \log n + k)$ using $O(n \log^{d-2+\epsilon} n)$ space.

The classical range-tree data structure can answer a 2D range-counting query in $O(\log n)$ time using $O(n \log n)$ space. Chazelle [Cha88] showed that the space can be improved to linear in the generalized pointer-machine or the RAM model by using a compressed range tree. For $d \geq 2$, by extending Chazelle's technique, a range-counting query in the rank space can be answered in $O(\mathrm{Lg}^{d-1} n)$ time using $O(n\mathrm{Lg}^{d-2} n)$ space [JMS04]. Such a data structure can be constructed in time $O(n \log^{d-2+1/d} n)$ [CP10]. Chan and Wilkinson [CW13] presented an adaptive data structure of $O(n \log \log n)$ size that can answer a 2D range-counting query in $O(\log \log n + \mathrm{Lg}k)$, where k is the number of points in the query range. Their technique can also be used to answer range-counting queries approximately: it returns a number k' with $k \leq k' \leq (1 + \delta)k$, for a given constant $\delta > 0$, in $O(\log \log n)$ time using $O(n \log \log n)$ space, or in $O(\log^\epsilon n)$ time using $O(n)$ space. If an additive error of ϵn is allowed, then an approximate range-counting query can be answered using $O(\frac{1}{\epsilon} \log(\frac{1}{\epsilon}) \log \log(\frac{1}{\epsilon}) \log n)$ space, and this bound is tight within $O(\log \log \frac{1}{\epsilon})$ factor [WY13].

In an off-line setting, where all queries are given in advance, n d-dimensional range-counting queries can be answered in $O(n \log^{d-2+1/d} n)$ time [CP10].

All the data structures discussed above can handle insertion/deletion of points using the standard partial-rebuilding technique [Ove83]. If the preprocessing time of the data structure is $P(n)$, then a point can be inserted into or deleted from the data structure in $O((P(n)/n) \log n)$ amortized time. The update time can be made worst-case using the known de-amortization techniques [DR91]. Faster dynamic data structures have been developed for a few cases, especially under the RAM model when $S \subseteq [0{:}U]^d$ and $d = 1, 2$. For example, Mortensen et al. [MPP05] dynamized the 1D range-reporting data structure with $O(\log \omega)$ update time and $O(\log \log \omega + k)$ query time.

TABLE 40.2.1 Orthogonal range reporting upper bounds; $h \in [1, \log^\epsilon n]$ and $\mathrm{Lg}n = \frac{\log n}{\log \log n}$.

d	Pointer Machine		RAM	
	$S(n)$	$Q(n)$	$S(n)$	$Q(n)$
$d = 1$	n	$\log n + k$	n	$1 + k$
$d = 2$	$n\mathrm{Lg}n$	$\log n + k$	$nh \log \log n$ $n \log_h \log n$	$\log \log n + k \log_h \log n$ $(1 + k)h \log \log n$
$d = 3$	$n\mathrm{Lg}^2 n$	$\log n + k$	$n \log^{1+\epsilon} n$	$\log \log n + k$
$d = 4$	$n\mathrm{Lg}^3 n$ $n\mathrm{Lg}^4 n$	$\log n\mathrm{Lg}n + k$ $\log n\mathrm{Lg}^{1/2} n + k$	$n \log^{2+\epsilon} n$	$\log n + k$
$d \geq 4$	$n\mathrm{Lg}^{d-1} n$ $n\mathrm{Lg}^d n$	$\log n\mathrm{Lg}^{d-3} n + k$ $\log n\mathrm{Lg}^\alpha n + k$ $\alpha = d - 4 + \frac{1}{d-2}$	$n \log^{d-2+\epsilon} n$	$\log n\mathrm{Lg}^{d-4} n + k$

LOWER BOUNDS

Semigroup model. Fredman [Fre80, Fre81a] was the first to prove nontrivial lower bounds on orthogonal range searching, but in a dynamic setting in which points can be inserted and deleted. He showed that a mixed sequence of n insertions, deletions, and queries takes $\Omega(n \log^d n)$ time under the semigroup model. Chazelle [Cha90b] proved lower bounds for the static version of orthogonal range searching, which almost match the best upper bounds known. He showed that there exists a set S of n weighted points in $\mathbb{R}^d$, with weights from a faithful semigroup, such that the worst-case query time, under the semigroup model, for an orthogonal range-searching data structure of size nh is $\Omega((\log_h n)^{d-1})$. If the data structure also supports insertions of points, the query time is $\Omega((\log_h n)^d)$. These lower bounds hold even if the queries are orthants instead of rectangles, i.e., for the so-called *dominance queries*. In fact, they apply to answering the dominance query for a randomly chosen query point. It should be pointed out that the bounds in [Cha90b] assume the weights of points in S to be a part of the input, i.e., the data structure is not tailored to a special set of weights. It is conceivable that a faster algorithm can be developed for answering orthogonal range-counting queries, exploiting the fact that the weight of each point is 1 in this case.

Group model. Pătraşcu [Păt07] proved that a dynamic data structure in the group model for 2D dominance counting queries that supports updates in expected time t_u requires $\Omega((\frac{\log n}{\log(t_u \log n)})^2)$ expected time to answer a query. Larsen [Lar14] proved lower bounds for dynamic range searching data structures in the group model in terms of combinatorial discrepancy of the underlying set system. For d-dimensional orthogonal range searching, any dynamic data structure with the worst-case t_u and t_q update and query time, respectively, requires $t_u t_q = \Omega(\log_\omega^{d-1} n)$ [LN12].

Cell-probe model. Pătraşcu [Păt07, Păt11] and Larsen [Lar12] also proved lower bounds on orthogonal range searching in the cell probe model. In particular, Pătraşcu proved that a data structure for 2D dominance counting queries that uses nh space requires $\Omega(\log_{h\omega} n)$ query time, where $\omega = \Omega(\log n)$ is the size of each memory cell. He also proved the same lower bound for 4D orthogonal range reporting. Note that for $h = n\text{polylog}n$ and $\omega = \log n$, the query time is $\Omega(\text{Lg}n)$ using $O(n\text{polylog}n)$ space, which almost matches the best known upper bound of $O(\log n + k)$ using $O(n\log^{2+\epsilon} n)$ space. In contrast, a 3D orthogonal range query can be answered in $O(\log\log n + k)$ time using $O(n\log^{1+\epsilon} n)$ space.

Larsen [Lar12] proved that a dynamic data structure for the 2D weighted range counting problem with t_u worst-case update time has $\Omega((\log_{\omega t_u} n)^2)$ expected query time; the weight of each point is an $O(\log n)$-bit integer. Note that for $t_u = \text{polylog}(n)$ and $\omega = \Theta(\log n)$, the expected query time is $\Omega(\text{Lg}^2 n)$.

Pointer-machine model. For range reporting queries, Chazelle [Cha90a] proved that the size of any data structure on a pointer machine that answers a d-dimensional range-reporting query in $O(\text{polylog}(n) + k)$ time is $\Omega(n\text{Lg}^{d-1}n)$. Notice that this lower bound is greater than the known upper bound for answering two-dimensional reporting queries on the RAM model. Afshani et al. [AAL10, AAL12] adapted Chazelle's technique to show that a data structure for range-reporting queries that uses nh space requires $\Omega(\log(n) \log_h^{\lfloor d/2 \rfloor - 2} n + k)$ time to answer a query.

Off-line searching. These lower bounds do not hold for off-line orthogonal range searching, where given a set of n weighted points in $\mathbb{R}^d$ and a set of n rectangles, one wants to compute the weight of points in each rectangle. Chazelle [Cha97] proved that the off-line version takes $\Omega(n\text{Lg}^{d-1}n)$ time in the semigroup model and $\Omega(n\log\log n)$ time in the group model. For $d = \Omega(\log n)$ (resp. $d = \Omega(\text{Lg}n)$), the lower bound for the off-line range-searching problem in the group model can be improved to $\Omega(n\log n)$ (resp. $\Omega(n\text{Lg}n)$) [CL01].

PRACTICAL DATA STRUCTURES

None of the data structures described above are used in practice, even in two dimensions, because of the polylogarithmic overhead in their size. For a data structure to be used in real applications, its size should be at most cn, where c is a very small constant, the time to answer a typical query should be small—the lower bounds mentioned earlier imply that we cannot hope for small worst-case bounds—and it should support insertions and deletions of points. Keeping these goals in mind, a plethora of data structures have been proposed.

Many practical data structures construct a recursive partition of space, typically into rectangles, and a tree induced by this partition. The simplest example of this type of data structure is the ***quad tree*** [Sam90]. A quad tree in $\mathbb{R}^2$ is a 4-way tree, each of whose nodes v is associated with a square R_v. R_v is partitioned into four equal-size squares, each of which is associated with one of the children of v. The squares are partitioned until at most one point is left inside a square. A range-search query can be answered by traversing the quad tree in a top-down fashion. Because of their simplicity, quad trees are one of the most widely used data structures for a variety of problems. One disadvantage of quad trees is that arbitrarily many levels of partitioning may be required to separate tightly clustered points. The ***compressed quad tree*** guarantees its size to be linear, though the depth can be linear in the worst case [Sam90].

Quad trees and their variants construct a grid on a square containing all the input points. One can instead partition the enclosing rectangle into two rectangles by drawing a horizontal or a vertical line and partitioning each of the two rectangles independently. This is the idea behind the *kd-**tree*** data structure of Bentley [Ben75]. In particular, a *kd*-tree is a binary tree, each of whose nodes v is associated with a rectangle R_v. If R_v does not contain any point in its interior, v is a leaf. Otherwise, R_v is partitioned into two rectangles by drawing a horizontal or vertical line so that each rectangle contains at most half of the points; splitting lines are alternately horizontal and vertical. A *kd*-tree answers a d-dimensional orthogonal range-reporting query in $O(n^{1-1/d} + k)$ using linear space; in fact, each point is stored only once.

Inserting/deleting points dynamically into a *kd*-tree is expensive, so a few variants of *kd*-trees have been proposed that can update the structure in $O(\text{polylog} n)$ time and can answer a query in $O(\sqrt{n} + k)$ time. On the practical side, many alternatives of *kd*-trees have been proposed to optimize space and query time, most notably ***buddy tree*** [SRF87] and ***hB-tree*** [LS90, ELS97]. A buddy tree is a combination of quad- and *kd*-trees in the sense that rectangles are split into sub-rectangles only at some specific locations, which simplifies the split procedure. In an hB-tree, the region associated with a node is allowed to be $R_1 \setminus R_2$ where R_1 and R_2 are rectangles. The same idea has been used in the ***BBd-tree***, a data structure developed for answering approximate nearest-neighbor and range queries [AMN+98].

Several extensions of *kd*-trees, such as random projection trees [DF08], randomized partition trees [DS15], randomly oriented *kd*-trees [Vem12], and cover trees [BKL06], have been proposed to answer queries on a set S of points in $\mathbb{R}^d$ when d is very high but the intrinsic dimension of S is low. A popular notion of intrinsic dimension of S is its *doubling dimension*, i.e., the smallest integer b such that for every $x \in \mathbb{R}^d$ and for every $r > 0$, the points of S within distance r from x can be covered by 2^b balls of radius $r/2$. The performance of these data structures on certain queries depends exponentially only on the doubling dimension of S.

PARTIAL-SUM QUERIES

Partial-sum queries require preprocessing a d-dimensional array A with n entries, in an additive semigroup, into a data structure, so that for a d-dimensional rectangle $\gamma = [a_1, b_1] \times \cdots \times [a_d, b_d]$, the sum

$$\sigma(A, \gamma) = \sum_{(k_1, k_2, \ldots, k_d) \in \gamma} A[k_1, k_2, \ldots, k_d]$$

can be computed efficiently. In the off-line version, given A and m rectangles $\gamma_1, \gamma_2, \ldots, \gamma_m$, we wish to compute $\sigma(A, \gamma_i)$ for each i. Partial-sum queries are a special case of orthogonal range queries, where the points lie on a regular d-dimensional lattice.

Partial-sum queries are widely used for on-line analytical processing (OLAP) of commercial databases. OLAP allows companies to analyze aggregate databases built from their data warehouses. A popular data model for OLAP applications is the multidimensional database called **data cube** [GBLP96] that represents the data as a d-dimensional array. An aggregate query on a data cube can be formulated as a partial-sum query. Driven by this application, several heuristics have been proposed to answer partial-sum queries on data cubes; see [HBA97, VW99] and the references therein.

If the sum is a group operation, then the query can be answered in $O(1)$ time in any fixed dimension by maintaining the prefix sums and using the inclusion-exclusion principle. Yao [Yao82] showed that, for $d = 1$, a partial-sum query where sum is a semigroup operation can be answered in $O(\alpha(n))$ time using $O(n)$ space; here $\alpha(n)$ is the inverse Ackermann function. If the sum operation is *max* or *min*, then a partial-sum query can be answered in $O(1)$ time under the RAM model [FH06, Vui80].

For $d > 1$, Chazelle and Rosenberg [CR89] developed a data structure of size $O(n \log^{d-1} n)$ that can answer a partial-sum query in time $O(\alpha(n) \log^{d-2} n)$. They also showed that the off-line version that answers m given partial-sum queries on n points takes $\Omega(n + m\alpha(m, n))$ time for any fixed $d \geq 1$. If the values in the array can be updated, then Fredman [Fre82] proved a lower bound of $\Omega(\log n / \log \log n)$ on the maximum of update and query time. The lower bound was later improved to $\Omega(\log n)$ by Pătrașcu and Demaine [PD06].

RANGE-STATISTICS QUERIES

The previous subsections focused on two versions of range searching—reporting and counting. In many applications, especially when the input data is large, neither of the two is quite satisfactory—there are too many points in the query range to report, and a simple count (or weighted sum) of the number of points gives too little information. There is recent work on computing some statistics on points lying in a query rectangle, mostly for $d = 1, 2$.

Let S be a set of weighted points. The simplest range statistics query is the *range min/max* query: return a point of the minimum/maximum weight in a query rectangle. If the coordinates of input points are real numbers, then the classical range tree can answer a range min/max query in time $O(\log^{d-1} n)$ using $O(n \log^{d-1} n)$ space. Faster data structures have been developed for answering range-min/max queries in the rank space. For $d = 1$, as mentioned above, a range min/max query can be answered in $O(1)$ time using $O(n)$ space in the RAM model. For $d = 2$, a range min/max query can be answered in $O(\log \log n)$ time using $O(n \log^\epsilon n)$ space [CLP11].

Recently, there has been some work on the *range-selection* query for $d = 1$: let $S = [1{:}n]$ and w_i be the weight of point i. Given an interval $[i{:}j]$ and an integer $r \in [1{:}j{-}i{+}1]$, return the r-th smallest value in $w_i, \ldots, w_j$. Chan and Wilkinson [CW13] have described a data structure of linear size that can answer the query in time $O(1 + \log_\omega r)$ in the ω-bit RAM model, and Jørgensen and Larsen [JL11] proved that this bound is tight in the worst case, i.e., any data structure that uses nh space, where $h \geq 1$ is a parameter, requires $\Omega(\log_{\omega h} n)$ time to answer a query. For $d \geq 2$, Chan and Zhou [CZ15] showed that a range-selection query can be answered in $O(\text{Lg}^d n)$ time using $O(n \text{Lg}^{d-1} n)$ space, where as before $\text{Lg} n = \log n / \log \log n$. A variant of range-selection query is the **top-*r* query** where the goal is to report the r points inside a query range with largest/smallest weights. By adapting the range-reporting structures, several data structures for answering top-r queries have been proposed [RT15, RT16]. For example, a 2D orthogonal top-r query can be answered in time $O(\log n + r)$ using $O(n \text{Lg} n)$ space.

A 1D *range-mode* query, i.e., given an interval $[i{:}j]$ return the mode of $w_i, \ldots, w_j$, can be computed in $O(\sqrt{n/\log n})$ time [CDL+14]. A close relationship between range-mode queries and Boolean matrix multiplication implies that any data structure for range-mode queries must have either $\Omega(n^{t/2})$ preprocessing time or $\Omega(n^{t/2-1})$ query time in the worst case, where t is the exponent in the running time of a matrix-multiplication algorithm; the best known matrix-multiplication algorithm has exponent 2.3727.

Rahul and Janardan [RJ12] considered the problem of reporting the set of maximal points (i.e., the points that are not dominated by any other point) in a query range, the so-called *skyline query*. They presented a data structure of size $O(n \log^{d+1} n)$ that can report all maximal points in a query rectangle in time $O(k \log^{d+2} n)$. For $d = 2$, Brodal and Larsen [BL14] show that a skyline-reporting query in the rank space can be answered in $O(\text{Lg} n + k)$ time using $O(n \log^\epsilon n)$ space or in $O(\text{Lg} n + k \log \log n)$ time using $O(n \log \log n)$ space. They also propose a linear-size data structure that answers a skyline-counting query in $O(\text{Lg} n)$ time, and show that this bound is optimal in the worst case.

OPEN PROBLEMS

1. For $d > 2$, prove a lower bound on the query time for orthogonal range counting that depends on d.

2. Can a range-reporting query for $d = 4, 5$ under the pointer-machine model be answered in $O(\log n + k)$ time using $n\text{polylog}(n)$ space?

3. How fast a skyline-counting query be answered for $d \geq 3$?

40.3 SIMPLEX RANGE SEARCHING

Unlike orthogonal range searching, no simplex range-searching data structure is known that can answer a query in polylogarithmic time using near-linear storage.

In fact, the lower bounds stated below indicate that there is little hope of obtaining such a data structure, since the query time of a linear-size data structure, under the semigroup model, is roughly at least $n^{1-1/d}$ (thus only saving a factor of $n^{1/d}$ over the naive approach). Because the size and query time of any data structure have to be at least linear and logarithmic, respectively, we consider these two ends of the spectrum: (i) What is the size of a data structure that answers simplex range queries in logarithmic time? (ii) How fast can a simplex range query be answered using a linear-size data structure?

GLOSSARY

Arrangement: The arrangement of a set H of hyperplanes in $\mathbb{R}^d$, denoted by $\mathsf{A}(H)$, is the subdivision of $\mathbb{R}^d$ into cells, each being a maximal connected set contained in the intersection of a fixed subset of H and not intersecting any other hyperplane of H.

Level: The level of a point p with respect to a set H of hyperplanes is the number of hyperplanes of H that lie below p.

Duality: The dual of a point $(a_1, \ldots, a_d) \in \mathbb{R}^d$ is the hyperplane $x_d = -a_1 x_1 - \cdots - a_{d-1} x_{d-1} + a_d$, and the dual of a hyperplane $x_d = b_1 x_1 + \cdots + b_d$ is the point $(b_1, \ldots, b_{d-1}, b_d)$.

1/r-cutting: Let H be a set of n hyperplanes in $\mathbb{R}^d$, and let $r \in [1, n]$ be a parameter. A $(1/r)$-cutting of H is a set Ξ of (relatively open) disjoint simplices covering $\mathbb{R}^d$ so that at most n/r hyperplanes of H cross (i.e., intersect but do not contain) each simplex of Ξ.

Shallow cutting: Let H be a set of n hyperplanes in $\mathbb{R}^d$, and let $r \in [1, n]$ and $q \in [0{:}n-1]$ be two parameters. A shallow $(1/r)$-cutting of H is a set Ξ of (relatively open) disjoint simplices covering all points with level (with respect to H) at most q so that at most n/r hyperplanes of H cross each simplex of Ξ.

DATA STRUCTURES WITH LOGARITHMIC QUERY TIME

For simplicity, consider the following *halfspace range-counting* problem: Preprocess a set S of n points in $\mathbb{R}^d$ into a data structure so that the number of points of S that lie above a query hyperplane can be counted quickly. Using the duality transform, the above halfspace range-counting problem can be reduced to the following point-location problem: *Given a set H of n hyperplanes in $\mathbb{R}^d$, determine the number of hyperplanes of H lying above a query point q.* Since the same subset of hyperplanes lies above all points of a single cell of $\mathsf{A}(H)$, the number of hyperplanes of H lying above q can be reported by locating the cell of $\mathsf{A}(H)$ that contains q. The following theorem of Chazelle [Cha93] leads to an $O(\log n)$ query-time data structure for halfspace range counting.

THEOREM 40.3.1 *Chazelle* [Cha93]

Let H be a set of n hyperplanes and $r \leq n$ a parameter. Set $s = \lceil \log_2 r \rceil$. There exist s cuttings $\Xi_1, \ldots, \Xi_s$ so that Ξ_i is a $(1/2^i)$-cutting of size $O(2^{id})$, each simplex of Ξ_i is contained in a simplex of Ξ_{i-1}, and each simplex of Ξ_{i-1} contains a constant number of simplices of Ξ_i. Moreover, $\Xi_1, \ldots, \Xi_s$ can be computed in time $O(nr^{d-1})$.

Choose $r = \lceil \frac{n}{\log_2 n} \rceil$. Construct the cuttings $\Xi_1, \ldots, \Xi_s$, for $s = \lceil \log_2 r \rceil$; for each simplex $\triangle \in \Xi_i$, for $i < s$, store pointers to the simplices of Ξ_{i+1} that are contained in $\triangle$; and for each simplex $\triangle \in \Xi_s$, store $H_\triangle \subseteq H$, the set of hyperplanes that intersect $\triangle$, and $k_\triangle$, the number of hyperplanes of H that lie above $\triangle$. Since $|\Xi_s| = O(r^d)$, the total size of the data structure is $O(nr^{d-1}) = O(n^d/\log^{d-1} n)$. For a query point $q \in \mathbb{R}^d$, by traversing the pointers, find the simplex $\triangle \in \Xi_s$ that contains q, count the number of hyperplanes of $H_\triangle$ that lie above q, and return $k_\triangle$ plus this quantity. The total query time is $O(\log n)$. The space can be reduced to $O(n^d/\log^d n)$ while keeping the query time to be $O(\log n)$ [Mat93].

The above approach can be extended to the *simplex range-counting* problem: store the solution of every combinatorially distinct simplex (two simplices are combinatorially distinct if they do not contain the same subset of S). Since there are $\Theta(n^{d(d+1)})$ combinatorially distinct simplices, such an approach will require $\Omega(n^{d(d+1)})$ storage. Chazelle et al. [CSW92] proposed a data structure of size $O(n^{d+\epsilon})$, for any $\epsilon > 0$, using a multi-level data structure (see Section 40.5), that can answer a simplex range-counting query in $O(\log n)$ time. The space bound can be reduced to $O(n^d)$ by increasing the query time to $O(\log^{d+1} n)$ [Mat93].

LINEAR-SIZE DATA STRUCTURES

Most of the linear-size data structures for simplex range searching are based on **partition trees**, originally introduced by Willard [Wil82] for a set of points in the plane. Roughly speaking, a partition tree is a hierarchical decomposition scheme (in the sense described at the end of Section 40.1) that recursively partitions the points into canonical subsets and encloses each canonical subset by a simple convex region (e.g., simplex), so that any hyperplane crosses only a fraction of the regions associated with the "children" of a canonical subset. A query is answered as described in Section 40.1. The query time depends on the maximum number of regions associated the with the children of a node that a hyperplane can cross. The partition tree proposed by Willard partitions each canonical subsets into four children, each contained in a wedge so that any line crosses at most three of them. As a result, the time spent in reporting all k points lying inside a triangle is $O(n^\alpha + k)$, where $\alpha = \log_4 3 \approx 0.793$. A major breakthrough in simplex range searching was made by Haussler and Welzl [HW87]. They formulated range searching in an abstract setting and, using elegant probabilistic methods, gave a randomized algorithm to construct a linear-size partition tree with $O(n^\alpha)$ query time, where $\alpha = 1 - \frac{1}{d(d-1)+1} + \epsilon$ for any $\epsilon > 0$. The best known linear-size data structure for simplex range searching, which almost matches the lower bounds mentioned below, was first given by Matoušek [Mat93] and subsequently simplified by Chan [Cha12]. These data structures answer a simplex range-counting (resp. range-reporting) query in $\mathbb{R}^d$ in time $O(n^{1-1/d})$ (resp. $O(n^{1-1/d} + k)$), and are based on the following theorem.

THEOREM 40.3.2 *Matoušek* [Mat92a]

Let S be a set of n points in $\mathbb{R}^d$, and let $1 < r \leq n/2$ be a given parameter. Then there exists a family of pairs $\Pi = \{(S_1, \triangle_1), \ldots, (S_m, \triangle_m)\}$ such that $S_i \subseteq S$ lies inside simplex $\triangle_i$, $n/r \leq |S_i| \leq 2n/r$, $S_i \cap S_j = \emptyset$ for $i \neq j$, and every hyperplane crosses at most $cr^{1-1/d}$ simplices of Π; here c is a constant. If r is a constant, then Π can be constructed in $O(n)$ time.

Using this theorem, a partition tree T can be constructed as follows. Each interior node v of T is associated with a subset $S_v \subseteq S$ and a simplex Δ_v containing S_v; the root of T is associated with S and $\mathbb{R}^d$. Choose r to be a sufficiently large constant. If $|S| \leq 4r$, T consists of a single node, and it stores all points of S. Otherwise, we construct a family of pairs $\Pi = \{(S_1, \Delta_1), \ldots, (S_m, \Delta_m)\}$ using Theorem 40.3.2. We recursively construct a partition tree T_i for each S_i and attach T_i as the ith subtree of u. The root of T_i also stores Δ_i. The total size of the data structure is linear, and it can be constructed in time $O(n \log n)$. Since any hyperplane crosses at most $cr^{1-1/d}$ simplices of Π, the query time of simplex range reporting is $O(n^{1-1/d+\log_r c} + k)$; the $\log_r c$ term in the exponent can be reduced to any arbitrarily small positive constant ϵ by choosing r sufficiently large. Although the query time can be improved to $O(n^{1-1/d}\text{polylog}(n) + k)$ by choosing r to be n^ϵ, a stronger version of Theorem 40.3.2 that builds a simplicial partition hierarchically analogous to Theorem 40.3.1, instead of building it at each level independently, leads to a linear-size data structure with $O(n^{1-1/d} + k)$ query time. Chan's (randomized) algorithm [Cha12] constructs a hierarchical simplicial partition in which the (relative) interiors of simplices at every level are pairwise-disjoint and they together induce a hierarchical partition of $\mathbb{R}^d$. A space/query-time trade-off for simplex range searching can be attained by combining the linear-size and logarithmic query-time data structures.

If the points in S lie on a b-dimensional algebraic surface of constant degree, a simplex range-counting query can be answered in time $O(n^{1-\gamma+\epsilon})$ using linear space, where $\gamma = 1/\lfloor (d+b)/2 \rfloor$ [AM94].

HALFSPACE RANGE REPORTING

A halfspace range-reporting query can be answered more quickly than a simplex range-reporting query using shallow cutting. For simplicity, throughout this subsection we assume the query halfspace to lie below its bounding hyperplane. We begin by discussing a simpler problem: the **halfspace-emptiness** query, which asks whether a query halfspace contains any input point. By the duality transform, the halfspace-emptiness query in $\mathbb{R}^d$ can be formulated as asking whether a query point $q \in \mathbb{R}^d$ lies below all hyperplanes in a given set H of hyperplanes in $\mathbb{R}^d$. This query is equivalent to asking whether q lies inside a convex polyhedron $\mathsf{P}(H)$, defined by the intersection of halfspaces lying below the hyperplanes of H. For $d \leq 3$, a point-location query in $\mathsf{P}(H)$ can be answered optimally in $O(\log n)$ time using $O(n)$ space and $O(n \log n)$ preprocessing since $\mathsf{P}(H)$ has linear size [BCKO08]. For $d \geq 4$, point-location query in $\mathsf{P}(H)$ becomes more challenging, and the query is answered using a shallow-cutting based data structure. The following theorem by Matoušek can be used to construct a point-location data structure for $\mathsf{P}(H)$:

THEOREM 40.3.3 *Matoušek [Mat92b]*

Let H be a set of n hyperplanes and $r \leq n$ a parameter. A shallow $(1/r)$-cutting of level 0 with respect to H of size $O(r^{\lfloor d/2 \rfloor})$ can be computed in time $O(nr^c)$, where c is a constant depending on d.

Choose r to be a sufficiently large constant, and compute a shallow $(1/r)$-cutting Ξ of level 0 of H using Theorem 40.3.3. For each simplex $\triangle \in \Xi$, let $H_\triangle \subseteq H$ be the set of hyperplanes that intersect $\triangle$. Recursively, construct the data structure for $H_\triangle$; the recursion stops when $|H_\triangle| \leq r$. The size of the data structure is $O(n^{\lfloor d/2 \rfloor + \epsilon})$, where $\epsilon > 0$ is an arbitrarily small constant, and it can be constructed in $O(n^{\lfloor d/2 \rfloor + \epsilon})$ time. If a query point q does not lie in a simplex of Ξ, then one can conclude that $q \notin P(H)$ and thus stop. Otherwise, if q lies in a simplex $\triangle \in \Xi$, recursively determine whether q lies below all the hyperplanes of $H_\triangle$. The query time is $O(\log n)$. Matoušek and Schwarzkopf [MS92] showed that the space can be reduced to $O(\frac{n^{\lfloor d/2 \rfloor}}{\log^{\lfloor d/2 \rfloor - \epsilon} n})$.

A linear-size data structure can be constructed for answering halfspace-emptiness queries by constructing a simplicial partition analogous to Theorem 40.3.2 but with the property that a hyperplane that has at most n/r points above it crosses $O(r^{1-1/\lfloor d/2 \rfloor})$ simplices. By choosing r appropriately, a linear-size data structure can be constructed in $O(n^{1+\epsilon})$ time that answers a query in $n^{1-1/\lfloor d/2 \rfloor} 2^{O(\log^* n)}$ time; the construction time can be reduced to $O(n \log n)$ at the cost of increasing the query time to $O(n^{1-1/\lfloor d/2 \rfloor} \text{polylog}(n))$. For even dimensions, a linear-size data structure with query time $O(n^{1-1/\lfloor d/2 \rfloor})$ can be constructed in $O(n \log n)$ randomized-expected time [Cha12].

The halfspace-emptiness data structure can be adapted to answer halfspace range-reporting queries. For $d = 2$, Chazelle *et al.* [CGL85] presented an optimal data structure with $O(\log n + k)$ query time, $O(n)$ space, and $O(n \log n)$ preprocessing time. For $d = 3$, after a series of papers with successive improvements, a linear-size data structure with $O(\log n + k)$ query times was proposed by Afshani and Chan [AC09]; this structure can be constructed in $O(n \log n)$ time [CT15]. Table 40.3.1 summarizes the best known bounds for halfspace range reporting in higher dimensions; halfspace-reporting data structures can also answer halfspace-emptiness queries without the output-size term in their query time.

TABLE 40.3.1   Near-linear-size data structures for half-space range reporting/emptiness.

d	$S(n)$	$Q(n)$	NOTES
$d = 2$	n	$\log n + k$	reporting
$d = 3$	n	$\log n + k$	reporting
$d > 3$	$n^{\lfloor d/2 \rfloor} \log^c n$	$\log n + k$	reporting
	$\frac{n^{\lfloor d/2 \rfloor}}{\log^{\lfloor d/2 \rfloor - \epsilon} n}$	$\log n$	emptiness
	n	$n^{1-1/\lfloor d/2 \rfloor} \log^c n + k$	reporting
	n	$n^{1-1/\lfloor d/2 \rfloor} 2^{O(\log^* n)}$	emptiness
even d	n	$n^{1-1/\lfloor d/2 \rfloor} + k$	reporting

Finally, we comment that halfspace-emptiness data structures have been adapted to answer halfspace range-counting queries approximately. For example, a set S of n points in $\mathbb{R}^3$ can be preprocessed, in $O(n \log n)$ time, into a linear-size data structure that for a query halfspace γ in $\mathbb{R}^3$, can report in $O(\log n)$ time a number t such that $|\gamma \cap S| \le t \le (1 + \delta)|\gamma \cap S|$, where $\delta > 0$ is a constant [AC09, AHZ10]. For $d > 3$, such a query can be answered in $O((\frac{n}{t})^{1-1/\lfloor d/2 \rfloor} \text{polylog}(n))$ time using linear space [Rah17].

LOWER BOUNDS

Fredman [Fre81b] showed that a sequence of n insertions, deletions, and halfplane queries on a set of points in the plane requires $\Omega(n^{4/3})$ time, under the semigroup model. His technique, however, does not extend to static data structures. In a seminal paper, Chazelle [Cha89] proved an almost matching lower bound on simplex range searching under the semigroup model. He showed that any data structure of size m, for $n \leq m \leq n^d$, for simplex range searching in the semigroup model requires a query time of $\Omega(n/\sqrt{m})$ for $d = 2$ and $\Omega(n/(m^{1/d} \log n))$ for $d \geq 3$ in the worst case. His lower bound holds even if the query ranges are wedges or strips. For halfspaces, Arya et al. [AMX12] proved a lower bound of $\Omega((\frac{n}{\log n}^{1-\frac{1}{d+1}} m^{-\frac{1}{d+1}}))$ on the query time under the semigroup model. They also showed that if the semigroup is integral (i.e., for all non-zero elements of the semigroup and for all $k \geq 2$, the k-fold sum $x + \cdots + x \neq x$), then the lower bound can be improved to $\Omega(\frac{n}{m^{1/d}} \log^{-1-\frac{2}{d}} n)$.

A few lower bounds on simplex range searching have been proved under the group model. Chazelle [Cha98] proved an $\Omega(n \log n)$ lower bound for off-line halfspace range searching under the group model. Exploiting a close-connection between range searching and discrepancy theory, Larsen [Lar14] showed that for any dynamic data structure for halfspace range searching with t_u and t_q update and query time, respectively, $t_u \cdot t_q = \Omega(n^{1-1/d})$.

The best-known lower bound for simplex range reporting in the pointer-machine model is by Afshani [Afs13] who proved that the size of any data structure that answers a simplex range reporting query in time $O(t_q + k)$ is $\Omega((\frac{n}{t_q})^d / 2^{O(\sqrt{\log t_q})})$. His technique also shows that the size of any halfspace range-reporting data structure in dimension $d(d+3)/2$ has size $\Omega((\frac{n}{t_q})^d / 2^{O(\sqrt{\log t_q})})$.

A series of papers by Erickson established the first nontrivial lower bounds for on-line and off-line emptiness query problems, in the partition-graph model of computation. He first considered this model for **Hopcroft's problem**—Given a set of n points and m lines, does any point lie on a line?—for which he obtained a lower bound of $\Omega(n \log m + n^{2/3}m^{2/3} + m \log n)$ [Eri96b], almost matching the best known upper bound $O(n \log m + n^{2/3}m^{2/3}2^{O(\log^*(n+m))} + m \log n)$, due to Matoušek [Mat93]. He later established lower bounds on a trade-off between space and query time, or preprocessing and query time, for on-line hyperplane emptiness queries [Eri00]. For d-dimensional hyperplane queries, $\Omega(n^d/\text{polylog}(n))$ preprocessing time is required to achieve polylogarithmic query time, and the best possible query time is $\Omega(n^{1/d}/\text{polylog}(n))$ if $O(n\text{polylog}(n))$ preprocessing time is allowed. For $d = 2$, if the preprocessing time is t_p, the query time is $\Omega(n/\sqrt{t_p})$.

OPEN PROBLEMS

1. Prove a near-optimal lower bound on static simplex range searching in the group model.

2. Prove an optimal lower bound on halfspace range reporting in the pointer-machine model.

3. Can a halfspace range counting query be answered more efficiently if query hyperplanes satisfy certain properties, e.g., they are tangent to $\mathbb{S}^{d-1}$?

40.4 SEMIALGEBRAIC RANGE SEARCHING

GLOSSARY

Semialgebraic set: A semialgebraic set is a subset of $\mathbb{R}^d$ obtained from a finite number of sets of the form $\{x \in \mathbb{R}^d \mid g(x) \geq 0\}$, where g is a d-variate polynomial with real coefficients, by Boolean operations (union, intersection, and complement). A semialgebraic set has *constant description complexity* if the dimension, the number of polynomials defining the set, as well as the maximum degree of these polynomials are all constants.

Tarski cell: A semialgebraic cell of constant description complexity.

Partitioning polynomial: For a set $S \subset \mathbb{R}^d$ of n points and a real parameter r, $1 < r \leq n$, an r-partitioning polynomial for S is a nonzero d-variate polynomial f such that each connected component of $\mathbb{R}^d \setminus Z(f)$ contains at most n/r points of S, where $Z(f) := \{x \in \mathbb{R}^d \mid f(x) = 0\}$ denotes the zero set of f. The decomposition of $\mathbb{R}^d$ into $Z(f)$ and the connected components of $\mathbb{R}^d \setminus Z(f)$ is called a *polynomial partition* (induced by f).

So far we have assumed the ranges to be bounded by hyperplanes, but many applications require ranges to be defined by non-linear functions. For example, a query of the form, *for a given point p and a real number r, find all points of S lying within distance r from p*, is a range-searching problem with balls as ranges. A more general class of ranges can be defined as follows.

Let $\Gamma_{d,\Delta,s}$ denote the family of all semialgebraic sets in $\mathbb{R}^d$ defined by at most s polynomial inequalities of degree at most Δ each. The range-searching problem in which query ranges belong to $\Gamma_{d,\Delta,s}$ for constants d, Δ, s, is referred to as *semialgebraic range searching*.

It suffices to consider the ranges bounded by a single polynomial because the ranges bounded by multiple polynomials can be handled using multi-level data structures. We therefore assume the ranges to be of the form

$$\Gamma_f(a) = \{x \in \mathbb{R}^d \mid f(x,a) \geq 0\},$$

where f is a $(d+p)$-variate polynomial specifying the type of ranges (disks, cylinders, cones, etc.), and a is a p-tuple specifying a specific range of the given type (e.g., a specific disk). We describe two approaches for answering such queries.

LINEARIZATION

One approach to answering Γ_f-range queries is the **linearization** method. The polynomial $f(x,a)$ is represented in the form

$$f(x,a) = \psi_0(a) + \psi_1(a)\varphi_1(x) + \cdots + \psi_\lambda(a)\varphi_\lambda(x),$$

where $\varphi_1, \ldots, \varphi_\lambda, \psi_0, \ldots, \psi_\lambda$ are real functions. A point $x \in \mathbb{R}^d$ is mapped to the point

$$\varphi(x) = (\varphi_1(x), \varphi_2(x), \ldots, \varphi_\lambda(x)) \in \mathbb{R}^\lambda.$$

TABLE 40.4.1   Semialgebraic range searching; λ is the dimension of linearization.

d	RANGE	$S(n)$	$Q(n)$	NOTES
$d = 2$	disk	n	$\dfrac{\sqrt{n}\log n}{\log n + k}$	Counting / Reporting
	Tarski cell	n	$\sqrt{n}\,\mathrm{polylog}(n)$	Counting
$d \geq 3$	ball	n	$n^{1-\frac{1}{d}+\epsilon}$ / $n^{1-\frac{1}{\lceil d/2 \rceil}}\,\mathrm{polylog}(n) + k$	Counting / Reporting
$d = 2t - 1$	ball	n	$n^{1-\frac{1}{t}} + k$	Reporting
$d \geq 3$	Tarski cell	n	$n^{1-1/d}\mathrm{polylog}(n)$ / $n^{1-\frac{1}{\lambda}+\epsilon}$	Counting

Then a range $\Gamma_f(a) = \{x \in \mathbb{R}^d \mid f(x,a) \geq 0\}$ maps to a halfspace

$$\varphi^{\#}(a) : \{y \in \mathbb{R}^\lambda \mid \psi_0(a) + \psi_1(a)y_1 + \cdots + \psi_\lambda(a)y_\lambda \geq 0\};$$

λ is called the **dimension** of linearization. For example, a set of spheres in $\mathbb{R}^d$ admit a linearization of dimension $d + 1$, using the well-known lifting transform. Agarwal and Matoušek [AM94] have described an algorithm for computing a linearization of the smallest dimension under certain assumptions on φ_i's and ψ_i's. If f admits a linearization of dimension λ, a Γ_f-range query can be answered using a λ-dimensional halfspace range-searching data structure.

ALGEBRAIC METHOD

Agarwal and Matoušek [AM94] had also proposed an approach for answering Γ_f-range queries, by extending Theorem 40.3.2 to Tarski cells and by constructing partition trees using this extension. The query time of this approach depends on the complexity of the so-called *vertical decomposition* of arrangements of surfaces, and it leads to suboptimal performance for $d > 4$. A better data structure has been proposed [AMS13, MP15] based on the *polynomial partitioning scheme* introduced by Guth and Katz [GK15]; see also [Gut16].

Let $S \subset \mathbb{R}^d$ be a set of n points, and let r, $1 < r \leq n$, be a real parameter. Guth and Katz show that an r-partitioning polynomial of degree $O(r^{1/d})$ for S always exists. Agarwal et al. [AMS13] described a randomized algorithm to compute such a polynomial in expected time $O(nr + r^3)$. A result by Barone and Basu [BB12] implies that an algebraic variety of dimension k defined by polynomials of constant-bounded degree crosses $O(r^{k/d})$ components of $\mathbb{R}^d \setminus Z(f)$, and that these components can be computed in time $r^{O(1)}$. Therefore, one can recursively construct the data structure for points lying in each component of $\mathbb{R}^d \setminus Z(f)$. The total time spent in recursively searching in the components crossed by a query range will be $n^{1-1/d}\mathrm{polylog}(n)$. However, this ignores the points in $S^* = S \cap Z(f)$. Agarwal et al. [AMS13] use a scheme based on the so-called cylindrical algebraic decomposition to handle S^*. A more elegant and simpler method was subsequently proposed by Matoušek and Patáková [MP15], which basically applies a generalized polynomial-partitioning scheme on S^* and $Z(f)$. Putting everything together, a semialgebraic range-counting query can be answered in $O(n^{1-1/d}\mathrm{polylog}(n))$ time using a linear-size data structure; all k points lying inside the query range can be reported by spending an additional $O(k)$ time.

Arya and Mount [AM00] have presented a linear-size data structure for approximate range-searching queries. Let γ be a constant-complexity semialgebraic set and $\epsilon > 0$ a parameter. Their data structure returns in $O(\frac{1}{\epsilon^d} \log n + k_\epsilon)$ time a subset S_ϵ of k_ϵ points such that $\gamma \cap S \subseteq S_\epsilon \subseteq \gamma_\epsilon \cap S$ where γ_ϵ is the set of points within distance $\epsilon \cdot \text{diam}(\gamma)$ of γ. If γ is convex, the query time improves to $O(\log n + \frac{1}{\epsilon^{d-1}} + k_\epsilon)$. A result by Larsen and Nguyen [LN12] implies that query time of a linear-size data structure is $\Omega(\log n + \epsilon^{-\frac{d}{1+\delta}-1})$ for any arbitrarily small constant $\delta > 0$. The data structure in [AM00] can also return a value k_ϵ, with $|S \cap \gamma| \le k_\epsilon \le |S \cap \gamma_\epsilon|$ in time $O(\frac{1}{\epsilon^d} \log n)$, or in $O(\log n + \frac{1}{\epsilon^{d-1}})$ time if γ is convex.

40.5 VARIANTS AND EXTENSIONS

In this section we review a few extensions of range-searching data structures: multi-level structures, secondary-memory structures, range searching in a streaming model, range searching on moving points, and coping with data uncertainty.

MULTI-LEVEL STRUCTURES

A powerful property of data structures based on decomposition schemes (described in Section 40.1) is that they can be cascaded together to answer more complex queries, at the increase of a logarithmic factor per level in their performance. The real power of the cascading property was first observed by Dobkin and Edelsbrunner [DE87], who used this property to answer several complex geometric queries. Since their result, several papers have exploited and extended this property to solve numerous geometric-searching problems. We briefly sketch the general cascading scheme.

Let S be a set of weighted objects. Recall that a geometric-searching problem P, with underlying relation $\Diamond$, requires computing $\sum_{p \Diamond \gamma} w(p)$ for a query range γ. Let P^1 and P^2 be two geometric-searching problems, and let $\Diamond^1$ and $\Diamond^2$ be the corresponding relations. Define $\mathsf{P}^1 \circ \mathsf{P}^2$ to be the conjunction of P^1 and P^2, whose relation is $\Diamond^1 \cap \Diamond^2$. For a query range γ, the goal is to compute $\sum_{p \Diamond^1 \gamma, p \Diamond^2 \gamma} w(p)$. Suppose there are hierarchical decomposition schemes D^1 and D^2 for problems P^1 and P^2. Let $\mathsf{F}^1 = \mathsf{F}^1(S)$ be the set of canonical subsets constructed by D^1, and for a range γ, let $\mathsf{C}^1_\gamma = \mathsf{C}^1(S, \gamma)$ be the corresponding partition of $\{p \in S \mid p \Diamond^1 \gamma\}$ into canonical subsets. For each canonical subset $C \in \mathsf{F}^1$, let $\mathsf{F}^2(C)$ be the collection of canonical subsets of C constructed by D^2, and let $\mathsf{C}^2(C, \gamma)$ be the corresponding partition of $\{p \in C \mid p \Diamond^2 \gamma\}$ into level-two canonical subsets. The decomposition scheme $\mathsf{D}^1 \circ \mathsf{D}^2$ for the problem $\mathsf{P}^1 \circ \mathsf{P}^2$ consists of the canonical subsets $\mathsf{F} = \bigcup_{C \in \mathsf{F}^1} \mathsf{F}^2(C)$. For a query range γ, the query output is $\mathsf{C}_\gamma = \bigcup_{C \in \mathsf{C}^1_\gamma} \mathsf{C}^2(C, \gamma)$. Any number of decomposition schemes can be cascaded in this manner.

Viewing D^1 and D^2 as tree data structures, cascading the two decomposition schemes can be regarded as constructing a two-level tree, as follows. First construct the tree induced by D^1 on S. Each node v of D^1 is associated with a canonical subset C_v. Next, construct a second-level tree D^2_v on C_v and store D^2_v at v as its secondary structure. A query is answered by first identifying the nodes that correspond to the canonical subsets $C_v \in \mathsf{C}_\gamma^1$ and then searching the corresponding secondary trees to compute the second-level canonical subsets $\mathsf{C}^2(C_v, \gamma)$.

Suppose the size and query time of each decomposition scheme are at most $S(n)$ and $Q(n)$, respectively, and D^1 is efficient and r-convergent (cf. Section 40.1), for some constant $r > 1$. Then the size and query time of the decomposition scheme D are $O(S(n)\log_r n)$ and $O(Q(n)\log_r n)$, respectively. If D^2 is also efficient and r-convergent, then D is efficient and r-convergent. In some cases, the logarithmic overhead in the query time or the space can be avoided.

The real power of multi-level data structures stems from the fact that there are no restrictions on the relations $\Diamond^1$ and $\Diamond^2$. Hence, any query that can be represented as a conjunction of a constant number of "primitive" queries, each of which admits an efficient, r-convergent decomposition scheme, can be answered by cascading individual decomposition schemes. Range trees for orthogonal range searching, logarithmic query-time data structures for simplex range searching, and data structures for semialgebraic range searching discussed above are a few examples of multi-level structures. More examples will be mentioned in the following sections.

COLORED RANGE SEARCHING

In *colored range searching* (or *categorical* range searching), each input point is associated with a color and the goal is to report or count the number of colors of points lying inside a query range. For $d = 1$, a colored (orthogonal) range-reporting query can be answered in $O(\log n + k)$ time using linear space, where k is now the number of colors reported [JL93]. For $d = 2$, a 3-sided-rectangle reporting query can be answered using $O(n)$ space in $O(\log n + k)$ time under the pointer-machine model, or in $O(\log\log n + k)$ time in the rank space under the RAM model [LW13]. These data structures extend to reporting the colors of points inside a (4-sided) rectangle in the same time but using $O(n \log n)$ space. Using the techniques in [ACY12], colored halfplane-reporting queries in the plane can be answered in $O(\log n + k)$ time using $O(n)$ space.[3] In general, a range-emptiness data structure with $S(n)$ space and $Q(n)$ query time can be extended to answer colored version of the range-reporting query for the same ranges in $O((1 + k)Q(n))$ time and $O(S(n)\log n)$ space; if $S(n) = \Omega(n^{1+\epsilon})$, the size remains $O(S(n))$.

It has been shown that a colored range counting query in $\mathbb{R}^1$ is equivalent to (uncolored) range counting in $\mathbb{R}^2$ [JL93, LW13], so a 1D colored range-counting query can be answered in $O(\log n)$ time under the pointer-machine model or in $O(\mathrm{Lg}n)$ time under the RAM model and rank space, using linear space. For $d = 2$, by establishing a connection between colored orthogonal range counting and Boolean matrix multiplication, Kaplan *et al.* [KRSV08] showed that the query time of a 2D colored range counting is $\Omega(n^{t/2-1})$ where n^t is the time taken by a Boolean matrix multiplication algorithm. They also showed that a d-dimensional colored orthogonal range counting query can be answered in $O(\log^{2d-1} n)$ time using $O(n^d \log^{2d-1} n)$ space; if the query ranges are d-dimensional orthants, then the query time and space can be improved to $O(n^{d-1} \log n)$ and $O(n^{\lfloor d/2 \rfloor} \log^{d-1} n)$, respectively. Finally, we note that an approximate counting query can be answered in $O(\log^{d+1} n)$ time using $O(n \log^{d+1} n)$ space [Rah17].

---

[3]The author thanks Saladi Rahul for pointing out this observation.

SECONDARY MEMORY STRUCTURES

If the input is too large to fit into main memory, then the data structure must be stored in secondary memory—on disk, for example—and portions of it must be moved into main memory when needed to answer a query. In this case the bottleneck in query and preprocessing time is the time spent in transferring data between main and secondary memory. A commonly used model is the standard *two-level I/O model*, in which main memory has finite size, secondary memory is unlimited, and data is stored in secondary memory in blocks of size B, where B is a parameter [AV88]. Each access to secondary memory transfers one block (i.e., B words), and we count this as one *input/output (I/O) operation*. The query (resp. preprocessing) time is defined as the number of I/O operations required to answer a query (resp. to construct the structure). Under this model, the range-reporting query time is $\Omega(\log_B n + k/B)$. There have been various extensions of this model, including the so-called *cache-oblivious model*, which provides a simple framework for designing algorithms for multi-level memory hierarchies. In this model, an algorithm is oblivious to the details of memory hierarchy but is analyzed in the I/O-model. In particular, the value of B is not known and the goal is to minimize the number of I/O operations as well as the total work performed.

I/O-efficient range-searching structures have received much attention because of large data sets in spatial databases. The main idea underlying these structures is to construct high-degree trees instead of binary trees. For example, B-trees and their variants are used to answer 1-dimensional range-reporting queries in $O(\log_B n + \kappa)$ I/Os [Sam90], where $\kappa = k/B$. Similarly, the kdB-tree, an I/O-efficient version of kd-tree, was proposed to answer high-dimensional orthogonal range queries [Rob81]. While storing each point only once, it can answer a d-dimensional range-reporting query using $O((n/B)^{1-1/d} + \kappa)$ I/Os.

Arge et al. [ASV99] developed an external priority search tree so that a 2D 3-sided rectangle-reporting query can be answered in $O(\log_B n + \kappa)$ I/Os using $O(n)$ space. The main ingredient of their algorithm is a data structure that can store B^2 points using $O(B)$ blocks and can report all points lying inside a 3-sided rectangle in $O(1 + \kappa)$ I/Os. In contrast, a data structure in the cache-oblivious model that answers a 3-sided query in $O(\text{polylog}(n) + \kappa)$ time needs $\Omega(n \log^\epsilon n)$ space, and the same holds for 3D halfspace range reporting queries [AZ11]. By extending the ideas proposed in [Cha90a], it can be shown that any I/O-efficient data structure that answers a range-reporting query using $O(\log_B^c n + \kappa)$ I/Os requires $\Omega(n \log_B n / \log \log_B n)$ storage. Table 40.5.1 summarizes the best known bounds on range reporting queries in the I/O and cache-oblivious models.

By extending the data structure in [AM00], Streppel and Yi [SY11] have presented a linear-size I/O-efficient data structure for approximate range reporting. For a constant complexity range γ, it returns using $O(\frac{1}{\epsilon^d} \log_B n + k_\epsilon/B)$ I/Os, a subset S_ϵ of k_ϵ points such that $S \cap \gamma \subseteq S_\epsilon \subseteq S \cap \gamma_\epsilon$, where γ_ϵ is the set of points in $\mathbb{R}^d$ within distance $\epsilon \cdot \text{diam}(\gamma)$ from γ.

Govindarajan et al. [GAA03] have shown that a two-dimensional orthogonal range-counting query can be answered in $O(\log_B n)$ I/Os using linear space, assuming that each word can store $\log n$ bits. As for internal memory data structures, I/O-range-emptiness data structures can be adapted to answer range-counting queries approximately. For example, a 3D halfspace or dominance range-counting query can be answered approximately in $O(\log_B n)$ I/Os using linear space in the cache-oblivious model [AHZ10]. Colored range searching also has been studied in the I/O model, and efficient data structures are known for $d \leq 2$ [LW13, Nek14]: a colored orthogonal range-reporting query in $\mathbb{R}^2$ can be answered in $O(\log_B n + k/B)$ I/Os using $O(n \log(n) \log^* n)$ space.

TABLE 40.5.1   Secondary-memory structures for range reporting queries; here $\mathrm{Lg}_B n = \log n / \log \log_B n$.

d	RANGE	MODEL	$Q(n)$	$S(n)$
$d = 1$	interval	C.O.	$\log_B n + \kappa$	n
$d = 2$	3-sided	I/O	$\log_B n + \kappa$	n
		C.O.	$\log_B n + \kappa$	$n\sqrt{\log n}$
	rectangle	I/O	$\log_B n + \kappa$	$n\mathrm{Lg}_B n$
		C.O.	$\log_B n + \kappa$	$n \log^{3/2} n$
	halfplane	I/O	$\log_B n + \kappa$	n
	triangle	I/O	$\sqrt{n/B} + \kappa$	n
$d = 3$	octant	I/O	$\log_B n + \kappa$	n
		C.O.	$\log_B n + \kappa$	$n \log n$
	box	I/O	$\log_B n + \kappa$	$n(\mathrm{Lg}_B n)^3$
		C.O	$\log_B n + \kappa$	$n \log^{7/2} n$
	halfspace	I/O	$\log_B n + \kappa$	$n \log^* n$
		C.O.	$\log_B n + \kappa$	$n \log n$
$d \geq 3$	box	I/O	$\log_B n (\mathrm{Lg}_B n)^{d-2} + \kappa$	$n(\mathrm{Lg}_B n)^{d-1}$
	simplex	I/O	$(n/B)^{1-1/d} + \kappa$	n

Perhaps the most widely used I/O-efficient data structure for range searching in higher dimensions is the **R-tree**, originally introduced by Guttman [Gut84]. An R-tree is a B-tree, each of whose nodes stores a set of rectangles. Each leaf stores a subset of input points, and each input point is stored at exactly one leaf. For each node v, let R_v be the smallest rectangle containing all the rectangles stored at v; R_v is stored at the parent of v (along with the pointer to v). R_v induces the subspace corresponding to the subtree rooted at v, in the sense that for any query rectangle intersecting R_v, the subtree rooted at v is searched. Rectangles stored at a node are allowed to overlap. Although allowing rectangles to overlap helps reduce the size of the data structure, answering a query becomes more expensive. Guttman suggests a few heuristics to construct an R-tree so that the overlap is minimized. Several heuristics for improving the performance, including R*- and Hilbert-R-trees, have been proposed though the query time is linear in the worst case. Agarwal et al. [ABG+02] showed how to construct a variant of the R-tree, called the *box tree*, on a set of n rectangles in $\mathbb{R}^d$ so that all k rectangles intersecting a query rectangle can be reported in $O(n^{1-1/d} + k)$ time. Arge et al. [ABHY04] adapted their method to define a version of R-tree with the same query time.

STREAMING MODEL

Motivated by a broad spectrum of applications, data streams have emerged as an important paradigm for processing data that arrives on-line. In many such applications, data is too large to be stored in its entirety or to even scan for real-time processing. The goal is therefore to construct a small-size summary of the data stream (arrived so far) that can be used to analyze or query the data. The monograph by Muthukrishnan [Mut06] gives a summary of algorithms developed in the streaming model.

In the context of range searching, a few algorithms have been proposed for constructing a "succinct" data structure so that a range-counting query can be answered approximately, roughly with additive error ϵn. All the proposed data structures are based on the notion of ϵ-approximation: For a parameter $\epsilon > 0$, a subset $A \subseteq S$ is an ϵ-approximation for a set R of ranges if for every $\gamma \in$ R,

$$\left| \frac{|S \cap \gamma|}{|S|} - \frac{|A \cap \gamma|}{|A|} \right| \leq \epsilon.$$

A more general result by Li *et al.* [LLS01] implies that a random subset of size $O(\frac{1}{\epsilon^2} \log \frac{1}{\delta})$ is an ϵ-approximation with probability at least $1 - \delta$ for geometric ranges of constant complexity. Better bounds are known for many cases. For example, an ϵ-approximation of size $O(\frac{1}{\epsilon} \text{polylog}(\frac{1}{\epsilon}))$ exists for rectangles in $\mathbb{R}^d$, and of size $O(\epsilon^{-\frac{2d}{d+1}})$ for halfspaces in $\mathbb{R}^d$.

Bagchi *et al.* [BCEG07] described a deterministic algorithm for maintaining an ϵ-approximation of size $O(\frac{1}{\epsilon^2} \log \frac{1}{\epsilon})$ for a large class of geometric ranges. It uses $O(\frac{1}{\epsilon^{2(d-1)}} \text{polylog}(n/\epsilon))$ space, and uses the same amount of time to update the ϵ-approximation when a new point arrives. Faster algorithms are known for special cases. For $d = 1$, an ϵ-approximation of size $O(\frac{1}{\epsilon} \log n)$ with respect to a set of intervals can be maintained efficiently by a deterministic algorithm in the streaming model [GK01]; the space can be improved to $O(\frac{1}{\epsilon} \log \frac{1}{\epsilon})$ using a randomized algorithm [FO15]. For $d \geq 2$, an ϵ-approximation of size $O(\frac{1}{\epsilon} \log^{2d+1} \frac{1}{\epsilon})$ for rectangles and of size $O(\epsilon^{-\frac{2d}{d+1}} \log^{d+1} \frac{1}{\epsilon})$ for halfspaces can be maintained efficiently [STZ04].

KINETIC RANGE SEARCHING

Let $S = \{p_1, \ldots, p_n\}$ be a set of n points in $\mathbb{R}^2$, each moving continuously with fixed velocity. Let $p_i(t) = a_i + b_i t$, for $a_i, b_i \in \mathbb{R}^2$, denote the position of p_i at time t, and let $S(t) = \{p_1(t), \ldots, p_n(t)\}$. The trajectory of a point p_i is a line $\bar{p}_i$. Let L denote the set of lines corresponding to the trajectories of points in S. We focus on the following two range-reporting queries:

Q1. Given an axis-aligned rectangle R in the xy-plane and a time value t_q, report all points of S that lie inside R at time t_q, i.e., report $S(t_q) \cap R$; t_q is called the *time stamp* of the query.

Q2. Given a rectangle R and two time values $t_1 \leq t_2$, report all points of S that lie inside R at any time between t_1 and t_2, i.e., report $\bigcup_{t=t_1}^{t_2} (S(t) \cap R)$.

Two general approaches have been proposed to preprocess moving points for range searching. The first approach, known as the *time-oblivious* approach, regards time as a new dimension and stores the trajectories $\bar{p}_i$ of input points p_i. An advantage of this scheme is that the data structure is updated only if the trajectory of a point changes or if a point is inserted into or deleted from the index. Since this approach preprocesses either curves in $\mathbb{R}^3$ or points in higher dimensions, the query time tends to be large. For example, if S is a set of points moving in $\mathbb{R}^1$, then the trajectory of each point is a line in $\mathbb{R}^2$ and a Q1 query corresponds to reporting all lines of L that intersect a query segment σ parallel to the x-axis. Using a 2D simplex range-reporting data structure, L can be preprocessed into a linear-size data structure so that all lines intersecting σ can be reported in $O(\sqrt{n} + k)$ time. A similar structure can answer Q2 queries within the same asymptotic time bound. The lower bounds on simplex range searching suggest that this approach will not lead to a near-linear-size data structure with $O(\log n + k)$ query time.

If S is a set of points moving in $\mathbb{R}^2$, then a Q1 query asks for reporting all lines of L that intersect a query rectangle R parallel to the xy-plane (in the xyt-space). A line ℓ in $\mathbb{R}^3$ (xyt-space) intersects R if and only if their projections onto the xt- and yt-planes both intersect. A two-level partition tree of size $O(n)$ can report, in $O(n^{1/2+\epsilon} + k)$ time, all k lines of L intersecting R [AAE00]. Again a Q2 query can be answered within the same time bound.

The second approach, based on the ***kinetic-data-structure*** framework, builds a dynamic data structure on the moving points (see Chapter 53). Roughly speaking, at any time it maintains a data structure on the current configuration of the points. As the points move, the data structure evolves and is updated at discrete time instances when certain *events* occur, e.g., when any of the coordinates of two points become equal. This approach leads to fast query time but at the cost of updating the structure periodically even if the trajectory of no point changes. Another disadvantage of this approach is that it can answer a query only at the current configurations of points, though it can be extended to handle queries arriving in chronological order, i.e., the time stamps of queries are in nondecreasing order. In particular, if S is a set of points moving in $\mathbb{R}^1$, using a kinetic balanced binary search tree, a one-dimensional Q1 query can be answered in $O(\log n + k)$ time. The data structure processes $O(n^2)$ events, each of which requires $O(\log n)$ time. Similarly, by kinetizing range trees, a two-dimensional Q1 query can be answered in $O(\log n + k)$ time; the data structure processes $O(n^2)$ events, each of which requires $O(\log^2 n / \log \log n)$ time [AAE00].

Since range trees are complicated, a more practical approach is to use the kinetic data structure framework on kd-trees, as proposed by Agarwal et al. [AGG02]. They propose two variants of kinetic kd-trees, each of which answers Q1 queries that arrive in chronological order in $O(n^{1/2+\epsilon})$ time, for any constant $\epsilon > 0$, process $O(n^2)$ kinetic events, and spend $O(\text{polylog}(n))$ time at each event. A variant of kd-tree with a slightly better performance was proposed in [ABS09]. Kinetic R-trees also have been proposed [ŠJLL00, PAHP02], which require weaker invariants than kinetic kd-trees and thus process fewer events.

RANGE SEARCHING UNDER UNCERTAINTY

Many applications call for answering range queries in the presence of uncertainty in data—the location of each point may be represented as a probability density function (pdf) or a discrete mass function, called *location uncertainty*, or each point may exist with certain probability, called *existential uncertainty*. In the presence of location uncertainty, the goal is to report the points that lie inside a query range with probability at least τ, for some parameter $\tau \in [0, 1]$, or count the number of such points. In the case of existential uncertainty, the goal is to return some statistics on the distribution of the points inside a query range, e.g., what is the probability distribution of the number of points inside a query range, or what is the expected/most-likely value of the maximum weight of a point inside a query range.

If the location of each point is given as a piecewise-constant pdf in $\mathbb{R}^1$, a data structure of Agarwal et. al. [ACY12] reports all k points that lie inside a query interval with probability at least τ. For fixed τ, the query time is $O(\log n + k)$ and the size of the data structure is $O(n)$. If τ is part of the query, then the query time and size are $O(\log^3 n + k)$ and $O(n \log^2 n)$, respectively. They also describe another data structure of near-linear size that can handle more general pdfs and answer fixed-τ queries in $O(\log n + k)$ time.

Agarwal *et al.* [AKSS16] have described a data structure for range-max/min queries under both existential and location uncertainty in $\mathbb{R}^d$. For $d = 1$, using $O(n^{3/2})$ space, their data structure can compute the most-likely or the expected value of the maximum weight of input points in a query interval in time $O(n^{1/2})$; their data structure extends to higher dimensions. They also present another data structure that can estimate the expected value of the maximum inside a query rectangle within factor 2 in $O(\text{polylog}(n))$ time using $O(n\text{polylog}(n))$ space.

OPEN PROBLEMS

1. Prove tight bounds on orthogonal range searching in higher dimensions in the I/O model.

2. Is there a simple, linear-size kinetic data structure that can answer Q1 queries in $O(\sqrt{n}+k)$ time and processes near-linear events, each requiring $O\text{polylog}(n))$ time?

3. How quickly can 2D range queries be answered under location uncertainty?

4. How quickly can 1D range-max query be answered under existential/location uncertainty in data using linear space?

40.6 INTERSECTION SEARCHING

A general intersection-searching problem can be formulated as follows: *Given a set S of objects in $\mathbb{R}^d$, a semigroup $(\mathbf{S}, +)$, and a weight function $w : S \to \mathbf{S}$; preprocess S into a data structure so that for a query object γ, the weighted sum $\sum w(p)$, taken over all objects of S that intersect γ, can be computed quickly.* Range searching is a special case of intersection-searching in which S is a set of points.

An intersection-searching problem can be formulated as a range-searching problem by mapping each object $p \in S$ to a point $\varphi(p)$ in a parametric space $\mathbb{R}^\lambda$ and every query range γ to a set $\psi(\gamma)$ so that p intersects γ if and only if $\varphi(p) \in \psi(\gamma)$. For example, suppose both S and the query ranges are sets of segments in $\mathbb{R}^2$. Each segment $e \in S$ with left and right endpoints (p_x, p_y) and (q_x, q_y), respectively, can be mapped to a point $\varphi(e) = (p_x, p_y, q_x, q_y)$ in $\mathbb{R}^4$, and a query segment γ can be mapped to a semialgebraic set $\psi(\gamma)$ so that γ intersects e if and only if $\psi(\gamma) \in \varphi(e)$. A shortcoming of this approach is that λ, the dimension of the parametric space, is typically much larger than d, thereby affecting the query time aversely. The efficiency can be significantly improved by expressing the intersection test as a conjunction of simple primitive tests (in low dimensions) and using a multi-level data structure (described in Section 40.5) to perform these tests. For example, a segment γ intersects another segment e if the endpoints of e lie on the opposite sides of the line containing γ and vice versa. A two-level data structure can be constructed to answer such a query—the first level sifts the subset $S_1 \subseteq S$ of all the segments that intersect the line supporting the query segment, and the second level reports those segments of S_1 whose supporting lines separate the endpoints of γ. Each level of this structure can be implemented using a two-dimensional simplex range-searching structure, and hence a query can be answered in $O(n/\sqrt{m}\,\mathrm{polylog}(n)+k)$ time using $O(m\log n)$ space, for $n \leq m \leq n^2$.

It is beyond the scope of this chapter to cover all intersection-searching problems. Instead, we discuss a selection of basic problems that have been studied extensively. All intersection-counting data structures described here can answer intersection-reporting queries at an additional cost proportional to the output size. In some cases an intersection-reporting query can be answered faster. Moreover, using intersection-reporting data structures, intersection-detection queries can be answered in time proportional to their query-search time.

POINT INTERSECTION SEARCHING

Preprocess a set S of objects (e.g., balls, halfspaces, simplices, Tarski cells) in $\mathbb{R}^d$ into a data structure so that the objects of S containing a query point can be reported (or counted) efficiently. This is the inverse of the range-searching problem, and it can also be viewed as locating a point in the subdivision induced by the objects in S. Table 40.6.1 gives some of the known results; polylogarithmic factors are omitted from the space and query-search time whenever the query time is of the form n/m^α.

SEGMENT INTERSECTION SEARCHING

Preprocess a set S of objects in $\mathbb{R}^d$ into a data structure so that the objects of S intersected by a query segment can be reported (or counted) efficiently. See Table 40.6.2 for some of the known results on segment intersection searching; polylogarithmic factors are omitted from the space and query-search time whenever the query time is of the form n/m^α.

COLORED INTERSECTION SEARCHING

Preprocess a given set S of colored objects in $\mathbb{R}^d$ (i.e., each object in S is assigned a color) so that we can report (or count) the colors of the objects that intersect the query range. This problem arises in many contexts in which one wants to answer intersection-searching queries for nonconstant-size input objects. For example, given a set $P = \{P_1, \ldots, P_m\}$ of m simple polygons, one may wish to report all polygons of P that intersect a query segment; the goal is to return the indices, and not the description, of these polygons. If the edges of P_i are colored with color i, the problem reduces to colored segment intersection searching in a set of segments.

A set S of n colored rectangles in the plane can be stored into a data structure

TABLE 40.6.1   Point intersection searching.

d	OBJECTS	$S(n)$	$Q(n)$	NOTES
	rectangles	n	$\log n + k$	reporting
	disks	m	$(n/\sqrt{m})^{4/3}$	counting
		n	$\log n + k$	reporting
$d = 2$	triangles	m	$\dfrac{n}{\sqrt{m}}$	counting
	fat triangles	$n \log^* n$	$\log n + k$	reporting
	Tarski cells	$n^{2+\epsilon}$	$\log n$	counting
$d = 3$	halfspaces	n	$\log n + k$	reporting
	Tarski cells	$n^{3+\epsilon}$	$\log n$	counting
	rectangles	$n \log^{d-2} n$ $n \log^{d-2+\epsilon} n$	$\log^{d-1} n + k$ $\log n (\frac{\log n}{\log \log n})^{d-2} + k$	reporting
$d \geq 3$	simplices	m	$\dfrac{n}{m^{1/d}}$	counting
	balls	$n^{d+\epsilon}$	$\log n$	counting
		m	$\dfrac{n}{m^{1/\lceil d/2 \rceil}} + k$	reporting
$d \geq 4$	Tarski cells	$n^{2d-4+\epsilon}$	$\log n$	counting

TABLE 40.6.2   Segment intersection counting queries.

d	OBJECTS	$S(n)$	$Q(n)$
	segments	m	$n/\sqrt{m}$
$d = 2$	circles	$n^{2+\epsilon}$	$\log n$
	circular arcs	m	$n/m^{1/3}$
	planes	m	$n/m^{1/3}$
$d = 3$	spheres	m	$n/m^{1/3}$
	triangles	m	$n/m^{1/4}$

of size $O(n \log n)$ so that the colors of all rectangles in S that contain a query point can be reported in time $O(\log n + k)$ [BKMT97]. If the vertices of the rectangles in S and all the query points lie on the grid $[0:U]^2$, the query time can be improved to $O(\log \log U + k)$ by increasing the storage to $O(n^{1+\epsilon})$.

Agarwal and van Kreveld [AK96] presented a linear-size data structure with $O(n^{1/2+\epsilon} + k)$ query time for colored segment intersection-reporting queries amid a set of segments in the plane, assuming that the segments of the same color form a connected planar graph or the boundary of a simple polygon.

40.7   RAY-SHOOTING QUERIES

Preprocess a set S of objects in $\mathbb{R}^d$ into a data structure so that the first object (if one exists) intersected by a query ray can be reported efficiently. Originally motivated by the ray-tracing problem in computer graphics, this problem has found many applications and has been studied extensively in computational geometry.

A general approach to the ray-shooting problem, using segment intersection-detection structures and Megiddo's parametric-searching technique, was proposed by Agarwal and Matoušek [AM93]. The basic idea of their approach is as follows: suppose there is a segment intersection-detection data structure for S, based on partition trees. Let ρ be a query ray. The query procedure maintains a segment $\vec{ab} \subseteq \rho$ that contains the first intersection point of ρ with S. If a lies on an object of S, it returns a. Otherwise, it picks a point $c \in ab$ and determines, using the segment intersection-detection data structure, whether the interior of the segment ac intersects any object of S. If the answer is YES, it recursively finds the first intersection point of $\vec{ac}$ with S; otherwise, it recursively finds the first intersection point of $\vec{cb}$ with S. Using parametric searching, the point c at each stage can be chosen so that the algorithm terminates after $O(\log n)$ steps.

In some cases the query time can be improved by a $\mathrm{polylog}(n)$ factor using a more direct approach. Table 40.7.1 gives a summary of known ray-shooting results; polylogarithmic factors are ignored in the space and query time whenever the query time is of the form n/m^{α}.

TABLE 40.7.1   Ray shooting.

d	OBJECTS	$S(n)$	$Q(n)$
$d=2$	simple polygon	n	$\log n$
	s disjoint polygons	n	$\sqrt{s}\log n$
	s disjoint polygons	$(s^2+n)\log s$	$\log s \log n$
	s convex polygons	$sn\log s$	$\log s \log n$
	segments	m	$n/\sqrt{m}$
	circlular arcs	m	$n/m^{1/3}$
	disjoint arcs	n	$\sqrt{n}$
$d=3$	convex polytope	n	$\log n$
	c-oriented polytopes	n	$\log n$
	s convex polytopes	$s^2 n^{2+\epsilon}$	$\log^2 n$
	fat convex polytopes	m	$n/\sqrt{m}$
	halfplanes	m	$n/\sqrt{m}$
	terrain	m	$n/\sqrt{m}$
	triangles	m	$n/m^{1/4}$
	spheres	m	$n/m^{1/3}$
$d>3$	hyperplanes	m	$n/m^{1/d}$
		$\dfrac{n^d}{\log^{d-\epsilon} n}$	$\log n$
	convex polytope	m	$n/m^{1/\lfloor d/2 \rfloor}$
		$\dfrac{n^{\lfloor d/2 \rfloor}}{\log^{\lfloor d/2 \rfloor -\epsilon} n}$	$\log n$

Practical data structures have been proposed that, notwithstanding poor worst-case performance, work well in practice. One common approach is to construct a subdivision of $\mathbb{R}^d$ into constant-size cells so that the interior of each cell does not intersect any object of S. A ray-shooting query can be answered by traversing the query ray through the subdivision until we find an object that intersects the ray. The worst-case query time is proportional to the maximum number of cells intersected by a segment that does not intersect any object in S. Hershberger and Suri [HS95] showed that a triangulation with $O(\log n)$ query time can be constructed when S is the boundary of a simple polygon in the plane. Agarwal et al. [AAS95] proved worst-case bounds for many cases on the number of cells in a subdivision in $\mathbb{R}^3$ that a line can intersect. Aronov and Fortune [AF99] obtained a bound on the expected number of cells in the subdivision in $\mathbb{R}^3$ that a line can intersect.

ARC-SHOOTING QUERIES

Arc shooting is a generalization of the ray-shooting problem, where given a set S of objects, one wishes to find the first object of S hit by an oriented arc. Cheong et al. [CCEO04] have shown that a simple polygon P can be preprocessed into a data structure of linear size so that the first point of P hit by a circular or parabolic arc can be computed in $O(\log^2 n)$ time. Sharir and Shaul [SS05] have described a linear-size data structure that given a set of triangles in $\mathbb{R}^3$ can compute in $O(n^{3/4+\epsilon})$ time the first triangle hit by a vertical parabolic arc.

LINEAR-PROGRAMMING (LP) QUERIES

Let H be a set of n halfspaces in $\mathbb{R}^d$. *Preprocess H into a data structure so that for a direction vector u, the first point of $P(H) := \bigcap_{h \in H} h$ in the direction u can be determined quickly.* LP queries are generalizations of ray-shooting queries in the sense that one wishes to compute the first point of $P(H)$ met by a hyperplane normal to u as it is translated in direction u.

For $d \leq 3$, an LP query can be answered in $O(\log n)$ time using $O(n)$ storage, by constructing the normal diagram of the convex polytope $P(S)$ and preprocessing it for point-location queries. For higher dimensions, Chan [Cha96] described a randomized technique that reduces LP queries to halfspace-emptiness queries. Using the best-known data structures for halfspace-emptiness queries, a linear-size data structure can be constructed that can answer LP-queries in expected time $O(n^{1-1/\lfloor d/2 \rfloor})$ for even values of d and in $n^{1-1/\lfloor d/2 \rfloor} 2^{O(\log^* n)}$ time for odd values of d. See also [Ram00].

40.8 SOURCES AND RELATED MATERIAL

RELATED READING

Books and Monographs

[BCKO08]: Basic topics in computational geometry.

[Mul93]: Randomized techniques in computational geometry. Chapters 6 and 8 cover range-searching, intersection-searching, and ray-shooting data structures.

[Cha01]: Lower bound techniques, ϵ-nets, cuttings, and simplex range searching.

[MTT99, Sam90]: Range-searching data structures in spatial database systems.

[Gut16]: Applications of the theory of polynomials and algebraic geometry to various problems in incidence geometry.

Survey Papers

[AE99, Mat94]: Range-searching data structures.

[GG98, NW00] Indexing techniques used in databases.

[AP02, ST99]: Range-searching data structures for moving points.

[Arg02]: Secondary-memory data structures.

RELATED CHAPTERS

Chapter 28: Arrangements
Chapter 38: Point location
Chapter 41: Ray shooting and lines in space
Chapter 43: Nearest neighbors in high-dimensional spaces
Chapter 47: Epsilon-approximations and epsilon-nets
Chapter 53: Modeling motion

REFERENCES

[AAE00]   P.K. Agarwal, L. Arge, and J. Erickson. Indexing moving points. In *Proc. 19th ACM Sympos. Principles Database Syst.*, pages 175–186, 2000.

[AAL09]   P. Afshani, L. Arge, and K.D. Larsen. Orthogonal range reporting in three and higher dimensions. In *Proc. 50th IEEE Sympos. Found. Comp. Sci.*, pages 149–158, 2009.

[AAL10]   P. Afshani, L. Arge, and K.D. Larsen. Orthogonal range reporting: Query lower bounds, optimal structures in 3-d, and higher-dimensional improvements. In *Proc. 26th Sympos. Comput. Geom.*, pages 240–246, 2010.

[AAL12]   P. Afshani, L. Arge, and K.G. Larsen. Higher-dimensional orthogonal range reporting and rectangle stabbing in the pointer machine model. In *Proc. 28th Sympos. Comput. Geom.*, pages 323–332, 2012.

[AAS95]   P.K. Agarwal, B. Aronov, and S. Suri. Stabbing triangulations by lines in 3d. In *Proc. 11th Sympos. Comput. Geom.*, pages 267–276, 1995.

[ABG+02]   P.K. Agarwal, M. de Berg, J. Gudmundsson, M. Hammar, and H.J. Haverkort. Box-trees and R-trees with near-optimal query time. *Discrete Comput. Geom.*, 26:291–312, 2002.

[ABHY04]   L. Arge, M. de Berg, H.J. Haverkort, and K. Yi. The priority R-tree: A practically efficient and worst-case optimal R-tree. In *Proc. ACM SIGMOD Intl. Conf. Manage. Data*, pages 347–358, 2004.

[ABR01]   S. Alstrup, G.S. Brodal, and T. Rauhe. Optimal static range reporting in one dimension. In *Proc. 33rd ACM Sympos. Theory Comput.*, pages 476–482, 2001.

[ABS09]   M.A. Abam, M. de Berg, and B. Speckmann. Kinetic kd-trees and longest-side kd-trees. *SIAM J. Comput.*, 39:1219–1232, 2009.

[AC09]   P. Afshani and T.M. Chan. Optimal halfspace range reporting in three dimensions. In *Proc. 20th ACM-SIAM Sympos. Discrete Algorithms*, pages 180–186, 2009.

[ACY12]   P.K. Agarwal, S.-W. Cheng, and K. Yi. Range searching on uncertain data. *ACM Trans. Algorithms*, 8:43, 2012.

[AE99]   P.K. Agarwal and J. Erickson. Geometric range searching and its relatives. In B. Chazelle, J. E. Goodman, and R. Pollack, editors, *Advances in Discrete and Computational Geometry*, vol. 223 of *Contemporary Mathematics*, pages 1–56, AMS, Providence, 1999.

[AF99]   B. Aronov and S. Fortune. Approximating minimum-weight triangulations in three dimensions. *Discrete Comput. Geom.*, 21:527–549, 1999.

[Afs13]   P. Afshani. Improved pointer machine and I/O lower bounds for simplex range reporting and related problems. *Internat. J. Comput. Geom. Appl.*, 23:233–252, 2013.

[AGG02]   P.K. Agarwal, J. Gao, and L.J. Guibas. Kinetic medians and kd-trees. In *Proc. 10th European Sympos. Algorithms*, vol. 2461 of *LNCS*, pages 15–26, Springer, Berlin, 2002.

[AHZ10]   P. Afshani, C.H. Hamilton, and N. Zeh. A general approach for cache-oblivious range reporting and approximate range counting. *Comput. Geom.*, 43:700–712, 2010.

[Ajt88]   M. Ajtai. A lower bound for finding predecessors in Yao's call probe model. *Combinatorica*, 8:235–247, 1988.

[AK96]   P.K. Agarwal and M. van Kreveld. Polygon and connected component intersection searching. *Algorithmica*, 15:626–660, 1996.

[AKSS16]   P.K. Agarwal, N. Kumar, S. Stavros, and S. Suri. Range-max queries on uncertain data. In *Proc. 35th ACM Sympos. Principles Database Syst.*, pages 465–476, 2016.

[AM93]   P.K. Agarwal and J. Matoušek. Ray shooting and parametric search. *SIAM J. Comput.*, 22:794–806, 1993.

[AM94]   P.K. Agarwal and J. Matoušek. On range searching with semialgebraic sets. *Discrete Comput. Geom.*, 11:393–418, 1994.

[AM00]   S. Arya and D.M. Mount. Approximate range searching. *Comput. Geom.*, 17:135–152, 2000.

[AMN⁺98]   S. Arya, D.M. Mount, N.S. Netanyahu, R. Silverman, and A.Y. Wu. An optimal algorithm for approximate nearest neighbor searching fixed dimensions. *J. ACM*, 45:891–923, 1998.

[AMS13]   P.K. Agarwal, J. Matoušek, and M. Sharir. On range searching with semialgebraic sets. II. *SIAM J. Comput.*, 42:2039–2062, 2013.

[AMX12]   S. Arya, D.M. Mount, and J. Xia. Tight lower bounds for halfspace range searching. *Discrete Comput. Geom.*, 47:711–730, 2012.

[AP02]   P.K. Agarwal and C.M. Procopiuc. Advances in indexing mobile objects. *IEEE Bulletin Data Engineering*, 25:25–34, 2002.

[Arg02]   L. Arge. External memory data structures. In J. Abello, P.M. Pardalos, and M.G.C. Resende, editors, *Handbook of Massive Data Sets*, pages 313–358. Kluwer Academic Publishers, Boston, 2002.

[ASV99]   L. Arge, V. Samoladas, and J.S. Vitter. On two-dimensional indexability and optimal range search indexing. In *Proc. 18th ACM Sympos. Principles Database Syst.*, pages 346–357, 1999.

[AT14]   P. Afshani and K. Tsakalidis. Optimal deterministic shallow cuttings for 3D dominance ranges. In *Proc. 25th ACM-SIAM Sympos. Discrete Algorithms*, pages 1389–1398, 2014.

[AV88]   A. Aggarwal and J.S. Vitter. The input/output complexity of sorting and related problems. *Commun. ACM*, 31:1116–1127, 1988.

[AZ11]   P. Afshani and N. Zeh. Improved space bounds for cache-oblivious range reporting. In *Proc. 22nd ACM-SIAM Sympos. Discrete Algorithms*, pages 1745–1758, 2011.

[BB12]   S. Barone and S. Basu. Refined bounds on the number of connected components of sign conditions on a variety. *Discrete Comput. Geom.*, 47:577–597, 2012.

[BCEG07]   A. Bagchi, A. Chaudhary, D. Eppstein, and M.T. Goodrich. Deterministic sampling and range counting in geometric data streams. *ACM Trans. Algorithms*, 3, 2007.

[BCKO08]   M. de Berg, O. Cheong, M. van Kreveld, and M. Overmars. *Computational Geometry*, 3rd edition. Springer, Berlin, 2008.

[Ben75]   J.L. Bentley. Multidimensional binary search trees used for associative searching. *Commun. ACM*, 18:509–517, 1975.

[Ben80]   J.L. Bentley. Multidimensional divide-and-conquer. *Commun. ACM*, 23:214–229, 1980.

[BKL06]   A. Beygelzimer, S. Kakade, and J. Langford. Cover trees for nearest neighbor. In *Proc. 23rd Intl. Conf. Machine Learning*, pages 97–104, 2006.

[BKMT97]   P. Bozanis, N. Ktsios, C. Makris, and A. Tsakalidis. New results on intersection query problems. *The Computer Journal*, 40:22–29, 1997.

[BL14]   G.S. Brodal and K.G. Larsen. Optimal planar orthogonal skyline counting queries. In *Proc. 14th Scand. Workshop Algo. Theory*, vol. 8503 of *LNCS*, pages 110–121, Springer, Berlin, 2014.

[CCEO04]   S.-W. Cheng, O. Cheong, H. Everett, and R. van Oostrum. Hierarchical decompositions and circular ray shooting in simple polygons. *Discrete Comput. Geom.*, 32:401–415, 2004.

[CDL+14]   T.M. Chan, S. Durocher, K.G. Larsen, J. Morrison, and B.T. Wilkinson. Linear-space data structures for range mode query in arrays. *Theory Comput. Syst.*, 55:719–741, 2014.

[CGL85]   B. Chazelle, L.J. Guibas, and D.T. Lee. The power of geometric duality. *BIT*, 25:76–90, 1985.

[Cha01]   B. Chazelle. *The Discrepancy Method: Randomness and Complexity*. Cambridge University Press, 2001.

[Cha12]   T.M. Chan. Optimal partition trees. *Discrete Comput. Geom.*, 47:661–690, 2012.

[Cha86]   B. Chazelle. Filtering search: A new approach to query-answering. *SIAM J. Comput.*, 15:703–724, 1986.

[Cha88]   B. Chazelle. A functional approach to data structures and its use in multidimensional searching. *SIAM J. Comput.*, 17:427–462, 1988.

[Cha89]   B. Chazelle. Lower bounds on the complexity of polytope range searching. *J. AMS*, 2:637–666, 1989.

[Cha90a]   B. Chazelle. Lower bounds for orthogonal range searching, I: The reporting case. *J. ACM*, 37:200–212, 1990.

[Cha90b]   B. Chazelle. Lower bounds for orthogonal range searching, II: The arithmetic model. *J. ACM*, 37:439–463, 1990.

[Cha93]   B. Chazelle. Cutting hyperplanes for divide-and-conquer. *Discrete Comput. Geom.*, 9:145–158, 1993.

[Cha96]   T.M. Chan. Fixed-dimensional linear programming queries made easy. in *Proc. 12th Sympos. Comput. Geom.*, pages 284–290, ACM Press, 1996.

[Cha97]   B. Chazelle. Lower bounds for off-line range searching. *Discrete Comput. Geom.*, 17:53–66, 1997.

[Cha98]   B. Chazelle. A spectral approach to lower bounds with applications to geometric searching. *SIAM J. Comput.*, 27:545–556, 1998.

[CL01]   B. Chazelle and A. Lvov. A trace bound for hereditary discrepancy. *Discrete Comput. Geom.*, 26:221–232, 2001.

[CLP11]   T.M. Chan, K. G. Larsen, and M. Pătraşcu. Orthogonal range searching on the RAM, revisited. In *Proc. 27th Sympos. Comput. Geom.*, pages 1–10, 2011.

[CP10]   T.M. Chan and M. Pătraşcu. Counting inversions, offline orthogonal range counting, and related problems. In *Proc. 21st ACM-SIAM Sympos. Discrete Algo.*, pages 161–173, 2010.

[CR89]   B. Chazelle and B. Rosenberg. Computing partial sums in multidimensional arrays. In *Proc. 5th Sympos. Comput. Geom.*, pages 131–139, ACM Press, 1989.

[CSW92]   B. Chazelle, M. Sharir, and E. Welzl. Quasi-optimal upper bounds for simplex range searching and new zone theorems. *Algorithmica*, 8:407–429, 1992.

[CT15]   T.M. Chan and K. Tsakalidis. Optimal deterministic algorithms for 2-d and 3-d shallow cuttings. In *Proc. 31st Sympos. Comput. Geom.*, vol. 34 of LIPIcs, pages 719–732, Schloss Dagstuhl, 2015.

[CW13]   T.M. Chan and B.T. Wilkinson. Adaptive and approximate orthogonal range counting. In *Proc. 24th ACM-SIAM Sympos. Discrete Algorithms*, pages 241–251, 2013.

[CZ15]   T.M. Chan and G. Zhou. Multidimensional range selection. In *Proc. 26th Intl. Sympos. Algo. Comput.*, pages 83–92, 2015.

[DE87] ✦ D.P. Dobkin and H. Edelsbrunner. Space searching for intersecting objects. *J. Algorithms*, 8:348–361, 1987.

[DF08]   S. Dasgupta and Y. Freund. Random projection trees and low dimensional manifolds. In *Proc. 40th ACM Sympos. Theory Comput.*, pages 537–546, 2008.

[DR91]   P.F. Dietz and R. Raman. Persistence, amortization and randomization. In *Proc. 2nd ACM-SIAM Sympos. Discrete Algorithms*, pages 78–88, 1991.

[DS15]   S. Dasgupta and K. Sinha. Randomized partition trees for nearest neighbor search. *Algorithmica*, 72:237–263, 2015.

[EBKZ77] P. van Emde Boas, R. Kaas, and E. Zijlstra. Design and implementation of an efficient priority queue. *Mathematical Systems Theory*, 10:99–127, 1977.

[ELS97]   G. Evangelidis, D.B. Lomet, and B. Salzberg. The hB$^\Pi$-tree: A multi-attribute index supporting concurrency, recovery and node consolidation. *VLDB Journal*, 6:1–25, 1997.

[Eri96a] J. Erickson. New lower bounds for halfspace emptiness. In *Proc. 37th IEEE Sympos. Found. Comput. Sci.*, pages 472–481, 1996.

[Eri96b] J. Erickson. New lower bounds for Hopcroft's problem. *Discrete Comput. Geom.*, 16:389–418, 1996.

[Eri00]   J. Erickson. Space-time tradeoffs for emptiness queries. *SIAM J. Comput.*, 19:1968–1996, 2000.

[FH06]   J. Fischer and V. Heun. Theoretical and practical improvements on the RMQ-problem, with applications to LCA and LCE. In *Proc. 17th Sympos. Combin. Pattern Matching*, vol. 4009 of *LNCS*, pages 36–48, Springer, Berlin, 2006.

[FO15]   D. Felber and R. Ostrovsky. A randomized online quantile summary in $o(1/\epsilon *$ $\log(1/\epsilon))$ words. In *Proc. APPROX/RANDOM*, vol. 40 of LIPIcs, pages 775–785, Schloss Dagstuhl, 2015.

[Fre80]   M.L. Fredman. The inherent complexity of dynamic data structures which accommodate range queries. In *Proc. 21st IEEE Sympos. Found. Comput. Sci.*, pages 191–199, 1980.

[Fre81a] M.L. Fredman. A lower bound on the complexity of orthogonal range queries. *J. ACM*, 28:696–705, 1981.

[Fre81b] M.L. Fredman. Lower bounds on the complexity of some optimal data structures. *SIAM J. Comput.*, 10:1–10, 1981.

[Fre82]   M.L. Fredman. The complexity of maintaining an array and computing its partial sums. *J. ACM*, 29:250–260, 1982.

[GAA03]   S. Govindarajan, P.K. Agarwal, and L. Arge. CRB-tree: An efficient indexing scheme for range aggregate queries. In *Proc. 9th Intl. Conf. Database Theory*, 2003.

[GBLP96] J. Gray, A. Bosworth, A. Layman, and H. Patel. Data cube: A relational aggregation operator generalizing group-by, cross-tab, and sub-totals. In *Proc. 12th IEEE Internat. Conf. Data Eng.*, pages 152–159, 1996.

[GG98]   V. Gaede and O. Günther. Multidimensional access methods. *ACM Comput. Surv.*, 30:170–231, 1998.

[GK01]   M. Greenwald and S. Khanna. Space-efficient online computation of quantile summaries. In *Proc. ACM SIGMOD Intl. Conf. Manage. Data*, pages 58–66, 2001.

[GK15]   L. Guth and N.H. Katz. On the Erdős distinct distances problem in the plane. *Annals Math.*, 181:155–190, 2015.

[Gut84]   A. Guttman. R-trees: A dynamic index structure for spatial searching. In *Proc. 3rd ACM Sympos. Principles Database Systems*, pages 47–57, 1984.

[Gut16]   L. Guth. *Polynomial Methods in Combinatorics*. Vol. 64 of *University Lecture Series*, AMS, Providence, 2016.

[HBA97]   C.-T. Ho, J. Bruck, and R. Agrawal. Partial-sum queries in OLAP data cubes using covering codes. In *Proc. 16th ACM Sympos. Principles Database Syst.*, pages 228–237, 1997.

[HS95]   J. Hershberger and S. Suri. A pedestrian approach to ray shooting: Shoot a ray, take a walk. *J. Algorithms*, 18:403–431, 1995.

[HW87]   D. Haussler and E. Welzl. Epsilon-nets and simplex range queries. *Discrete Comput. Geom.*, 2:127–151, 1987.

[JL93]   R. Janardan and M. Lopez. Generalized intersection searching problems. *Internat. J. Comput. Geom. Appl.*, 3:39–69, 1993.

[JL11]   A.G. Jørgensen and K.G. Larsen. Range selection and median: Tight cell probe lower bounds and adaptive data structures. In *Proc. 22nd ACM-SIAM Sympos. Discrete Algo.*, pages 805–813, 2011.

[JMS04]   J. JáJá, C.W. Mortensen, and Q. Shi. Space-efficient and fast algorithms for multidimensional dominance reporting and counting. In *Proc. 15th Intl. Sympos. Algorithms Computation*, vol. 3341 of *LNCS*, pages 558–568, Springer, Berlin, 2004.

[KRSV08] H. Kaplan, N. Rubin, M. Sharir, and E. Verbin. Efficient colored orthogonal range counting. *SIAM J. Comput.*, 38:982–1011, 2008.

[Lar12]   K.G. Larsen. The cell probe complexity of dynamic range counting. In *Proc. 44th ACM Sympos. Theory Comput.*, pages 85–94, 2012.

[Lar14]   K.G. Larsen. On range searching in the group model and combinatorial discrepancy. *SIAM J. Comput.*, 43:673–686, 2014.

[LLS01]   Y. Li, P.M. Long, and A. Srinivasan. Improved bounds on the sample complexity of learning. *J. Comp. Syst. Sci.*, 62:516–527, 2001.

[LN12]   K.G. Larsen and H.L. Nguyen. Improved range searching lower bounds. In *Proc. 28th Sympos. Comput. Geom.*, pages 171–178, 2012.

[LS90]   D.B. Lomet and B. Salzberg. The hB-tree: A multiattribute indexing method with good guaranteed performance. *ACM Trans. Database Syst.*, 15:625–658, 1990.

[LW13]   K.G. Larsen and F. van Walderveen. Near-optimal range reporting structures for categorical data. In *Proc. 24th ACM-SIAM Sympos. Discrete Algorithms*, pages 265–276, 2013.

[Mat92a]   J. Matoušek. Efficient partition trees. *Discrete Comput. Geom.*, 8:315–334, 1992.

[Mat92b]   J. Matoušek. Reporting points in halfspaces. *Comput. Geom.*, 2:169–186, 1992.

[Mat93]   J. Matoušek. Range searching with efficient hierarchical cuttings. *Discrete Comput. Geom.*, 10:157–182, 1993.

[Mat94]   J. Matoušek. Geometric range searching. *ACM Comput. Surv.*, 26:421–461, 1994.

[McC85]   E.M. McCreight. Priority search trees. *SIAM J. Comput.*, 14(2):257–276, 1985.

[MP15]   J. Matoušek and Z. Patáková. Multilevel polynomial partitions and simplified range searching. *Discrete Comput. Geom.*, 54:22–41, 2015.

[MPP05]   C.W. Mortensen, R. Pagh, and M. Pǎtraşcu. On dynamic range reporting in one dimension. In *Proc. 37th ACM Sympos. Theory Comput.*, pages 104–111, 2005.

[MS92]   J. Matoušek and O. Schwarzkopf. Linear optimization queries. In *Proc. 8th Sympos. Comput. Geom.*, pages 16–25, 1992.

[MTT99]   Y. Manolopoulos, Y. Theodoridis, and V.J. Tsotras. *Advanced Database Indexing*. Kluwer Academic Publishers, Boston, 1999.

[Mul93]   K. Mulmuley. *Computational Geometry: An Introduction through Randomized Algorithms*. Prentice Hall, Englewood Cliffs, 1993.

[Mut06]   S. Muthukrishnan. Data streams: Algorithms and applications. *Foundations and Trends in Theoretical Computer Science*, 1, 2006.

[Nek14]   Y. Nekrich. Efficient range searching for categorical and plain data. *ACM Trans. Database Syst.*, 39:9, 2014.

[NW00]   J. Nievergelt and P. Widmayer. Spatial data structures: Concepts and design choices. In J.-R. Sack and J. Urrutia, editors, *Handbook of Computational Geometry*, pages 725–764, Elsevier, Amsterdam, 2000.

[Ove83]   M.H. Overmars. *The Design of Dynamic Data Structures*. Vol. 156 of *LNCS*, Springer, Berlin, 1983.

[PAHP02]   C.M. Procopiuc, P.K. Agarwal, and S. Har-Peled. Star-tree: An efficient self-adjusting index for moving points. In *Proc. 4th Workshop on Algorithm Engineering and Experiments*, vol. 2409 *LNCS*, pages 178–193, Springer, Berlin, 2002.

[PD06]   M. Pǎtraşcu and E.D. Demaine. Logarithmic lower bounds in the cell-probe model. *SIAM J. Comput.*, 35:932–963, 2006.

[Pǎt07]   M. Pǎtraşcu. Lower bounds for 2-dimensional range counting. In *Proc. 39th ACM Sympos. Theory Comput.*, pages 40–46, 2007.

[Pǎt11]   M. Pǎtraşcu. Unifying the landscape of cell-probe lower bounds. *SIAM J. Comput.*, 40:827–847, 2011.

[Rah17]   S. Rahul. Approximate range counting revisited. In *Proc. 33rd Sympos. Comput. Geom.*, vol. 77 of *LIPIcs*, article 55, Schloss Dagstuhl, 2017.

[Ram00]   E.A. Ramos. Linear programming queries revisited. In *Proc. 16th Sympos. Comput. Geom.*, pages 176–181, ACM Press, 2000.

[RJ12]   S. Rahul and R. Janardan. Algorithms for range-skyline queries. In *Proc. Intl. Conf. Adv. Geog. Inf. Sys.*, pages 526–529, 2012.

[Rob81]   J.T. Robinson. The k-d-B-tree: A search structure for large multidimensional dynamic indexes. Report CMU-CS-81-106, Dept. Comput. Sci., Carnegie-Mellon Univ., Pittsburgh, 1981.

[RT15]   S. Rahul and Y. Tao. On top-k range reporting in $2D$ space. In *Proc. 34th ACM Sympos. Principles Database Systems*, pages 265–275, 2015.

[RT16]   S. Rahul and Y. Tao. Efficient top-k indexing via general reductions. In *Proc. 35th ACM Sympos. Principles Database Systems*, pages 277–288, 2016.

[Sam90]   H. Samet. *The Design and Analysis of Spatial Data Structures*. Addison-Wesley, Reading, 1990.

[SRF87]   T. Sellis, N. Roussopoulos, and C. Faloutsos. The R^+-tree: A dynamic index for multi-dimensional objects. In *Proc. 13th VLDB Conference*, pages 507–517, 1987.

[SS05]   M. Sharir and H. Shaul. Ray shooting and stone throwing with near-linear storage. *Comput. Geom.*, 30:239–252, 2005.

[ST99]   B. Salzberg and V.J. Tsotras. A comparison of access methods for time evolving data. *ACM Comput. Surv.*, 31:158–221, 1999.

[STZ04]   S. Suri, C.D. Tóth, and Y. Zhou. Range counting over multidimensional data streams. In *Proc. 20th Sympos. Comput. Geom.*, pages 160–169, 2004.

[SY11]   M. Streppel and K. Yi. Approximate range searching in external memory. *Algorithmica*, 59:115–128, 2011.

[Vem12]   S. Vempala. Randomly-oriented k-d trees adapt to intrinsic dimension. In *Proc. IARCS Conf. Found. Soft. Tech. Theo. Comp. Sci.*, pages 48–57, 2012.

[ŠJLL00]   S. Šaltenis, C.S. Jensen, S.T. Leutenegger, and M.A. Lopez. Indexing the positions of continuously moving objects. In *Proc. ACM SIGMOD Internat. Conf. Management Data*, pages 331–342, 2000.

[Vui80]   J. Vuillemin. A unifying look at data structures. *Commun. ACM*, 23:229–239, 1980.

[VW99]   J.S. Vitter and M. Wang. Approximate computation of multidimensional aggregates of sparse data using wavelets. In *Proc. ACM SIGMOD Intl. Conf. Management Data*, pages 193–204, 1999.

[Wil82]   D.E. Willard. Polygon retrieval. *SIAM J. Comput.*, 11:149–165, 1982.

[WY13]   Z. Wei and K. Yi. The space complexity of 2-dimensional approximate range counting. In *Proc. 24th ACM-SIAM Sympos. Discrete Algorithms*, pages 252–264, 2013.

[Yao81]   A.C.-C. Yao. Should tables be sorted? *J. ACM*, 28:615–628, 1981.

[Yao82]   A.C.-C. Yao. Space-time trade-off for answering range queries. In *Proc. 14th ACM Sympos. Theory Comput.*, pages 128–136, 1982.

[Yao85]   A.C.-C. Yao. On the complexity of maintaining partial sums. *SIAM J. Comput.*, 14:277–288, 1985.

41 RAY SHOOTING AND LINES IN SPACE

Marco Pellegrini

INTRODUCTION

The geometry of lines in 3-space has been a part of the body of classical algebraic geometry since the pioneering work of Plücker. Interest in this branch of geometry has been revived by several converging trends in computer science. The discipline of computer graphics (Chapter 52) has pursued the task of rendering realistic images by simulating the flow of light within a scene according to the laws of elementary optical physics. In these models light moves along straight lines in 3-space and a computational challenge is to find efficiently the intersections of a very large number of rays with the objects comprising the scene. In robotics (Chapters 50 and 51) the chief problem is that of moving 3D objects without collisions. Effects due to the edges of objects have been studied as a special case of the more general problem of representing and manipulating lines in 3-space. Computational geometry (whose core is better termed "design and analysis of geometric algorithms") has moved in the nineties from the realm of planar problems to tackling directly problems that are specifically 3D. The new and sometimes unexpected computational phenomena generated by lines (and segments) in 3-space have emerged as a main focus of research.

In this chapter we will survey the present state of the art on lines and ray shooting in 3-space from the point of view of computational geometry. The emphasis is on provable nontrivial bounds for the time and storage used by algorithms for solving natural problems on lines, rays, and polyhedra in 3-space. We start by mentioning different possible choices of coordinates for lines (Section 41.1. This is an essential initial step because different coordinates highlight different properties of the lines in their interaction with other geometric objects. Here a special role is played by *Plücker coordinates* [Plu65], which represent the starting point for many results. Then we consider how lines interact with each other (Section 41.2). We are given a finite set of lines L that act as obstacles and we will define other (infinite) sets of lines induced by L that capture some of the important properties of visibility and motion problems. We show bounds on the storage required for a complete description of such sets. Then we move a step forward by considering the same sets of lines when the obstacles are polyhedral sets, more commonly encountered in applications. We arrive in Section 41.3 at the ray-shooting problem and its variants (on-line, off-line, arbitrary direction, fixed direction, and shooting with objects other than rays). Again, the obstacles are usually polyhedral objects, but in one case we are able to report a ray-shooting result on spheres.

Section 41.4 is devoted to the problem of collision-free movements (arbitrary or translation only) of lines among obstacles. This problem arises, for example, when lines are used to model radiation or light beams (e.g., lasers). In Section 41.5 we define a few notions of distance among lines, and as a consequence we have several natural proximity problems for lines in 3-space. Finding the closest pair in a set of lines is the most basic of such problems.

In Section 41.6 we survey what is known about the "dominance" relation among lines. This relation is central for many visibility problems in graphics. It is used, for example, in the painter's algorithm for hidden surface removal (Chapter 52). Another direction of research has explored the relation between lines in 3-space and their orthogonal projections. A central topic here is realizability: Given a set of planar lines together with a relation, does there exist a corresponding set of lines in 3-space whose dominance is consistent with the given relation?

41.1 COORDINATES OF LINES

GLOSSARY

Homogeneous coordinates: A point (x, y, z) in Cartesian coordinates has homogeneous coordinates (x_0, x_1, x_2, x_3), where $x = x_1/x_0$, $y = x_2/x_0$, and $z = x_3/x_0$.

Oriented lines: A line may have two distinct orientations. A line and an orientation form an oriented line.

Unoriented line: A line for which an orientation is not distinguished.

(I) Canonical coordinates by pairs of planes. The intersection of two planes with equations $y = az+b$ and $x = cz+d$ is a nonhorizontal line in 3-space, uniquely defined by the four parameters (a, b, c, d). Thus these parameters can be taken as coordinates of such lines. In fact, the space of nonhorizontal lines is homeomorphic to $\mathbb{R}^4$. Results on ray shooting among boxes and some lower bounds on stabbing are obtained using these coordinates.

(II) Canonical coordinates by pairs of points. Given two parallel horizontal planes, $z = 1$ and $z = 0$, the intersection points of a nonhorizontal line l with the two planes uniquely define that line. If $(x_0, y_0, 0)$ and $(x_1, y_1, 1)$ are two such points for l, then the quadruple (x_0, y_0, x_1, y_1) can be used as coordinates of l. Results on sets of horizontal polygons are obtained using these coordinates.

Although four is the minimum number of coordinates needed to represent an *unoriented* line, such parametrizations have proved useful only in special cases. Many interesting results have been derived using instead a five-dimensional parametrization for *oriented* lines, called **Plücker coordinates**.

(III) Plücker coordinates of lines. An oriented line in 3-space can be given by the homogeneous coordinates of two of its points. Let l be a line in 3-space and let $a = (a_0, a_1, a_2, a_3)$ and $b = (b_0, b_1, b_2, b_3)$ be two distinct points in homogeneous coordinates on l. We can represent the line l, oriented from a to b, by the matrix

$$l = \begin{pmatrix} a_0 & a_1 & a_2 & a_3 \\ b_0 & b_1 & b_2 & b_3 \end{pmatrix}, \quad \text{with } a_0, b_0 > 0.$$

By taking the determinants of the six 2×2 submatrices of the above 2×4 matrix we obtain the **homogeneous Plücker coordinates** of the line:

$$p(l) = (\xi_{01}, \xi_{02}, \xi_{03}, \xi_{12}, \xi_{31}, \xi_{23}), \quad \text{with } \xi_{ij} = det \begin{pmatrix} a_i & a_j \\ b_i & b_j \end{pmatrix}.$$

The six numbers ξ_{ij} are interpreted as homogeneous coordinates of a point in 5-space. For a given line l the six numbers are unique modulo a positive multiplicative factor, and they do not depend on the particular distinct points a and b that we have chosen on l. We call $p(l)$ the **Plücker point** of l in real projective 5-dimensional space $\mathbb{P}^5$.

We also define the **Plücker hyperplane** of the line l to be the hyperplane in $\mathbb{P}^5$ with vector of coefficients $v(l) = (\xi_{23}, \xi_{31}, \xi_{12}, \xi_{03}, \xi_{02}, \xi_{01})$. So the Plücker hyperplane is:

$$h(l) = \{p \in \mathbb{P}^5 \mid v(l) \cdot p = 0\}.$$

For each Plücker hyperplane we have a positive and a negative halfspace given by $h^+(l) = \{p \in \mathbb{P}^5 \mid v(l) \cdot p \geq 0\}$ and $h^-(l) = \{p \in \mathbb{P}^5 \mid v(l) \cdot p \leq 0\}$. Not every tuple of 6 real numbers corresponds to a line in 3-space since the Plücker coordinates must satisfy the condition

$$\xi_{01}\xi_{23} + \xi_{02}\xi_{31} + \xi_{03}\xi_{12} = 0. \tag{41.1.1}$$

The set of points in $\mathbb{P}^5$ satisfying Equation 41.1.1 forms the so-called **Plücker hypersurface** Π; it is also called the **Klein quadric** or the **Grassmannian** (manifold). The converse is also true: every tuple of six real numbers satisfying Equation 41.1.1 is the Plücker point of some line in 3-space. Given two lines l and l', they intersect or are parallel (i.e., they intersect at infinity) when the four defining points are coplanar. In this case the determinant of the 4×4 matrix formed by the 16 homogeneous coordinates of the four points is zero. In terms of Plücker coordinates we have the following basic lemmas.

LEMMA 41.1.1

Lines l and l' intersect or are parallel (meet at infinity) if and only if $p(l) \in h(l')$.

Note that Equation 41.1.1 states in terms of Plücker coordinates the fact that any line always meets itself.

LEMMA 41.1.2

Let l be an oriented line and t a triangle in Cartesian 3-space with vertices (p_0, p_1, p_2). Let l_i be the oriented line through (p_i, p_{i+1}) (indices mod 3). Then l intersects t if and only if either $p(l) \in h^+(l_0) \cap h^+(l_1) \cap h^+(l_2)$ or $p(l) \in h^-(l_0) \cap h^-(l_1) \cap h^-(l_2)$.

These two lemmas allow us to map combinatorial and algorithmic problems involving lines (and polyhedral sets) in 3-space into problems involving sets of hyperplanes and points in projective 5-space (Plücker space). The main advantage is that we can use the rich collection of results on the combinatorics of high dimensional arrangements of hyperplanes (see Chapter 28). The main drawback is that we are using five (nonhomogeneous) parameters, instead of four which is the minimum number necessary. This choice has a potential for increasing the time bounds of line algorithms. We are rescued by the following theorem:

THEOREM 41.1.3   [APS93]

Given a set H of n hyperplanes in 5-dimensional space, the complexity of the cells of the arrangement $\mathcal{A}(H)$ intersected by the Plücker hypersurface Π (also called the **zone** *of Π in $\mathcal{A}(H)$) is $O(n^4 \log n)$.*

Although the entire arrangement $\mathcal{A}(H)$ can be of complexity $\Theta(n^5)$, if we are working only with Plücker points we can limit our constructions to the zone of Π, the complexity of which is one order of magnitude smaller. Theorem 41.1.3 is especially useful for deriving ray-shooting results.

The list of coordinatizations discussed in this section is by no means exhaustive. Other parametrizations are used, for example, in [Ame92] and [AAS97].

A TYPICAL EXAMPLE

A typical example of the use of Plücker coordinates in 3D problems is the result for fast ray shooting among polyhedra (see Table 41.3.1). We triangulate the faces of the polyhedra and extend each edge to a full line. Each such line is mapped to a Plücker hyperplane. Lemma 41.1.2 guarantees that each cell in the resulting arrangement of Plücker hyperplanes contains Plücker points that pass through the same set of triangles. Thus to answer a ray-shooting query, we first locate the query Plücker point in the arrangement, and then search the list of triangles associated with the retrieved cell. This final step is accomplished using a binary search strategy when the polyhedra are disjoint. Theorem 41.1.3 guarantees that we need to build a point-location structure only for the zone of the Plücker hypersurface, thus saving an order of magnitude over general point-location methods for arrangements (see Sections 28.7 and 38.3).

41.2   SETS OF LINES IN 3-SPACE

With Plücker coordinates (III) to represent oriented lines, we can use the topology induced by the standard topology of 5-dimensional projective space $\mathbb{P}^5$ on Π as a natural topology on sets of oriented lines. Using the four-dimensional coordinatizations (I) or (II), we can impose the standard topology of $\mathbb{R}^4$ on the set of nonhorizontal unoriented lines. Thus we can define the concepts of "neighborhood," "continuous path," "open set," "closed set," "boundary," "path-connected component," and so on, for the set $\mathcal{L}$ of lines in 3-space. The distinction between oriented lines and unoriented lines is mainly technical and the complexity bounds hold in either case.

FAMILIES OF LINES INDUCED BY A FINITE SET OF LINES

GLOSSARY

Semialgebraic set: The set of all points that satisfy a Boolean combination of a finite number of algebraic constraints (equalities and inequalities) in the Cartesian coordinates of $\mathbb{R}^d$. See Chapter 37.

Path-connected component: A maximal set of lines that can be connected by a path of lines, a continuous function from the interval $[0, 1]$ to the space of lines.

Positively-oriented lines: Oriented lines l'_1 and l'_2 on the xy-plane are positively-oriented if the triple scalar product of vectors parallel to l_1', l_2', and the positive z-axis is positive.

Consistently-oriented lines: An oriented line l in 3-space is oriented consistently with a 3D set L of oriented lines if the projection l' of l onto the xy-plane is positively-oriented with the projection of every line in L.

A finite set L of n lines in 3-space can be viewed as an obstacle to the free movement of other lines in 3-space. Many applications lead to defining families of lines with some special properties with respect to the fixed lines L. The resources used by algorithms for these applications are often bounded by the "complexity" of such families.

The boundary of a semialgebraic set in $\mathbb{R}^4$ is partitioned into a finite number of faces of dimension 0, 1, 2, and 3, each of which is also a semialgebraic set. The number of faces on the boundary of a semialgebraic set is the ***complexity*** of that set. The families of lines that we consider are represented in $\mathbb{R}^4$ by semialgebraic sets, with the coefficients of the corresponding algebraic constraints a function of the given finite set of lines L.

The set Miss(L) consists of lines that do not meet any line in L. The sets Vert(L) and Free(L) consists of lines that may be translated to infinity without collision with lines in L. The basic complexities displayed in Table 41.2.1 are derived from [CEG$^+$96, Pel94b, Aga94].

TABLE 41.2.1   Complexity of families of lines defined by lines.

SET OF LINES	DEFINITION	COMPLEXITY
Miss(L)	do not meet any line in L	$\Theta(n^4)$
1 component of Miss(L)	1 path-connected component	$\Theta(n^2)$
Vert(L)	can be translated vertically to ∞	$\Theta(n^3)$
Free(L)	can be translated to ∞ in some direction	$\Omega(n^3), O(n^3 \log n)$
Vert$CO(L)$	above L and oriented consistently with L	$\Theta(n^2)$

MEMBERSHIP TESTS

Given L, we can build a data structure during a preprocessing phase so that when presented with a new (query) line l, we can decide efficiently whether l is in one of the sets defined in the previous section. Such an algorithm implements a membership test for a group of lines. Table 41.2.2 shows the main results.

TABLE 41.2.2   Membership tests for families of lines defined by lines.

SET OF LINES	QUERY TIME	PREPROC/STORAGE	SOURCE
Miss(L)	$O(\log n)$	$O(n^{4+\epsilon})$	[Pel93b, AM93]
1 component of Miss(L)	$O(\log n)$	$O(n^{2+\epsilon})$	[Pel91]
Vert(L), VertCO(L)	$O(\log n)$	$O(n^{2+\epsilon})$	[CEG$^+$96]
Free(L)	$O(\log n)$	$O(n^{3+\epsilon})$	[Pel94b]

FAMILIES OF LINES INDUCED BY POLYHEDRA

GLOSSARY

ϵ: A positive real number, which we may choose arbitrarily close to zero for each algorithm or data structure. A caveat is that the multiplicative constant implicit in the big-O notation depends on ϵ and its value increases when ϵ tends to zero.

$\alpha(\,\cdot\,)$:   The inverse of Ackermann's function. $\alpha(n)$ grows very slowly and is at most 4 for any practical value of n. See Section 28.10.

$\beta(\,\cdot\,)$: $\beta(n) = 2^{c\sqrt{\log n}}$ for a constant c. $\beta(\cdot)$ is a function that is smaller than any polynomial but larger than any polylogarithmic factor. Formally we have that for every $a, b > 0$, $\log^a n \leq \beta(n) \leq n^b$ for any $n \geq n_0(a, b)$.

Polyhedral set P: A region of 3-space bounded by a collection of interior-disjoint vertices, segments, and planar polygons. We denote by n the total number of vertices, edges, and faces.

Star-shaped polyhedron: A polyhedron P for which there exists a point $o \in P$ such that for every point $p \in P$, the open segment op is contained in P.

Terrain: When the star-shaped polyhedron is unbounded and o is at infinity we obtain a terrain, a monotone surface (cf. Section 30.1).

Horizontal polygons: Convex polygons contained in planes parallel to the xy-plane.

A collection of polyhedra in 3-space may act as obstacles limiting the collision-free movements of lines. Following the blueprint of the previous section, the complexity of some interesting families of lines induced by polyhedra are displayed in Table 41.2.3 (see [HS94, Pel94b, Aga94] and [AAKS04, GL10, Rub12]).

TABLE 41.2.3   Complexity of families of lines defined by polyhedra and spheres.

SET OF LINES	DEFINITION	COMPLEXITY
Miss(P)	do not meet polyhedron P	$\Theta(n^4)$
Vert(P)	can be translated vertically to ∞	$\Omega(n^3)$, $O(n^3\beta(n))$
Free(P)	can be translated to ∞ in some direction	$\Omega(n^3)$
Miss(Q), Free(Q)	Q star-shaped polyhedron or a terrain	$\Omega(n^2\alpha(n))$, $O(n^3\log n)$
Miss(U)	U a set of n unit balls	$O(n^{3+\epsilon})$
Miss(B)	B a set of n balls	$\Omega(n^3)$, $O(n^{3+\epsilon})$
Miss(H)	H a set of horizontal polygons with n edges	$\Omega(n^2)$, $O(n^4)$

Similarly, we can define families of *3D segments* defined by polyhedra in 3D. The set of relatively open segments that miss P is also a semialgebraic set, known as the **3D Visibility skeleton** (see [DDP97, Dur99] and [DDP02, Zha09]). Its combinatorial complexity is $\Theta(n^4)$. The visibility skeleton induced by k possibly intersecting convex polyhedra of total size n has $\Theta(n^2k^2)$ connected components [BDD+07]. The Visibility skeleton induced by n uniformly distributed unit spheres has linear expected complexity [DDE+03].

OPEN PROBLEMS

1. Find an almost cubic upper bound on the complexity of the group of lines Free(P) for a polyhedron P.

2. Close the gap between the quadratic lower and the cubic upper bound for the group Free(T) induced by a terrain T (Table 41.2.3).

SETS OF STABBING LINES

GLOSSARY

Stabber: A line l that intersects every member of a collection $\mathcal{P} = \{P_1, ..., P_k\}$ of polyhedral sets. The sum of the sizes of the polyhedral sets in $\mathcal{P}$ is n. The set of lines stabbing $\mathcal{P}$ is denoted $S(\mathcal{P})$. Stabbers are also called *line transversals*.

Box: A parallelepiped each of whose faces is orthogonal to one of the three Cartesian axes.

c-oriented: Convex polyhedra whose face normals come from a set of c fixed directions.

Table 41.2.4 lists the worst-case complexity of the set $S(\mathcal{P})$ and the time to find a witness stabbing line.

TABLE 41.2.4   Complexity of the set of stabbing lines and detection time.

OBJECTS	COMPLEXITY OF $S(\mathcal{P})$	FIND TIME	SOURCES
Convex polyhedra	$\Omega(n^3)$, $O(n^3 \log n)$	$O(n^3 \beta(n))$	[PS92, Pel93a, Aga94]
k polyhedra	$O(n^2 k^{1+\epsilon})$	$O(n^2 k^{1+\epsilon})$	[KRS10]
Boxes	$O(n^2)$	$O(n)$	[Ame92, Meg91]
c-oriented polyhedra	$O(n^2)$	$O(n^2)$	[Pel91]
Horiz. polygons	$\Theta(n^2)$	$O(n)$	[Pel91]

Note that in some cases (boxes, parallel polygons) a stabbing line can be found in linear time, even though the best bound known for the complexity of the stabbing set is quadratic. These results are obtained using linear programming techniques (Chapter 49).

We can determine whether a given line l is a stabber for a preprocessed set $\mathcal{P}$ of convex polyhedra in time $O(\log n)$, using data structures of size $O(n^{2+\epsilon})$ that can be constructed in time $O(n^{2+\epsilon})$ [PS92].

For an *oriented* stabber l and a set $\mathcal{O}$ of k *disjoint* convex bodies in $\mathbb{R}^d$, the order of the intersection of the objects along l is called a **geometric permutation** (cf. Chapter 4). A result of Cheong et al. [CGN05] shows that for k disjoint balls of unit radius in $\mathbb{R}^3$ and $k > 9$, there are at most two geometric permutations, while for $3 \leq k \leq 9$ there are at most three geometric permutations. For k disjoint balls of any radius in $\mathbb{R}^3$ there are $\Theta(k^2)$ geometric permutations [SMS00]. If the ratio of the largest radius to the smallest radius of the spheres in the collection is a constant c, then there are at most $O(c^{\log c})$ geometric permutations. For k rectangular boxes in $\mathbb{R}^d$ there are at most 2^{d-1} geometric permutations, which is tight (see also [O'R01]). There are at most n geometric permutations of n line segments in $\mathbb{R}^3$ [BEL+05]; while for arbitrary disjoint convex sets there are $O(n^3 \log n)$ geometric permutations [RKS12].

OPEN PROBLEMS

1. Can linear programming techniques yield a linear-time algorithm for c-oriented polyhedra?

2. The lower bound for $S(\mathcal{P})$ for a set of pairwise *disjoint* convex polyhedra is only $\Omega(n^2)$ [PS92]. Close the gap between this and the cubic upper bound.

41.3 RAY SHOOTING

Ray shooting is an important operation in computer graphics and a primitive operation useful in several geometric computations (e.g., hidden surface removal, and detecting and computing intersections of polyhedra). The problem is defined as follows. Given a large collection $\mathcal{P}$ of simple polyhedral objects, we want to know, for a given point p and direction $\vec{d}$, the first object in $\mathcal{P}$ intersected by the ray defined by the pair $p, \vec{d}$. A single polyhedron with many faces can be represented without loss of generality by the collection of its faces, each treated as a separate polygon.

ON-LINE RAY SHOOTING IN AN ARBITRARY DIRECTION

Here we consider the on-line model in which the set $\mathcal{P}$ is given in advance and a data structure is produced and stored. Afterward we are given the query rays one-by-one and the answer to one query must be produced before the next query is asked.

Table 41.3.1 summarizes the known complexity bounds on this problem. For a given class of objects we report the query time, the storage, and the preprocessing time of the method with the best bound. In this table and in the following ones we omit the big-O symbols. Again, n denotes the sum of the sizes of all the polyhedra in $\mathcal{P}$. The main references on ray shooting (Table 41.3.1) are in [Pel93b, BHO+94] (boxes), [AM93, AM94, Pel93b, BHO+94, AS93] (polyhedra), [Pel96] (horizontal polygons), [AAS97, MS94] (spheres), [DK85, AS96] (convex polyhedra), [ABG08] (fat convex polyhedra), and [Kol04] (semialgebraic sets of constant complexity).

GLOSSARY

Fat horizontal polygons: Convex polygons contained in planes parallel to the xy-plane, with a constant lower bound on the ratio of the radius of the maximum inscribed circle over the radius of the minimum enclosing circle.

Curtains: Polygons in 3-space bounded by one segment and by two vertical rays from the endpoints of the segment.

Axis-oriented curtains: Curtains hanging from a segment parallel to the x- or y-axis.

c-fat polyhedra: A convex polyhedron P in $\mathbb{R}^3$ is c-fat, for $0 \leq c \leq 1$ if, for any ball b whose center lies in P and which does not completely contain P, we have: $vol(b \cap P) \geq c \cdot vol(b)$.

TABLE 41.3.1   On-line ray shooting in an arbitrary direction.

OBJECTS	QUERY	STORAGE	PREPROCESSING
Boxes, terrains, curtains	$\log n$	$n^{2+\epsilon}$	$n^{2+\epsilon}$
Boxes	$n^{1+\epsilon}/m^{1/2}$	$n \leq m \leq n^2$	$m^{1+\epsilon}$
Polyhedra	$\log n$	$n^{4+\epsilon}$	$n^{4+\epsilon}$
Polyhedra	$n^{1+\epsilon}/m^{1/4}$	$n \leq m \leq n^4$	$m^{1+\epsilon}$
Fat horiz. polygons	$\log n$	$n^{2+\epsilon}$	$n^{2+\epsilon}$
Horiz. polygons	$\log^3 n$	$n^{3+\epsilon} + K$	$n^{3+\epsilon} + K \log n$
Spheres	$\log^4 n$	$n^{3+\epsilon}$	$n^{3+\epsilon}$
1 convex polyhedron	$\log n$	n	$n \log n$
s convex polyhedra	$\log^2 n$	$n^{2+\epsilon} s^2$	$n^{2+\epsilon} s^2$
c-fat convex polyhedra	$(n/m^{1/2}) \log^2 n$	$n^{1+\epsilon} \leq m \leq n^{2+\epsilon}$	$m^{1+\epsilon}$
semialgebraic sets of $O(1)$ complexity	$n \log n$	$n^{4+\epsilon}$	$n^{4+\epsilon}$

When we drop the fatness assumption for horizontal polygons we obtain bounds that depend on K, the actual complexity of the set of lines missing the *edges* of the polygons (see Section 41.2).

Most of the data structures for ray shooting mentioned in Table 41.3.1 can be made dynamic (under insertion and deletion of objects in the scene) by using general dynamization techniques (see [Meh84, AEM92]).

ON-LINE RAY SHOOTING IN A FIXED DIRECTION

We can usually improve on the general case if the direction of the rays is fixed a priori, while the source of the ray can lie anywhere in $\mathbb{R}^3$. See Table 41.3.2; here k is the number of vertices, edges, faces, and cells of the arrangement of the (possibly intersecting) polyhedra. References for ray shooting in a fixed direction (Table 41.3.2) are [Ber93, BGH94] and [BG08, KRS09].

OFF-LINE RAY SHOOTING IN AN ARBITRARY DIRECTION

In the previous section we considered the on-line situation when the answer to the query must be generated before the next question is asked. In many situations we do not need such strict requirements. For example, we might know all the queries from the start and are interested in minimizing the total time needed to answer all of the queries (the *off-line* situation). In this case there are simpler algorithms that improve on the storage bounds of on-line algorithms:

TABLE 41.3.2   On-line ray shooting in a fixed direction.

OBJECTS	QUERY TIME	STORAGE	PREPROCESSING
Boxes	$\log n$	$n^{1+\epsilon}$	$n^{1+\epsilon}$
Boxes	$\log n (\log \log n)^2$	$n \log n$	$n \log^2 n$
Axis-oriented curtains	$\log n$	$n \log n$	$n \log n$
Polyhedra	$\log^2 n$	$n^{2+\epsilon} + k$	$n^{2+\epsilon} + k \log n$
Polyhedra	$n^{1+\epsilon}/m^{1/3}$	$n \leq m \leq n^3$	$m^{1+\epsilon}$
c-fat convex constant size polyhedra	$\log^2 n$	$n \log^2 n$	$n \log^2 n$
h polyhedra	$\log^2 n$	$nh^2 \log^2 n$	$nh^2 \log^2 n$

THEOREM 41.3.1

Given a polyhedral set $\mathcal{P}$ with n vertices, edges, and faces, and given m rays off-line, we can answer the m ray-shooting queries in time $O(m^{0.8}n^{0.8+\epsilon} + m \log^2 n + n \log n \log m)$ using $O(n+m)$ storage.

One of the most interesting applications of this result is the current asymptotically fastest algorithm for detecting whether two nonconvex polyhedra in 3-space intersect, and to compute their intersection. See Table 41.3.3; here k is the size of the intersection.

TABLE 41.3.3   Detection and computation of intersection among polyhedra.

OBJECTS	DETECTION	COMPUTATION	SOURCES
Polyhedra	$n^{1.6+\epsilon}$	$n^{1.6+\epsilon} + k \log^2 n$	[Pel93b]
Terrains	$n^{4/3+\epsilon}$	$n^{4/3+\epsilon} + k^{1/3}n^{1+\epsilon} + k \log^2 n$	[CEGS94, Pel94b]

For a set of *convex* polyhedra in 3D with a total of n vertices, the number h of intersecting pairs of polyhedra can be computed in time $O(n^{1.6+\epsilon} + h)$ [ABHP+02]. Given s convex polyhedra of total complexity n, we can pre-process them in linear time and storage so that any pair of them can be tested for intersection in time $O(\log n)$ per pair [BL15]. Lower bounds on off-line ray-shooting and intersection problems in 3D are difficult to prove. It has been shown in [Eri95] that many such problems are at least as hard as Hopcroft's incidence problem in the appropriate ambient space (see Chapter 40).

RAY-SHOOTING IN SIMPLICIAL COMPLEXES

If we have a subdivision of the free space $\mathbb{R}^3 \setminus \mathcal{P}$ into a simplicial complex we can answer ray-shooting queries by locating the tetrahedron containing the source of the ray and tracing the ray in the complex at cost $O(1)$ for each visited face of the complex. There are scenes $\mathcal{P}$ for which any simplicial complex has some line meeting $\Omega(n)$ faces of the complex. The average time for tracing a ray in a simplicial complex is proportional to the sum of the areas of all faces in the complex. It is possible to find a complex of total surface area within a constant multiplicative factor of the minimum, with $O(n^3 \log n)$ simplices in time $O(n^3 \log n)$ for general $\mathcal{P}$. For $\mathcal{P}$, a point set or a single polyhedron $O(n^2 \log n)$ time suffices (see [AAS95, AF99, CD99]). These results are obtained via a generalization of Eppstein's method for two-dimensional Minimum Weighted Steiner Triangulation (2D-MWST) of a point set [Epp94]. In the 3D context the weight is the surface area of the 2D faces of the complex. Starting from the set $\mathcal{P}$ of polyhedral obstacles in $\mathbb{R}^3$, an oct-tree-based decomposition of $\mathbb{R}^3$ is produced which is "balanced" and "smooth." It is then proved, via a charging argument, that the sum of the surface areas of all the boxes in the decomposition is within a constant factor of the surface area of any Minimum Surface Steiner Simplicial Complex compatible with $\mathcal{P}$. From the oct-tree the final complex is derived within just a constant factor increase in the total surface.

EXTENSIONS AND ALTERNATIVE METHODS

Some ray-shooting results of Agarwal and Matoušek are obtained from the observation that a ray is traced by a family of segments $\rho(t)$, where one endpoint is the ray source and the second endpoint lies on the ray at distance t from the source. Using *parametric search* techniques Agarwal and Matoušek compute the first value of t for which $\rho(t)$ intersects $\mathcal{P}$, and thus answer the ray-shooting query.

An interesting extension of the concept of shooting rays against obstacles is obtained by shooting triangles and more generally simplices. We consider a family of simplices $s(t)$, indexed by real parameter $t \in \mathbb{R}^+$, where t is the volume of the simplex $s(t)$, such that the simplices form a chain of inclusions: $t_1 \leq t_2 \Rightarrow s(t_1) \subset s(t_2)$, Intuitively we grow a simplex until it first meets one of the obstacles. Surprisingly, when the obstacles are general polyhedra, shooting simplices is not harder than shooting rays.

THEOREM 41.3.2   [Pel94a]

Given a set of polyhedra $\mathcal{P}$ with n edges we can preprocess it in time $O(m^{1+\epsilon})$ into a data structure of size m, such that the following queries can be answered in time $O(n^{1+\epsilon}/m^{1/4})$: Given a simplex s, does s avoid $\mathcal{P}$? Given a family of simplices $s(t)$ as above, which is the first value of t^* for which $s(t^*)$ intersects $\mathcal{P}$?

When the polyhedra are convex and c-fat, fixed simplex intersection queries can be answered in time $O((n/m^{1/3}) \log n)$ [ABG08] with $O(m^{1+\epsilon})$ storage. Thus by applying parametric search to this data structure one can also solve the shooting simplices problem within the same time/storage bound for this class of polyhedra. Computing the interaction between beams and polyhedral objects is a central problem in radio-therapy and radio-surgery (see, e.g., [SAL93, For99, CHX00]).

Other popular methods for solving ray-shooting problems are based binary space partitions, kD-trees, solid modeling schemes, etc. These methods, although important in practice, are usually not fully analyzable a priori using algorithmic analysis. In [ABCC06] Aronov et al. propose techniques that give a posteriori estimates of the cost of ray shooting.

OPEN PROBLEMS

1. Find time and storage bounds for ray-shooting among general polyhedra that

are sensitive to the actual complexity of a group of lines (as opposed to the worst-case bound on such a complexity).

2. For a collection of s convex polyhedra there is a wide gap in storage and preprocessing requirements for ray-shooting between the special case $s = 1$ and the case for general s. It would be interesting to obtain a bound that depends smoothly on s.

3. No lower bound on time or storage required for ray shooting is known.

41.4 MOVING LINES AMONG OBSTACLES

ARBITRARY MOTIONS

So far we have treated lines as static objects. In this section we consider moving lines. A laser beam in manufacturing or a radiation beam in radiation therapy can be modeled as lines in 3-space moving among obstacles. The main computational problem is to decide whether a source line l_1 can be moved continuously until it coincides with a target line l_2 while avoiding a set of obstacles $\mathcal{P}$. We consider the following situation where the set of obstacles $\mathcal{P}$ is given in advance and preprocessed to obtain a data structure. When the query lines l_1 and l_2 are given the answer is produced before a new query is accepted. We have the results shown in Table 41.4.1, where K is the complexity of the set of lines missing the edges of a set of horizontal polygons (cf. Section 41.2). The result on moving lines among polyhedral obstacles is extended in [Kol05] to moving a line segment, with the same complexity.

TABLE 41.4.1   On-line collision-free movement of lines among obstacles.

OBJECTS	QUERY TIME	STORAGE	PREPROC	SOURCES
Polyhedra	$\log n$	$n^{4+\epsilon}$	$n^{4+\epsilon}$	[Pel93b]
Horiz. polygons	$\log^3 n$	$n^{3+\epsilon} + K$	$n^{3+\epsilon} + K \log n$	[Pel96]

OPEN PROBLEMS

It is not known how to trade off storage and query time, or whether better bounds can be obtained in an off-line situation.

TRANSLATIONS

We now restrict the type of motion and consider only translations of lines. The first result is negative: there are sets of lines which cannot be split by any collision-free translation. There exists a set L of 9 lines such that, for all directions v and all subsets $L_1 \subset L$, L_1 cannot be translated continuously in direction v without collisions with $L \setminus L_1$ [SS93]. Positive results are displayed in Table 41.4.2.

GLOSSARY

Towering property: Two sets of lines L_1 and L_2 are said to satisfy the towering property if we can translate simultaneously all lines in L_1 in the vertical direction without any collision with any lines in L_2.

Separation property: Two sets of lines satisfy the separation property if they satisfy the towering property in some direction (not necessarily vertical).

TABLE 41.4.2   Separating lines by translations.

PROPERTY	TIME TO CHECK PROPERTY	SOURCES
Towering	$O(n^{4/3+\epsilon})$	[CEG$^+$96]
Separation	$O(n^{3/2+\epsilon})$	[Pel94b]

41.5 CLOSEST PAIR OF LINES

GLOSSARY

Distance between lines: The Euclidean distance between two lines l_1 and l_2 in 3-space is the length of the shortest segment with one endpoint on l_1 and the other on l_2.

Vertical distance between lines (segments):   The length of the shortest vertical segment with one endpoint on line l_1 (resp. segment s_1) and one endpoint on line l_2 (resp. segment s_2). If no vertical segment joins two lines (resp. segments) the vertical distance is undefined.

TABLE 41.5.1   Closest and farthest pair of lines and segments.

PROBLEM	OBJECTS	TIME	SOURCES
Smallest distance	lines	$O(n^{8/5+\epsilon})$	[CEGS93]
Smallest vertical distance	lines, segments	$O(n^{8/5+\epsilon})$	[Pel94a]
Largest vertical distance	lines, segments	$O(n^{4/3+\epsilon})$	[Pel94a]

Note that when some of the lines/segments are co-planar on a vertical plane, the problem of finding the smallest/largest vertical distance among them degenerates to a simpler distance problem relative to a planar arrangements of lines (or segments). Any centrally symmetric convex polyhedron C in 3D defines a metric. If C has constant combinatorial complexity, then the complexity of the Voronoi diagram of n lines in 3-space is $O(n^2\alpha(n)\log n)$ [CKS$^+$98]. For Euclidean distance the best bound is $O(n^{3+\epsilon})$.

OPEN PROBLEM

1. Finding an algorithm with subquadratic time complexity for the smallest distance among segments (and more generally, among polyhedra) is a notable open question.

2. Close the gap between the complexity of Voronoi diagrams of lines induced by polyhedral metrics and the Euclidean metric.

41.6 DOMINANCE RELATION AND WEAVINGS

GLOSSARY

Dominance relation: Given a finite set L of nonvertical disjoint lines in $\mathbb{R}^3$, define a dominance relation $\prec$ among lines in L as follows: $l_1 \prec l_2$ if l_2 lies above l_1, i.e., if, on the vertical line intersecting l_1 and l_2, the intersection with l_1 has a smaller z-coordinate than does the intersection with l_2.

Weaving: A weaving is a pair $(L', \prec')$ where L' is a set of lines on the plane and $\prec'$ is an anti-symmetric nonreflexive binary relation $\prec' \subset L' \times L'$ among the lines in L'.

Realizable: A weaving is realizable if there exists a set of lines L in 3-space such that L' is the projection of L and $\prec'$ is the image of the dominance relation $\prec$ for L.

Elementary cycle: A cycle in the dominance relation such that the projections of the lines in the cycle bound a cell of the arrangement of projected lines.

Perfect: A weaving $(L', \prec')$ is perfect if each line l alternates below and above the other lines in the order they cross l (see Figure 41.6.1a).

Bipartite weaving: Two families of segments in 3-space such that, when projecting on the xy-plane, each segment does not meet segments from its own family and meets all the segments from the other family in the same order. (A bipartite weaving of size 4×4 is shown in Figure 41.6.1b.)

Perfect bipartite weaving: Every segment alternates above and below the segments of the other family (see Figure 41.6.1b).

The dominance relation is possibly cyclic, that is, there may be three lines such that $l_1 \prec l_2 \prec l_3 \prec l_1$. Some results in [CEG+92, PPW93, BOS94, Sol98] [HPS01, BDP05, ABGM08, AS16] related to dominance are the following:

1. *How fast can we generate a consistent linear extension if the relation $\prec$ is acyclic?* $O(n^{4/3+\epsilon})$ time is sufficient for the case of lines. This result has been extended to the case of segments and of polyhedra. If an ordering is given as input, it is possible to *verify* that it is a linear extension of $\prec$ in time $O(n^{4/3+\epsilon})$.

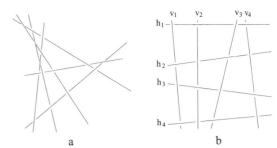

FIGURE 41.6.1

(a) *A perfect weaving;*

(b) *a perfect bipartite weaving.*

2. *How many elementary cycles in the dominance relation can n lines define?* In the case of bipartite weavings, the dominance relation has $O(n^{3/2})$ elementary cycles and there is a family of bipartite weavings attaining the lower bound $\Omega(n^{4/3})$. For general weavings there is a construction attaining $\Omega(n^{3/2})$.

3. *If we cut the segments to eliminate cycles, how many "cuts" are necessary to eliminate all cycles?* From the previous result we have that sometimes $\Omega(n^{4/3})$ cuts are necessary since a single cut can eliminate only one elementary cycle. In order to eliminate all cycles (including the nonelementary ones) in any weaving, $O(n^{3/2}\text{polylog}(n))$ cuts are always sufficient.

4. *How fast can we find those cuts?* There are algorithms to find cuts in bipartite weavings in time $O(n^{9/5}\log n)$, and in time $O(n^{11/6+\epsilon})$ for general weavings. In a general weaving, calling μ is the minimum number of cuts, there is an algorithm to cut all cycles in time $O(n^{4/3+\epsilon}\mu^{1/3})$ that produces $O(n^{1+\epsilon}\mu^{1/3})$ cuts. Finding the minimum number μ of cuts is an NP-complete problem, and there is a polynomial-time approximation algorithm producing a set of cuts of size within a factor $O(\log \mu \log \log \mu)$ of the optimal.

5. The fraction of realizable weavings over all possible weavings of n lines tends to 0 exponentially as n tends to ∞. This result holds also when we generalize lines into semi-algebraic curves defined coordinate-wise by polynomials of constant degree.

6. A perfect weaving of $n \geq 4$ lines is not realizable.

7. Perfect bipartite weavings are realizable if and only if one of the families has fewer than four segments.

41.7 SOURCES AND RELATED MATERIAL

FURTHER READING

[Som51, HP52, Jes69]: Extensive book-length treatments of the geometry of lines in space.

[Sto89, Sto91]: Algorithmic aspects of computing in projective spaces.

[BR79, Shi78]: Uses of the geometry of lines in robotics. For uses in graphics see [FDFH90].

[Ber93] [PS09, Chapter 7]: A detailed description of many ray-shooting results.

[Spe92, Dur99, Hav00]: Pointers to the vast related literature on pragmatic aspects of ray shooting.

[Goa10]: A survey on line transversals for sets of balls.

RELATED CHAPTERS

REFERENCES

[AAKS04]   P.K. Agarwal, B. Aronov, V. Koltun, and M. Sharir. On lines avoiding unit balls in three dimensions. *Discrete Comput. Geom.*, 34:231–250, 2005.

[AAS95]   P.K. Agarwal, B. Aronov, and S. Suri. Stabbing triangulations by lines in 3D. In *Proc. 11th Sympos. Comput. Geom.*, pages 267–276, ACM Press, 1995.

[AAS97]   P.K. Agarwal, B. Aronov, and M. Sharir. Computing envelopes in four dimensions with applications. *SIAM J. Computing*, 26:1714–1732, 1997.

[ABCC06]   B. Aronov, H. Brönnimann, A.Y. Chang, and Y.-J. Chiang. Cost prediction for ray shooting in octrees. *Comput. Geom.*, 34:159–181, 2006.

[ABG08]   B. Aronov, M. de Berg, and C. Gray. Ray shooting and intersection searching amidst fat convex polyhedra in 3-space. *Comput. Geom.*, 41:68–76, 2008.

[ABGM08]   B. Aronov, M. de Berg, C. Gray, and E. Mumford. Cutting cycles of rods in space: hardness and approximation. In *Proc. 19th ACM-SIAM Sympos. Discrete Algorithms*, pages 1241–1248, 2008.

[ABHP+02] P.K. Agarwal, M. de Berg, S. Har-Peled, M.H. Overmars, M. Sharir, and J. Vahrenhold. Reporting intersecting pairs of convex polytopes in two and three dimensions. *Comput. Geom.*, 23:195–207, 2002.

[AEM92]   P.K. Agarwal, D. Eppstein, and J. Matoušek. Dynamic half-space range searching, geometric optimization and minimum spanning trees. In *Proc. 33rd IEEE Sympos. Found. Comp. Sci.*, pages 80–89, 1992.

[AF99]   B. Aronov and S. Fortune. Approximating minimum-weight triangulations in three dimensions. *Discrete Comput. Geom.*, 21:527–549, 1999.

[Aga94]   P.K. Agarwal. On stabbing lines for convex polyhedra in 3D. *Comput. Geom.*, 4:177–189, 1994.

[AM93]   P.K. Agarwal and J. Matoušek. Ray shooting and parametric search. *SIAM J. Comput.*, 22:794–806, 1993.

[AM94]   P.K. Agarwal and J. Matoušek. Range searching with semialgebraic sets. *Discrete Comput. Geom.*, 11:393–418, 1994.

[Ame92]   N. Amenta. Finding a line transversal of axial objects in three dimensions. In *Proc. 3rd ACM-SIAM Sympos. Discrete Algorithms*, pages 66–71, 1992.

[APS93]   B. Aronov, M. Pellegrini, and M. Sharir. On the zone of an algebraic surface in a hyperplane arrangement. *Discrete Comput. Geom.*, 9:177–188, 1993.

[AS93]   P.K. Agarwal and M. Sharir. Applications of a new space partitioning technique. *Discrete Comput. Geom.*, 9:11–38, 1993.

[AS96]   P.K. Agarwal and M. Sharir. Ray shooting amidst convex polyhedra and polyhedral terrains in three dimensions. *SIAM J. Comput.*, 25:100–116, 1996.

[AS16]   B. Aronov and M. Sharir. Almost tight bounds for eliminating depth cycles in three dimensions. In *Proc. 48th ACM Sympos. Theory Comput.*, pages 1–8, 2016.

[BDD+07]   H. Brönnimann, O. Devillers, V. Dujmović, H. Everett, M. Glisse, X. Goaoc, S. Lazard, H.-S. Na, and S. Whitesides. Lines and free line segments tangent to arbitrary three-dimensional convex polyhedra. *SIAM J. Comput.*, 37:522–551, 2007.

[BDP05]   S. Basu, R. Dhandapani, and R. Pollack. On the realizable weaving patterns of polynomial curves in $\mathbb{R}^3$. In *Proc. 12th Sympos. Graph Drawing*, pages 36–42, vol. 3383 of *LNCS*, Springer, Berlin, 2005.

[BEL+05]   H. Brönnimann, H. Everett, S. Lazard, F. Sottile, and S. Whitesides. Transversals to line segments in three-dimensional space. *Discrete Comput. Geom.*, 34:381–390, 2005.

[Ber93]   M. de Berg. *Ray Shooting, Depth Orders and Hidden Surface Removal.* Vol. 703 of *LNCS*, Springer, Berlin, 1993.

[BG08]   M. de Berg and C. Gray. Vertical ray shooting and computing depth orders for fat objects. *SIAM J. Comput.*, 38:257–275, 2008.

[BGH94]   M. de Berg, L. Guibas, and D. Halperin. Vertical decompositions for triangles in 3-space. *Discrete Comput. Geom.*, 15:35–61, 1996.

[BHO+94]   M. de Berg, D. Halperin, M. Overmars, J. Snoeyink, and M. van Kreveld. Efficient ray-shooting and hidden surface removal. *Algorithmica*, 12:31–53, 1994.

[BL15]   L. Barba and S. Langerman. Optimal detection of intersections between convex polyhedra. In *Proc. 26th ACM-SIAM Sympos. Discrete Algorithms*, pages 1641–1654, 2015.

[BOS94]   M. de Berg, M. Overmars, and O. Schwarzkopf. Computing and verifying depth orders. *SIAM J. Comput.*, 23:437–446, 1994.

[BR79]   O. Bottima and B. Roth. *Theoretical Kinematics.* North-Holland, Amsterdam, 1979.

[CD99]   S.-W. Cheng and T.K. Dey. Approximate minimum weight Steiner triangulation in three dimensions. In *Proc. ACM-SIAM Sympos. Discrete Algorithms*, pages 205–214, 1999.

[CEG+92]   B. Chazelle, H. Edelsbrunner, L.J. Guibas, R. Pollack, R. Seidel, M. Sharir, and J. Snoeyink. Counting and cutting cycles of lines and rods in space. *Comput. Geom.*, 1:305–323, 1992.

[CEG+96]   B. Chazelle, H. Edelsbrunner, L.J. Guibas, M. Sharir, and J. Stolfi. Lines in space: Combinatorics and algorithms. *Algorithmica*, 15:428–447, 1996.

[CEGS93]   B. Chazelle, H. Edelsbrunner, L. Guibas, and M. Sharir. Diameter, width, closest line pair and parametric search. *Discrete Comput. Geom.*, 10:183–196, 1993.

[CEGS94]   B. Chazelle, H. Edelsbrunner, L. Guibas, and M. Sharir. Algorithms for bichromatic line segment problems and polyhedra terrains. *Algorithmica*, 11:116–132, 1994.

[CGN05]   O. Cheong, X. Goaoc, and H.-S. Na. Geometric permutations of disjoint unit spheres. *Comput. Geom.*, 30:253–270, 2005.

[CHX00]   D.Z. Chen, X. Hu, and J. Xu. Optimal beam penetrations in two and three dimensions. *J. Comb. Optim.*, 7:111–136, 2003.

[CKS⁺98]   L.P. Chew, K. Kedem, M. Sharir, B. Tagansky, and E. Welzl. Voronoi diagrams of lines in 3-space under polyhedral convex distance functions. *J. Algorithms*, 29:238–255, 1998.

[DDE⁺03]   O. Devillers, V. Dujmović, H. Everett, X. Goaoc, S. Lazard, H.-S. Na, and S. Petitjean. The expected number of 3D visibility events is linear. *SIAM J. Comput.*, 32:1586–1620, 2003.

[DDP97]   F. Durand, G. Drettakis, and C. Puech. The visibility skeleton: a powerful and efficient multi-purpose global visibility tool. In *Proc 24th Conf. Comp. Graphics Interactive Techniques*, pages 89–100, ACM Press, 1997.

[DDP02]   F. Durand, G. Drettakis, and C. Puech. The 3D visibility complex. *ACM Trans. Graph.*, 21:176–206, 2002.

[DK85]   D.P. Dobkin and D.G. Kirkpatrick. A linear algorithm for determining the separation of convex polyhedra. *J. Algorithms*, 6:381–392, 1985.

[Dur99]   F. Durand. *3D Visibility: Analytical Study and Applications*. PhD thesis, Université Joseph Fourier, Grenoble, 1999.

[Epp94]   D. Eppstein. Approximating the minimum weight Steiner triangulation. *Discrete Comput. Geom.*, 11:163–191, 1994.

[Eri95]   J. Erickson. On the relative complexities of some geometric problems. In *Proc. Canad. Conf. Comput. Geom.*, pages 85–90, 1995.

[FDFH90]   J.D. Foley, A. van Dam, S.K. Feiner, and J.F. Hughes. *Computer Graphics: Principles and Practice*. Addison-Wesley, Reading, 1990.

[For99]   S. Fortune. Topological beam tracing. In *Proc. 15th Sympos. Comput. Geom.*, pages 59–68, ACM Press, 1999.

[GL10]   M. Glisse and S. Lazard. On the complexity of sets of free lines and line segments among balls in three dimensions. *Discrete Comput. Geom.*, 47:756–772, 2012.

[Goa10]   X. Goaoc. Some discrete properties of the space of line transversals to disjoint balls. In *Nonlinear Computational Geometry*, pages 51–83. Springer, Berlin, 2010.

[Hav00]   V. Havran. *Heuristic Ray Shooting Algorithms*. PhD thesis, Faculty of Electrical Engineering, Czech Technical University, Prague, 2000.

[HP52]   W.V.D. Hodge and D. Pedoe. *Methods of Algebraic Geometry*. Cambridge University Press, 1952.

[HPS01]   S. Har-Peled and M. Sharir. Online point location in planar arrangements and its applications. *Discrete Comput. Geom.*, 26:19–40, 2001.

[HS94]   D. Halperin and M. Sharir. New bounds for lower envelopes in three dimensions, with applications to visibility in terrains. *Discrete Comput. Geom.*, 12:313–326, 1994.

[Jes69]   C.M. Jessop. *A Treatise on the Line Complex*. Chelsea, 1969.

[Kol04]   V. Koltun. Almost tight upper bounds for vertical decompositions in four dimensions. *J. ACM*, 51:699–730, 2004.

[Kol05]   V. Koltun. Pianos are not flat: Rigid motion planning in three dimensions. In *Proc. 16th ACM-SIAM Sympos. Discrete Algorithms*, pages 505–514, 2005.

[KRS09]   H. Kaplan, N. Rubin, and M. Sharir. Linear data structures for fast ray-shooting amidst convex polyhedra. *Algorithmica*, 55:283–310, 2009.

[KRS10]   H. Kaplan, N. Rubin, and M. Sharir. Line transversals of convex polyhedra in $\mathbb{R}^3$. *SIAM J. Comput.*, 39:3283–3310, 2010.

[Meg91]   N. Megiddo. Personal communication. 1991.

[Meh84]   K. Mehlhorn. *Multidimensional Searching and Computational Geometry.* Springer, Berlin, 1984.

[MS94]   S. Mohaban and M. Sharir. Ray-shooting amidst spheres in three-dimensions and related problems. *SIAM J. Comput.*, 26:654–674, 1997.

[O'R01]   J. O'Rourke. Computational geometry column 41. *Internat. J. Comput. Geom. Appl.*, 11:239–242, 2001.

[Pel91]   M. Pellegrini. *Combinatorial and Algorithmic Analysis of Stabbing and Visibility Problems in 3-Dimensional Space.* PhD thesis, New York University—Courant Institute of Mathematical Sciences, 1991.

[Pel93a] M. Pellegrini. Lower bounds on stabbing lines in 3-space. *Comput. Geom. Theory Appl.*, 3:53–58, 1993.

[Pel93b] M. Pellegrini. Ray shooting on triangles in 3-space. *Algorithmica*, 9:471–494, 1993.

[Pel94a] M. Pellegrini. On collision-free placements of simplices and the closest pair of lines in 3-space. *SIAM J. Comput.*, 23:133–153, 1994.

[Pel94b] M. Pellegrini. On lines missing polyhedral sets in 3-space. *Discrete Comput. Geom.*, 12:203–221, 1994.

[Pel96]   M. Pellegrini. On point location and motion planning among simplices. *SIAM J. Comput.*, 25:1061–1081, 1996.

[Plu65]   J. Plücker. On a new geometry of space. *Phil. Trans. Royal Soc. London*, 155:725–791, 1865.

[PPW93]   J. Pach, R. Pollack, and E. Welzl. Weaving patterns of lines segments in space. *Algorithmica*, 9:561–571, 1993.

[PS92]   M. Pellegrini and P.W. Shor. Finding stabbing lines in 3-space. *Discrete Comput. Geom.*, 8:191–208, 1992.

[PS09]   J. Pach and M. Sharir. *Combinatorial Geometry and Its Algorithmic Applications.* Vol. 152 of *Mathematical Surveys and Monographs*, AMS, Providence, 2009.

[RKS12]   N. Rubin, H. Kaplan, and M. Sharir. Improved bounds for geometric permutations. *SIAM J. Comput.*, 41:367–390, 2012.

[Rub12]   N. Rubin. Lines avoiding balls in three dimensions revisited. *Discrete Comput. Geom.*, 48:65–93, 2012.

[SAL93]   A. Schweikard, J.R. Adler, and J.-C. Latombe. Motion planning in stereotaxic radiosurgery. *IEEE Trans. Robot. Autom.*, 9:764–774, 1993.

[Shi78]   B.E. Shimano. *The Kinematic Design and Force Control of Computer Controlled Manipulators.* PhD thesis, Dept. of Mechanical Engineering, Stanford University, 1978.

[SMS00]   S. Smorodinsky, J.S.B. Mitchell, and M. Sharir. Sharp bounds on geometric permutations of pairwise disjoint balls in $\mathbb{R}^d$. *Discrete Comput. Geom.*, 23:247–259, 2000.

[Sol98]   A. Solan. Cutting cycles of rods in space. In *Proc. 14th Sympos. Comput. Geom.*, pages 135–142, ACM Press, 1998.

[Som51]   D.M.Y. Sommerville. *Analytical Geometry of Three Dimensions.* Cambridge University Press, 1951.

[Spe92]   R. Speer. An updated cross-indexed guide to the ray-tracing literature. *Comput. Graph.*, 26:41–72, 1992.

[SS93]   J. Snoeyink and J. Stolfi. Objects that cannot be taken apart with two hands. *Discrete Comput. Geom.*, 12:367–384, 1994.

[Sto89]   J. Stolfi. Primitives for computational geometry. Technical Report 36, Digital SRC, 1989.

[Sto91]   J. Stolfi. *Oriented Projective Geometry: A Framework for Geometric Computations.* Academic Press, San Diego, 1991.

[Zha09]   L. Zhang. *On the Three-Dimensional Visibility Skeleton: Implementation and Analysis.* PhD thesis, McGill University, School of Computer Science, Montréal, 2009.

42 GEOMETRIC INTERSECTION

David M. Mount

INTRODUCTION

Detecting whether two geometric objects intersect and computing the region of intersection are fundamental problems in computational geometry. Geometric intersection problems arise naturally in a number of applications. Examples include geometric packing and covering, wire and component layout in VLSI, map overlay in geographic information systems, motion planning, and collision detection. In solid modeling, computing the volume of intersection of two shapes is an important step in defining complex solids. In computer graphics, detecting the objects that overlap a viewing window is an example of an intersection problem, as is computing the first intersection of a ray and a collection of geometric solids.

Intersection problems are fundamental to many aspects of geometric computing. It is beyond the scope of this chapter to completely survey this area. Instead we illustrate a number of the principal techniques used in efficient intersection algorithms. This chapter is organized as follows. Section 42.1 discusses intersection primitives, the low-level issues of computing intersections that are common to high-level algorithms. Section 42.2 discusses detecting the existence of intersections. Section 42.3 focuses on issues related to counting the number of intersections and reporting intersections. Section 42.4 deals with problems related to constructing the actual region of intersection. Section 42.5 considers methods for geometric intersections based on spatial subdivisions.

42.1 INTERSECTION PREDICATES

GLOSSARY

Geometric predicate: A function that computes a discrete relationship between basic geometric objects.

Boundary elements: The vertices, edges, and faces of various dimensions that make up the boundary of an object.

Complex geometric objects are typically constructed from a number of primitive objects. Intersection algorithms that operate on complex objects often work by breaking the problem into a series of primitive geometric predicates acting on basic elements, such as points, lines and curves, that form the boundary of the objects involved. Examples of geometric predicates include determining whether two line segments intersect each other or whether a point lies above, below, or on a given line. Computing these predicates can be reduced to computing the sign of a polynomial, ideally of low degree. In many instances the polynomial arises as the determinant of a symbolic matrix.

Computing geometric predicates in a manner that is efficient, accurate, and robust can be quite challenging. Floating-point computations are fast but suffer from round-off errors, which can result in erroneous decisions. These errors in turn can lead to topological inconsistencies in object representations, and these inconsistencies can cause the run-time failures. Some of the approaches used to address robustness in geometric predicates include approximation algorithm that are robust to floating-point errors [SI94], computing geometric predicates exactly using adaptive floating-point arithmetic [Cla92, ABD+97], exact arithmetic combined with fast floating-point filters [BKM+95, FW96], and designing algorithms that are based on a restricted set of geometric predicates [BS00, BV02].

When the points of intersection are themselves used to as inputs in the construction of other discrete geometric structures, they are typically first rounded to finite precision. The rounding process needs to be performed with care, for otherwise topological inconsistencies may result. **Snap rounding** is a method for converting an arrangement of line segments in the plane into a fixed-precision representation by rounding segment intersection points to the vertices of a square grid [Hob99] (see also [GY86]). Various methods have been proposed and analyzed for implementing this concept (see, e.g., [HP02, BHO07, Her13]). For further information, see Chapter 28.

We will concentrate on geometric intersections involving flat objects (line segments, polygons, polyhedra), but there is considerable interest in computing intersections of curves and surfaces (see, e.g., [AKO93, APS93, BO79, CEGS91, JG91, LP79b]). Predicates for curve and surface intersections are particularly challenging, because the intersection of surfaces of a given algebraic degree generally results in a curve of a significantly higher degree. Computing intersection primitives typically involves solving an algebraic system equations, which can be performed either exactly by algebraic and symbolic methods [Yap93] or approximately by numerical methods [Hof89, MC91]. See Chapter 45.

42.2 INTERSECTION DETECTION

GLOSSARY

Polygonal chain: A sequence of line segments joined end-to-end.

Self-intersecting: Said of a polygonal chain if any pair of nonadjacent edges intersects one another.

Bounding box: A rectangular box surrounding an object, usually axis-aligned (*isothetic*).

Intersection detection, the easiest of all intersection tasks, requires merely determining the existence of an intersection. Nonetheless, detecting intersections efficiently in the absence of simplifying geometric structure can be challenging. As an example, consider the following fundamental intersection problem, posed by John Hopcroft in the early 1980s. Given a set of n points and n lines in the plane, does any point lie on any line? A series of efforts to solve **Hopcroft's problem** culminated in the best algorithm known for this problem to date, due to Matoušek [Mat93], which runs in $O(n^{4/3})2^{O(\log^* n)}$. There is reason to believe that this may be close to optimal; Erickson [Eri96] has shown that, in certain models of computation, $\Omega(n^{4/3})$ is a lower bound. Agarwal and Sharir [AS90] have shown that, given two sets of line segments denoted red and blue, it is possible to determine whether there is any red-blue intersection in $O(n^{4/3+\epsilon})$ time, for any positive constant ϵ.

Another example of a detection problem is that of determining whether a set of line segments intersect using the information-theoretic minimum number of operations, or close to this. Chan and Lee [CL15] showed that it is possible to determine whether there are any intersections among a set of n axis-parallel line segments in the plane with $n \log_2 n + O(n\sqrt{\log n})$.

The types of objects considered in this section are polygons, polyhedra, and line segments. Let P and Q denote the two objects to be tested for intersection. Throughout, n_p and n_q denote the combinatorial complexity of P and Q, respectively, that is, the number of vertices, edges, and faces (for polyhedra). Let $n = n_p + n_q$ denote the total complexity.

Table 42.2.1 summarizes a number of results on intersection detection, which will be discussed further in this section. In the table, the terms *convex* and *simple* refer to convex and simple polygons, respectively. The notation $(s(n), q(n))$ in the "Time" column means that the solution involves preprocessing, where a data structure of size $O(s(n))$ is constructed so that intersection detection queries can be answered in $O(q(n))$ time.

TABLE 42.2.1   Intersection detection.

DIM	OBJECTS	TIME	SOURCE
2	convex-convex	$\log n$	[DK83]
	simple-simple	n	[Cha91]
	simple-simple	$(n, s \log^2 n)$	[Mou92]
	line segments	$n \log n$	[SH76]
	Hopcroft's problem	$n^{4/3} 2^{O(\log^* n)}$	[Mat93]
3	convex-convex	n	[DK85]
	convex-convex	$(n, \log n_p \log n_q)$	[DK90]

INTERSECTION DETECTION OF CONVEX POLYGONS

Perhaps the most easily understood example of how the structure of geometric objects can be exploited to yield an efficient intersection test is that of detecting the intersection of two convex polygons. There are a number of solutions to this problem that run in $O(\log n)$ time. We present one due to Dobkin and Kirkpatrick [DK83].

Assume that each polygon is given by an array of vertex coordinates, sorted in counterclockwise order. The first step of the algorithm is to find the vertices of each of P and Q with the highest and lowest y-coordinates. This can be done in $O(\log n)$ time by an appropriate modification of binary search and consideration of the direction of the edges incident to each vertex [O'R94, Section 7.6]. After these vertices are located, the boundary of each polygon is split into two semi-infinite convex chains, denoted P_L, P_R and Q_L, Q_R (see Figure 42.2.1(a)). P and Q intersect if and only if P_L and Q_R intersect, and P_R and Q_L intersect.

Consider the case of P_L and Q_R. The algorithm applies a variant of binary

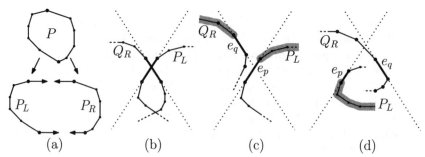

FIGURE 42.2.1
Intersection detection for two convex polygons.

search. Consider the median edge e_p of P_L and the median edge e_q of Q_R (shown as heavy lines in the figure). By a simple analysis of the relative positions of these edges and the intersection point of the two lines on which they lie, it is possible to determine in constant time either that the polygons intersect, or that half of at least one of the two boundary chains can be eliminated from further consideration. The cases that arise are illustrated in Figure 42.2.1(b)-(d). The shaded regions indicate the portion of the boundary that can be eliminated from consideration.

SIMPLE POLYGONS

Without convexity, it is generally not possible to detect intersections in sublinear time without preprocessing; but efficient tests do exist.

One of the important intersection questions is whether a closed polygonal chain defines the edges of a simple polygon. The problem reduces to detecting whether the chain is self-intersecting. This problem can be solved efficiently by supposing that the polygonal chain is a simple polygon, attempting to triangulate the polygon, and seeing whether anything goes wrong in the process. Some triangulation algorithms can be modified to detect self-intersections. In particular, the problem can be solved in $O(n)$ time by modifying Chazelle's linear-time triangulation algorithm [Cha91]. See Section 29.2.

Another variation is that of determining the intersection of two simple polygons. Chazelle observed that this can also be reduced to testing self-intersections in $O(n)$ time by joining the polygons into a single closed chain by a narrow channel as shown in Figure 42.2.2.

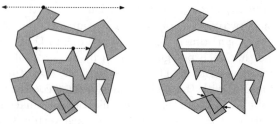

FIGURE 42.2.2
Intersection detection for two simple polygons.

DETECTING INTERSECTIONS OF MULTIPLE OBJECTS

In many applications, it is important to know whether any pair of a set of objects intersects one another. Shamos and Hoey showed that the problem of detecting whether a set of n line segments in the plane have an intersecting pair can be solved in $O(n \log n)$ time [SH76]. This is done by plane sweep, which will be discussed below. They also showed that the same can be done for a set of circles. Reichling showed that this can be generalized to detecting whether any pair of m convex n-gons intersects in $O(m \log m \log n)$ time, and whether they all share a common intersection point in $O(m \log^2 n)$ time [Rei88]. Hopcroft, Schwartz, and Sharir [HSS83] showed how to detect the intersection of any pair of n spheres in 3-space in $O(n \log^2 n)$ time and $O(n \log n)$ space by applying a 3D plane sweep.

INTERSECTION DETECTION WITH PREPROCESSING

If preprocessing is allowed, then significant improvements in intersection detection time may be possible. One of the best-known techniques is to filter complex intersection tests is to compute an axis-aligned bounding box for each object. Two objects need to be tested for intersection only if their bounding boxes intersect. It is very easy to test whether two such boxes intersect by comparing their projections on each coordinate axis. For example, in Figure 42.2.3, of the 15 possible pairs of object intersections, all but three may be eliminated by the bounding box filter.

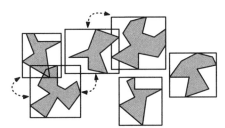

FIGURE 42.2.3
Using bounding boxes as an intersection filter.

It is hard to prove good worst-case bounds for the bounding-box filter since it is possible to create instances of n disjoint objects in which all $O(n^2)$ pairs of bounding boxes intersect. Nonetheless, this popular heuristic tends to perform well in practice. Suri and others [SHH99, ZS99] provided an explanation for this. They proved that if the boxes have bounded aspect ratio and the relative object sizes are within a constant factor each other, then (up to an additive linear term) the number of intersecting boxes is proportional to the number of intersecting object pairs. Combining this with Dobkin and Kirkpatrick's results leads to an algorithm, which given n convex polytopes in dimension d, reports all k intersecting pairs in time $O(n \log^{d-1} n + k \log^{d-1} m)$, where m is the maximum number of vertices in any polytope.

Another example is that of ray shooting in a simple polygon. This is a planar version of a well-known 3D problem in computer graphics. The problem is to preprocess a simple polygon so that given a query ray, the first intersection of the ray with the boundary of the polygon can be determined. After $O(n)$ preprocessing it is possible to answer ray-shooting queries in $O(\log n)$ time. A particularly elegant solution was given by Hershberger and Suri [HS95]. The polygon is triangulated in a special way, called a *geodesic triangulation*, so that any line segment that does not intersect the boundary of the polygon crosses at most $O(\log n)$ triangles. Ray-shooting queries are answered by locating the triangle that contains the origin of the ray, and "walking" the ray through the triangulation. See also Section 31.2.

Mount showed how the geodesic triangulation can be used to generalize the bounding box test for the intersection of simple polygons. Each polygon is preprocessed by computing a geodesic triangulation of its exterior. From this it is possible to determine whether they intersect in $O(s \log^2 n)$ time, where s is the minimum number of edges in a polygonal chain that separates the two polygons [Mou92]. Separation sensitive intersections of polygons has been studied in the context of kinetic algorithms for collision detection. See Chapter 50.

CONVEX POLYHEDRA IN HIGHER DIMENSIONS

Extending a problem from the plane to 3-space and higher often involves in a significant increase in difficulty. Computing the intersection of convex polyhedra is among the first problems studied in the field of computational geometry [MP78]. Dobkin and Kirkpatrick showed that detecting the intersection of convex polyhedra can be performed efficiently by adapting Kirkpatrick's hierarchical decomposition of planar triangulations. Consider convex polyhedra P and Q in 3-space having combinatorial boundary complexities n_p and n_q, respectively. They showed that each can be preprocessed in linear time and space so that it is possible to determine the intersection of any translation and rotation of the two in time $O(\log n_p \cdot \log n_q)$ [DK90].

DOBKIN-KIRKPATRICK DECOMPOSITION

Before describing the intersection algorithm, let us discuss the hierarchical representation. Let $P = P_0$ be the initial polyhedron. Assume that P's faces have been triangulated. The vertices, edges, and faces of P's boundary define a planar graph with triangular faces. Let n denote the number of vertices in this graph. An important fact is that every planar graph has an independent set (a subset of pairwise nonadjacent vertices) that contains a constant fraction of the vertices formed entirely from vertices of bounded degree. Such an independent set is computed and is removed along with any incident edges and faces from P. Then any resulting "holes" in the boundary of P are filled in with triangles, resulting in a convex polyhedron with fewer vertices (cf. Section 38.3).

These holes can be triangulated independently of one another, each in constant time. The resulting convex polyhedron is denoted P_1. The process is repeated until reaching a polyhedron having a constant number of vertices. The result is a sequence of polyhedra, $\langle P_0, P_1, \ldots, P_k \rangle$, called the **Dobkin-Kirkpatrick hierarchy**. Because a constant fraction of vertices are eliminated at each stage, the depth k of the hierarchy is $O(\log n)$. The hierarchical decomposition is illustrated in Figure 42.2.4. The vertices that are eliminated at each stage, which form an independent set, are highlighted in the figure.

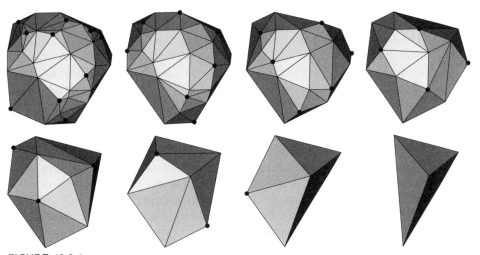

FIGURE 42.2.4
Dobkin-Kirkpatrick decomposition of a convex polyhedron.

INTERSECTION DETECTION ALGORITHM

Suppose that the hierarchical representations of P and Q have already been computed. The intersection detection algorithm actually computes the *separation*, that is, the minimum distance between the two polyhedra. First consider the task of determining the separation between P and a triangle T in 3-space. We start with the top of the hierarchy, P_k. Because P_k and T are both of constant complexity, the separation between P_k and T can be computed in constant time. Given the separation between P_i and T, it is possible to determine the separation between P_{i-1} and T in constant time. This is done by a consideration of the newly added boundary elements of P_{i-1} that lie in the neighborhood of the two closest points.

Given the hierarchical decompositions of two polyhedra P and Q, the Dobkin-Kirkpatrick intersection algorithm begins by computing the separation at the highest common level of the two hierarchies (so that at least one of the decomposed polyhedra is of bounded complexity). They show that in $O(\log n_p + \log n_q)$ time it is possible to determine the separation of the polyhedra at the next lower level of the hierarchies. This leads to a total running time of $O(\log n_p \cdot \log n_q)$.

IMPROVEMENTS AND HIGHER DIMENSIONS

Barba and Langerman revisited this problem over 30 years after Dobkin and Kirkpatrick's first result, both improving and extending it. Given convex polyhedra P and Q in d-dimensional space, for any fixed constant d, let n_p and n_q denote their respective combinatorial complexities, that is, the total number of faces of all dimensions on their respective boundaries. They show that it is possible to preprocess such polyhedra such that after any translation and rotation, it is possible to determine whether they intersect in time $O(\log n_p + \log n_q)$ [BL15]. In 3-space, the preprocessing time and space are linear in the combinatorial complexity. In general in d-dimensional space the preprocessing time and space are of the form $O(n_p^{\lfloor d/2 \rfloor + \varepsilon})$, for any $\varepsilon > 0$. Their improvement arises by considering intersection detection from both a primal and polar perspective and applying ε-nets for sampling.

SUBLINEAR INTERSECTION DETECTION

The aforementioned approaches assume that the input polyhedra have been preprocessed into a data structure. Without such preprocessing, it would seem impossible to detect the presence of an intersection in time that is sublinear in the input size. Remarkably, Chazelle, Liu and Magen [CLM05] showed that there exists a randomized algorithm that, without any preprocessing, detects whether two 3-dimensional n-vertex convex polyhedra intersect in $O(\sqrt{n})$ expected time. Algorithms like this whose running time is sublinear in the input size are called ***sublinear algorithms***.

It is assumed that each polyhedron is presented in memory using any standard boundary representation, such as a DCEL or winged-edge data structure (see Chapter 67.2.3), and that it is possible to access a random edge or vertex of the polyhedron in constant time. The algorithm randomly samples $O(\sqrt{n})$ vertices from each polyhedron and applies low-dimensional linear programming to test whether the convex hulls of the two sampled sets intersect. If so, the original polyhedra intersect. If they do not intersect, the region of possible intersection can be localized to a portion of the boundary of expected size $O(\sqrt{n})$. An efficient algorithm for identifying and searching this region is presented.

42.3 INTERSECTION COUNTING AND REPORTING

GLOSSARY

Plane sweep: An algorithm paradigm based on simulating the left-to-right sweep of the plane with a vertical ***sweepline***. See Figure 42.3.1.

Bichromatic intersection: Segment intersection between segments of two colors, where only intersections between segments of different colors are to be reported (also called ***red-blue intersection***).

In many applications, geometric intersections can be viewed as a discrete set of entities to be counted or reported. The problems of intersection counting and reporting have been heavily studied in computational geometry from the perspective of *intersection searching*, employing preprocessing and subsequent queries (Chapter 40). We limit our discussion here to batch problems, where the geometric objects are all given at once. In many instances, the best algorithms known for batch counting and reporting reduce the problem to intersection searching.

Table 42.3.1 summarizes a number of results on intersection counting and reporting. The quantity n denotes the combinatorial complexity of the objects, d denotes the dimension of the space, and k denotes the number of intersections. Because every pair of elements might intersect, the number of intersections k may generally be as large as $O(n^2)$, but it is frequently much smaller.

Computing the intersection of line segments is among the most fundamental problems in computational geometry [Cha86]. It is often an initial step in computing the intersection of more complex objects. In such cases particular properties of the class of objects involved may influence the algorithm used for computing the underlying intersections. For example, if it is known that the objects involved are fat or if only certain faces of the resulting arrangement are of interest, then more efficient approaches may be possible (see, e.g., [AMS98, MPS+94, Vig03]).

TABLE 42.3.1   Intersection counting and reporting.

PROBLEM	DIM	OBJECTS	TIME	SOURCE
Reporting	2	line segments	$n \log n + k$	[CE92][Bal95]
	2	bichromatic segments (general)	$n^{4/3} \log^{O(1)} n + k$	[Aga90][Cha93]
	2	bichromatic segments (disjoint)	$n + k$	[FH95]
	d	orthogonal segments	$n \log^{d-1} n + k$	[EM81]
Counting	2	line segments	$n^{4/3} \log^{O(1)} n$	[Aga90][Cha93]
	2	bichromatic segments (general)	$n^{4/3} \log^{O(1)} n$	[Aga90][Cha93]
	2	bichromatic segments (disjoint)	$n \log n$	[CEGS94]
	d	orthogonal segments	$n \log^{d-1} n$	[EM81, Cha88]

A related problem is that of computing properties of the set of intersection points of a collection of objects. Atallah showed that it is possible to compute the convex hull of the (quadratic sized) set of intersection points of a collection of n lines in the plane in time $O(n \log n)$ [?]. Arkin et al. [AMS08] showed that it is possible to achieve the same running time for the more general case of intersection points of line segments in the plane.

REPORTING LINE SEGMENT INTERSECTIONS

Consider the problem of reporting the intersections of n line segments in the plane. This problem is an excellent vehicle for introducing the powerful technique of plane sweep (Figure 42.3.1). The plane-sweep algorithm maintains an active list of segments that intersect the current sweepline, sorted from bottom to top by intersection point. If two line segments intersect, then at some point prior to this intersection they must be consecutive in the sweep list. Thus, we need only test consecutive pairs in this list for intersection, rather than testing all $O(n^2)$ pairs.

At each step the algorithm advances the sweepline to the next event: a line segment endpoint or an intersection point between two segments. Events are stored in a *priority queue* by their x-coordinates. After advancing the sweepline to the next event point, the algorithm updates the contents of the active list, tests new consecutive pairs for intersection, and inserts any newly-discovered events in the priority queue. For example, in Figure 42.3.1 the locations of the sweepline are shown with dashed lines.

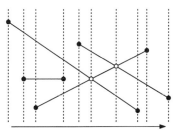

FIGURE 42.3.1
Plane sweep for line segment intersection.

Bentley and Ottmann [BO79] showed that by using plane sweep it is possible to report all k intersecting pairs of n line segments in $O((n+k)\log n)$ time. If the number of intersections k is much less than the $O(n^2)$ worst-case bound, then this is great savings over a brute-force test of all pairs.

For many years the question of whether this could be improved to $O(n\log n+k)$ was open, until Edelsbrunner and Chazelle presented such an algorithm [CE92]. This algorithm is optimal with respect to running time because at least $\Omega(k)$ time is needed to report the result, and it can be shown that $\Omega(n\log n)$ time is needed to detect whether there is any intersection at all. However, their algorithm uses $O(n+k)$ space. Balaban [Bal95] showed how to achieve the same running time using only $O(n)$ space. Vahrenhold [Vah07] further improved the space bound by showing that there is an ***in-place algorithm*** that runs in time $O(n\log^2 n + k)$. This means that the input is assumed to be stored in an array of size n so that random access is possible and only $O(1)$ additional working space is needed.

Clarkson and Shor [CS89] and later Mulmuley [Mul91] presented simple, randomized algorithms that achieve running time $O(n\log n + k)$ with $O(n)$ space. Mulmuley's algorithm is particularly elegant. It involves maintaining a ***trapezoidal decomposition***, a subdivision which results by shooting a vertical ray up and down from each segment endpoint and intersection point until it hits another segment. The algorithm inserts the segments one by one in random order by "walking" each segment through the subdivision and updating the decomposition as it goes. (This is shown in Figure 42.3.2, where the broken horizontal line on the left is being inserted and the shaded regions on the right are the newly created trapezoids.) Chan and Chen [CC10] improved Vahrenhold's result by showing there is a randomized in-place algorithm for line-segment intersection that achieves a running time of $O(n\log n + k)$.

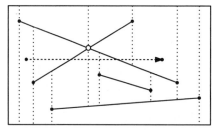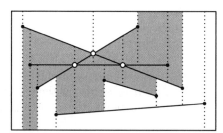

FIGURE 42.3.2
Incremental construction of a trapezoidal decomposition.

BICHROMATIC INTERSECTION PROBLEMS

Among the numerous variations of the segment intersection problem, the most widely studied is the problem of computing intersections that arise between two sets of segments, say red and blue, whose total size is n. The goal is to compute all ***bichromatic intersections***, that is, intersections that arise when a red segment intersects a blue segments. Let k denote the number of such intersections.

The case where there are no monochromatic (blue-blue or red-red) intersections is particularly important. It arises, for example, when two planar subdivisions are overlaid, called the ***map overlay*** problem in GIS applications, as well as in many intersection algorithms based on divide-and-conquer. (See Figure 42.3.3.) In this case the problem can be solved by in $O(n \log n + k)$ time by any optimal monochromatic line-segment intersection algorithm. This problem seems to be somewhat simpler than the monochromatic case, because Mairson and Stolfi [MS88] showed the existence of an $O(n \log n + k)$ algorithm prior to the discovery of these optimal monochromatic algorithms. Chazelle et al. [CEGS94] presented an algorithm based on a simple but powerful data structure, called the *hereditary segment tree*. Chan [Cha94] presented a practical approach based on a plane sweep of the trapezoidal decomposition of the two sets. Mantler and Snoeyink [MS01] presented an algorithm that is not only optimal with respect to running time but is also optimal with respect to the arithmetic precision needed.

Guibas and Seidel [GS87] showed that, if the segments form a simple connected convex subdivision of the plane, the problem can be solved more efficiently in $O(n + k)$ time. This was extended to simply connected subdivisions that are not necessarily convex by Finke and Hinrichs [FH95].

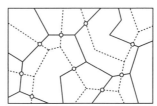

FIGURE 42.3.3
Overlaying planar subdivisions.

The problem is considerably more difficult if monochromatic intersections exist. This is because there may be quadratically many monochromatic intersections, even if there are no bichromatic intersections. Agarwal [Aga90] and Chazelle [Cha93] showed that the k bichromatic intersections can be reported in $O(n^{4/3} \log^{O(1)} n + k)$ time through the use of a partitioning technique called *cuttings*. Basch et al. [BGR96] showed that if the set of red segments forms a connected set and the blue set does as well, then it is possible to report all bichromatic intersections in $O((n + k) \log^{O(1)} n)$ time. Agarwal et al. [ABH+02] and Gupta et al. [GJS99] considered a multi-chromatic variant in which the input consists of m convex polygons and the objective is to report all intersections between pairs of polygons. They show that many of the same techniques can be applied to this problem and present algorithms with similar running times.

COUNTING LINE-SEGMENT INTERSECTIONS

Efficient intersection counting often requires quite different techniques from reporting because it is not possible to rely on the lower bound of k needed to report the results. Nonetheless, a number of the efficient intersection reporting algorithms can be modified to count intersections efficiently. For example, methods based on cuttings [Aga90, Cha93] can be used to count the number of intersections among n planar line segments and bichromatic intersections between n red and blue segments in $O(n^{4/3} \log^{O(1)} n)$ time. If there are no monochromatic intersections then the hereditary segment tree [CEGS94] can be used to count the number bichromatic intersections in $O(n \log n)$ time. Chan and Wilkinson showed that in the word RAM computational model it is possible to count bichromatic intersections even faster in $O(n\sqrt{\log n})$ time [CW11].

Many of the algorithms for performing segment intersection exploit the observation that if the line segments span a closed region, it is possible to infer the number of segment intersections within the region simply by knowing the order in which the lines intersect the boundary of the region. Consider, for example, the problem of counting the number of line intersections that occur within a vertical strip in the plane. This problem can be solved in $O(n \log n)$ time by sorting the points according to their intersections on the left side of the strip, computing the associated permutation of indices on the right side, and then counting the number *inversions* in the resulting sequence [DMN92, Mat91]. An inversion is any pair of values that are not in sorted order. See Figure. 42.3.4. Inversion counting can be performed by a simple modification of the Mergesort algorithm. It is possible to generalization this idea to regions whose boundary is not simply connected [Asa94, MN01].

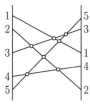
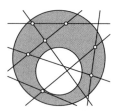

FIGURE 42.3.4
Intersections and inversion counting.

Intersection counting of other varieties have also been considered. Pellegrini [Pel97] considered an offline variant of simplicial range searching where, given a set of points and a set of triangles, the objective is to count the number of points lying within each of the triangles. Ezra and Sharir [ES05] solved an open problem posed in Pellegrini's paper, in which the input consists of triangles in 3-space, and the problem is to report all triples of intersecting triangles. Their algorithm runs in nearly quadratic time. They show that the (potentially cubic-sized) output can be expressed concisely as the disjoint union of complete tripartite hypergraphs.

INTERSECTION SEARCHING AND RANGE SEARCHING

Range and intersection searching are powerful tools that can be applied to more complex intersection counting and reporting problems. This fact was first observed by Dobkin and Edelsbrunner [DE87], and has been applied to many other intersection searching problems since.

As an illustration, consider the problem of counting all intersecting pairs from a set of n rectangles. Edelsbrunner and Maurer [EM81] observed that intersections among orthogonal objects can be broken down to a set of orthogonal search queries (see Figure 42.3.5). For each rectangle x we can count all the intersecting rectangles of the set satisfying each of these conditions and sum them. Each of these counting queries can be answered in $O(\log n)$ time after $O(n \log n)$ preprocessing time [Cha88], leading to an overall $O(n \log n)$ time algorithm. This counts every intersection twice and counts self-intersections, but these are easy to factor out from the final result. Generalizations to hyperrectangle intersection counting in higher dimensions are straightforward, with an additional factor of $\log n$ in time and space for each increase in dimension. We refer the reader to Chapter 40 for more information on intersection searching and its relationship to range searching.

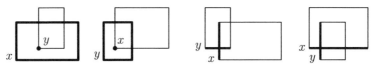

FIGURE 42.3.5
Types of intersections between rectangles x and y.

42.4 INTERSECTION CONSTRUCTION

GLOSSARY

Regularization: Discarding measure-zero parts of the result of an operation by taking the closure of the interior.

Clipping: Computing the intersection of each of many polygons with an axis-aligned rectangular viewing **window**.

Kernel of a polygon: The set of points that can see every point of the polygon. (See Section 30.1.)

Intersection construction involves determining the region of intersection between geometric objects. Many of the same techniques that are used for computing geometric intersections are used for computing Boolean operations in general (e.g., union and difference). Many of the results presented here can be applied to these other problems as well. Typically intersection construction reduces to the following tasks: (1) compute the intersection between the boundaries of the objects; (2) if the boundaries do not intersect then determine whether one object is nested within the other; and (3) if the boundaries do intersect then classify the resulting boundary fragments and piece together the final intersection region.

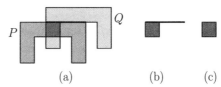

(a)   (b)   (c)

FIGURE 42.4.1
Regularized intersection: (a) *Polygons P and Q;* (b) *$P \cap Q$;* (c) *$P \cap^* Q$.*

When Boolean operations are computed on solid geometric objects, it is possible that lower-dimensional "dangling" components may result. It is common to eliminate these lower-dimensional components by a process called **regularization** [RV85] (see Section 57.1.1). The regularized intersection of P and Q, denoted $P \cap^* Q$, is defined formally to be the closure of the interior of the standard intersection $P \cap Q$ (see Figure 42.4.1).

Some results on intersection construction are summarized in Table 42.4.1, where n is the total complexity of the objects being intersected, and k is the number of pairs of intersecting edges.

TABLE 42.4.1   Intersection construction.

DIM	OBJECTS	TIME	SOURCE
2	convex-convex	n	[SH76, OCON82]
2	simple-simple	$n \log n + k$	[CE92]
2	kernel	n	[LP79]
3	convex-convex	n	[Cha92]

CONVEX POLYGONS

Determining the intersection of two convex polygons is illustrative of many intersection construction algorithms. Observe that the intersection of two convex polygons having a total of n edges is either empty or a convex polygon with at most n edges. O'Rourke et al. present an $O(n)$ time algorithm, which given two convex polygons P and Q determines their intersection [OCON82].

The algorithm can be viewed as a geometric generalization of merging two sorted lists. It performs a counterclockwise traversal of the boundaries of the two polygons. The algorithm maintains a pair of edges, one from each polygon. From a consideration of the relative positions of these edges the algorithm advances one of them to the next edge in counterclockwise order around its polygon. Intuitively, this is done in such a way that these two edges effectively "chase" each other around the boundary of the intersection polygon (see Figure 42.4.2(a)-(i)).

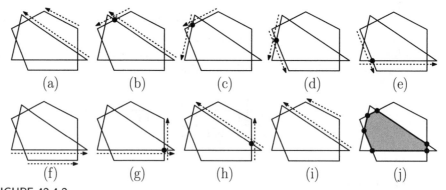

(a)	(b)	(c)	(d)	(e)
(f)	(g)	(h)	(i)	(j)

FIGURE 42.4.2
Convex polygon intersection construction.

OPEN PROBLEM

Reichling has shown that it is possible to detect whether m convex n-gons share a common point in $O(m \log^2 n)$ time [Rei88]. Is there an output-sensitive algorithm of similar complexity for constructing the intersection region?

SIMPLE POLYGONS AND CLIPPING

As with convex polygons, computing the intersection of two simple polygons reduces to first computing the points at which the two boundaries intersect and then classifying the resulting edge fragments. Computing the edge intersections and edge fragments can be performed by any algorithm for reporting line segment intersections. Classifying the edge fragments is a simple task. Margalit and Knott describe a method for edge classification that works not only for intersection, but for any Boolean operation on the polygons [MK89].

Clipping a set of polygons to a rectangular window is a special case of simple polygon intersection that is particularly important in computer graphics (see Section 52.3). One popular algorithm for this problem is the Sutherland-Hodgman algorithm [FDFH90]. It works by intersecting each polygon with each of the four halfplanes that bound the clipping window. The algorithm traverses the boundary of the polygon, and classifies each edge as lying either entirely inside, entirely outside, or crossing each such halfplane.

An elegant feature of the algorithm is that it effectively "pipelines" the clipping process by clipping each edge against one of the window's four sides and then passing the clipped edge, if it is nonempty, to the next side to be clipped. This makes the algorithm easy to implement in hardware. An unusual consequence, however, is that if a polygon's intersection with the window has multiple connected components (as can happen with a nonconvex polygon), then the resulting clipped polygon consists of a single component connected by one or more "invisible" channels that run along the boundary of the window (see Figure 42.4.3).

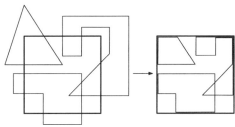

FIGURE 42.4.3
Clipping using the Sutherland-Hodgman algorithm.

INTERSECTION CONSTRUCTION IN HIGHER DIMENSIONS

Intersection construction in higher dimensions, and particularly in dimension 3, is important to many applications such as solid modeling. The basic paradigm of computing boundary intersections and classifying boundary fragments applies here as well. Muller and Preparata gave an $O(n \log n)$ algorithm that computes the intersection of two convex polyhedra in 3-space (see [PS85]). The existence of a linear-time algorithm remained open for years until Chazelle discovered such an algorithm [Cha92]. He showed that the Dobkin-Kirkpatrick hierarchical representation of polyhedra can be applied to the problem. A particularly interesting element of his algorithm is the use of the hierarchy for representing the interior of each polyhedron, and a dual hierarchy for representing the exterior of each polyhedron. Many years later Chan presented a significantly simpler algorithm, which also uses the Dobkin-Kirkpatrick hierarchy [Cha16]. Dobrindt, Mehlhorn, and Yvinec [DMY93] presented an output-sensitive algorithm for intersecting two polyhedra, one of which is convex.

Another class of problems can be solved efficiently are those involving *polyhedral terrains*, that is, a polyhedral surface that intersects every vertical line in at most one point. Chazelle et al. [CEGS94] show that the hereditary segment tree can be applied to compute the smallest vertical distance between two polyhedral terrains in roughly $O(n^{4/3})$ time. They also show that the upper envelope of two polyhedral terrains can be computed in $O(n^{3/2+\epsilon} + k \log^2 n)$ time, where ϵ is an arbitrary constant and k is the number of edges in the upper envelope.

KERNELS AND THE INTERSECTION OF HALFSPACES

Because of the highly structured nature of convex polygons, algorithms for convex polygons can often avoid additional $O(\log n)$ factors that seem to be necessary when dealing with less structured objects. An example of this structure arises in computing the kernel of a simple polygon: the (possibly empty) locus of points that can see every point in the polygon (the shaded region of Figure 42.4.4). Put another way, the kernel is the intersection of inner halfplanes defined by all the sides of P. The kernel of P is a convex polygon having at most n sides. Lee and Preparata gave an $O(n)$ time algorithm for constructing it [LP79] (see also Table 30.3.1). Their algorithm operates by traversing the boundary of the polygon, and incrementally updating the boundary of the kernel as each new edge is encountered.

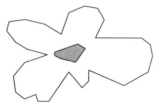

FIGURE 42.4.4
The kernel of a simple polygon.

The general problem of computing the intersection of halfplanes, when the halfplanes do not necessarily arise from the sides of a simple polygon, requires $\Omega(n \log n)$ time. See Chapter 26 for more information on this problem.

42.5 METHODS BASED ON SPATIAL SUBDIVISIONS

So far we have considered methods with proven worst-case asymptotic efficiency. However, there are numerous approaches to intersection problems for which worst-case efficiency is hard to establish, but that practical experience has shown to be quite efficient on the types of inputs that often arise in practice. Most of these methods are based on subdividing space into disjoint regions, or *cells*. Intersections can be computed by determining which objects overlap each cell, and then performing primitive intersection tests between objects that overlap the same cell.

GRIDS

Perhaps the simplest spatial subdivision is based on "bucketing" with square grids. Space is subdivided into a regular grid of squares (or generally hypercubes) of equal side length. The side length is typically chosen so that either the total number of cells is bounded, or the expected number of objects overlapping each cell is bounded. Edahiro et al. [ETHA89] showed that this method is competitive with and often performs much better than more sophisticated data structures for reporting intersections between randomly generated line segments in the plane. Conventional wisdom is that grids perform well as long as the objects are small on average and their distribution is roughly uniform.

HIERARCHICAL SUBDIVISIONS

The principle shortcoming of grids is their inability to deal with nonuniformly distributed objects. Hierarchical subdivisions of space are designed to overcome this weakness. There is quite a variety of different data structures based on hierarchical subdivisions, but almost all are based on the principal of recursively subdividing space into successively smaller regions, until each region is sufficiently simple in the sense that it overlaps only a small number of objects. When a region is subdivided, the resulting subregions are its *children* in the hierarchy. Well-known examples of hierarchical subdivisions for storing geometric objects include quadtrees and k-d trees, R-trees, and binary space partition (BSP) trees. See [Sam90b] for a discussion of all of these.

Intersection construction with hierarchical subdivisions can be performed by a process of *merging* the two hierarchical spatial subdivisions. This method is described by Samet for quadtrees [Sam90a] and Naylor et al. [NAT90] for BSP trees. To illustrate the idea on a simple example, consider a quadtree representation of two black-and-white images. The problem is to compute the intersection of the two black regions. For example, in Figure 42.5.1 the two images on the left are intersected, resulting in the image on the right.

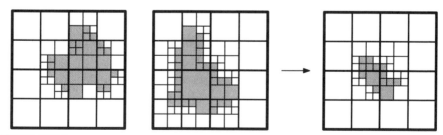

FIGURE 42.5.1
Intersection of images using quadtrees.

The algorithm recursively considers two overlapping square regions from each quadtree. A region of the quadtree is *black* if the entire region is black, *white* if the entire region is white, and *gray* otherwise. If either region is white, then the result is white. If either region is black, then the result is the other region. Otherwise both regions are gray, and we apply the procedure recursively to each of the four pairs of overlapping children.

42.6 SOURCES

Geometric intersections and related topics are covered in general sources on computational geometry [BCKO00, O'R94, Mul93, Ede87, PS85, Meh84]. A good source of information on the complexity of the lower envelopes and faces in arrangements are the books by Agarwal [Aga91] and Sharir and Agarwal [SA95]. Intersections of convex objects are discussed in the paper by Chazelle and Dobkin [CD87]. For information on data structures useful for geometric intersections see Samet's books [Sam90a, Sam90b, Sam06]. Sources on computing intersection primitives include O'Rourke's book on computational geometry [O'R94], Yap's book [Yap93] on algebraic algorithms, and most texts on computer graphics, for example [FDFH90]. For 3D surface intersections, consult books on solid modeling, including those by Hoffmann [Hof89] and Mäntylä [Män88]. The *Graphics Gems* series (e.g., [Pae95]) contains a number of excellent tips and techniques for computing geometric operations including intersection primitives.

RELATED CHAPTERS

Chapter 26: Convex hull computations
Chapter 28: Arrangements
Chapter 29: Triangulations and mesh generation
Chapter 40: Range searching
Chapter 41: Ray shooting and lines in space
Chapter 52: Computer graphics
Chapter 56: Splines and geometric modeling

REFERENCES

[ABD⁺97]   F. Avnaim, J.-D. Boissonnat, O. Devillers, F.P. Preparata, and M. Yvinec. Evaluating signs of determinants using single-precision arithmetic. *Algorithmica*, 17:111–132, 1997.

[ABH⁺02]   P.K. Agarwal, M. de Berg, S. Har-Peled, M.H. Overmars, M. Sharir, and J. Vahrenhold. Reporting intersecting pairs of convex polytopes in two and three dimensions. *Comput. Geom.*, 23:197–207, 2002.

[Aga90]   P.K. Agarwal. Partitioning arrangements of lines: II. Applications. *Discrete Comput. Geom.*, 5:533–573, 1990.

[Aga91]   P.K. Agarwal. *Intersection and Decomposition Algorithms for Planar Arrangements.* Cambridge University Press, New York, 1991.

[AKO93]   P.K. Agarwal, M. van Kreveld, and M. Overmars. Intersection queries in curved objects. *J. Algorithms*, 15:229–266, 1993.

[AMS98]   P.K. Agarwal, J. Matoušek, and O. Schwarzkopf. Computing many faces in arrangements of lines and segments. *SIAM J. Comput.*, 27:491–505, 1998.

[AMS08]   E.M. Arkin, J.S.B. Mitchell, and J. Snoeyink. Capturing crossings: Convex hulls of segment and plane intersections. *Inform. Process. Lett.*, 107:194–197, 2008.

[APS93]   P.K. Agarwal, M. Pellegrini, and M. Sharir. Counting circular arc intersections. *SIAM J. Comput.*, 22:778–793, 1993.

[AS90]   P.K. Agarwal and M. Sharir. Red-blue intersection detection algorithms, with applications to motion planning and collision detection. *SIAM J. Comput.*, 19:297–321, 1990.

[Asa94]   T. Asano. Reporting and counting intersections of lines within a polygon. In *Proc. 5th Internat. Sympos. Algorithms Comput.*, vol. 834 of *LNCS*, pages 652–659, Springer, Berlin, 1994.

[Bal95]   I.J. Balaban. An optimal algorithm for finding segment intersections. In *Proc. 11th Sympos. Comput. Geom.*, pages 211–219, ACM Press, 1995.

[BCKO00]   M. de Berg, O. Cheong, M. van Kreveld, and M. Overmars. *Computational Geometry: Algorithms and Applications.* Springer-Verlag, Berlin, 3rd edition, 2008.

[BGR96]   J. Basch, L.J. Guibas, and G.D. Ramkumar. Reporting red-blue intersections between two sets of connected line segments. In *Proc. 4th European Sympos. Algorithms*, vol. 1136 of *LNCS*, pages 302–319, Springer, Berlin, 1996.

[BHO07]   M. de Berg, D. Halperin, and M. Overmars. An intersection-sensitive algorithm for snap rounding. *Comput. Geom.*, 36:159–165, 2007.

[BKM+95]   C. Burnikel, J. Könnemann, K. Mehlhorn, S. Näher, S. Schirra, and C. Uhrig. Exact geometric computation in LEDA. In *Proc. 11th Sympos. Comput. Geom.*, pages C18–C19, 1995.

[BL15]   L. Barba and S. Langerman. Optimal detection of intersections between convex polyhedra. In *Proc. 26th ACM-SIAM Sympos. Discrete Algorithms*, pages 1641–1654, 2015.

[BO79]   J.L. Bentley and T.A. Ottmann. Algorithms for reporting and counting geometric intersections. *IEEE Trans. Comput.*, C-28:643–647, 1979.

[BS00]   J.-D. Boissonnat and J. Snoeyink. Efficient algorithms for line and curve segment intersection using restricted predicates. *Comput. Geom*, 16:35–52, 2000.

[BV02]   J.-D. Boissonnat and A. Vigneron. An elementary algorithm for reporting intersections of red/blue curve segments. *Comput. Geom.*, 21:167–175, 2002.

[CC10]   T.M. Chan and E.Y. Chen. Optimal in-place and cache-oblivious algorithms for 3-d convex hulls and 2-d segment intersection. *Comput. Geom.*, 43:636–646, 2010.

[CD87]   B. Chazelle and D.P. Dobkin. Intersection of convex objects in two and three dimensions. *J. ACM*, 34:1–27, 1987.

[CE92]   B. Chazelle and H. Edelsbrunner. An optimal algorithm for intersecting line segments in the plane. *J. ACM*, 39:1–54, 1992.

[CEGS91]   B. Chazelle, H. Edelsbrunner, L.J. Guibas, and M. Sharir. A singly exponential stratification scheme for real semi-algebraic varieties and its applications. *Theoret. Comput. Sci.*, 84:77–105, 1991.

[CEGS94]   B. Chazelle, H. Edelsbrunner, L.J. Guibas, and M. Sharir. Algorithms for bichromatic line segment problems and polyhedral terrains. *Algorithmica*, 11:116–132, 1994.

[Cha86]   B. Chazelle. Reporting and counting segment intersections. *J. Comput. Syst. Sci.*, 32:156–182, 1986.

[Cha88]   B. Chazelle. A functional approach to data structures and its use in multidimensional searching. *SIAM J. Comput.*, 17:427–462, 1988.

[Cha91]   B. Chazelle. Triangulating a simple polygon in linear time. *Discrete Comput. Geom.*, 6:485–524, 1991.

[Cha92]   B. Chazelle. An optimal algorithm for intersecting three-dimensional convex polyhedra. *SIAM J. Comput.*, 21:671–696, 1992.

[Cha93]   B. Chazelle. Cutting hyperplanes for divide-and-conquer. *Discrete Comput. Geom.*, 9:145–158, 1993.

[Cha94]   T.M. Chan. A simple trapezoid sweep algorithm for reporting red/blue segment intersections. In *Proc. 6th Canad. Conf. Comput. Geom.*, pages 263–268, 1994.

[Cha16]   T.M. Chan. A simpler linear-time algorithm for intersecting two convex polyhedra in three dimensions. *Discrete Comput. Geom.*, 56:860–865, 2016.

[CL15]   T.M. Chan and P. Lee. On constant factors in comparison-based geometric algorithms and data structures. *Discrete Comput. Geom.*, 53:489–513, 2015.

[Cla92]   K.L. Clarkson. Safe and effective determinant evaluation. In *Proc. 33rd IEEE Sympos. Found. Comput. Sci.*, pages 387–395, 1992.

[CLM05]   B. Chazelle, D. Liu, and A. Magen. Sublinear geometric algorithms. *SIAM J. Comput.*, 35:627–646, 2005.

[CS89]   K.L. Clarkson and P.W. Shor. Applications of random sampling in computational geometry, II. *Discrete Comput. Geom.*, 4:387–421, 1989.

[CW11]   T.M. Chan and B.T. Wilkinson. Bichromatic line segment intersection counting in $o(n\sqrt{\log n})$ time. In *Proc. 23rd Canadian Conf. Comput. Geom.*, Toronto, 2011.

[DE87]   D.P. Dobkin and H. Edelsbrunner. Space searching for intersecting objects. *J. Algorithms*, 8:348–361, 1987.

[DK83]   D.P. Dobkin and D.G. Kirkpatrick. Fast detection of polyhedral intersection. *Theoret. Comput. Sci.*, 27:241–253, 1983.

[DK85]   D.P. Dobkin and D.G. Kirkpatrick. A linear algorithm for determining the separation of convex polyhedra. *J. Algorithms*, 6:381–392, 1985.

[DK90]   D.P. Dobkin and D.G. Kirkpatrick. Determining the separation of preprocessed polyhedra—a unified approach. In *Proc. 17th Internat. Colloq. Automata Lang. Program.*, vol. 443 of *LNCS*, pages 400–413, Springer, Berlin, 1990.

[DMN92]   M.B. Dillencourt, D.M. Mount, and N.S. Netanyahu. A randomized algorithm for slope selection. *Internat. J. Comput. Geom. Appl.*, 2:1–27, 1992.

[DMY93]   K. Dobrindt, K. Mehlhorn, and M. Yvinec. A complete and efficient algorithm for the intersection of a general and a convex polyhedron. In *Proc. 3rd Workshop Algorithms Data Struct.*, vol. 709 of *LNCS*, pages 314–324, Springer, Berlin, 1993.

[Ede87]   H. Edelsbrunner. *Algorithms in Combinatorial Geometry*, vol. 10 of *EATCS Monogr. Theoret. Comput. Sci.* Springer-Verlag, Heidelberg, 1987.

[EM81]   H. Edelsbrunner and H.A. Maurer. On the intersection of orthogonal objects. *Inform. Process. Lett.*, 13:177–181, 1981.

[Eri96]   J. Erickson. New lower bounds for Hopcroft's problem. *Discrete Comput. Geom.*, 16:389–418, 1996.

[ES05]   E. Ezra and M. Sharir. Counting and representing intersections among triangles in three dimensions. *Comput. Geom.*, 32:196–215, 2005.

[ETHA89]   M. Edahiro, K. Tanaka, R. Hoshino, and Ta. Asano. A bucketing algorithm for the orthogonal segment intersection search problem and its practical efficiency. *Algorithmica*, 4:61–76, 1989.

[FDFH90]   J.D. Foley, A. van Dam, S.K. Feiner, and J.F. Hughes. *Computer Graphics: Principles and Practice*. Addison-Wesley, Reading, 1990.

[FH95]   U. Finke and K.H. Hinrichs. Overlaying simply connected planar subdivisions in linear time. In *Proc. 11th Sympos. Comput. Geom.*, pages 119–126, ACM Press, 1995.

[FW96]   S. Fortune and C.J. van Wyk. Static analysis yields efficient exact integer arithmetic for computational geometry. *ACM Trans. Graph.*, 15:223–248, 1996.

[GJS99]   P. Gupta, R. Janardan, and M. Smid. Efficient algorithms for counting and reporting pairwise intersections between convex polygons. *Inform. Process. Lett.*, 69:7–13, 1999.

[GS87]   L.J. Guibas and R. Seidel. Computing convolutions by reciprocal search. *Discrete Comput. Geom.*, 2:175–193, 1987.

[GY86]   D.H. Greene and F.F. Yao. Finite-resolution computational geometry. In *Proc. 27th IEEE Sympos. Found. Comp. Sci.*, pages 143–152, 1986.

[Her13]   J. Hershberger. Stable snap rounding. *Comput. Geom.*, 46:403–416, 2013.

[Hob99]   J.D. Hobby. Practical segment intersection with finite precision output. *Comput. Geom.*, 13:199–214, 1999.

[Hof89]   C.M. Hoffmann. *Geometric and Solid Modeling*. Morgan Kaufmann, San Francisco, 1989.

[HP02]   D. Halperin and E. Packer. Iterated snap rounding. *Comput. Geom.*, 23:209–225, 2002.

[HS95]   J. Hershberger and S. Suri. A pedestrian approach to ray shooting: Shoot a ray, take a walk. *J. Algorithms*, 18:403–431, 1995.

[HSS83]   J.E. Hopcroft, J.T. Schwartz, and M. Sharir. Efficient detection of intersections among spheres. *Internat. J. Robot. Res.*, 2:77–80, 1983.

[JG91]   J.K. Johnstone and M.T. Goodrich. A localized method for intersecting plane algebraic curve segments. *Visual Comput.*, 7:60–71, 1991.

[LP79]   D.T. Lee and F.P. Preparata. An optimal algorithm for finding the kernel of a polygon. *J. ACM*, 26:415–421, 1979.

[LP79b]   J.J. Little and T.K. Peucker. A recursive procedure for finding the intersection of two digital curves. *Comput. Graph. Image Process.*, 10:159–171, 1979.

[Män88]   M. Mäntylä. *An Introduction to Solid Modeling*. Computer Science Press, Rockville, 1988.

[Mat91]   J. Matoušek. Randomized optimal algorithm for slope selection. *Inform. Process. Lett.*, 39:183–187, 1991.

[Mat93]   J. Matoušek. Range searching with efficient hierarchical cuttings. *Discrete Comput. Geom.*, 10:157–182, 1993.

[MC91]   D. Manocha and J.F. Canny. A new approach for surface intersection. *Internat. J. Comput. Geom. Appl.*, 1:491–516, 1991.

[Meh84]   K. Mehlhorn. *Multi-dimensional Searching and Computational Geometry*, vol. 3 of *Data Structures and Algorithms*, Springer-Verlag, Berlin, 1984.

[MK89]   A. Margalit and G.D. Knott. An algorithm for computing the union, intersection or difference of two polygons. *Comput. Graph.*, 13:167–183, 1989.

[MN01]   D.M. Mount and N.S. Netanyahu. Efficient randomized algorithms for robust estimation of circular arcs and aligned ellipses. *Comput. Geom.*, 19:1–33, 2001.

[Mou92]   D.M. Mount. Intersection detection and separators for simple polygons. In *Proc. 8th Sympos. Comput. Geom.*, pages 303–311, ACM Press, 1992.

[MP78]   D.E. Muller and F.P. Preparata. Finding the intersection of two convex polyhedra. *Theoret. Comput. Sci.*, 7:217–236, 1978.

[MPS⁺94]   J. Matoušek, J. Pach, M. Sharir, S. Sifrony, and E. Welzl. Fat triangles determine linearly many holes. *SIAM J. Comput.*, 23:154–169, 1994.

[MS88]   H.G. Mairson and J. Stolfi. Reporting and counting intersections between two sets of line segments. In R.A. Earnshaw, editor, *Theoretical Foundations of Computer Graphics and CAD*, vol. F40 of *NATO ASI*, pages 307–325, Springer, Berlin, 1988.

[MS01]   A. Mantler and J. Snoeyink. Intersecting red and blue line segments in optimal time and precision. In *Japan. Conf. Discrete Comput. Geom.*, vol. 2098 of *LNCS*, pages 244–251, Springer, Berlin, 2001.

[Mul91]   K. Mulmuley. A fast planar partition algorithm, II. *J. ACM*, 38:74–103, 1991.

[Mul93]   K. Mulmuley. *Computational Geometry: An Introduction through Randomized Algorithms*. Prentice Hall, Englewood Cliffs, 1993.

[NAT90]   B. Naylor, J.A. Amatodes, and W. Thibault. Merging BSP trees yields polyhedral set operations. *Comput. Graph.*, 24:115–124, 1990.

[OCON82]   J. O'Rourke, C.-B. Chien, T. Olson, and D. Naddor. A new linear algorithm for intersecting convex polygons. *Comput. Graph. Image Process.*, 19:384–391, 1982.

[O'R94]   J. O'Rourke. *Computational Geometry in C*. Cambridge University Press, 1994.

[Pae95]   A.W. Paeth, editor. *Graphics Gems V*. Academic Press, Boston, 1995.

[Pel97]   M. Pellegrini. On counting pairs of intersecting segments and off-line triangle range searching. *Algorithmica*, 17:380–398, 1997.

[PS85]   F.P. Preparata and M.I. Shamos. *Computational Geometry: An Introduction*. Springer-Verlag, New York, 1985.

[Rei88]   M. Reichling. On the detection of a common intersection of k convex objects in the plane. *Inform. Process. Lett.*, 29:25–29, 1988.

[RV85]   A.A.G. Requicha and H.B. Voelcker. Boolean operations in solid modelling: Boundary evaluation and merging algorithms. *Proc. IEEE*, 73:30–44, 1985.

[SA95]   M. Sharir and P.K. Agarwal. *Davenport-Schinzel Sequences and Their Geometric Applications*. Cambridge University Press, 1995.

[Sam90a]   H. Samet. *Applications of Spatial Data Structures*. Addison-Wesley, Reading, 1990.

[Sam90b]   H. Samet. *The Design and Analysis of Spatial Data Structures*. Addison-Wesley, Reading, 1990.

[Sam06]   H. Samet. *Foundations of Multidimensional and Metric Data Structures*. Morgan Kaufmann, San Francisco, 2006.

[SH76]   M.I. Shamos and D. Hoey. Geometric intersection problems. In *Proc. 17th IEEE Sympos. Found. Comput. Sci.*, pages 208–215, 1976.

[SHH99]   S. Suri, P.M. Hubbard, and J.F. Hughes. Analyzing bounding boxes for object intersection. *ACM Trans. Graph.*, 18:257–277, 1999.

[SI94]   K. Sugihara and M. Iri. A robust topology-oriented incremental algorithm for Voronoi diagrams. *Internat. J. Comput. Geom. Appl.*, 4:179–228, 1994.

[Vah07]   J. Vahrenhold. Line-segment intersection made in-place. *Comput. Geom.*, 38:213–230, 2007.

[Vig03]   A. Vigneron. Reporting intersections among thick objects. *Inform. Process. Lett.*, 85:87–92, 2003.

[Yap93]   C.K. Yap. *Fundamental Problems in Algorithmic Algebra*. Princeton University Press, 1993.

[ZS99]   Y. Zhou and S. Suri. Analysis of a bounding box heuristic for object intersection. *J. ACM*, 46:833–857, 1999.

43 NEAREST NEIGHBORS IN HIGH-DIMENSIONAL SPACES

Alexandr Andoni and Piotr Indyk

INTRODUCTION

In this chapter we consider the following problem: given a set P of points in a high-dimensional space, construct a data structure that given any *query* point q finds the point in P closest to q. This problem, called **nearest neighbor search**[1], is of significant importance to several areas of computer science, including pattern recognition, searching in multimedia data, vector compression [GG91], computational statistics [DW82], and data mining. Many of these applications involve data sets that are very large (e.g., a database containing Web documents could contain over one billion documents). Moreover, the dimensionality of the points is usually large as well (e.g., in the order of a few hundred). Therefore, it is crucial to design algorithms that scale well with the database size as well as with the dimension.

The nearest neighbor problem is an example of a large class of **proximity problems**, which, roughly speaking, are problems whose definitions involve the notion of distance between the input points. Apart from nearest neighbor search, the class contains problems like closest pair, diameter, minimum spanning tree and variants of clustering problems.

Many of these problems were among the first investigated in the field of computational geometry. As a result of this research effort, many efficient solutions have been discovered for the case when the points lie in a space of *constant* dimension. For example, if the points lie in the plane, the nearest neighbor problem can be solved with $O(\log n)$ time per query, using only $O(n)$ storage [SH15b, LT80]. Similar results can be obtained for other problems as well. Unfortunately, as the dimension grows, the algorithms become less and less efficient. More specifically, their space or time requirements grow *exponentially* in the dimension. In particular, the nearest neighbor problem has a solution with $O(d^{O(1)} \log n)$ query time, but using roughly $n^{O(d)}$ space [Cla88, Mei93]. Alternatively, if one insists on linear or near-linear storage, the best known running time bound for *random* input is of the form $\min(2^{O(d)}, dn)$, which is essentially linear in n even for moderate d. Worse still, the exponential dependence of space and/or time on the dimension (called the "curse of dimensionality") has been observed in applied settings as well. Specifically, it is known that many popular data structures (using linear or near-linear storage), exhibit query time linear in n when the dimension exceeds a certain threshold (usually 10–20, depending on the number of points), e.g., see [WSB98] for more information.

The lack of success in removing the exponential dependence on the dimension led many researchers to conjecture that no efficient solutions exists for these problems when the dimension is sufficiently large (e.g., see [MP69]). At the same time, it raised the question: Is it possible to remove the exponential dependence on d,

---

[1]Many other names occur in literature, including *best match*, *post office problem* and *nearest neighbor*.

if we allow the answers to be **approximate**. The notion of approximation is best explained for the nearest neighbor search: instead of reporting a point p closest to q, the algorithm is allowed to report *any* point within distance $(1 + \epsilon)$ times the distance from q to p. Similar definitions can be naturally applied to other problems. Note that this approach is similar to designing efficient approximation algorithms for NP-hard problems.

Over the years, a number of researchers have shown that indeed in many cases approximation enables a reduction of the dependence on dimension from exponential to polynomial. In this chapter we will survey these results. In addition, we will discuss the issue of *proving* that the curse of dimensionality is inevitable if one insists on exact answers, and survey the known results in this direction.

Although this chapter is devoted almost entirely to approximation algorithms with running times polynomial in the dimension, the notion of approximate nearest neighbor was first formulated in the context of algorithms with exponential query times, leading to a large number of results including [Cla94, AMN+98, DGK01, HP01, Cha02a, SSS06, AMM09, ADFM11]. Chapter 32 of this Handbook covers such results in more detail.

43.1 APPROXIMATE NEAR NEIGHBOR

Almost all algorithms for proximity problems in high-dimensional spaces proceed by reducing the problem to the problem of finding an *approximate near neighbor*, which is the decision version of the approximate near<u>est</u>-neighbor problem. Thus, we start from describing the results for the former problem.

All the NNS algorithms are based on space partitions (even if not always framed this way). We distinguish two broad classes of partitions: 1) data-independent approaches, where the partition is independent of the given dataset P, and 2) data-dependent approaches, where the partition depends on the dataset P.

For the definitions of metric spaces and normed spaces, see Chapter 8.

GLOSSARY

Approximate Near Neighbor, or (r,c)-NN: Given a set P on n points in a metric space $M = (X, D)$, design a data structure that supports the following operation: For any query $q \in X$, if there exists $p \in P$ such that $D(p, q) \le r$, find a point $p' \in P$ such that $D(q, p') \le cr$

Dynamic problems: Problems that involve designing a data structure for a set of points (e.g., approximate near neighbor) and support insertions and deletions of points. We distinguish dynamic problems from their **static** versions by adding the word "Dynamic" (or letter "D") in front of their names (or acronyms). E.g., the dynamic version of the approximate near-neighbor problem is denoted by (r,c)-DNN.

Hamming metric: A metric (Σ^d, D) where Σ is a set of *symbols*, and for any $p, q \in \Sigma^d$, $D(p, q)$ is equal to the number of $i \in \{1 \ldots d\}$ such that $p_i \ne q_i$.

DATA-INDEPENDENT APPROACH

The first algorithms for (r,c)-NN in high dimensions were obtained by using the technique of random projections, as introduced in the paper by Kleinberg [Kle97]. Although his algorithms still suffered from the curse of dimensionality (i.e., used exponential storage or had $\Omega(n)$ query time), his ideas provided inspiration for designing improved algorithms.

Later algorithms are based on the technique of *Locality-Sensitive Hashing*. We describe both approaches next.

TABLE 43.1.1   Approximate Near Neighbors under the Hamming distance.

#	APPROX.	QUERY TIME	SPACE	UPDATE TIME
Very low query time	Source: [KOR00] (see also [HIM12])			
	$c = 1 + \epsilon$	$d \log n / \min(\epsilon^2, 1)$	$n^{O(1/\epsilon^2 + \log c/c)}$	$n^{O(1/\epsilon^2 + \log c/c)}$
Low query time	Source: [Pan06, Kap15, Laa15, ALRW17]			
	c	$dn^{o(1)}$	$n^{\left(\frac{c}{c-1}\right)^2}$	$n^{\left(\frac{c}{c-1}\right)^2}$
Balanced	Source: [HIM12, AINR14, AR15]			
	c	$dn^{\frac{1}{2c-1}+o(1)}$	$n^{1+\frac{1}{2c-1}+o(1)} + dn$	$dn^{\frac{1}{2c-1}+o(1)}$
Low space	Source: [Ind01a, Pan06, AI06, Kap15, Laa15, ALRW17]			
	c	$n^{\frac{2c-1}{c^2}}$	$dn^{1+o(1)}$	$dn^{o(1)}$

Algorithms via dimensionality reduction.   We first focus on the case where all input and query points are binary vectors from $\{0,1\}^d$, and D is the Hamming distance. *Dimensionality reduction* is a randomized procedure that reduces the dimension of Hamming space from d to $k = O(\log n/\epsilon^2)$, while preserving a certain range of distances between the input points and the query up to a factor of $1 + \epsilon$. This notion has been introduced earlier in Chapter 8 in the context of Euclidean space. In the case of the Hamming space, the following holds.

THEOREM 43.1.1   *[KOR00] (see also [HIM12])*

For any given $r \in \{1 \dots d\}$, $\epsilon \in (0,1]$ and $\delta \in (0,1)$, one can construct a distribution over mappings $A : \{0,1\}^d \to \{0,1\}^k$, $k = O(\log(1/\delta)/\epsilon^2)$, and a "scaling factor" S, so that for any $p,q \in \{0,1\}^d$, if $D(p,q) \in [r, 10r]$, then $D(A(p), A(q)) = S \cdot D(p,q)(1 \pm \epsilon)$ with probability at least $1 - \delta$.

The factor 10 can be replaced by any constant. As in the case of Euclidean norm, the mapping A is linear (over the field $GF(2)$). The $k \times n$ matrix A is obtained by choosing each entry of A independently at random from the set $\{0,1\}$. The probability that an entry is equal to 1 is roughly r/d.

The first algorithm in Table 43.1.1 is an immediate consequence of Theorem 43.1.1. Specifically, it allows us to reduce the $(r, c + \epsilon)$-NN problem in d-dimensional space to (r, c)-NN problem in k-dimensional space. Since the *exact* nearest neighbor problem in k-dimensional space can be solved by storing the answers to all 2^k queries q, the bound follows.

This resulting algorithm is randomized and has a Monte Carlo guarantee of correctness. One can also obtain stronger guarantees of correctness: Las Vegas guarantee with similar performance and a deterministic algorithm with $3 + \epsilon$ approximation are described in [Ind00].

We note that one can apply the same approach to solve the near-neighbor problem in the Euclidean space. In particular, it is fairly easy to solve the $(r, 1 + \epsilon)$-NN problem in l_2^d using $n(1/\epsilon)^{O(d)}$ space [HIM12]. Applying the Johnson-Lindenstrauss lemma leads to an algorithm with storage bound similar to (although slightly worse than) the bound of algorithm (1a) [HIM12].

Algorithms via Locality-Sensitive Hashing.   The storage bound for the first algorithm in Table 43.1.1 is high and, oftentimes, one needs space to be much closer to the linear in the dataset size. The next two algorithms in the table obtain a better trade-off between space and the query time.

These algorithms are based on the concept of ***Locality-Sensitive Hashing***, or *LSH* [HIM12] (see also [KWZ95, Bro98]). A family of hash functions $h : \{0,1\}^d \to U$ is called (r, cr, P_1, P_2)-sensitive (for $c > 1$ and $P_1 > P_2$) if for any $q, p \in \{0,1\}^d$

- if $D(p, q) \leq r$, then $\Pr[h(q) = h(p)] \geq P_1$, and
- if $D(p, q) > cr$, then $\Pr[h(q) = h(p)] \leq P_2$,

where $\Pr[\cdot]$ is defined over the random choice of h. We note that the notion of locality-sensitive hashing can be defined for any metric space M in a natural way (see [Cha02b] for sufficient and necessary conditions for existence of LSH for M).

For Hamming space, there are particularly simple LSH families: it is sufficient to take all functions h_i, $i = 1 \dots d$, such that $h_i(p) = p_i$ for $p \in \{0,1\}^d$. The resulting family is sensitive due to the fact that $\Pr[h(p) = h(q)] = 1 - D(p, q)/d$.

TABLE 43.1.2   Approximate Near Neighbors via LSH and data-dependent hashing approaches.

METRIC	TYPE	QUERY TIME	EXPONENT ρ	FOR $c = 2$	REFERENCE
ℓ_1	LSH	$n^\rho d$	$\rho = 1/c$	$\rho = 1/2$	[HIM12]
			$\rho \geq 1/c - o(1)$	$\rho \geq 1/2$	[MNP07, OWZ14]
	Data dependent hashing	$n^\rho d$	$\rho = \frac{1}{2c-1} + o(1)$	$\rho = 1/3$	[AINR14, AR15]
			$\rho \geq \frac{1}{2c-1} - o(1)$	$\rho \geq 1/3$	[AR16]
ℓ_2	LSH	$n^\rho d$	$\rho \leq 1/c$	$\rho \leq 1/2$	[HIM12, DIIM04]
			$\rho = 1/c^2 + o(1)$	$\rho = 1/4$	[AI06]
			$\rho \geq 1/c^2 - o(1)$	$\rho \geq 1/4$	[MNP07, OWZ14]
	Data dependent hashing	$n^\rho d$	$\rho = \frac{1}{2c^2-1} + o(1)$	$\rho = 1/7$	[AINR14, AR15]
			$\rho \geq \frac{1}{2c^2-1} - o(1)$	$\rho \geq 1/7$	[AR16]

An LSH family with a "large" gap between P_1 and P_2 immediately yields a solution to the (c, r)-NN problem. During preprocessing, all input points p are hashed to the bucket $h(p)$. In order to answer the query q, the algorithm retrieves the points in the bucket $h(q)$ and checks if any one of them is close to q. If the gap between P_1 and P_2 is sufficiently large, this approach can be shown to result in sublinear query time. Unfortunately, the P_1/P_2 gap guaranteed by the above LSH family is not large enough. However, the gap can be amplified by concatenating several independently chosen hash functions $h_1 \ldots h_l$ (i.e., hashing the points using functions h' such that $h'(p) = (h_1(p), \ldots, h_l(p))$). This decreases the probability of finding a near neighbor in one hash tables, and therefore we build a few hash tables, and, at query time, look-up one bucket of each of the hash tables. Details can be found in [HIM12].

The overall LSH algorithm uses $O(n^\rho)$ hash tables, where $\rho = \frac{\log P_1}{\log P_2} \in (0, 1)$ turns out to be the key measure of quality of the hash function (equivalently, the space partition). In particular, the query time becomes $O(n^\rho d)$, and the space becomes $O(n^{1+\rho} + nd)$. For the above LSH family for the Hamming space, one can prove that $\rho = 1/c$, resulting in the [HIM12] algorithm with $O(n^{1/c}d)$ time and $O(n^{1+1/c} + nd)$ space.

A somewhat similar hashing-based algorithm (for the closest-pair problem) was earlier proposed in [KWZ95], and also in [Bro98]. Due to different problem formulation and analysis, comparing their performance with the guarantees of the LSH approach seems difficult.

Time-space trade-offs.   While the LSH approach achieves sub-quadratic space, one may hope to obtain even better guarantee: a near-linear space, $\tilde{O}(nd)$. Indeed, such near-linear space algorithms have been proposed in [Ind01a, Pan06, AI06, Kap15, Laa15]. For the smallest possible space of $O(nd)$, [Kap15] obtains query time $O(n^{4/(c+1)}d)$. For other algorithms, see the "low space" regime in Table 43.1.1, as well as the "(very) low query time" regime for the opposite extreme, of the lowest possible query time.

Algorithms for l_2 and other ℓ_p norms.   To solve the problem under the ℓ_1 norm, we can reduce it to the Hamming case. If we assume that all points of interest p have coordinates in the range $\{1, \ldots, M\}$, we can define $U(p) = (U(p_1), \ldots, U(p_d))$ where $U(x)$ is a string of x ones followed by $M - x$ zeros. Then we get $\|p - q\|_1 = D(U(p), U(q))$. In general, M could be quite large, but can be reduced to $d^{O(1)}$ in the context of approximate near neighbor [HIM12]. Thus we can reduce (r, c)-NN under l_1 to (r, c)-NN in the Hamming space.

For the Euclidean space ℓ_2, one can design more efficient algorithms. Specifically, [AI06] obtain $\rho = 1/c^2 + o(1)$, improving over the (optimal) $\rho = 1/c$ exponent for the Hamming space. For ℓ_p's, for $p \in [1, 2)$, one can then reduce the problem to the ℓ_2 case by embedding the $(p/2)$-th root of ℓ_p into ℓ_2 [Kal08, Theorem 4.1] (see also a quantitative version in [Ngu14]).

For ℓ_p with $p < 1$, [DIIM04] obtain an LSH exponent $\rho = 1/c^p + o(1)$.

Las Vegas algorithms.   The standard LSH scheme guarantees 90% success probability of recovering an (approximate) near neighbor (Monte Carlo randomness). There are algorithms that guarantee to return an (approximate) near neighbor, albeit the runtime is in expectation only (Las Vegas randomness) [GPY94, Ind00, AGK06, Pag16]. Such algorithms proceed by constructing a number of space partitions, such that any pair of close points will collide in at least on of the space partitions. The query time of such algorithms is usually (polynomially) higher than the query time of the Monte Carlo ones.

Lower bounds.   For algorithms based on the LSH concept, we can prove tight lower bounds, ruling out better exponents ρ [MNP07, OWZ14]. In particular, for $p \in (0, 2]$, the tight exponent is $\rho = 1/c^p \pm o(1)$. Such lower bounds assume that P_1 is not too small, namely inversely exponential in $d^{\Omega(1)}$. LSH schemes with P_1 exponentially small in d is not useful for the high dimensional spaces with $d \gg \log n$ — in this situation, $1/P_1 \ll 1/n$. We note however that in the "moderate dimension regime," when $d = \Theta(\log n)$, it may be tolerable to have $P_1 = 2^{-\Theta(d)}$ and indeed, [BDGL16, Laa15] obtain somewhat better exponents ρ in this case.)

For $p > 2$, it is natural to conjecture that any LSH scheme must incur either a super-constant factor approximation, or must have P_1 exponentially small in $d^{\Omega(1)}$; see [AN11] for partial progress towards the conjecture.

DATA-DEPENDENT APPROACH

The data-independent techniques from the previous section have natural limitations. It turns out that sometimes vastly better algorithms are possible using *data-dependent* approaches. In particular, these are the methods where the hash function h itself depends on the entire dataset P. Note that an important requirement for a data-dependent hashing function is to have *efficient evaluation procedure* on a new (query) point q. Without such condition, the obvious best data-dependent partition would be the Voronoi diagram—i.e., $h(q)$ returns the identity of the closest point from the dataset—which is obviously useless (computing the hash function is as hard as the original problem!). Indeed, the space partitions mentioned below are (provably) better than the data-independent variants while being efficient to compute as well.

As before, the resulting algorithms give an improvement for worst-case datasets.

Algorithm for the ℓ_∞ norm.   The first algorithm from this category is for near-neighbor problem under the ℓ_∞ norm, where data-independent methods are otherwise powerless. In particular, [Ind01b] solves (r, c)-NN under the l_∞^d norm with the following guarantees, for any $\rho > 0$:

- Approximation factor: $c = O(\lfloor \log_{1+\rho} \log 4d \rfloor)$; if $\rho = \log d$ then $c = 3$.
- Space: $dn^{1+\rho}$.
- Query time: $O(d \log n)$ for the static, or $(d + \log n)^{O(1)}$ for the dynamic case.
- Update time: $d^{O(1)} n^\rho$ (described in [Ind01a]).

The basic idea of the algorithm is to use a divide and conquer approach. In particular, consider hyperplanes H consisting of all points with one (say the ith) coordinate equal to the same value. The algorithm tries to find a hyperplane H having the property that the set of points $P_L \subset P$ that are on the left side of H and at distance $\geq r$ from H, is not "much smaller" than the set P_M of points at distance r from H. Moreover, a similar condition has to be satisfied for an analogously defined set P_R of points on the right side of H. If such H exists, we divide P into $P_{LM} = P_L \cup P_M$ and set $P_{RM} = P \setminus P_L$ and build the data structure recursively on P_{LM} and P_{RM}. It is easy to see that while processing a query q, it suffices to recurse on either P_{LM} or P_{RM}, depending on the side of H the query q lies on. Also, one can prove that the increase in storage caused by duplicating P_M is moderate. On the other hand, if H does not exist, one can prove that a large subset $P' \subset P$ has $O(cr)$ diameter. In such a case we can pick any point from P' as its representative, and apply the algorithm recursively on $P \setminus P'$.

Algorithms for the ℓ_p norms.   Later work defined data-dependent hashing more formally and showed that one can obtain better runtime exponents ρ than LSH, for the exponent ρ defined similarly to the one from LSH. In particular, data-dependent hashing gives the following performance: for the Hamming space, the exponent is $\rho = \frac{1}{2c-1}$, and for the Euclidean space, the exponent is $\rho = \frac{1}{2c^2-1}$. See Table 43.2.1 for a comparison with LSH and lower bounds.

The data-dependent hashing algorithms in [AINR14, AR15] have two major components. First, if the given dataset has a certain "canonical geometric configuration," then one can design a data-*independent* hashing (LSH) scheme with better parameters than for datasets in a general position. This canonical setting essentially corresponds to a dataset which is distributed (pseudo-)randomly on a sphere (i.e., points are at $\approx \pi/2$ angle with respect to the origin), and the query is planted to be at $\theta < \pi/2$ angle from some point in the dataset. Second, there is a procedure to decompose any *worst-case* dataset and reduce it to this canonical case.

For ℓ_p, where $0 < p < 2$, all algorithms for ℓ_2 apply as well (with c^2 replaced by c^p in the exponent ρ); see [Ngu14].

For ℓ_p, where $p > 2$, efficient algorithms are possible via data-dependent hashing, however optimal bounds are not presently known. [And09] shows how to obtain $O(\log \log d)$ approximation, and [BG15, NR06] obtain approximation $2^{O(p)}$.[2]

Lower Bounds.   One can prove that, for $p \leq 2$ and $p = \infty$, the above exponents ρ are optimal within the class of data-dependent hashing schemes. To prove such lower bounds, one has to formalize the class of data-dependent hashing schemes (which in particular would rule out the aforementioned Voronoi diagram solution). For ℓ_∞, matching lower bounds were shown in [ACP08, KP12], and for ℓ_p in [AR16]. Both type of results formalize the class by assuming that the hash function has description complexity of $n^{1-\Omega(1)}$, as well as that $d = \log^{1+\Omega(1)} n$ and P_1 is not too small. These lower bounds also have implication for (unconditional) cell probe lower bounds; see Section 43.3.

Time-space trade-offs.   As for LSH, one can obtain other time-space trade-offs with the data-dependent approach as well. In particular, [Laa15, ALRW17] obtain trade-offs for the ℓ_2 space. One can obtain an algorithm with query time $n^{\rho_q+o(1)}d$ and space $n^{1+\rho_s+o(1)} + O(nd)$ for any $\rho_s, \rho_q > 0$ that satisfy the following equality:

$$c^2\sqrt{\rho_q} + (c^2 - 1)\sqrt{\rho_s} = \sqrt{2c^2 - 1}.$$

This trade-off is also optimal (in the right formalization) [ALRW17, Chr17]. For $\rho_q = 0$, there are also cell-probe lower bounds; see Section 43.3.

GLOSSARY

Product metrics: An f-product of metrics $X_1, \ldots, X_k$ with distance functions $D_1, \ldots, D_k$ is a metric over $X_1 \times \ldots \times X_k$ with distance function D such that $D((p_1, \ldots, p_k), (q_1, \ldots, q_k)) = f(D_1(p_1, q_1), \ldots, D_k(p_k, q_k))$.

---

[2]At the moment of writing of this document, [NR06] does not seem to be available but is referenced in [Nao14, Remark 4.12].

Although the l_∞ data structure seems to rely on the geometry of the l_∞ norm, it turns out that it can be used in a much more general setting. In particular, assume that we are given k metrics $M_1 \dots M_k$ such that for each metric M_i we have a data structure for (a variant of) (r, c)-NN in metric M_i, with $Q(n)$ query time and $S(n)$ space. In this setting, it is possible to construct a data structure solving $(r, O(c \log \log n))$-NN in the max-product metric M of $M_1, \dots, M_k$ (i.e., an f-product with f computing the maximum of its arguments) [Ind02]. The data structure for M achieves query time roughly $O(Q(n) \log n + k \log n)$ and space $O(kS(n)n^{1+\delta})$, for any constant $\delta > 0$. The data structure could be viewed as an abstract version of the data structure for the l_∞ norm (note that the l_∞^d norm is a max-product of l_p^1 norms). For the particular case of the l_∞^d norm, it is easy to verify that the result of [Ind02] provides a $O(\log \log n)$-approximate algorithm using space polynomial in n. At the same time, the algorithm of [Ind01b] has $O(\log \log d)$-approximation guarantee when using the same amount of space. Interestingly, the former data structure gives an approximation bound comparable to the latter one, while being applicable in a much more general setting.

The above result can also be used for developing NNS under product spaces, defined as follows. For a vector $x \in \mathbb{R}^{d_1 \cdot d_2}$, its $\ell_p^{d_1}(\ell_q^{d_2})$ norm is computed by taking the ℓ_q norm of each of the d_1 rows and then taking the ℓ_p norm of these d_1 values. For such product norms (in fact for any fixed iterated product norm), [AIK09, And09] showed how to obtain efficient NNS with $O(\log \log n)^{O(1)}$ approximation.

EXTENSIONS VIA EMBEDDINGS

Most of the algorithms described so far work only for l_p norms. However, they can be used for other metric spaces M, by using low-distortion embeddings of M into l_p norms. See Chapter 8 for more information.

Similarly, one can use low-distortion embedding of M into a product space to obtain efficient NNS under M. This has been used for the Ulam metric, which is the edit (Levenshtein) distance on nonrepetitive strings. In particular, [AIK09] showed a $O(1)$-distortion embedding of Ulam distance into an iterated product space, which gives the currently best known NNS algorithms for Ulam.

43.2 REDUCTIONS TO APPROXIMATE NEAR NEIGHBOR

GLOSSARY

We define the following problems, for a given set of points P in a metric space $M = (X, D)$:

Approximate Closest Pair, or c-CP: Find a pair of points $p', q' \in P$ such that $D(p', q') \le c \min_{p,q \in P, p \ne q} D(p, q)$

Approximate Close Pair, or (r, c)-CP: If there exists $p, q \in P, p \ne q$, such that $D(p, q) \le r$, find a pair $p', q' \in P, p' \ne q'$, such that $D(q', p') \le cr$.

Approximate Chromatic Closest Pair, or c-CCP: Assume that each point $p \in P$ is labeled with a color $c(p)$. Find a pair of points p, q such that $c(p) \ne c(q)$ and $D(p, q)$ is approximately minimal (as in the definition of c-CP).

Approximate Bichromatic Closest Pair, or c-BCP: As above, but $c(p)$ assumes only two values.

Approximate Chromatic/Bichromatic Close Pair, or (r, c)-CCP/(r, c)-BCP: Decision versions of c-CCP or c-BCP (as in the definition of (r, c)-CP).

Approximate Farthest Pair, or Diameter, or c-FP: Find $p, q \in P$ such that $D(p, q) \geq \max_{p', q' \in P} D(p', q')/c$. The decision problem, called ***Approximate Far Pair***, or (r, c)-FP, is defined in the natural way.

Approximate Farthest Neighbor, or c-FN: A maximization version of the Approximate Near Neighbor. The decision problem, called Approximate Far Neighbor or (r, c)-FN, is defined in a natural way.

Approximate Minimum Spanning Tree, or c-MST: Find a tree T spanning all points in P whose weight $w(T) = \sum_{(p, q) \in T} D(p, q)$ is at most c times larger than the weight of any tree spanning P.

Approximate Bottleneck Matching, or c-BM: Assuming $|P|$ is even, find a set of $|P|/2$ nonadjacent edges E joining points in P (i.e., a matching), such that the following function is minimized (up to factor of c)

$$\max_{\{p, q\} \in E} D(p, q)$$

Approximate Facility Location, or c-FL: Find a set $F \subset P$ such that the following function is minimized (up to factor of c), given the cost function $c : P \to \Re^+$

$$\sum_{p \in F} c(p) + \sum_{p \in P} \min_{f \in F} D(p, f)$$

In general, we could have two sets: P_c of *cities* and P_f of *facilities*; in this case we require that $F \subset P_f$ and we are only interested in the cost of P_c.

Spread (of a point set): The ratio between the diameter of the set to the distance between its closest pair of points.

In this section we show that the problems defined above can be efficiently reduced to the approximate near-neighbor problem discussed in the previous section.

First, we observe that any problem from the above list, say $c(1 + \delta)$-P for some $\delta > 0$, can be easily reduced to its decision version (say (r, c)-P), if we assume that the spread of $P \cup \{q\}$ is always bounded by some value, say Δ. For simplicity, assume that the minimum distance between the points in P is 1. The reduction proceeds by building (or maintaining) $O(\log_{1+\delta} \Delta)$ data structures for (r, c)-P, where r takes values $(1 + \delta)^i/2$ for $i = 0, 1 \ldots$. It is not difficult to see that a query to $c(1 + \delta)$-P can be answered by $O(\log \log_{1+\delta} \Delta)$ calls to these structures for (r, c)-P, via binary search.

In general, the spread of P could be unbounded. However, in many cases it is easy to reduce it to $n^{O(1)}$. This can be accomplished, for example, by "discretizing" the input to c-MST or c-FL. In those cases, the above reduction is very efficient.

Reductions from other problems are specified in the following table. The bounds for the time and space used by the algorithm in the "To" column are denoted by $T(n)$ and $S(n)$, respectively.

We mention that a few other reductions have been given in [KOR00, BOR04]. For the problems discussed in this section, they are less efficient than the reductions in the above table. Additionally, [BOR04] reduces the problems of computing

---

TABLE 43.2.1   Reductions to Approximate Near Neighbors.

#	FROM	TO	TIME	SPACE
1	Source: [HIM12].			
	$c(1+\delta)$-NN	(r,c)-NN	$T(n)\log^{O(1)} n$	$S(n)\log^{O(1)} n$
2	Source: [Epp95]; amortized time.			
	c-DBCP	c-DNN	$T(n)\log^{O(1)} n$	$S(n)\log^{O(1)} n$
	(r,c)-DBCP	(r,c)-DNN	$T(n)\log^{O(1)} n$	$S(n)\log^{O(1)} n$
3	Source: [HIM12]; via Kruskal alg.			
	$c(1+\delta)$-MST	(r,c)-DBCP	$nT(n)\log^{O(1)} n$	
4	Source: [GIV01, Ind01a]; via Primal-Dual			
	$3c^3(1+\delta)$-FL	(r,c)-DBCP	$nT(n)\log^{O(1)} n$	
5	Source: [GIV01, Ind01a].			
	$2c$-BM	c-DBCP	$nT(n)\log^{O(1)} n$	

approximate agglomerative clustering and *sparse partitions* to $O(n\log^{O(1)} n)$ calls to a dynamic approximate nearest neighbor data structure. See [BOR04] for the definitions and algorithms.

Also, we mention that a reduction from $(1+\epsilon)$-approximate farthest neighbor to $(1+\epsilon)$-approximate nearest neighbor (for the static case and under the l_2 norm) has been given in [GIV01]. However, a direct (and dynamic) algorithm for the approximate farthest neighbor in l_2^d, achieving a better query and update times of $dn^{1/(1+\epsilon)^2}$, has been given in [Ind03]. The former paper also presents an algorithm for computing a $(\sqrt{2}+\epsilon)$-approximate diameter (for any $\epsilon > 0$) of a given pointset in $dn\log^{O(1)} n$ time.

We now describe briefly the main techniques used to achieve the above results.

Nearest neighbor.   We start from the reduction of c-NN to (r,c)-NN. As we have seen already, the reduction is easy if the spread of P is small. Otherwise, it is shown that the data set can be clustered into $n/2$ clusters, in such a way that:

- If the query point q is "close" to one of the clusters, it must be far away from a constant fraction of points in P; thus, we can ignore these points in the search for an approximate nearest neighbor.

- If the query point q is "far" from a cluster, then all points in the clusters are equally good candidates for the *approximate* nearest neighbor; thus we can replace the cluster by its representative point.

See the details of the construction in [HIM12].

Bichromatic closest pair.   A very powerful reduction from various variants of c-DBCP to c-DNN was given by Eppstein [Epp95]. His algorithm was originally designed for the case $c = 1$, but it can be verified to work also for general $c \geq 1$ [Epp99]. Moreover, as mentioned in the original paper, the reduction does not require the distance function D be a metric.

The basic idea of the algorithm is to try to maintain a graph that contains an edge connecting the two closest bichromatic points. A natural candidate for such a graph is the graph formed by connecting each point to its nearest neighbor. This, however, does not work, because a vertex in such a graph can have very high degree, leading to high update cost. Another option would be to maintain a single path, such that the ith vertex points to its nearest neighbor of the opposite color, chosen from points not yet included in the path. This graph has low degree, but its rigid structure makes it difficult to update it at each step. So the actual data structure is based on the path idea but allows its structure to degrade in a controlled way, and only rebuilds it when it gets too far degraded, so that the rebuilding work is spread over many updates. Then, however, one needs to keep track of the information from the degraded parts of the path, which can be done using a second shorter path, and so on. The constant factor reduction in the lengths of each successive path means the total number of paths is only logarithmic.

We note that recent research [Val15, KKK16] showed an alternative approach to the bichromatic closest pair problem, which, in certain settings, vastly outperforms the approach via the NNS algorithms. This line of work is best described using the following parameterization. Suppose the two color classes of points $A, B \subset \mathbb{R}^d$ satisfy the following: 1) all points are of unit norm, and 2) for each $a \in A, b \in B$, we have that $\langle a, b \rangle < \alpha$ except for a single (close) pair that satisfies $\langle a, b \rangle \geq \beta$, for some $0 < \alpha < \beta < 1$. Then the algorithm of [Val15] obtains a runtime of $n^{\frac{5-\omega}{4-\omega} + \omega \frac{\log \beta}{\log \alpha}}$. $d^{O(1)}$, where $2 \leq \omega < 2.373$ is the exponent of the fast matrix multiplication algorithm. An algorithm in [KKK16] runs in time $n^{2\omega/3 + O(\log \beta / \log \alpha)} \cdot d^{O(1)}$. The most interesting case is where the point sets A and B are random, except for a pair that has inner product ϵ (termed the *light bulb problem* [Val88]). In this case the runtime of these algorithms is of the form $n^{2 - \Omega(1)} (d/\epsilon)^{O(1)}$, improving over $n^{2 - O(\epsilon)}$ time obtainable via LSH methods. The algorithm of [Val15] also obtains $n^{2 - \Omega(\sqrt{\epsilon})} \cdot d^{O(1)}$ time for the standard $(1 + \epsilon)$-BCP.

Finally, there is recent work on *exact* BCP for medium dimensions, which also relies on faster matrix multiplication. In particular, the algorithm of [AW15] achieves a runtime of $n^{2 - 1/O(c \log^2 c)}$ for Hamming space of dimension $d = c \log n$ for $c > 1$.

Minimum spanning tree. Many existing algorithms for computing MST (e.g., Kruskal's algorithm) can be expressed as a sequence of operations on a CCP data structure. For example, Kruskal's algorithm repetitively seeks the lightest edge whose endpoints belong to different components, and then merges the components. These operations can be easily expressed as operations on a CCP data structure, where each component has a different color. The contribution of [HIM12] was to show that in case of Kruskal's algorithm, using an *approximate* c-CCP data structure enables one to compute an *approximate* c-MST. Also, note that c-CCP can be implemented by $\log n$ c-BCP data structures [HIM12]. Other reductions from c-MST to c-BCP are given in [BOR04, IST99].

Minimum bottleneck matching. The main observation here is that a matching is also a spanning forest with the property that any connected component has even cardinality (call it an *even* forest). At the same time, it is possible to convert *any* even forest to a matching, in a way that increases the length of the longest edge by at most a factor of 2. Thus, it suffices to find an even forest with minimum edge length. This can be done by including longer and longer edges to the graph, and stopping at the moment when all components have even cardinality. It is not difficult to implement this procedure as a sequence of c-CCP (or c-BCP) calls.

Other algorithms.   The algorithm for the remaining problem (c-FL) is obtained by implementing the primal-dual approximation algorithm [JV01]. Intuitively, the algorithm proceeds by maintaining a set of balls of increasing radii. The latter process can be implemented by resorting to c-CCP. The approximation factor follows from the analysis of the original algorithm.

43.3 LOWER BOUNDS

In the previous sections we presented many algorithms solving approximate versions of proximity problems. The main motivation for designing approximation algorithms was the "curse of dimensionality" conjecture, i.e., the conjecture that finding exact solutions to those problems requires either superpolynomial (in d) query time, or superpolynomial (in n) space. In this section we state the conjecture more rigorously, and describe the progress toward proving it. We also describe the lower bounds for the approximation algorithms as well, which sometimes match the algorithmic results presented earlier.

Curse of dimensionality.   We start from the exact near-neighbor problem, where the curse of dimensionality can be formalized as follows.

CONJECTURE 43.3.1

Assume that $d = n^{o(1)}$ but $d = \omega(\log n)$. Any data structure for $(r, 1)$-NN in Hamming space over $\{0, 1\}^d$, with $d^{O(1)}$ query time, must use $n^{\omega(1)}$ space.

The conjecture as stated above is probably the weakest version of the "curse of dimensionality" phenomenon for the near-neighbor problem. It is plausible that other (stronger) versions of the conjecture could hold. In particular, at present, we do not know any data structure that simultaneously achieves $o(dn)$ query time and $2^{o(d)}$ space for the above range of d. At the same time, achieving $O(dn)$ query time with space dn, or $O(d)$ query time with space 2^d is quite simple (via linear scan or using exhaustive storage).

Also note that if $d = O(\log n)$, achieving $2^{o(d)} = o(n)$ space is impossible via a simple incompressibility argument.

Below we describe the work toward proving the conjecture. The first results addressed the complexity of a simpler problem, namely the *partial match* problem. This problem is of importance in databases and other areas and has been long investigated (e.g., see [Riv74]). Thus, the lower bounds for this problem are interesting in their own right.

GLOSSARY

Partial match: Given a set P of n vectors from $\{0, 1\}^d$, design a data structure that supports the following operation: For any query $q \in \{0, 1, *\}^d$, check if there exists $p \in P$ such that for all $i = 1 \ldots d$, if $q_i \neq *$ then $p_i = q_i$.

It is not difficult to see that any data structure solving $(r, 1)$-NN in the Hamming metric $\{0, 1\}^d$, can be used to solve the partial match problem using essentially the same space and query time. Thus, any lower bound for the partial match problem implies a corresponding lower bound for the near-neighbor problem. The best currently known lower bound for the partial match has been established in [Păt10], following earlier work of [JKKR04, BOR03, MNSW98]. Their lower bound holds in the *cell-probe* model, a very general model of computation, capturing e.g., the standard Random Access Machine model. The lower bound implies that any (possibly randomized) cell-probe algorithm for the partial match problem, in which the algorithm is allowed to retrieve at most $O(n^{1-\epsilon})$ bits from any memory cell in one step for $\epsilon > 0$, must use space $2^{\Omega(d/t)}$ for t cell probes (memory accesses, and hence query time).

Similar space lower bounds have been proven for *exact* (possibly randomized) near-neighbor [BR02], and for *deterministic* $O(1)$-approximate near neighbor [Liu04]. Note that allowing both approximation and randomization allows for much better upper bounds (see the "very low query time" regime in Table 43.1.1).

All the aforementioned lower bounds are proved in a very general model, using the tools of *asymmetric communication complexity*. As a result, they cannot yield lower bounds of $\omega(d/\log n)$ query time, for $n^{O(1)}$ space, as we now explain.

The communication complexity approach interprets the data structure as a communication channel between Alice (holding the query point q) and Bob (holding the database P). The goal of the communication is to determine the nearest neighbor of q. Since the data structure has polynomial size, each access to one of its memory cell is equivalent to Alice sending $O(\log n)$ bits of information to Bob. If we show that Alice needs to send at least a bits to Bob to solve the problem, we obtain $\Omega(a/\log n)$ lower bound for the query time. However, $b \leq d$, since Alice can always choose to transmit the whole input vector q. Thus, $\Omega(d/\log n)$ lower bound is the best result one can achieve using the communication complexity approach.

A partial step toward removing this obstacle was made in [BV02], employing the *branching programs* model of computation. In particular, they focused on randomized algorithms that have very small (inversely polynomial in n) probability of error. They showed that any algorithm for the $(r, 1)$-NN problem in the Hamming metric over $\{1 \ldots d^6\}^d$ has either $\Omega(d \log(d \log d/S))$ query time or uses $\Omega(S)$ space. This holds for $n = \Omega(d^6)$. Thus, if the query time is $o(d \log d)$, then the data structure must use $2^{d^{\Omega(1)}}$ space.

Another such step was made in [PT09], who show higher lower bound for *near-linear* space. In particular, they show that number of cell probes must be $t = \Omega(d \cdot \log \frac{Sd}{n})$ for space S.

Lower bounds for approximate randomized algorithms.   More recent efforts focused on lower bounds for approximate randomized problem, where most of the improved algorithms were obtained. Even if it is harder to prove lower bounds in this regime, researchers often managed to prove *tight* lower bounds, matching the known algorithms.

First of all, if we consider the *nearest neighbor* problem (in contrast to the *near* neighbor, which is considered in this chapter), [CR10, CCGL03] show that any data structure for the $O(1)$-approximate nearest neighbor on $\{0, 1\}^d$ requires either $\Omega(\log \log d/\log \log \log d)$ query time or $n^{\omega(1)}$ space. [CR10] also show a matching upper bound.

For the *near* neighbor problem, the authors of [AIP06] show that a $(1 + \epsilon)$-approximation requires $\left(n^{\Omega(1/\epsilon^2)}\right)^{1/t}$ space lower bound for t cell-probe data structures for the Hamming space. This lower bound matches the upper bound of

[KOR00, HIM12] (the "very low query time" regime in Table 43.1.1), which uses just 1 cell probe. We note that while the lower bound of [AIP06] is tight up to a constant in the exponent, a tight constant was shown in [ALRW17], although the matching upper bound uses $n^{o(1)}$ cell probes.

Restricting the attention to *near-linear space* algorithms, higher lower bounds were shown. In particular, the authors of [PTW08, PTW10] show how to obtain the following lower bounds: any t cell probe data structure for (r, c)-NNS under ℓ_1 requires space $n^{1+\Omega(1/c)/t}$. Note that this lower bound goes beyond the simple communication complexity framework described above. The lower bounds from [PTW10] were tightened in [ALRW17], which, when setting $t \in \{1, 2\}$, match the space bound of the data structure of [Laa15, ALRW17] for $n^{o(1)}$ query time.

Finally, for NNS under the ℓ_∞ metric, there are similar lower bounds, which are also based on the tools of asymmetric communication complexity. The authors of [ACP08] show that any deterministic decision tree for ℓ_∞ must incur an approximation of $O(\log_{1+\rho} \log n)$ if using space $n^{1+\rho}$ (unless the query time is polynomially large). This matches the upper bound of [Ind01b] described above. This lower bound was also extended to randomized decision trees in [PTW10, KP12].

REDUCTIONS

Despite the progress toward resolving the "curse of dimensionality" conjecture and the widespread belief in its validity, proving it seems currently beyond reach. Nevertheless, it is natural to assume the validity of the conjecture (or its variants), and see what conclusions can be derived from this assumption. Below we survey a few results of this type.

In order to describe the results, we need to state another conjecture.

CONJECTURE 43.3.2

Let $d = n^{o(1)}$ but $d = \log^{\omega(1)} n$. Any data structure for the partial match problem with parameters d and n that provides $d^{O(1)}$ query time must use $2^{d^{\Omega(1)}}$ space.

Note that, for the same ranges[3] of d, Conjecture 43.3.2 is analogous to Conjecture 43.3.1, but much stronger: it considers an easier problem, and states stronger bounds. However, since the partial match problem was extensively investigated on its own, and no algorithm with bounds remotely resembling the above have been discovered (cf. [CIP02] for a survey), Conjecture 43.3.2 is believed to be true.

Assuming Conjecture 43.3.2, it is possible to show lower bounds for some of the approximate nearest neighbor problems discussed in Section 43.1. In particular, it was shown [Ind01b] that any data structure for (r, c)-NN under l_∞^d for $c < 3$ can be used to solve the partial match problem with parameter d, using essentially the same query time and storage (the number of points in the database is the same in both cases). Thus, unless Conjecture 43.3.2 is false, the 3-approximation algorithm from Section 43.1 is optimal, in the sense that it provides the smallest approximation factor possible while preserving polynomial (in d) query time and subexponential (in d) storage. Note that this result resembles the nonapproximability results based on the P $\neq$ NP conjecture.

On the other hand, it was shown [CIP02] that the exact near-neighbor prob-

---

[3]For $d = \log^{\omega(1)} n$, Conjecture 43.3.2 is true by a simple incompressibility argument. At the same time, the status of Conjecture 43.3.1 for $d \in [\omega(\log n), \log^{O(1)} n]$ is still unresolved.

lem under the l_∞^k norm can be reduced to solving the partial match problem with the parameter $d = (k + \log n)^{O(1)}$; the number of points n is the same for both problems. In fact, the same holds for a more general problem of *orthogonal range queries*. Thus, Conjecture 43.3.2, and its variant for the $(r, 1)$-NN under l_∞^d (or for orthogonal range queries), are equivalent. This strengthens the belief in the validity of Conjecture 43.3.2, since the exact nearest neighbor under l_∞ norm and the orthogonal range query problem have received additional attention in the Computational Geometry community.

43.4 OTHER TOPICS

There are many other research directions on high-dimensional NNS; we briefly mention some of them below.

Average-case algorithms.   The approximate algorithms described so far are designed to work for any (i.e., worst-case) input. However, researchers have also investigated *exact* algorithms for the NN problem that achieve fast query times for *average* input.

In fact, the first such average-case NNS solutions have been proposed for its offline version (the bichromatic closest pair from above). In particular, [Val88] introduced the *light-bulb problem*, defined as follows: the point set P is chosen at random i.i.d. from $\{0, 1\}^d$, and a point q is inserted at random within distance r from some point $p \in P$; the goal is to find q given $P \cup \{q\}$. Some results were obtained in [PRR95, GPY94] (with performance comparable to LSH [HIM12]), and [Dub10] (with performance comparable to data-dependent LSH [AR15]), as well as in [Val15, KKK16] which use fast matrix multiplication (see Section 43.2).

For ℓ_∞^d, there also are average-case algorithms for the NNS problem. Consider a point set where each point, including the query, is chosen randomly i.i.d. from $[0, 1]^d$. In this setting, it was shown [AHL01, HL01] that there is a nearest neighbor data structure using $O(dn)$ space, with query time $O(n \log d)$. Note that a naive algorithm would suffer from $O(nd)$ query time. The algorithm uses a clever pruning approach to quickly eliminate points that *cannot* be nearest neighbors of the query point.

Low-intrinsic dimension.   Another way to depart from the worst-case analysis is to assume further structure in the dataset P. A particularly lucrative approach is to assume that the dataset has "low intrinsic dimension." This is justified by the fact that the high-dimensional data is often actually explained by a few free parameters, and the coordinate values are functions of these parameters. One such notion is the *doubling dimension* of a pointset P, defined as follows. Let λ be the smallest number such that any ball of radius r in P can be covered by λ balls of radius $r/2$ (with centers in P), for any $r > 0$. Then we say that the pointset P has doubling dimension $\log \lambda$. This definition was introduced in [Cla99, GKL03], and is related to the Assouad constant [Ass83]. Note that it can be defined for any metric space on P.

Algorithms for datasets P with intrinsic dimension k typically achieve performance comparable to NNS algorithms for k-dimensional spaces. Some algorithms designed for low doubling dimensional spaces include [KL04, BKL06, IN07].

There are now a few notions of "intrinsic dimension," including KR-dimension [KR02], smooth manifolds [BW09, Cla08], and others [CNBYM01, FK97, IN07, Cla06, DF08, AAKK14].

NNS for higher dimensional flats.   Finally, a further line of work investigated NNS in a setting where the dataset points or the queries are in fact larger objects, such as lines, or, more generally, k-dimensional flats. The new goal becomes to find, say, the dataset point closest to a given *query line* (up to some approximation). For the case when the dataset is composed of points in $\mathbb{R}^d$ and the queries are lines, the results of [AIKN09, MNSS15] provide efficient algorithms, with polynomial space and sublinear query time. Such algorithms often use vanilla approximate NNS algorithms as subroutines. For the dual case, where the dataset is composed of lines and the queries are points, the work of [Mah15] provides an efficient algorithm (improving over the work of [Mag07, BHZ07]). Some of the cited results generalize lines to k-dimensional flats, with performance degrading (rapidly) with the parameter k.

We note that the problem is intrinsically related to the partial match problem, since a query with k "don't care" symbols can be represented as a k-dimensional flat query.

43.5 SOURCES AND RELATED MATERIAL

RELATED CHAPTERS

Chapter 8: Low-distortion embeddings of discrete metric spaces
Chapter 28: Arrangements
Chapter 32: Proximity algorithms
Chapter 40: Range searching

REFERENCES

[AAKK14]   A. Abdullah, A. Andoni, R. Kannan, and R. Krauthgamer. Spectral approaches to nearest neighbor search. In *Proc. 55th IEEE Sympos. Found. Comp. Sci.*, pages 581–590, 2014. Full version at arXiv:1408.0751.

[ACP08]   A. Andoni, D. Croitoru, and M. Pătraşcu. Hardness of nearest neighbor under L-infinity. In *Proc. 49th IEEE Sympos. Found. Comp. Sci.*, pages 424–433, 2008.

[ADFM11]   S. Arya, G.D. Da Fonseca, and D.M. Mount. Approximate polytope membership queries. In *Proc. 43rd ACM Sympos. Theory Comput.*, pages 579–586, 2011. Full version at arXiv:1604.01183.

[AGK06]   A. Arasu, V. Ganti, and R. Kaushik. Efficient exact set-similarity joins. In *Proc. 32nd Conf. Very Large Data Bases*, pages 918–929, 2006.

[AHL01]   H. Alt and L. Heinrich-Litan. Exact l_∞-nearest neighbor search in high dimensions. *Proc. 17th Sympos. Comput. Geom.*, pages 157–163, ACM Press, 2001.

[AI06]   A. Andoni and P. Indyk. Near-optimal hashing algorithms for approximate nearest neighbor in high dimensions. In *Proc. 47th IEEE Sympos. Found. Comp. Sci.*, pages 459–468, 2006.

[AIK09]   A. Andoni, P. Indyk, and R. Krauthgamer. Overcoming the ℓ_1 non-embeddability barrier: Algorithms for product metrics. In *Proc. 20th ACM-SIAM Sympos. Discrete Algorithms*, pages 865–874, 2009.

[AIKN09]   A. Andoni, P. Indyk, R. Krauthgamer, and H.L. Nguyen. Approximate line nearest neigbor in high dimensions. In *Proc. 20th ACM-SIAM Sympos. Discrete Algorithms*, pages 293–301, 2009.

[AINR14]   A. Andoni, P. Indyk, H.L. Nguyen, and I. Razenshteyn. Beyond locality-sensitive hashing. In *Proc. 25th ACM-SIAM Sympos. Discrete Algorithms*, pages 1018–1028, 2014.

[AIP06]   A. Andoni, P. Indyk, and M. Pătraşcu. On the optimality of the dimensionality reduction method. In *Proc. 47th IEEE Sympos. Found. Comp. Sci.*, pages 449–458, 2006.

[ALRW17]   A. Andoni, T. Laarhoven, I.P. Razenshteyn, and E. Waingarten. Optimal hashing-based time-space trade-offs for approximate near neighbors. In *Proc. 28th ACM-SIAM Sympos. Discrete Algorithms*, pages, 47–66, 2017.

[AMM09]   S. Arya, T. Malamatos, and D.M. Mount. Space-time tradeoffs for approximate nearest neighbor searching. *J. ACM*, 57:1, 2009.

[AMN+98]   S. Arya, D.M. Mount, N.S. Netanyahu, R. Silverman, and A.Y. Wu. An optimal algorithm for approximate nearest neighbor searching. *J. ACM*, 6:891–923, 1998.

[AN11]   A. Andoni and H. Nguyen. Lower bounds for locality sensitive hashing. Manuscript, 2011.

[And09]   A. Andoni. *Nearest Neighbor Search: the Old, the New, and the Impossible.* PhD thesis, MIT, Cambridge, 2009.

[AR15]   A. Andoni and I. Razenshteyn. Optimal data-dependent hashing for approximate near neighbors. In *Proc. 47th ACM Sympos. Theory Comput.*, pages 793–801, 2015. Full version available at `arXiv:1501.01062`.

[AR16]   A. Andoni and I. Razenshteyn. Tight lower bounds for data-dependent locality-sensitive hashing. In *Proc. Sympos. Comput. Geom.*, vol 51 of *LIPIcs*, pages 9:1-9:11, Schloss Dagstuhl, 2016.

[Ass83]   P. Assouad. Plongements lipschitziens dans $\mathbb{R}^n$. *Bull. Soc. Math. France*, 111:429–448, 1983.

[AW15]   J. Alman and R. Williams. Probabilistic polynomials and Hamming nearest neighbors. In *Proc. 56th IEEE Sympos. Found. Comp. Sci.*, pages 136–150, 2015.

[BDGL16]   A. Becker, L. Ducas, N. Gama, and T. Laarhoven. New directions in nearest neighbor searching with applications to lattice sieving. In *Proc. 27th ACM-SIAM Sympos. Discrete Algorithms*, pages 10–24, 2016.

[BG15]   Y. Bartal and L.-A. Gottlieb. Approximate nearest neighbor search for ℓ_p-spaces $(2 < p < \infty)$ via embeddings. Preprint, `arXiv:1512.01775`, 2015.

[BHZ07]   R. Basri, T. Hassner, and L. Zelnik-Manor. Approximate nearest subspace search with applications to pattern recognition. In *Proc. IEEE Conf. Computer Vision Pattern Recogn.*, pages 1–8, 2007.

[BKL06]   A. Beygelzimer, S. Kakade, and J. Langford. Cover trees for nearest neighbor. In *Proc. 23rd Internat. Conf. Machine Learning*, pages 97–104, ACM Press, 2006.

[BOR03]   A. Borodin, R. Ostrovsky, and Y. Rabani. Lower bounds for high dimensional nearest neighbor search and related problems. In *Discrete and Computational Geometry: The Goodman-Pollack Festschrift*, pages 253–274, Springer, Berlin, 2003.

[BOR04]   A. Borodin, R. Ostrovsky, and Y. Rabani. Subquadratic approximation algorithms for clustering problems in high dimensional spaces. *Mach. Learn.*, 56:153–167, 2004.

[BR02]   O. Barkol and Y. Rabani. Tighter bounds for nearest neighbor search and related problems in the cell probe model. *J. Comput. Syst. Sci.*, 64:873–896, 2002.

[Bro98]   A. Broder. Filtering near-duplicate documents. *Proc. Fun with Algorithms*, Carleton University, 1998.

[BV02]   P. Beame and E. Vee. Time-space tradeoffs, multiparty communication complexity, and nearest-neighbor problems. *Proc. 34th ACM Sympos. Theory Comput.*, pages 688–697, 2002.

[BW09]   R.G. Baraniuk and M.B. Wakin. Random projections of smooth manifolds. *Found. Comput. Math.*, 9:51–77, 2009.

[CCGL03]   A. Chakrabarti, B. Chazelle, B. Gum, and A. Lvov. A lower bound on the complexity of approximate nearest-neighbor searching on the Hamming cube. In *Discrete and Computational Geometry: The Goodman-Pollack Festschrift*, pages 313–328, Springer, Berlin, 2003.

[Cha02a]   T.M. Chan. Closest-point problems simplified on the RAM. In *Proc. 13th ACM-SIAM Sympos. Discrete Algorithms*, pages 472–473, 2002.

[Cha02b]   M.S. Charikar. Similarity estimation techniques from rounding. In *Proc. 34th ACM Sympos. Theory Comput.*, pages 380–388, 2002.

[Chr17]   T. Christiani. A framework for similarity search with space-time tradeoffs using locality-sensitive filtering. In *Proc. 20th ACM-SIAM Sympos. Discrete Algorithms*, pages 31–46, 2017.

[CIP02]   M. Charikar, P. Indyk, and R. Panigrahy. New algorithms for subset query, partial match, orthogonal range searching and related problems. *Proc. Internat. Coll. Automata, Languages, Progr.*, vol. 2380 of *LNCS*, pages 451–462, Springer, Berlin, 2002.

[Cla88]   K.L. Clarkson. A randomized algorithm for closest-point queries. *SIAM J. Comput.*, 17:830–847, 1988.

[Cla94]   K.L. Clarkson. An algorithm for approximate closest-point queries. *Proc. 10th Sympos. Comput. Geom.*, pages 160–164, ACM Press, 1994.

[Cla99]   K.L. Clarkson. Nearest neighbor queries in metric spaces. *Discrete Comput. Geom.*, 22:63–93, 1999.

[Cla06]   K.L. Clarkson. Nearest-neighbor searching and metric space dimensions. In G. Shakhnarovich, T. Darrell, and P. Indyk, editors, *Nearest-Neighbor Methods for Learning and Vision: Theory and Practice*, pages 15–59, MIT Press, Cambridge, 2006.

[Cla08]   K.L. Clarkson. Tighter bounds for random projections of manifolds. In *Proc. 24th Sympos. Comput. Geom.*, pages 39–48, ACM Press, 2008.

[CNBYM01] E. Chávez, G. Navarro, R. Baeza-Yates, and J. Marroquin. Searching in metric spaces. *ACM Comput. Surv*, 33:273–321, 2001.

[CR10]   A. Chakrabarti and O. Regev. An optimal randomised cell probe lower bound for approximate nearest neighbor searching. *SIAM J. Comput.*, 39:1919–1940, 2010.

[DF08]   S. Dasgupta and Y. Freund. Random projection trees and low dimensional manifolds. In *Proc. 40th ACM Sympos. Theory Comput.*, pages 537–546, 2008.

[DGK01]   C.A. Duncan, M.T. Goodrich, and S. Kobourov. Balanced aspect ratio trees: Combining the advantages of *k*-d trees and octrees. *J. Algorithms*, 38:303–333, 2001.

[DIIM04]   M. Datar, N. Immorlica, P. Indyk, and V. Mirrokni. Locality-sensitive hashing scheme based on *p*-stable distributions. In *Proc. 20th Sympos. Comput. Geom.*, pages 253–262, ACM Press, 2004.

[Dub10]   M. Dubiner. Bucketing coding and information theory for the statistical high dimensional nearest neighbor problem. *IEEE Trans. Inform. Theory*, 56:4166–4179, 2010.

[DW82]   L. Devroye and T. Wagner. Nearest neighbor methods in discrimination. P.R. Krishnaiah and L.N. Kanal, editors, *Handbook of Statistics*, vol. 2, North-Holland, Amsterdam, 1982.

[Epp95]   D. Eppstein. Dynamic Euclidean minimum spanning trees and extrema of binary functions. *Discrete Comput. Geom.*, 13:111–122, 1995.

[Epp99]   D. Eppstein. Personal communication. 1999.

[FK97]   C. Faloutsos and I. Kamel. Relaxing the uniformity and independence assumptions using the concept of fractal dimension. *J. Comput. System Sci.*, 55:229–240, 1997.

[GG91]   A. Gersho and R. Gray. *Vector Quantization and Data Compression*. Kluwer, Norwell, 1991.

[GIV01]   A. Goel, P. Indyk, and K. Varadarajan. Reductions among high-dimensional geometric problems. *Proc. 12th ACM-SIAM Sympos. Discrete Algorithms*, pages 769–778, 2001.

[GKL03]   A. Gupta, R. Krauthgamer, and J.R. Lee. Bounded geometries, fractals, and low-distortion embeddings. In *Proc. 44th IEEE Sympos. Found. Comp. Sci.*, pages 534–543, 2003.

[GPY94]   D. Greene, M. Parnas, and F. Yao. Multi-index hashing for information retrieval. In *Proc. 35th IEEE Sympos. Found. Comp. Sci.*, pages 722–731, 1994.

[HIM12]   S. Har-Peled, P. Indyk, and R. Motwani. Approximate nearest neighbor: Towards removing the curse of dimensionality. *Theory Comput.*, 8:321–350, 2012.

[HL01]   L. Heinrich-Litan. Exact l_∞-nearest neighbor search in high dimensions. Personal communication, 2001.

[HP01]   S. Har-Peled. A replacement for Voronoi diagrams of near linear size. In *Proc. 42nd IEEE Sympos. Found. Comp. Sci.*, pages 94–103, 2001.

[IN07]   P. Indyk and A. Naor. Nearest neighbor preserving embeddings. *ACM Trans. Algorithms*, 3:31, 2007.

[Ind00]   P. Indyk. Dimensionality reduction techniques for proximity problems. *Proc. 11th ACM-SIAM Sympos. Discrete Algorithms*, pages 371–378, 2000.

[Ind01a]   P. Indyk. *High-Dimensional Computational Geometry*. Ph.D. Thesis. Department of Computer Science, Stanford University, 2001.

[Ind01b]   P. Indyk. On approximate nearest neighbors in ℓ_∞ norm. *J. Comput. System Sci.*, 63:627–638, 2001.

[Ind02]   P. Indyk. Approximate nearest neighbor algorithms for Frechet metric via product metrics. *Proc. 18th Sympos. Comput. Geom.*, pages 102–106, ACM Press, 2002.

[Ind03]   P. Indyk. Better algorithms for high-dimensional proximity problems via asymmetric embeddings. *Proc. 14th ACM-SIAM Sympos. Discrete Algorithms*, pages 539–545, 2003.

[IST99]   P. Indyk, S.E. Schmidt, and M. Thorup. On reducing approximate MST to closest pair problems in high dimensions. Manuscript, 1999.

[JKKR04]   T.S. Jayram, S. Khot, R. Kumar, and Y. Rabani. Cell-probe lower bounds for the partial match problem. *J. Comput. Syst. Sci.*, 69:435–447, 2004.

[JV01]   K. Jain and V.V. Vazirani. Primal-dual approximation algorithms for metric facility location and k-median problems. *J. ACM*, 48:274–296, 2001.

[Kal08]   N.J. Kalton. The nonlinear geometry of Banach spaces. *Rev. Mat. Complut.*, 21:7–60, 2008.

[Kap15]   M. Kapralov. Smooth tradeoffs between insert and query complexity in nearest neighbor search. In *Proc. 34th ACM Sympos. Principles Database Syst.*, pages 329–342, 2015.

[KKK16]   M. Karppa, P. Kaski, and J. Kohonen. A faster subquadratic algorithm for finding outlier correlations. In *Proc. 27th ACM-SIAM Sympos. Discrete Algorithms*, pages 1288–1305, 2016.

[KL04]   R. Krauthgamer and J.R. Lee. Navigating nets: Simple algorithms for proximity search. *Proc. 15th ACM-SIAM Sympos. Discrete Algorithms*, pages 798–807, 2004.

[Kle97]   J. Kleinberg. Two algorithms for nearest-neighbor search in high dimensions. *Proc. 29th ACM Sympos. Theory Comput.*, pages 599–608, 1997.

[KOR00]   E. Kushilevitz, R. Ostrovsky, and Y. Rabani. Efficient search for approximate nearest neighbor in high dimensional spaces. *SIAM J. Comput.*, 30:457–474, 2000.

[KP12]   M. Kapralov and R. Panigrahy. NNS lower bounds via metric expansion for ℓ_∞ and EMD. In *Proc. Internat. Coll. Automata, Languages and Progr.*, vol. 7391 of *LNCS*, pages 545–556, Springer, Berlin, 2012.

[KR02]   D. Karger and M. Ruhl. Finding nearest neighbors in growth-restricted metrics. *Proc. 34th ACM Sympos. Theory Comput.*, pages 741–750, 2002.

[KWZ95]   R. Karp, O. Waarts, and G. Zweig. The bit vector intersection problem. In *Proc. 36th IEEE Sympos. Found. Comp. Sci.*, pages 621–630, 1995.

[Laa15]   T. Laarhoven. Tradeoffs for nearest neighbors on the sphere. Preprint, `arXiv: 1511.07527`, 2015.

[Liu04]   D. Liu. A strong lower bound for approximate nearest neighbor searching in the cell probe model. *Inform. Process. Lett*, 92:23–29, 2004.

[LT80]   R.J. Lipton and R.E. Tarjan. Applications of a planar separator theorem. *SIAM J. Comput.*, 9:615–627, 1980.

[Mag07]   A. Magen. Dimensionality reductions in ℓ_2 that preserve volumes and distance to affine spaces. *Discrete Comput. Geom.*, 38:139–153, 2007.

[Mah15]   S. Mahabadi. Approximate nearest line search in high dimensions. In *Proc. 26th ACM-SIAM Sympos. Discrete Algorithms*, pages 337–354, 2015.

[Mei93]   S. Meiser. Point location in arrangements of hyperplanes. *Inform. and Comput.*, 106:286–303, 1993.

[MNP07]   R. Motwani, A. Naor, and R. Panigrahy. Lower bounds on locality sensitive hashing. *SIAM J. Discrete Math.*, 21:930–935, 2007.

[MNSS15]   W. Mulzer, H.L. Nguyên, P. Seiferth, and Y. Stein. Approximate k-flat nearest neighbor search. In *Proc. 47th ACM Sympos. Theory Comput.*, pages 783–792, 2015.

[MNSW98] P. Miltersen, N. Nisan, S. Safra, and A. Wigderson. Data structures and asymmetric communication complexity. *J. Comput. Syst. Sci.*, 57:37–49, 1998.

[MP69] M. Minsky and S. Papert. *Perceptrons*. MIT Press, Cambridge, 1969.

[Nao14] A. Naor. Comparison of metric spectral gaps. *Anal. Geom. Metr. Spaces*, 2(1), 2014.

[Ngu14] H.L. Nguyen. *Algorithms for High Dimensional Data*. PhD thesis, Princeton University, 2014.

[NR06] A. Naor and Y. Rabani. On approximate nearest neighbor search in ℓ_p, $p > 2$. Manuscript, 2006.

[OWZ14] R. O'Donnell, Y. Wu, and Y. Zhou. Optimal lower bounds for locality sensitive hashing (except when q is tiny). *ACM Trans. Comput. Theory*, 6:5, 2014.

[Pag16] R. Pagh. Locality-sensitive hashing without false negatives. In *Proc. 27th ACM-SIAM Sympos. Discrete Algorithms*, pages 1–9, 2016.

[Pan06] R. Panigrahy. Entropy-based nearest neighbor algorithm in high dimensions. In *Proc. 17th ACM-SIAM Sympos. Discrete Algorithms*, pages 1186–1195, 2006.

[Pǎt10] M. Pǎtraşcu. Unifying the landscape of cell-probe lower bounds. In *SIAM J. Comput.*, 40:827–847, 2010.

[PRR95] R. Paturi, S. Rajasekaran, and J. Reif. The light bulb problem. *Inform. and Comput.*, 117:187–192, 1995.

[PT09] M. Pǎtraşcu and M. Thorup. Higher lower bounds for near-neighbor and further rich problems. *SIAM J. Comput.*, 39:730–741, 2009.

[PTW08] R. Panigrahy, K. Talwar, and U. Wieder. A geometric approach to lower bounds for approximate near-neighbor search and partial match. In *Proc. 49th IEEE Sympos. Found. Comp. Sci.*, pages 414–423, 2008.

[PTW10] R. Panigrahy, K. Talwar, and U. Wieder. Lower bounds on near neighbor search via metric expansion. In *Proc. 51st IEEE Sympos. Found. Comp. Sci.*, pages 805–814, 2010.

[Riv74] R.L. Rivest. *Analysis of Associative Retrieval Algorithms*. Ph.D. thesis, Stanford University, 1974.

[SH75] M.I. Shamos and D. Hoey. Closest point problems. *Proc. 16th IEEE Sympos. Found. Comp. Sci.*, pages 152–162, 1975.

[SSS06] Y. Sabharwal, N. Sharma, and S. Sen. Nearest neighbors search using point location in balls with applications to approximate Voronoi decompositions. *J. Comput. Syst. Sci.*, 72:955–977, 2006.

[Val88] L.G. Valiant. Functionality in neural nets. In *Proc. 1st Workshop Comput. Learning Theory*, pages 28–39, Morgan Kaufmann, San Francisco, 1988.

[Val15] G. Valiant. Finding correlations in subquadratic time, with applications to learning parities and the closest pair problem. *J. ACM*, 62:13, 2015.

[WSB98] R. Weber, H.J. Schek, and S. Blott. A quantitative analysis and performance study for similarity-search methods in high-dimensional spaces. *Proc. 24th Conf. Very Large Data Bases*, pages 194 205, Morgan Kaufman, San Francisco, 1998.

Part VI

COMPUTATIONAL TECHNIQUES

44 RANDOMIZATION AND DERANDOMIZATION
Otfried Cheong, Ketan Mulmuley, and Edgar Ramos

INTRODUCTION

Randomization as an algorithmic technique goes back at least to the seventies. In a seminal paper published in 1976, Rabin uses randomization to solve a geometric problem, giving an algorithm for the closest-pair problem with expected linear running time [Rab76]. Randomized and probabilistic algorithms and constructions were applied successfully in many areas of theoretical computer science. Following influential work in the mid-1980s, a significant proportion of published research in computational geometry employed randomized algorithms or proof techniques. Even when both randomized and deterministic algorithms of comparable asymptotic complexity are available, the randomized algorithms are often much simpler and more efficient in an actual implementation. In some cases, the best deterministic algorithm known for a problem has been obtained by "derandomizing" a randomized algorithm.

This chapter focuses on the randomized algorithmic *techniques* being used in computational geometry, and not so much on particular results obtained using these techniques. Efficient randomized algorithms for specific problems are discussed in the relevant chapters throughout this Handbook.

GLOSSARY

Probabilistic or "Monte Carlo" algorithm: Traditionally, any algorithm that uses random bits. Now often used in contrast to *randomized algorithm* to denote an algorithm that is allowed to return an incorrect or inaccurate result, or fail completely, but with small probability. Monte Carlo methods for numerical integration provide an example. Algorithms of this kind have started to play a larger role in computational geometry in the 21st century (Section 44.8).

Randomized or "Las Vegas" algorithm: An algorithm that uses random bits and is guaranteed to produce a correct answer; its running time and space requirements may depend on random choices. Typically, one tries to bound the expected running time (or other resource requirements) of the algorithm. In this chapter, we will concentrate on randomized algorithms in this sense.

Expected running time: The expected value of the running time of the algorithm, that is, the average running time over all possible choices of the random bits used by the algorithm. No assumptions are made about the distribution of input objects in space. When expressing bounds as a function of the input size, the worst case over all inputs of that size is given. Normally the random choices made by the algorithm are hidden from the outside, in contrast with average running time.

Average running time: The average of the running time, over all possible inputs. Some suitable distribution of inputs is assumed.

To illustrate the difference between expected running time and average running time, consider the *Quicksort* algorithm. If it is implemented so that the pivot element is the first element of the list (and the assumed input distribution is the set of all possible permutations of the input set), then it has $O(n \log n)$ *average* running time. By providing a suitable input (here, a sorted list), an adversary can force the algorithm to perform worse than the average. If, however, Quicksort is implemented so that the pivot element is chosen at random, then it has $O(n \log n)$ *expected* running time, for any possible input. Since the random choices are hidden, an adversary cannot force the algorithm to behave badly, although it may perform poorly with some positive probability.

Randomized divide-and-conquer: A divide-and-conquer algorithm that uses a random sample to partition the original problem into subproblems (Section 44.1).

Randomized incremental algorithm: An incremental algorithm where the order in which the objects are examined is a random permutation (Section 44.2).

Tail estimate: A bound on the probability that a random variable deviates from its expected value. Tail estimates for the running time of randomized algorithms are useful but seldom available (Section 44.9).

High-probability bound: A strong tail estimate, where the probability of deviating from the expected value decreases as a fast-growing function of the input size n. The exact definition varies between authors, but a typical example would be to ask that for any $\alpha > 0$, there exists a $\beta > 0$ such that the probability that the random variable $X(n)$ exceeds $\alpha E[X(n)]$ be at most $n^{-\beta}$.

Derandomization: Obtaining a deterministic algorithm by "simulating" a randomized one (Section 44.6).

Coreset: A data set of small size that can be used as a proxy for a large data set. Algorithms can be run on the coreset to obtain a good approximation of the result for the full data set. Since a random sample captures many characteristics of a given data set, a coreset can be considered a stronger form of a random sample. Chapter 48 discusses coresets in detail, and mentions randomization frequently.

Trapezoidal map: A planar subdivision $\mathcal{T}(S)$ induced by a set S of line segments with disjoint interiors in the plane (cf. Section 33.3). $\mathcal{T}(S)$ can be obtained by passing vertical attachments through every endpoint of the given segments, extending upward and downward until each hits another segment, or extending to infinity; see Figure 44.0.1. Every face of the subdivision is a trapezoid (possibly degenerated to a triangle, or with a missing top or bottom side), hence the name.

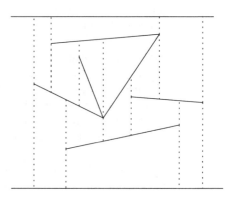

FIGURE 44.0.1
The trapezoidal map of a set of 6 line segments.

We will use the problem of computing the trapezoidal map of a set of line segments with disjoint interiors as a running example throughout this chapter. We assume for presentation simplicity that no two distinct endpoints have the same x-coordinate, so that every trapezoid is adjacent to at most four segments. (This can be achieved by a slight rotation of the vertical direction.)

The trapezoidal map can also be defined for intersecting line segments. In that situation, vertical attachments must be added to intersection points as well, and the map may consist of a quadratic number of trapezoids. The trapezoidal map is also called the **vertical decomposition** of the set of line segments. Decompositions similar to this play an important role in randomized algorithms, because most algorithms assume that the structure to be computed has been subdivided into elementary objects. (Section 44.5 explains why this assumption is necessary.)

44.1 RANDOMIZED DIVIDE-AND-CONQUER

GLOSSARY

Top-down sampling: Sampling with small, usually constant-size random samples, and recursing on the subproblems.

Cutting: A subdivision Ξ of space into simple cells Δ (of constant description complexity, most often simplices). The **size** of a cutting is the number of cells.

ϵ-cutting Ξ: For a set X of n geometric objects, a cutting such that every cell $\Delta \in \Xi$ intersects at most n/r of the objects in X (also called a $1/r$-cutting with $\epsilon = 1/r$ when convenient). See also Chapter 47.

Bottom-up sampling: Sampling with random samples large enough that the subproblems may be solved directly (without recursion).

Bernoulli sampling: The "standard" way of obtaining a random sample of size r from a given n-element set uses a random number generator to choose among all the possible subsets of size r, with equal probability for each subset (also obtained as the first r elements in a random permutation of n elements). In Bernoulli sampling, we instead toss a coin for each element of the set independently, and accept it as part of the sample with probability r/n. While the size of the sample may vary, its expected size is r, and essentially all the bounds and results of this chapter hold for both sampling models. Sharir showed that in fact this model can be analysed more easily than the standard one [Sha03].

Gradation: A hierarchy of samples for a set X of objects obtained by bottom-up sampling:
$$X = X_1 \supset X_2 \supset X_3 \supset \cdots \supset X_{r-1} \supset X_r = \emptyset.$$

With Bernoulli sampling, a new element can be inserted into the gradation by flipping a coin at most r times, leading to efficient dynamic data structures (Section 44.3).

Geometric problems lend themselves to solution by divide-and-conquer algorithms. It is natural to solve a geometric problem by dividing space into regions (perhaps with a grid), and solving the problem in every region separately. When the geometric objects under consideration are distributed uniformly over the space, then gridding or "slice-and-dice" techniques seem to work well. However, when object density varies widely throughout the environment, then the decomposition has to be fine in the areas where objects are abundant, while it may be coarse in places with low object density. Random sampling can help achieve this: the density of a random sample R of the set of objects will approach that of the original set. Therefore dividing space according to the sample R will create small regions where the geometric objects are dense, and larger regions that are sparsely populated.

We can distinguish two main types of randomized divide-and-conquer algorithm, depending on whether the size of the sample is rather small or quite large.

TOP-DOWN SAMPLING

Top-down sampling is the most common form of random sampling in computational geometry. It uses a random sample of small, usually constant, size to partition the problem into subproblems. We sketch the technique by giving an algorithm for the computation of the trapezoidal map of a set of segments in the plane.

Given a set S of n line segments with disjoint (relative) interiors, we take a sample $R \subset S$ consisting of r segments, where r is a constant. We compute the trapezoidal map $\mathcal{T}(R)$ of R. It consists of $O(r)$ trapezoids. For every trapezoid $\Delta \in \mathcal{T}(R)$, we determine the **conflict list** S_Δ, the list of segments in S intersecting Δ. We construct the trapezoidal map of every set S_Δ recursively, clip it to the trapezoid Δ, and finally glue all these maps together to obtain $\mathcal{T}(S)$.

The running time of this algorithm can be analyzed as follows. Because r is a constant, we can afford to compute $\mathcal{T}(R)$ and the lists S_Δ naively, in time $O(r^2)$ and $O(nr)$ respectively. Gluing together the small maps can be done in time $O(n)$. But what about the recursive calls? If we denote the size of S_Δ by n_Δ, then bounding the n_Δ becomes the key issue here. It turns out that the right intuition is to assume that the n_Δ are about n/r. Assuming this, we get the recursion

$$T(n) \leq O(r^2 + nr) + O(r)T(n/r),$$

which solves to $T(n) = O(n^{1+\epsilon})$, where $\epsilon > 0$ is a constant depending on r. By increasing the value of r, ϵ can be made arbitrarily small, but at the same time the constant of proportionality hidden in the O-notation increases.

The truth is that one cannot really assume that $n_\Delta = O(n/r)$ holds for every trapezoid Δ at the same time. Valid bounds are as follows. For randomly chosen R of size r, we have:

■ The **pointwise bound:** With probability increasing with r,

$$n_\Delta \leq C \frac{n}{r} \log r, \tag{44.1.1}$$

for all $\Delta \in \mathcal{T}(R)$, where the constant C does not depend on r and n.

■ The **higher-moments bound:** For any constant $c \geq 1$, there is a constant $C(c)$ (independent of r and n) such that

$$\sum_{\Delta \in \mathcal{T}(R)} (n_\Delta)^c = C(c) \left(\frac{n}{r}\right)^c |\mathcal{T}(R)|. \tag{44.1.2}$$

In other words, while the maximum n_Δ can be as much as $O((n/r)\log r)$, on the average the n_Δ behave as if they indeed were $O(n/r)$.

Both bounds can be used to prove that $T(n) = O(n^{1+\epsilon})$, with the dependence on ϵ being somewhat better using the latter bound. The difference between the two bounds becomes more marked for larger values of r, as will be detailed below. (For a more general result that subsumes these two bounds, see Theorem 44.5.2.)

The same scheme used to compute $\mathcal{T}(S)$ will also give a data structure for point location in the trapezoidal map. This data structure is a tree, constructed as follows. If the set S is small enough, simply store $\mathcal{T}(S)$ explicitly. Otherwise, take a random sample R, and store $\mathcal{T}(R)$ in the root node. Subtrees are created for every $\Delta \in \mathcal{T}(R)$. These subtrees are constructed recursively, using the sets S_Δ.

By the pointwise bound, the depth of the tree is $O(\log n)$ with high probability, and therefore the query time is also $O(\log n)$. The storage requirement is easily seen to be $O(n^{1+\epsilon})$ as above.

The algorithmic technique described in this section is surprisingly robust. It works for a large number of problems in computational geometry, and for many problems it is the only known approach to solve the problem. It does have two major drawbacks, however.

First, it seems to be difficult to remove the ϵ-term in the exponent, and truly optimal random-sampling algorithms are scarce. If the size r of the sample is a function of n, say $r = n^\delta$, then the extra factor can often be reduced to $\log^c n$. To entirely eliminate this extra factor, one needs to control the total conflict list size using problem-specific insights. Some examples where this has been achieved are listed below as applications.

Second, the practicality of this method remains to be established. If the size of the random sample is chosen too small, then the problem size may not decrease fast enough to guarantee a fast-running algorithm, or even termination. Few papers in the literature calculate this size constant, and so for most applications it remains unclear whether the size of the random sample can be chosen considerably smaller than the problem size in practice.

CUTTINGS

The only use of randomization in the above algorithm was to subdivide the plane into a number of simply-shaped regions Δ, such that every region is intersected by only a few line segments. Such a subdivision is called a *cutting* Ξ for the set X of n segments; if every $\Delta \in \Xi$ intersects at most n/r of the objects in X, it is a $1/r$-*cutting*. Cuttings are interesting in their own right, and have been studied intensively. See Section 47.5 for results on the deterministic construction of efficient cuttings, with useful properties that go beyond those of the simple cutting based on a random sample discussed above. Cuttings form the basis for many algorithms and search structures in computational geometry; see Chapter 40. As a result, many recent geometric divide-and-conquer algorithms no longer explicitly use randomization, and randomized divide-and-conquer has to some extent been replaced by divide-and-conquer based on cuttings.

In practice, however, cuttings may still be constructed most efficiently using random sampling. There are two basic techniques, which we illustrate again using a set X of n line segments with disjoint interiors in the plane.

- **ε-net based cuttings:** The easiest way to obtain a $1/r$-cutting is to take a random sample $N \subset X$ of size $O(r \log r)$. If N is a $1/r$-net for the range space (X, Γ) (defined in Section 44.4 and Chapter 47), then the trapezoidal map of N is a $1/r$-cutting of size $O(r \log r)$. If not, we try a different sample.

- **Splitting the excess:** The construction based on ε-nets can be improved as follows. First take a random sample N of X of size $O(r)$, and compute its trapezoidal map. Every trapezoid Δ may be intersected by $O((n/r) \log r)$ segments. If we take a random sample of these segments, and form their trapezoidal map again (restricted to Δ), the pieces obtained are intersected by at most n/r segments. The size of this cutting is only $O(r)$, which is optimal.

Har-Peled [HP00] investigates the constants achievable for cuttings of lines in the plane.

BOTTOM-UP SAMPLING

In bottom-up sampling, the random sample is so large that the resulting subproblems are small enough to be solved directly. However, it is no longer trivial to compute the auxiliary structures needed to subdivide the problem. We again illustrate with the trapezoidal map.

Given a set S of n line segments, we take a sample R of size $n/2$, and compute the trapezoidal map of R recursively. For every $\Delta \in \mathcal{T}(R)$, we compute the list S_Δ of segments in $S \setminus R$ intersecting Δ. This can be done by locating an endpoint of every segment in $S \setminus R$ in $\mathcal{T}(R)$ and traversing $\mathcal{T}(R)$ from there. If we use a planar point location structure (Section 38.3), this takes time $O(n \log n + \sum_{\Delta \in \mathcal{T}(R)} n_\Delta)$. For every Δ, we then compute the trapezoidal map $\mathcal{T}(S_\Delta)$, and clip it to Δ. This can be done naively in time $O(n_\Delta^2)$. Finally, we glue together all the little maps.

The running time of the algorithm is bounded by the recursion

$$T(n) \leq T(n/2) + O(n \log n) + \sum_{\Delta \in \mathcal{T}(R)} O(n_\Delta^2).$$

The pointwise bound shows that with high probability, $n_\Delta = O(\log n)$ for all Δ. That would imply that the last term in the recursion is $O(n \log^2 n)$. Here, the higher-moments bound turns out to give a strictly better result, as it shows that the expected value of that term is only $O(n)$. The recursion therefore solves to $O(n \log^2 n)$.

Bottom-up sampling has the potential to lead to more efficient algorithms than top-down sampling, because it avoids the blow-up in problem size that manifests itself in the n^ϵ-term in top-down sampling. However, it needs more refined ingredients—as the constructions of $\mathcal{T}(R)$ and the lists S_Δ demonstrate—and therefore seems to apply to fewer problems.

As with top-down sampling, bottom-up sampling can be used for point location. These search structures have the advantage that they can often easily be made dynamic (Section 44.3).

APPLICATIONS

Proofs for the theorems above can be found in the surveys and books cited in Section 44.11, in particular [Cla92, Mul00, BCKO08, Mul93].

In the following we list a few advanced applications of geometric divide-and-conquer.

- Computing the diameter of a point set in $\mathbb{R}^3$ [Ram01], achieves optimality by clustering subproblems together to achieve a small boundary between subproblems.

- An optimal data structure for simplex range searching [Cha12], builds a partition tree by refining a cutting on each level.

- A *shallow cutting* is a cutting for only the "shallow" part of a structure, that is, the region of space that lies in few of the objects. Shallow cuttings are used in range searching data structures and can be computed in optimal time [CT16].

44.2 RANDOMIZED INCREMENTAL ALGORITHMS

GLOSSARY

Backwards analysis: Analyzing the time complexity of an algorithm by viewing it as running backwards in time [Sei93].

Conflict graph: A bipartite graph whose arcs represent conflicts (usually intersections) between objects to be added and objects already constructed.

History graph: A directed, acyclic graph that records the history of changes in the geometric structure being maintained. Also known as an **influence graph**.

Many problems in computational geometry permit a natural computation by an incremental algorithm. Incremental algorithms need only process one new object at a time, which often implies that changes in the geometric data structure remain localized in the neighborhood of the new object.

As an example, consider the computation of the trapezoidal map of a set of line segments (cf. Fig. 38.3.2; for another example, see Section 26.3). To add a new line segment s to the map, one would first identify the trapezoids of the map intersected by s. Those trapezoids must be split, creating new trapezoids, some of which then must be merged along the segment s. All these update operations can be accomplished in time linear in the sum of the number of old trapezoids that are destroyed and the number of new trapezoids that are created during the insertion of s. This quantity is called the **structural change**.

This results in a rather simple algorithm to compute the trapezoidal map of a set of line segments. Starting with the empty set, we treat the line segments one-by-one, maintaining the trapezoidal map of the set of line segments inserted so far.

However, a general disadvantage of incremental algorithms is that the total structural change during the insertions of n objects, and hence the running time of the algorithm, depends strongly on the order in which the objects are processed. In our case, it is not difficult to devise a sequence of n line segments leading to a total structural change of $\Theta(n^2)$. Even if we know that a good order of insertion exists (one that implies a small structural change), it seems difficult to determine this order beforehand. And this is exactly where randomization can help: we simply treat the n objects in random order. In the case of the trapezoidal map, we will show below that if the n segments are processed in random order, the *expected* structural change in every step of the algorithm is only constant.

BACKWARDS ANALYSIS

An easy way to see this is via *backwards analysis*. We first observe that it suffices to bound the number of trapezoids created in each stage of the algorithm. All these trapezoids are incident to the segment inserted in that stage. We imagine the algorithm removing the line segments from the final map one-by-one. In each step, we must bound the number of trapezoids incident to the segment s removed. Now we make two observations:

- The trapezoidal map is a planar graph, with every trapezoid incident to at most 4 segments. Hence, if there are m segments in the current set, the total number of trapezoid-segment incidences is $O(m)$.

- Since the order of the segments is a random permutation of the set of segments, each of the m segments is equally likely to be removed.

These two facts suffice to show that the expected number of trapezoids incident to s is constant. In fact, this number is bounded by the average degree of a segment in a trapezoidal map.

It follows that the expected total structural change during the course of the algorithm is $O(n)$. To obtain an efficient algorithm, however, we need a second ingredient: whenever a new segment s is inserted, we need to identify the trapezoids of the old map intersected by s. Two basic approaches are known to solve this problem: the conflict graph and the history graph.

CONFLICT GRAPH

A conflict graph is a bipartite graph whose nodes are the not-yet-added segments on one side and the trapezoids of the current map on the other side. There is an arc between a segment s and a trapezoid Δ if and only if s intersects Δ, in which case we say that *s is in conflict with* Δ.

It is possible to maintain the conflict graph during the course of the incremental algorithm. Whenever a new segment is inserted, all the conflicts of the newly-created trapezoids are found. This is not difficult, because a segment can only conflict with a newly-created trapezoid if it was previously in conflict with the old trapezoids at the same place. Thus the trapezoids intersected by the new segment s are just the neighbors of s in the conflict graph.

The time necessary to maintain the conflict graph can be bounded by summing the number of conflicts of all trapezoids created during the course of the algorithm. It follows from the higher-moments bound (Eq. 44.1.2) that the average number of conflicts of the trapezoids present after inserting the first r segments—note that these segments form a random sample of size r of S—is $O(n/r)$. Intuitively, we can assume that this is also correct if we look only at the trapezoids that are *created* by the insertion of the rth segment. Since the expected number of trapezoids created in every step of the algorithm is constant, the expected total time is $\sum_{i=1}^{n} O(n/r) = O(n \log n)$.

Note that an algorithm using a conflict graph needs to know the entire set of objects (segments in our example) in advance.

HISTORY GRAPH

A different approach uses a history graph, which records the history of changes in the maintained structure.

In our example, we can maintain a directed acyclic graph whose nodes correspond to trapezoids constructed during the course of the algorithm. The leaves are the trapezoids of the current map; all inner nodes correspond to trapezoids that have already been destroyed (with the root corresponding to the entire plane). When we insert a segment s, we create new nodes for the newly-created trapezoids, and create a pointer from an old trapezoid to every new one that overlaps it. Hence, there are at most four outgoing pointers for every inner node of the history graph.

We can now find the trapezoids intersected by a new segment s by performing a graph search from the root, using say, depth-first search on the connected subgraph consisting of all trapezoids intersecting s. Note that this search performs precisely the same computations that would have been necessary to maintain the conflict graph during the sequence of updates, but at a different time. We can therefore consider a history graph as a lazy implementation of a conflict graph: it postpones each computation to the moment it is actually needed. Consequently, the analysis is exactly the same as for conflict graphs.

Algorithms using a history graph are ***on-line*** or ***semidynamic*** in the sense that they do not need to know about a point until the moment it is inserted.

ABSTRACT FRAMEWORK AND ANALYSIS

Most randomized incremental algorithms in the literature follow the framework sketched here for the computation of the trapezoidal map: the structure to be computed is maintained while the objects defining it are inserted in random order. To insert a new object, one first has to find a "conflict" of that object (the *location step*), then local updates in the structure are sufficient to bring it up to date (the *update step*). The cost of the update is usually linear in the size of the change in the combinatorial structure being maintained, and can often be bounded using backwards analysis. The location step can be implemented using either a conflict graph or a history graph. In both cases, the analysis is the same (since the actual computations performed are also often identical). To avoid having to prove the same bounds repeatedly for different problems, researchers have defined an axiomatic framework that captures the combinatorial essence of most randomized incremental algorithms. This framework, which uses *configuration spaces*, provides ready-to-use bounds for the expected running time of most randomized incremental algorithms. See Section 44.5.

POINT LOCATION THROUGH HISTORY GRAPH

In our trapezoidal map example, the history graph may be used as a point location structure for the trapezoidal map: given a query point q, find the trapezoid containing q by following a path from the root to a leaf node of the history graph. At each step, we continue to the child node corresponding to the trapezoid containing q.

The search time is clearly proportional to the length of the path. Backwards analysis shows that the expected length of this path is $O(\log n)$ for any fixed query point. Even stronger, one can show that the maximum length of any search path in the history graph is $O(\log n)$ with high probability.

If point location is the goal, the history graph can be simplified: instead of storing trapezoids, the inner nodes of the graph can denote two different kinds of elementary tests ("Does a point lie to the left or right of another point?" and "Does a point lie above or below a line?"). The final result is then an efficient and practical planar point location structure [Sei91].

This observation can also lead to a somewhat different location step inside the randomized incremental algorithm. Instead of performing a graph search with the whole segment s, point location can be used to find the trapezoid containing one endpoint of s. From there, a traversal of the trapezoidal map allows locating all trapezoids intersected by s.

APPLICATIONS

The randomized incremental framework has been successfully applied to a large variety of problems. We list a number of important such applications. Details on the results can be found in the chapters dealing with the respective area, or in one of the surveys cited in Section 44.11.

- Trapezoidal decomposition formed by segments in the plane, and point location structures for this decomposition (Section 38.3).

- Triangulation of simple polygons: an optimal randomized algorithm with linear running time, and a simple algorithm with running time $O(n \log^* n)$ (Section 30.3).

- Convex hulls of points in d-dimensional space, output-sensitive convex hulls in $\mathbb{R}^3$ (Section 26.3).

- Voronoi diagrams in different metrics, including higher order and abstract Voronoi diagrams (Section 27.3).

- Linear programming in finite-dimensional space (Chapter 49).

- Generalized linear programming: optimization problems that are combinatorially similar to linear programming (Section 49.6).

- Hidden surface removal (Section 33.8 and Chapter 52).

- Constructing a single face in an arrangement of (curved) segments in the plane, or in an arrangement of triangles or surface patches in $\mathbb{R}^3$ (Sections 28.5 and 50.2); computing zones in an arrangement of hyperplanes in $\mathbb{R}^d$ (Section 28.4).

44.3 DYNAMIC ALGORITHMS

DYNAMIC RANDOMIZED INCREMENTAL

Any on-line randomized incremental algorithm can be used as a semidynamic algorithm, a dynamic algorithm that can only perform insertions of objects. The bound on the expected running time of the randomized incremental algorithm then turns into a bound on the *average* running time, under the assumption that every permutation of the input is equally likely. (The relation between the two uses of the algorithms is similar to that between randomized and ordinary Quicksort as mentioned in the Introduction.)

This observation has motivated researchers to extend randomized incremental algorithms so that they can also manage deletions of objects. Then bounds on the average running time of the algorithm are given, under the assumption that the input sequence is a *random update sequence*. In essence, one assumes that for an addition, every object currently not in the structure is equally likely to be inserted, while for a deletion every object currently present is equally likely to be removed (the precise definition varies between authors; see, e.g., [Mul91c, Sch91, Cla92, DMT92]).

Two approaches have been suggested to handle deletions in history-graph based incremental algorithms. The first adds new nodes at the leaf level of the history graph for every deletion. This works for a wide variety of problems and is relatively easy to implement, but after a number of updates the history graph will become "dirty": it will contain elements that are no longer part of the current structure but which still must be traversed by the point-location steps. Therefore, the history graph needs periodic "cleaning." This can be accomplished by discarding the current graph, and reconstructing it from scratch using the elements currently present.

In the second approach, for every deletion the history graph is transformed to the state it would have been had the object never been inserted. The history graph is therefore always "clean." However, in this model deletions are more complicated, and it therefore seems to apply to fewer problems.

DYNAMIC SAMPLING AND GRADATIONS

A rather different approach permits a number of search structures based on bottom-up sampling to be dynamized surprisingly easily. Such a search structure consists of a *gradation* using Bernoulli sampling (Section 44.1): The gradation is a hierarchy of $O(\log n)$ levels. Every object is included in the first level, and is chosen independently to be in the second level with probability $\frac{1}{2}$. Every object in the second level is propagated to level 3 with probability $\frac{1}{2}$, and so forth. Whenever an object is added to or removed from the current set, the search structure is updated to the proper state. When adding an object, it suffices to flip a coin at most $\log n$ times to determine where to place the object. Using this technique, it is possible to give high-probability bounds on the search time and sometimes also on the update time [Mul91a, Mul91b, Mul93].

44.4 RANGE SPACES

"Pointwise bounds" of the form in Equation 44.1.1 can be proved in the axiomatic framework of range spaces, which then leads to immediate application to a wide variety of geometric settings. Chapter 47 also discusses range spaces, but calls them (abstract) set systems.

GLOSSARY

Range space: A pair (X, Γ), with X a universe (possibly infinite), and Γ a family of subsets of X. The elements of Γ are called **ranges**. Typical examples of range spaces are of the form $(\mathbb{R}^d, \Gamma)$, where Γ is a set of geometric figures, such as all line segments, halfspaces, simplices, balls, etc.

Shattered: A set $A \subseteq X$ is shattered if every subset A' of A can be expressed as $A' = A \cap \gamma$, for some range $\gamma \in \Gamma$.

In the range space $(\mathbb{R}^2, \mathcal{H})$, where $\mathcal{H}$ is the set of all closed halfplanes, a set of three points in convex position is shattered. However, no set of four points is shattered. See Figure 44.4.1: whether the point set is in convex position or not, there always is a subset (encircled) that cannot be expressed as $A \cap h$ for any halfplane h.

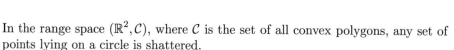

FIGURE 44.4.1
No set of four points can be shattered by halfplanes.

In the range space $(\mathbb{R}^2, \mathcal{C})$, where $\mathcal{C}$ is the set of all convex polygons, any set of points lying on a circle is shattered.

Vapnik-Chervonenkis dimension (VC-dimension):   The VC-dimension of a range space (X, Γ) is the smallest integer d such that there is no shattered subset $A \subseteq X$ of size $d + 1$. If no such d exists, the VC-dimension is said to be infinite.

Range spaces $(\mathbb{R}^d, \Gamma)$, where Γ is the set of line segments, of simplices, of balls, or of halfspaces, have finite VC-dimension. For example, the range space $(\mathbb{R}^2, \mathcal{H})$ has VC-dimension 3. The range space $(\mathbb{R}^2, \mathcal{C})$, however, has infinite VC-dimension.

Shatter function: For a range space (X, Γ), the shatter function $\pi_\Gamma(m)$ is defined as

$$\pi_\Gamma(m) = \max_{A \subset X, |A| = m} |\{A \cap \gamma \mid \gamma \in \Gamma\}|.$$

If the VC-dimension of the range space is infinite, then $\pi_\Gamma(m) = 2^m$. Otherwise the shatter function is bounded by $O(m^d)$, where d is the VC-dimension. (So the shatter function of any range space is either exponential or polynomially bounded.) If the shatter function is polynomial, the VC-dimension is finite. The order of magnitude of the shatter function is not necessarily the same as the VC-dimension; for instance, the range space $(\mathbb{R}^2, \mathcal{H})$ has VC-dimension 3 and shatter function $O(m^2)$. Since the VC-dimension is often difficult to compute, some authors have defined the *VC-exponent* as the order of magnitude of the shatter-function.

ϵ-net: A subset $N \subseteq X$ is called an ϵ-net for the range space (X, Γ) if $N \cap \gamma \neq \emptyset$ for every $\gamma \in \Gamma$ with $|\gamma|/|X| > \epsilon$ (here, $\epsilon \in [0, 1)$ and X is finite). It is often more convenient to write $1/r$ for ϵ, with $r > 1$.

ϵ-approximation: A subset $A \subseteq X$ is called an ϵ-approximation for the range space (X, Γ) if, for every $\gamma \in \Gamma$, we have

$$\left| \frac{|A \cap \gamma|}{|A|} - \frac{|\gamma|}{|X|} \right| \leq \epsilon \, .$$

An ϵ-approximation is also an ϵ-net, but not necessarily vice versa.

Relative (p, ϵ)-approximation: An ϵ-approximation provides an absolute error on the term $|\gamma|/|X|$. In many applications we would prefer a relative error instead. This cannot be achieved by a small approximation, so a relative (p, ϵ)-approximation guarantees a relative error only when $|\gamma|/|X| \geq p$, otherwise an absolute error of ϵp [HPS11].

ϵ-NETS AND ϵ-APPROXIMATIONS

The pointwise bound translates into the abstract framework of range spaces as follows:

THEOREM 44.4.1

Let (X, Γ) be a range space with X finite and of finite VC-dimension d. Then a random sample $R \subset X$ of size $C(d)r \log r$ is a $1/r$-net for (X, Γ) with probability whose complement to 1 is polynomially small in r, where $C(d)$ is a constant that depends only on d.

This theorem forms the basis for "traditional" randomized divide-and-conquer algorithms, such as the one for the trapezoidal map of line segments sketched in Section 44.1. The pointwise bound used there follows from the theorem. Consider the range space (S, Γ), where $\Gamma := \{\gamma(\Delta) \mid \Delta$ an open trapezoid$\}$, and $\gamma(\Delta)$ is the set of all segments in S intersecting Δ. The VC-dimension of this range space is finite. The easiest way to see this is by looking at the shatter function. Consider a set of m line segments. Extend them to full lines, pass $2m$ vertical lines through all endpoints, and look at the arrangement of these $3m$ lines. Clearly, for any two trapezoids Δ and Δ' whose corners lie in the same faces of this arrangement we have $\gamma(\Delta) = \gamma(\Delta')$. Consequently, there are at most $O(m^8)$ different ranges, and that crudely bounds the shatter function as $O(m^8)$. Thus the VC-dimension is finite and Theorem 44.4.1 applies: with probability increasing rapidly with r, the sample R of size r is an ϵ-net for S with $\epsilon = \Omega((1/r) \log r)$. Assume this is the case, and consider some trapezoid $\Delta \in \mathcal{T}(R)$. The interior of Δ does not intersect any segment in R, so by the property of ϵ-nets, the range $\gamma(\Delta)$ can intersect at most ϵn segments of S. And so we have $n_\Delta = O((n/r) \log r)$.

The construction of ϵ-nets has been so successfully derandomized that ϵ-nets now are used routinely in deterministic algorithms. Indeed, Chapter 47, which covers ϵ-nets and ϵ-approximations in detail, hardly mentions randomization.

For general range spaces, the bound $O(r \log r)$ in Theorem 44.4.1 is the best possible. For some geometrically defined spaces the bound has been improved, see Section 47.4. For others, including the case of halfspaces in more than three dimensions, a lower bound of $\Omega(r \log r)$ has been shown [PT13].

An ϵ-approximation serves as a coreset for density estimation, see Section 48.2.

44.5 CONFIGURATION SPACES

The framework of configuration spaces is somewhat more complicated than range spaces, but facilitates proving higher-moment bounds as in Equation 44.1.2. Terminology, axiomatics, and notation vary widely between authors. Note that the term "configuration space" is used in robotics with a different meaning (see Chapters 50 and 51).

GLOSSARY

Configuration space: A four-tuple $(X, \mathcal{T}, D, K)$. X is a finite set of geometric objects (the universe of size n). $\mathcal{T}$ is a mapping that assigns to every subset $S \subseteq X$ a set $\mathcal{T}(S)$; the elements of $\mathcal{T}(S)$ are called **configurations**. $\Pi(X) :=$ $\bigcup_{S \subseteq X} \mathcal{T}(S)$ is the set of all configurations occurring over some subset of X. D and K assign to every configuration $\Delta \in \Pi(X)$ subsets $D(\Delta)$ and $K(\Delta)$ of X. Elements of the set $D(\Delta)$ are said to **define** the configuration (they are also called **triggers**) and the elements of the set $K(\Delta)$ are said to **kill** the configuration (they are also said to be in **conflict** with the configuration and are sometimes called **stoppers**).

Conflict size of Δ: The number of elements of $K(\Delta)$.

We will require the following axioms:

(i) The number $d = \max\{|D(\Delta)| \mid \Delta \in \Pi(X)\}$ is a constant (called the **maximum degree** or the **dimension** of the configuration space). Moreover, the number of configurations sharing the same defining set is bounded by a constant.

(ii) For any $\Delta \in \mathcal{T}(S)$, $D(\Delta) \subseteq S$ and $S \cap K(\Delta) = \emptyset$.

(iii) If $\Delta \in \mathcal{T}(S)$ and $D(\Delta) \subseteq S' \subseteq S$, then $\Delta \in \mathcal{T}(S')$.

(iii′) If $D(\Delta) \subseteq S$ and $K(\Delta) \cap S = \emptyset$, then $\Delta \in \mathcal{T}(S)$.

Note that axiom (iii) follows from (iii′); see below.

EXAMPLES

1. *Trapezoidal map.* The universe X is a set of segments in the plane, and $\mathcal{T}(S)$ is the set of trapezoids in the trapezoidal map of S. The defining set $D(\Delta)$ is the set of segments that are necessary to define Δ (at most four segments suffice, so $d = 4$), and the killing set $K(\Delta)$ is the set of segments that intersect the trapezoid. It is easy to verify that conditions (i), (ii), (iii), (iii′) all hold.

2. *Delaunay triangulation.* X is a set of points in the plane (assume that no four points lie on a circle), and $\mathcal{T}(S)$ is the set of triangles of the Delaunay triangulation of S. $D(\Delta)$ consists of the vertices of triangle Δ (so $d = 3$), while $K(\Delta)$ is the set of points lying inside the circumcircle of the triangle. Again, axioms (i), (ii), (iii), (iii′) all hold.

3. *Convex hulls in 3D.* The universe X is a set of points in 3D (assume that no four points are coplanar), and $\mathcal{T}(S)$ is the set of facets of the convex hull of S. The defining set of a facet Δ is the set of its vertices ($d = 3$), and the killing set is the set of points lying in the outer open halfspace defined by Δ. Note that there can be two configurations sharing the same defining set. Again, axioms (i)–(iii′) all hold.

4. *Single cell.* The universe X is a set of possibly intersecting segments in the plane, and $\mathcal{T}(S)$ is the set of trapezoids in the trapezoidal map of S that belongs to the cell of the line segment arrangement containing the origin (Figure 44.5.1). The defining and killing sets are defined as in the case of the trapezoidal map of the whole arrangement above. In this situation, axiom (iii′) does not hold. Whether or not a given trapezoid appears in $\mathcal{T}(S)$ depends on segments other than the ones in $D(\Delta) \cup K(\Delta)$. Axioms (i), (ii), (iii) are nevertheless valid.

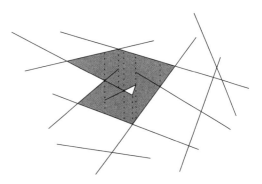

FIGURE 44.5.1
A single cell in an arrangement of line segments.

5. *LP-type problems.* LP-type problems and *violator spaces* are generalizations of linear programming. Consider as an example the problem of finding the smallest enclosing disk for a set of points in the plane. The universe X is the set of points, and $\mathcal{T}(S)$ is the unique smallest enclosing disk. If we assume general position, then a disk is defined by two or three points, its killing set is the set of points lying outside the disk. Axioms (i)–(iii′) hold.

   LP-type problems in general are defined as a set H of *constraints* and a mapping assigning a "value" $w(S)$ to each subset $S \subset H$, the goal is to find a minimal subset $B \subset H$ (a *basis*) such that $w(B) = w(H)$, see Section 49.6 for details. LP-type problems of constant combinatorial dimension (that is, the size of bases is bounded by a constant) satisfy our axioms, but degeneracies (many bases with the same value) need to be handled with care.

6. *Counterexample.* Let X be a set of line segments, and let $\mathcal{T}(S)$ be a decomposition of the arrangement that is obtained by drawing vertical extensions for faces with an even number of edges, and horizontal extensions for faces with an odd number of edges. Axioms (i) and (ii) hold, but neither (iii) nor (iii′) is satisfied.

Note that when (ii) and (iii′) both hold, then $\Delta \in \mathcal{T}(S)$ if and only if $D(\Delta) \subseteq S$ and $K(\Delta) \cap S = \emptyset$. In other words, the mapping $\mathcal{T}$ is then completely defined by the functions D and K. In fact, in the first three examples we can decide from local information alone whether or not a configuration appears in $\mathcal{T}(S)$. For instance, a triangle Δ is in the Delaunay triangulation of S if and only if the vertices of Δ are in S, and no point of S lies in the circumcircle of Δ.

As mentioned above, axiom (iii) follows from (iii'), but not conversely. Axiom (iii) requires a kind of monotonicity: if Δ occurs in $\mathcal{T}(S)$ for some S, then we cannot destroy it by removing elements from S unless we remove some element in $D(\Delta)$.

We may say that the configuration spaces of the first three examples are defined *locally* and *canonically*. The fourth example is *canonical*, but *nonlocal*. The last example is not canonical and cannot be treated with the methods described here. (Fortunately, this is an artificial example with no practical use—but see the open problems below.)

HIGHER-MOMENTS AND EXPONENTIAL DECAY LEMMA

The higher-moments bound for configuration spaces generalizes the bound for trapezoidal maps, Equation 44.1.2:

THEOREM 44.5.1 *Higher-moments bound*

Let $(X, \mathcal{T}, D, K)$ be a configuration space satisfying axioms (i), (ii), (iii), and let R be a random sample of X of size r. For any constant c, we have

$$E\left[\sum_{\Delta \in \mathcal{T}(R)} |K(\Delta)|^c\right] = O((n/r)^c E[|\mathcal{T}(R)|]).$$

(Technically, rather than R, a sample R' of size $\lfloor r/2 \rfloor$ should appear on the right, but $E[|\mathcal{T}(R')|] = O(E[|\mathcal{T}(R)|])$ in all cases of interest). In other words, as far as the cth-degree average is concerned, the conflict size behaves as if it were $O(n/r)$, instead of $O((n/r)\log r)$ from the pointwise bound.

Let $(X, \mathcal{T}, D, K)$ and R be as in Theorem 44.5.1. For any natural number t, we define $\mathcal{T}_t(R)$ to be the subset of configurations of $\mathcal{T}(R)$ whose conflict size exceeds the "natural" value n/r by at least the factor t:

$$\mathcal{T}_t(R) := \{\Delta \in \mathcal{T}(S) \mid |K(\Delta)| \geq tn/r\}.$$

The following exponential-decay lemma [AMS98] states that the number of such configurations decreases exponentially with t:

THEOREM 44.5.2 *Exponential decay lemma*

Let $(X, \mathcal{T}, D, K)$ be a configuration space satisfying axioms (i), (ii), (iii), and let R be a random sample of X of size r. For any t with $1 \leq t \leq r/d$ (where d is as in axiom (i)), we have

$$E[|\mathcal{T}_t(R)|] = O(2^{-t}) \cdot E[|\mathcal{T}(R')|],$$

where $R' \subseteq X$ denotes a random sample of size $\lfloor r/t \rfloor$.

The exponential decay lemma implies both the higher-moments bound, by adding over t, and the pointwise bound, by Markov's inequality.

RANDOMIZED INCREMENTAL CONSTRUCTION

Many, if not most, randomized incremental algorithms in the literature can be analyzed using the configuration space framework. Given the set X, the goal of the randomized incremental algorithm is to compute $\mathcal{T}(X)$. This is done by maintaining $\mathcal{T}(X^i)$, for $1 \leq i \leq n$, where $X^i = \{x_1, x_2, \ldots, x_i\}$ and the x_i form a random permutation of X.

To bound the number of configurations created during the insertion of x_i into X^{i-1}, we observe that by axiom (iii) these configurations are exactly those $\Delta \in \mathcal{T}(X^i)$ with $x_i \in D(\Delta)$. The expected number of these can be bounded by

$$\frac{d}{i} E[|\mathcal{T}(X^i)|]$$

using backwards analysis. Here, d is the maximum degree of the configuration space.

The expected total change in the conflict graph or history graph can be bounded by summing $|K(\Delta)|$ over all Δ created during the course of the algorithm. Using axioms (i) to (iii′), we can derive the following bound:

$$\sum_{i=1}^{n} d^2 \frac{n-i}{i} \frac{E[|\mathcal{T}(X^i)|]}{i}.$$

(The exact form of this expression depends on the model used.) The book [Mul93] treats randomized incremental algorithms systematically using the configuration space framework (assuming axiom (iii′)).

LAZY RANDOMIZED INCREMENTAL CONSTRUCTION

In problems that have nonlocal definition, such as the computation of a single cell in an arrangement of segments, single cells in arrangements of surface patches, or zones in arrangements, the update step of a randomized incremental construction becomes more difficult. Besides the local updates in the neighborhood of the newly inserted object, there may also be global changes. For instance, when a line segment is inserted into an arrangement of line segments, it may cut the single cell being computed into several pieces, only one of which is still interesting. The technique of lazy randomized incremental construction [BDS95] deals with these problems by simply postponing the global changes to a few "clean-up" stages. Since the setting of all these problems is nonlocal, the analysis uses only axioms (i), (ii), (iii).

OPEN PROBLEM

The canonical framework of randomized incremental algorithms sketched above is sometimes too restrictive. For instance, to make a problem fit into the framework, one often has to assume that objects are in general position. While many algorithms could deal with special cases (e.g., four points on a circle in the case of Delaunay triangulations) directly, the analysis does not hold for those situations, and one has to resort to a symbolic perturbation scheme to save the analysis. Can a more relaxed framework for randomized incremental construction be given [Sei93]?

44.6 DERANDOMIZATION TECHNIQUES

Even when an efficient randomized algorithm for a problem is known, researchers
still find it worthwhile to obtain a deterministic algorithm of the same efficiency.
The reasons for doing this are varied, from scientific curiosity (what is the real power
of randomness?), to practical reasons (truly random bits are quite expensive), to
a preference for "deterministicity" that may not be strictly rational. Sometimes
a deterministic algorithm for a given problem may be obtained by "simulating"
or "derandomizing" a randomized algorithm. Derandomization has turned out to
be a powerful theoretical tool: for several problems the only known worst-case
optimal deterministic algorithm has been obtained by derandomization. The most
famous example is computing the convex hull of n points in d-dimensional space
(Section 26.3).

General derandomization techniques can be used to produce a deterministic
counterpart of random sampling in both configuration spaces and range spaces.
As a result, it is possible to obtain in polynomial time a sample that satisfies the
higher-moment bound, or that is a net or an approximation. Taking advantage
of separability and composition properties of approximations, these constructions
can be made efficient. In most applications, deterministic sampling is the base of
a deterministic divide-and-conquer algorithm or data structure, which is almost as
efficient as the randomized counterpart.

On the other hand, incremental algorithms are considerably harder to derandomize: the convex hull algorithm mentioned above is essentially the only successful case. The problem is that in an incremental algorithm each insertion must be
"globally good," while in the divide-and-conquer case, items are chosen locally in
a neighborhood that shrinks as the algorithm progresses. In some cases, such as
linear programming, a derandomized divide-and-conquer approach leads to a deterministic algorithm with better dependency on the dimension than previously known
methods (prune-and-search), but there still remains a large gap with respect to the
best randomized algorithm (which is an incremental one).

METHOD OF CONDITIONAL PROBABILITIES

The **method of conditional probabilities** (also called the **Raghavan-Spencer
method**) [Spe87, Rag88] implements a binary search of the probability space to
determine an event with the desired properties (guaranteed by a probabilistic analysis). Given a configuration space $(X, \mathcal{T}, D, K)$, the goal is to obtain a random
sample of size (approximately) r that satisfies the higher-moments bound. Let
$X = \{x_1, \ldots, x_n\}$ and Ω be the probability space on $\{0, 1\}^n$, and consider the
probability distribution on Ω induced by selecting each component equal to 1 independently with probability $p = r/n$ (for convenience, we use Bernoulli sampling).
Let $F : \Omega \to \mathbb{R}$ be the random variable that assigns to the vector $(q_1, \ldots, q_n)$ the
value $\sum_{\Delta \in \mathcal{T}(R)} f(|K(\Delta) \cap X|)$, where $x_i \in R$ iff $q_i = 1$, and $f(x) = x^k$ (for the
kth moment; using $f(x) = e^{c(r/n)x}$ with an appropriate constant c, one can achieve
the exponential decay bound). We know that $\mathrm{E}[F] \leq M$ with $M = Cf(n/r)t(r)$,
where $t(r)$ is an upper bound for $\mathrm{E}[|\mathcal{T}(R)|]$. The method is based on the following

relation, for $0 \leq i < n$:

$$
\begin{aligned}
\mathrm{E}[F|q_1 &= v_1, \ldots q_i = v_i] \\
&= p \cdot \mathrm{E}[F|q_1 = v_1, \ldots q_i = v_i, q_{i+1} = 1] + \\
&\quad (1-p) \cdot \mathrm{E}[F|q_1 = v_1, \ldots q_i = v_i, q_{i+1} = 0] \\
&\geq \min\{\mathrm{E}[F|q_1 = v_1, \ldots q_i = v_i, q_{i+1} = 1], \mathrm{E}[F|q_1 = v_1, \ldots q_i = v_i, q_{i+1} = 0]\}
\end{aligned}
$$

If these conditional expectations can be computed *efficiently*, then this implies an efficient procedure to select v_{i+1} so that

$$
\mathrm{E}[F|q_1 = v_1, \ldots q_i = v_i] \geq \mathrm{E}[F|q_1 = v_1, \ldots q_i = v_i, q_{i+1} = v_{i+1}].
$$

Iterating this procedure, one finally obtains a solution $(v_1, \ldots, v_n)$ that satisfies the probabilistic bound

$$
M \geq \mathrm{E}[F] \geq \mathrm{E}[F|q_1 = v_1, \ldots q_n = v_n].
$$

If the locality property holds, the conditional probabilities involved can indeed be computed in polynomial time: Let $X_i = \{x_1, x_2, \ldots, x_i\}$ and $R_i = \{x_j \in X : q_j = 1, j \leq i\}$, then $\mathrm{E}[F|q_1 = v_1, \ldots q_i = v_i]$ is equal to

$$
\sum_{\Delta \in \Pi(X)} \Pr\{\Delta \in \mathcal{T}(R)|q_1 = v_1, \ldots q_i = v_i\} f(|K(\Delta) \cap X|)
$$

$$
= \sum_{\Delta \in \Pi(X): D(\Delta) \cap X_i \subseteq R_i, K(\Delta) \cap R_i = \emptyset} p^{|D(\Delta) \setminus S_i|} (1-p)^{|K(\Delta) \cap (X \setminus X_i)|} f(|K(\Delta) \cap X|),
$$

which can be approximated with sufficient accuracy. Similarly, $(1/r)$-nets and $(1/r)$-approximations of sizes $O(r \log r)$ and $O(r^2 \log r)$ can be computed in polynomial time, see Chapter 47.

k-WISE INDEPENDENT DISTRIBUTIONS

The method of conditional probabilities is highly sequential. An approach that is more suitable for parallel algorithms is to construct a probability space of polynomial size, and to execute the algorithm on each vector of this space. This is possible, for example, when the variables q_i need only be k-wise independent rather than being fully independent: for any indices $i_1, \ldots, i_k$, and 0-1 values, $v_1, \ldots, v_k$,

$$
\Pr\{q_{i_1} = v_1, \ldots, q_{i_k} = v_k\} = \Pi_{j=1}^{k} \Pr\{q_{i_j} = v_{i_j}\} = \Pi_{j=1}^{k} p^{v_j} (1-p)^{1-v_j}.
$$

A probability space and distribution of size $O(n^k)$ with such k-wise independence can be computed effectively [Jof74, KM97]. Let $\rho \geq n$ be a prime number and suppose that $p_1, \ldots, p_n \in [0, 1]$ satisfy $p_i = j_i/\rho$, for some integers j_i. Define a probability space with at most n^k points, as follows. For each $\langle a_0, a_1, \ldots, a_{k-1} \rangle$ in $\{0, 1, \ldots, \rho - 1\}^k$, let

$$
X_i = a_0 + a_1 i + a_2 i^2 + \cdots + a_{k-1} i^{k-1} \bmod \rho,
$$

for $1 \leq i \leq n$, assign probability $1/\rho^k$ and associate the vector $\langle Y_1, \ldots, Y_n \rangle$ where $Y_i = 1$ if $X_i \in \{0, 1, \ldots, j_i - 1\}$ and $Y_i = 0$ otherwise. The 0-1 probability space defined by the vectors $\langle Y_1, \ldots, Y_n \rangle$ is a k-wise independent 0-1 probability space for $p_1, \ldots, p_n$. With this construction, arbitrary probabilities can be approximated (within a factor of 2) by an appropriate choice of ρ. Using a larger space of size $O(n^{2k})$, arbitrary probabilities can be achieved exactly [KM97].

For some randomized algorithms one can show that they still work under k-wise independency for an appropriate k. For example, a quasi-random permutation with k-wise independence suffices for the randomized incremental approach to work [Mul96] (thus $O(\log n)$ random bits suffice rather than the $\Omega(n \log n)$ bits needed to define a fully random permutation). To verify that k-wise independence suffices, a tail inequality under k-independence is used [SSS95, BR94]. Let $q_1, \ldots, q_n$ be a sequence of k-wise independent random variables in $\{0, 1\}$, with $k \geq 2$ even, let $Q = \sum_{i=1}^{n} q_i$, $\mu = \mathrm{E}[Q]$ and assume that $\mu \geq k$, then

$$\Pr\{Q = 0\} < \frac{C_k}{\mu^{k/2}} \qquad (44.6.1)$$

where C_k is a constant depending on k. Let R be a $2k$-wise independent random sample from X with uniform probability p. For $\Delta \in \Pi(X)$, note that (no independence assumption needed)

$$\Pr\{D(\Delta) \subseteq R, K(\Delta) \cap R = \emptyset\} = \Pr\{D(\Delta) \subseteq R\} \cdot \Pr\{K(\Delta) \cap R = \emptyset | D(\Delta) \subseteq R\}.$$

Let d be an upper bound on $|D(\Delta)|$. The first factor can be computed using $2k$-wise independence assuming $2k \geq d$:

$$\Pr\{D(\Delta) \subseteq R\} = p^{|D(\Delta)|}.$$

To upper bound the second factor, let $t_\Delta = p|K(\Delta)|$; then using the tail bound above, for $t_\Delta \geq k$:

$$\Pr\{K(\Delta) \cap R = \emptyset | D(\Delta) \subseteq R\} \leq \frac{C}{t_\Delta^{k-d/2}},$$

since after fixing $D(\Delta) \subseteq R$, the remaining random variables are still $(2k - d)$-wise independent. Choosing k so that $c \leq k - d/2 + 2$, one can verify that the cth moment bound holds. Similarly, $1/r$-nets and $1/r$-approximations with sizes $O(rn^\delta)$ and $O(r^2 n^\delta)$ can be computed in polynomial time, where $\delta = O(1/k)$. It does not seem possible, however, to achieve the exponential decay bound with a limited-independence space of polynomial size.

For fixed k, the size of the space can be reduced if a certain deviation from k-wise independence is allowed [NN93]. Furthermore, the approach of testing all the vectors in the probability space can be combined with the approach of performing a binary search so that even a space of superpolynomial size is usable [MNN94, BRS94]. Still, these approaches do not lead to the exponential decay bound, or to nets or approximations of size matching the probabilistic analysis.

CONSTRAINT-BASED PROBABILITY SPACES

An alternative approach that is implementable in parallel constructs a probability distribution tailored to a particular algorithm and even to a specific input [Nis92, KM94, KK97, MRS01], leading to smaller probability spaces. The approach models the sampling process using ***randomized finite automata*** (RFA), and fools the automaton using a probability distribution D_n with support of size E_0 that depends polynomially on the error and on the size of the problem. Once the probability distribution has been constructed, it is only a matter of testing the algorithm for each point in D_n.

For each configuration Δ we construct an RFA M_Δ as follows: It consists of $n + 1$ levels $N_{\Delta,j}$, $0 \leq j \leq n$, each with two states $\langle j, \text{Yes} \rangle$ and $\langle j, \text{No} \rangle$, with transitions that reflect whether $\Delta \in \mathcal{T}(R)$: $\langle j - 1, \text{No} \rangle$ is always connected to $\langle j, \text{No} \rangle$; if $x_j \in D(\Delta)$ then $\langle j - 1, \text{Yes} \rangle$ is connected to $\langle j, \text{Yes} \rangle$ under $q_j = 1$ and to $\langle j, \text{No} \rangle$ under $q_j = 0$; if $x_j \in K(\Delta)$ then $\langle j - 1, \text{Yes} \rangle$ is connected to $\langle j, \text{Yes} \rangle$ under $q_j = 0$ and to $\langle j, \text{No} \rangle$ under $q_j = 1$; if $x_j \notin D(\Delta) \cup K(\Delta)$ then $\langle j - 1, \text{Yes} \rangle$ is connected to $\langle j, \text{Yes} \rangle$, and $\langle j - 1, \text{No} \rangle$ is connected to $\langle j, \text{No} \rangle$, in either case. D_n is determined by a recursive approach in which the generic procedure $\text{fool}(l, l')$ constructs a distribution that fools the transition probabilities between level l and l' in *all* the RFAs as follows. It computes, using $\text{fool}(l, l'')$ and $\text{fool}(l'', l')$ recursively, distributions D_1 and D_2, each of size at most $E_0(1 + o(1))$, that fool the transitions between states in levels l and $l'' = \lfloor (l + l')/2 \rfloor$, and between states in levels l'' and l; a procedure $\text{reduce}(D_1 \times D_2)$ then combines D_1 and D_2 into a distribution D of size at most $E_0(1 + o(1))$ that fools the transitions between states in levels l and l' in all the RFAs. Let $\tilde{D} = D_1 \times D_2$ be the product distribution with $\text{support}(\tilde{D}) = \{w_1 w_2 : w_i \in \text{support}(D_i)\}$ and $\text{Pr}_{\tilde{D}}\{w_1 w_2\} = \text{Pr}_{D_1}\{w_1\}\text{Pr}_{D_2}\{w_2\}$, a randomized version of reduce is to retain each $w \in \tilde{D}$ with probability $q(w) = E_0/|\text{support}(\tilde{D})|$ into $\text{support}(D)$ with $\text{Pr}_D\{w\} = \text{Pr}_{\tilde{D}}\{w\}/q(w)$. Thus, for all pairs of states s, t in the RFAs the transition probabilities are preserved in expectation:

$$\text{E}[\text{Pr}_D\{s \to t\}] = \sum_{w \, : \, s \overset{w}{\to} t} \frac{\text{Pr}_{\tilde{D}}\{w\}}{q(w)} q(w) = \text{Pr}_{\tilde{D}}\{s \to t\}, \qquad (44.6.2)$$

where the sum is over all the strings w that lead from state s to state t. This selection also implies that the expected size of $\text{support}(D)$ is $\sum_w q(w) = E_0$. This randomized combining can be derandomized using a 2-wise independent probability space. The bottom of the recursion is reached when the number of levels between l and l' is at most $\log E_0$, and then the distribution (of size E_0) is implemented by $\log E_0$ unbiased bits. E_0 depends polynomially on $1/\delta$, where δ is the relative error that is allowed for the transition probabilities. Taking δ as a small constant suffices to obtain a constant approximation of the moment bounds.

APPLICATIONS

Some interesting examples for which optimal deterministic algorithms have been obtained using derandomization are the following:

- *ϵ-nets and ϵ-approximations:* See Chapter 47.

- *Convex hulls:* The only optimal deterministic algorithm for the computation of the convex hull of n points in $\mathbb{R}^d$ space is the derandomization of a randomized-incremental algorithm. The reader is referred to [BCM99] for the details (this reference is much more readable than the original paper [Cha93]) (Chapter 26).

- *Output-sensitive convex hull in $\mathbb{R}^d$:* An optimal algorithm for $d = 3$ was obtained using derandomization [CM95]; afterward a surprisingly simple solution avoiding derandomization was found [Cha96].

- *Diameter of a point set in $\mathbb{R}^3$:* After a sequence of improvements, an optimal algorithm using derandomization was found [Ram01]. Currently, the best solution that avoids derandomization has a running time with an extra $\log n$ factor [Bes01].

- *Linear programming:* In $\mathbb{R}^d$, the best deterministic solution is achieved through derandomization and has running time $O(C_d n)$ with $C_d = \exp(O(d \log d))$ [CM96, Cha16]; in contrast, it is possible to achieve $C_d \approx \exp(\sqrt{d})$ with randomization [MSW96] (Chapter 49).

- *Segment intersection:* A first algorithm for reporting intersections between n line segments in $O(n)$ space and optimal time used derandomization [AGR95], followed shortly by a relatively simple algorithm that avoids derandomization [Bal95]. Optimal parallel algorithms have been obtained with derandomization and have not been matched by other approaches.

OPEN PROBLEMS

Is derandomization truly necessary to obtain an optimal algorithm in cases such as the convex hull in d dimensions or the diameter in dimension 3?

44.7 OPTIMIZATION

In geometric optimization we seek to optimize some measure on a geometric structure that satisfies given constraints, such as the length of a tour visiting given points, or the area of a convex container into which given shapes can be translated.

PARAMETRIC SEARCH AND RANDOMIZATION

Parametric search is a technique that allows us to take a decision algorithm for the problem (that is, an algorithm that takes a parameter r and tells us if a solution of quality better than r exists), and to convert it into an algorithm solving the optimization problem. Several geometric examples can be found in [CEGS93, AST94].

In many applications of parametric searching, it can be considered as a search among the vertices of an arrangement, where each vertex determines a critical value of the parameter r. If we could compute these vertices efficiently, we could use the decision algorithm to perform a binary search—but the arrangement is too large to be built explicitly.

If we can randomly sample vertices of some range of this (implicit) arrangement, then parametric search can be replaced by randomization to obtain a much simpler algorithm. A classic example is the *slope selection* problem [Mat91].

Alternatively, we may be able to guide the search using an appropriate cutting of this arrangement, obtaining a deterministic algorithm that is much simpler than using parametric search [BC98].

CHAN'S TECHNIQUE

Another randomized alternative to parametric search is a simple technique by Chan [Cha99]. Again, it assumes that we have an algorithm to solve the decision version of the problem, and in addition requires that we can split the problem into a constant number k of subproblems smaller by a constant fraction. Then the optimization problem can be solved as follows: Split the problem P into k subproblems $P_1, P_2, \ldots, P_k$, pick a random permutation of the P_i, and set r to the solution for P_1. For $i \in \{2, \ldots, k\}$, use the decision algorithm to determine if the solution for P_i is better than r. If it is, recursively compute the solution for P_i and set r to this solution.

The final value of r is the optimal solution for P, the algorithm also determines a constant-size subset that determines this optimum. If the running time of the decision algorithm is at least polynomial, then the expected running time of this procedure asymptotically matches the running time of the decision algorithm.

A simple example is determining the closest pair, or more generally the b-tuple minimizing some measure, in a given set of points P. Indeed, we can arbitrarily partition P into $b + 1$ sets $Q_1, \ldots, Q_{b+1}$ of roughly equal size, and define the subproblem $P_i = P \setminus Q_i$. Then $k = b + 1$ and each subproblem's size is $b/(b+1)$ of the original problem's size.

This technique was generalized for certain families of implicit LP-type problems [Cha04].

44.8 MONTE CARLO ALGORITHMS

Monte Carlo algorithms, that is, algorithms that are allowed to fail or report an incorrect answer with a small probability, used to be uncommon in computational geometry, but started to appear more frequently since the early 2000s (Chapter 48 also mentions several Monte Carlo algorithms).

A typical Monte Carlo algorithm in computational geometry takes a random sample R of some input set S, and proceeds under the assumption that R is an ϵ-approximation for S. This is true with high probability, but if the assumption fails, an incorrect result can be returned.

As an example, consider the problem of computing, given a simple polygon P, a point p inside P that maximizes the area of P visible from p. A naive approach would simply try many random points $p \in P$, compute their visibility region, and choose the best one. However, the probability of finding a good point can be arbitrarily small, and we can do better by using an ϵ-approximation to estimate the visibility region of a point [CEHP07]: We uniformly sample a set S of points from P. We compute the visibility region $V(s)$ of each point $s \in S$, take the arrangement of all these regions, and choose a point p^* in the most heavily covered cell of this arrangement.

This point p maximizes the number of points in S visible from p. Since with high probability, S is an ϵ-approximation for the points in P, that is, the number of points of S visible from a point p is an estimate for the area of P visible from p, this implies that the area seen by p^* is a good approximation for the largest area that any point in P can see.

The technique can be improved by observing that we do not actually need an

ϵ-approximation—we need that S is a good approximation only for the vertices of the arrangement of the visibility regions $V(s)$. We can also make the size of S dependent on the value of the optimal solution, see [CEHP07].

Similar ideas have been used in shape matching [CEHP07, ASS10, AS12]. *Relative (p, ϵ)-approximations* help to achieve a guaranteed relative error with a sample of small size [HPS11].

44.9 BETTER GUARANTEES

Bounds for the expected performance of randomized algorithms are usually available. Sometimes stronger results are desired. If the analysis of the algorithm cannot be extended to provide such bounds, then some techniques may help to achieve them:

Randomized space vs. deterministic space. Any randomized algorithm using expected space S and expected time T can be converted to an algorithm that uses deterministic space $2S$, and whose expected running time is at most $2T$. We simply need to maintain a count of the memory allocated by the algorithm. Whenever it exceeds $2S$, we stop the computation and restart it again with fresh choices for the random variables. The expected number of retrials is one.

Tail estimates. The knowledge that the expected running time of a given program is one second does not exclude the possibility that it sometimes takes one hour. Markov's inequality implies that the probability that this happens is at most $1/3600$. While this seems innocuous, it implies that it is likely to occur if we repeat this particular computation, say, 10000 times.

For randomized incremental construction, better tail estimates can sometimes be proven [CHPR16, Lemma 3.4]. In particular, bounds are known for segment intersection in the plane [MSW93a] (see also [BCKO08, Section 6.4]) and for LP-type optimization [GW00]. Tail estimates are also available for the *space complexity* of randomized incremental construction [CMS93, MSW93b].

In all other cases, one can still apply a simple modification to the algorithm to yield a stronger bound. We run it for two seconds. If it does not finish the computation within two seconds, then we abandon the computation and restart with fresh choices for the random variables. Clearly, the probability that the algorithm does not terminate within one hour is at most 2^{-1800}. Alt et al. [AGM+96] work out this technique, which is a special case of success amplification [Hro05, Chapter 5], in detail.

44.10 PROBABILISTIC PROOF TECHNIQUES

Randomized algorithms are related to probabilistic proofs and constructions in combinatorics, which precede them historically. Conversely, the concepts developed to design and analyze randomized algorithms in computational geometry can be used as tools in proving purely combinatorial results. Many of these results are based on the following theorem:

THEOREM 44.10.1

*Let $(X, \mathcal{T}, D, K)$ be a configuration space satisfying axioms (i), (ii), (iii), and (iii')
of Section 44.5. For $S \subseteq X$ and $0 \le k \le n$, let*

$$\Pi^k(S) := \{ \Delta \in \Pi(X) \mid |K(\Delta) \cap S| \le k \}$$

denote the set of configurations with at most k conflicts in S.

*Then $|\Pi^k(S)| = O(k^d)E[|\mathcal{T}(R)|]$, where R is a random sample of S of size n/k,
and d is as in axiom (i).*

Note that $\Pi^0(S) = \mathcal{T}(S)$. The theorem relates the number of configurations
with at most k conflicts to those without conflict.

An immediate application is to prove a bound on the number of vertices of
level at most k in an arrangement of lines in the plane (the level of a vertex is the
number of lines lying above it; see Section 20.2). We define a configuration space
$(X, \mathcal{T}, D, K)$ where X is the set of lines, $\mathcal{T}(S)$ is the set of vertices of the upper
envelope of the lines, $D(\Delta)$ are the two lines forming the vertex Δ (so $d = 2$), and
$K(\Delta)$ is the set of lines lying above Δ. Theorem 44.10.1 implies that the number
of vertices of level up to k is bounded by $O(nk)$. The same argument works in any
dimension.

Sharir and others have proved a number of combinatorial results using this
technique [Sha94, AES99, ASS96, SS97]. They define a configuration space and
need to bound $|\mathcal{T}(S)|$. They do this by proving a geometric relationship between
the configurations with zero conflicts (the ones appearing in $\mathcal{T}(S)$) and the configurations with at most k conflicts. Applying Theorem 44.10.1 yields a recursion
that bounds $|\mathcal{T}(S)|$ in terms of $|\mathcal{T}(R)|$. A refined approach that uses a sample of
size $n - 1$ (instead of n/k) has been suggested by Tagansky [Tag96].

Sharir [Sha03] reviews this technique and gives a new proof for Theorem 44.10.1
based on the *Crossing lemma* (Chapter 28).

44.11 SOURCES AND RELATED MATERIAL

SURVEYS AND BOOKS

[BCKO08]: This textbook on computational geometry contains a gentle introduction to randomized incremental construction for several problems.

[HP11]: This monograph covers approximation algorithms in computational geometry and includes many randomized algorithms.

[Cla92, Mul00]: General surveys of randomized algorithms in computational geometry.

[Sei93]: An introduction to randomized incremental algorithms using backwards
analysis.

[GS93]: Surveys computations with arrangements, including randomized algorithms.

[AS01] Surveys randomized techniques in geometric optimization problems.

[Mul93]: This monograph is an extensive treatment of randomized algorithms in
computational geometry.

[Mat00]: An introduction to derandomization for geometric algorithms, with many references.

[MR95]: A book on randomized algorithms and their analysis in computer science, including derandomization techniques.

[AS16]: This monograph covers probabilistic proof techniques mostly in combinatorics, and includes some algorithmic aspects, geometric results, and derandomization.

[HP09]: A survey of ϵ-samples, approximations, and relative (p, ϵ)-approximations.

RELATED CHAPTERS

Because randomized algorithms have been used successfully in nearly all areas of computational geometry, they are mentioned throughout Parts C and D of this Handbook. Areas where randomization plays a particularly important role include:

Chapter 26: Convex hull computations
Chapter 28: Arrangements
Chapter 40: Range searching
Chapter 47: Epsilon-approximations and epsilon-nets
Chapter 48: Coresets and sketches
Chapter 49: Linear programming

REFERENCES

[AES99]   P.K. Agarwal, A. Efrat, and M. Sharir. Vertical decomposition of shallow levels in 3-dimensional arrangements and its applications. *SIAM J. Comput.*, 29:912–953, 1999.

[AGM+96]   H. Alt, L. Guibas, K. Mehlhorn, R. Karp, and A. Wigderson. A method for obtaining randomized algorithms with small tail probabilities. *Algorithmica*, 16:543–547, 1996.

[AGR95]   N.M. Amato, M.T. Goodrich, and E.A. Ramos. Computing faces in segment and simplex arrangements. In *Proc. 27th ACM Sympos. Theory Comput.*, pages 672–682, 1995.

[AMS98]   P.K. Agarwal, J. Matoušek, and O. Schwarzkopf. Computing many faces in arrangements of lines and segments. *SIAM J. Comput.*, 27:491–505, 1998.

[AS01]   P.K. Agarwal and S. Sen. Randomized algorithms for geometric optimization problems. In S. Rajasekaran, P.M. Pardalos, J.H. Reif, and J. Rolim, editors, *Handbook of Randomized Computing*, pages 151–201, Kluwer Academic, Boston, 2001.

[AS12]   H. Alt and L. Scharf. Shape matching by random sampling. *Theoret. Comput. Sci.*, 442:2–12, 2012.

[AS16]   N. Alon and J.H. Spencer. *The Probabilistic Method*, 4th edition. John Wiley & Sons, New York, 2016.

[ASS96]   P.K. Agarwal, O. Schwarzkopf, and M. Sharir. The overlay of lower envelopes and its applications. *Discrete Comput. Geom.*, 15:1–13, 1996.

[ASS10]   H. Alt, L. Scharf, and D. Schymura. Probabilistic matching of planar regions. *Comput. Geom.*, 43:99–114, 2010.

[AST94]   P.K. Agarwal, M. Sharir, and S. Toledo. Applications of parametric searching in geometric optimization. *J. Algorithms*, 17:292–318, 1994.

[Bal95]   I.J. Balaban. An optimal algorithm for finding segment intersections. In *Proc. 11th Sympos. Comput. Geom.*, pages 211–219, ACM Press, 1995.

[BC98]   H. Brönnimann and B. Chazelle. Optimal slope selection via cuttings. *Comput. Geom.*, 10:23–29, 1998.

[BCKO08]   M. de Berg, O. Cheong, M. van Kreveld, and M. Overmars. *Computational Geometry: Algorithms and Applications*, 3rd edition. Springer, Berlin, 2008.

[BCM99]   H. Brönnimann, B. Chazelle, and J. Matoušek. Product range spaces, sensitive sampling, and derandomization. *SIAM J. Comput.*, 28:1552–1575, 1999.

[BDS95]   M. de Berg, K. Dobrindt, and O. Schwarzkopf. On lazy randomized incremental construction. *Discrete Comput. Geom.*, 14:261–286, 1995.

[Bes01]   S. Bespamyatnikh. An efficient algorithm for the three-dimensional diameter problem. *Discrete Comput. Geom.*, 25:235–255, 2001.

[BR94]   M. Bellare and J. Rompel. Randomness-efficient oblivious sampling. In *Proc. 35th IEEE Sympos. Found. Comp. Sci.*, pages 276–287, 1994.

[BRS94]   B. Berger, J. Rompel, and P.W. Shor. Efficient NC algorithms for set cover with applications to learning and geometry. *J. Comput. Syst. Sci.*, 49:454–477, 1994.

[CEGS93]   B. Chazelle, H. Edelsbrunner, L.J. Guibas, and M. Sharir. Diameter, width, closest line pair and parametric searching. *Discrete Comput. Geom.*, 10:183–196, 1993.

[CEHP07]   O. Cheong, A. Efrat, and S. Har-Peled. Finding a guard that sees most and a shop that sells most. *Discrete Comput. Geom.*, 37:545–563, 2007.

[Cha93]   B. Chazelle. An optimal convex hull algorithm in any fixed dimension. *Discrete Comput. Geom.*, 10:377–409, 1993.

[Cha96]   T.M. Chan. Optimal output-sensitive convex hull algorithms in two and three dimensions. *Discrete Comput. Geom.*, 16:361–368, 1996.

[Cha99]   T.M. Chan. Geometric applications of a randomized optimization technique. *Discrete Comput. Geom.*, 22:547–567, 1999.

[Cha04]   T.M. Chan. An optimal randomized algorithm for maximum Tukey depth. In *Proc. 15th ACM-SIAM Sympos. Discrete Algorithms*, pages 430–436, 2004.

[Cha12]   T.M. Chan. Optimal partition trees. *Discrete Comput. Geom.*, 47:661–690, 2012.

[Cha16]   T.M. Chan. Improved deterministic algorithms for linear programming in low dimensions. In *Proc. 27th ACM-SIAM Sympos. Discrete Algorithms*, pages 1213–1219, 2016.

[CHPR16]   H.-C. Chang, S. Har-Peled, and B. Raichel. From proximity to utility: A Voronoi partition of Pareto optima. *Discrete Comput. Geom.*, 56:631–656, 2016.

[Cla92]   K.L. Clarkson. Randomized geometric algorithms. In D.-Z. Du and F.K. Hwang, editors, *Computing in Euclidean Geometry*, vol. 1 of *Lecture Notes Series on Computing*, pages 117–162, World Scientific, Singapore, 1992.

[CM95]   B. Chazelle and J. Matoušek. Derandomizing an output-sensitive convex hull algorithm in three dimensions. *Comput. Geom.*, 5:27–32, 1995.

[CM96]   B. Chazelle and J. Matoušck. On linear-time deterministic algorithms for optimization problems in fixed dimension. *J. Algorithms*, 21:579–597, 1996.

[CMS93]   K.L. Clarkson, K. Mehlhorn, and R. Seidel. Four results on randomized incremental constructions. *Comput. Geom.*, 3:185–212, 1993.

[CT16]   T.M. Chan and K. Tsakalidis. Optimal deterministic algorithms for 2-d and 3-d shallow cuttings. *Discrete Comput. Geom.*, 2016.

[DMT92]   O. Devillers, S. Meiser, and M. Teillaud. Fully dynamic Delaunay triangulation in logarithmic expected time per operation. *Comput. Geom.*, 2:55–80, 1992.

[GS93]   L.J. Guibas and M. Sharir. Combinatorics and algorithms of arrangements. In J. Pach, editor, *New Trends in Discrete and Computational Geometry*, vol. 10 of *Algorithms and Combin.*, pages 9–36. Springer-Verlag, Berlin, 1993.

[GW00]   B. Gärtner and E. Welzl. Random sampling in geometric optimization: New insights and applications. In *Proc. 16th Sympos. Comput. Geom.*, pages 91–99, ACM Press, 2000.

[HP00]   S. Har-Peled. Constructing planar cuttings in theory and practice. *SIAM J. Comput.*, 29:2016–2039, 2000.

[HP09]   S. Har-Peled. Carnival of samplings: Nets, approximations, relative and sensitive. Preprint, `arXiv:0908.3716`, 2009.

[HP11]   S. Har-Peled. *Geometric Approximation Algorithms*. AMS, Providence, 2011.

[HPS11]   S. Har-Peled and M. Sharir. Relative (p, ϵ)-approximations in geometry. *Discrete Comput. Geom.*, 45:462–496, 2011.

[Hro05]   J. Hromkovič. *Design and Analysis of Randomized Algorithms*. Springer, Berlin, 2005.

[Jof74]   A. Joffe. On a set of almost deterministic k-independent random variables. *Ann. Probab.*, 2:161–162, 1974.

[KK97]   D.R. Karger and D. Koller. (De)randomized construction of small sample spaces in NC. *J. Comput. Syst. Sci.*, 55:402–413, 1997.

[KM94]   D. Koller and N. Megiddo. Constructing small sample spaces satisfying given constraints. *SIAM J. Discrete Math.*, 7:260–274, 1994.

[KM97]   H. Karloff and Y. Mansour. On construction of k–wise independent random variables. *Combinatorica*, 17:91–107, 1997.

[Mat91]   J. Matoušek. Randomized optimal algorithm for slope selection. *Inform. Process. Lett.*, 39:183–187, 1991.

[Mat00]   J. Matoušek. Derandomization in computational geometry. In J.-R. Sack and J. Urrutia, editors, *Handbook of Computational Geometry*, pages 559–595, North-Holland, Amsterdam, 2000.

[MNN94]   R. Motwani, J. Naor, and M. Naor. The probabilistic method yields deterministic parallel algorithms. *J. Comput. Syst. Sci.*, 49:478–516, 1994.

[MR95]   R. Motwani and P. Raghavan. *Randomized Algorithms*. Cambridge University Press, 1995.

[MRS01]   S. Mahajan, E.A. Ramos, and K.V. Subrahmanyam. Solving some discrepancy problems in NC. *Algorithmica*, 29:371–395, 2001.

[MSW93a] K. Mehlhorn, M. Sharir, and E. Welzl. Tail estimates for the efficiency of randomized incremental algorithms for line segment intersection. *Comput. Geom.*, 3:235–246, 1993.

[MSW93b] K. Mehlhorn, M. Sharir, and E. Welzl. Tail estimates for the space complexity of randomized incremental algorithms. *Comput. Geom.*, 4:185–246, 1993.

[MSW96]   J. Matoušek, M. Sharir, and E. Welzl. A subexponential bound for linear programming. *Algorithmica*, 16:498–516, 1996.

[Mul91a] K. Mulmuley. Randomized multidimensional search trees: Dynamic sampling. In *7th Sympos. Comput. Geom.*, pages 121–131, ACM Press, 1991.

[Mul91b] K. Mulmuley. Randomized multidimensional search trees: Further results in dynamic sampling. In *32nd IEEE Sympos. Found. Comp. Sci.*, pages 216–227, 1991.

[Mul91c]   K. Mulmuley. Randomized multidimensional search trees: Lazy and dynamic shuffling. In *32nd IEEE Sympos. Found. Comp. Sci.*, pages 180–196, 1991.

[Mul93]   K. Mulmuley. *Computational Geometry: An Introduction through Randomized Algorithms*. Prentice Hall, Englewood Cliffs, 1993.

[Mul96]   K. Mulmuley. Randomized geometric algorithms and pseudorandom generators. *Algorithmica*, 16:450–463, 1996.

[Mul00]   K. Mulmuley. Randomized algorithms in computational geometry. In J.-R. Sack and J. Urrutia, editors, *Handbook of Computational Geometry*, pages 703–724, North-Holland, Amsterdam, 2000.

[Nis92]   N. Nisan. Pseudorandom generators for space-bounded computation. *Combinatorica*, 12:449–461, 1992.

[NN93]   J. Naor and M. Naor. Small-bias probability spaces: Efficient constructions and applications. *SIAM J. Comput.*, 22:838–856, 1993.

[PT13]   J. Pach and G. Tardos. Tight lower bounds for the size of epsilon-nets. *J. Amer. Math. Soc.*, 26:645–658, 2013.

[Rab76]   M.O. Rabin. Probabilistic algorithms. In J. Traub, editor, *Algorithms and Complexity*, pages 21–39. Academic Press, New York, 1976.

[Rag88]   P. Raghavan. Probabilistic construction of deterministic algorithms: Approximating packing integer programs. *J. Comput. Syst. Sci.*, 37:130–143, 1988.

[Ram01]   E.A. Ramos. An optimal deterministic algorithm for computing the diameter of a three-dimensional point set. *Discrete Comput. Geom.*, 26:233–244, 2001.

[Sch91]   O. Schwarzkopf. Dynamic maintenance of geometric structures made easy. In *Proc. 32nd IEEE Sympos. Found. Comp. Sci.*, pages 197–206, 1991.

[Sei91]   R. Seidel. A simple and fast incremental randomized algorithm for computing trapezoidal decompositions and for triangulating polygons. *Comput. Geom.*, 1:51–64, 1991.

[Sei93]   R. Seidel. Backwards analysis of randomized geometric algorithms. In J. Pach, editor, *New Trends in Discrete and Computational Geometry*, vol. 10 of *Algorithms and Combin.*, pages 37–68. Springer-Verlag, Berlin, 1993.

[Sha94]   M. Sharir. Almost tight upper bounds for lower envelopes in higher dimensions. *Discrete Comput. Geom.*, 12:327–345, 1994.

[Sha03]   M. Sharir. The Clarkson-Shor technique revisited and extended. *Combin. Probab. Comput.*, 12:191–201, 2003.

[Spe87]   J. Spencer. *Ten Lectures on the Probabilistic Method*. CBMS-NSF. SIAM, 1987.

[SS97]   O. Schwarzkopf and M. Sharir. Vertical decomposition of a single cell in a three-dimensional arrangement of surfaces and its applications. *Discrete Comput. Geom.*, 18:269–288, 1997.

[SSS95]   J.P. Schmidt, A. Siegel, and A. Srinivasan. Chernoff-Hoeffding bounds for applications with limited independence. *SIAM J. Discrete Math.*, 8:223–250, 1995.

[Tag96]   B. Tagansky. A new technique for analyzing substructures in arrangements of piecewise linear surfaces. *Discrete Comput. Geom.*, 16:455–479, 1996.

45 ROBUST GEOMETRIC COMPUTATION
Vikram Sharma and Chee K. Yap

INTRODUCTION

Nonrobustness refers to qualitative or catastrophic failures in geometric algorithms arising from numerical errors. Section 45.1 provides background on these problems. Although nonrobustness is already an issue in "purely numerical" computation, the problem is compounded in "geometric computation." In Section 45.2 we characterize such computations. Researchers trying to create robust geometric software have tried two approaches: making fixed-precision computation robust (Section 45.3), and making the exact approach viable (Section 45.4). Another source of nonrobustness is the phenomenon of degenerate inputs. General methods for treating degenerate inputs are described in Section 45.5. For some problems the exact approach may be expensive or infeasible. To ensure robustness in this setting, a recent extension of exact computation, the so-called "soft exact approach," has been proposed. This is described in Section 45.6.

45.1 NUMERICAL NONROBUSTNESS ISSUES

Numerical nonrobustness in scientific computing is a well-known and widespread phenomenon. The root cause is the use of **fixed-precision numbers** to represent real numbers, with precision usually fixed by the machine word size (e.g., 32 bits). The unpredictability of floating-point code across architectural platforms in the 1980s was resolved through a general adoption of the IEEE standard 754-1985. But this standard only makes program behavior predictable and consistent across platforms; the errors are still present. Ad hoc methods for fixing these errors (such as treating numbers smaller than some ε as zero) cannot guarantee their elimination.

Nonrobustness is already problematic in purely numerical computation: this is well documented in numerous papers in numerical analysis with the key word "pitfalls" in their title. But nonrobustness apparently becomes intractable in "geometric" computation. In Section 45.2, we elucidate the concept of geometric computations. Based on this understanding, we conclude that nonrobustness problems within fixed-precision computation cannot be solved by purely arithmetic solutions (better arithmetic packages, etc.). Rather, some suitable *fixed-precision geometry* is needed to substitute for the original geometry (which is usually Euclidean). We describe such approaches in Section 45.3. But in Section 45.4, we describe the alternative *exact approach* that requires arbitrary precision. Naively, the exact approach suggests that each numerical predicate must be computed exactly. But the formulation of **Exact Geometric Computation** (EGC) asks only for error-free evaluation of predicates. The simplicity and generality of the EGC solution has made it the dominant nonrobustness approach among computational geometers.

In Section 45.5, we address a different but common cause of numerical non-robustness, namely, *data degeneracy*. Geometric inputs can be degenerate: an input triangle might degenerate into a line segment, or an input set of points might contain collinear triples, etc. This can cause geometric algorithms to fail. But if geometric algorithms must detect such special situations, the number of such special cases can be formidable, especially in higher dimensions. For nonlinear problems, such analysis usually requires new and nontrivial facts of algebraic geometry. This section looks at general techniques to avoid explicit enumeration of degeneracies.

In Section 45.6, we note some formidable barriers to extending the EGC approach for nonlinear and nonalgebraic problems. This motivates some current research directions that might be described as *soft exact computation*. It goes beyond "simply exact" computation by giving a formal role to numerical approximations in our computing concepts and notions of correctness.

GLOSSARY

Fixed-precision computation: A mode of computation in which every number is represented using some fixed number L of bits, usually 32 or 64. The representation of floating point numbers using these L bits is dictated by the IEEE Floating Point standard. **Double-precision mode** is a relaxation of fixed precision: the intermediate values are represented in $2L$ bits, but these are finally truncated back to L bits.

Nonrobustness: The property of code failing on certain kinds of inputs. Here we are mainly interested in nonrobustness that has a numerical origin: the code fails on inputs containing certain patterns of numerical values. Degenerate inputs are just extreme cases of these "bad patterns."

Benign vs. catastrophic errors: Fixed-precision numerical errors are fully expected and so are normally considered to be "benign." In purely numerical computations, errors become "catastrophic" when there is a severe loss of precision. In geometric computations, errors are "catastrophic" when the computed results are qualitatively different from the true answer (e.g., the combinatorial structure is wrong) or when they lead to unexpected or *inconsistent* states of the programs.

Big number packages: They refer to software packages for representing and computing with arbitrary precision numbers. There are three main types of such number packages, called *BigIntegers*, *BigRationals* and *BigFloats*, representing (respectively) integers, rational numbers, and floating point numbers. For instance, $+$, $-$, and $\times$ are implemented exactly with *BigIntegers*. With *BigRationals*, division can also be exact. Beyond rational numbers, *BigFloats* become essential, and operations such as $\sqrt{\ }$ can be approximated to any desired precision in *BigFloat*. But ensuring that *BigFloats* achieve the correct rounding is a highly nontrivial issue especially for transcendental function. The MPFR [mpf16] package is the only Big number package to address this.

45.2 THE NATURE OF GEOMETRIC COMPUTATION

It is well known that numerical approximations may cause an algorithm to crash or enter an infinite loop; but there is a persistent belief that when the algorithm halts, then the output is a reasonably close approximation (up to the machine precision) to the true output. The paper [KMP$^+$07, §4.3] wishes to "refute this myth" by considering a simple algorithm for computing the convex hull of a planar point set. They constructed inputs such that the output "convex hull" might (1) miss a point far from the interior, (2) contain a large concave corner, or (3) be a self-intersecting polygon. The paper provides "classroom examples" based on a systematic analysis of floating point errors and their effects on common predicates in computational geometry. If the root cause of nonrobustness is arithmetic, then it may appear that the problem can be solved with the right kind of arithmetic package. We may roughly divide the approaches into two camps, depending on whether one uses finite precision arithmetic or insists on exactness (or at least the possibility of computing to arbitrary precision). While arithmetic is an important topic in its own right, our focus here will be on geometric rather than purely arithmetic approaches for achieving robustness.

To understand why nonrobustness is especially problematic for geometric computation, we need to understand what makes a computation "geometric." Indeed, we are revisiting the age-old question *"What is Geometry?"* that has been asked and answered many times in mathematical history, by Euclid, Descartes, Hilbert, Dieudonné and others. But as in many other topics, the perspective stemming from a modern computational viewpoint sheds new light. Geometric computation clearly involves numerical computation, but there is something more. We use the aphorism GEOMETRIC = NUMERIC + COMBINATORIAL to capture this. Instead of "combinatorial" we could have substituted "discrete" or sometimes "topological." What is important is that this combinatorial part is concerned with discrete relations among geometric objects. Examples of discrete relations are "a point is on a line," "a point is inside a simplex," or "two disks intersect." The geometric objects here are points, lines, simplices, and disks. Following Descartes, each object is defined by numerical parameters. Each discrete relation is reduced to the truth of suitable numerical inequalities involving these parameters. Geometry arises when such discrete relations are used to characterize configurations of geometric objects.

The mere presence of combinatorial structures in a numerical computation does not make a computation "geometric." There must be some nontrivial *consistency condition* holding between the numerical data and the combinatorial data. Thus, we would not consider the classical shortest-path problems on graphs to be geometric: the numerical weights assigned to edges of the graphs are not restricted by any consistency condition. Note that common restrictions on the weights (positivity, integrality, etc.) are not consistency restrictions. But the related *Euclidean shortest-path problem* (Chapter 31) is geometric. See Table 45.2.1 for further examples from well-known problems.

Alternatively, we can characterize a computation as "geometric" if it involves constructing or searching a geometric structure (which may only be implicit). The incidence graph of an arrangement of hyperplanes (Chapter 28), with suitable additional labels and constraints, is a primary example of such a structure. The reader may keep this example in mind in the following definition. A *geometric structure*

TABLE 45.2.1   Examples of geometric and non-geometric problems.

PROBLEM	GEOMETRIC?
Matrix multiplication, determinant	no
Hyperplane arrangements	yes
Shortest paths on graphs	no
Euclidean shortest paths	yes
Point location	yes
Convex hulls, linear programming	yes
Minimum circumscribing circles	yes

---

is comprised of four components:

$$D = (G, \lambda, \Phi(\mathbf{z}), I), \qquad (45.2.1)$$

where $G = (V, E)$ is a directed graph, λ is a labeling function on the vertices and edges of G, Φ is the consistency predicate, and I the input assignment. Intuitively, G is the combinatorial part, λ the geometric part, and Φ constrains λ based on the structure of G. The **input assignment**, for an input of size n, is $I : \{z_1, \ldots, z_n\} \to \mathbb{R}$ where the z_i's are called **structural variables**. We informally identify I with the sequence "$\mathbf{c} = (c_1, \ldots, c_n)$" where $I(z_i) = c_i$. The c_i's are called **(structural) parameters**. For each $u \in V \cup E$, the label $\lambda(u)$ is a Tarski formula of the form $\xi(\mathbf{x}, \mathbf{z})$, where $\mathbf{z} = (z_1, \ldots, z_n)$ are the structural variables and $\mathbf{x} = (x_1, \ldots, x_d)$ for some $d \geq 1$. This d is fixed and determines the ambient space $\mathbb{R}^d$ containing the geometric object. This formula defines a **semialgebraic set** (Chapter 37) parameterized by the structural variables. For a given $\mathbf{c}$, the semialgebraic set is $f_{\mathbf{c}}(v) = \{\mathbf{a} \in \mathbb{R}^d \mid \xi(\mathbf{a}, \mathbf{c}) \text{ holds}\}$. Following Tarski, we are identifying semialgebraic sets in $\mathbb{R}^d$ with d-dimensional geometric objects. The consistency relation $\Phi(\mathbf{z})$ is another Tarski formula of the form $\Phi(\mathbf{z}) = (\forall x_1, \ldots, x_d)\phi(\lambda(u_1), \ldots, \lambda(u_m), \mathbf{x}, \mathbf{z})$ where $u_1, \ldots, u_m$ ranges over elements of $V \cup E$. For each class of geometric structures, e.g., hyperplane arrangements, or Voronoi diagrams of points, the formula $\Phi(\mathbf{z})$ can be systematically constructed from G. Note that if D is to be computed in an output, we need not explicitly specify $\Phi(\mathbf{z})$, as this is understood (or implicit in our understanding of the geometric structure). The definition above can be contrasted with Fortune's definition, where a geometric problem on input of size n is a map from $\mathbb{R}^n$ to a discrete set, e.g., the set of cyclic permutations for convex hulls or the incidence graph for arrangements [For89]; this definition, however, fails to capture discrete relations on the input.

An example of the notation in (45.2.1), is an arrangement S of hyperplanes in $\mathbb{R}^d$. The combinatorial structure $D(S)$ is the incidence graph $G = (V, E)$ of the arrangement and V is the set of faces of the arrangement. The parameter $\mathbf{c}$ consists of the coefficients of the input hyperplanes. If $\mathbf{z}$ is the corresponding structural parameters then the input assignment is $I(\mathbf{z}) = \mathbf{c}$. The geometric data associates to each node v of the graph the Tarski formula $\lambda(v)$ involving $\mathbf{x}, \mathbf{z}$. When $\mathbf{c}$ is substituted for $\mathbf{z}$, then the formula $\lambda(v)$ defines a face $f_{\mathbf{c}}(v)$ (or $f(v)$ for short) of the arrangement. We use the convention that an edge $(u, v) \in E$ represents an "incidence" from $f(u)$ to $f(v)$, where the dimension of $f(u)$ is one more than that of $f(v)$. So $f(v)$ is contained in the closure of $f(u)$. Let aff(X) denote the affine span of a set $X \subseteq \mathbb{R}^d$. Then $(u, v) \in E$ implies aff$(f(v)) \subseteq$ aff$(f(u))$ and $f(u)$ lies on one of the two open halfspaces defined by aff$(f(u)) \setminus$ aff$(f(v))$. We let $\lambda(u, v)$ be the Tarski formula $\xi(\mathbf{x}, \mathbf{z})$ that defines the open halfspace in aff$(f(u))$ that contains $f(u)$. Again, let $f(u, v) = f_{\mathbf{c}}(u, v)$ denote this open halfspace. The consistency requirement is that (a) the set $\{f(v) : v \in V\}$ is a partition of $\mathbb{R}^d$, and (b) for each $u \in V$, the set $f(u)$ is nonempty with an irredundant representation of the form

$$f(u) = \bigcap \{f(u, v) \mid (u, v) \in E\}.$$

Although the above definition appears complicated, all its elements are necessary in order to capture the following additional concepts. We can suppress the input assignment I, so there are only structural variables $\mathbf{z}$ (which is implicit in λ and Φ) but no parameters $\mathbf{c}$. The triple

$$\widehat{D} = (G, \lambda, \Phi(\mathbf{z})) \tag{45.2.2}$$

becomes an ***abstract geometric structure***, and $D = (G, \lambda, \Phi(\mathbf{z}), I)$ is an ***instance*** of $\widehat{D}$. The structure D in (45.2.1) is ***consistent*** if the predicate $\Phi(\mathbf{c})$ holds. An abstract geometric structure $\widehat{D}$ is ***realizable*** if it has some consistent instance. Two geometric structures D, D' are ***structurally similar*** if they are instances of a common abstract geometric structure. We can also introduce metrics on structurally similar geometric structures: if $\mathbf{c}$ and $\mathbf{c}'$ are the parameters of D, D' then define $d(D, D')$ to be the Euclidean norm of $\mathbf{c} - \mathbf{c}'$.

The graph $G = (V, E)$ in (45.2.1) is an abstract graph where each $v \in V$ is a symbol that represents the semi-algebraic set $f(v) \subseteq \mathbb{R}^d$. The Tarski formula $\lambda(v)$ is an exact but symbolic representation of $f(v)$. For most applications of geometric algorithms, such a symbolic representation is alone insufficient. We need an approximate "embedding" of the underlying semi-algebraic sets in $\mathbb{R}^d$. Consider the problem of meshing curves and surfaces (see Boissonnat et al. [BCSM+06] for a survey). In meshing, the set $\{f(v) : v \in V\}$ is typically a simplicial complex (i.e., triangulation). For each $v \in V$, if $f(v)$ is a k-simplex $(k = 0, \ldots, d)$, then we want to compute a piecewise linear set $\widetilde{f}(v) \subseteq \mathbb{R}^d$ that is homeomorphic to a k-ball. Moreover, $(u, v) \in E$ iff $\widetilde{f}(v) \subseteq \partial \widetilde{f}(u)$ (∂ is the boundary operator). Then the set $\widetilde{V} = \{\widetilde{f}(v) : v \in V\}$ is a topological simplicial complex that captures all information in the symbolic graph $G = (V, E)$. The algorithms in [PV04, LY11, LYY12] (for $d \leq 3$) can even ensure that for all $v \in V$, we have $d_H(f(v), \widetilde{f}(v)) < \varepsilon$ (for any given ε) where d_H denotes the Hausdorff distance on Euclidean sets. In that case, $\widetilde{V}$ is an ε-***approximation*** of the simplicial complex $\{f(v) : v \in V\}$. This is the "explicit" geometric representation of D, which applications need.

45.3 FIXED-PRECISION APPROACHES

This section surveys the various approaches within the fixed-precision paradigm. Such approaches have strong motivation in the modern computing environment where fast floating point hardware has become a de facto standard in every computer. If we can make our geometric algorithms robust within machine arithmetic, we are assured of the fastest possible implementation. We may classify the approaches into several basic groups. We first illustrate our classification by considering the simple question: "What is the concept of a line in fixed-precision geometry?" Four basic answers to this question are illustrated in Figure 45.3.1 and in Table 45.3.1.

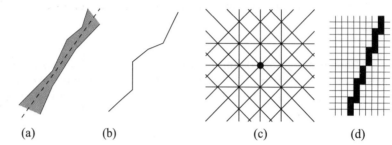

(a)   (b)   (c)   (d)

FIGURE 45.3.1

Four concepts of "finite-precision" lines.

WHAT IS A FINITE-PRECISION LINE?

We call the first approach ***interval geometry*** because it is the geometric analogue of interval arithmetic. Segal and Sequin [SS85] and others define a zone surrounding the line composed of all points within some ϵ distance from the actual line.

The second approach is called ***topologically consistent distortion***. Greene and Yao [GY86] distorted their lines into polylines, where the vertices of these polylines are constrained to be at grid points. Note that although the "fixed-precision representation" is preserved, the number of bits used to represent these polylines can have arbitrary complexity.

TABLE 45.3.1   Concepts of a finite-precision line.

	APPROACH	SUBSTITUTE FOR IDEAL LINE	SOURCE
(a)	Interval geometry	a line fattened into a tubular region	[SS85]
(b)	Topological distortion	a polyline	[GY86, Mil89]
(c)	Rounded geometry	a line whose equation has bounded coefficients	[Sug89]
(d)	Discretization	a suitable set of pixels	computer graphics

The third approach follows a tack of Sugihara [Sug89]. An ideal line is specified by a linear equation, $ax + by + c = 0$. Sugihara interprets a "fixed-precision line" to mean that the coefficients in this equation are integer and bounded: $|a|, |b| < K, |c| < K^2$ for some constant K. Call such lines *representable* (see Figure 45.3.1(c) for the case $K = 2$). There are $O(K^4)$ representable lines. An arbitrary line must be "rounded" to the closest (or some nearby) representable line in our algorithms. Hence we call this ***rounded geometry***.

The last approach is based on *discretization*: in traditional computer graphics and in the pattern recognition community, a "line" is just a suitable collection of pixels. This is natural in areas where pixel images are the central objects of study, but less applicable in computational geometry, where compact line representations are desired. This approach will not be considered further in this chapter.

INTERVAL GEOMETRY

In interval geometry, we thicken a geometric object into a zone containing the object. Thus a point may become a disk, and a line becomes a strip between two parallel lines: this is the simplest case and is treated by Segal and Sequin [SS85, Seg90]. They called these "toleranced objects," and in order to obtain correct predicates, they enforce *minimum feature separations*. To do this, features that are too close must be merged (or pushed apart).

Guibas, Salesin, and Stolfi [GSS89] treat essentially the same class of thick objects as Segal and Sequin, although their analysis is mostly confined to geometric data based on points. Instead of insisting on minimum feature separations, their predicates are allowed to return the DON'T KNOW truth value. Geometric predicates (called ϵ-predicates) for objects are systematically treated in this paper.

In general we can consider zones with nonconstant descriptive complexity, e.g., a planar zone with polygonal boundaries. As with interval arithmetic, a zone is generally a conservative estimate because the precise region of uncertainty may be too complicated to compute or to maintain. In applications where zones expand rapidly, there is danger of the zone becoming catastrophically large: Segal [Seg90] reports that a sequence of duplicate-rotate-union operations repeated eleven times to a cube eventually collapsed it to a single vertex.

TOPOLOGICALLY CONSISTENT DISTORTION

Sugihara and Iri [SI89b, SIII00] advocate an approach based on preserving topological consistency. These ideas have been applied to several problems, including geometric modeling [SI89a] and Voronoi diagrams for point sets [SI92]. In their approach, one first chooses some topological property (e.g., planarity of the underlying graph) and constructs geometric algorithms that preserve the chosen property in the following sense: the algorithm will always terminate with an output whose topological properties match the topological properties for the output corresponding to some input but not necessarily a nearby instance. It is not clear in this prescription how to choose appropriate topological properties. Greene and Yao [GY86] consider the problem of maintaining certain "topological properties" of an arrangement of finite-precision line segments. They introduce polylines as substitutes for ideal line segments in order to preserve certain properties of ideal arrangements (e.g., two line segments intersect in a *connected* subset). Each polyline is a distortion of an ideal segment σ when constrained to pass through the "hooks" of σ (i.e., grid points nearest to the intersections of σ with other line segments). But this may generate new intersections (derived hooks) and the cascaded effects must be carefully controlled. The grid model of Greene-Yao has been taken up by several other authors [Hob99, GM95, GGHT97]. Extension to higher dimensions is harder: there is a solution of Fortune [For99] in 3 dimensions. Further developments include the numerically stable algorithms in [FM91]. The interesting twist here is the use of pseudolines rather than polylines.

Hoffmann, Hopcroft, and Karasick [HHK88] address the problem of intersecting polygons in a consistent way. Phrased in terms of our notion of "geometric structure" (Section 45.2) their goal is to compute a combinatorial structure G that is *consistent* in the sense that G is the structure underlying a consistent geometric structure $D = (G, \lambda, \Phi, \mathbf{c}')$. Here, $\mathbf{c}'$ need not equal the actual input parameter vector $\mathbf{c}$. They show that the intersection of two polygons R_1, R_2 can be efficiently computed, i.e., a consistent G representing $R_1 \cap R_2$ can be computed. However, in their framework, $R_1 \cap (R_2 \cap R_3) \neq (R_1 \cap R_2) \cap R_3$. Hence they need to consider the triple intersection $R_1 \cap R_2 \cap R_3$. Unfortunately, this operation seems to require a nontrivial amount of geometric theorem proving ability.

This suggests that the problem of verifying consistency of combinatorial structures (the "reasoning paradigm" [HHK88]) is generally hard. Indeed, the NP-hard existential theory of reals can be reduced to such problems. In some sense, the ultimate approach to ensuring consistency is to design "parsimonious algorithms" in the sense of Fortune [For89]. This also amounts to theorem proving as it entails deducing the consequences of all previous decisions along a computation path. A similar approach has been proposed more recently by Sugihara [Sug11].

STABILITY

This is a metric form of topological distortion where we place a priori bounds on the amount of distortion. It is analogous to backward error analysis in numerical analysis. Framed as the problem of computing the graph G underlying some geometric structure D (as above, for [HHK88]), we could say, following Fortune [For89], that an algorithm is ϵ-*stable* if there is a consistent geometric structure $D = (G, \lambda, \Phi, \mathbf{c}')$ such that $\|\mathbf{c} - \mathbf{c}'\| < \epsilon$ where $\mathbf{c}$ is the input parameter vector. We say an algorithm has **strong** (resp. *linear*) stability if ϵ is a constant (resp., $O(n)$) where n is the input size. Fortune and Milenkovic [FM91] provide both linearly stable and strongly stable algorithms for line arrangements. Stable algorithms have been achieved for two other problems on planar point sets: maintaining a triangulation of a point set [For89], and Delaunay triangulations [For92, For95]. The latter problem can be solved stably using either an incremental or a diagonal-flipping algorithm that is $O(n^2)$ in the worst case. Jaromczk and Wasilkowski [JW94] presented stable algorithms for convex hulls. Stability is a stronger requirement than topological consistency, e.g., the topological algorithms in [SI92] have not been proved stable.

ROUNDED GEOMETRY

Sugihara [Sug89] shows that the above problem of "rounding a line" can be reduced to the classical problem of *simultaneous approximation by rationals*: given real numbers $a_1, \ldots, a_n$, find integers $p_1, \ldots, p_n$ and q such that $\max_{1 \leq i \leq n} |a_i q - p_i|$ is minimized. There are no efficient algorithms to solve this exactly, although lattice reduction techniques yield good approximations. The above approach of Greene and Yao can also be viewed as a geometric rounding problem. The "rounded lines" in the Greene-Yao sense are polylines with unbounded combinatorial complexity; but rounded lines in the Sugihara sense still have constant complexity. Milenkovic and Nackman [MN90] show that rounding a collection of disjoint simple polygons while preserving their combinatorial structure is NP-complete. In Section 45.5, rounded geometry is seen in a different light.

ARITHMETICAL APPROACHES

Certain approaches might be described as mainly based on arithmetic considerations (as opposed to geometric considerations). Ottmann, Thiemt, and Ullrich [OTU87] show that the use of an accurate scalar product operator leads to improved robustness in segment intersection algorithms; that is, the onset of qualitative errors is delayed. A case study of Dobkin and Silver [DS88] shows that permutation of operations combined with random rounding (up or down) can give accurate predictions of the total round-off error. By coupling this with a multiprecision arithmetic package that is invoked when the loss in significance is too severe, they are able to improve the robustness of their code. There is a large literature on computation under the interval arithmetic model (e.g., [Ull90]). It is related to what we call interval geometry above. There are also systems providing programming language support for interval analysis.

45.4 EXACT APPROACH

As the name suggests, this approach proposes to compute without any error. The initial interpretation is that every numerical quantity is computed exactly. While this has a natural meaning when all numerical quantities are rational, it is not obvious what this means for values such as $\sqrt{2}$ which cannot be exactly represented "explicitly." Informally, a number representation is explicit if it facilitates efficient comparison operations. In practice, this amounts to representing numbers by one or more integers in some positional notation (this covers the usual representation of rational numbers as well as floating point numbers). Although we could achieve numerical exactness in some modified sense, this turns out to be unnecessary. The solution to the nonrobustness only requires a weaker notion of exactness: it is enough to ensure "geometric exactness." In the GEOMETRIC = NUMERIC + COMBINATORIAL formulation, the exactness is not to be found in the numeric part, but in the combinatorial part, as this encodes the geometric relations. Hence this approach is called ***Exact Geometric Computation*** (EGC), and it entails the following:

Input is exact. We cannot speak of exact geometry unless this is true. This assumption can be an issue if the input is inherently approximate. Sometimes we can simply treat the approximate inputs as ***nominally*** exact, as in the case of an input set of points without any constraints. Otherwise, there are two options: (1) "clean up" the inexact input, by transforming it to data that is exact; or (2) formulate a related problem in which the inexact input can be treated as exact (e.g., inexact input points can be viewed as the *exact* centers of small balls). So the convex hull of a set of points becomes the convex hull of a set of balls. The cleaning-up process in (1) may be nontrivial as it may require perturbing the data to achieve some consistency property and lies outside our present scope. The transformation

(2) typically introduces a computationally harder problem. Not much research is currently available for such transformed problems. In any case, (1) and (2) still end up with exact inputs for a well-defined computational problem.

Numerical quantities may be implicitly represented.   This is necessary if we want to represent irrational values exactly. In practice, we will still need explicit numbers for various purposes (e.g., comparison, output, display, etc.). So a corollary is that numerical approximations will be important, a remark that was not obvious in the early days of EGC.

All branching decisions in a computation are errorless.   At the heart of EGC is the idea that all "critical" phenomena in geometric computations are determined by the particular sequence branches taken in a *computation tree*. The key observation is that the sequence of branching decisions completely decides the combinatorial nature of the output. Hence if we make only errorless branches, the combinatorial part of a geometric structure D (see Section 45.2) will be correctly computed. To ensure this, we only need to evaluate *test values* to one bit of relative precision, i.e., enough to determine the sign correctly.

For problems (such as convex hulls) requiring only rational numbers, exact computation is possible. In other applications rational arithmetic is not enough. The most general setting in which exact computation is known to be possible is the framework of *algebraic problems* [Yap97].

GLOSSARY

Computation tree: A geometric algorithm in the algebraic framework can be viewed as an infinite sequence $T_1, T_2, T_3, \ldots$ of computation trees. Each T_n is restricted to inputs of size n, and is a finite tree with two kinds of nodes: (a) nonbranching nodes, (b) branching nodes. Assume the input to T_n is a sequence of n real parameters $c_1, \ldots, c_n$ (this is I in (45.2.1)). A nonbranching node at depth i computes a value v_i, say $v_i \leftarrow f_i(v_1, \ldots, v_{i-1}, c_1, \ldots, c_n)$. A branching node tests a previous computed value v_i and makes a 3-way branch depending on the sign of v_i. In case v_i is a complex value, we simply take the sign of the real part of v_i. Call any v_i that is used solely in a branching node a ***test value***. The branch corresponding to a zero test value is the ***degenerate branch***.

Exact Geometric Computation (EGC): Preferred name for the general approach of "exact computation," as it accurately identifies the goal of determining geometric relations exactly. The exactness of the computed numbers is either unnecessary, or should be avoided if possible.

Composite Precision Bound: This is specified by a pair $[r, a]$ where $r, a \in \mathbb{R} \cup \{\infty\}$. For any $z \in \mathbb{C}$, let $z[r, a]$ denote the set of all $\widetilde{z} \in \mathbb{C}$ such that $|z - \widetilde{z}| \leq \max\{2^{-a}, |z| 2^{-r}\}$. When $r = \infty$, then $z[\infty, a]$ comprises all the numbers $\widetilde{z}$ that approximate z with an absolute error of 2^{-a}; we say this approximation $\widetilde{z}$ has a ***absolute bits***. Similarly, $z[r, \infty]$ comprises all numbers $\widetilde{z}$ that approximate z with a relative error of 2^{-r}; we say this approximation $\widetilde{z}$ has r ***relative bits***.

Constant Expressions: Let Ω be a set of complex algebraic operators; each operator $\omega \in \Omega$ is a partial function $\omega : \mathbb{C}^{a(\omega)} \to \mathbb{C}$ where $a(\omega) \in \mathbb{N}$ is the arity of ω. If $a(\omega) = 0$, then ω is identified with a complex number. Let $\mathcal{E}(\Omega)$ be the set of expressions over Ω where an expression E is a rooted DAG (directed acyclic graph) and each node with outdegree $n \in \mathbb{N}$ is labeled with an operator of Ω of arity n. There is a natural **evaluation function** val : $\mathcal{E}(\Omega) \to \mathbb{R}$. If Ω has partial functions, then val() is also partial. If val(E) is undefined, we write val(E) $=\uparrow$ and say E is **invalid**. When $\Omega = \Omega_2 = \{+, -, \times, \div, \sqrt{\;}\} \cup \mathbb{Z}$ we get the important class of **constructible expressions**, so called because their values are precisely the constructible reals.

Constant Zero Problem, ZERO(Ω):   Given $E \in \mathcal{E}(\Omega)$, decide if val($E$) $=\uparrow$; if not, decide if val(E) $= 0$.

Guaranteed Precision Evaluation Problem, GVAL(Ω):   Given $E \in \mathcal{E}(\Omega)$ and $a, r \in \mathbb{Z} \cup \{\infty\}$, $(a, r) \neq (\infty, \infty)$, compute some approximate value in val(E)[r, a].

Schanuel's Conjecture: If $z_1, \ldots, z_n \in \mathbb{C}$ are linearly independent over $\mathbb{Q}$, then the set $\{z_1, \ldots, z_n, e^{z_1}, \ldots, e^{z_n}\}$ contains a subset $B = \{b_1, \ldots, b_n\}$ that is algebraically independent, i.e., there is no polynomial $P(X_1, \ldots, X_n) \in \mathbb{Q}[X_1, \ldots, X_n]$ such that $P(b_1, \ldots, b_n) = 0$. This conjecture generalizes several deep results in transcendental number theory, and implies many other conjectures.

NAIVE APPROACH

For lack of a better term, we call the approach to exact computation in which every numerical quantity is computed exactly (explicitly if possible) the *naive approach.* Thus an exact algorithm that relies solely on the use of a big number package is probably naive. This approach, even for rational problems, faces the "bugbear of exact computation," namely, high numerical precision. Using an off-the-shelf big number package does not appear to be a practical option [FW93a, KLN91, Yu92]. There is evidence (surveyed in [YD95]) that just improving current big number packages alone is unlikely to gain a factor of more than 10.

BIG EXPRESSION PACKAGES

The most common examples of expressions are determinants and the distance $\sqrt{\sum_{i=1}^{n}(p_i - q_i)^2}$ between two points p, q. A big expression package allows a user to construct and evaluate expressions with big number values. They represent the next logical step after big number packages, and are motivated by the observation that the numerical part of a geometric computation is invariably reduced to repeated evaluations of a few variable[1] expressions (each time with different constants substituted for the variables). When these expressions are test values, then it is sufficient to compute them to one bit of relative precision. Some implementation efforts are shown in Table 45.4.1.

One of LN's goals [FW96] is to remove all overhead associated with function calls or dynamic allocation of space for numbers with unknown sizes. It incorporates an effective floating-point filter based on static error analysis. The experience in [CM93] suggests that LN's approach is too aggressive as it leads to code bloat. The **LEA** system philosophy [BJMM93] is to delay evaluating an expression until forced to, and to maintain intervals of uncertainty for values. Upon complete evaluation, the expression is discarded. It uses root bounds to achieve exactness

---

[1] These expressions involve variables, unlike the constant expressions in $\mathcal{E}(\Omega)$.

TABLE 45.4.1   Expression packages.

SYSTEM	DESCRIPTION	REFERENCES
LN	Little Numbers	[FW96]
LEA	Lazy ExAct Numbers	[BJMM93]
Real/Expr	Precision-driven exact expressions	[YD95]
LEDA Real	Exact numbers of Library of Efficient Data structures and Algorithms	[BFMS99, BKM+95]
Core Library	Package with Numerical Accuracy API and C++ interface	[KLPY99, YYD+10]

and floating point filters for speed. The Real/Expr Package [YD95] was the first package to achieve guaranteed precision for a general class of nonrational expressions. It introduces the "precision-driven mechanism" whereby a user-specified precision at the root of the expression is transformed and downward-propagated toward the leaves, while approximate values generated at the leaves are evaluated and error bounds up-propagated up to the root. This up-down process may be iterated. See [LPY04] for a general description of this evaluation mechanism, and [MS15] for optimizations issues.   LEDA Real [BFMS99, BKM+95] is a number type with a similar mechanism. It is part of a much more ambitious system of data structures for combinatorial and geometric computing (see Chapter 68). The semantics of Real/Expr expression assignment is akin to constraint propagation in the constraint programming paradigm.   The Core Library (CORE) is derived from Real/Expr with the goal of making the system as easy to use as possible. The two pillars of this transformation are the adoption of conventional assignment semantics and the introduction of a simple *Numerical Accuracy API* [Yap98].

The CGAL Library (Chapter 68) is a major library of geometric algorithms which are designed according to the EGC principles. While it has some native number types supporting rational expressions, the current distribution relies on LEDA Real or CORE for more general algebraic expressions. Shewchuk [She96] implements an arithmetic package that uses adaptive-precision floating-point representations. While not a big expression package, it has been used to implement polynomial predicates and shown to be extremely efficient.

THEORY

The class of algebraic computational problems encompasses most problems in contemporary computational geometry. Such problems can be solved exactly in singly exponential space [Yap97]. This general result is based on a solution to the decision problem for Tarski's language, on the associated cell decomposition problems, as well as cell adjacency computation (Chapter 37). However, general EGC libraries such as the Core Library and LEDA Real depend directly on the algorithms for the guaranteed precision evaluation problem GVAL(Ω) (see Glossary), where Ω is the set of operators in the computation model. The possibility of such algorithms can be reduced to the recursiveness of a constellation of problems that might be called the *Fundamental Problems of EGC*. The first is the Constant Zero Problem ZERO(Ω). But there are two closely related problems. In the *Constant Va-*

lidity Problem VALID(Ω), we are to decide if a given $E \in \mathcal{E}(\Omega)$ is valid, i.e., val(E) $\neq \uparrow$. The *Constant Sign Problem* SIGN(Ω) is to compute $\text{sign}(E)$ for any given $E \in \mathcal{E}(\Omega)$, where $\text{sign}(E) \in \{\uparrow, -1, 0, +1\}$. In case val($E$) is complex, define $\text{sign}(E)$ to be the sign of the real part of val(E).

TABLE 45.4.2   Expression hierarchy.

OPERATORS	NUMBER CLASS	EXTENSIONS
$\Omega_0 = \{+, -, \times\} \cup \mathbb{Z}$	Integers	
$\Omega_1 = \Omega_0 \cup \{\div\}$	Rational Numbers	$\Omega_1^+ = \Omega_1 \cup \mathbb{Q}$
$\Omega_2 = \Omega_1 \cup \{\sqrt{\cdot}\}$	Constructible Numbers	$\Omega_2^+ = \Omega_2 \cup \{\sqrt[k]{\cdot} : k \geq 3\}$
$\Omega_3 = \Omega_2 \cup \{\text{RootOf}(P(X), I)\}$	Algebraic Numbers	Use of $\diamond(E_1, \ldots, E_d, i)$, [BFM$^+$09]
$\Omega_4 = \Omega_3 \cup \{\exp(\cdot), \ln(\cdot)\}$	Elementary Numbers (cf. [Cho99])	

There is a natural hierarchy of the expression classes, each corresponding to a class of complex numbers as shown in Table 45.4.2. In Ω_3, $P(X)$ is any polynomial with integer coefficients and I is some means of identifying a unique root of $P(X)$: I may be a complex interval bounding a unique root of $P(X)$, or an integer i to indicate the ith largest real root of $P(X)$. The operator RootOf(P, I) can be generalized to allow expressions as coefficients of $P(X)$ as in Burnikel et al. [BFM$^+$09], or by introducing systems of polynomial equations as in Richardson [Ric97]. The problem of isolating the roots of such a polynomial in the univariate case has been recently addressed in [BSS$^+$16, BSSY15]. Although Ω_4 can be treated as a set of real operators, it is more natural to treat Ω_4 (and sometimes Ω_3) as complex operators. Thus the elementary functions $\sin x, \cos x, \arctan x$, etc., are available as expressions in Ω_4.

It is clear that ZERO(Ω) and VALID(Ω) are reducible to SIGN(Ω). For Ω_4, all three problems are recursively equivalent. The fundamental problems related to Ω_i are decidable for $i \leq 3$. It is a major open question whether the fundamental problems for Ω_4 are decidable. These questions have been studied by Richardson and others [Ric97, Cho99, MW96]. The most general positive result is that SIGN(Ω_3) is decidable. An intriguing conditional result of Richardson [Ric07] is that ZERO(Ω_4) is decidable, conditional on the truth of Schanuel's conjecture in Transcendental Number Theory. The most general unconditional result known about ZERO(Ω_4) is (essentially) Baker's theory of linear form in logarithms [Bak75].

CONSTRUCTIVE ROOT BOUNDS

In practice, algorithms for the guaranteed precision problem GVAL(Ω_3) can exploit the fact that algebraic numbers have computable root bounds. A *root bound* for Ω is a total function $\beta : \mathcal{E}(\Omega) \to \mathbb{R}_{\geq 0}$ such that for all $E \in \mathcal{E}(\Omega)$, if E is valid and val(E) $\neq 0$ then $|\text{val}(E)| \geq \beta(E)$. More precisely, β is called an *exclusion* root bound; it is an *inclusion root bound* when the inequality becomes "$|\text{val}(E)| \leq \beta(E)$." We use the (exclusion) root bound β to solve ZERO(Ω) as follows: to test if an expression E evaluates to zero, we compute an approximation α to val(E) such that $|\alpha - \text{val}(E)| < \beta(E)/2$. While computing α, we can recursively verify the validity of E. If E is valid, we compare α with $\beta/2$. It is easy to conclude that $\text{val}(E) = 0$ if $|\alpha| \leq \beta/2$. Otherwise $|\alpha| > \beta/2$, and the sign of $\text{val}(E)$ is that of α. An important remark is that the root bound β determines the worst-case complexity. This is unavoidable if $\text{val}(E) = 0$. But if $\text{val}(E) \neq 0$, the worst case may be avoided by iteratively computing α_i with increasing absolute precision ε_i. If for any $i \geq 1$, $|\alpha_i| > \varepsilon_i$, we stop and conclude $\text{sign}(\text{val}(E)) = \text{sign}(\alpha_i) \neq 0$.

There is an extensive classical mathematical literature on root bounds, but they are usually not suitable for computation. Recently, new root bounds have been introduced that explicitly depend on the structure of expressions $E \in \mathcal{E}(E)$. In [LY01], such bounds are called **constructive** in the following sense: (i) There are easy-to-compute recursive rules for maintaining a set of numerical parameters $u_1(E), \ldots, u_m(E)$ based on the structure of E, and (ii) $\beta(E)$ is given by an explicit formula in terms of these parameters. The first constructive bounds in EGC were the degree-length and degree-height bounds of Yap and Dubé [YD95, Yap00] in their implementation of `Real/Expr`. The (Mahler) Measure Bound was introduced even earlier by Mignotte [Mig82, BFMS00] for the problem of "identifying algebraic numbers." A major improvement was achieved with the introduction of the BFMS Bound [BFMS00]. Li-Yap [LY01] introduced another bound aimed at improving the BFMS Bound in the presence of division. Comparison of these bounds is not easy: but let us say a bound β **dominates** another bound β' if for every $E \in \mathcal{E}(\Omega_2)$, $\beta(E) \leq \beta'(E)$. Burnikel et al. [BFM+09] generalized the BFMS Bound to the BFMSS Bound. Yap noted that if we incorporate a symmetrizing trick for the $\sqrt{x/y}$ transformation, then BFMSS will dominate BFMS. Among current constructive root bounds, three are not dominated by other bounds: BFMSS, Measure, and Li-Yap Bounds. In general, BFMSS seems to be the best. Scheinerman [Sch00] provides an interesting bound based on eigenvalues. A factoring technique of Pion and Yap [PY06] can be combined with the above methods (e.g., BFMSS) to yield sharper bounds; it exploits the presence of k-ary input numbers (such as binary or decimal numbers) which appear in realistic inputs. In general, we need multivariate root bounds such as **Canny's bound** [Can88], the **Brownawell-Yap bound** [BY09], and the **DMM bound** [EMT10].

FILTERS

An extremely effective technique for speeding up predicate evaluation is based on the filter concept. Since evaluating the predicate amounts to determining the sign of an expression E, we can first use machine arithmetic to quickly compute an approximate value α of E. For a small overhead, we can simultaneously determine an error bound ε where $|\text{val}(E) - \alpha| \leq \varepsilon$. If $|\alpha| > \varepsilon$, then the sign of α is the correct one and we are done. Otherwise, we evaluate the sign of E again, this time using a sure-fire if slow evaluation method. The algorithm used in the first evaluation is called a (floating-point) **filter**. The expected cost of the two-stage evaluation is small if the filter is efficient with a high probability of success. This idea was first used by Fortune and van Wyk [FW96]. Floating-point filters can be classified along the static-to-dynamic dimension: **static filters** compute the bound ε solely from information that is known at compile time while **dynamic filters** depend on information available at run time. There is an **efficiency-efficacy tradeoff**: static filters (e.g., FvW Filter [FW96]) are more efficient, but dynamic filters (e.g., BFS Filter [BFS98]) are more accurate (efficacious). Interval arithmetic has been shown to be an effective way to implement dynamic filters

[BBP01]. Automatic tools for generating filter code are treated in [FW93b, Fun97]. Filters can be elaborated in several ways. First, we can use a cascade of filters [BFS98]. The "steps" of an algorithm which are being filtered can be defined at different levels of granularity. One extreme is to consider an entire algorithm as one step [MNS⁺96, KW98]. A general formulation "structural filtering" is proposed in [FMN05]. Probabilistic analysis [DP99] shows the efficacy of arithmetic filters. The filtering of determinants is treated in several papers [Cla92, BBP01, PY01, BY00].

Filtering is related to program checking [BK95, BLR93]. View a computational problem P as an input-output relation, $P \subseteq I \times O$ where I, O is the input and output spaces, respectively. Let A be a (standard) **algorithm** for P which, viewed as a total function $A : I \to O \cup \{\uparrow\}$, has the property that for all $i \in I$, we have $A(i) \neq \uparrow$ iff $(i, A(i)) \in P$. Let $H : I \to O \cup \{\uparrow\}$ be another algorithm with no restrictions; call H a **heuristic algorithm** for P. Let $F : I \times O \to \{true, false\}$. Then F is a **checker** for P if F computes the characteristic function for P, i.e., $F(i, o) = true$ iff $(i, o) \in P$. Note that F is a checker for the problem P, and not for any purported program for P. Unlike program checking literature, we do not require any special properties of P such as self-reducibility. We call F a **filter** for P if $F(i, o) = true$ implies $(i, o) \in P$. So filters are less restricted than checkers. Another filter F' is **more efficacious** than F if $F(i, o) = true$ implies $F'(i, o) = true$. A **filtered program** for P is therefore a triple (H, F, A) where H is a heuristic algorithm, A a standard algorithm and F a filter. To run this program on input i, we first compute $H(i)$ and check if $F(i, H(i))$ is true. If so, we output $H(i)$; otherwise compute and output $A(i)$. Filtered programs can be extremely effective when H, F are both efficient and efficacious. Usually H is easy—it is just a machine arithmetic implementation of an exact algorithm. The filter F can be more subtle, but it is still more readily constructed than any checker. In illustration, let P be the problem *SDET* of computing the sign of determinants. The only checker we know here is trivial, amounting to computing the determinant itself. On the other hand, effective filters for *SDET* are known [BBP01, PY01].

PRECISION COMPLEXITY

An important goal of EGC is to control the cost of high-precision computation. For instance, consider the sum of square roots problem: given $2k$ numbers $a_1, \ldots, a_k$ and $b_1, \ldots, b_k$ such that $0 < a_i, b_i \leq N$, derive a bound on the precision required to compute a 1-relative bit approximation to $(\sum_{i=1}^{k} \sqrt{a_i} - \sum_{i=1}^{k} \sqrt{b_i})$. The best known bounds on the precision are of the form $O(2^k \log N)$; for instance, we can compute the constructive root bound for the expression denoting the difference and take its logarithm to obtain such a bound. In practice, it has been observed that $O(k \log N)$ precision is sufficient. The current best upper bound implies that the problem is in PSPACE. We next describe two approaches to address the issue of precision complexity based on modifying the algorithmic specification.

In predicate evaluation, there is an in-built precision of 1-relative bit (this precision guarantees the correct sign in the predicate evaluation). But in construction steps, any precision guarantees must be explicitly requested by the user. For optimization problems, a standard method to specify precision is to incorporate an extra input parameter $\epsilon > 0$. Assume the problem is to produce an output x to minimizes the function $\mu(x)$. An **ϵ-approximation algorithm** will output a solution x such that $\mu(x) \leq (1 + \varepsilon)\mu(x^*)$ for some optimum x^*. An example is the **Euclidean Shortest-path Problem in 3-space** (3ESP). Since this problem is NP-hard (Section 31.5), we seek an ϵ-approximation algorithm. A simple way to implement an ϵ-approximation algorithm is to directly implement any *exact* algorithm in which the underlying arithmetic has guaranteed precision evaluation (using, e.g., `Core Library`). However, the bit complexity of such an algorithm may not be obvious. The more conventional approach is to explicitly build the necessary approximation scheme directly into the algorithm. The first such scheme from Papadimitriou [Pap85] is polynomial time in n and $1/\varepsilon$. Choi et al. [CSY97] give an improved scheme, and perform a rare bit-complexity analysis.

Another way to control precision is to consider output complexity. In geometric problems, the input and output **sizes** are measured in two independent ways: combinatorial size and bit sizes. Let the input combinatorial and input bit sizes be n and L, respectively. By an L-bit input, we mean each of the numerical parameters in the description of the geometric object (see Section 45.2) is an L-bit number. Now an extremely fruitful concept in algorithmic design is this: an algorithm is said to be **output-sensitive** if the complexity of the algorithm can be made a function of the output size as well as of the input size parameters. In the usual view of output-sensitivity, only the output combinatorial size is exploited. Choi et al. [SCY00] introduced the concept of **precision-sensitivity** to remedy this gap. They presented the first precision-sensitive algorithm for 3ESP, and gave some experimental results. Using the framework of **pseudo-approximation algorithms**, Asano et al. [AKY04] gave new and more efficient precision-sensitive algorithms for 3ESP, as well as for an optimal d_1-motion for a rod.

GEOMETRIC ROUNDING

We saw rounded geometry as one of the fixed-precision approaches (Section 45.3) to robustness. But geometric rounding is also important in EGC, with a difference. The EGC problem is to "round" a geometric structure (Section 45.2) D to a geometric structure D' with lower precision. In fixed-precision computation, one is typically asked to construct D' from some input S that *implicitly* defines D. In EGC, D is explicitly given (e.g., D may be computed from S by an EGC algorithm). The EGC view should be more tractable since we have separated the two tasks: (a) computing D and (b) rounding D. We are only concerned with (b), the **pure rounding problem**. For instance, if S is a set of lines that are specified by linear equations with L-bit coefficients, then the arrangement $D(S)$ of S would have vertices with $2L + O(1)$-bit coordinates. We would like to round the arrangement, say, back to L bits. Such a situation, where the output bit precision is larger than the input bit precision, is typical. If we pipeline several of these computations in a sequence, the final result could have a very high bit precision unless we perform rounding.

If D rounds to D', we could call D' a **simplification** of D. This viewpoint connects to a larger literature on simplification of geometry (e.g., simplifying geometric models in computer graphics and visualization (Chapter 52). Two distinct goals in simplification are **combinatorial** versus **precision simplification**. For example, a problem that has been studied in a variety of contexts (e.g., Douglas-Peucker algorithm in computational cartography) is that of simplifying a polygonal line P. We can use **decimation** to reduce the combinatorial complexity (i.e., number of vertices $\#(P)$), for example, by omitting every other vertex in P. Or we can use ***clustering*** to reduce the bit-complexity of P to L-bits, e.g., we collapse all vertices that lie within the same grid cell, assuming grid points are L-bit numbers. Let $d(P, P')$ be the Hausdorff distance between P and another polyline P'; other similar measures of distance may be used. In any simplification P' of P, we want to keep $d(P, P')$ small. In [BCD$^+$02], two optimization problems are studied: in the ***Min-# Problem***, given P and ε, find P' to minimize $\#(P)$, subject to $d(P, P') \leq \varepsilon$. In the ***Min-$\varepsilon$ Problem***, the roles of $\#(P)$ and $d(P, P')$ are reversed. For EGC applications, optimality can often be relaxed to simple feasibility. Path simplification can be generalized to the simplification of any cell complexes.

BEYOND ALGEBRAIC

Non-algebraic computation over Ω_4 is important in practice. This includes the use of elementary functions such as $\exp x, \ln x, \sin x$, etc., which are found in standard libraries (`math.h` in C/C++). Elementary functions can be implemented via their representation as ***hypergeometric functions***, an approach taken by Du et al. [DEMY02]. They described solutions for fundamental issues such as automatic error analysis, hypergeometric parameter processing and argument reduction. If f is a hypergeometric function and x is an explicit number, one can compute $f(x)$ to any desired absolute accuracy. But in the absence of root bounds for Ω_4, we cannot solve the guaranteed precision problem GVAL(Ω_4). In `Core Library` 2.1, transcendental functions are provided, but they are computed up to some user-specified "Escape Bound." Numbers that are smaller than this bound are declared zero, and a record is made. Thus our computation is correct subject to these records being actual zeros. Another intriguing approach is to invoke some strong version of the uniformity conjecture [RES06], which provides a bound implemented in an EGC library like [YYD$^+$10], or to use a decision procedure conditioned on the truth of Schanuel's conjecture [Ric07]. If our library ever led to an error, we would have produced a counterexample to the underlying conjectures. Thus, we are continually testing some highly nontrivial conjectures when we use the transcendental parts of the library.

A rare example of a simple geometric problem that is provably transcendental and yet decidable was shown in [CCK$^+$06]. This is the problem of shortest path between two points amidst a set of circular obstacles. Although a direct argument can be used to show the decidability of this problem, to get a complexity bound, it was necessary to invoke Baker's theory of linear form in logarithms [Bak75] to derive a single-exponential time bound.

The need for transcendental functions may be only apparent: e.g., path planning in Euclidean space only appears to need trigonometric functions, but they can be replaced by algebraic relations. Moreover, we can get arbitrarily good approximations by using *rational rigid transformations* where the sines or cosines are rational. Solutions in 2 and 3 dimensions are given by Canny et al. [CDR92] and Milenkovic and Milenkovic [MM93], respectively.

APPLICATIONS

We now consider issues in implementing specific algorithms under the EGC paradigm. The rapid growth in the number of such algorithms means the following list is quite partial. We attempt to illustrate the range of activities in several groups: **(i)** The early EGC algorithms are easily reduced to integer arithmetic and polynomial predicates, such convex hulls or Delaunay triangulations. The goal was to demonstrate that such algorithms are implementable and relatively efficient (e.g., [FW96]). To treat irrational predicates, the careful analysis of root bounds were needed to ensure efficiency. Thus, Burnikel, Mehlhorn, and Schirra [BMS94, Bur96] gave sharp bounds in the case of Voronoi diagrams for line segments. Similarly, Dubé and Yap [DY93] analyzed the root bounds in Fortune's sweepline algorithm, and first identified the usefulness of floating point approximations in EGC. Another approach is to introduce algorithms that use new predicates with low algebraic degrees. This line of work was initiated by Liotta, Preparata, and Tamassia [LPT98, BS00]. **(ii)** Polyhedral modeling is a natural domain for EGC techniques. Two efforts are [CM93, For97]. The most general viewpoint here uses Nef polyhedra [See01] in which open, closed or half-open polyhedral sets are represented. This is a radical departure from the traditional solid modeling based on ***regularized sets*** and the associated ***regularized operators***; see Chapter 57. The regularization of a set $S \subseteq \mathbb{R}^d$ is obtained as the closure of the interior of S; regularized sets do not allow lower dimensional features, e.g., a line sticking out of a solid is not permitted. Treatment of Nef polyhedra was previously impossible outside the EGC framework. **(iii)** An interesting domain is optimization problems such as linear and quadratic programming [Gae99, GS00] and the smallest enclosing cylinder problem [SSTY00]. In linear programming, there is a tradition of using benchmark problems for evaluating algorithms and their implementations. But what is lacking in the benchmarks is ***reference solutions*** with guaranteed accuracy to (say) 16 digits. One application of EGC algorithms is to produce such solutions. **(iv)** An area of major challenge is computation of algebraic curves and surfaces. Algorithms for low degree curves and surfaces can be efficiently solved today e.g., [BEH$^+$02, GHS01, Wei02, EKSW06]. Arbitrary precision libraries for algebraic curves have been implemented (see Krishnan et al. [KFC$^+$01]) but without zero bounds, there are no topology guarantees. More recently, there has been a lot of progress in determining the topology of curves, which may be given implicitly, e.g., as the Voronoi diagram of ellipses [ETT06], or are given explicitly and are of arbitrary degree [EKW07, BEKS13]. The more general problems of computing the topology of arrangements of curves [BEKS11], and the topology of surfaces [BKS08, Ber14] are also being addressed under the EGC paradigm. These algorithms have been implemented under the EXACUS project [BEH$^+$05]. **(v)** The development of general geometric libraries such as CGAL [HHK$^+$07] or LEDA [MN95] exposes a range of issues peculiar to EGC (cf. Chapter 68). For instance, in EGC we want a framework where various number kernels and filters can be used for a single algorithm.

45.5 TREATMENT OF DEGENERACIES

Suppose the input to an algorithm is a set of planar points. Depending on the context, any of the following scenarios might be considered "degenerate": two covertical points, three collinear points, four cocircular points. Intuitively, these are degenerate because arbitrarily small perturbations can result in qualitatively different geometric structures. Degeneracy is basically a discontinuity [Yap90b, Sei98]. Sedgewick [Sed02] calls degeneracies the "bugbear of geometric algorithms." Degeneracy is a major cause of nonrobustness for two reasons. First, it presents severe difficulties for approximate arithmetic. Second, even under the EGC paradigm, implementers are faced with a large number of special degenerate cases that must be treated (this number grows exponentially in the dimension of the underlying space). For instance, the Voronoi diagram of the arrangements of three lines in $\mathbb{R}^3$ is characterized by five cases [EGL$^+$09]. In general, the types of Voronoi cells grow exponentially in the dimension [YSL12]. Thus there is a need to develop general techniques for handling degeneracies.

GLOSSARY

Inherent and induced degeneracy: This is illustrated by the planar convex hull problem: an input set S with three collinear points p, q, r is inherently degenerate if S lies entirely in one halfplane determined by the line through p, q, r. If p, q, r are collinear but S does not lie on one side of the line through p, q, r, then we may have an induced degeneracy for a divide-and-conquer algorithm. This happens when the algorithm solves a subproblem $S' \subseteq S$ containing p, q, r with all the remaining points on one side. Induced degeneracy is algorithm-dependent. In this chapter, we simply say "degeneracy" for induced degeneracy. More precisely, an input is **degenerate** if it leads to a path containing a vanishing test value in the computation tree [Yap90b]. A nondegenerate input is also said to be **generic**.

Generic versus General algorithm: A generic algorithm is one that is only guaranteed to be correct on generic inputs. A general algorithm is one that works correctly for all (legal) inputs. Beware that "general" and "generic" are used synonymously in the literature (e.g., "generic inputs" often means inputs in "general position").

THE BASIC ISSUES

1. One basic goal of this field is to provide a *systematic transformation* of a generic algorithm A into a general algorithm A'. Since generic algorithms are widespread in the literature, the availability of general tools for this $A \mapsto A'$ transformation is useful for implementing robust algorithms.

2. Underlying any transformations $A \mapsto A'$ is some kind of *perturbation* of the inputs. There are two kinds of perturbations: symbolic (a.k.a. infinitesimal) or numeric. Informally, perturbation has the connotation of being random. But there are applications where we want to "control" these perturbations to achieve some properties: e.g., for a convex polytope A, we may like this to be an "outward perturbation" so that A' contains A. However, the term *controlled perturbation* has now taken a rather specific meaning in the framework introduced by Halperin et al. [HS98, Raa99]. The goal of controlled perturbation is to guarantee that a random δ-perturbation achieves a *sufficiently nondegenerate state* with high probability. Sufficient nondegeneracy here means that the underlying predicates can be computed with (say) IEEE floating point arithmetic, with a certification of nondegeneracy.

3. There is a *postprocessing issue*: although A' is "correct" in some technical sense, it may not necessarily produce the same outputs as an ideal algorithm A^*. This issue arises in symbolic rather than numeric perturbation. For example, suppose A computes the Voronoi diagram of a set of points in the plane. Four cocircular points are a degeneracy and are not treated by A. The transformed A' can handle four cocircular points but it may output two Voronoi vertices that have identical coordinates and are connected by a Voronoi edge of length 0. This may arise if we use infinitesimal perturbations. The postprocessing problem amounts to cleaning up the output of A' (removing the length-0 edges in this example) so that it conforms to the ideal output of A^*.

CONVERTING GENERIC TO GENERAL ALGORITHMS

There are three main methods for converting a generic algorithm to a general one:

Symbolic perturbation schemes (Blackbox sign evaluation)   We postulate a *sign blackbox* that takes as input a function $f(\mathbf{x}) = f(x_1, \ldots, x_n)$ and parameters $\mathbf{a} = (a_1, \ldots, a_n) \in \mathbb{R}^n$, and outputs a nonzero sign (either $+$ or $-$). In case $f(\mathbf{a}) \neq 0$, this sign is guaranteed to be the sign of $f(\mathbf{a})$, but the interesting fact is that we get a nonzero sign even if $f(\mathbf{a}) = 0$. We can formulate a consistency property for the blackbox, both in an algebraic setting [Yap90b] and in a geometric setting [Yap90a]. The transformation $A \mapsto A'$ amounts to replacing all evaluations of test values by calls to this blackbox. In [Yap90b], a family of *admissible schemes* for blackboxes is given in case the functions $f(\mathbf{x})$ are polynomials. This method of Simulation of Simplicity (SoS) is a special case of this scheme.

Numerical and controlled perturbation.   In contrast to symbolic perturbation, we can make a random numerical perturbation to the input. Intuitively, we expect this results in nondegeneracy. But controlled perturbation [HS98, Raa99] in the sense of Halperin assumes other ingredients: one idea is that the perturbation must be "sufficiently nondegenerate" so that it could be "efficiently certified." To formalize this, consider this problem: *given any $\varepsilon > 0$ and input z, compute $\delta = \delta(\varepsilon) > 0$ such that a random δ-perturbation of z has probability $> 1/2$ of being ε-nondegenerate.*   The set of δ-perturbations of z is $U_\delta(z) := \{z' : \|z' - z\| \leq \delta\}$, where $\|\cdot\|$ is any norm. Usually the infinity-norm is chosen (so $U_\delta(z)$ is a box). Assume that z is degenerate if $f(z) = 0$, where f is a given real function. Note that it is not enough to be "simply nondegenerate" (the probability of this is 1, trivially). We define ε-nondegenerate to mean $|f(u)| > \varepsilon$. How to choose this ε? It is dictated by a pragmatic assumption: for efficiency, assume we want to use the IEEE floating point arithmetic. Let $\widetilde{f}(z)$ denote the result of computing $f(z)$ in IEEE floating point. There are techniques (see below and [MOS11, appendix]) to compute a bound $B(f, M)$ such that for all $\|z\| \leq M$, if $|\widetilde{f}(z)| > B(f, M)$ then $|f(z)| > 0$, i.e., z is nondegenerate. Moreover, $B(f, M)$ can be computed in floating point, taking no more than big-Oh of the time to evaluate $f(z)$. Therefore, if we choose ε to be $B(f, M)$, we can verify that z is nondegenerate by checking that $|\widetilde{f}(z)| > B(f, M)$. If this check succeeds, we say z is *certifiably nondegenerate* and stop. If it fails, we repeat the process with another random δ-perturbation z'; but we are assured that the expected number of such repeats is at most one. In actual algorithms, we do not seek a monolithic perturbation of z, but view $z = (z_1, z_2, \ldots, z_n)$ to be a sequence of points where each z_i is an m-vector. We perturb the z_i's sequentially (in the order $i = 1, \ldots, n$). Inductively, suppose $(z'_1, \ldots, z'_{i-1})$ is certifiably nondegenerate. The perturbed z'_i must be chosen to extend this inductive hypothesis.

Perturbation toward a nondegenerate instance.   A third approach is provided by Seidel [Sei98], based on the following idea. For any problem, if we know one nondegenerate input $\mathbf{a}^*$ for the problem, then every other input $\mathbf{a}$ can be made nondegenerate by perturbing it in the direction of $\mathbf{a}^*$. We can take the perturbed input to be $\mathbf{a} + \epsilon \mathbf{a}^*$ for some small ϵ. For example, for the convex hull of points in $\mathbb{R}^n$, we can choose $\mathbf{a}^*$ to be distinct points on the moment curve $(t, t^2, \ldots, t^n)$. This method can be regarded as symbolic or numeric, depending on whether ϵ is an infinitesimal or an actual numerical value.

We now elaborate on these methods. We currently only have blackbox schemes for rational functions, while Seidel's method would apply even in nonalgebraic settings. Blackbox schemes are independent of particular problems, while the nondegenerate instances $\mathbf{a}^*$ depend on the problem (and on the input size); no systematic method to choose $\mathbf{a}^*$ is known. An early work in this area is the Simulation of Simplicity (SoS) technique of Edelsbrunner and Mücke [EM90]. The method amounts to adding powers of an indeterminate ϵ to each input parameter. Such ϵ-methods were first used in linear programming in the 1950s. The SoS scheme (for determinants) turns out to be an admissible scheme [Yap90b]. Intuitively, sign blackbox invocations should be almost as fast as the actual evaluations with high probability [Yap90b]. But the worst-case exponential behavior led Emiris and Canny to propose more efficient numerical approaches [EC95]. To each input parameter a_i in $\mathbf{a}$, they add a perturbation $b_i \epsilon$ (where $b_i \in \mathbb{Z}$ and ϵ is again an infinitesimal): these are called ***linear perturbations***. In case the test values are determinants, they show that a simple choice of the b_i's will ensure nondegeneracy and efficient computation. For general rational function tests, a lemma of Schwartz shows that a random choice of the b_i's is likely to yield nondegeneracy. Emiris, Canny, and Seidel [ECS97, Sei98] give a general result on the validity of linear perturbations, and apply it to common test polynomials.

Controlled perturbation was developed explicitly to be practical and to exploit fast floating point arithmetic. Our above assumption of the specific IEEE standard is only for simplicity: in general, we will need to consider floating point systems with L-bit mantissa as discussed in [MOS11]. By choosing L arbitrarily large, we can make the error bound $\varepsilon = B(f, M)$ arbitrarily close to 0. The idea of computing $B(f, M)$ goes back to Fortune and van Wyk [FW93a, FW96]. Given an expression E (typically a polynomial perhaps also with square-root), certain parameters such as m_E, ind_E, $\deg_E$, etc., can be recursively defined from the structure of E (e.g., [MOS11, Table 1]). They can even be computed in the chosen floating point system, costing no more than evaluating E in floating point. If $\widetilde{E}$ is the floating point value of E then [MOS11, Theorem 16] says that the error $|\widetilde{E} - E|$ is at most $(\text{ind}_E + 1) \cdot \mathbf{u} \cdot m_E$, where $\mathbf{u} = 2^{-L-1}$ is the unit round-off error. If E is an expression for evaluating $f(z)$, then the bound $B(f, M)$ may be defined as $(\text{int}_E + 1) \cdot \mathbf{u} \cdot m_E$.

We note two additional issues in controlled perturbation: (1) The above perturbation analysis operates at a predicate level. An algorithm execution will call these predicates many times on different inputs. We must translate the bounds and probabilities from predicate level to algorithm level. (2) The analysis assumes the input and perturbation are real numbers. Since the actual perturbations must be floating point numbers, we must convert our probability estimates based on real numbers into corresponding estimates for a discrete set of floating point numbers. Say u is a δ-perturbation of z. Let the $w = f\ell(u)$ be the floating point number closest to u (breaking ties arbitrarily). In our algorithm, we must be able to generate w with the probability p_w of the set $\{u \in U : f\ell(u) = w\}$. This issue was first addressed explicitly in Mehlhorn et al. [MOS11].

APPLICATIONS AND PRACTICE

Michelucci [Mic95] describes implementations of blackbox schemes, based on the concept of "ϵ-arithmetic." One advantage of his approach is the possibility of controlling the perturbations. Experiences with the use of perturbation in the beneath-beyond convex hull algorithm in arbitrary dimensions are reported in [ECS97]. Neuhauser [Neu97] improved and implemented the rational blackbox scheme of Yap. He also considered controlled perturbation techniques. Comes and Ziegelmann [CZ99] implemented the linear perturbation ideas of Seidel in CGAL.

Controlled perturbation was initially applied by Halperin et al. to arrangements for spheres, polyhedra and circles [HS98, Raa99, HL04]. This was extended to controlled perturbations for Delaunay triangulations [FKMS05], convex hulls in all dimensions [Kle04], and Voronoi diagrams of line segments [Car07]. Finally, Mehlhorn et al. [MOS11] provided the general analysis for all degeneracy predicates defined by multivariate polynomials, subsuming the previously studied cases.

In solid modeling systems, it is very useful to systematically avoid degenerate cases (numerous in this setting). Fortune [For97] uses symbolic perturbation to allow an "exact manifold representation" of nonregularized polyhedral solids (see Section 57.1). The idea is that a dangling rectangular face (for instance) can be perturbed to look like a very flat rectangular solid, which has a manifold representation. Hertling and Weihrauch [HW94] define "levels of degeneracy" and use this to obtain lower bounds on the size of decision computation trees.

In contrast to our general goal of *avoiding* degeneracies, there are some papers that propose to directly handle degeneracies. Burnikel, Mehlhorn, and Schirra [BMS95] describe the implementation of a line segment intersection algorithm and semidynamic convex hull maintenance in arbitrary dimensions. Based on this experience, they question the usefulness of perturbation methods using three observations: (i) perturbations may increase the running time of an algorithm by an arbitrary amount; (ii) the postprocessing problem can be significant; and (iii) it is not hard to handle degeneracies directly. But the probability of (i) occurring in a drastic way (e.g., for a degenerate input of n identical points) is so negligible that it may not deter users when they have the option of writing a generic algorithm, especially when the general algorithm is very complex or not readily available. Unfortunately, property (iii) is the exception rather than the rule, especially in nonlinear and nonplanar settings. In illustration, consider the mildly nonlinear problem of computing the Voronoi diagram of convex polyhedra in $\mathbb{R}^3$. An explicit exact algorithm for this "Voronoi quest" is still elusive at the moment of this writing. Solution of a special case (Voronoi diagram of arbitrary lines in $\mathbb{R}^3$) was called a major milestone in Hemmer et al. [HSH10]. The fundamental barrier here is traced by [YSL12] to the lack of a complete degeneracy analysis, requiring 10

cases (with subcases) and nontrivial facts of algebraic geometry (the requisite algebraic geometry lemma for lines in $\mathbb{R}^3$ was provided by Everett et al. [ELLD09]). In
short, users must weigh these opposing considerations for their particular problem
(cf. [Sch94]).

45.6 SOFT EXACT COMPUTATION

EGC has emerged as a successful general solution to numerical nonrobustness for
basic geometric problems, mainly planar and linear problems. In principle, this
solution extends to nonlinear algebraic problems. But its complexity is a barrier in
practice. For transcendental problems, the specter of noncomputable Zero problems
(Section 45.4) must be faced. These considerations alone press us to weaken the
notion of exactness. On top of this, the input is inherently inexact in the vast ma-
jority of geometric applications (e.g., in robotics or biology). All common physical
constants, save the speed of light which is exact by definition, is known to at most 8
digits of accuracy. In such settings, exact computation could be regarded as a device
to ensure robustness (treating the inexact input as nominally exact, Section 45.4).
But this device is no longer reasonable when addressing regimes where exact computation is expensive or impossible. Numerical approaches [Yap09] promise to open
up vast new domains in Computational Science and Engineering (CS&E) that are
inaccessible to current techniques of Computational Geometry. This section describes current attempts to exploit numerics and achieve some weak notion (or soft
notion) of exactness in regimes challenging for the current exact approach.

GLOSSARY

Traditional Numerical Algorithm: An algorithm described in the Real RAM
model and then implemented in the *Standard Model* of Numerical Analysis
[TB97]: namely, each primitive numerical operation returns a value with relative
error at most **u**, where $\mathbf{u} > 0$ is called the *unit round-off error*. Of course,
these errors will accumulate in subsequent steps.

Subdivision Model of Computation: A popular model for practitioners and
can take many forms. Here we describe a fully adaptive version [LY11, Yap15]
that supports rigorous soft algorithms using interval methods. Computational
problems in a metric space X can often be reduced to "local computation"
in small neighborhoods of X. Concretely, let $X = \mathbb{R}^d$, and $\square X$ denote the
set of d-boxes in X, representing neighborhoods. A *subdivision* of any set
$S \subseteq X$ is a finite collection C of sets with pairwise-disjoint interiors, whose
union $\bigcup C$ equals S. Assume an operator that replaces any box B by a finite
subdivision of B called split(B); each subbox in split(B) is a *child* of B. In
this way, we form a *subdivision tree* $T(B)$ rooted at B where the parent-child
relationship is determined by split operations. This is a dynamic tree that grows
by splitting at leaves. The leaves of $T(B)$ form a subdivision of B. For example,
for $d = 2$, split(B) is the set of four congruent subboxes sharing the midpoint of
B, and $T(B)$ is called a *quadtree*. To solve a problem in $B \subseteq X$, we grow the
subdivision tree rooted at B by splitting at any leaf. Termination is determined
by some numerical predicate on boxes. These predicates need only be evaluated

to a precision determined by the depth, thus side-stepping exact computation. In general, we can replace boxes by other well-structured sets such as simplices; see [Yap15] for a treatment that includes subdivision in non-Euclidean spaces.

RESOLUTION EXACTNESS

Our return to numerics needs new approaches: after all, we were driven to the exact approach in the first place because the **Traditional Numerical Algorithms** led to insoluble nonrobustness issues. Such algorithms do not have an a priori guarantee of correctness, but depend on an a posteriori error analysis. What we seek are algorithms with a priori correctness guarantees, and which can dynamically adjust their arbitrary precision and steps depending on the input instance. Informally, we call this mode of numerical computation "Soft Exact Computation."

The approximate version of an arbitrary problem is obtained by introducing a new **resolution parameter** $\varepsilon > 0$, in addition to its normal parameters. We must not identify ε with the unit round-off error **u** of the **Standard Model** of numerical analysis; intuitively, we want ε to bound the output error. In approximation algorithms, this has a standard interpretation called ε-**approximation** for optimization problems. It says that the output value should be within a relative error of ε of the optimal. For example, for shortest path problems, the output path length is $\leq (1 + \varepsilon)$ times the shortest length. If no path exists, the optimal length is ∞. Observe that such an ε-approximation algorithm implicitly solves the same path-existence predicate as the original algorithm: it outputs ∞ iff there is no path. Since this predicate requires exact computation, it is not really what we want. In fact, we want to approximate even the nonoptimization problems; in particular, what are "ε-approximate predicates"? In general, this is not a meaningful thought. However we can define a sensible notion in case of **numerical predicates**: if $f : X \to \mathbb{R}$ is a continuous function over some metric space X, then it defines a numerical predicate $C_f : X \to \{\texttt{true}, \texttt{false}\}$ given by $C_f(x) = \texttt{true}$ iff $f(x) > 0$.

To be concrete, consider the robot motion planning problem (Chapter 50) for some fixed robot in $\mathbb{R}^3$: given $x = (A, B, \Omega)$, we must find a path from start configuration A to goal configuration B while avoiding some polygonal set $\Omega \subseteq \mathbb{R}^3$. Let X be the set of all such inputs x. There is[2] a nontrivial path-existence predicate here: does input $x = (A, B, \Omega)$ admit a path? Let $f(x)$ be defined as the minimum clearance of an Ω-avoiding path from A to B if one exists; $f(x) = 0$ if there is no such path. Thus C_f is the path-existence predicate. Following [WCY15, LCLY15, Yap15], we call a function of the form $\widetilde{C}_f : X \times \mathbb{R}_{>0} \to \{\texttt{true}, \texttt{false}\}$ **resolution-exact** (or ε-exact) version of C_f if there exists a $K > 1$ such that for all $(x, \varepsilon) \in X \times \mathbb{R}_{>0}$, we have the following:

(T) If $f(x) > K\varepsilon$, then $\widetilde{C}_f(x, \varepsilon) = \texttt{true}$.

(F) If $f(x) < \varepsilon/K$, then $\widetilde{C}_f(x, \varepsilon) = \texttt{false}$.

The conditions (T) and (F) are nonexhaustive: in case $f(x) \in [\varepsilon/K, K\varepsilon]$, $\widetilde{C}_f(x)$ can output either answer. Because of this, we say that the ε-exactness problem is

---

[2] Current robotics literature ignores the path-existence predicate and formulates their correctness criteria for path planning, but *only* for inputs that admit a path; two such criteria are "resolution completeness" and "probabilistic completeness" [LaV06].

indeterminate. Indeterminacy allows us to bypass the specter of the Zero problem because the computability of $\tilde{C}_f$ is not in question as long as we can arbitrarily approximate f.

Another crucial step in transitioning to numerical approximation is to abandon the Real RAM model implicit in Traditional Numerical Algorithms. The alternative is conveniently captured by the **Subdivision Model**. We need techniques from interval arithmetic [Moo66]: assume $X = \mathbb{R}^d$ and $\square\mathbb{R}^d$ is the set of full-bodied axes-parallel boxes. If $f : X \to \mathbb{R}$ and $\square f : \square X \to \square\mathbb{R}$, we call $\square f$ a **soft version** of f if $\square f$ is **conservative** (i.e., $f(B) \subseteq \square f(B)$ for $B \in \square\mathbb{R}^d$) and **convergent** (i.e., $\lim_{i\to\infty} \square f(B_i) = f(p)$ for any sequence $\{B_i : i \geq 0\}$ of boxes that monotonically converges to a point $p \in X$). Such soft functions are relatively easy to design and implement for motion planning problems, leading to efficient and practical ε-exact planners [WCY15]. Soft exact algorithms have been developed for Voronoi diagrams [BPY16], meshing curves and surfaces [PV04, LY11, LYY12], and arrangement of curves [LSVY14].

The approach can be systematically extended to the first-order theory of reals as in Gao et al. [GAC12b]: given a Tarski formula φ, there are standard ways to bring it into a positive prenex form whose matrix is a Boolean combination of atomic predicates of the form $t(\mathbf{x}) > 0$ or $t(\mathbf{x}) \geq 0$ where $t(\mathbf{x})$ are terms (i.e., polynomials). Positive means that there are no negations in φ. We can enrich Tarski's language by allowing terms that include functions from some subset F of real functions that are computable in the sense of computable analysis [Wei00]. Call these F-**formulas**. For any $\delta \geq 0$, the δ-**strengthened form** of φ, denoted by $\varphi^{+\delta}$, is obtained by replacing the atomic predicate $t(\mathbf{x}) > 0$ by $t(\mathbf{x}) > \delta$ and $t(\mathbf{x}) \geq 0$ by $t(\mathbf{x}) \geq \delta$ in the positive form. Clearly, $\varphi^{+\delta}$ implies φ. Next, an F-formula φ is said to be **bounded** if each occurrence of a quantifier Q_i ($i = 1, 2, \ldots$) appears in the form $(Q_i x_i \in [u_i, v_i])[\ldots]$ where $u_i \leq v_i$ are both terms involving variables x_j, for $j < i$. The δ-**decision problem** is this: *given a bounded F-sentence φ and $\delta > 0$, decide whether φ is true or $\varphi^{+\delta}$ is false.* Their main result in [GAC12b] is that this problem is decidable. Prima facie, this is surprising since the decision problem for F-sentences is already undecidable if the $\sin(\cdot)$ function is included in F [Ric68]. It is less surprising when we realize that δ-decision is indeterminate, not the usual "hard" decision involving the Zero problem for F-terms. The δ-strengthening is one of several ways to strengthen (or weaken) numerical predicates; the above notion of ε-exactness is a variant better suited to nonnegative functions like clearance $f(x)$ and to the physical interpretations of robot uncertainty. It is also possible to strengthen using relative errors. Gao et al. [GAC12a] address the more practical case of δ-decision for bounded existential F-sentences. They further introduce a subdivision algorithm with interval evaluations (in the terminology of verification literature, this is called the **DPLL-Interval Constraint Propagation** (or DPLL/ICP)). It is proved that the DPLL/ICP algorithm can δ-decide bounded existential F-sentences within the complexity class NP^P assuming that the functions in F are P-computable.

45.7 OPEN PROBLEMS

1. The main theoretical question in EGC is whether the Constant Zero Problem for Ω_4 is decidable. This is decidable if Schanuel's conjecture is true. Baker's

theory gives a positive answer to a very special subproblem. This is expected to be very deep, so deciding any nontrivial subproblem would be of interest. A simpler question is whether $\mathrm{ZERO}(\Omega_3 \cup \{\sin(\cdot), \pi\})$ is decidable.

2. In constructive root bounds, it is unknown if there exists a root bound $\beta : \mathcal{E}(\Omega_2) \to \mathbb{R}_{\geq 0}$ where $-\lg(\beta(E)) = O(D(E))$ and $D(E)$ is the degree of E. In current bounds, we only know a quadratic bound, $-\lg(\beta(E)) = O(D(E)^2)$. The Revised Uniformity Conjecture of Richardson and Sonbaty [RES06] is a useful starting point.

3. Give an optimal algorithm for the guaranteed precision evaluation problem $\mathrm{GVAL}(\Omega)$ for, say, $\Omega = \Omega_2$. The solution includes a reasonable cost model.

4. In geometric rounding, we pose two problems: (a) Extend the Greene-Yao rounding problem to nonuniform grids (e.g., the grid points are L-bit floating point numbers). (b) Round simplicial complexes. The preferred notion of rounding here should not increase combinatorial complexity (unlike Greene-Yao), but rather allow features to collapse (triangles can degenerate to a vertex), but disallow inversion (triangles cannot flip its orientation).

5. Extend the control perturbation technique to more general classes of expressions, including rational functions, square-root functions, analytic functions [MOS11].

6. Give a systematic treatment of inexact (dirty) data. Held [Hel01a, Hel01b] describes the engineering of reliable algorithms to handle such inputs.

7. Design soft exact algorithms for kinodynamic planning and for nonholonomic planning in robotics. Known theoretical algorithms are far from practical (and practical ones are nonrigorous). Even when subdivision is used, such algorithms use exact ("hard") predicates.

8. Develop techniques for nontrivial complexity analysis of subdivision algorithms in computation geometry. Since such algorithms are adaptive, we would like to use complexity parameters based on the geometry of the input instance such as feature size or separation bounds.

45.8 SOURCES AND RELATED MATERIAL

SURVEYS

Forrest [For87] is an influential overview of the field of computational geometry. He deplores the gap between theory and practice and describes the open problem of robust intersection of line segments (expressing a belief that robust solutions do not exist). Other surveys of robustness issues in geometric computation are Schirra [Sch00], Yap and Dubé [YD95] and Fortune [For93]. Robust geometric modelers are surveyed in [PCH+95].

RELATED CHAPTERS

Chapter 26: Convex hull computations
Chapter 28: Arrangements
Chapter 31: Shortest paths and networks
Chapter 37: Computational and quantitative real algebraic geometry
Chapter 41: Ray shooting and lines in space
Chapter 50: Algorithmic motion planning
Chapter 51: Robotics
Chapter 57: Solid modeling
Chapter 67: Software
Chapter 68: Two computational geometry libraries: LEDA and CGAL

REFERENCES

[AKY04]   T. Asano, D. Kirkpatrick, and C. Yap. Pseudo approximation algorithms, with applications to optimal motion planning. *Discrete Comput. Geom.*, 31:139–171, 2004.

[Bak75]   A. Baker. *Transcendental Number Theory*. Cambridge University Press, 1975.

[BBP01]   H. Brönnimann, C. Burnikel, and S. Pion. Interval arithmetic yields efficient dynamic filters for computational geometry. *Discrete Appl. Math.*, 109:25–47, 2001.

[BCD⁺02]   G. Barequet, D.Z. Chen, O. Daescu, M.T. Goodrich, and J. Snoeyink. Efficiently approximating polygonal paths in three and higher dimensions. *Algorithmica*, 33:150–167, 2002.

[BCSM⁺06]   J.-D. Boissonnat, D. Cohen-Steiner, B. Mourrain, G. Rote, and G. Vegter. Meshing of surfaces. Chap. 5 in J.-D. Boissonnat and M. Teillaud, editors, *Effective Computational Geometry for Curves and Surfaces*, Springer, Berlin, 2006.

[BEH⁺02]   E. Berberich, A. Eigenwillig, M. Hemmer, S. Hert, K. Mehlhorn, and E. Schömer. A computational basis for conic arcs and Boolean operations on conic polygons. In *Proc. 10th European Sympos. Algorithms*, vol. 2461 of *LNCS*, pages 174–186, Springer, Berlin, 2002.

[BEH⁺05]   E. Berberich, A. Eigenwillig, M. Hemmer, S. Hert, L. Kettner, K. Mehlhorn, J. Reichel, S. Schmitt, E. Schömer, and N. Wolpert. EXACUS: Efficient and Exact Algorithms for Curves and Surfaces. In *Proc. 13th European Sympos. Algorithms*, vol. 3669 of *LNCS*, pages 155–166, Springer, Berlin, 2005.

[BEKS11]   E. Berberich, P. Emeliyanenko, A. Kobel, and M. Sagraloff. Arrangement computation for planar algebraic curves. In *Proc. Int. Workshop Symbolic-Numeric Comput.*, pages 88–98, ACM Press, 2011.

[BEKS13]   E. Berberich, P. Emeliyanenko, A. Kobel, and M. Sagraloff. Exact symbolic-numeric computation of planar algebraic curves. *Theor. Comput. Sci.*, 491:1–32, 2013.

[Ber14]   E. Berberich. Robustly and efficiently computing algebraic curves and surfaces. In *Proc. 4th Int. Congr. Math. Software*, vol. 8592 of *LNCS*, pages 253–260, Springer, Berlin, 2014.

[BFM⁺09]   C. Burnikel, S. Funke, K. Mehlhorn, S. Schirra, and S. Schmitt. A separation bound for real algebraic expressions. *Algorithmica*, 55:14–28, 2009.

[BFMS99]   C. Burnikel, R. Fleischer, K. Mehlhorn, and S. Schirra. Exact efficient geometric computation made easy. In *Proc. 15th ACM Sympos. Comput. Geom.*, pages 341–450, ACM Press, 1999.

[BFMS00]   C. Burnikel, R. Fleischer, K. Mehlhorn, and S. Schirra. A strong and easily computable separation bound for arithmetic expressions involving radicals. *Algorithmica*, 27:87–99, 2000.

[BFS98]   C. Burnikel, S. Funke, and M. Seel. Exact geometric predicates using cascaded computation. In *Proc. 14th Sympos. Comput. Geom.*, pages 175–183, ACM Press, 1998.

[BJMM93]   M.O. Benouamer, P. Jaillon, D. Michelucci, and J.-M. Moreau. A lazy arithmetic library. In *Proc. 11th IEEE Sympos. Comp. Arithmetic*, pages 242–269, 1993.

[BK95]   M. Blum and S. Kannan. Designing programs that check their work. *J. ACM*, 42:269–291, 1995.

[BKM⁺95]   C. Burnikel, J. Könnemann, K. Mehlhorn, S. Näher, S. Schirra, and C. Uhrig. Exact geometric computation in LEDA. In *Proc. 11th Sympos. Comput. Geom.*, pages C18–C19, ACM Press, 1995.

[BKS08]   E. Berberich, M. Kerber, and M. Sagraloff. Exact geometric-topological analysis of algebraic surfaces. In *Proc. 24th Sympos. Comput. Geom.*, pages 164–173, ACM Press, 2008.

[BLR93]   M. Blum, M. Luby, and R. Rubinfeld. Self-testing and self-correcting programs, with applications to numerical programs. *J. Comp. Sys. Sci.*, 47:549–595, 1993.

[BMS94]   C. Burnikel, K. Mehlhorn, and S. Schirra. How to compute the Voronoi diagram of line segments: Theoretical and experimental results. In *Proc. 2nd Eur. Sympos. Algo.*, vol. 855 of *LNCS*, pages 227–239, Springer, Berlin, 1994.

[BMS95]   C. Burnikel, K. Mehlhorn, and S. Schirra. On degeneracy in geometric computations. In *Proc. 5th ACM-SIAM Sympos. Discrete Algorithms*, pages 16–23, 1995.

[BPY16]   H. Bennett, E. Papadopoulou, and C. Yap. Planar minimization diagrams via subdivision with applications to anisotropic Voronoi diagrams. In *Proc. Eurographics Sympos. Geom. Process.*, 2016.

[BS00]   J.-D. Boissonnat and J. Snoeyink. Efficient algorithms for line and curve segment intersection using restricted predicates. *Comput. Geom.*, 16:35–52, 2000.

[BSS⁺16]   R. Becker, M. Sagraloff, V. Sharma, J. Xu, and C. Yap. Complexity analysis of root clustering for a complex polynomial. In *Proc. Int. Sympos. Symb. Algebr. Comput.*, pages 71–78, ACM Press, 2016.

[BSSY15]   R. Becker, M. Sagraloff, V. Sharma, and C. Yap. A simple near-optimal subdivision algorithm for complex root isolation based on the pellet test and Newton iteration. Preprint, arXiv:1509.06231, 2015.

[Bur96]   C. Burnikel. *Exact Computation of Voronoi Diagrams and Line Segment Intersections*. Ph.D thesis, Universität des Saarlandes, 1996.

[BY00]   H. Brönnimann and M. Yvinec. Efficient exact evaluation of signs of determinants. *Algorithmica*, 27:21–56, 2000.

[BY09]   W.D. Brownawell and Chee K. Yap. Lower bounds for zero-dimensional projections. In *Proc. 34th Int. Sympos. Symb. Algebr. Comput.*, pages 79–86, ACM Press, 2009.

[Can88]   J.F. Canny. *The Complexity of Robot Motion Planning*. ACM Doctoral Dissertion Award Series, MIT Press, Cambridge, 1988.

[Car07]   M. Caroli. Evaluation of a generic method for analyzing controlled-perturbation algorithms. Master's thesis, University of Saarland, 2007.

[CCK⁺06]   E.-C. Chang, S.W. Choi, D. Kwon, H. Park, and C.K. Yap. Shortest paths for disc obstacles is computable. *Internat. J. Comput. Geom. Appl.*, 16:567–590, 2006.

[CDR92]   J.F. Canny, B. Donald, and E.K. Ressler. A rational rotation method for robust geometric algorithms. *Proc. 8th Sympos. Comp. Geom.*, pages 251–160, ACM Press, 1992.

[Cho99]   T.Y. Chow. What is a closed-form number? *Amer. Math. Monthly*, 106:440–448, 1999.

[Cla92]   K.L. Clarkson. Safe and effective determinant evaluation. In *Proc. 33rd IEEE Sympos. Found. Comp. Sci.*, pages 387–395, 1992.

[CM93]   J.D. Chang and V.J. Milenkovic. An experiment using LN for exact geometric computations. *Proc. 5th Canad. Conf. Comput. Geom.*, pages 67–72, 1993.

[CSY97]   J. Choi, J. Sellen, and C. Yap. Approximate Euclidean shortest paths in 3-space. *Internat. J. Comput. Geom. Appl.*, 7:271–295, 1997.

[CZ99]   J. Comes and M. Ziegelmann. An easy to use implementation of linear perturbations within CGAL. In *Proc. 3rd Workshop Algorithms Eng.*, vol. 1668 of *LNCS*, pages 169–182, Springer, Berlin, 1999.

[DEMY02]   Z. Du, M. Eleftheriou, J. Moreira, and C. Yap. Hypergeometric functions in exact geometric computation. *Electron. Notes Theor. Comput. Sci.*, 66:53–64, 2002.

[DP99]   O. Devillers and F.P. Preparata. Further results on arithmetic filters for geometric predicates. *Comput. Geom.*, 13:141–148, 1999.

[DS88]   D.P. Dobkin and D. Silver. Recipes for Geometry & Numerical Analysis – Part I: An empirical study. In *Proc. 4th Sympos. Comput. Geom.*, pages 93–105, ACM Press, 1988.

[DY93]   T. Dubé and C.K. Yap. A basis for implementing exact geometric algorithms (extended abstract). Unpublished manuscript, 1993. `http://cs.nyu.edu/exact/doc/basis.ps.gz`.

[EC95]   I.Z. Emiris and J.F. Canny. A general approach to removing degeneracies. *SIAM J. Comput.*, 24:650–664, 1995.

[ECS97]   I.Z. Emiris, J.F. Canny, and R. Seidel. Efficient perturbations for handling geometric degeneracies. *Algorithmica*, 19:219–242, 1997.

[EGL⁺09]   H. Everett, C. Gillot, D. Lazard, S. Lazard, and M. Pouget. The Voronoi diagram of three arbitrary lines in $\mathbb{R}^3$. In *Abstracts of 25th Eur. Workshop Comput. Geom.*, Bruxelles, 2009.

[EKSW06]   A. Eigenwillig, L. Kettner, E. Schömer, and N. Wolpert. Complete, exact, and efficient computations with cubic curves. *Comput. Geom.*, 35:36–73, 2006.

[EKW07]   A. Eigenwillig, M. Kerber, and N. Wolpert. Fast and exact geometric analysis of real algebraic plane curves. In *Int. Sympos. Symb. Alge. Comp.*, pages 151–158, ACM Press, 2007.

[ELLD09]   H. Everett, D. Lazard, S. Lazard, and M.S. El Din. The Voronoi diagram of three lines. *Discrete Comput. Geom.*, 42:94–130, 2009.

[EM90]   H. Edelsbrunner and E.P. Mücke. Simulation of simplicity: A technique to cope with degenerate cases in geometric algorithms. *ACM Trans. Graph.*, 9:66–104, 1990.

[EMT10]   I.Z. Emiris, B. Mourrain, and E.P. Tsigaridas. The DMM bound: Multivariate (aggregate) separation bounds. In *35th Int. Sympos. Symb. Algebr. Comput.*, pages 243–250, ACM Press, 2010.

[ETT06]   I.Z. Emiris, E.P. Tsigaridas, and G.M. Tzoumas. The predicates for the Voronoi diagram of ellipses. In *Proc. 22nd Sympos. Comput. Geom.*, pages 227–236, ACM Press, 2006.

[FHL⁺07]   L. Fousse, G. Hanrot, V. Lefèvre, P. Pélissier, and P. Zimmermann. MPFR: A multiple-precision binary floating-point library with correct rounding. *ACM Trans. Math. Software*, 33: article 13, 2007.

[FKMS05]   S. Funke, C. Klein, K. Mehlhorn, and S. Schmitt. Controlled perturbation for Delaunay triangulations. In *Proc. 16th ACM-SIAM Sympos. Discrete Algorithms*, pages 1047–1056, 2005.

[FM91]   S.J. Fortune and V.J. Milenkovic. Numerical stability of algorithms for line arrangements. In *Proc. 7th Sympos. Comput. Geom.*, pages 334–341, ACM Press, 1991.

[FMN05]   S. Funke, K. Mehlhorn, and S. Näher. Structural filtering: A paradigm for efficient and exact geometric programs. *Comput. Geom.*, 31:179–194, 2005.

[For87]   A.R. Forrest. Computational geometry and software engineering: Towards a geometric computing environment. In D.F. Rogers and R.A. Earnshaw, editors, *Techniques for Computer Graphics*, pages 23–37, Springer, New York, 1987.

[For89]   S.J. Fortune. Stable maintenance of point-set triangulations in two dimensions. In *Proc. 30th IEEE Sympos. Found. Comp. Sci.*, pages 494–499, 1989.

[For92]   S.J. Fortune. Numerical stability of algorithms for 2-d Delaunay triangulations. In *Proc. 8th Sympos. Comput. Geom.*, pages 83–92, ACM Press, 1992.

[For93]   S.J. Fortune. Progress in computational geometrey. In R. Martin, editor, *Directions in Geometric Computing*, pages 81–127, Information Geometers Ltd., Winchester, 1993.

[For95]   S.J. Fortune. Numerical stability of algorithms for 2-d Delaunay triangulations. *Internat. J. Comput. Geom. Appl.*, 5:193–213, 1995.

[For97]   S.J. Fortune. Polyhedral modeling with multiprecision integer arithmetic. *Comput.-Aided Des.*, 29:123–133, 1997.

[For99]   S.J. Fortune. Vertex-rounding a three-dimensional polyhedral subdivision. *Discrete Comput. Geom.* 22:593–618, 1999.

[Fun97]   S. Funke. *Exact Arithmetic Using Cascaded Computation.* Master's thesis, Max Planck Institute for Computer Science, Saarbrücken, 1997.

[FW93a]   S.J. Fortune and C.J. van Wyk. Efficient exact arithmetic for computational geometry. In *Proc. 9th Sympos. Comput. Geom.*, pages 163–172, ACM Press, 1993.

[FW93b]   S.J. Fortune and C.J. van Wyk. *LN User Manual.* AT&T Bell Laboratories, 1993.

[FW96]   S.J. Fortune and C.J. van Wyk. Static analysis yields efficient exact integer arithmetic for computational geometry. *ACM Trans. Graph.*, 15:223–248, 1996.

[GAC12a]   S. Gao, J. Avigad, and E.M. Clarke. Delta-complete decision procedures for satisfiability over the reals. In *Porc. 6th Int. Joint Conf. Automated Reasoning*, vol. 7364 of *LNCS*, pages 286–300, Springer, Berlin 2012.

[GAC12b]   S. Gao, J. Avigad, and E.M. Clarke. Delta-decidability over the reals. In *Proc. 27th IEEE Sympos. Logic in Comp. Sci.*, pages 305–314, 2012.

[Gae99]   B. Gärtner. Exact arithmetic at low cost—a case study in linear programming. *Comput. Geom.*, 13:121–139, 1999.

[GGHT97]   M.T. Goodrich, L.J. Guibas, J. Hershberger, and P. Tanenbaum. Snap rounding line segments efficiently in two and three dimensions. In *Proc. 13th Sympos. Comput. Geom.*, pages 284–293, ACM Press, 1997.

[GHS01]   N. Geismann, M. Hemmer, and E. Schömer. Computing a 3-dimensional cell in an arrangement of quadrics: Exactly and actually! In *Proc. 17th Sympos. Comput. Geom.*, pages 264–273, ACM Press, 2001.

[GM95]   L.J. Guibas and D. Marimont. Rounding arrangements dynamically. In *Proc. 11th Sympos. Comput. Geom.*, pages 190–199, ACM Press, 1995.

[GS00]   B. Gärtner and S. Schönherr. An efficient, exact, and generic quadratic programming solver for geometric optimization. In *Proc. 16th Sympos. Comput. Geom.*, pages 110–118, ACM Press, 2000.

[GSS89]   L.J. Guibas, D. Salesin, and J. Stolfi. Epsilon geometry: Building robust algorithms from imprecise computations. In *Proc. 5th Sympos. Comput. Geom.*, pages 208–217, ACM Press, 1989.

[GY86]   D.H. Greene and F.F. Yao. Finite-resolution computational geometry. In *Proc. 27th IEEE Sympos. Found. Comp. Sci.*, pages 143–152, 1986.

[Hel01a]   M. Held. FIST: Fast industrial-strength triangulation of polygons. *Algorithmica*, 30:563–596, 2001.

[Hel01b]   M. Held. VRONI: An engineering approach to the reliable and efficient computation of Voronoi diagrams of points and line segments. *Comput. Geom.*, 18:95–123, 2001.

[HHK88]   C.M. Hoffmann, J.E. Hopcroft, and M.S. Karasick. Towards implementing robust geometric computations. In *Proc. 4th Sympos. Comput. Geom.*, pages 106–117, ACM Press, 1988.

[HHK+07]   S. Hert, M. Hoffmann, L. Kettner, S. Pion, and M. Seel. An adaptable and extensible geometry Kernel. *Comput. Geom.*, 38:16–36, 2007.

[HL04]   D. Halperin and E. Leiserowitz. Controlled perturbation for arrangements of circles. *Internat. J. Comput. Geom. Appl.*, 14:277–310, 2004.

[Hob99]   J.D. Hobby. Practical segment intersection with finite precision output. *Comput. Geom.*, 13:199–214, 1999.

[HS98]   D. Halperin and C.R. Shelton. A perturbation scheme for spherical arrangements with applications to molecular modeling. *Comput. Geom.*, 10:273–288, 1998.

[HSH10]   M. Hemmer, O. Setter, and D. Halperin. Constructing the exact Voronoi diagram of arbitrary lines in three-dimensional space. In *Proc. 18th Eur. Sympos. Algorithms* vol. 6346 of *LNCS*, pages 398–409, Springer, Berlin, 2010.

[HW94]   P. Hertling and K. Weihrauch. Levels of degeneracy and exact lower complexity bounds for geometric algorithms. In *Proc. 6th Canad. Conf. Comput. Geom.*, pages 237–242, 1994.

[JW94]   J.W. Jaromczyk and G.W. Wasilkowski. Computing convex hull in a floating point arithmetic. *Comput. Geom.*, 4:283–292, 1994.

[KFC+01]   S. Krishnan, M. Foskey, T. Culver, J. Keyser, and D. Manocha. PRECISE: Efficient multiprecision evaluation of algebraic roots and predicates for reliable geometric computation. In *17th Sympos. Comput. Geom.*, pages 274–283, ACM Press, 2001.

[Kle04]   C. Klein. *Controlled Perturbation for Voronoi Diagrams*. Master's thesis, University of Saarland, 2004.

[KLN91]   M. Karasick, D. Lieber, and L.R. Nackman. Efficient Delaunay triangulation using rational arithmetic. *ACM Trans. Graph.*, 10:71–91, 1991.

[KLPY99]   V. Karamcheti, C. Li, I. Pechtchanski, and C. Yap. A Core library for robust numerical and geometric computation. In *15th Sympos. Comput. Geom.*, p. 351–359, ACM Press, 1999.

[KMP⁺07]   L. Kettner, K. Mehlhorn, S. Pion, S. Schirra, and C. Yap. Classroom examples of robustness problems in geometric computation. *Comput. Geom.*, 40:61–78, 2007.

[KW98]   L. Kettner and E. Welzl. One sided error predicates in geometric computing. In *Proc. 15th IFIP World Computer Congress*, pages 13–26, 1998.

[LaV06]   S.M. LaValle. *Planning Algorithms*. Cambridge University Press, 2006.

[LCLY15]   Z. Luo, Y.-J. Chiang, J.-M. Lien, and C. Yap. Resolution exact algorithms for link robots. In *Proc. 11th Workshop Algo. Found. Robot.*, vol. 107 of *Springer Tracts in Advanced Robotics*, pages 353–370, 2015.

[LPT98]   G. Liotta, F.P. Preparata, and R. Tamassia. Robust proximity queries: An illustration of degree-driven algorithm design. *SIAM J. Comput.*, 28:864–889, 1998.

[LPY04]   C. Li, S. Pion, and C.K. Yap. Recent progress in Exact Geometric Computation. *J. Log. Algebr. Program.*, 64:85–111, 2004.

[LSVY14]   J.-M. Lien, V. Sharma, G. Vegter, and C. Yap. Isotopic arrangement of simple curves: An exact numerical approach based on subdivision. In *Proc. 4th Int. Congr. Math. Software*, vol. 8592 of *LNCS*, pages 277–282, Springer, Berlin, 2014.

[LY01]   C. Li and C.K. Yap. A new constructive root bound for algebraic expressions. In *Proc. 12th ACM-SIAM Sympos. Discrete Algorithms*, pages 496–505, 2001.

[LY11]   L. Lin and C. Yap. Adaptive isotopic approximation of nonsingular curves: The parameterizability and nonlocal isotopy approach. *Discrete Comput. Geom.*, 45:760–795, 2011.

[LYY12]   L. Lin, C. Yap, and J. Yu. Non-local isotopic approximation of nonsingular surfaces. *Comput.-Aided Des.*, 45:451–462, 2012.

[Mic95]   D. Michelucci. An epsilon-arithmetic for removing degeneracies. In *Proc. 12th IEEE Sympos. Comp. Arith.*, pages 230–237, 1995.

[Mig82]   M. Mignotte. Identification of algebraic numbers. *J. Algorithms*, 3:197–204, 1982.

[Mil89]   V. Milenkovic. Double precision geometry: A general technique for calculating line and segment intersections using rounded arithmetic. In *Proc. 30th IEEE Sympos. Found. Comp. Sci.*, pages 500–506, 1989.

[MM93]   V.J. Milenkovic and Ve. Milenkovic. Rational orthogonal approximations to orthogonal matrices. *Proc. 5th Canad. Conf. on Comp. Geom.*, pages 485–490, Waterloo, 1993.

[MN90]   V. Milenkovic and L.R. Nackman. Finding compact coordinate representations for polygons and polyhedra. In *Proc. 6th Sympos. Comp. Geom.*, pages 244–252, ACM Press, 1990.

[MN95]   K. Mehlhorn and S. Näher. LEDA: A platform for combinatorial and geometric computing. *Comm. ACM*, 38:96–102, 1995.

[MNS⁺96]   K. Mehlhorn, S. Näher, T. Schilz, R. Seidel, M. Seel, and C. Uhrig. Checking geometric programs or verification of geometric structures. In *Proc. 12th Sympos. Comput. Geom.*, pages 159–165, ACM Press, 1996.

[Moo66]   R.E. Moore. *Interval Analysis*. Prentice Hall, Englewood Cliffs, 1966.

[MOS11]   K. Mehlhorn, R. Osbild, and M. Sagraloff. A general approach to the analysis of controlled perturbation algorithms. *Comput. Geom.*, 44:507–528, 2011.

[MS15]   M. Mörig and S. Schirra. Precision-driven computation in the evaluation of expression-dags with common subexpressions: Problems and solutions. In *Proc. 6th Math. Aspects Comp. Info. Sci.*, vol. 9582 of *LNCS*, pages 451–465, Springer, Berlin, 2015.

[MW96]   A. Macintyre and A. Wilkie. On the decidability of the real exponential field. In *Kreiseliana, about and around Georg Kreisel*, pages 441–467, A.K. Peters, Wellesley, 1996.

[Neu97]   M.A. Neuhauser. Symbolic perturbation and special arithmetics for controlled handling of geometric degeneracies. In *Proc. 5th Int. Conf. Central Europe Comput. Graphics Visualization*, pages 386–395, 1997.

[OTU87]   T. Ottmann, G. Thiemt, and C. Ullrich. Numerical stability of geometric algorithms. In *Proc. 3rd Sympos. Comput. Geom.*, pages 119–125, ACM Press, 1987.

[Pap85]   C.H. Papadimitriou. An algorithm for shortest-path motion in three dimensions. *Inform. Process. Lett.*, 20:259–263, 1985.

[PCH$^+$95] N.M. Patrikalakis, W. Cho, C.-Y. Hu, T. Maekawa, E.C. Sherbrooke, and J. Zhou. Towards robust geometric modelers, 1994 progress report. In *Proc. NSF Design and Manufacturing Grantees Conference*, pages 139–140, 1995.

[PV04]   S. Plantinga and G. Vegter. Isotopic approximation of implicit curves and surfaces. In *Proc. Eurographics Sympos. Geom. Process.*, pages 245–254, ACM Press, 2004.

[PY01]   V.Y. Pan and Y. Yu. Certification of numerical computation of the sign of the determinant of a matrix. *Algorithmica*, 30:708–724, 2001.

[PY06]   S. Pion and C. K. Yap. Constructive root bound method for k-ary rational input numbers. *Theoret. Comput. Sci.* 369:361–376, 2006.

[Raa99]   S. Raab. Controlled perturbation for arrangements of polyhedral surfaces with application to swept volumes. In *Proc. 15th Sympos. Comput. Geom.*, pages 163–172, ACM Press, 1999.

[RES06]   D. Richardson and A. El-Sonbaty. Counterexamples to the uniformity conjecture. *Comput. Geom.*, 33:58–64, 2006.

[Ric68]   D. Richardson. Some undecidable problems involving elementary functions of a real variable. *J. Symb. Log.*, 33:514–520, 1968.

[Ric97]   D. Richardson. How to recognize zero. *J. Symbolic Comput.*, 24:627–645, 1997.

[Ric07]   D. Richardson. Zero tests for constants in simple scientific computation. *Math. Comput. Sci.*, 1(:21–37, 2007.

[Sch94]   P. Schorn. Degeneracy in geometric computation and the perturbation approach. *The Computer J.*, 37:35–42, 1994.

[Sch00]   S. Schirra. Robustness and precision issues in geometric computation. In J.R. Sack and J. Urrutia, editors, *Handbook of Computational Geometry*, pages 597–632, Elsevier, Amsterdam, 2000.

[Sch00]   E.R. Scheinerman. When close enough is close enough. *Amer. Math. Monthly*, 107:489–499, 2000.

[SCY00]   J. Sellen, J. Choi, and C. Yap. Precision-sensitive Euclidean shortest path in 3-Space. *SIAM J. Comput.*, 29:1577–1595, 2000.

[Sed02]   R. Sedgewick. *Algorithms in C: Part 5: Graph Algorithms*, 3rd edition. Addison-Wesley, Boston, 2002.

[See01]   M. Seel. *Planar Nef Polyhedra and Generic High-dimensional Geometry*. Ph.D. thesis, Universität des Saarlandes, 2001.

[Seg90]   M. Segal. Using tolerances to guarantee valid polyhedral modeling results. *Comput. Graph.*, 24:105–114, 1990.

[Sei98]   R. Seidel. The nature and meaning of perturbations in geometric computing. *Discrete Comput. Geom.*, 19:1–17, 1998.

[She96]   J.R. Shewchuk. Robust adaptive floating-point geometric predicates. In *Proc. 12th Sympos. Comput. Geom.*, pages 141–150, ACM Press, 1996.

[SI89a]   K. Sugihara and M. Iri. A solid modeling system free from topological inconsistency. *J. Inf. Process.*, 12:380–393, 1989.

[SI89b]   K. Sugihara and M. Iri. Two design principles of geometric algorithms in finite precision arithmetic. *Appl. Math. Lett.*, 2:203–206, 1989.

[SI92]   K. Sugihara and M. Iri. Construction of the Voronoi diagram for 'one million' generators in single-precision arithmetic. *Proc. IEEE*, 80:1471–1484, 1992.

[SIII00]   K. Sugihara, M. Iri, H. Inagaki, and T. Imai. Topology-oriented implementation—an approach to robust geometric algorithms. *Algorithmica*, 27:5–20, 2000.

[SS85]   M.G. Segal and C.H. Séquin. Consistent calculations for solids modelling. In *Proc. 1st Sympos. Comput. Geom.*, pages 29–38, ACM Press, 1985.

[SSTY00]   E. Schömer, J. Sellen, M. Teichmann, and C. Yap. Smallest enclosing cylinders. *Algorithmica*, 27:170–186, 2000.

[Sug89]   K. Sugihara. On finite-precision representations of geometric objects. *J. Comput. Syst. Sci.*, 39:236–247, 1989.

[Sug11]   K. Sugihara. Robust geometric computation based on the principle of independence. *Nonlinear Theory Appl., IEICE*, 2:32–42, 2011.

[TB97]   L.N. Trefethen and D. Bau. *Numerical Linear Algebra*. SIAM, Philadelphia, 1997.

[Ull90]   C. Ullrich, editor. *Computer Arithmetic and Self-validating Numerical Methods*. Academic Press, Boston, 1990.

[WCY15]   C. Wang, Y.J. Chiang, and C. Yap. On soft predicates in subdivision motion planning. *Comput. Geom.*, 48:589–605, 2015.

[Wei00]   K. Weihrauch. *Computable Analysis*. Springer, Berlin, 2000.

[Wei02]   R. Wein. High level filtering for arrangements of conic arcs. In *Proc. 10th Eur. Sympos. Algorithms*, vol. 2461 of *LNCS*, pages 884–895, Springer, Berlin, 2002.

[Yap90a]   C.K. Yap. A geometric consistency theorem for a symbolic perturbation scheme. *J. Comput. Syst. Sci.*, 40:2–18, 1990.

[Yap90b]   C.K. Yap. Symbolic treatment of geometric degeneracies. *J. Symbolic Comput.*, 10:349–370, 1990.

[Yap97]   C.K. Yap. Towards exact geometric computation. *Comput. Geom.*, 7:3–23, 1997.

[Yap98]   C.K. Yap. A new number core for robust numerical and geometric libraries. In *Abstract 3rd CGC Workshop Comput. Geom.*, Brown University, Providence, 1998. http://www.cs.brown.edu/cgc/cgc98/home.html.

[Yap00]   C.K. Yap. *Fundamental Problems of Algorithmic Algebra*. Oxford Univ. Press, 2000.

[Yap09]   C.K. Yap. In praise of numerical computation. In S. Albers, H. Alt, and S. Näher, editors, *Efficient Algorithms: Essays Dedicated to Kurt Mehlhorn on the Occasion of His 60th Birthday*, vol. 5760 of *LNCS*, pages 308–407. Springer, Berlin, 2009.

[Yap15]   C.K. Yap. Soft subdivision search and motion planning, II: Axiomatics. In *Proc. 9th Conf. Frontiers in Algorithmics*, vol 9130 of *LNCS*, pages 7–22, Springer, Berlin, 2015.

[YD95]   C.K. Yap and T. Dubé. The exact computation paradigm. In D.-Z. Du and F.K. Hwang, editors, *Computing in Euclidean Geometry*, 2nd edition, pages 452–492, World Scientific, Singapore, 1995.

[YSL12]   C. Yap, V. Sharma, and J.M. Lien. Towards exact numerical Voronoi diagrams. In
            Proc. 9th Int. Sympos. Voronoi Diagrams in Sci. Eng., pages 2–16, 2012.

[Yu92]   J. Yu. *Exact Arithmetic Solid Modeling.* Ph.D. dissertation, Department of Computer
            Science, Purdue University, West Lafayette, 1992.

[YYD$^+$10] J. Yu, C. Yap, Z. Du, S. Pion, and H. Brönnimann. The design of Core 2: A library
            for exact numeric computation in geometry and algebra. In *Proc. 3rd Int. Congr.
            Math. Software*, vol. 6327 of *LNCS*, pages 121–141, Springer, Berlin, 2010.

46 PARALLEL ALGORITHMS IN GEOMETRY
Michael T. Goodrich and Nodari Sitchinava

INTRODUCTION

The goal of parallel algorithm design is to develop parallel computational methods that run very fast with as few processors as possible, and there is an extensive literature of such algorithms for computational geometry problems. There are several different parallel computing models, and in order to maintain a focus in this chapter, we will describe results in the Parallel Random Access Machine (PRAM) model, which is a synchronous parallel machine model in which processors share a common memory address space (and all inter-processor communication takes place through this shared memory). Although it does not capture all aspects of parallel computing, it does model the essential properties of parallelism. Moreover, it is a widely accepted model of parallel computation, and all other reasonable models of parallel computation can easily simulate a PRAM.

Interestingly, parallel algorithms can have a direct impact on efficient sequential algorithms, using a technique called *parametric search*. This technique involves the use of a parallel algorithm to direct searches in a parameterized geometric space so as to find a critical location (e.g., where an important parameter changes sign or achieves a maximum or minimum value).

The PRAM model is subdivided into submodels based on how one wishes to handle concurrent memory access to the same location. The Exclusive-Read, Exclusive-Write (EREW) variant does not allow for concurrent access. The Concurrent-Read, Exclusive-Write (CREW) variant permits concurrent memory reads, but memory writes must be exclusive. Finally, the Concurrent-Read, Concurrent-Write (CRCW) variant allows for both concurrent memory reading and writing, with concurrent writes being resolved by some simple rule, such as having an arbitrary member of a collection of conflicting writes succeed. One can also define randomized versions of each of these models (e.g., an rCRCW PRAM), where in addition to the usual arithmetic and comparison operations, each processor can generate a random number from 1 to n in one step.

Early work in parallel computational geometry, in the way we define it here, began with the work of Chow [Cho80], who designed several parallel algorithms with polylogarithmic running times using a linear number of processors. Subsequent to this work, several researchers initiated a systematic study of work-efficient parallel algorithms for geometric problems, including Aggarwal *et al.* [ACG⁺88], Akl [Akl82, Akl84, Akl85], Amato and Preparata [AP92, AP95], Atallah and Goodrich [AG86, Goo87], and Reif and Sen [RS92, Sen89].

In Section 46.1 we give a brief discussion of general techniques for parallel geometric algorithm design. We then partition the research in parallel computational geometry into problems dealing with convexity (Section 46.2), arrangements and decompositions (Section 46.3), proximity (Section 46.4), geometric searching (Section 46.5), and visibility, envelopes, and geometric optimization (Section 46.6).

46.1 SOME PARALLEL TECHNIQUES

The design of efficient parallel algorithms for computational geometry problems often depends upon the use of powerful general parallel techniques (e.g., see [AL93, Já92, KR90, Rei93]). We review some of these techniques below.

PARALLEL DIVIDE-AND-CONQUER

Possibly the most general technique is parallel divide-and-conquer. In applying this technique one divides a problem into two or more subproblems, solves the subproblems recursively in parallel, and then merges the subproblem solutions to solve the entire problem. As an example application of this technique, consider the problem of constructing the upper convex hull of a set S of n points in the plane presorted by x-coordinates. Divide the list S into $\lceil \sqrt{n} \rceil$ contiguous sublists of size $\lfloor \sqrt{n} \rfloor$ each and recursively construct the upper convex hull of the points in each list. Assign a processor to each pair of sublists and compute the common upper tangent line for the two upper convex hulls for these two lists, which can be done in $O(\log n)$ time using a well-known "binary search" computation [Ede87, O'R94, PS85]. By maximum computations on the left and right common tangents, respectively, for each subproblem S_i, one can determine which vertices on the upper convex hull of S_i belong to the upper convex hull of S. Compressing all the vertices identified to be on the upper convex hull of S constructs an array representation of this hull, completing the construction.

The running time of this method is characterized by the recurrence relation $T(n) \leq T(\sqrt{n}) + O(\log n)$, which implies that $T(n)$ is $O(\log n)$. It is important to note that the coefficient for the $T(\sqrt{n})$ term is 1 even though we had $\lceil \sqrt{n} \rceil$ subproblems, for all these subproblems were processed simultaneously in parallel. The number of processors needed for this computation can be characterized by the recurrence relation $P(n) = \lceil \sqrt{n} \rceil P(\sqrt{n}) + O(1)$, which implies that $P(n) = O(n)$. Thus, the *work* needed for this computation is $O(n \log n)$, which is not quite optimal. Still, this method can be adapted to achieve work-optimal algorithms [BSV96, Che95, GG97].

BUILD-AND-SEARCH

Another important technique in parallel computational geometry is the build-and-search technique. It is a paradigm that often yields efficient parallel adaptations of sequential algorithms designed using the powerful plane sweeping technique. In the build-and-search technique, the solution to a problem is partitioned into a *build* phase, where one constructs in parallel a data structure built from the geometric data present in the problem, and a *search* phase, where one searches this data structure in parallel to solve the problem at hand. An example of an application of this technique is for the trapezoidal decomposition problem: given a collection of nonintersecting line segments in the plane, determine the first segments intersected by vertical rays emanating from each segment endpoint (cf. Figure 44.0.1). The existing efficient parallel algorithm for this problem is based upon first building in parallel a data structure on the input set of segments that allows for such vertical ray-shooting queries to be answered in $O(\log n)$ time by a single processor, and then querying this structure for each segment endpoint in parallel. This results in a parallel algorithm with an efficient $O(n \log n)$ work bound and fast $O(\log n)$ query time.

46.2 CONVEXITY

Results on the problem of constructing the convex hull of n points in $\mathbb{R}^d$ are summarized in Table 46.2.1, for various fixed values of d, and, in the case of $d = 2$, under assumptions about whether the input is presorted. We restrict our attention to parallel algorithms with efficient work bounds, where we use the term **work** of an algorithm here to refer to the product of its running time and the number of processors used by the algorithm. A parallel algorithm has an **optimal** work bound if the work used asymptotically matches the sequential lower bound for the problem. In the table, h denotes the size of the hull, and c is some fixed constant. Also, throughout this chapter we use $\bar{O}(f(n))$ to denote an asymptotic bound that holds with high probability.

TABLE 46.2.1   Parallel convex hull algorithms.

PROBLEM	MODEL	TIME	WORK	REF
2D presorted	rand-CRCW	$\bar{O}(\log^* n)$	$\bar{O}(n)$	[GG91]
2D presorted	CRCW	$O(\log \log n)$	$O(n)$	[BSV96]
2D presorted	EREW	$O(\log n)$	$O(n)$	[Che95]
2D polygon	EREW	$O(\log n)$	$O(n)$	[Che95]
2D	rand-CRCW	$\bar{O}(\log n)$	$\bar{O}(n \log h)$	[GG91]
2D	rand-CRCW	$O(\log h \log \log n)$	$\bar{O}(n \log h)$	[GS97]
2D	EREW	$O(\log n)$	$O(n \log n)$	[MS88]
2D	EREW	$O(\log^2 n)$	$O(n \log h)$	[GG91]
3D	rand-CRCW	$\bar{O}(\log n)$	$\bar{O}(n \log n)$	[RS92]
3D	rand-CRCW	$O(\log h \log^2 \log n)$	$\bar{O}(n \log h)$	[GS03]
3D	CREW	$O(\log n)$	$O(n^{1+1/c})$	[AP93]
3D	EREW	$O(\log^2 n)$	$O(n \log n)$	[AGR94]
3D	EREW	$O(\log^3 n)$	$O(n \log h)$	[AGR94]
Fixed $d \geq 4$	rand-EREW	$\bar{O}(\log^2 n)$	$\bar{O}(n^{\lfloor d/2 \rfloor})$	[AGR94]
Even $d \geq 4$	EREW	$O(\log^2 n)$	$O(n^{\lfloor d/2 \rfloor})$	[AGR94]
Odd $d > 4$	EREW	$O(\log^2 n)$	$O(n^{\lfloor d/2 \rfloor} \log^c n)$	[AGR94]

We discuss a few of these algorithms to illustrate their flavor.

2-DIMENSIONAL CONVEX HULLS

The two-dimensional convex hull algorithm of Miller and Stout [MS88] is based upon a parallel divide-and-conquer scheme where one presorts the input and then divides it into many subproblems ($O(n^{1/4})$ in their case), solves each subproblem independently in parallel, and then merges all the subproblem solutions together in $O(\log n)$ parallel time. Of course, the difficult step is the merge of all the subproblems, with the principal difficulty being the computation of common tangents between hulls. The total running time is characterized by the recurrence

$$T(n) \leq T(n^{1/4}) + O(\log n),$$

which solves to $T(n) = O(\log n)$.

3-DIMENSIONAL CONVEX HULLS

All of the 3D convex hull algorithms listed in Table 46.2.1 are also based upon this many-way, divide-and-conquer paradigm, except that there is no notion of presorting in three dimensions, so the subdivision step also becomes nontrivial. Reif and Sen [RS92] use a random sample to perform the division, and the methods of Amato, Goodrich, and Ramos [AGR94] derandomize this approach. Amato and Preparata [AP93] use parallel separating planes, an approach extended to higher dimensions in [AGR94].

LINEAR PROGRAMMING

A problem strongly related to convex hull construction, which has also been addressed in a parallel setting, is d-dimensional linear programming, for fixed dimensions d (see Chapter 49). Of course, one could solve this problem by transforming it to its dual problem, constructing a convex hull in this dual space, and then evaluating each vertex in the simplex that is dual to this convex hull. This would be quite inefficient, however, for $d \geq 4$. The best parallel bounds for this problem are listed in Table 46.2.2. See Section 49.6 for a detailed discussion.

TABLE 46.2.2   Fixed d-dimensional parallel linear programming.

MODEL	TIME	WORK	REF
Rand-CRCW	$\bar{O}(1)$	$\bar{O}(n)$	[AM94]
CRCW	$O((\log \log n)^{d-1})$	$O(n)$	[GR97]
EREW	$O(\log n (\log \log n)^{d-1})$	$O(n)$	[Goo96]
EREW (d=2)	$O(\log n (\log \log n)^*)$	$O(n)$	[CX02]

OPEN PROBLEMS

There are a number of interesting open problems regarding convexity:

1. Can d-dimensional linear programming be solved (deterministically) in $O(\log n)$ time using $O(n)$ work in the CREW PRAM model?

2. Is there an efficient output-sensitive parallel convex hull algorithm for $d \geq 4$?

3. Is there a work-optimal $O(\log^2 n)$-time CREW PRAM convex hull algorithm for odd dimensions greater than 4?

46.3 ARRANGEMENTS AND DECOMPOSITIONS

Another important class of geometric problems that has been addressed in the parallel setting are arrangement and decomposition problems, which deal with ways of partitioning space. We review the best parallel bounds for such problems in Table 46.3.1.

GLOSSARY

Arrangement: The partition of space determined by the intersections of a collection of geometric objects, such as lines, line segments, or (in higher dimensions) hyperplanes. In this chapter, algorithms for constructing arrangements produce the *incidence graph*, which stores all adjacency information between the various primitive topological entities determined by the partition, such as intersection points, edges, faces, etc. See Section 28.3.1.

Red-blue arrangement: An arrangement defined by two sets of objects A and B such that the objects in A (resp. B) are nonintersecting.

Axis-parallel: All segments/lines are parallel to one of the coordinate axes.

Monotone polygon: A polygon, which is intersected by any line parallel to some fixed direction at most twice. See Section 30.1.

Polygon triangulation: A decomposition of the interior of a polygon into triangles by adding non-crossing diagonals between vertices. See Section 30.2.

Trapezoidal decomposition: A decomposition of the plane into trapezoids (and possibly triangles) by adding appropriate vertical line segments incident to vertices. See Section 38.3.

Star-shaped polygon: A (simple) polygon that is completely visible from a single point. A polygon with nonempty kernel. See Section 30.1.

$1/r$-cutting: Given n hyperplanes in $\mathbb{R}^d$ and a parameter $1 \leq r \leq n$, a $1/r$-cutting is a partition of $\mathbb{R}^d$ into (relatively open) simplices such that each simplex intersects at most n/r hyperplanes. See Sections 40.2 and 44.1.

We sketch one randomized algorithm in Table 46.3.1 to illustrate how randomization and parallel computation can be mixed. Let S be a set of segments in the plane with k intersecting pairs. The goal is to construct $\mathcal{A}(S)$, the arrangement induced by S. First, an estimate $\hat{k}$ for k is obtained from a random sample. Then a random subset $R \subset S$ of a size r dependent on $\hat{k}$ is selected. $\mathcal{A}(R)$ is constructed using a suboptimal parallel algorithm, and processed (in parallel) for point location. Next the segments intersecting each cell of $\mathcal{A}(R)$ are found using a parallel point-location algorithm, together with some ad hoc techniques. Visibility information among the segments meeting each cell is computed using another suboptimal parallel algorithm. Finally, the resulting cells are merged in parallel. Because various key parameters in the suboptimal algorithms are kept small by the sampling, optimal expected work is achieved.

All of the algorithms for computing segment arrangements are **output-sensitive**, in that their work bounds depend upon both the input size and the output size. In these cases we must slightly extend our computational model to allow for

TABLE 46.3.1   Parallel arrangement and decomposition algorithms.

PROBLEM	MODEL	TIME	WORK	REF
d-dim hyperplane arr	EREW	$O(\log n)$	$O(n^d)$	[AGR94]
2D seg arr	rand-CRCW	$\bar{O}(\log n)$	$\bar{O}(n \log n + k)$	[CCT92a, CCT92b]
2D axis-par seg arr	CREW	$O(\log n)$	$O(n \log n + k)$	[Goo91a]
2D red-blue seg arr	CREW	$O(\log n)$	$O(n \log n + k)$	[GSG92, GSG93, Rüb92]
2D seg arr	EREW	$O(\log^2 n)$	$O(n \log n + k)$	[AGR95]
Monotone polygon triangulation	EREW	$O(\log n)$	$O(n)$	[Che95]
Polygon triangulation	CRCW	$O(\log n)$	$O(n)$	[Goo95]
Polygon triangulation	CREW	$O(\log n)$	$O(n \log n)$	[Goo89, Yap88]
2D nonint seg trap decomp	CREW	$O(\log n)$	$O(n \log n)$	[ACG89]
2D quadtree decomp	EREW	$O(\log n)$	$O(n \log n + k)$	[BET99]

the machine to request additional processors if necessary. In all these algorithms, this request may originate only from a single "master" processor, however, so this modification is not that different from our assumption that the number of processors assigned to a problem can be a function of the input size. Of course, to solve a problem on a real parallel computer, one would simulate one of these efficient parallel algorithms to achieve an optimal speed-up over what would be possible using a sequential method.

A class of related problems deals with methods for detecting intersections. Testing whether a collection of objects contains at least one intersecting pair is frequently easier than finding all such intersections. Table 46.3.2 reviews such results in the parallel domain.

TABLE 46.3.2   Parallel intersection detection algorithms.

PROBLEM	MODEL	TIME	WORK	REF
2 convex polygons	CREW	$O(1)$	$O(n^{1/c})$	[DK89a]
2 star-shaped polygons	CREW	$O(\log n)$	$O(n)$	[GM91]
2 convex polyhedra	CREW	$O(\log n)$	$O(n)$	[DK89a]

Given a collection of n hyperplanes in $\mathbb{R}^d$, another important decomposition problem is the construction of a $(1/r)$-cutting. For this problem there exists an EREW algorithm running in $O(\log n \log r)$ time using $O(nr^{d-1})$ work [Goo93].

OPEN PROBLEMS

1. Is there a work-optimal $O(\log n)$-time polygon triangulation algorithm that does not use concurrent writes?

2. Can a line segment arrangement be constructed in $O(\log n)$ time using $O(n \log n + k)$ work in the CREW PRAM model?

46.4 PROXIMITY

An important property of Euclidean space is that it is a metric space, and distance plays an important role in many computational geometry applications. For example, computing a closest pair of points can be used in collision detection, as can the more general problem of computing the nearest neighbor of each point in a set S, a problem we will call the ***all-nearest neighbors (ANN)*** problem. Perhaps the most fundamental problem in this domain is the subdivision of space into regions where each region $V(s)$ is defined by a *site* s in a set S of geometric objects such that each point in $V(s)$ is closer to s than to any other object in S. This subdivision is the ***Voronoi diagram*** (Chapter 27); its graph-theoretic dual, which is also an important geometric structure, is the ***Delaunay triangulation*** (Section 29.1). For a set of points S in $\mathbb{R}^d$, there is a simple "lifting" transformation that takes each point $(x_1, x_2, \ldots, x_d) \in S$ to the point $(x_1, x_2, \ldots, x_d, x_1^2 + x_2^2 + \ldots + x_d^2)$, forming a set of points S' in $\mathbb{R}^{d+1}$ (Section 27.1). Each simplex on the convex hull of S' with a negative $(d+1)$-st component in its normal vector projects back to a simplex of the Delaunay triangulation in $\mathbb{R}^d$. Thus, any $(d+1)$-dimensional convex hull algorithm immediately implies a d-dimensional Voronoi diagram (VD) algorithm. Table 46.4.1 summarizes the bounds of efficient parallel algorithms for constructing Voronoi diagrams in this way, as well as methods that are designed particularly for Voronoi diagram construction or other specific proximity problems. (In the table, the underlying objects are points unless stated otherwise.)

GLOSSARY

Convex position: A set of points are in convex position if they are all on the boundary of their convex hull.

Voronoi diagram for line segments: A Voronoi diagram that is defined by a set of nonintersecting line segments, with distance from a point p to a segment s being defined as the distance from p to a closest point on s. See Section 27.3.

TABLE 46.4.1 Parallel proximity algorithms.

PROBLEM	MODEL	TIME	WORK	REF
2D ANN in convex pos	EREW	$O(\log n)$	$O(n)$	[CG92]
2D ANN	EREW	$O(\log n)$	$O(n \log n)$	[CG92]
d-dim ANN	CREW	$O(\log n)$	$O(n \log n)$	[Cal93]
2D VD in L_1 metric	CREW	$O(\log n)$	$O(n \log n)$	[WC90]
2D VD for points & segments	rand-CREW	$\bar{O}(\log n)$	$\bar{O}(n \log n)$	[Ram97]
2D VD for points & segments	CRCW	$O(\log n \log \log n)$	$O(n \log n)$	[Ram97]
2D VD	CREW	$O(\log n \log \log n)$	$O(n \log^2 n)$	[CGÓ96]
2D VD	EREW	$O(\log^2 n)$	$O(n \log n)$	[AGR94]
2D VD for segments	EREW	$O(\log^2 n)$	$O(n \log n)$	[Ram97]
3D VD	EREW	$O(\log^2 n)$	$O(n^2)$	[AGR94]

OPEN PROBLEMS

1. Can a 2D Voronoi diagram be constructed deterministically in $O(\log n)$ time using $O(n \log n)$ work in either of the PRAM models?

2. Is there an efficient output-sensitive parallel algorithm for constructing 3D Voronoi diagrams?

46.5 GEOMETRIC SEARCHING

Given a subdivision of space by a collection S of geometric objects, such as line segments, the point location problem is to build a data structure for this set that can quickly answer *vertical ray-shooting queries*, where one is given a point p and asked to report the first object in S hit by a vertical ray from p. We summarize efficient parallel algorithms for planar point location in Table 46.5.1. The time and work bounds listed, as well as the computational model, are for building the data structure to achieve an $O(\log n)$ query time. We do not list the space bounds for any of these methods in the table since, in every case, they are equal to the preprocessing work bounds.

GLOSSARY

Arbitrary planar subdivision: A subdivision of the plane (not necessarily connected), defined by a set of line segments that intersect only at their endpoints.

Monotone subdivision: A connected subdivision of the plane in which each face is intersected by a vertical line in a single segment.

Triangulated subdivision: A connected subdivision of the plane into triangles whose corners are vertices of the subdivision (see Chapter 29).

Shortest path in a polygon: The shortest path between two points that does not go outside of the polygon (see Section 30.4).

Ray-shooting query: A query whose answer is the first object hit by a ray oriented in a specified direction from a specified point.

TABLE 46.5.1   Parallel geometric searching algorithms.

QUERY PROBLEM	MODEL	TIME	WORK	REF
Point loc in arb subdivision	CREW	$O(\log n)$	$O(n \log n)$	[ACG89]
Point loc in monotone subdivision	EREW	$O(\log n)$	$O(n)$	[TV91]
Point loc in triangulated subdivision	CREW	$O(\log n)$	$O(n)$	[CZ90]
Point loc in d-dim hyp arr	EREW	$O(\log n)$	$O(n^d)$	[AGR94]
Shortest path in triangulated polygon	CREW	$O(\log n)$	$O(n)$	[GSG92]
Ray shooting in triangulated polygon	CREW	$O(\log n)$	$O(n)$	[HS95]
Line & convex polyhedra intersection	CREW	$O(\log n)$	$O(n)$	[DK89b, CZ90]

OPEN PROBLEMS

1. Is there an efficient data structure that allows n simultaneous point locations to be performed in $O(\log n)$ time using $O(n)$ processors in the EREW PRAM model?

2. Is there an efficient data structure for 3-dimensional point location in convex subdivisions that can be constructed in $O(n \log n)$ work and at most $O(\log^2 n)$ time and which allows for a query time that is at most $O(\log^2 n)$?

46.6 VISIBILITY, ENVELOPES, AND OPTIMIZATION

We summarize efficient parallel methods for various visibility and lower envelope problems for a simple polygon with n vertices in Table 46.6.1. In the table, m denotes the number of edges in a visibility graph. For definitions see Chapter 33.

TABLE 46.6.1   Parallel visibility algorithms for a simple polygon.

PROBLEM	MODEL	TIME	WORK	REF
Kernel	EREW	$O(\log n)$	$O(n)$	[Che95]
Vis from a point	EREW	$O(\log n)$	$O(n)$	[ACW91]
Vis from an edge	CRCW	$O(\log n)$	$O(n)$	[Her95]
Vis from an edge	CREW	$O(\log n)$	$O(n \log n)$	[GSG92, GSG93]
Vis graph	CREW	$O(\log n)$	$O(n \log^2 n + m)$	[GSG92, GSG93]

We sketch the algorithm for computing the point visibility polygon [ACW91], which is notable for two reasons: first, it is employed as a subprogram in many other algorithms; and second, it requires much more intricate processing and analysis than the relatively simple optimal sequential algorithm (Section 33.3). The parallel algorithm is recursive, partitioning the boundary into $n^{1/4}$ subchains, and computing *visibility chains* from the source point of visibility x. Each of these chains is star-shaped with respect to x, i.e., effectively "monotone" (see Section 30.1). This monotonicity property is, however, insufficient to intersect the visibility chains quickly enough in the merge step to obtain optimal bounds. Rather, the fact that the chains are subchains of the boundary of a simple polygon must be exploited to achieve logarithmic-time computation of the intersection of two chains. This then leads to the optimal bounds quoted in Table 46.6.1.

The bounds of efficient parallel methods for visibility problems on general sets of segments and curves in the plane are summarized in Table 46.6.2.

GLOSSARY

Lower envelope: The function $F(x)$ defined as the pointwise minimum of a collection of functions $\{f_1, f_2, \ldots, f_n\}$: $F(x) = \min_i f_i(x)$ (see Section 28.2).

k-intersecting curves: A set of curves, every two of which intersect at most k times (where they cross).

$\lambda_s(n):$ The maximum length of a Davenport-Schinzel sequence [SA95, AS00] of order s on n symbols. If s is a constant, $\lambda_s(n)$ is $o(n \log^* n)$. See Section 28.10.

TABLE 46.6.2   General parallel visibility and enveloping algorithms.

PROBLEM	MODEL	TIME	WORK	REF
Lower env for segments	EREW	$O(\log n)$	$O(n \log n)$	[CW02]
Lower env for k-int curves	rand-CRCW	$\bar{O}(\log n \log^* n)$	$\bar{O}(\lambda_k(n) \log n)$	[Goo91b]
Lower env for k-int curves	EREW	$O(\log^{1+\epsilon} n)$	$O(\lambda_{k+1}(n) \log n \log^* n)$	[CW02]

Finally, we summarize some efficient parallel algorithms for solving several geometric optimization problems in Table 46.6.3.

GLOSSARY

Largest-area empty rectangle: For a collection S of n points in the plane, the largest-area rectangle that does not contain any point of S in its interior.

All-farthest neighbors problem in a simple polygon: Determine for each vertex p of a simple polygon the vertex q such that the shortest path from p is longest.

Closest visible-pair between polygons: A closest pair of mutually-visible vertices between two nonintersecting simple polygons in the plane.

Minimum circular-arc cover: For a collection of n arcs of a given circle C, a minimum-cardinality subset that covers C.

Optimal-area inscribed/circumscribed triangle: For a convex polygon P, the largest-area triangle inscribed in P, or, respectively, the smallest-area triangle circumscribing P.

Min-link path in a polygon: A piecewise-linear path of fewest "links" inside a simple polygon between two given points p and q; see Sections 30.4.

TABLE 46.6.3   Parallel geometric optimization algorithms.

PROBLEM	MODEL	TIME	WORK	REF
Largest-area empty rectangle	CREW	$O(\log^2 n)$	$O(n \log^3 n)$	[AKPS90]
All-farthest neighbors in polygon	CREW	$O(\log^2 n)$	$O(n \log^2 n)$	[Guh92]
Closest visible-pair btw polygons	CREW	$O(\log n)$	$O(n \log n)$	[HCL92]
Min circular-arc cover	EREW	$O(\log n)$	$O(n \log n)$	[AC89]
Opt-area inscr/circum triangle	CRCW	$O(\log \log n)$	$O(n)$	[CM92]
Opt-area inscr/circum triangle	CREW	$O(\log n)$	$O(n)$	[CM92]
Min-link path in a polygon	CREW	$\bar{O}(\log n \log \log n)$	$O(n \log n \log \log n)$	[CGM+95]

OPEN PROBLEMS

1. Can the visibility graph of a set of n nonintersecting line segments be constructed using $O(n \log n + m)$ work in time at most $O(\log^2 n)$ in the CREW model, where m is the size of the graph?

2. Can the visibility graph of a triangulated polygon be computed in $O(\log n)$ time using $O(n + m)$ work in the CREW model?

46.7 SOURCES AND RELATED MATERIAL

FURTHER READING

Our presentation has been results oriented and has not provided much problem intuition or algorithmic techniques. There are several excellent surveys available in the literature [Ata92, AC94, AC00, AG93, RS93, RS00] that are more techniques oriented. Another good location for related material is the book by Akl and Lyons [AL93].

Finally, while we have focused on the classical PRAM model, recently, several extensions have been proposed that consider the effects of the hierarchical memory design of modern multicore processors on the runtime [BG04, CGK+07, BCG+08, AGNS08, CRSB13]. In these models, the goal is to organize data and schedule computations in such a way as to minimize the number of accesses to the slow shared memory by utilizing fast private and/or shared caches. This is still a relatively new research direction with many problems remaining open. However, some results have already been obtained in these models for 2D axis-parallel line segment and rectangle intersection reporting [ASZ10, ASZ11], 1D and 2D range reporting [ASZ10, SZ12], 2D lower envelope and convex hull computation [ASZ10] and 3D maxima reporting [ASZ10].

RELATED CHAPTERS

Chapter 26: Convex hull computations
Chapter 27: Voronoi diagrams and Delaunay triangulations
Chapter 28: Arrangements
Chapter 30: Polygons
Chapter 38: Point location
Chapter 42: Geometric intersection
Chapter 44: Randomization and derandomization
Chapter 49: Linear programming

REFERENCES

[AC89]   M.J. Atallah and D.Z. Chen. An optimal parallel algorithm for the minimum circle-cover problem. *Inform. Process. Lett.*, 34:159–165, 1989.

[AC94]   M.J. Atallah and D.Z. Chen. Parallel computational geometry. In A.Y. Zomaya, editor, *Parallel Computations: Paradigms and Applications*, World Scientific, Singapore, 1994.

[AC00]   M.J. Atallah and D.Z. Chen. Deterministic parallel computational geometry. In J.-R. Sack and J. Urrutia, editors, *Handbook of Computational Geometry*, pages 155–200, Elsevier, Amsterdam, 2000.

[ACG⁺88]   A. Aggarwal, B. Chazelle, L.J. Guibas, C. Ó'Dúnlaing, and C. Yap. Parallel computational geometry. *Algorithmica*, 3:293–327, 1988.

[ACG89]   M.J. Atallah, R. Cole, and M.T. Goodrich. Cascading divide-and-conquer: A technique for designing parallel algorithms. *SIAM J. Comput.*, 18:499–532, 1989.

[ACW91]   M.J. Atallah, D.Z. Chen, and H. Wagener. Optimal parallel algorithm for visibility of a simple polygon from a point. *J. ACM*, 38:516–553, 1991.

[AG86]   M.J. Atallah and M.T. Goodrich. Efficient parallel solutions to some geometric problems. *J. Parallel Distrib. Comput.*, 3:492–507, 1986.

[AG93]   M.J. Atallah and M.T. Goodrich. Deterministic parallel computational geometry. In J.H. Reif, editor, *Synthesis of Parallel Algorithms*, pages 497–536. Morgan Kaufmann, San Mateo, 1993.

[AGNS08]   L. Arge, M.T. Goodrich, M. Nelson, and N. Sitchinava. Fundamental parallel algorithms for private-cache chip multiprocessors. In *Proc. 20th ACM Sympos. Parallel. Algorithms Architect.*, pages 197–206, 2008.

[AGR94]   N.M. Amato, M.T. Goodrich, and E.A. Ramos. Parallel algorithms for higher-dimensional convex hulls. In *Proc. 35th IEEE Sympos. Found. Comp. Sci.*, pages 683–694, 1994.

[AGR95]   N.M. Amato, M.T. Goodrich, and E.A. Ramos. Computing faces in segment and simplex arrangements. In *Proc. 27th ACM Sympos. Theory Comput.*, pages 672–682, 1995.

[Akl82]   S.G. Akl. A constant-time parallel algorithm for computing convex hulls. *BIT*, 22:130–134, 1982.

[Akl84]   S.G. Akl. Optimal parallel algorithms for computing convex hulls and for sorting. *Computing*, 33:1–11, 1984.

[Akl85]   S.G. Akl. Optimal parallel algorithms for selection, sorting and computing convex hulls. In G.T. Toussaint, editor, *Computational Geometry*, vol. 2 of *Machine Intelligence and Pattern Recognition*, pages 1–22. North-Holland, Amsterdam, 1985.

[AKPS90]   A. Aggarwal, D. Kravets, J.K. Park, and S. Sen. Parallel searching in generalized Monge arrays with applications. In *Proc. 2nd ACM Sympos. Parallel Algorithms Architect.*, pages 259–268, 1990.

[AL93]   S.G. Akl and K.A. Lyons. *Parallel Computational Geometry*. Prentice Hall, Englewood Cliffs, 1993.

[AM94]   N. Alon and N. Megiddo. Parallel linear programming in fixed dimension almost surely in constant time. *J. ACM*, 41:422–434, 1994.

[AP92]   N.M. Amato and F.P. Preparata. The parallel 3D convex hull problem revisited. *Internat. J. Comput. Geom. Appl.*, 2:163–173, 1992.

[AP93]   N.M. Amato and F.P. Preparata. An NC¹ parallel 3D convex hull algorithm. In *Proc. 9th Sympos. Comput. Geom.*, pages 289–297, ACM Press, 1993.

[AP95]   N.M. Amato and F.P. Preparata. A time-optimal parallel algorithm for three-dimensional convex hulls. *Algorithmica*, 14:169–182, 1995.

[AS00]   P.K. Agarwal and M. Sharir. Davenport-Schinzel sequences and their geometric applications. In J.-R. Sack and J. Urrutia, editors, *Handbook of Computational Geometry*, pages 1–47, Elsevier, Amsterdam, 2000.

[ASZ10]   D. Ajwani, N. Sitchinava, and N. Zeh. Geometric algorithms for private-cache chip multiprocessors. In *Proc. 18th Europ. Sympos. Alg.*, part II, vol. 6347 of *LNCS*, pages 75–86, Springer, Berlin, 2010.

[ASZ11]   D. Ajwani, N. Sitchinava, and N. Zeh. I/O-optimal distribution sweeping on private-cache chip multiprocessors. In *Proc. 25th IEEE Int. Parallel Distrib. Proc. Sympos.*, 'pages 1114–1123, 2011.

[Ata92]   M.J. Atallah. Parallel techniques for computational geometry. *Proc. IEEE*, 80:1435–1448, 1992.

[BCG⁺08]   G.E. Blelloch, R.A. Chowdhury, P.B. Gibbons, V. Ramachandran, S. Chen, and M. Kozuch. Provably good multicore cache performance for divide-and-conquer algorithms. In *Proc. 19th ACM-SIAM Sympos. Discrete Algorithms*, pages 501–510, 2008.

[BET99]   M. Bern, D. Eppstein, and S.-H. Teng. Parallel construction of quadtrees and quality triangulations. *Internat. J. Comput. Geom. Appl.*, 9:517–532, 1999.

[BG04]   G.E. Blelloch and P.B. Gibbons. Effectively sharing a cache among threads. In *Proc. 16th ACM Sympos. Parallel. Algorithms Architect.*, pages 235–244, 2004.

[BSV96]   O. Berkman, B. Schieber, and U. Vishkin. A fast parallel algorithm for finding the convex hull of a sorted point set. *Internat. J. Comput. Geom. Appl.*, 6:231–242, 1996.

[Cal93]   P.B. Callahan. Optimal parallel all-nearest-neighbors using the well-separated pair decomposition. In *Proc. 34th IEEE Sympos. Found. Comp. Sci.*, pages 332–340, 1993.

[CCT92a]   K.L. Clarkson, R. Cole, and R.E. Tarjan. Erratum: Randomized parallel algorithms for trapezoidal diagrams. *Internat. J. Comput. Geom. Appl.*, 2:341–343, 1992.

[CCT92b]   K.L. Clarkson, R. Cole, and R.E. Tarjan. Randomized parallel algorithms for trapezoidal diagrams. *Internat. J. Comput. Geom. Appl.*, 2:117–133, 1992.

[CG92]   R. Cole and M.T. Goodrich. Optimal parallel algorithms for polygon and point-set problems. *Algorithmica*, 7:3–23, 1992.

[CGK⁺07]   S. Chen, P.B. Gibbons, M. Kozuch, V. Liaskovitis, A. Ailamaki, G.E. Blelloch, B. Falsafi, L. Fix, N. Hardavellas, T.C. Mowry, and C. Wilkerson. Scheduling threads for constructive cache sharing on CMPs. In *Proc. 19th ACM Sympos. Parallel. Algorithms Architect.*, pages 105–115, 2007.

[CGM⁺95]   V. Chandru, S.K. Ghosh, A. Maheshwari, V.T. Rajan, and S. Saluja. *NC*-algorithms for minimum link path and related problems. *J. Algorithms*, 19:173–203, 1995.

[CGÓ96]   R. Cole, M.T. Goodrich, and C. Ó'Dúnlaing. A nearly optimal deterministic parallel Voronoi diagram algorithm. *Algorithmica*, 16:569–617, 1996.

[Che95]   D.Z. Chen. Efficient geometric algorithms on the EREW PRAM. *IEEE Trans. Parallel Distrib Syst.*, 6:41–47, 1995.

[Cho80]   A.L. Chow. *Parallel Algorithms for Geometric Problems*. Ph.D. thesis, Dept. Comp. Sci., Univ. Illinois, Urbana, IL, 1980.

[CM92]   S. Chandran and D.M. Mount. A parallel algorithm for enclosed and enclosing triangles. *Internat. J. Comput. Geom. Appl.*, 2:191–214, 1992.

[CRSB13]   R.A. Chowdhury, V. Ramachandran, F. Silvestri, and B. Blakeley. Oblivious algorithms for multicores and networks of processors. *J. Parallel Distrib. Comput.*, 73:911–925, 2013.

[CW02]   W. Chen and K. Wada. On computing the upper envelope of segments in parallel. *IEEE Trans. Parallel Distrib. Syst.*, 13:5–13, 2002.

[CX02]   D.Z. Chen and J. Xu. Two-variable linear programming in parallel. *Computat Geom*, 21:155–165, 2002.

[CZ90]   R. Cole and O. Zajicek. An optimal parallel algorithm for building a data structure for planar point location. *J. Parallel Distrib. Comput.*, 8:280–285, 1990.

[DK89a] N. Dadoun and D.G. Kirkpatrick. Cooperative subdivision search algorithms with applications. In *Proc. 27th Allerton Conf. Commun. Control Comput.*, pages 538–547, 1989.

[DK89b] N. Dadoun and D.G. Kirkpatrick. Parallel construction of subdivision hierarchies. *J. Comp. Syst. Sci.*, 39:153–165, 1989.

[Ede87] H. Edelsbrunner. *Algorithms in Combinatorial Geometry.* Volume 10 of *EATCS Monogr. Theoret. Comp. Sci.*. Springer-Verlag, Heidelberg, 1987.

[GG91]   M.R. Ghouse and M.T. Goodrich. In-place techniques for parallel convex hull algorithms. In *Proc. 3rd ACM Sympos. Parallel Algorithms Architect.*, pages 192–203, 1991.

[GG97]   M.R. Ghouse and M.T. Goodrich. Fast randomized parallel methods for planar convex hull construction. *Comput. Geom.*, 7:219–235, 1997.

[GM91]   S.K. Ghosh and A. Maheshwari. An optimal parallel algorithm for determining the intersection type of two star-shaped polygons. In *Proc. 3rd Canad. Conf. Comput. Geom.*, pages 2–6, 1991.

[Goo87] M.T. Goodrich. *Efficient Parallel Techniques for Computational Geometry.* Ph.D. thesis, Dept. Comp. Sci., Purdue Univ., West Lafayette, 1987.

[Goo89] M.T. Goodrich. Triangulating a polygon in parallel. *J. Algorithms*, 10:327–351, 1989.

[Goo91a] M.T. Goodrich. Intersecting line segments in parallel with an output-sensitive number of processors. *SIAM J. Comput.*, 20:737–755, 1991.

[Goo91b] M.T. Goodrich. Using approximation algorithms to design parallel algorithms that may ignore processor allocation. In *Proc. 32nd IEEE Sympos. Found. Comp. Sci.*, pages 711–722, 1991.

[Goo93] M.T. Goodrich. Geometric partitioning made easier, even in parallel. In *Proc. 9th Sympos. Comput. Geom.*, pages 73–82, ACM Press, 1993.

[Goo95] M.T. Goodrich. Planar separators and parallel polygon triangulation. *J. Comp. Syst. Sci.*, 51:374–389, 1995.

[Goo96] M.T. Goodrich. Fixed-dimensional parallel linear programming via relative epsilon-approximations. In *Proc. 7th ACM-SIAM Sympos. Discrete Algorithms*, pages 132–141, 1996.

[GR97]   M.T. Goodrich and E.A. Ramos. Bounded-independence derandomization of geometric partitioning with applications to parallel fixed-dimensional linear programming. *Discrete Comput. Geom.*, 18:397–420, 1997.

[GS97]   N. Gupta and S. Sen. Optimal, output-sensitive algorithms for constructing planar hulls in parallel. *Comput. Geom.*, 8:151–166, 1997.

[GS03]   N. Gupta and S. Sen. Faster output-sensitive parallel algorithms for 3d convex hulls and vector maxima. *J. Parallel Distrib. Comput.*, 63:488–500, 2003.

[GSG92] M.T. Goodrich, S. Shauck, and S. Guha. Parallel methods for visibility and shortest path problems in simple polygons. *Algorithmica*, 8:461–486, 1992.

[GSG93]   M.T. Goodrich, S. Shauck, and S. Guha. Addendum to "parallel methods for visibility and shortest path problems in simple polygons." *Algorithmica*, 9:515–516, 1993.

[Guh92]   S. Guha. Parallel computation of internal and external farthest neighbours in simple polygons. *Internat. J. Comput. Geom. Appl.*, 2:175–190, 1992.

[HCL92]   F.R. Hsu, R.C. Chang, and R.C.T. Lee. Parallel algorithms for computing the closest visible vertex pair between two polygons. *Internat J. Comput. Geom. Appl.*, 2:135–162, 1992.

[Her95]   J. Hershberger. Optimal parallel algorithms for triangulated simple polygons. *Internat. J. Comput. Geom. Appl.*, 5:145–170, 1995.

[HS95]   J. Hershberger and S. Suri. A pedestrian approach to ray shooting: Shoot a ray, take a walk. *J. Algorithms*, 18:403–431, 1995.

[Já92]   J. JáJá. *An Introduction to Parallel Algorithms*. Addison-Wesley, Reading, 1992.

[KR90]   R.M. Karp and V. Ramachandran. Parallel algorithms for shared memory machines. In J. van Leeuwen, editor, *Handbook of Theoretical Computer Science*, pages 869–941, Elsevier/The MIT Press, Amsterdam, 1990.

[MS88]   R. Miller and Q.F. Stout. Efficient parallel convex hull algorithms. *IEEE Trans. Comp.*, 37:1605–1618, 1988.

[O'R94]   J. O'Rourke. *Computational Geometry in C*. Cambridge University Press, 1st edition, 1994.

[PS85]   F.P. Preparata and M.I. Shamos. *Computational Geometry: An Introduction*. Springer-Verlag, New York, 1985.

[Ram97]   E.A. Ramos. Construction of 1-d lower envelopes and applications. In *Proc. 13th Sympos. Comput. Geom.*, pages 57–66, ACM Press, 1997.

[Rei93]   J.H. Reif. *Synthesis of Parallel Algorithms*. Morgan Kaufmann, San Mateo, 1993.

[RS92]   J.H. Reif and S. Sen. Optimal parallel randomized algorithms for three-dimensional convex hulls and related problems. *SIAM J. Comput.*, 21:466–485, 1992.

[RS93]   S. Rajasekaran and S. Sen. Random sampling techniques and parallel algorithms design. In J.H. Reif, editor, *Synthesis of Parallel Algorithms*, pages 411–452, Morgan Kaufmann, San Mateo, 1993.

[RS00]   J.H. Reif and S. Sen. Parallel computational geometry: An approach using randomization. In J.-R. Sack and J. Urrutia, editors, *Handbook of Computational Geometry*, pages 765–828, Elsevier, Amsterdam, 2000.

[Rüb92]   C. Rüb. Computing intersections and arrangements for red-blue curve segments in parallel. In *Proc 4th Canad. Conf. Comput. Geom.*, pages 115–120, 1992.

[SA95]   M. Sharir and P.K. Agarwal. *Davenport-Schinzel Sequences and Their Geometric Applications*. Cambridge University Press, 1995.

[Sen89]   S. Sen. *Random Sampling Techniques for Efficient Parallel Algorithms in Computational Geometry*. Ph.D. thesis, Dept. Comp. Sci., Duke Univ., 1989.

[SZ12]   N. Sitchinava and N. Zeh. A parallel buffer tree. In *Proc. 24th ACM Sympos. Parallel. Algorithms Architect.*, pages 214–223, 2012.

[TV91]   R. Tamassia and J.S. Vitter. Parallel transitive closure and point location in planar structures. *SIAM J. Comput.*, 20:708–725, 1991.

[WC90]   Y.C. Wee and S. Chaiken. An optimal parallel L_1-metric Voronoi diagram algorithm. In *Proc. 2nd Canadian Conf. Comput. Geom.*, pages 60–65, 1990.

[Yap88]   C.K. Yap. Parallel triangulation of a polygon in two calls to the trapezoidal map. *Algorithmica*, 3:279–288, 1988.

47 EPSILON-APPROXIMATIONS AND EPSILON-NETS

Nabil H. Mustafa and Kasturi Varadarajan

INTRODUCTION

The use of random samples to approximate properties of geometric configurations has been an influential idea for both combinatorial and algorithmic purposes. This chapter considers two related notions—ϵ-approximations and ϵ-nets—that capture the most important quantitative properties that one would expect from a random sample with respect to an underlying geometric configuration. An example problem: given a set P of points in the plane and a parameter $\epsilon > 0$, is it possible to choose a set N of $O(\frac{1}{\epsilon})$ points of P such that N contains at least one point from *each* disk containing $\epsilon|P|$ points of P? More generally, what is the smallest non-empty set $A \subseteq P$ that can be chosen such that for any disk D in the plane, the proportion of points of P contained in D is within ϵ to the proportion of points of A contained in D? In both these cases, a random sample provides an answer "in expectation," establishing worst-case guarantees is the topic of this chapter.

47.1 SET SYSTEMS DERIVED FROM GEOMETRIC CONFIGURATIONS

Before we present work on ϵ-approximations and ϵ-nets for geometric set systems, we briefly survey different types of set systems that can be derived from geometric configurations and study the combinatorial properties of these set systems due to the constraints induced by geometry. For example, consider the fact that for any set P of points in the plane, there are only $O(|P|^3)$ subsets of P induced by containment by disks. This is an immediate consequence of the property that three points of P are sufficient to "anchor" a disk. This property will be abstracted to a purely combinatorial one, called the *VC-dimension* of a set system, from which can be derived many analogous properties for abstract set systems.

GLOSSARY

Set systems: A pair $\Sigma = (X, \mathcal{R})$, where X is a set of base elements and $\mathcal{R}$ is a collection of subsets of X, is called a set system. The *dual set system* to $(X, \mathcal{R})$ is the system $\Sigma^* = (X^*, \mathcal{R}^*)$, where $X^* = \mathcal{R}$, and for each $x \in X$, the set $\mathcal{R}_x := \{R \in \mathcal{R} : x \in R\}$ belongs to $\mathcal{R}^*$.

VC-dimension: For any set system $(X, \mathcal{R})$ and $Y \subseteq X$, the projection of $\mathcal{R}$ on Y is the set system $\mathcal{R}|_Y := \{Y \cap R : R \in \mathcal{R}\}$. The Vapnik-Chervonenkis dimension (or VC-dimension) of $(X, \mathcal{R})$, denoted as VC-dim$(\mathcal{R})$, is the minimum integer d such that $|\mathcal{R}|_Y| < 2^{|Y|}$ for any finite subset $Y \subseteq X$ with $|Y| > d$.

Shatter function: A set Y is *shattered* by $\mathcal{R}$ if $|\mathcal{R}|_Y| = 2^{|Y|}$. The *shatter function*, $\pi_\mathcal{R} : \mathbb{N} \to \mathbb{N}$, of a set system $(X, \mathcal{R})$ is obtained by letting $\pi_\mathcal{R}(m)$ be the maximum number of subsets in $\mathcal{R}|_Y$ for any set $Y \subseteq X$ of size m.

Shallow-cell complexity: A set system $(X, \mathcal{R})$ has *shallow-cell complexity* $\varphi_\mathcal{R} : \mathbb{N} \times \mathbb{N} \to \mathbb{N}$, if for every $Y \subseteq X$, the number of sets of size at most l in the set system $\mathcal{R}|_Y$ is $O\big(|Y| \cdot \varphi_\mathcal{R}(|Y|, l)\big)$. For convenience, dropping the second argument of $\varphi_\mathcal{R}$, we say that $(X, \mathcal{R})$ has *shallow-cell complexity* $\varphi_\mathcal{R} : \mathbb{N} \to \mathbb{N}$, if there exists a constant $c(\mathcal{R}) > 0$ such that for every $Y \subseteq X$ and for every positive integer l, the number of sets of size at most l in $\mathcal{R}|_Y$ is $O\big(|Y| \cdot \varphi_\mathcal{R}(|Y|) \cdot l^{c(\mathcal{R})}\big)$.

Geometric set systems: Let $\mathcal{R}$ be a family of (possibly unbounded) geometric objects in $\mathbb{R}^d$, and X be a finite set of points in $\mathbb{R}^d$. Then the set system $(X, \mathcal{R}|_X)$ is called a *primal set system induced by* $\mathcal{R}$. Given a finite set $\mathcal{S} \subseteq \mathcal{R}$, the *dual set system induced by* $\mathcal{S}$ is the set system $(\mathcal{S}, \mathcal{S}^*)$, where $\mathcal{S}^* = \{S_x \; : \; x \in \mathbb{R}^d\}$ and $S_x := \{S \in \mathcal{S} \; : \; x \in S\}$.

Union complexity of geometric objects: The *union complexity*, $\kappa_\mathcal{R} : \mathbb{N} \to \mathbb{N}$, of a family of objects $\mathcal{R}$ is obtained by letting $\kappa_\mathcal{R}(m)$ be the maximum number of faces of all dimensions that the union of any m members of $\mathcal{R}$ can have.

δ-Separated set systems: The symmetric difference of two sets R, R' is denoted as $\Delta(R, R')$, where $\Delta(R, R') = (R \setminus R') \cup (R' \setminus R)$. Call a set system $(X, \mathcal{R})$ δ-*separated* if for every pair of sets $R, R' \in \mathcal{R}$, $|\Delta(R, R')| \geq \delta$.

VC-DIMENSION

First defined by Vapnik and Chervonenkis [VC71], a crucial property of VC-dimension is that it is *hereditary*—if a set system $(X, \mathcal{R})$ has VC-dimension d, then for any $Y \subseteq X$, the VC-dimension of the set system $(Y, \mathcal{R}|_Y)$ is at most d.

LEMMA 47.1.1 [VC71, Sau72, She72]

Let $(X, \mathcal{R})$ be a set system with VC-dim$(\mathcal{R}) \leq d$ for a fixed constant d. Then for all positive integers m,

$$\pi_\mathcal{R}(m) \leq \sum_{i=0}^{d} \binom{m}{i} = O\left(\left(\frac{em}{d}\right)^d\right).$$

Conversely, if $\pi_\mathcal{R}(m) \leq cm^d$ for some constant c, then VC-dim$(\mathcal{R}) \leq 4d\log(cd)$.

Throughout this chapter, we usually state the results in terms of shatter functions of set systems; the first part of Lemma 47.1.1 implies that these results carry over for set systems with bounded VC-dimension as well. Geometric set systems often have bounded VC-dimension, a key case being the primal set system induced by half-spaces in $\mathbb{R}^d$, for which Radon's lemma [Rad21] implies the following.

LEMMA 47.1.2

Let $\mathcal{H}$ be the family of all half-spaces in $\mathbb{R}^d$. Then VC-dim$(\mathcal{H}) = d + 1$. Consequently, $\pi_\mathcal{H}(m) = O(m^{d+1})$.

Lemma 47.1.2 is the starting point for bounding the VC-dimension of a large category of geometric set systems. For example, it implies that the VC-dimension of the primal set system induced by balls in $\mathbb{R}^d$ is $d + 1$, since if a set of points is shattered by the primal set system induced by balls, then it is also shattered by the primal set system induced by half-spaces[1]. More generally, sets defined by polynomial inequalities can be lifted to half-spaces in some higher dimension by Veronese maps and so also have bounded VC-dimension. Specifically, identify each d-variate polynomial $f(x_1, \ldots, x_d)$ with its induced set $S_f := \{p \in \mathbb{R}^d : f(p) \geq 0\}$. Then Veronese maps—i.e., identifying the $d' = \binom{D+d}{d}$ coefficients of a d-variate polynomial of degree at most D with distinct coordinates of $\mathbb{R}^{d'}$—together with Lemma 47.1.2 immediately imply the following.

LEMMA 47.1.3 [Mat02a]

Let $\mathcal{R}_{d,D}$ be the primal set system induced by all d-variate polynomials over $\mathbb{R}^d$ of degree at most D. Then VC-dim$(\mathcal{R}_{d,D}) \leq \binom{D+d}{d}$.

Set systems derived from other bounded VC-dimension set systems using a finite sequence of set operations can be shown to also have bounded VC-dimension. The number of sets in this derived set system can be computed by a direct combinatorial argument, which together with the second part of Lemma 47.1.1 implies the following.

LEMMA 47.1.4 [HW87]

Let $(X, \mathcal{R})$ be a set system with VC-dim$(\mathcal{R}) \leq d$, and $k \geq 1$ an integer. Define the set system

$$F_k(\mathcal{R}) := \{F(R_1, \ldots, R_k) : R_1, \ldots, R_k \in \mathcal{R}\},$$

where $F(S_1, \ldots, S_k)$ denotes the set derived from the input sets $S_1, \ldots, S_k$ from a fixed finite sequence of union, intersection and difference operations. Then we have VC-dim$(F_k(\mathcal{R})) = O(kd \log k)$.

LEMMA 47.1.5 [Ass83]

Given a set system $\Sigma = (X, \mathcal{R})$ and its dual system $\Sigma^* = (X^*, \mathcal{R}^*)$, VC-dim$(\mathcal{R}^*) < 2^{\text{VC-dim}(\mathcal{R})+1}$.

Note that if $\pi_{\mathcal{R}^*}(m) = O(m^d)$ for some constant d, then the second part of Lemma 47.1.1 implies that VC-dim$(\mathcal{R}^*) = O(d \log d)$, and Lemma 47.1.5 then implies that VC-dim$(\mathcal{R}) = 2^{O(d \log d)} = d^{O(d)}$.

On the other hand, the primal set system induced by convex objects in $\mathbb{R}^2$ has unbounded VC-dimension, as it shatters any set of points in convex position.

SHALLOW-CELL COMPLEXITY

A key realization following from the work of Clarkson and Varadarajan [CV07] and Varadarajan [Var10] was to consider a finer classification of set systems than just based on VC-dimension, namely its shallow-cell complexity, first defined explicitly in Chan *et al.* [CGKS12]. Note that if $(X, \mathcal{R})$ has shallow-cell complexity $\varphi_{\mathcal{R}}(m) = O(m^t)$ for some constant t, then $\pi_{\mathcal{R}}(m) = O(m^{1+t+c(\mathcal{R})})$ for an absolute constant $c(\mathcal{R})$, and so $\mathcal{R}$ has bounded VC-dimension. On the other hand, while the shatter function bounds the total number of sets in the projection of $\mathcal{R}$ onto a subset Y, it does not give any information on the *distribution* of the set sizes, which has turned

---

[1]Assume that a set X of points in $\mathbb{R}^d$ is shattered by the primal set system induced by balls. Then for any $Y \subseteq X$, there exists a ball B with $Y = B \cap X$, and a ball B' with $X \setminus Y = B' \cap X$. Then any hyperplane that separates $B \setminus B'$ from $B' \setminus B$ also separates Y from $X \setminus Y$.

TABLE 47.1.1   Combinatorial properties of some primal (P) and dual (D) geometric set systems.

OBJECTS	SETS	$\varphi(m)$	VC-dim	$\pi(m)$
Intervals	P/D	$O(1)$	2	$\Theta(m^2)$
Lines in $\mathbb{R}^2$	P/D	$O(m)$	2	$\Theta(m^2)$
Pseudo-disks in $\mathbb{R}^2$	P	$O(1)$	3	$O(m^3)$
Pseudo-disks in $\mathbb{R}^2$	D	$O(1)$	$O(1)$	$O(m^2)$
Half-spaces in $\mathbb{R}^d$	P/D	$O(m^{\lfloor d/2 \rfloor - 1})$	$d+1$	$\Theta(m^d)$
Balls in $\mathbb{R}^d$	P	$O(m^{\lceil d/2 \rceil - 1})$	$d+1$	$\Theta(m^{d+1})$
Balls in $\mathbb{R}^d$	D	$O(m^{\lceil d/2 \rceil - 1})$	$d+1$	$\Theta(m^d)$
Triangles in $\mathbb{R}^2$	D	$O(m)$	7	$O(m^7)$
Fat triangles in $\mathbb{R}^2$	D	$O(\log^* m)$	7	$O(m^7)$
Axis-par. rect. in $\mathbb{R}^2$	P	$O(m)$	4	$\Theta(m^4)$
Axis-par. rect. in $\mathbb{R}^2$	D	$O(m)$	4	$\Theta(m^2)$
Convex sets in $\mathbb{R}^d$	P	$O(2^m/m)$	∞	$\Theta(2^m)$
Translates of a convex set in $\mathbb{R}^d, d \geq 3$	P	$O(2^m/m)$	∞	$\Theta(2^m)$

out to be a key parameter (as we will see later in, e.g., Theorem 47.4.5). Tight bounds on shatter functions and shallow-cell complexity are known for many basic geometric set systems.

LEMMA 47.1.6 [CS89]

Let $\mathcal{H}$ be the family of all half-spaces in $\mathbb{R}^d$. Then $\varphi_{\mathcal{H}}(m) = O(m^{\lfloor d/2 \rfloor - 1})$. Furthermore, this bound is tight, in the sense that for any integer $m \geq 1$, there exist m points for which the above bound can be attained.

The following lemma, a consequence of a probabilistic technique by Clarkson and Shor [CS89], bounds the shallow-cell complexity of the dual set system induced by a set of objects in $\mathbb{R}^2$.

LEMMA 47.1.7 [Sha91]

Let $\mathcal{R}$ be a finite set of objects in $\mathbb{R}^2$, each bounded by a closed Jordan curve, and with union complexity $\kappa_{\mathcal{R}}(\cdot)$. Further, each intersection point in the arrangement of $\mathcal{R}$ is defined by a constant number of objects of $\mathcal{R}$. Then the shallow-cell complexity of the dual set system induced by $\mathcal{R}$ is bounded by $\varphi_{\mathcal{R}^*}(m) = O\left(\frac{\kappa_{\mathcal{R}}(m)}{m}\right)$.

Table 47.1.1 states the shatter function as well as the shallow-cell complexity of some commonly used set systems. Some of these bounds are derived from the above two lemmas using known bounds on union complexity of geometric objects (e.g., pseudo-disks [BPR13], fat triangles [ABES14]) while others are via explicit constructions [NT10].

A packing lemma. A key combinatorial statement at the heart of many of the results in this chapter is inspired by packing properties of geometric objects. It was first proved for the primal set system induced by half-spaces in $\mathbb{R}^d$ by geometric techniques [CW89]; the following more general form was first shown by Haussler [Hau95][2] (see [Mat99, Chapter 5.3] for a nice exposition of this result).

---

[2]The theorem as stated in [Hau95] originally required that VC-dim$(\mathcal{P}) \leq d$. It was later verified that the proof also works with the assumption of polynomially bounded shatter functions;

LEMMA 47.1.8 [Hau95]

Let $(X, \mathcal{P})$, $|X| = n$, be a δ-separated set system with $\delta \geq 1$ and $\pi_{\mathcal{P}}(m) = O(m^d)$ for some constant $d > 1$. Then $|\mathcal{P}| \leq e(d+1)\left(\frac{2en}{\delta}\right)^d = O\left(\left(\frac{n}{\delta}\right)^d\right)$. Furthermore, this bound is asymptotically tight.

A strengthening of this statement, for specific values of δ, was studied for some geometric set systems in [PR08, MR14], and for any $\delta \geq 1$ for the so-called Clarkson-Shor set systems in [Ezr16, DEG16]. This was then generalized in terms of the shallow-cell complexity of a set system to give the following statement.

LEMMA 47.1.9 [Mus16]

Let $(X, \mathcal{P})$, $|X| = n$, be a δ-separated set system with $\pi_{\mathcal{P}}(m) = O(m^d)$ for some constant $d > 1$, and with shallow-cell complexity $\varphi_{\mathcal{P}}(\cdot, \cdot)$. If $|P| \leq k$ for all $P \in \mathcal{P}$, then $|\mathcal{P}| \leq O\left(\frac{n}{\delta} \cdot \varphi_{\mathcal{P}}\left(\frac{4dn}{\delta}, \frac{24dk}{\delta}\right)\right)$.

A matching lower-bound for Clarkson-Shor set systems was given in [DGJM17].

47.2 EPSILON-APPROXIMATIONS

Given a set system $(X, \mathcal{R})$ and a set $A \subseteq X$, a set $R \in \mathcal{R}$ is *well-represented* in A if $\frac{|R|}{|X|} \approx \frac{|R \cap A|}{|A|}$. Intuitively, a set $A \subseteq X$ is an ϵ-approximation for $\mathcal{R}$ if *every* $R \in \mathcal{R}$ is well-represented in A; the parameter ϵ captures quantitatively the additive error between these two quantities. In this case the value $\frac{|R \cap A|}{|A|} \cdot |X|$ is a good estimate for $|R|$. As an example, suppose that X is a finite set of points in the plane, and let A be an ϵ-approximation for the primal set system on X induced by half-spaces. Then given a query half-space h, one can return $\frac{|h \cap A|}{|A|} \cdot |X|$ as an estimate for $|h \cap X|$. If $|A| \ll |X|$, computing this estimate is more efficient than computing $|h \cap X|$.

GLOSSARY

ϵ-Approximation: Given a finite set system $(X, \mathcal{R})$, and a parameter $0 \leq \epsilon \leq 1$, a set $A \subseteq X$ is called an ϵ-approximation if, for each $R \in \mathcal{R}$,

$$\left| \frac{|R|}{|X|} - \frac{|R \cap A|}{|A|} \right| \leq \epsilon.$$

Sensitive ϵ-approximation: Given a set system $(X, \mathcal{R})$ and a parameter $0 < \epsilon \leq 1$, a set $A \subseteq X$ is a sensitive ϵ-approximation if for each $R \in \mathcal{R}$,

$$\left| \frac{|R|}{|X|} - \frac{|R \cap A|}{|A|} \right| \leq \frac{\epsilon}{2}\left(\sqrt{\frac{|R|}{|X|}} + \epsilon\right).$$

Relative (ϵ, δ)-approximation: Given a set system $(X, \mathcal{R})$ and parameters $0 < \delta, \epsilon \leq 1$, a set $A \subseteq X$ is a relative (ϵ, δ)-approximation if for each $R \in \mathcal{R}$,

$$\left| \frac{|R|}{|X|} - \frac{|R \cap A|}{|A|} \right| \leq \max\left\{\delta \cdot \frac{|R|}{|X|}, \ \delta \cdot \epsilon\right\}$$

see [Mat95] for details.

Discrepancy: Given a set system $(X, \mathcal{R})$, and a two-coloring $\chi : X \to \{-1, 1\}$, define the discrepancy of $R \in \mathcal{R}$ with respect to χ as $\text{disc}_\chi(R) = \left| \sum_{p \in R} \chi(p) \right|$, and the discrepancy of $\mathcal{R}$ with respect to χ as $\text{disc}_\chi(\mathcal{R}) = \max_{R \in \mathcal{R}} \text{disc}_\chi(R)$. The discrepancy of $(X, \mathcal{R})$ is $\text{disc}(\mathcal{R}) = \min_{\chi: X \to \{-1, 1\}} \text{disc}_\chi(\mathcal{R})$.

EPSILON-APPROXIMATIONS AND DISCREPANCY

When no other constraints are known for a given set system $(X, \mathcal{R})$, the following is the currently best bound on the sizes of ϵ-approximations for $\mathcal{R}$.

THEOREM 47.2.1 [Cha00]

Given a finite set system $(X, \mathcal{R})$ and a parameter $0 < \epsilon \leq 1$, an ϵ-approximation for $(X, \mathcal{R})$ of size $O\left(\frac{1}{\epsilon^2} \log |\mathcal{R}|\right)$ can be found in deterministic $O\left(|X| \cdot |\mathcal{R}|\right)$ time.

If VC-dim$(\mathcal{R}) = d$, the shatter function $\pi_\mathcal{R}(m)$ for $(X, \mathcal{R})$ is bounded by $O(m^d)$ (Lemma 47.1.1). In this case, $|\mathcal{R}| = O(|X|^d)$, and Theorem 47.2.1 guarantees an ϵ-approximation of size at most $O\left(\frac{d}{\epsilon^2} \log |X|\right)$. An influential idea originating in the work of Vapnik and Chervonenkis [VC71] is that for any set system $(X, \mathcal{R})$ with VC-dim$(\mathcal{R}) \leq d$, one can construct an ϵ-approximation of $\mathcal{R}$ by uniformly sampling a subset $A \subseteq X$ of size $O\left(\frac{d \log \frac{1}{\epsilon}}{\epsilon^2}\right)$. Remarkably, this gives a bound on sizes of ϵ-approximations which are independent of $|X|$ or $|\mathcal{R}|$. To get an idea behind the proof, it should be first noted that the factor of $\log |\mathcal{R}|$ in Theorem 47.2.1 comes from applying union bound to a number of failure events, one for each set in $\mathcal{R}$. The key idea in the proof of [VC71], called *symmetrization*, is to "cluster" failure events based on comparing the random sample A with a second sample (sometimes called a *ghost sample* in learning theory literature; see [DGL96]). Together with later work which removed the logarithmic factor, one arrives at the following.

THEOREM 47.2.2 [VC71, Tal94, LLS01]

Let $(X, \mathcal{R})$ be a finite set system with $\pi_\mathcal{R}(m) = O(m^d)$ for a constant $d \geq 1$, and $0 < \epsilon, \gamma < 1$ be given parameters. Let $A \subseteq X$ be a subset of size

$$c \cdot \left(\frac{d}{\epsilon^2} + \frac{\log \frac{1}{\gamma}}{\epsilon^2} \right)$$

chosen uniformly at random, where c is a sufficiently large constant. Then A is an ϵ-approximation for $(X, \mathcal{R})$ with probability at least $1 - \gamma$.

The above theorem immediately implies a randomized algorithm for computing approximations. There exist near-linear time deterministic algorithms for constructing ϵ-approximations of size slightly worse than the above bound; see [STZ06] for algorithms for computing ϵ-approximations in data streams.

THEOREM 47.2.3 [CM96]

Let $(X, \mathcal{R})$ be a set system with VC-dim$(\mathcal{R}) = d$, and $0 < \epsilon \leq \frac{1}{2}$ be a given parameter. Assume that given any finite $Y \subseteq X$, all the sets in $\mathcal{R}|_Y$ can be computed explicitly in time $O\left(|Y|^{d+1}\right)$. Then an ϵ-approximation for $(X, \mathcal{R})$ of size $O\left(\frac{d}{\epsilon^2} \log \frac{d}{\epsilon}\right)$ can be computed deterministically in $O\left(d^{3d}\right)\left(\frac{1}{\epsilon^2} \log \frac{d}{\epsilon}\right)^d |X|$ time.

Somewhat surprisingly, it is possible to show the existence of ϵ-approximations of size smaller than that guaranteed by Theorem 47.2.2. Such results are usually established using a fundamental relation between the notions of approximations and discrepancy: assume $|X|$ is even and let $\chi : X \to \{-1, +1\}$ be any two-coloring of X. For any $R \subseteq X$, let R^+ and R^- denote the subsets of R of the two colors, and w.l.o.g., assume that $|X^+| = \frac{|X|}{2} + t$ and $|X^-| = \frac{|X|}{2} - t$ for some integer $t \geq 0$. Assuming that $X \in \mathcal{R}$, we have $\big||X^+| - |X^-|\big| \leq \mathrm{disc}_\chi(\mathcal{R})$, and so $t \leq \frac{\mathrm{disc}_\chi(\mathcal{R})}{2}$. Take A to be any subset of X^+ of size $\frac{|X|}{2}$. Then for any $R \in \mathcal{R}$,

$$\big||R^+| - |R^-|\big| = \big||R^+| - (|R| - |R^+|)\big| \leq \mathrm{disc}_\chi(\mathcal{R}) \implies \left||R^+| - \frac{|R|}{2}\right| \leq \frac{\mathrm{disc}_\chi(\mathcal{R})}{2}.$$

As $|R \cap A| \geq |R^+| - t$, this implies that $\left||R \cap A| - \frac{|R|}{2}\right| \leq \mathrm{disc}_\chi(\mathcal{R})$. Thus

$$\left|\frac{|R|}{|X|} - \frac{|R \cap A|}{|A|}\right| \leq \left|\frac{|R|}{|X|} - \frac{\frac{|R|}{2} \pm \mathrm{disc}_\chi(\mathcal{R})}{\frac{|X|}{2}}\right| \leq \frac{2 \cdot \mathrm{disc}_\chi(\mathcal{R})}{|X|},$$

and we arrive at the following.

LEMMA 47.2.4   [MWW93]

Let $(X, \mathcal{R})$ be a set system with $X \in \mathcal{R}$, and let $\chi : X \to \{+1, -1\}$ be any two-coloring of X. Then there exists a set $A \subset X$, with $|A| = \lceil \frac{|X|}{2} \rceil$, such that A is an ϵ-approximation for $(X, \mathcal{R})$, with $\epsilon = \frac{2 \cdot \mathrm{disc}_\chi(\mathcal{R})}{|X|}$.

The following simple observation on ϵ-approximations is quite useful.

OBSERVATION 47.2.5   [MWW93]

If A is an ϵ-approximation for $(X, \mathcal{R})$, then any ϵ'-approximation for $(A, \mathcal{R}|_A)$ is an $(\epsilon + \epsilon')$-approximation for $(X, \mathcal{R})$.

Given a finite set system $(X, \mathcal{R})$ with $X \in \mathcal{R}$, put $X_0 = X$, and compute a sequence $X_1, X_2, \ldots, X_t$, where $X_i \subseteq X_{i-1}$ satisfies $|X_i| = \left\lceil \frac{|X_{i-1}|}{2} \right\rceil$, and is computed from a two-coloring of $(X_{i-1}, \mathcal{R}|_{X_{i-1}})$ derived from Lemma 47.2.4. Assume that X_i is an ϵ_i-approximation for $(X_{i-1}, \mathcal{R}|_{X_{i-1}})$. Then Observation 47.2.5 implies that X_t is a ϵ-approximation for $(X, \mathcal{R})$ with $\epsilon = \sum_{i=1}^t \epsilon_i$. The next statement follows by setting the parameter t to be as large as possible while ensuring that $\sum_{i=1}^t \epsilon_i \leq \epsilon$.

LEMMA 47.2.6   [MWW93]

Let $(X, \mathcal{R})$ be a finite set system with $X \in \mathcal{R}$, and let $f(\cdot)$ be a function such that $\mathrm{disc}(\mathcal{R}|_Y) \leq f(|Y|)$ for all $Y \subseteq X$. Then, for every integer $t \geq 0$, there exists an ϵ-approximation A for $(X, \mathcal{R})$ with $|A| = \lceil \frac{n}{2^t} \rceil$ and

$$\epsilon \leq \frac{2}{n}\left(f(n) + 2f\left(\left\lceil \frac{n}{2} \right\rceil\right) + \cdots + 2^t f\left(\left\lceil \frac{n}{2^t} \right\rceil\right)\right).$$

In particular, if there exists a constant $c > 1$ such that we have $f(2m) \leq \frac{2}{c} f(m)$ for all $m \geq \lceil \frac{n}{2^t} \rceil$, then $\epsilon = O\left(\frac{f(\lceil \frac{n}{2^t} \rceil) 2^t}{n}\right)$.

Many of the currently best bounds on ϵ-approximations follow from applications of Lemma 47.2.6; e.g., the existence of ϵ-approximations of size $O\left(\frac{1}{\epsilon^2} \log \frac{1}{\epsilon}\right)$ for set systems $(X, \mathcal{R})$ with $\pi_{\mathcal{R}}(m) = O(m^d)$ (for some constant $d > 1$) follows immediately from the fact that for such $\mathcal{R}$, we have $\mathrm{disc}(\mathcal{R}|_Y) = O\big(\sqrt{|Y|}\log|Y|\big)$. The next two theorems, from a seminal paper of Matoušek, Welzl, and Wernisch [MWW93], were established by deriving improved discrepancy bounds (which turn out to be based on Lemma 47.1.8), and then applying Lemma 47.2.6.

THEOREM 47.2.7   [MWW93, Mat95]

Let $(X, \mathcal{R})$ be a finite set system with the shatter function $\pi_{\mathcal{R}}(m) = O(m^d)$, where $d > 1$ is a fixed constant. For any $0 < \epsilon \le 1$, there exists an ϵ-approximation for $\mathcal{R}$ of size $O\left(\dfrac{1}{\epsilon^{2-\frac{2}{d+1}}}\right)$.

The above theorem relies on the existence of low discrepancy colorings, whose initial proof was non-algorithmic (using the "entropy method"). However, recent work by Bansal [Ban12] and Lovett and Meka [LM15] implies polynomial time algorithms for constructing such low discrepancy colorings and consequently ϵ-approximations whose sizes are given by Theorem 47.2.7; see [Ezr16, DEG16].

Improved bounds on approximations are also known in terms of the shatter function of the set system dual to $(X, \mathcal{R})$.

THEOREM 47.2.8   [MWW93]

Let $(X, \mathcal{R})$ be a finite set system and $0 < \epsilon \le 1$ be a given parameter. Suppose that for the set system $(X^*, \mathcal{R}^*)$ dual to $(X, \mathcal{R})$, we have $\pi_{\mathcal{R}^*}(m) = O(m^d)$, where $d > 1$ is a constant independent of m. Then there exists an ϵ-approximation for $\mathcal{R}$ of size $O\left(\dfrac{1}{\epsilon^{2-\frac{2}{d+1}}}\Big(\log\dfrac{1}{\epsilon}\Big)^{1-\frac{1}{d+1}}\right)$.

Theorems 47.2.7 and 47.2.8 yield the best known bounds for several geometric set systems. For example, the shatter function (see Table 47.1.1) of the primal set system induced by half-spaces in $\mathbb{R}^2$ is $O(m^2)$, and thus one obtains ϵ-approximations for it of size $O\big(\frac{1}{\epsilon^{4/3}}\big)$ from Theorem 47.2.7. For the primal set system induced by disks in $\mathbb{R}^2$, the shatter function is bounded by $\Theta(m^3)$; Theorem 47.2.7 then implies the existence of ϵ-approximations of size $O\big(\frac{1}{\epsilon^{3/2}}\big)$. In this case, it turns out that Theorem 47.2.8 gives a better bound: the shatter function of the dual set system is bounded by $O(m^2)$, and thus there exist ϵ-approximations of size $O\big(\frac{1}{\epsilon^{4/3}}(\log\frac{1}{\epsilon})^{\frac{2}{3}}\big)$.

Table 47.2.1 states the best known bounds for some common geometric set systems. Observe that for the primal set system induced by axis-parallel rectangles in $\mathbb{R}^d$, there exist ϵ-approximations of size near-linear in $\frac{1}{\epsilon}$.

RELATIVES OF EPSILON-APPROXIMATIONS

It is easy to see that a sensitive ϵ-approximation is an ϵ-approximation and an ϵ'-net, for $\epsilon' > \epsilon^2$ (see the subsequent section for the definition of ϵ-nets) simultaneously. This notion was first studied by Brönnimann *et al.* [BCM99]. The following result improves slightly on their bounds.

TABLE 47.2.1   Sizes of ϵ-approximations for geometric set systems (multiplicative constants omitted for clarity).

Objects	SETS	UPPER-BOUND
Intervals	Primal	$\frac{1}{\epsilon}$
Half-spaces in $\mathbb{R}^d$	Primal/Dual	$\frac{1}{\epsilon^{2-\frac{2}{d+1}}}$ [MWW93, Mat95]
Balls in $\mathbb{R}^d$	Primal	$\frac{1}{\epsilon^{2-\frac{2}{d+1}}}(\log\frac{1}{\epsilon})^{1-\frac{1}{d+1}}$ [MWW93]
Balls in $\mathbb{R}^d$	Dual	$\frac{1}{\epsilon^{2-\frac{2}{d+1}}}$ [MWW93, Mat95]
Axis-par. rect. in $\mathbb{R}^d$	Primal	$\frac{1}{\epsilon}\cdot(\log^{2d}\frac{1}{\epsilon})\cdot\log^{c_d}(\log\frac{1}{\epsilon})$ [Phi08]

THEOREM 47.2.9 [BCM99, HP11]

Let $(X,\mathcal{R})$ be a finite system with $\mathrm{VC\text{-}dim}(\mathcal{R})\leq d$, where d is a fixed constant. For a given parameter $0<\epsilon\leq 1$, let $A\subseteq X$ be a subset of size

$$\frac{c\cdot d}{\epsilon^2}\log\frac{d}{\epsilon}$$

chosen uniformly at random, where $c>0$ is an absolute constant. Then A is a sensitive ϵ-approximation for $(X,\mathcal{R})$ with probability at least $\frac{1}{2}$. Furthermore, assuming that given any $Y\subseteq X$, all the sets in $\mathcal{R}|_Y$ can be computed explicitly in time $O(|Y|^{d+1})$, a sensitive ϵ-approximation of size $O(\frac{d}{\epsilon^2}\log\frac{d}{\epsilon})$ can be computed deterministically in time $O(d^{3d})\cdot\frac{1}{\epsilon^{2d}}(\log\frac{d}{\epsilon})^d\cdot|X|$.

On the other hand, a relative (ϵ,δ)-approximation is both a δ-approximation and an ϵ'-net, for any $\epsilon'>\epsilon$. It is easy to see that a $(\epsilon\cdot\delta)$-approximation is a relative (ϵ,δ)-approximation. Thus, using Theorem 47.2.2, one obtains a relative (ϵ,δ)-approximation of size $O(\frac{d}{\epsilon^2\cdot\delta^2})$. This bound can be improved to the following.

THEOREM 47.2.10 [LLS01, HPS11]

Let $(X,\mathcal{R})$ be a finite set system with shatter function $\pi_{\mathcal{R}}(m)=O(m^d)$ for some constant d, and $0<\delta,\epsilon,\gamma\leq 1$ be given parameters. Let $A\subseteq X$ be a subset of size

$$c\cdot\left(\frac{d\log\frac{1}{\epsilon}}{\epsilon\delta^2}+\frac{\log\frac{1}{\gamma}}{\epsilon\delta^2}\right)$$

chosen uniformly at random, where $c>0$ is an absolute constant. Then A is a relative (ϵ,δ)-approximation for $(X,\mathcal{R})$ with probability at least $1-\gamma$.

A further improvement is possible on the size of relative (ϵ,δ)-approximations for the primal set system induced by half-spaces in $\mathbb{R}^2$ [HPS11] and $\mathbb{R}^3$ [Ezr16], as well as other bounds with a better dependency on $\frac{1}{\delta}$ (at the cost of a worse dependence on $\frac{1}{\epsilon}$) for systems with small shallow-cell complexity [Ezr16, DEG16].

47.3 APPLICATIONS OF EPSILON-APPROXIMATIONS

One of the main uses of ϵ-approximations is in constructing a small-sized representation or "sketch" A of a potentially large set of elements X with respect to an underlying set system $\mathcal{R}$. Then data queries from $\mathcal{R}$ on X can instead be performed on A to get provably approximate answers. Suppose that we aim to preprocess a finite set X of points in the plane, so that given a query half-space h, we can efficiently return an approximation to $|h \cap X|$. For this data structure, one could use an ϵ-approximation $A \subseteq X$ for the set system $(X, \mathcal{R})$ induced by the set of all half-spaces in $\mathbb{R}^2$. Then given a query half-space h, simply return $\frac{|h \cap A|}{|A|} \cdot |X|$; this answer differs from $|h \cap X|$ by at most $\epsilon \cdot |X|$. If instead A is a relative (δ, ϵ)-approximation, then our answer differs from the true answer by at most $\delta \cdot |h \cap X|$, provided $|h \cap X| \geq \epsilon|X|$. Two key properties of approximations useful in applications are (a) $\frac{|R \cap A|}{|A|}$ approximates $\frac{|R \cap X|}{|X|}$ *simultaneously* for each $R \in \mathcal{R}$, and (b) ϵ-approximations exist of size independent of $|X|$ or $|\mathcal{R}|$. This enables the use of ϵ-approximations for computing certain estimators on geometric data; e.g., a combinatorial median $q \in \mathbb{R}^d$ for a point set X can be approximated by the one for an ϵ-approximation, which can then be computed in near-linear time.

GLOSSARY

Product set systems: Given finite set systems $\Sigma_1 = (X_1, \mathcal{R}_1)$ and $\Sigma_2 = (X_2, \mathcal{R}_2)$, the product system $\Sigma_1 \otimes \Sigma_2$ is defined as the system $(X_1 \times X_2, \mathcal{T})$, where $\mathcal{T}$ consists of all subsets $T \subseteq X_1 \times X_2$ for which the following hold: (a) for any $x_2 \in X_2$, $\{x \in X_1 \ : \ (x, x_2) \in T\} \in \mathcal{R}_1$, and (b) for any $x_1 \in X_1$, $\{x \in X_2 \ : \ (x_1, x) \in T\} \in \mathcal{R}_2$.

Centerpoints: Given a set X of n points in $\mathbb{R}^d$, a point $q \in \mathbb{R}^d$ is said to be a centerpoint for X if any half-space containing q contains at least $\frac{n}{d+1}$ points of X; for $\epsilon > 0$, q is said to be an ϵ-centerpoint if any half-space containing q contains at least $(1-\epsilon)\frac{n}{d+1}$ points of X. By Helly's theorem, a centerpoint exists for all point sets.

Shape fitting: A shape fitting problem consists of the triple $(\mathbb{R}^d, \mathcal{F}, \text{dist})$, where $\mathcal{F}$ is a family of non-empty closed subsets (*shapes*) in $\mathbb{R}^d$ and dist : $\mathbb{R}^d \times \mathbb{R}^d \to \mathbb{R}^+$ is a continuous, symmetric, positive-definite (*distance*) function. The *distance* of a point $p \in \mathbb{R}^d$ from the shape $F \in \mathcal{F}$ is defined as $\text{dist}(p, F) = \min_{q \in F} \text{dist}(p, q)$. A finite subset $P \subset \mathbb{R}^d$ defines an instance of the shape fitting problem, where the goal is to find a shape $F^* = \arg\min_{F \in \mathcal{F}} \sum_{p \in P} \text{dist}(p, F)$.

ϵ-Coreset: Given an instance $P \subset \mathbb{R}^d$ of a shape fitting problem $(\mathbb{R}^d, \mathcal{F}, \text{dist})$, and an $\epsilon \in (0, 1)$, an ϵ-coreset of size s is a pair (S, w), where $S \subseteq P$, $|S| = s$, and $w \ : \ S \to \mathbb{R}$ is a weight function such that for any $F \in \mathcal{F}$:

$$\left| \sum_{p \in P} \text{dist}(p, F) - \sum_{q \in S} w(q) \cdot \text{dist}(q, F) \right| \leq \epsilon \sum_{p \in P} \text{dist}(p, F).$$

APPROXIMATING GEOMETRIC INFORMATION

One of the main uses of ϵ-approximations is in the design of efficient approximation algorithms for combinatorial queries on geometric data. An illustrative example is that of computing a centerpoint of a finite point set $X \subset \mathbb{R}^d$; the proof of the following lemma is immediate.

LEMMA 47.3.1 [Mat91a]

Let $X \subset \mathbb{R}^d$ be a finite point set, $0 \le \epsilon < 1$ be a given parameter, and A be an ϵ-approximation for the primal set system induced by half-spaces in $\mathbb{R}^d$ on X. Then any centerpoint for A is an ϵ-centerpoint for X.

We now describe a more subtle application in the same spirit, in fact one of the motivations for considering products of set systems, first considered in Brönnimann, Chazelle, and Matoušek [BCM99]. For $i = 1, 2$, let X_i be a finite set of lines in $\mathbb{R}^2$ such that $X_1 \cup X_2$ is in general position, and let $\mathcal{R}_i$ be the family of subsets of X_i that contains every subset $X' \subseteq X_i$ such that X' is precisely the subset of lines intersected by some line segment. The VC-dimension of the set system $\Sigma_i = (X_i, \mathcal{R}_i)$ is bounded by some constant. We can identify $(r, b) \in X_1 \times X_2$ with the intersection point of r and b.

Considering the product set system $\Sigma_1 \otimes \Sigma_2 = (X_1 \times X_2, \mathcal{T})$, it is easy to see that for any convex set C, the set of intersection points between lines of X_1 and X_2 that lie within C is an element of $\mathcal{T}$. The VC-dimension of $\Sigma_1 \otimes \Sigma_2$ is in fact unbounded. Indeed, notice that any matching $\{(r_1, b_1), (r_2, b_2), \ldots, (r_k, b_k)\} \subset X_1 \times X_2$ is shattered by $\Sigma_1 \otimes \Sigma_2$. Nevertheless, it is possible to construct small ϵ-approximations for this set system:

LEMMA 47.3.2 [Cha93, BCM99]

For $i = 1, 2$ and $0 \le \epsilon_i \le 1$, let A_i be an ϵ_i-approximation for the finite set system $\Sigma_i = (X_i, \mathcal{R}_i)$. Then $A_1 \times A_2$ is an $(\epsilon_1 + \epsilon_2)$-approximation for $\Sigma_1 \otimes \Sigma_2$.

We can apply this general result on $\Sigma_1 \otimes \Sigma_2$ to estimate $V(X_1 \times X_2, C)$—defined to be the number of intersections between lines in X_1 and X_2 that are contained in a query convex set C—by $\frac{|V(A_1 \times A_2, C)| \cdot |X_1| \cdot |X_2|}{|A_1| \cdot |A_2|}$. Lemma 47.3.2 implies that the error of this estimate can be bounded by

$$\left| \frac{V(X_1 \times X_2, C)}{|X_1| \cdot |X_2|} - \frac{V(A_1 \times A_2, C)}{|A_1| \cdot |A_2|} \right| \le \epsilon_1 + \epsilon_2.$$

The notion of a product of set systems and Lemma 47.3.2 can be generalized to more than two set systems [BCM99, Cha00].

SHAPE FITTING AND CORESETS

Consider the scenario where the shape family $\mathcal{F}$ contains, as its elements, all possible k-point subsets of $\mathbb{R}^d$; that is, each $F \in \mathcal{F}$ is a subset of $\mathbb{R}^d$ consisting of k points. If the function $\text{dist}(\cdot, \cdot)$ is the Euclidean distance, then the corresponding shape fitting problem $(\mathbb{R}^d, \mathcal{F}, \text{dist})$ is the well-known k-median problem. If $\text{dist}(\cdot, \cdot)$ is the square of the Euclidean distance, then the shape fitting problem is the k-means problem. If the shape family $\mathcal{F}$ contains as its elements all hyperplanes in $\mathbb{R}^d$, and $\text{dist}(\cdot, \cdot)$ is the Euclidean distance, then the corresponding shape fitting problem asks for a hyperplane that minimizes the sum of the Euclidean distances from points in the given instance $P \subset \mathbb{R}^d$. The shape fitting problem as defined is just one of many versions that have been considered. In another well-studied version, given an instance $P \subset \mathbb{R}^d$, the goal is to find a shape that minimizes $\max_{p \in P} \text{dist}(p, F)$.

Given an instance P, and a parameter $0 < \epsilon < 1$, an ϵ-coreset (S, w) "approximates" P with respect to every shape F in the given family $\mathcal{F}$. Such an ϵ-coreset can be used to find a shape that approximately minimizes $\sum_{p \in P} \text{dist}(p, F)$: one simply finds a shape that minimizes $\sum_{q \in S} w(q) \cdot \mathrm{dist}(q, F)$. For this approach to be useful, the size of the coreset needs to be small as well as efficiently computable. Building on a long sequence of works, Feldman and Langberg [FL11] (see also Langberg and Schulman [LS10]) showed the existence of a function $f : \mathbb{R} \to \mathbb{R}$ such that an ϵ-approximation for a carefully constructed set system associated with the shape fitting problem $(\mathbb{R}^d, \mathcal{F}, \mathrm{dist})$ and instance P yields an $f(\epsilon)$-coreset for the instance P. For many shape fitting problems, this method often yields coresets with size guarantees that are not too much worse than bounds via more specialized arguments. We refer the reader to the survey [BLK17] for further details.

47.4 EPSILON-NETS

While an ϵ-approximation of a set system $(X, \mathcal{R})$ aims to achieve equality in the *proportion* of points picked from each set, often only a weaker threshold property is needed. A set $N \subseteq X$ is called an ϵ-net for $\mathcal{R}$ if it has a non-empty intersection with each set of $\mathcal{R}$ of cardinality at least $\epsilon|X|$. For all natural geometric set systems, trivial considerations imply that any such N must have size $\Omega(\frac{1}{\epsilon})$: one can always arrange the elements of X into disjoint $\lfloor \frac{1}{\epsilon} \rfloor$ groups, each with at least $\epsilon|X|$ elements, such that the set consisting of the elements in each group is induced by the given geometric family. While ϵ-nets form the basis of many algorithmic and combinatorial tools in discrete and computational geometry, here we present only two applications, one combinatorial and one algorithmic.

GLOSSARY

ϵ-Nets: Given a finite set system $(X, \mathcal{R})$ and a parameter $0 \leq \epsilon \leq 1$, a set $N \subseteq X$ is an ϵ-net for $\mathcal{R}$ if $N \cap R \neq \emptyset$ for all sets $R \in \mathcal{R}$ with $|R| \geq \epsilon|X|$.

Weak ϵ-nets: Given a set X of points in $\mathbb{R}^d$ and family of objects $\mathcal{R}$, a set $Q \subseteq \mathbb{R}^d$ is a *weak ϵ-net* with respect to $\mathcal{R}$ if $Q \cap R \neq \emptyset$ for all $R \in \mathcal{R}$ containing at least $\epsilon|X|$ points of X. Note that in contrast to ϵ-nets, we do not require Q to be a subset of X.

Semialgebraic sets: Semialgebraic sets are subsets of $\mathbb{R}^d$ obtained by taking Boolean operations such as unions, intersections, and complements of sets of the form $\{x \in \mathbb{R}^d \mid g(x) \geq 0\}$, where g is a d-variate polynomial in $\mathbb{R}[x_1, \ldots, x_d]$.

ϵ-Mnets: Given a set system $(X, \mathcal{R})$ and a parameter $0 \leq \epsilon \leq 1$, a collection of sets $\mathcal{M} = \{X_1, \ldots, X_t\}$ on X is an ϵ-Mnet of size t if $|X_i| = \Theta(\epsilon|X|)$ for all i, and for any set $R \in \mathcal{R}$ with $|R| \geq \epsilon|X|$, there exists an index $j \in \{1, \ldots, t\}$ such that $X_j \subseteq R$.

EPSILON-NETS FOR ABSTRACT SET SYSTEMS

The systematic study of ϵ-nets started with the breakthrough result of Haussler and Welzl [HW87], who first showed the existence of ϵ-nets whose size was a function of the parameter ϵ and the VC-dimension. A different framework, with somewhat similar ideas and consequences, was independently introduced by Clarkson [?]. The result of Haussler and Welzl was later improved upon and extended in several ways: the precise dependency on VC-dim($\mathcal{R}$) was improved, the probabilistic proof in [HW87] was de-randomized to give a deterministic algorithm, and finer probability estimates were derived for randomized constructions of ε-nets.

THEOREM 47.4.1 [HW87, KPW92]

Let $(X, \mathcal{R})$ be a finite set system, such that $\pi_{\mathcal{R}}(m) = O(m^d)$ for a fixed constant d, and let $\epsilon > 0$ be a sufficiently small parameter. Then there exists an ε-net for $\mathcal{R}$ of size $(1 + o(1)) \frac{d}{\epsilon} \log \frac{1}{\epsilon}$. Furthermore, a uniformly chosen random sample of X of the above size is an ε-net with constant probability.

An alternate proof, though with worse constants, follows immediately from ε-approximations: use Theorem 47.2.2 to compute an $\frac{\epsilon}{2}$-approximation A for $(X, \mathcal{R})$, where $|A| = O(\frac{d}{\epsilon^2})$. Observe that an $\frac{\epsilon}{2}$-net for $(A, \mathcal{R}|_A)$ is an ε-net for $(X, \mathcal{R})$, as for each $R \in \mathcal{R}$ with $|R| \geq \epsilon|X|$, we have $\left| \frac{|R|}{|X|} - \frac{|R \cap A|}{|A|} \right| \leq \frac{\epsilon}{2}$ and so $\frac{|R \cap A|}{|A|} \geq \frac{\epsilon}{2}$. Now a straightforward random sampling argument with union bound (or an iterative greedy construction) gives an $\frac{\epsilon}{2}$-net for $\mathcal{R}|_A$, of total size $O(\frac{1}{\epsilon} \log |\mathcal{R}|_A|) = O(\frac{d}{\epsilon} \log \frac{d}{\epsilon})$.

THEOREM 47.4.2 [AS08]

Let $(X, \mathcal{R})$ be a finite set system with $\pi_{\mathcal{R}}(m) = O(m^d)$ for a constant d, and $0 < \epsilon, \gamma \leq 1$ be given parameters. Let $N \subseteq X$ be a set of size

$$\max \left\{ \frac{4}{\epsilon} \log \frac{2}{\gamma}, \ \frac{8d}{\epsilon} \log \frac{8d}{\epsilon} \right\}$$

chosen uniformly at random. Then N is an ε-net with probability at least $1 - \gamma$.

THEOREM 47.4.3 [BCM99]

Let $(X, \mathcal{R})$ be a finite set system such that VC-dim($\mathcal{R}$) = d, and $\epsilon > 0$ a given parameter. Assume that for any $Y \subseteq X$, all sets in $\mathcal{R}|_Y$ can be computed explicitly in time $O(|Y|^{d+1})$. Then an ε-net of size $O(\frac{d}{\epsilon} \log \frac{d}{\epsilon})$ can be computed deterministically in time $O(d^{3d}) \cdot (\frac{1}{\epsilon} \log \frac{1}{\epsilon})^d \cdot |X|$.

It was shown in [KPW92] that for any $0 < \epsilon \leq 1$, there exist ε-nets of size $\max \left\{ 2, \lceil \frac{1}{\epsilon} \rceil - 1 \right\}$ for any set system $(X, \mathcal{R})$ with VC-dim($\mathcal{R}$) = 1. For the case when VC-dim($\mathcal{R}$) ≥ 2, the quantitative bounds of Theorem 47.4.1 are near-optimal, as the following construction shows. For a given integer $d \geq 2$ and a real $\epsilon > 0$, set $n = \Theta(\frac{1}{\epsilon} \log \frac{1}{\epsilon})$ and construct a random ϵn-uniform set system by choosing $\Theta(\frac{1}{\epsilon^{d+\gamma-1}})$ sets uniformly from all possible sets of size ϵn, where γ is sufficiently small. It can be shown that, with constant probability, this set system has VC-dimension at most d and any ε-net for it must have large size.

THEOREM 47.4.4 [KPW92]

Given any $\epsilon > 0$ and integer $d \geq 2$, there exists a set system $(X, \mathcal{R})$ such that VC-dim($\mathcal{R}$) $\leq d$ and any ε-net for $\mathcal{R}$ has size at least $\left(1 - \frac{2}{d} + \frac{1}{d(d+2)} + o(1) \right) \frac{d}{\epsilon} \log \frac{1}{\epsilon}$.

Over the years it was realized that the shatter function of a set system is too crude a characterization for purposes of ε-nets, and that the existence of smaller sized ε-nets can be shown if one further knows the distribution of sets of any fixed size in the set system. This was first understood for the case of geometric dual set systems in $\mathbb{R}^2$ using spatial partitioning techniques, initially in the work of Clarkson and

Varadarajan [CV07] and then in its improvements by Aronov *et al.* [AES10]. Later it was realized by Varadarajan [Var09, Var10] and in its improvement by Chan *et al.* [CGKS12] that one could avoid spatial partitioning altogether, and get improved bounds on sizes of ϵ-nets in terms of the shallow-cell complexity of a set system.

THEOREM 47.4.5   [Var10, CGKS12]

Let $(X, \mathcal{R})$ be a set system with shallow-cell complexity $\varphi_{\mathcal{R}}(\cdot)$, where $\varphi_{\mathcal{R}}(n) = O(n^d)$ for some constant d. Let $\epsilon > 0$ be a given parameter. Then there exists an ϵ-net[3] for $\mathcal{R}$ of size $O\bigl(\frac{1}{\epsilon} \log \varphi_{\mathcal{R}}(\frac{1}{\epsilon})\bigr)$. Furthermore, such an ϵ-net can be computed in deterministic polynomial time.

We sketch a simple proof of the above theorem due to Mustafa *et al.* [MDG17]. For simplicity, assume that $|R| = \Theta(\epsilon n)$ for all $R \in \mathcal{R}$. Let $\mathcal{P} \subseteq \mathcal{R}$ be a maximal $\frac{\epsilon n}{2}$-separated system, of size $|\mathcal{P}| = O\bigl(\frac{1}{\epsilon} \varphi_{\mathcal{R}}(\frac{1}{\epsilon})\bigr)$ by Lemma 47.1.9. By the maximality of $\mathcal{P}$, for each $R \in \mathcal{R}$ there exists a $P_R \in \mathcal{P}$ such that $|R \cap P_R| \geq \frac{\epsilon n}{2}$, and thus a set N which is a $\frac{1}{2}$-net for each of the $|\mathcal{P}|$ set systems $(P, \mathcal{R}|_P)$, $P \in \mathcal{P}$, is an ϵ-net for $\mathcal{R}$. Construct the set N by picking each point of X uniformly with probability $\Theta\bigl(\frac{1}{\epsilon n} \log \varphi_{\mathcal{R}}(\frac{1}{\epsilon})\bigr)$. For each $P \in \mathcal{P}$, $P \cap N$ is essentially a random subset of size $\Theta\bigl(\log \varphi_{\mathcal{R}}(\frac{1}{\epsilon})\bigr)$, and so by Theorem 47.4.2, N fails to be a $\frac{1}{2}$-net for $\mathcal{R}|_P$ with probability $O\bigl(\frac{1}{\varphi_{\mathcal{R}}(1/\epsilon)}\bigr)$. By linearity of expectation, N is a $\frac{1}{2}$-net for all but expected $O\bigl(\frac{1}{\varphi_{\mathcal{R}}(1/\epsilon)}\bigr) \cdot |\mathcal{P}| = O(\frac{1}{\epsilon})$ sets of $\mathcal{P}$, and for those a $O(1)$-size $\frac{1}{2}$-net can be constructed individually (again by Theorem 47.4.2) and added to N, resulting in an ϵ-net of expected size $\Theta\bigl(\frac{1}{\epsilon} \log \varphi_{\mathcal{R}}(\frac{1}{\epsilon})\bigr)$.

Furthermore, this bound can be shown to be near-optimal by generalizing the random construction used in Theorem 47.4.4.

THEOREM 47.4.6   [KMP16]

Let d be a fixed positive integer and let $\varphi : \mathbb{N} \to \mathbb{R}^+$ be any submultiplicative function[4] with $\varphi(n) = O(n^d)$ for some constant d. Then, for any $\epsilon > 0$, there exists a set system $(X, \mathcal{R})$ with shallow-cell complexity $\varphi(\cdot)$, and for which any ϵ-net has size $\Omega\bigl(\frac{1}{\epsilon} \log \varphi(\frac{1}{\epsilon})\bigr)$.

On the other hand, there are examples of natural set systems with high shallow-cell complexity and yet with small ϵ-nets [Mat16]: for a planar undirected graph $G = (V, E)$, let $\mathcal{R}$ be the set system on V induced by shortest paths in G; i.e., for every pair of vertices $v_i, v_j \in V$, the set $R_{i,j} \in \mathcal{R}$ consists of the set of vertices on the shortest path between v_i and v_j. Further, assume that these shortest paths are unique for every pair of vertices. Then $(V, \mathcal{R})$ has ϵ-nets of size $O(\frac{1}{\epsilon})$ [KPR93], and yet $\varphi_{\mathcal{R}}(n) = \Omega(n)$ can be seen, e.g., by considering the star graph. As we will see in the next part, the primal set system induced by axis-parallel rectangles is another example with high shallow-cell complexity and yet small ϵ-nets.

The proof in [Var10, CGKS12] presents a randomized method to construct an ϵ-net N such that each element $x \in X$ belongs to N with probability $O\bigl(\frac{1}{\epsilon|X|} \log \varphi_{\mathcal{R}}(\frac{1}{\epsilon})\bigr)$. This implies the following more general result.

---

[3]The bound in these papers is stated as $O\bigl(\frac{1}{\epsilon} \log \varphi_{\mathcal{R}}(|X|)\bigr)$, which does not require the assumption that $\varphi_{\mathcal{R}}(n) = O(n^d)$ for some constant d. However, standard techniques using ϵ-approximations imply the stated bound; see [Var09, KMP16] for details.

[4]A function $\varphi : \mathbb{R}^+ \to \mathbb{R}^+$ is called *submultiplicative* if (a) $\varphi^\alpha(n) \leq \varphi(n^\alpha)$ for any $0 < \alpha < 1$ and a sufficiently large positive n, and (b) $\varphi(x)\varphi(y) \geq \varphi(xy)$ for any sufficiently large $x, y \in \mathbb{R}^+$.

COROLLARY 47.4.7   [Var10, CGKS12]

Let $(X, \mathcal{R})$ be a set system with shallow-cell complexity $\varphi_{\mathcal{R}}(\cdot)$, and $\epsilon > 0$ be a given parameter. Further let $w : X \to \mathbb{R}^+$ be weights on the elements of X, with $W = \sum_{x \in X} w(x)$. Then there exists an ϵ-net for $\mathcal{R}$ of total weight $O\left(\frac{W}{\epsilon|X|} \log \varphi_{\mathcal{R}}(\frac{1}{\epsilon})\right)$.

The notion of ϵ-Mnets of a set system $(X, \mathcal{R})$, first defined explicitly and studied in Mustafa and Ray [MR14], is related to both ϵ-nets (any transversal of the sets in an ϵ-Mnet is an ϵ-net for $\mathcal{R}$) as well as the so-called *Macbeath regions* in convex geometry (we refer the reader to the surveys [BL88, Bár07] for more details on Macbeath regions, and to Mount *et al.* [AFM17] for some recent applications). The following theorem concerns ϵ-Mnets with respect to volume for the primal set system induced by half-spaces.

THEOREM 47.4.8   [BCP93]

Given a compact convex body K in $\mathbb{R}^d$ and a parameter $0 < \epsilon < \frac{1}{(2d)^{2d}}$, let $\mathcal{R}$ be the primal set system on K induced by half-spaces in $\mathbb{R}^d$, equipped with Lebesgue measure. There exists an ϵ-Mnet for $\mathcal{R}$ of size $O\left(\frac{1}{\epsilon^{1-\frac{2}{d+1}}}\right)$. Furthermore, the sets in the ϵ-Mnet are pairwise-disjoint convex bodies lying in K.

The role of shallow-cell complexity carries over to the bounds on ϵ-Mnets; the proof of the following theorem uses the packing lemma (Lemma 47.1.9).

THEOREM 47.4.9   [DGJM17]

Given a set X of points in $\mathbb{R}^d$, let $\mathcal{R}$ be the primal set system on X induced by a family of semialgebraic sets in $\mathbb{R}^d$ with shallow-cell complexity $\varphi_{\mathcal{R}}(\cdot)$, where $\varphi_{\mathcal{R}}(n) = O(n^t)$ for some constant t. Let $\epsilon > 0$ be a given parameter. Then there exists an ϵ-Mnet for $\mathcal{R}$ of size $O\left(\frac{1}{\epsilon} \varphi_{\mathcal{R}}(\frac{1}{\epsilon})\right)$, where the constants in the asymptotic notation depend on the degree and number of inequalities defining the semialgebraic sets.

Together with bounds on shallow-cell complexity for half-spaces (Lemma 47.1.6), this implies the existence of ϵ-Mnets of size $O\left(\frac{1}{\epsilon^{\lfloor d/2 \rfloor}}\right)$ for the primal set system induced by half-spaces on a finite set of points in $\mathbb{R}^d$. Further, as observed in [DGJM17], Theorem 47.4.9 implies Theorem 47.4.5 for semialgebraic set systems by a straightforward use of random sampling and the union bound.

EPSILON-NETS FOR GEOMETRIC SET SYSTEMS

We now turn to set systems, both primal and dual, induced by geometric objects in $\mathbb{R}^d$. The existence of ϵ-nets of size $O(\frac{1}{\epsilon} \log \frac{1}{\epsilon})$ for several geometric set systems follow from the early breakthroughs of Clarkson [?] and Clarkson and Shor [CS89] via the use of random sampling together with spatial partitioning. For the case of primal and dual set systems, it turns out that all known asymptotic bounds on sizes of ϵ-nets follow from Theorem 47.4.5 and bounds on shallow-cell complexity (Table 47.1.1). The relevance of shallow-cell complexity for ϵ-nets was realized after considerable effort was spent on inventing a variety of specialized techniques for constructing ϵ-nets for geometric set systems. These techniques and ideas have their own advantages, often yielding algorithms with low running times and low constants hidden in the asymptotic notation. Table 47.4.1 lists the most precise upper bounds known for many natural geometric set systems; all except one are, asymptotically, direct consequences of Theorem 47.4.5. The exception is the case of the primal set system induced by the family $\mathcal{R}$ of axis-parallel rectangles in the plane, which have shallow-cell complexity $\varphi_{\mathcal{R}}(n) = n$, as for any integer n there exist a set X of n points in $\mathbb{R}^2$ such that the number of subsets of X of size at most two induced by $\mathcal{R}$ is $\Theta(n^2)$. However, Aronov et al. [AES10] showed that there exists another family of objects[5] $\mathcal{R}'$ with $\varphi_{\mathcal{R}'}(n) = O(\log n)$, such that an $\frac{\epsilon}{2}$-net for the primal set system on X induced by $\mathcal{R}'$ is an ϵ-net for the one induced by $\mathcal{R}$; now ϵ-nets of size $O\left(\frac{1}{\epsilon} \log \log \frac{1}{\epsilon}\right)$ for the primal set system induced by $\mathcal{R}$ follow by applying Theorem 47.4.5 on $\mathcal{R}'$.

Precise sizes of ϵ-nets for some constant values of ϵ have been studied for the primal set system induced by axis-parallel rectangles and disks in $\mathbb{R}^2$ [AAG14]. It is also known that the visibility set system for a simple polygon P and a finite set of guards G—consisting of all sets S_p, where S_p is the set of points of G visible from $p \in P$—admits ϵ-nets of size $O\left(\frac{1}{\epsilon} \log \log \frac{1}{\epsilon}\right)$ [KK11]. In the case where the underlying base set is $\mathbb{R}^d$, bounds better than those following from Theorem 47.4.5 are known from the theory of geometric coverings.

THEOREM 47.4.10   [Rog57]

Let $K \subset \mathbb{R}^d$ be a bounded convex body, and let $Q = [-r, r]^d$ be a cube of side-length $2r$, where $r \in \mathbb{R}^+$. Let $\mathcal{R}$ be the primal set system induced by translates of K completely contained in Q. Then there exists a hitting set $P \subset Q$ for $\mathcal{R}$ of size at most

$$\frac{r^d}{\mathrm{vol}(K)} \cdot (d \ln d + d \ln \ln d + 5d).$$

Note that Theorem 47.4.5 cannot be used here, as translates of a convex set have unbounded VC-dimension and exponential shallow-cell complexity. Furthermore, even for the case where K is a unit ball in $\mathbb{R}^d$, Theorem 47.4.5 would give a worse bound of $O\left(\frac{r^d}{\mathrm{vol}(K)} \cdot d^2 \log r\right)$.

Lower bounds for ϵ-nets for geometric set systems are implied by the following connection, first observed by Alon [Alo12], between ϵ-nets and density version of statements in Ramsey theory. Given a function $f : \mathbb{N}^+ \to \mathbb{N}^+$, let $(X, \mathcal{R})$, $|X| = n$, be a set system with the Ramsey-theoretic property that for any $X' \subset X$ of size $\frac{n}{2}$, there exists a set $R \in \mathcal{R}$ such that $|R| \geq f(n)$ and $R \subseteq X'$. Then any $\frac{f(n)}{n}$-net N for $(X, \mathcal{R})$ must have size at least $\frac{n}{2}$, as otherwise the set $X \setminus N$ of size at least $\frac{n}{2}$ would violate the Ramsey property. As $\frac{n}{2} = \omega\left(\frac{n}{f(n)}\right)$ for any monotonically increasing function $f(\cdot)$ with $f(n) \to \infty$ as $n \to \infty$, this gives a super-linear lower bound on the size of any $\frac{f(n)}{n}$-net; the precise lower bound will depend on the function $f(\cdot)$. Using this relation, Alon [Alo12] showed a super-linear lower bound for ϵ-nets for the primal set system induced by lines, for which the corresponding Ramsey-theoretic statement is the density version of the Hales-Jewett theorem. By Veronese maps[6], this implies a nonlinear bound for ϵ-nets for the primal set system induced by half-spaces in $\mathbb{R}^5$. Next, Pach and Tardos [PT13] showed that, for any

---

[5]Constructed as follows: let l be a vertical line that divides X into two equal-sized subsets, say X_1 and X_2; then add to $\mathcal{R}'$ all subsets of X induced by axis-parallel rectangles with one vertical boundary edges lying on l. Add recursively subsets to $\mathcal{R}'$ for X_1 and X_2.

[6]Map each point $p : (p_x, p_y) \in \mathbb{R}^2$ to the point $f(p) = (p_x, p_y, p_x p_y, p_x^2, p_y^2) \in \mathbb{R}^5$, and each line $l : ax + by = c$ to the half-space $f(l) : (-2ac) \cdot x_1 + (-2bc) \cdot x_2 + (2ab) \cdot x_3 + a^2 \cdot x_4 + b^2 \cdot x_5 \leq -c^2$. Then it can be verified by a simple calculation that a point $p \in \mathbb{R}^2$ lies on a line l if and only if the point $f(p) \in \mathbb{R}^5$ lies in the half-space $f(l)$.

TABLE 47.4.1   Sizes of ε-nets for both primal (P) and dual (D) set systems (ceilings/floors and lower-order terms are omitted for clarity).

Objects	SETS	UPPER BOUND		LOWER BOUND	
Intervals	P/D	$\frac{1}{\epsilon}$		$\frac{1}{\epsilon}$	
Lines, $\mathbb{R}^2$	P/D	$\frac{2}{\epsilon}\log\frac{1}{\epsilon}$	[HW87]	$\frac{1}{2\epsilon}\frac{\log^{1/3}\frac{1}{\epsilon}}{\log\log\frac{1}{\epsilon}}$	[BS17]
Half-spaces, $\mathbb{R}^2$	P/D	$\frac{2}{\epsilon}-1$	[KPW92]	$\frac{2}{\epsilon}-2$	[KPW92]
Half-spaces, $\mathbb{R}^3$	P/D	$O(\frac{1}{\epsilon})$	[MSW90]	$\Omega(\frac{1}{\epsilon})$	
Half-spaces, $\mathbb{R}^d$, $d \geq 4$	P/D	$\frac{d}{\epsilon}\log\frac{1}{\epsilon}$	[KPW92]	$\frac{\lfloor d/2\rfloor-1}{9}\frac{1}{\epsilon}\log\frac{1}{\epsilon}$	[PT13] [KMP16]
Disks, $\mathbb{R}^2$	P	$\frac{13.4}{\epsilon}$	[BGMR16]	$\frac{2}{\epsilon}-2$	[KPW92]
Balls, $\mathbb{R}^3$	P	$\frac{2}{\epsilon}\log\frac{1}{\epsilon}$		$\Omega(\frac{1}{\epsilon})$	
Balls, $\mathbb{R}^d$, $d \geq 4$	P	$\frac{d+1}{\epsilon}\log\frac{1}{\epsilon}$	[KPW92]	$\frac{\lfloor d/2\rfloor-1}{9}\frac{1}{\epsilon}\log\frac{1}{\epsilon}$	[KMP16]
Pseudo-disks, $\mathbb{R}^2$	P/D	$O(\frac{1}{\epsilon})$	[PR08]	$\Omega(\frac{1}{\epsilon})$	
Fat triangles, $\mathbb{R}^2$	D	$O(\frac{1}{\epsilon}\log\log^*\frac{1}{\epsilon})$	[AES10]	$\Omega(\frac{1}{\epsilon})$	
Axis-par. rect., $\mathbb{R}^2$	D	$\frac{5}{\epsilon}\log\frac{1}{\epsilon}$	[HW87]	$\frac{1}{9}\frac{1}{\epsilon}\log\frac{1}{\epsilon}$	[PT13]
Axis-par. rect., $\mathbb{R}^2$	P	$O(\frac{1}{\epsilon}\log\log\frac{1}{\epsilon})$	[AES10]	$\frac{1}{16}\frac{1}{\epsilon}\log\log\frac{1}{\epsilon}$	[PT13]
Union $\kappa_{\mathcal{R}}(\cdot)$, $\mathbb{R}^2$	D	$O(\frac{\log(\epsilon\cdot\kappa_{\mathcal{R}}(1/\epsilon))}{\epsilon})$	[AES10]	$\Omega(\frac{1}{\epsilon})$	
Convex sets, $\mathbb{R}^d$, $d \geq 2$	P	$|X|-\epsilon|X|$		$|X|-\epsilon|X|$	

$\epsilon > 0$ and large enough integer n, there exists a set X of n points in $\mathbb{R}^4$ such that any ε-net for the primal set system on X induced by half-spaces must have size at least $\frac{1}{9\epsilon}\log\frac{1}{\epsilon}$; when $\frac{1}{\epsilon}$ is a power of two, then it improves to the lower bound of $\frac{1}{8\epsilon}\log\frac{1}{\epsilon}$. See Table 47.4.1 for all known lower bounds.

Weak ε-nets. When the net for a given primal geometric set system $(X, \mathcal{R})$ need not be a subset of X—i.e., the case of weak ε-nets—one can sometimes get smaller bounds. For example, $O(\frac{1}{\epsilon})$ size weak ε-nets exist for the primal set system induced by balls in $\mathbb{R}^d$ [MSW90]. We outline a different construction than the one in [MSW90], as follows. Let B be the smallest radius ball containing a set X' of at least $\epsilon|X|$ points of X and no point of the current weak ε-net Q (initially $Q = \emptyset$). Now add a set $Q' \subseteq \mathbb{R}^d$ of $O(1)$ points to Q such that any ball, of radius at least that of B, intersecting B must contain a point of Q', and compute a weak ε-net for $X \setminus X'$. Weak ε-nets of size $O(\frac{1}{\epsilon}\log\log\frac{1}{\epsilon})$ exist for the primal set system induced by axis-parallel rectangles in $\mathbb{R}^d$, for $d \geq 4$ [Ezr10].

The main open question at this time on weak ε-nets is for the primal set system induced on a set X of n points by the family $\mathcal{C}$ of all convex objects in $\mathbb{R}^d$. Note that if X is in convex position, then any ε-net for this set system must have size at least $(1-\epsilon)n$. All currently known upper bounds depend exponentially on the dimension d. In Alon *et al.* [ABFK92], a bound of $O(\frac{1}{\epsilon^2})$ was shown for this problem for $d = 2$ and $O(\frac{1}{\epsilon^{d+1}})$ for $d \geq 3$. This was improved by Chazelle *et al.* [CEG+95], and then slightly further via an elegant proof by Matoušek and Wagner [MW04].

THEOREM 47.4.11   [MW04]

Let X be a finite set of points in $\mathbb{R}^d$, and let $0 < \epsilon \leq 1$ be a given parameter. Then there exists a weak ε-net for the primal set system induced by convex objects of size $O\big(\frac{1}{\epsilon^d}\log^a(\frac{1}{\epsilon})\big)$, where $a = \Theta\big(d^2\ln(d+1)\big)$. Furthermore, such a net can be computed in time $O\big(n\log\frac{1}{\epsilon}\big)$.

The above theorem—indeed many of the weak ϵ-net constructions—are based on the following two ideas. First, for a parameter t that is chosen carefully, construct a partition $\mathcal{P} = \{X_1, \ldots, X_t\}$ of X such that (a) $|X_i| \leq \lceil \frac{n}{t} \rceil$ for all i, and (b) for any integer $k \geq 1$, there exists a point set Q_k of small size such that any convex object having non-empty intersection with at least ϵk sets of $\mathcal{P}$ must contain a point of Q_k. Note that Q_t is a weak ϵ-net, as any convex set containing ϵn points must intersect at least $\frac{\epsilon n}{(n/t)} = \epsilon t$ sets. Second, compute recursively a weak ϵ'-net Q_i' for each X_i, for a suitably determined value of ϵ'. If a convex set C is not hit by $\bigcup Q_i'$, it contains at most $\frac{\epsilon' n}{t}$ points from each set of $\mathcal{P}$, and so has non-empty intersection with at least $\frac{\epsilon n}{(\epsilon' n/t)} = \frac{t\epsilon}{\epsilon'}$ sets of $\mathcal{P}$. Then $\bigcup Q_i'$ together with $Q_{\frac{t}{\epsilon'}}$ is a weak ϵ-net; fixing the trade-off parameters t, ϵ' gives the final bound. Theorem 47.4.11 uses simplicial partitions for $\mathcal{P}$, and centerpoints of some representative points from each set of $\mathcal{P}$ as the set Q_k.

There is a wide gap between the best known upper and lower bounds. Matoušek [Mat02b] showed the existence of a set X of points in $\mathbb{R}^d$ such that any weak $\frac{1}{50}$-net for the set system induced by convex objects on X has size $\Omega\big(e^{\frac{\sqrt{d}}{2}}\big)$. For arbitrary values of ϵ, the current best lower bound is the following.

THEOREM 47.4.12 [BMN11]

For every $d \geq 2$ and every $\epsilon > 0$, there exists a set X of points in $\mathbb{R}^d$ such that any weak ϵ-net for the primal set system induced on X by convex objects has size $\Omega\big(\frac{1}{\epsilon} \log^{d-1} \frac{1}{\epsilon}\big)$.

There is a relation between weak ϵ-nets induced by convex sets and ϵ-nets for the primal set system induced by intersections of half-spaces, though the resulting size of the weak ϵ-net is still exponential in the dimension [MR08]. The weak ϵ-net problem is closely related to an old (and still open) problem of Danzer and Rogers, which asks for the area of the largest convex region avoiding a given set of n points in a unit square (see [PT12] for a history of the problem). Better bounds for weak ϵ-nets for primal set systems induced by convex objects are known for special cases: an upper bound of $O\big(\frac{1}{\epsilon}\alpha(\frac{1}{\epsilon})\big)$ when X is a set of points in $\mathbb{R}^2$ in convex position [AKN+08]; optimal bounds when ϵ is a large constant [MR09]; a bound of $O\big(\frac{1}{\epsilon}(\log \frac{1}{\epsilon})^{\Theta(d^2 \ln d)}\big)$ when the points lie on a moment curve in $\mathbb{R}^d$ [MW04].

47.5 APPLICATIONS OF EPSILON-NETS

As ϵ-nets capture some properties of random samples with respect to a set system, a natural use of ϵ-nets has been for derandomization; the best deterministic combinatorial algorithms for linear programming [CM96, Cha16] are derived via derandomization using ϵ-nets. Another thematic use originates from the fact that an ϵ-net of a set system $(X, \mathcal{R})$ can be viewed as a hitting set for sets in $\mathcal{R}$ of size at least $\epsilon|X|$, and so is relevant for many types of covering optimization problems; a recent example is the beautiful work of Arya *et al.* [AFM12] in approximating a convex body by a polytope with few vertices. At first glance, the restriction that an ϵ-net only guarantees to hit sets of size at least $\epsilon|X|$ narrows its applicability. A breakthrough idea, with countless applications, has been to first assign *multiplicities* (or *weights*) to the elements of X such that all multisets have large size; then ϵ-nets can be used to "round" this to get a solution. Lastly, ϵ-nets can be used for constructing spatial partitions that enable the use of divide-and-conquer methods; indeed, one of the earliest applications introducing ε-nets was by Clarkson [Cla88] to construct a spatial partitioning data-structure for answering nearest-neighbor queries.

SPATIAL PARTITIONING

Consider the set system $(\mathcal{H}, \mathcal{R})$ where the base set $\mathcal{H}$ is a set of n hyperplanes in $\mathbb{R}^d$, and $\mathcal{R}$ is the set system induced by intersection of simplices in $\mathbb{R}^d$ with $\mathcal{H}$. An ε-net for $\mathcal{R}$ consists of a subset $\mathcal{H}'$ such that any simplex intersecting at least ϵn hyperplanes of $\mathcal{H}$ intersects a hyperplane in $\mathcal{H}'$. This implies that for any simplex Δ lying in the interior of a cell in the arrangement of $\mathcal{H}'$, the number of hyperplanes of $\mathcal{H}$ intersecting Δ is less than ϵn. One can further partition each cell in the arrangement of $\mathcal{H}'$ into simplices, leading to the powerful concept of *cuttings*. After a series of papers in the 1980s and early 1990s [CF90, Mat91b], the following is the best result in terms of both combinatorial and algorithmic bounds.

THEOREM 47.5.1 [Cha93]

Let $\mathcal{H}$ be a set of n hyperplanes in $\mathbb{R}^d$, and $r \geq 1$ a given parameter. Then there exists a partition of $\mathbb{R}^d$ into $O(r^d)$ interior-disjoint simplices, such that the interior of each simplex intersects at most $\frac{n}{r}$ hyperplanes of $\mathcal{H}$. These simplices, together with the list of hyperplanes intersecting the interior of each simplex, can be found deterministically in time $O(nr^{d-1})$.

There are many extensions of such a partition, called a $\frac{1}{r}$-cutting, known for objects other than hyperplanes; see Chapter 28. Here we state just one such result.

THEOREM 47.5.2 [BS95, Pel97]

Let $\mathcal{S}$ be a set of n $(d-1)$-dimensional simplices in $\mathbb{R}^d$ and let $m = m(\mathcal{S})$ denote the number of d-tuples of $\mathcal{S}$ having a point in common. Then, for any $\epsilon > 0$ and any given parameter $r \geq 1$, there exists a $\frac{1}{r}$-cutting of $\mathcal{S}$ with the number of simplices at most $O\left(r + \frac{mr^2}{n^2}\right)$ for $d = 2$, and $O\left(r^{d-1+\epsilon} + \frac{mr^d}{n^d}\right)$ for $d \geq 3$.

Cuttings have found countless applications, both combinatorial and algorithmic, for their role in divide-and-conquer arguments. A paradigmatic combinatorial use for upper-bounding purposes, initiated in a seminal paper by Clarkson *et al.* [CEG+90], is using cuttings to partition $\mathbb{R}^d$ into simplices, each of which forms an independent sub-problem where one can apply a worse—and often purely combinatorial—bound. The sum of this bound over all simplices together with accounting for interaction on the boundaries of the simplices gives an upper bound. This remains a key technique for bounding incidences between points and various geometric objects (see the book [Gut16]), as well as for many Turán-type problems on geometric configurations (see [MP16] for a recent example). Algorithmically, cuttings have proven invaluable for divide-and-conquer based methods for point location, convex hulls, Voronoi diagrams, combinatorial optimization problems, clustering, range reporting and range searching. An early use was for the half-space range searching problem, which asks for pre-processing a finite set X of points in $\mathbb{R}^d$ such that one can efficiently count the set of points of X contained in any query half-space [Mat93b]. The current best data structure [AC09] for the related problem of reporting points contained in a query half-space is also based on cuttings; see Chapter 40.

Finally, we state one consequence of a beautiful result of Guth [Gut15] which achieves spatial partitioning for more general objects, with a topological approach replacing the use of ϵ-nets: given a set $\mathcal{H}$ of n k-dimensional flats in $\mathbb{R}^d$ and a parameter $r \geq 1$, there exists a nonzero d-variate polynomial P, of degree at most r, such that each of the $O(r^d)$ cells induced by the zero set $Z(P)$ of P (i.e., each component of $\mathbb{R}^d \setminus Z(P)$) intersects $O(r^{k-d}n)$ flats of $\mathcal{H}$. Note that for the case $k = d - 1$, this is a "polynomial partitioning" version of Theorem 47.5.1.

ROUNDING FRACTIONAL SYSTEMS

We now present two uses of ϵ-nets in rounding fractional systems to integral ones— as before, one will be algorithmic and the other combinatorial. Given a set system $(X, \mathcal{R})$, the *hitting set problem* asks for the smallest set $Y \subseteq X$ that intersects all sets in $\mathcal{R}$. Let $\mathrm{OPT}_{\mathcal{R}}$ be the size of a minimum hitting set for $\mathcal{R}$. Given a weight function $w : X \to \mathbb{R}^+$ with $w(x) > 0$ for at least one $x \in X$, we say that $N \subseteq X$ is an ϵ-net *with respect to* $w(\cdot)$ if $N \cap R \neq \emptyset$ for any $R \in \mathcal{R}$ such that $w(R) \geq \epsilon \cdot w(X)$. The construction of an ϵ-net with respect to weight function $w(\cdot)$ can be reduced to the construction of a regular ϵ-net for a different set system $(X', \mathcal{R}')$; the main idea is that for each $x \in X$ we include multiple "copies" of x in the base set X', with the number of copies being proportional to $w(x)$. Using this reduction, many of the results on ϵ-nets carry over to ϵ-nets with respect to a weight function.

THEOREM 47.5.3   [BG95, Lon01, ERS05]

Given $(X, \mathcal{R})$, assume there is a function $f : \mathbb{R}^+ \to \mathbb{N}^+$ such that for any $\epsilon > 0$ and weight function $w : X \to \mathbb{R}^+$, an ϵ-net of size at most $\frac{1}{\epsilon} \cdot f(\frac{1}{\epsilon})$ exists with respect to $w(\cdot)$. Further assume a net of this size can be computed in polynomial time. Then one can compute a $f(\mathrm{OPT}_{\mathcal{R}})$-approximation to the minimum hitting set for $\mathcal{R}$ in polynomial time, where $\mathrm{OPT}_{\mathcal{R}}$ is the size of a minimum hitting set for $\mathcal{R}$.

The proof proceeds as follows: to each $p \in X$ assign a weight $w(p) \in [0, 1]$ such that the total weight $W = \sum_{p \in X} w(p)$ is minimized, under the constraint that $w(R) = \sum_{p \in R} w(p) \geq 1$ for each $R \in \mathcal{R}$. Such weights can be computed in polynomial time using linear programming. Now a $\frac{1}{W}$-net (with respect to the weight function $w(\cdot)$) is a hitting set for $\mathcal{R}$; crucially, as $W \leq \mathrm{OPT}_{\mathcal{R}}$, this net is of size at most $W f(W) \leq \mathrm{OPT}_{\mathcal{R}} \cdot f(\mathrm{OPT}_{\mathcal{R}})$. In particular, when the set system has ϵ-nets of size $O(\frac{1}{\epsilon})$, one can compute a constant-factor approximation to the minimum hitting set problem; e.g., for the geometric minimum hitting set problem for points and disks in the plane. Furthermore these algorithms can be implemented in near-linear time [AP14, BMR15]. When the elements of X have costs, and the goal is to minimize the cost of the hitting set, Varadarajan [Var10] showed that ϵ-nets imply the corresponding approximation factor.

THEOREM 47.5.4   [Var10]

Given $(X, \mathcal{R})$ with a cost function $c : X \to \mathbb{R}^+$, assume that there exists a function $f : \mathbb{N} \to \mathbb{N}$ such that for any $\epsilon > 0$ and weight function $w : X \to \mathbb{R}^+$, there is an ϵ-net with respect to $w(\cdot)$ of cost at most $\frac{c(X)}{\epsilon n} \cdot f(\frac{1}{\epsilon})$. Further assume such a net can be computed in polynomial time. Then one can compute a $f(\mathrm{OPT}_{\mathcal{R}})$-approximation to the minimum cost hitting set for $\mathcal{R}$ in polynomial time.

We now turn to a combinatorial use of ϵ-nets in rounding. A set $\mathcal{C}$ of n convex objects in $\mathbb{R}^d$ is said to satisfy the $HD(p,q)$ property if for any set $\mathcal{C}' \subseteq \mathcal{C}$ of size p, there exists a point common to at least q objects in $\mathcal{C}'$ (see Chapter 4). Answering a long-standing open question, Alon and Kleitman [AK92] showed that then there exists a hitting set for $\mathcal{C}$ whose size is a function of only p, q and d— in particular, independent of n. The resulting function was improved to give the following statement.

THEOREM 47.5.5 [AK92, KST17]

Let $\mathcal{C}$ be a finite set of convex objects in $\mathbb{R}^d$, and p, q be two integers, where $p \geq q \geq d + 1$, such that for any set $\mathcal{C}' \subseteq \mathcal{C}$ of size p, there exists a point in $\mathbb{R}^d$ common to at least q objects in $\mathcal{C}'$. Then there exists a hitting set for $\mathcal{C}$ of size $O\left(p^{d\frac{q-1}{q-d}} \log^{c'd^3 \log d} p\right)$, where c' is an absolute constant.

We present a sketch of the proof. Let P be a point set consisting of a point from each cell of the arrangement of $\mathcal{C}$. For each $p \in P$, let $w(p)$ be the weight assigned to p such that the total weight $W = \sum_p w(p)$ is minimized, while satisfying the constraint that each $C \in \mathcal{C}$ contains points of total weight at least 1. Similarly, let $w^*(C)$ be the weight assigned to each $C \in \mathcal{C}$ such that the total weight $W^* = \sum_C w^*(C)$ is maximized, while satisfying the constraint that each $p \in P$ lies in objects of total weight at most 1. Now linear programming duality implies that $W = W^*$, and crucially, we have $c \cdot W^* \leq 1$ for some constant $c > 0$: using the $HD(p,q)$ property, a straightforward counting argument shows that there exists a point $p \in P$ hitting objects in $\mathcal{C}$ of total weight at least $c \cdot W^*$, where $c > 0$ is a constant depending only on p, q and d. Thus $W = W^* \leq \frac{1}{c}$, and so a weak c-net for P (with respect to the weight function $w(\cdot)$) induced by convex objects hits all objects in $\mathcal{C}$, and has size $O\left(\frac{1}{c^d} \log^{\Theta(d^2 \log d)} \frac{1}{c}\right)$ by Theorem 47.4.11. This idea was later used in proving combinatorial bounds for a variety of geometric problems; see [AK95, Alo98, AKMM02, MR16] for a few examples.

47.6 OPEN PROBLEMS

We conclude with some open problems.

1. Show a lower bound of $\Omega(\frac{1}{\epsilon} \log \frac{1}{\epsilon})$ on the size of any ϵ-net for the primal set system induced by lines in the plane.

2. Prove a tight bound on the size of weak ϵ-nets for the primal set system induced by convex objects in $\mathbb{R}^d$. An achievable goal may be to prove the existence of weak ϵ-nets of size $O\left(\frac{1}{\epsilon^{\lceil d/2 \rceil}}\right)$.

3. Improve the current best bound of $O\left(\frac{1}{\epsilon} \log \log \frac{1}{\epsilon}\right)$ for weak ϵ-nets for the primal set system induced by axis-parallel rectangles in $\mathbb{R}^2$.

4. Show a lower bound of $\left(\frac{d}{2} - o(1)\right)\frac{1}{\epsilon} \log \frac{1}{\epsilon}$ for the size of any ϵ-net for the primal set system induced by half-spaces in $\mathbb{R}^d$.

5. Show a lower bound of $\Omega\left(\frac{1}{\epsilon} \log \frac{1}{\epsilon}\right)$ for ϵ-nets for the primal set system induced by balls in $\mathbb{R}^3$.

6. An unsatisfactory property of many lower bound constructions for ϵ-nets is that the construction of the set system depends on the value of ϵ—typically

the number of elements in the construction is only $\Theta(\frac{1}{\epsilon} \log \frac{1}{\epsilon})$; each element is then "duplicated" to derive the statement for arbitrary values of n. Do constructions exist that give a lower bound on the ϵ-net size for every value of ϵ?

7. Improve the slightly sub-optimal bound of Theorem 47.5.2 to show the following. Let $\mathcal{S}$ be a set of n $(d-1)$-dimensional simplices in $\mathbb{R}^d$, $d \geq 3$, and let $m = m(\mathcal{S})$ denote the number of d-tuples of $\mathcal{S}$ having a point in common. Then for any $r \leq n$, there is a $\frac{1}{r}$-cutting of $\mathcal{S}$ with size at most $O\left(r^{d-1} + \dfrac{mr^d}{n^d}\right)$.

8. Improve the current bounds for ϵ-approximations for the primal set system induced by balls in $\mathbb{R}^d$ to $O\left(\dfrac{1}{\epsilon^{2 - \frac{2}{d+1}}}\right)$.

9. Let $(X, \mathcal{R})$ be a set system with $\varphi_{\mathcal{R}}(m, k) = O\left(m^{d_1} k^{d - d_1}\right)$, where $1 < d_1 \leq d$ are constants (with $\varphi_{\mathcal{R}}(m, k)$ as defined in the first section). Do there exist relative (ϵ, δ)-approximations of size $O\left(\dfrac{1}{\epsilon^{\frac{d+d_1}{d+1}} \delta^{\frac{2d}{d+1}}}\right)$ for $(X, \mathcal{R})$?

47.7 SOURCES AND RELATED MATERIALS

READING MATERIAL

See Matoušek [Mat98] for a survey on VC-dimension, and its relation to discrepancy, sampling and approximations of geometric set systems. An early survey on ϵ-nets was by Matoušek [Mat93a], and a more general one on randomized algorithms by Clarkson [Cla92]. Introductory expositions to ϵ-approximations and ϵ-nets can be found in the books by Pach and Agarwal [PA95], Matoušek [Mat02a], and Har-Peled [HP11]. The monograph of Har-Peled [HP11] also discusses sensitive approximations and relative approximations. The books by Matoušek [Mat99] on geometric discrepancy and by Chazelle [Cha00] on the discrepancy method give a detailed account of some of the material in this chapter. From the point of view of learning theory, a useful survey on approximations is Boucheron *et al.* [BBL05], while the books by Devroye, Györfi, and Lugosi [DGL96] and Anthony and Bartlett [AB09] contain detailed proofs on random sampling for set systems with bounded VC-dimension. For spatial partitioning and its many applications, we refer the reader to the book by Guth [Gut16].

RELATED CHAPTERS

Chapter 13: Geometric discrepancy theory and uniform distribution
Chapter 40: Range searching
Chapter 44: Randomization and derandomization
Chapter 48: Coresets and sketches

REFERENCES

[AAG14]   P. Ashok, U. Azmi, and S. Govindarajan. Small strong epsilon nets. *Comput. Geom.*, 47:899–909, 2014.

[AB09]   M. Anthony and P.L. Bartlett. *Neural Network Learning: Theoretical Foundations.* Cambridge University Press, 2009.

[ABES14] B. Aronov, M. de Berg, E. Ezra, and M. Sharir. Improved bounds for the union of locally fat objects in the plane. *SIAM J. Comput.*, 43:543–572, 2014.

[ABFK92] N. Alon, I. Bárány, Z. Füredi, and D.J. Kleitman. Point selections and weak ε-nets for convex hulls. *Combin. Probab. Comput.*, 1:189–200, 1992.

[AC09]   P. Afshani and T.M. Chan. Optimal halfspace range reporting in three dimensions. In *Proc. 20th ACM-SIAM Sympos. Discrete Algorithms*, pages 180–186, 2009.

[AES10]   B. Aronov, E. Ezra, and M. Sharir. Small-size ε-nets for axis-parallel rectangles and boxes. *SIAM J. Comput.*, 39:3248–3282, 2010.

[AFM12]   S. Arya, G.D. da Fonseca, and D.M. Mount. Polytope approximation and the Mahler volume. In *Proc. 23rd ACM-SIAM Sympos. Discrete Algorithms*, pages 29–42, 2012.

[AFM17]   S. Arya, G.D. da Fonseca, and D.M. Mount. Near-optimal epsilon-kernel construction and related problems. In *Proc. 33rd Sympos. Comput. Geom.*, article 10, vol. 77 of *LIPIcs*, Schloss Dagstuhl, 2017.

[AK92]   N. Alon and D.J. Kleitman. Piercing convex sets and the Hadwiger-Debrunner (p, q)-problem. *Adv. Math.*, 96:103–112, 1992.

[AK95]   N. Alon and G. Kalai. Bounding the piercing number. *Discrete Comput. Geom.*, 13:245–256, 1995.

[AKMM02] N. Alon, G. Kalai, J. Matoušek, and R. Meshulam. Transversal numbers for hypergraphs arising in geometry. *Adv. Appl. Math.*, 29:79–101, 2002.

[AKN+08] N. Alon, H. Kaplan, G. Nivasch, M. Sharir, and S. Smorodinsky. Weak ε-nets and interval chains. *J. ACM*, 55(6), 2008.

[Alo98]   N. Alon. Piercing d-intervals. *Discrete Comput. Geom.*, 19:333–334, 1998.

[Alo12]   N. Alon. A non-linear lower bound for planar epsilon-nets. *Discrete Comput. Geom.*, 47:235–244, 2012.

[AP14]   P.K. Agarwal and J. Pan. Near-linear algorithms for geometric hitting sets and set covers. In *Proc. 30th Sympos. Comput. Geom.*, pages 271–279, ACM Press, 2014.

[AS08]   N. Alon and J.H. Spencer. *The Probabilistic Method*, 3rd edition. John Wiley & Sons, New York, 2008.

[Ass83]   P. Assouad. Density and dimension. *Ann. Inst. Fourier*, 33:233–282, 1983.

[Ban12]   N. Bansal. Semidefinite optimization in discrepancy theory. *Math. Program.*, 134:5–22, 2012.

[Bár07]   I. Bárány. Random polytopes, convex bodies, and approximation. In W. Weil, editor, *Stochastic Geometry*, pages 77–118, Springer, Berlin, 2007.

[BL88]   I. Bárány and D.G. Larman. Convex bodies, economic cap coverings, random polytopes. *Mathematika*, 35:274–291, 1988.

[BBL05]   S. Boucheron, O. Bousquet, and G. Lugosi. Theory of classification: A survey of some recent advances. *ESAIM: Probab. Stat.*, 9:323–375, 2005.

[BCM99]   H. Brönnimann, B. Chazelle, and J. Matoušek. Product range spaces, sensitive sampling, and derandomization. *SIAM J. Comput.*, 28:1552–1575, 1999.

[BCP93]   H. Brönnimann, B. Chazelle, and J. Pach. How hard is halfspace range searching? *Discrete Comput. Geom.*, 10:143–155, 1993.

[BG95]   H. Brönnimann and M.T. Goodrich. Almost optimal set covers in finite VC-dimension. *Discrete Comput. Geom.*, 14:463–479, 1995.

[BGMR16]   N. Bus, S. Garg, N.H. Mustafa, and S. Ray. Tighter estimates for ϵ-nets for disks. *Comput. Geom.*, 53:27–35, 2016.

[BLK17]   O. Bachem, M. Lucic, and A. Krause. Practical coreset constructions for machine learning. Preprint, `arXiv:1703.06476`, 2017.

[BMN11]   B. Bukh, J. Matoušek, and G. Nivasch. Lower bounds for weak epsilon-nets and stair-convexity. *Israel J. Math.*, 182:199–228, 2011.

[BMR15]   N. Bus, N.H. Mustafa, and S. Ray. Geometric hitting sets for disks: Theory and practice. In *Proc. 23rd European Sympos. Algorithms*, vol. 9294 of *LNCS*, pages 903–914, Springer, Berlin, 2015.

[BPR13]   S. Buzaglo, R. Pinchasi, and G. Rote. Topological hyper-graphs. In J. Pach, editor, *Thirty Essays on Geometric Graph Theory*, pages 71–81, Springer, New York, 2013.

[BS95]   M. de Berg and O. Schwarzkopf. Cuttings and applications. *Internat. J. Comput. Geom. Appl.*, 5:343–355, 1995.

[BS17]   J. Balogh and J. Solymosi. On the number of points in general position in the plane. Preprint, `arXiv:1704.05089`, 2017.

[CEG+90]   K.L. Clarkson, H. Edelsbrunner, L.J. Guibas, M. Sharir, and E. Welzl. Combinatorial complexity bounds for arrangement of curves and spheres. *Discrete Comput. Geom.*, 5:99–160, 1990.

[CEG+95]   B. Chazelle, H. Edelsbrunner, M. Grigni, L. Guibas, M. Sharir, and E. Welzl. Improved bounds on weak ε-nets for convex sets. *Discrete Comput. Geom.*, 13:1–15, 1995.

[CGKS12]   T.M. Chan, E. Grant, J. Könemann, and M. Sharpe. Weighted capacitated, priority, and geometric set cover via improved quasi-uniform sampling. In *Proc. 23rd ACM-SIAM Sympos. Discrete Algorithms*, pages 1576–1585, 2012.

[CF90]   B. Chazelle and J. Friedman. A deterministic view of random sampling and its use in geometry. *Combinatorica*, 10:229–249, 1990.

[Cha00]   B. Chazelle. *The Discrepancy Method: Randomness and Complexity*. Cambridge University Press, 2000.

[Cha16]   T.M. Chan. Improved deterministic algorithms for linear programming in low dimensions. In *Proc. 27th ACM-SIAM Sympos. Discrete Algorithms*, pages 1213–1219, 2016.

[Cha93]   B. Chazelle. Cutting hyperplanes for divide-and-conquer. *Discrete Comput. Geom.*, 9:145–158, 1993.

bibitem[Cla87]C87 K.L. Clarkson. New applications of random sampling in computational geometry. *Discrete Comput. Geom.*, 2:195–222, 1987.

[Cla88]   K.L. Clarkson. A randomized algorithm for closest-point queries. *SIAM J. Comput.*, 17:830–847, 1988.

[Cla92]   K.L. Clarkson. Randomized geometric algorithms. In F.K. Hwang and D.Z. Hu, editors, *Computers and Euclidean Geometry*, World Scientific Publishing, 1992.

[CM96]   B. Chazelle and J. Matoušek. On linear-time deterministic algorithms for optimization problems in fixed dimension. *J. Algorithms*, 21:579–597, 1996.

[CS89]   K.L. Clarkson and P.W. Shor. Application of random sampling in computational geometry, II. *Discrete Comput. Geom.*, 4:387–421, 1989.

[CV07]   K.L. Clarkson and K. Varadarajan. Improved approximation algorithms for geometric set cover. *Discrete Comput. Geom.*, 37:43–58, 2007.

[CW89]   B. Chazelle and E. Welzl. Quasi-optimal range searching in space of finite VC-dimension. *Discrete Comput. Geom.*, 4:467–489, 1989.

[DEG16]   K. Dutta, E. Ezra, and A. Ghosh. Two proofs for shallow packings. *Discrete Comput. Geom.*, 56:910–939, 2016.

[DGJM17]   K. Dutta, A. Ghosh, B. Jartoux, and N.H. Mustafa. Shallow packings, semialgebraic set systems, Macbeath regions, and polynomial partitioning. In *Proc. 33rd Sympos. Comput. Geom.*, article 38, vol. 77 of *LIPIcs*, Schloss Dagstuhl, 2017.

[DGL96]   L. Devroye, L. Györfi, and G. Lugosi. *A Probabilistic Theory of Pattern Recognition.* Springer, Berlin, 1996.

[ERS05]   G. Even, D. Rawitz, and S. Shahar. Hitting sets when the VC-dimension is small. *Inform. Process. Lett.*, 95:358–362, 2005.

[Ezr10]   E. Ezra. A note about weak epsilon-nets for axis-parallel boxes in d-space. *Inform. Process. Lett.*, 110:835–840, 2010.

[Ezr16]   E. Ezra. A size-sensitive discrepancy bound for set systems of bounded primal shatter dimension. *SIAM J. Comput.*, 45:84–101, 2016.

[FL11]   D. Feldman and M. Langberg. A unified framework for approximating and clustering data. In *Proc. 43rd ACM Sympos. Theory Comput.*, pages 569–578, 2011.

[Gut15]   L. Guth. Polynomial partitioning for a set of varieties. *Math. Proc. Cambridge Philos. Soc.*, 159:459–469, 2015.

[Gut16]   L. Guth. *Polynomial Methods in Combinatorics.* University Lecture Series, AMS, Providence, 2016.

[Hau95]   D. Haussler. Sphere packing numbers for subsets of the Boolean n-cube with bounded Vapnik-Chervonenkis dimension. *J. Combin. Theory Ser. A*, 69:217–232, 1995.

[HP11]   S. Har-Peled. *Geometric Approximation Algorithms.* AMS, Providence, 2011.

[HPS11]   S. Har-Peled and M. Sharir. Relative (p, ε)-approximations in geometry. *Discrete Comput. Geom.*, 45:462–496, 2011.

[HW87]   D. Haussler and E. Welzl. Epsilon-nets and simplex range queries. *Discrete Comput. Geom.*, 2:127–151, 1987.

[KK11]   J. King and D. Kirkpatrick. Improved approximation for guarding simple galleries from the perimeter. *Discrete Comput. Geom.*, 46:252–269, 2011.

[KPR93]   P. Klein, S.A. Plotkin, and S. Rao. Excluded minors, network decomposition, and multicommodity flow. In *Proc. 25th ACM Sympos. Theory Comput.*, pages 682–690, 1993.

[KMP16]   A. Kupavskii, N.H. Mustafa, and J. Pach. New lower bounds for epsilon-nets. In *Proc. 32nd Sympos. Comput. Geom.*, vol. 51 of *LIPIcs*, article 54, Schloss Dagstuhl, 2016.

[KPW92]   J. Komlós, J. Pach, and G.J. Woeginger. Almost tight bounds for epsilon-nets. *Discrete Comput. Geom.*, 7:163–173, 1992.

[KST17]   C. Keller, S. Smorodinsky, and G. Tardos. On MAX-CLIQUE for intersection graphs of sets and the Hadwiger-Debrunner numbers. In *Proc. ACM-SIAM Sympos. Discrete Algorithms*, pages 2254–2263, 2017.

[LLS01]   Y. Li, P.M. Long, and A. Srinivasan. Improved bounds on the sample complexity of learning. *J. Comput. Syst. Sci.*, 62:516–527, 2001.

[Lon01]   P.M. Long. Using the pseudo-dimension to analyze approximation algorithms for integer programming. In *Proc. 7th Workshop on Algorithms and Data Structures*, vol. 2125 of *LNCS*, pages 26–37, Springer, Berlin, 2001.

[LM15]   S. Lovett and R. Meka. Constructive discrepancy minimization by walking on the edges. *SIAM J. Comput.*, 44:1573–1582, 2015.

[LS10]   M. Langberg and L.J. Schulman. Universal ε-approximators for integrals. In *Proc. 21st ACM-SIAM Sympos. Discrete Algorithms*, pages 598–607, 2010.

[Mat16]   C. Mathieu. Personal communication, 2016.

[Mat91a]   J. Matoušek. Computing the center of planar point sets. In J.E. Goodman, R. Pollack, and W.L. Steiger, editors, *Discrete and Computational Geometry: Papers from the DIMACS Special Year*, vol. 6 of *DIMACS Ser. Discrete Math. Theor. Comp. Sci.*, pages 221–230, AMS, Providence, 1991.

[Mat91b]   J. Matoušek. Cutting hyperplane arrangements. *Discrete Comput. Geom.*, 6:385–406, 1991.

[Mat93a]   J. Matoušek. Epsilon-nets and computational geometry. In J. Pach, editor, *New Trends in Discrete and Computational Geometry*, vol. 10 of *Algorithms and Combinatorics*, pages 69–89, Springer, Berlin, 1993.

[Mat93b]   J. Matoušek. Range searching with efficient hierarchical cuttings. *Discrete Comput. Geom.*, 10:157–182, 1993.

[Mat95]   J. Matoušek. Tight upper bounds for the discrepancy of half-spaces. *Discrete Comput. Geom.*, 13:593–601, 1995.

[Mat98]   J. Matoušek. Geometric set systems. In A. Balog et al., editors, *European Congress of Mathematics*. vol. 169 of *Progress in Math.*, pages 1–27, Birkhäuser, Basel, 1998.

[Mat99]   J. Matoušek. *Geometric Discrepancy: An Illustrated Guide*. Vol. 18 of *Algorithms and Combinatorics*, Springer, Berlin, 1999.

[Mat02a]   J. Matoušek. *Lectures on Discrete Geometry*. Springer, Berlin, 2002.

[Mat02b]   J. Matoušek. A lower bound for weak epsilon-nets in high dimension. *Discrete Comput. Geom.*, 28:45–48, 2002.

[MDG17]   N.H. Mustafa, K. Dutta, and A. Ghosh. A simple proof of optimal epsilon-nets. *Combinatorica*, in press, 2017.

[MP16]   N.H. Mustafa and J. Pach. On the Zarankiewicz problem for intersection hypergraphs. *J. Combin. Theory Ser. A*, 141:1–7, 2016.

[MR08]   N.H. Mustafa and S. Ray. Weak ϵ-nets have a basis of size $O(1/\epsilon \log 1/\epsilon)$. *Comput. Geom.*, 40:84–91, 2008.

[MR09]   N.H. Mustafa and S. Ray. An optimal extension of the centerpoint theorem. *Comput. Geom.*, 42:505–510, 2009.

[MR14]   N.H. Mustafa and S. Ray. ε-Mnets: Hitting geometric set systems with subsets. *Discrete Comput. Geom.*, 57:625–640, 2017.

[MR16]   N.H. Mustafa and S. Ray. An optimal generalization of the colorful Carathéodory theorem. *Discrete Math.*, 339:1300–1305, 2016.

[MSW90]   J. Matoušek, R. Seidel, and E. Welzl. How to net a lot with little: Small epsilon-nets for disks and halfspaces. In *Proc. 6th Sympos. Comput. Geom.*, pages 16–22, ACM Press, 1990.

[Mus16]   N.H. Mustafa. A simple proof of the shallow packing lemma. *Discrete Comput. Geom.*, 55:739–743, 2016.

[MW04]   J. Matoušek and U. Wagner. New constructions of weak epsilon-nets. *Discrete Comput. Geom.*, 32:195–206, 2004.

[MWW93]   J. Matoušek, E. Welzl, and L. Wernisch. Discrepancy and approximations for bounded VC-dimension. *Combinatorica*, 13:455–466, 1993.

[NT10]   M. Naszódi and S. Taschuk. On the transversal number and VC-dimension of families of positive homothets of a convex body. *Discrete Math.*, 310:77–82, 2010.

[PA95]   J. Pach and P.K. Agarwal. *Combinatorial Geometry.* John Wiley & Sons, New York, 1995.

[Pel97]   M. Pellegrini. On counting pairs of intersecting segments and off-line triangle range searching. *Algorithmica*, 17:380–398, 1997.

[Phi08]   J.M. Phillips. Algorithms for ε-approximations of terrains. In *Proc. 35th Internat. Coll. Automata, Languages, and Prog.*, part 1, vol. 5125 of *LNCS*, pages 447–458, Springer, Berlin, 2008.

[PR08]   E. Pyrga and S. Ray. New existence proofs for ε-nets. In *Proc. 24th Sympos. Comput. Geom.*, pages 199–207, ACM Press, 2008.

[PT12]   J. Pach and G. Tardos. Piercing quasi-rectangles—on a problem of Danzer and Rogers. *J. Combin. Theory Ser. A*, 119:1391–1397, 2012.

[PT13]   J. Pach and G. Tardos. Tight lower bounds for the size of epsilon-nets. *J. Amer. Math. Soc.*, 26:645–658, 2013.

[Rad21]   J. Radon. Mengen konvexer Körper, die einen gemeinsamen Punkt enthalten. *Math. Ann.*, 83:113–115, 1921.

[Rog57]   C.A. Rogers. A note on coverings. *Mathematika*, 4:1–6, 1957.

[Sau72]   N. Sauer. On the density of families of sets. *J. Combin. Theory Ser. A*, 13:145–147, 1972.

[Sha91]   M. Sharir. On k-sets in arrangement of curves and surfaces. *Discrete Comput. Geom.*, 6:593–613, 1991.

[She72]   S. Shelah. A combinatorial problem; stability and order for models and theories in infinitary languages. *Pacific J. Math.*, 41:247–261, 1972.

[STZ06]   S. Suri, C.D. Tóth, and Y. Zhou. Range counting over multidimensional data streams. *Discrete Comput. Geom.*, 36:633–655, 2006.

[Tal94]   M. Talagrand. Sharper bounds for Gaussian and empirical processes. *Ann. Prob.*, 22:28–76, 1994.

[Var09]   K. Varadarajan. Epsilon nets and union complexity. In *Proc. 25th Sympos. Comput. Geom.*, pages 11–16, ACM Press, 2009.

[Var10]   K. Varadarajan. Weighted geometric set cover via quasi uniform sampling. In *Proc. 42nd ACM Sympos. Theory Comput.*, pages 641–648, 2010.

[VC71]   V.N. Vapnik and A.Y. Chervonenkis. On the uniform convergence of relative frequencies of events to their probabilities. *Theory Probab. Appl.*, 16:264–280, 1971.

48 CORESETS AND SKETCHES

Jeff M. Phillips

INTRODUCTION

Geometric data summarization has become an essential tool in both geometric approximation algorithms and where geometry intersects with big data problems. In linear or near-linear time, large data sets can be compressed into a summary, and then more intricate algorithms can be run on the summaries whose results approximate those of the full data set. Coresets and sketches are the two most important classes of these summaries.

A *coreset* is a reduced data set which can be used as proxy for the full data set; the same algorithm can be run on the coreset as the full data set, and the result on the coreset approximates that on the full data set. It is often required or desired that the coreset is a subset of the original data set, but in some cases this is relaxed. A *weighted coreset* is one where each point is assigned a weight, perhaps different than it had in the original set. A *weak coreset* associated with a set of queries is one where the error guarantee holds for a query which (nearly) optimizes some criteria, but not necessarily all queries; a *strong coreset* provides error guarantees for all queries.

A *sketch* is a compressed mapping of the full data set onto a data structure which is easy to update with new or changed data, and allows certain queries whose results approximate queries on the full data set. A *linear sketch* is one where the mapping is a linear function of each data point, thus making it easy for data to be added, subtracted, or modified.

These definitions can blend together, and some summaries can be classified as either or both. The overarching connection is that the summary size will ideally depend only on the approximation guarantee but not the size of the original data set, although in some cases logarithmic dependence is acceptable.

We focus on five types of coresets and sketches: shape-fitting (Section 48.1), density estimation (Section 48.2), high-dimensional vectors (Section 48.3), high-dimensional point sets / matrices (Section 48.4), and clustering (Section 48.5). There are many other types of coresets and sketches (e.g., for graphs [AGM12] or Fourier transforms [IKP14]) which we do not cover due to space limitations or because they are less geometric.

COMPUTATIONAL MODELS AND PRIMATIVES

Often the challenge is not simply to bound the size of a coreset or sketch as a function of the error tolerance, but to also do so efficiently and in a restricted model. So before we discuss the specifics of the summaries, it will be useful to outline some basic computational models and techniques.

The most natural fit is a *streaming model* that allows limited space (e.g., the size of the coreset or sketch) and where the algorithm can only make a single scan over the data, that is, one can read each data element once. There are several other relevant models which are beyond the scope of this chapter to describe precisely. Many of these consider settings where data is distributed across or streaming into different locations and it is useful to compress or maintain data as coresets and sketches at each location before communicating only these summaries to a central coordinator. The **mergeable model** distills the core step of many of these distributed models to a single task: given two summaries S_1 and S_2 of disjoint data sets, with error bounds ε_1 and ε_2, the model requires a process to create a single summary S of all of the data, of size $\max\{\text{size}(S_1), \text{size}(S_2)\}$, and with error bound $\varepsilon = \max\{\varepsilon_1, \varepsilon_2\}$. Specific error bound definitions will vary widely, and will be discussed subsequently. We will denote any such merge operation as $\oplus$, and a summary where these size and error constraints can be satisfied is called *mergeable* [ACH+13].

A more general **merge-reduce framework** [CM96, BS80] is also often used, including within the streaming model. Here we may consider less sophisticated merge $\oplus$ operations, such as the union where the size of S is $\text{size}(S_1) + \text{size}(S_2)$, and then a reduce operation to shrink the size of S, but resulting in an increased error, for instance as $\varepsilon = \varepsilon_1 + \varepsilon_2$. Combining these operations together into an efficient framework can obtain a summary of size g (asymptotically, perhaps up to log factors) from a dataset of size n as follows. First, arbitrarily divide the data into n/g subsets, each of size g (assume n/g is a power of 2, otherwise pad the data with dummy points). Think of organizing these subsets in a binary tree. Then in $\log(n/g)$ rounds until there is one remaining set, perform each of the next two steps. First, pair up all remaining sets, and merge each pair using an $\oplus$ operator. Second, reduce each remaining set to be a summary of size g. If the summary follows the mergeable model, the reduce step is unnecessary.

Even if the merge or reduce step requires some polynomial m^c time to process m data points, this is only applied to sets of size at most $2g$, hence the full runtime is dominated by the first round as $(n/g) \cdot (2g)^c = O(n \cdot g^{c-1})$. The log factor increase in error (for that many merge-reduce steps) can be folded into the size g, or in many cases removed by delaying some reduce steps and careful bookkeeping [CM96].

In a streaming model this framework is applied by mapping data points to the n/g subsets in the order they arrive, and then always completing as much of the merge-reduce process as possible given the data seen; e.g., scanning the binary tree over the initial subsets from left to right. Another $\log(n/g)$ space factor is incurred for those many summaries which can be active at any given time.

48.1 SHAPE FITTING

In this section we will discuss problems where, given an input point set P, the goal is to find the best fitting shape from some class to P. The two central problems in this area are the minimum (or smallest) enclosing ball, which has useful solutions in high dimensions, and the ε-kernel coreset for directional width which approximates the convex hull but also can be transformed to solve many other problems.

GLOSSARY

Minimum enclosing ball (MEB): Given a point set $P \subset \mathbb{R}^d$, it is the smallest ball B which contains P.

ε-Approximate minimum enclosing ball problem: Given a point set $P \subset \mathbb{R}^d$, and a parameter $\varepsilon > 0$, the problem is to find a ball B whose radius is no larger than $(1 + \varepsilon)$ times the radius of the MEB of P.

Directional width: Given a point set $P \subset \mathbb{R}^d$ and a unit vector $u \in \mathbb{R}^d$, then the *directional width* of P in direction u is $\omega(P, u) = \max_{p \in P} \langle p, u \rangle - \min_{p \in P} \langle p, u \rangle$.

ε-Kernel coreset: An ε-*kernel coreset* of a point set $P \in \mathbb{R}^d$ is subset $Q \subset P$ so that for all unit vectors $u \in \mathbb{R}^d$,

$$0 \leq \omega(P, u) - \omega(Q, u) \leq \varepsilon \omega(P, u).$$

Functional width: Given a set $\mathcal{F} = \{f_1, \ldots, f_n\}$ of functions each from $\mathbb{R}^d$ to $\mathbb{R}$, the width at a point $x \in \mathbb{R}^d$ is defined $\omega_{\mathcal{F}}(x) = \max_{f_i \in \mathcal{F}} f_i(x) - \min_{f_i \in \mathcal{F}} f_i(x)$.

ε-Kernel for functional width: Given a set $\mathcal{F} = \{f_1, \ldots, f_n\}$ of functions each from $\mathbb{R}^d$ to $\mathbb{R}$, an ε-kernel coreset is a subset $\mathcal{G} \subset \mathcal{F}$ such that for all $x \in \mathbb{R}^d$ the functional width $\omega_{\mathcal{G}}(x) \geq (1 - \varepsilon)\omega_{\mathcal{F}}(x)$.

Faithful measure: A measure μ is faithful if there exists a constant c, depending on μ, such that for any point set $P \subset \mathbb{R}^d$ any ε-kernel coreset Q of P is a coreset for μ with approximation parameter $c\varepsilon$.

Diameter: The diameter of a point set P is $\max_{p,p' \in P} \|p - p'\|$.

Width: The width of a point set P is $\min_{u \in \mathbb{R}^d, \|u\|=1} \omega(P, u)$.

Spherical shell: For a point $c \in \mathbb{R}^d$ and real numbers $0 \leq r \leq R$, it is the closed region $\sigma(c, r, R) = \{x \in \mathbb{R}^d \mid r \leq \|x - c\| \leq R\}$ between two concentric spheres of radius r and R centered at c. Its *width* is defined $R - r$.

SMALLEST ENCLOSING BALL CORESET

Given a point set $P \subset \mathbb{R}^d$ of size n, there exists a ε-coreset for the smallest enclosing ball problem of size $\lceil 2/\varepsilon \rceil$ that runs in time $O(nd/\varepsilon + 1/\varepsilon^5)$ [BC03]. Precisely, this finds a subset $S \subset P$ with the smallest enclosing ball $B(S)$ described by center point c and radius r; it holds that if r is expanded to $(1 + \varepsilon)r$, then the ball with the same center would contain P.

The algorithm is very simple and iterative: At each step, maintain the center c_i of the current set S_i, add to S_i the point $p_i \in P$ farthest from c_i, and finally update $S_{i+1} = S_i \cup \{p_i\}$ and c_{i+1} as the center of the smallest enclosing ball of S_{i+1}. Clarkson [Cla10] discusses the connection to the Frank-Wolfe [FW56] algorithm, and the generalizations towards several sparse optimization problems relevant for machine learning, for instance support vector machines [TKC05], polytope distance [GJ09], uncertain data [MSF14], and general Riemannian manifolds [AN12].

These algorithms do not work in the streaming model, as they require $\Omega(1/\varepsilon)$ passes over the data, but the runtime can be improved to $O((d/\varepsilon + n/\varepsilon^2) \log(n/\varepsilon))$ with high probability [CHW12]. Another approach [AS15] maintains a set of $O((1/\varepsilon^3) \log(1/\varepsilon))$ points in a stream that handles updates in $O((d/\varepsilon^2) \log(1/\varepsilon))$

time. But it is not a coreset (a true proxy for P) since in order to handle updates, it needs to maintain these points as $O((1/\varepsilon^2)\log(1/\varepsilon))$ different groups.

EPSILON-KERNEL CORESET FOR WIDTH

Given point sets $P \subset \mathbb{R}^d$ of size n, an ε-kernel coreset for directional width exists of size $O(1/\varepsilon^{(d-1)/2})$ [AHPV04] and can be constructed in $O(n + 1/\varepsilon^{d-(3/2)})$ time [Cha06, YAPV08]. These algorithms are quite different from those for MEB, and the constants have heavy dependence on d (in addition to it being in the exponent of $1/\varepsilon$). They first estimate the rough shape of the points so that they can be made fat (so width and diameter are $\Theta(1)$) through an affine transform that does not change which points form a coreset. Then they carefully choose a small set of points in the extremal directions.

In the streaming model in $\mathbb{R}^d$, the ε-kernel coreset can be computed using $O((1/\varepsilon^{(d-1)/2})\cdot\log(1/\varepsilon))$ space with $O(1 + (1/\varepsilon^{(d-3)/2})\log(1/\varepsilon))$ update time, which can be amortized to $O(1)$ update time [ZZ11]. In $\mathbb{R}^2$ this can be reduced to $O(1/\sqrt{\varepsilon})$ space and $O(1)$ update time [AY07].

Similar to ε-kernels for directional width, given a set of n d-variate linear functions $\mathcal{F}$ and a parameter ε, then an ε-kernel for functional width can be computed of size $O(1/\varepsilon^{d/2})$ in time $O(n + 1/\varepsilon^{d-(1/2)})$ [AHPV04, Cha06].

Many other measures can be shown to have ε-approximate coresets by showing they are *faithful*; this includes diameter, width, minimum enclosing cylinder, and minimum enclosing box. Still other problems can be given ε-approximate coresets by linearizing the inputs so they represent a set of n linear functions in higher dimensions. Most naturally this works for creating an ε-kernel for the width of polynomial functions. Similar linearization is possible for a slew of other shape-fitting problems including the minimum width spherical shell problem, overviewed nicely in a survey by Agarwal, Har-Peled and Varadarajan [AHPV07].

These coresets can be extended to handle a small number of outliers [HPW04, AHPY08] or uncertainty in the input [HLPW16]. A few approaches also extend to high dimensions, such as fitting a k-dimensional subspace [HPV04, BHPR16].

48.2 DENSITY ESTIMATION

Here we consider a point set $P \subset \mathbb{R}^d$ which represents a discrete density function. A coreset is then a subset $Q \subset P$ such that Q represents a similar density function to P under a restricted family of ways to measure the density on subsets of the domain, e.g., defined by a range space.

GLOSSARY

Range space: A range space $(P, \mathcal{A})$ consists of a ground set P and a family of ranges $\mathcal{R}$ of subsets from P. In this chapter we consider ranges that are defined geometrically, for instance when P is a point set and $\mathcal{R}$ is the collection of all subsets defined by a ball, that is, any subset of P which coincides with $P \cap B$ for any ball B.

ε-Net: Given a range space $(P, \mathcal{R})$, it is a subset $Q \subset P$, so for any $R \in \mathcal{R}$ such that $|R \cap P| \geq \varepsilon |P|$, then $R \cap Q \neq \emptyset$.

ε-Approximation (or ε-sample): Given a range space $(P, \mathcal{R})$, it is a subset $Q \subset P$, so for all $R \in \mathcal{R}$ it implies $\left| \frac{|R \cap P|}{|P|} - \frac{|R \cap Q|}{|Q|} \right| \leq \varepsilon$.

VC-dimension: For a range space $(P, \mathcal{R})$ it is the size of the largest subset $Y \subset P$ such that for each subset $Z \subset Y$ it holds that $Z = Y \cap R$ for some $R \in \mathcal{R}$.

RANDOM SAMPLING BOUNDS

Unlike the shape fitting coresets, these density estimate coresets can be constructed by simply selecting a large enough random sample of P. The best such size bounds typically depend on VC-dimension ν [VC71] (or shattering dimension σ), which for many geometrically defined ranges (e.g., by balls, halfspaces, rectangles) is $\Theta(d)$. A random subset $Q \subset P$ of size $O((1/\varepsilon^2)(\nu + \log(1/\delta)))$ [LLS01] is an ε-approximation of any range space $(P, \mathcal{R})$ with VC-dimension ν, with probability at least $1 - \delta$. A subset $Q \subset P$ of size $O((\nu/\varepsilon) \log(1/\varepsilon\delta))$ [HW87] is an ε-net of any range space $(P, \mathcal{R})$ with VC-dimension ν, with probability at least $1 - \delta$.

These bounds are of broad interest to learning theory, because they describe how many samples are sufficient to learn various sorts of classifiers. In machine learning, it is typical to assume each data point $q \in Q$ is drawn iid from some unknown distribution, and since the above bounds have no dependence on n, we can replace P by any probability distribution with domain $\mathbb{R}^d$. Consider that each point in Q has a value from $\{-, +\}$, and a separator range (e.g., a halfspace) should ideally have all $+$ points inside, and all $-$ points outside. Then for an ε-approximation Q of a range space $(P, \mathcal{A})$, the range $R \in \mathcal{R}$ which misclassifies the fewest points on Q, misclassifies at most an ε-fraction of points in P more than the optimal separator does. An ε-net (which requires far fewer samples) can make the same claim as long as there exists a separator in $\mathcal{A}$ that has zero misclassified points on P; it was recently shown [Han16] that a weak coreset for this problem only requires $\Theta((1/\varepsilon)(\nu + \log(1/\delta)))$ samples.

The typical ε-approximation bound provides an additive error of ε in estimating $|R \cap P|/|P|$ with $|R \cap Q|/|Q|$. One can achieve a stronger *relative* (ρ, ε)-*approximation* such that

$$\max_{R \in \mathcal{R}} \left| \frac{|R \cap P|}{|P|} - \frac{|R \cap Q|}{|Q|} \right| \leq \varepsilon \max \left\{ \rho, \frac{|R \cap P|}{|P|} \right\}.$$

This requires $O((1/\rho\varepsilon^2)(\nu \log(1/\rho) + \log(1/\delta)))$ samples [LLS01, HPS11] to succeed with probability at least $1 - \delta$.

DISCREPANCY-BASED RESULTS

Tighter bounds for density estimation coresets arise through discrepancy. The basic idea is to build a coloring on the ground set $\chi : X \to \{-1, +1\}$ to minimize $\sum_{x \in R} \chi(x)$ over all ranges (the ***discrepancy***). Then we can plug this into the merge-reduce framework where merging takes the union and reducing discards the points colored -1. Chazelle and Matoušek [CM96] showed how slight modifications of the merge-reduce framework can remove extra log factors in the approximation.

Based on discrepancy results (see Chapters 13 and 47) we can achieve the following bounds. These assume $d \geq 2$ is a fixed constant, and is absorbed in $O(\cdot)$ notation. For any range space $(P, \mathcal{R})$ with VC-dimension ν (a fixed constant) we can construct an ε-approximation of size $g = O(1/\varepsilon^{2-\nu/(\nu+1)})$ in $O(n \cdot g^{w-1})$ time. This is tight for range spaces $\mathcal{H}_d$ defined by halfspaces in $\mathbb{R}^d$, where $\nu = d$. For range spaces $\mathcal{B}_d$ defined by balls in $\mathbb{R}^d$, where $\nu = d+1$ this can be improved slightly to $g = O(1/\varepsilon^{2-\nu/(\nu+1)}\sqrt{\log(1/\varepsilon)})$; it is unknown if the log factor can be removed. For range spaces $\mathcal{T}_d$ defined by axis-aligned rectangles in $\mathbb{R}^d$, where $\nu = 2d$, this can be greatly improved to $g = O((1/\varepsilon)\log^{d+1/2}(1/\varepsilon))$ with the best lower bound as $g = \Omega((1/\varepsilon)\log^{d-1}(1/\varepsilon))$ for $d \geq 2$ [Lar14, MNT15]. These colorings can be constructed by adapting techniques from Bansal [Ban10, BS13]. Various generalizations (typically following one of these patterns) can be found in books by Matoušek [Mat10] and Chazelle [Cha00]. Similar bounds exist in the streaming and mergeable models, adapting the merge-reduce framework [BCEG07, STZ06, ACH+13].

Discrepancy based results also exist for constructing ε-nets. However, often the improvement over the random sampling bounds are not as dramatic. For halfspaces in $\mathbb{R}^3$ and balls in $\mathbb{R}^2$ we can construct ε-nets of size $O(1/\varepsilon)$ [MSW90, CV07, PR08]. For axis-aligned rectangles and fat objects we can construct ε-nets of size $O((1/\varepsilon)\log\log(1/\varepsilon))$ [AES10]. Pach and Tardos [PT13] then showed these results are tight, and that similar improvements cannot exist in higher dimensions.

GENERALIZATIONS

One can replace the set of ranges $\mathcal{R}$ with a family of functions $\mathcal{F}$ so that $f \in \mathcal{F}$ has range $f : \mathbb{R}^d \to [0,1]$, or scaled to other ranges, including $[0,\infty)$. For some $\mathcal{F}$ we can interpret this as replacing a binary inclusion map $R : \mathbb{R}^d \to \{0,1\}$ for $R \in \mathcal{R}$, with a continuous one $f : \mathbb{R}^d \to [0,1]$ for $f \in \mathcal{F}$. A family of functions $\mathcal{F}$ is *linked* to a range space $(P, \mathcal{R})$ if for every value $\tau > 0$ and every function $f \in \mathcal{F}$, the points $\{p \in P \mid f(p) \geq \tau\} = R \cap P$ for some $R \in \mathcal{R}$. When $\mathcal{F}$ is linked to $(P, \mathcal{R})$, then an ε-approximation Q for $(P, \mathcal{R})$ also ε-approximates $(P, \mathcal{F})$ [JKPV11] (see also [HP06, LS10] for similar statements) as

$$\max_{f \in \mathcal{F}} \left| \frac{\sum_{p \in P} f(p)}{|P|} - \frac{\sum_{q \in Q} f(q)}{|Q|} \right| \leq \varepsilon.$$

One can also show ε-net type results. An (τ, ε)-net for $(P, \mathcal{F})$ has for all $f \in \mathcal{F}$ such that $\frac{\sum_{p \in P} (p)}{|P|} \geq \varepsilon$, then there exists some $q \in Q$ such at $f(q) \geq \tau$. Then an $(\varepsilon - \tau)$-net Q for $(P, \mathcal{R})$ is an (τ, ε)-net for $(P, \mathcal{F})$ if they are linked [PZ15].

A concrete example is for centrally-symmetric shift-invariant kernels $\mathcal{K}$ (e.g., Gaussians $K(x, p) = \exp(-\|x - p\|^2)$) then we can set $f_x(p) = K(x, p)$. Then the above ε-approximation corresponds with an approximate kernel density estimate [JKPV11]. Surprisingly, there exist discrepancy-based ε-approximation constructions that are smaller for many kernels (including Gaussians) than for the linked ball range space; for instance in $\mathbb{R}^2$ with $|Q| = O((1/\varepsilon)\sqrt{\log(1/\varepsilon)})$ [Phi13].

One can also consider the minimum cost from a set $\{f_1, \ldots, f_k\} \subset \mathcal{F}$ of functions [LS10], then the size of the coreset often only increases by a factor k. This setting will, for instance, be important for k-means clustering when $f(p) = \|x - p\|^2$ for some center $x \in \mathbb{R}^d$ [FL11]. And it can be generalized to robust error functions [FS12] and Gaussian mixture models [FFK11].

QUANTILES SKETCH

Define the **rank** of v for set $X \in \mathbb{R}$ as $\text{rank}(X, v) = |\{x \in X \mid x \leq v\}|$. A quantiles sketch S over a data set X of size n allows for queries such that $|S(v) - \text{rank}(X, v)| \leq \varepsilon n$ for all $v \in \mathbb{R}$. This is equivalent to an ε-approximation of a one-dimensional range space $(X, \mathcal{I})$ where $\mathcal{I}$ is defined by half-open intervals of the form $(-\infty, a]$.

A ε-approximation coreset of size $1/\varepsilon$ can be found by sorting X and taking evenly spaced points in that sorted ordering. Streaming sketches are also known; most famously the Greenwald-Khanna sketch [GK01] which takes $O((1/\varepsilon) \log(\varepsilon n))$ space, where X is size n. Recently, combining this sketch with others [ACH+13, FO15], Karnin, Lang, and Liberty [KLL16] provided new sketches which require $O((1/\varepsilon) \log \log(1/\varepsilon))$ space in the streaming model and $O((1/\varepsilon) \log^2 \log(1/\varepsilon))$ space in the mergeable model.

48.3 HIGH-DIMENSIONAL VECTORS

In this section we will consider high-dimensional vectors $v = (v_1, v_2, \ldots, v_d)$. When each v_i is a positive integer, we can imagine these as the counts of a labeled set (the d dimensions); a subset of the set elements or the labels is a coreset approximating the relative frequencies. Even more generally, a sketch will compactly represent another vector u which behaves similarly to v under various norms.

GLOSSARY

ℓ_p-**Norm:** For a vector $v \in \mathbb{R}^d$ the ℓ_p norm, for $p \in [1, \infty)$, is defined $\|v\|_p = (\sum_{i=1}^{d} |v_i|^p)^{1/p}$; if clear we use $\|v\| = \|v\|_2$. For $p = 0$ define $\|v\|_0 = |\{i \mid v_i \neq 0\}|$, the number of nonzero coordinates, and for $p = \infty$ define $\|v\|_\infty = \max_{i=1}^{d} |v_i|$.

k-**Sparse:** A vector is k-sparse if $\|v\|_0 \leq k$.

Additive ℓ_p/ℓ_q approximation: A vector v has an additive ε-(ℓ_p/ℓ_q) approximation with vector u if $\|v - u\|_p \leq \varepsilon \|v\|_q$.

k-**Sparse ℓ_p/ℓ_q approximation:** A vector v has a k-sparse ε-(ℓ_p/ℓ_q) approximation with vector u if u is k-sparse and $\|v - u\|_p \leq \varepsilon \|v - u\|_q$.

Frequency count: For a vector $v = (v_1, v_2, \ldots v_d)$ the value v_i is called the ith frequency count of v.

Frequency moment: For a vector $v = (v_1, v_2, \ldots v_d)$ the value $\|v\|_p$ is called the pth frequency moment of v.

FREQUENCY APPROXIMATION

There are several types of coresets and sketches for frequency counts. Derived by ε-approximation and ε-net bounds, we can create the following coresets over dimensions. Assume v has positive integer coordinates, and each coordinate's count v_i represents v_i distinct objects. Then let S be a random sample of size k of these objects and $u(S)$ be an approximate vector defined so $u(S)_i = (\|v\|_1/k) \cdot |\{s \in S \mid s = i\}|$. Then with $k = O((1/\varepsilon^2) \log(1/\delta))$ we have $\|v - u(S)\|_\infty \leq \varepsilon \|v\|_1$ (an additive ε-(ℓ_∞/ℓ_1) approximation) with probability at least $1 - \delta$. Moreover, if $k = O((1/\varepsilon)\log(1/\varepsilon\delta))$ then for all i such that $v_i \geq \varepsilon\|v\|_1$, then $u(S)_i \neq 0$, and we can then measure the true count to attain a weighted coreset which is again an additive ε-(ℓ_∞/ℓ_1) approximation. And in fact, there can be at most $1/\varepsilon$ dimensions i with $v_i \geq \varepsilon\|v\|_1$, so there always exists a weighted coreset of size $1/\varepsilon$.

Such a weighted coreset for additive ε-(ℓ_∞/ℓ_1) approximations that is $(1/\varepsilon)$-sparse can be found deterministically in the streaming model via the Misra-Gries sketch [MG82] (or other variants [MAA06, DLOM02, KSP03]). This approach keeps $1/\varepsilon$ counters with associated labels. For a new item, if it matches a label, the counter is incremented, else if any counter is 0 it takes over that counter/label, and otherwise, (perhaps unintuitively) all counters are decremented.

The count-min sketch [CM05] also provides an additive ε-(ℓ_∞/ℓ_1) approximation with space $O((1/\varepsilon)\log(1/\delta))$ and is successful with probability $1 - \delta$. A count-sketch [CCFC04] provides an additive ε-(ℓ_∞/ℓ_2) approximation with space $O((1/\varepsilon^2)\log(1/\delta))$, and is successful with probability at least $1 - \delta$. Both of these linear sketches operate by using $O(\log 1/\delta)$ hash functions, each mapping $[d]$ to one of $O(1/\varepsilon)$ or $O(1/\varepsilon^2)$ counters. The counters are incremented or decremented with the value v_i. Then an estimate for v_i can be recovered by examining all cells where i hashes; the effect of other dimensions which hash there can be shown bounded.

Frequency moments. Another common task is to approximate the frequency moments $\|v\|_p$. For $p = 1$, this is the count and can be done exactly in a stream. The AMS Sketch [AMS99] maintains a sketch of size $O((1/\varepsilon^2)\log(1/\delta))$ that can derive a value $\widehat{F_2}$ so that $|\|v\|_2 - \widehat{F_2}| \leq \varepsilon\|v\|_2$ with probability at least $1 - \delta$.

The FM Sketch [FM85] (and its extensions [AMS99, DF03]) show how to create a sketch of size $O((1/\varepsilon^2)\log(1/\delta))$ which can derive an estimate $\widehat{F_0}$ so that $|\|v\|_0 - \widehat{F_0}| \leq \varepsilon\|v\|_1$ with probability at least $1 - \delta$. This works when v_i are positive counts, and those counts are incremented one at a time in a stream. Usually sketches and coresets have implicit assumptions that a "word" can fit $\log n$ bits where the stream is of size n, and is sufficient for each counter. Interestingly and in contrast, these ℓ_0 sketches operate with bits, and only have a hidden $\log\log n$ factor for bits.

k-sparse tail approximation. Some sketches can achieve k-sparse approximations (which are akin to coresets of size k) and have stronger error bounds that depend only on the "tail" of the matrix; this is the class of k-sparse ε-(ℓ_p/ℓ_q) approximations. See the survey by Gilbert and Indyk for more details [GI10].

These bounds are typically achieved by increasing the sketch size by a factor k, and then the k-sparse vector is the top k of those elements. The main recurring argument is roughly as follows: If you maintain the top $1/\varepsilon$ counters, then the largest counter not maintained is of size at most $\varepsilon\|v\|$. Similarly, if you first remove the top k counters (a set $K = \{i_1, i_2, \ldots, i_k\} \subset [d]$, let their collective norm be $\|v_K\|$), then maintain $1/\varepsilon$ more, the largest not-maintained counter is at most $\varepsilon(\|v\| - \|v_K\|)$. The goal is then to sketch a k-sparse vector which approximates v_K; for instance the Misra-Gries Sketch [MG82] and Count-Min sketch [CM05] achieve k-sparse ε-(ℓ_∞/ℓ_1)-approximations with $O(k/\varepsilon)$ counters, and the Count sketch [CCFC04] achieves k-sparse ε-(ℓ_∞/ℓ_2)-approximations with $O(k^2/\varepsilon^2)$ counters [BCIS10].

48.4 HIGH-DIMENSIONAL POINT SETS (MATRICES)

Matrix sketching has gained a lot of interest due to its close connection to scalability issues in machine learning and data mining. The goal is often to replace a matrix with a small space and low-rank approximation. However, given a $n \times d$ matrix A, it can also be imagined as n points each in $\mathbb{R}^d$, and the span of a rank-k approximation is a k-dimensional subspace that approximately includes all of the points.

Many of the techniques build on approaches from vector sketching, and again, many of the sketches are naturally interpretable as weighted coresets. Sometimes it is natural to represent the result as a reduced set of rows in a $\ell \times d$ matrix B. Other times it is more natural to consider the dimensionality reduction problem where the goal is an $n \times c$ matrix, and sometimes you do both! But since these problems are typically phrased in terms of matrices, the difference comes down to simply transposing the input matrix. We will write all results as approximating an $n \times d$ matrix A using fewer rows, for instance, with an $\ell \times d$ matrix B.

Notoriously this problem can be solved optimally using the numerical linear algebra technique, the *singular value decomposition*, in $O(nd^2)$ time. The challenges are then to compute this more efficiently in streaming and other related settings.

We will describe three basic approaches (row sampling, random projections, and iterative SVD variants), and then some extensions and applications [FT15]. The first two approaches are mainly randomized, and we will describe results with constant probability, and for the most part these bounds can be made to succeed with any probability $1 - \delta$ by increasing the size by a factor $\log(1/\delta)$.

GLOSSARY

Matrix rank: The *rank* of an $n \times d$ matrix A, denoted $\mathsf{rank}(A)$, is the smallest k such that all rows (or columns) lie in a k-dimensional subspace of $\mathbb{R}^d$ (or $\mathbb{R}^n$).

Singular value decomposition: Given an $n \times d$ matrix A, the singular value decomposition is a product $U\Sigma V^T$ where U and V are orthogonal, and Σ is diagonal. U is $n \times n$, and V is $d \times d$, and $\Sigma = \mathsf{diag}(\sigma_1, \sigma_2, \ldots, \sigma_{\min\{n,d\}})$ (padded with either $n - d$ rows or $d - n$ columns of all 0s, so Σ is $n \times d$) where $\sigma_1 \geq \sigma_2 \geq \ldots \geq \sigma_{\min\{n,d\}} \geq 0$, and $\sigma_i = 0$ for all $i > \mathsf{rank}(A)$.
The ith column of U (resp. column of V) is called the ith left (resp. right) *singular vector*; and σ_i is the ith *singular value*.

Spectral norm: The spectral norm of matrix A is denoted $\|A\|_2 = \max_{x \neq 0} \|Ax\|/\|x\|$.

Frobenius norm: The Frobenius norm of a matrix A is $\|A\|_F = \sqrt{\sum_{i=1}^n \|a_i\|^2}$ where a_i is the ith row of A.

Low rank approximation of a matrix: The best rank k approximation of a matrix A is denoted $[A]_k$. Let Σ_k be the matrix Σ (the singular values from the SVD of A) where the singular values σ_i are set to 0 for $i > k$. Then $[A]_k = U\Sigma_k V^T$. Note we can also ignore the columns of U and V after k; these are implicitly set to 0 by multiplication with $\sigma_i = 0$. The $n \times d$ matrix $[A]_k$ is optimal in that over all rank k matrices B it minimizes $\|A - B\|_2$ and $\|A - B\|_F$.

Projection: For a subspace $F \subset \mathbb{R}^d$ and point $x \in \mathbb{R}^d$, define the projection $\pi_F(x) = \arg\min_{y \in F} \|x - y\|$. For an $n \times d$ matrix A, then $\pi_F(A)$ defines the $n \times d$ matrix where each row is individually projected on to F.

ROW SUBSET SELECTION

The first approach towards these matrix sketches is to choose a careful subset of the rows (note: the literature in this area usually discusses selecting columns). An early analysis of these techniques considered sampling $\ell = O((1/\varepsilon^2)k \log k)$ rows proportional to their squared norm as $\ell \times d$ matrix B, and showed [FKV04, DFK+04, DKM06] one could describe a rank-k matrix $P = [\pi_B(A)]_k$ so that

$$\|A - P\|_F^2 \le \|A - [A]_k\|_F^2 + \varepsilon\|A\|_F^2 \quad \text{and} \quad \|A - P\|_2^2 \le \|A - [A]_k\|_2^2 + \varepsilon\|A\|_F^2.$$

This result can be extended for sampling columns in addition to rows.

This bound was then improved by sampling proportional to the *leverage scores*; If U_k is the $n \times k$ matrix of the first k left singular vectors of A, then the leverage score of row i is $\|U_k(i)\|^2$, the norm of the ith row of U_k. In this case $O((1/\varepsilon^2)k \log k)$ rows achieve a relative error bound [DMM08]

$$\|A - \pi_B(A)\|_F \le (1 + \varepsilon)\|A - [A]_k\|_F^2.$$

These relative error results can be extended to sample rows and columns, generating a so-called CUR decomposition of A. Similar relative error bounds can be achieved through volume sampling [DV06]. Computing these leverage scores exactly can be as slow as the SVD; instead one can approximate the leverage scores [DMI+12, CLM+15], for instance in a stream in $O((kd/\varepsilon^2) \log^4 n)$ bits of space [DMI+12].

Better algorithms exist outside the streaming model [FVR16]. These can, for instance, achieve the strong relative error bounds with only $O(k/\varepsilon)$ rows (and $O(k/\varepsilon)$ columns) and only require time $O(\mathsf{nnz}(A) \log n + n\mathsf{poly}(\log n, k, 1/\varepsilon))$ time where $\mathsf{nnz}(A)$ is the number of nonzero entries in A [BDMI14, BW17]. Or Batson, Spielman and Srivastava [BSS14, Sri10, Nao12] showed that $O(d/\varepsilon^2)$ reweighted rows are sufficient and necessary to achieve bounds as below in (48.4.1).

RANDOM PROJECTIONS

The second approach to matrix sketching is based on the Johnson-Lindenstrauss (JL) Lemma [JL84], which says that projecting any vector x (independent of its dimension, for which it will be useful here to denote as n) onto a random subspace F of dimension $\ell = O(1/\varepsilon^2)$ preserves, with constant probability, its norm up to $(1 + \varepsilon)$ relative error, after rescaling: $(1 - \varepsilon)\|x\| \le \sqrt{n/\ell}\|\pi_F(x)\| \le (1 + \varepsilon)\|x\|$. Follow-up work has shown that the projection operator π_F can be realized as an $\ell \times n$ matrix S so that $(\sqrt{n/\ell})\pi_F(x) = Sx$. And in particular, we can fill the entries of S with iid Gaussian random variables [DG03], uniform $\{-1, +1\}$ or $\{-1, 0, +1\}$ random variables [Ach03], or any sub-Gaussian random variable [Mat08], rescaled; see Chapter 8 for more details. Alternatively, we can make S all 0s except for one uniform $\{-1, +1\}$ random variable in each column of S [CW17]. This latter construction essentially "hashes" each element of A to one of the elements in $\pi_F(x)$ (see also a variant [NN13]); basically an extension of the count-sketch [CCFC04].

To apply these results to matrix sketching, we simply apply the sketch matrix S to A instead of just a single "dimension" of A. Then $B = SA$ is our resulting $\ell \times d$ matrix. However, unlike in typical uses of the JL Lemma on a point set of size m, where it can be shown to preserve all distances using $\ell = O((1/\varepsilon^2) \log m)$ target dimensions, we will strive to preserve the norm over all d dimensions. As such we use $\ell = O(d/\varepsilon^2)$ for iid JL results [Sar06], or $\ell = O(d^2/\varepsilon^2)$ for hashing-based approaches [CW17, NN13]. As was first observed by Sarlos [Sar06] this allows one to create an oblivious subspace embedding so that for *all* $x \in \mathbb{R}^d$ guarantees

$$(1 - \varepsilon)\|Ax\|_2^2 \leq \|Bx\|_2^2 \leq (1 + \varepsilon)\|Ax\|_2^2. \tag{48.4.1}$$

The obliviousness of this linear projection matrix S (it is created independent of A) is very powerful. It means this result can not only be performed in the update-only streaming model, but also one that allows merges, deletions, or arbitrary updates to an individual entry in a matrix. Moreover, given a matrix A with only $\mathsf{nnz}(A)$ nonzero entries, it can be applied in roughly $O(\mathsf{nnz}(A))$ time [CW17, NN13]. It also implies bounds for matrix multiplication, and as we will discuss, linear regression.

FREQUENT DIRECTIONS

This third class of matrix sketching algorithms tries to more directly replicate those properties of the SVD, and can be deterministic. So why not just use the SVD? These methods, while approximate, are faster than SVD, and work in the streaming and mergeable models.

The Frequent Directions algorithm [Lib13, GLPW16] essentially processes each row (or $O(\ell)$ rows) of A at a time, always maintaining the best rank-ℓ approximation as the sketch. But this can suffer from data drift, so crucially after each such update, it also shrinks all squared singular values of $[B]_\ell$ by s_ℓ^2; this ensures that the additive error is never more than $\varepsilon\|A\|_F^2$, precisely as in the Misra-Gries [MG82] sketch for frequency approximation. Setting $\ell = k + 1/\varepsilon$ and $\ell = k + k/\varepsilon$, respectively, the following bounds have been shown [GP14, GLPW16] for any unit vector x:

$$0 \leq \|Ax\|^2 - \|Bx\|^2 \leq \varepsilon\|A - [A]_k\|_F^2 \quad \text{and} \quad \|A - \pi_{[B]_k}(A)\|_F^2 \leq (1+\varepsilon)\|A - [A]_k\|_F^2.$$

Operating in a batch to process $\Theta(\ell)$ rows at a time, this takes $O(nd\ell)$ time. A similar approach by Feldman et al. [FSS13] provides a more general bound, and will be discussed in the context of subspace clustering below.

LINEAR REGRESSION AND ROBUST VARIANTS

The regression problem takes as input again an $n \times d$ matrix A and also an $n \times w$ matrix T (most commonly $w = 1$ so T is a vector); the goal is to find the $d \times w$ matrix $X^* = \arg\min_X \|AX - T\|_F$. One can create a coreset of ℓ rows (or weighted linear combination of rows): the $\ell \times d$ matrix $\widehat{A}$ and $\ell \times w$ matrix $\widehat{T}$ imply a matrix $\widehat{X} = \arg\min_X \|\widehat{A}X - \widehat{T}\|_2$ that satisfies

$$(1 - \varepsilon)\|AX^* - T\|_F^2 \leq \|A\widehat{X} - T\|_F^2 \leq (1 + \varepsilon)\|AX^* - T\|_F^2.$$

Using the random projection techniques described above, one can sketch $\widehat{A} = SA$ and $\widehat{T} = SA$ with $\ell = O(d^2/\varepsilon^2)$ for hashing approaches or $\ell = O(d/\varepsilon^2)$ for iid approaches. Moreover, Sarlos [Sar06] observed that for the $w = 1$ case, since only a single direction (the optimal one) is required to be preserved (see *weak coresets* below), one can also use just $\ell = O(d^2/\varepsilon)$ rows. Using row-sampling, one can deterministically select $\ell = O(d/\varepsilon^2)$ rows [BDMI13]. The above works also provide bounds for approximating the multiple-regression spectral norm $\|AX^* - T\|_2$.

Mainly considering the single-regression problem when $w = 1$ (in this case spectral and Frobenius norms bounds are equivalent $p = 2$ norms), there also exist bounds for approximating $\|AX - T\|_p$ for $p \in [1, \infty)$ using random projection approaches and row sampling [CDMI+13, Woo14]. The main idea is to replace iid Gaussian random variables which are 2-stable with iid p-stable random variables. These results are improved using max-norm stability results [WZ13] embedding into ℓ_∞, or for other robust error functions like the Huber loss [CW15b, CW15a].

48.5 CLUSTERING

An assignment-based clustering of a data set $X \subset \mathbb{R}^d$ is defined by a set of k centers $C \subset \mathbb{R}^d$ and a function $\phi_C : \mathbb{R}^d \to C$, so $\phi_C(x) = \arg\min_{c \in C} \|x - c\|$. The function ϕ_C maps to the closest center in C, and it assigns each point $x \in X$ to a center and an associated cluster. It will be useful to consider a weight $w : X \to \mathbb{R}^+$. Then a clustering is evaluated by a cost function

$$\text{cost}_p(X, w, C) = \sum_{x \in X} w(x) \cdot \|x - \phi_C(x)\|^p.$$

For uniform weights (i.e., $w(x) = 1/|X|$, which we assume as default), then we simply write $\text{cost}_p(X, C)$. We also define $\text{cost}_\infty(X, C) = \max_{x \in X} \|x - \phi_C(x)\|$. These techniques extend to when the centers of the clusters are not just points, but can also be higher-dimensional subspaces.

GLOSSARY

k-Means / k-median / k-center clustering problem: Given a set $X \subset \mathbb{R}^d$, find a point set C of size k that minimizes $\text{cost}_2(X, C)$ (respectively, $\text{cost}_1(X, C)$ and $\text{cost}_\infty(X, C)$).

(k, ε)-Coreset for k-means / k-median / k-center: Given a point set $X \subset \mathbb{R}^d$, then a subset $S \subset X$ is a (k, ε)-coreset for k-means (respectively, k-median and k-center) if for all center sets C of size k and parameter $p = 2$ (respectively $p = 1$ and $p = \infty$) that

$$(1 - \varepsilon)\text{cost}_p(X, C) \le \text{cost}_p(S, C) \le (1 + \varepsilon)\text{cost}_p(X, C).$$

Projective distance: Consider a set $C = (C_1, C_2, \ldots, C_k)$ of k affine subspaces of dimension j in $\mathbb{R}^d$, and a power $p \in [1, \infty)$. Then for a point $x \in \mathbb{R}^d$ the projective distance is defined as $\text{dist}_p(C, x) = \min_{C_i \in C} \|x - \pi_{C_i}(x)\|^p$, recalling that $\pi_{C_i}(x) = \arg\min_{y \in C_i} \|x - y\|$.

Projective (k, j, p)-clustering problem: Given a set $X \subset \mathbb{R}^d$, find a set C of k j-dimensional affine subspaces that minimizes $\text{cost}_p(X, C) = \sum_{x \in X} \text{dist}_p(C, x)$.

(k, j, ε)-Coreset for projective (k, j, p)-clustering: Given a point set $X \subset \mathbb{R}^d$, then a subset $S \subset X$, a weight function $w : S \to \mathbb{R}^+$, and a constant γ, is a (k, j, ε)-coreset for projective (k, j, p)-clustering if for all j-dimensional center sets C of size k that

$$(1 - \varepsilon)\text{cost}_p(X, C) \le \text{cost}_p(X, w, C) + \gamma \le (1 + \varepsilon)\text{cost}_p(X, C).$$

In many cases the constant γ may not be needed; it will be 0 unless stated.

Strong coreset: Given a point set $X \subset \mathbb{R}^d$, it is a subset $S \subset X$ that approximates the distance to any k-tuple of j-flats up to a multiplicative $(1 + \varepsilon)$ factor.

Weak coreset: Given a point set $X \subset \mathbb{R}^d$, it is a subset $S \subset X$ such that the cost of the optimal solution (or one close to the optimal solution) of (k, j)-clustering on S, approximates the optimal solution on X up to a $(1+\varepsilon)$ factor. So a strong coreset is also a weak coreset, but not the other way around.

k-MEANS AND k-MEDIAN CLUSTERING CORESETS

(k, ε)-Coresets for k-means and for k-median are closely related. The best bounds on the size of these coresets are independent of n and sometimes also d, the number and dimension of points in the original point set X. Feldman and Langberg [FL11] showed how to construct a strong (k, ε)-coreset for k-median and k-means clustering of size $O(dk/\varepsilon^2)$. They also show how to construct a weak (k, ε)-coreset [FMS07] of size $O(k \log(1/\varepsilon)/\varepsilon^3)$. These bounds can generalize for any cost_p for $p \geq 1$. However, note that for any fixed X and C that $\mathsf{cost}_p(X, C) > \mathsf{cost}_{p'}(X, C)$ for $p > p'$, hence these bounds are not meaningful for the $p = \infty$ special case associated with the k-center problem. Using the merge-reduce framework, the weak coreset constructions work in the streaming model with $O((k/\varepsilon^3) \log(1/\varepsilon) \log^4 n)$ space.

Interestingly, in contrast to earlier work on these problems [HPK07, HPM04, Che09] which applied various forms of geometric discretization of $\mathbb{R}^d$, the above results make an explicit connection with VC-dimension-type results and density approximation [LS10, VX12]. The idea is each point $x \in X$ is associated with a function $f_x(\cdot) = \mathsf{cost}_p(x, \cdot)$, and the total cost $\mathsf{cost}_p(X, \cdot)$ is a sum of these. Then the mapping of these functions onto k centers results in a generalized notion of dimension, similar to the VC-dimension of a dual range space, with dimension $O(kd)$, and then standard sampling arguments can be applied.

k-CENTER CLUSTERING CORESETS

The k-center clustering problem is harder than the k-means and k-median ones. It is NP-hard to find a set of centers $\tilde{C}$ such that $\mathsf{cost}_\infty(X, \tilde{C}) \leq (2 - \eta)\mathsf{cost}_\infty(X, C^*)$ where C^* is the optimal center set and for any $\eta > 0$ [Hoc97]. Yet, famously the Gonzalez algorithm [Gon85], which always greedily chooses the point farthest away from any of the points already chosen, finds a set $\widehat{C}$ of size k, so $\mathsf{cost}_\infty(X, \widehat{C}) \leq 2 \cdot \mathsf{cost}_\infty(X, C^*)$. This set $\widehat{C}$, plus the farthest point from any of these points (i.e., run the algorithm for $k+1$ steps instead of k) is a $(k, 1)$ coreset (yielding the above stated 2 approximation) of size $k + 1$. In a streaming setting, McCutchen and Khuller [MK08] describe a $O(k \log k \cdot (1/\varepsilon) \log(1/\varepsilon))$ space algorithm that provides $(2 + \varepsilon)$ approximation for the k-center clustering problem, and although not stated, can be interpreted as a streaming $(k, 1 + \varepsilon)$-coreset for k-center clustering.

To get a (k, ε)-coreset, in low dimensions, one can use the result of the Gonzalez algorithm to define a grid of size $O(k/\varepsilon^d)$, keeping one point from each grid cell as a coreset of the same size [AP02], in time $O(n + k/\varepsilon^d)$ [HP04a]. In high dimensions one can run $O(k^{O(k/\varepsilon)})$ sequences of k parallel MEB algorithms to find a k-center coreset of size $O(k/\varepsilon)$ in $O(dnk^{O(k/\varepsilon)})$ time [BC08].

PROJECTIVE CLUSTERING CORESETS

Projective clustering seeks to find a set of k subspaces of dimension j which approximate a large, high-dimensional data set. This can be seen as the combination of the subspace (matrix) approximations and clustering coresets.

Perhaps surprisingly, not all shape-fitting problems admit coresets, and in particular, subspace clustering ones pose a problem. Har-Peled showed [HP04b] that no coreset exists for the 2-line-center clustering problem of size sublinear in the dataset. This result can be interpreted so that for $j = 1$ (and extended to $j = 2$), $k = 2$, and $d = 3$, there is no coreset for the projective (k, j)-center clustering problem sublinear in n. Moreover, a result of Meggido and Tamir [MT83] can be interpreted to say for $j \geq 2$ and $k > \log n$, the solution cannot be approximated in polynomial time, for any approximation factor, unless $P = NP$.

This motivates the study of bicriteria approximations, where the solution to the projective (j, k, p)-clustering problem can be approximated using a solution for larger values of j and/or k. Feldman and Langberg [FL11] describe a strong coreset for projective (j, k)-clustering of size $O(djk/\varepsilon^2)$ or a weak coreset of size $O(kj^2 \log(1/\varepsilon)/\varepsilon^3)$, which approximated k subspaces of dimension j using $O(k \log n)$ subspaces of dimensions j. This technique yields stronger bounds in the $j = 0$ and $p = \infty$ case (the k-center clustering problem) where a set of $O(k \log n)$ cluster centers can be shown to achieve error no more than the optimal set of k centers: a $(k, 0)$-coreset for k-center clustering with an extra $O(\log n)$ factor in the number of centers. Other tradeoffs are also described in their paper where the size or approximation factor varies as the required number of subspaces changes. These approaches work in the streaming model with an extra factor $\log^4 n$ in space.

Feldman, Schmidt, and Sohler [FSS13] consider the specific case of cost_2 and crucially make use of a nonzero γ value in the definition of a (k, j, ε)-coreset for projective $(k, j, 2)$-clustering. They show strong coresets of size $O(j/\varepsilon)$ for $k = 1$ (subspace approximation), of size $O(k^2/\varepsilon^4)$ for $j = 0$ (k-means clustering), of size $\mathsf{poly}(2^k, \log n, 1/\varepsilon)$ if $j = 1$ (k-lines clustering), and under the assumption that the coordinates of all points are integers between 1 and $n^{O(1)}$, of size $\mathsf{poly}(2^{kj}, 1/\varepsilon)$ if $j, k > 1$. These results are improved slightly in efficiency [CEM+15], and these constructions also extend to the streaming model with extra $\log n$ factors in space.

48.6 SOURCES AND RELATED MATERIAL

SURVEYS

[AHPV07]: A slightly dated, but excellent survey on coresets in geometry.

[HP11]: Book on geometric approximation that covers many of the above topics, for instance Chapter 5 (ε-approximations and ε-nets), Chapter 19 (dimensionality reduction), and Chapter 23 (ε-kernel coresets).

[Mut05]: On streaming, including history, puzzles, applications, and sketching.

[CGHJ11]: Nice introduction to sketching and its variations.

[GI10]: Survey on k-sparse ε-(ℓ_p/ℓ_q) approximations.

[Mah11, Woo14]: Surveys of randomized algorithms for matrix sketching.

RELATED CHAPTERS

Chapter 8: Low-distortion embeddings of finite metric spaces
Chapter 13: Geometric discrepancy theory and uniform distribution
Chapter 47: Epsilon-approximations and epsilon-nets

REFERENCES

[Ach03]   D. Achlioptas. Database-friendly random projections: Johnson-Lindenstrauss with binary coins. *J. Comp. Syst. Sci.*, 66:671–687, 2003.

[ACH$^+$13]   P.K. Agarwal, G. Cormode, Z. Huang, J. Phillips, Z. Wei, and K. Yi. Mergeable summaries. *ACM Trans. Database Syst.*, 38:26, 2013.

[AES10]   B. Aronov, E. Ezra, and M. Sharir. Small-size epsilon-nets for axis-parallel rectangles and boxes. *SIAM J. Comput.*, 39:3248–3282, 2010.

[AGM12]   K.J. Ahn, S. Guha, and A. McGregor. Graph sketches: Sparsification, spanners, and subgraphs. In *Proc. 27th ACM Sympos. Principles Database Syst.*, pages 253–262, 2012.

[AHPV04]   P.K. Agarwal, S. Har-Peled, and K.R. Varadarajan. Approximating extent measure of points. *J. ACM*, 51:606–635, 2004.

[AHPV07]   P.K. Agarwal, S. Har-Peled, and K. Varadarajan. Geometric approximations via coresets. In J.E. Goodman, J. Pach and E. Welzl, editor, *Combinatorial and Computational Geometry*. Vol. 52 of *MSRI Publications*, pages 1–30, Cambridge University Press, 2007.

[AHPY08]   P.K. Agarwal, S. Har-Peled, and H. Yu. Robust shape fitting via peeling and grating coresets. *Discrete Comput. Geom.*, 39:38–58, 2008.

[AMS99]   N. Alon, Y. Matias, and M. Szegedy. The space complexity of approximating the frequency moments. *J. Comp. Syst. Sci.*, 58:137–147, 1999.

[AN12]   M. Arnaudon and F. Nielsen. On approximating the Riemannian 1-center. *Comput. Geom.*, 1:93–104, 2012.

[AP02]   P.K. Agarwal and C.M. Procopiuc. Exact and approximate algorithms for clustering. *Algorithmica*, 33:201–226, 2002.

[AS15]   P.K. Agarwal and R. Sharathkumar. Streaming algorithms for extent problems in high dimensions. *Algorithmica*, 72:83–98, 2015.

[AY07]   P.K. Agarwal and H. Yu. A space-optimal data-stream algorithm for coresets in the plane. In *Proc. 23rd Sympos. Comput. Geom.*, pages 1–10, ACM Press, 2007.

[Ban10]   N. Bansal. Constructive algorithms for discrepancy minimization. In *Proc. 51st IEEE Sympos. Found. Comp. Sci.*, pages 3–10, 2010.

[BC08]   M. Bădoiu and K.L. Clarkson. Optimal coresets for balls. *Comput. Geom.*, 40:14–22, 2008.

[BC03]   M. Bădoiu and K.L. Clarkson. Smaller core-sets for balls. In *Proc. 14th ACM-SIAM Sympos. Discrete Algorithms*, pages 801–802, 2003.

[BCEG07]   A. Bagchi, A. Chaudhary, D. Eppstein, and M.T. Goodrich. Deterministic sampling and range counting in geometric data streams. *ACM Trans. Algorithms*, 3:2, 2007.

[BCIS10]   R. Berinde, G. Cormode, P. Indyk, and M. Strauss. Space-optimal heavy hitters with strong error bounds. *ACM Trans. Database Syst.*, 35:4, 2010.

[BDMI13]   C. Boutsidis, P. Drineas, and M. Magdon-Ismail. Near-optimal coresets for least-squares regression. *IEEE Trans. Inform. Theory*, 59:6880–6892, 2013.

[BDMI14]   C. Boutsidis, P. Drineas, and M. Magdon-Ismail. Near optimal column-based matrix reconstruction. *SIAM J. Comput.*, 43:687–717, 2014.

[BHPR16]   A. Blum, S. Har-Peled, and B. Raichel. Sparse approximation via generating point sets. In *Proc. ACM-SIAM Sympos. Discrete Algorithms*, pages 548–557, 2016.

[BS80]   J.L. Bentley and J.B. Saxe. Decomposable searching problems I: Static-to-dynamic transformations. *J. Algorithms*, 1:301–358, 1980.

[BS13]   N. Bansal and J. Spencer. Deterministic discrepancy minimization. *Algorithmica*, 67:451–471, 2013.

[BSS14]   J.D. Batson, D.A. Spielman, and N. Srivastava. Twice-Ramanujan sparsifiers. *SIAM Review*, 56:315–334, 2014.

[BW17]   C. Boutsidis and D.P. Woodruff. Optimal CUR decompositions. *SIAM J. Comput.*, 46:543–589, 2017.

[CCFC04]   M. Charikar, K. Chen, and M. Farach-Colton. Finding frequent items in data streams. *Theoret. Comput. Sci.*, 312:3–15, 2004.

[CDMI+13] K.L. Clarkson, P. Drineas, M. Magdon-Ismail, M.W. Mahoney, X. Meng, and D.P. Woodruff. The fast Cauchy transform and faster robust linear regression. In *Proc. 24th ACM-SIAM Sympos. Discrete Algorithms*, pages 466–477, 2013. (Extended version: arXiv:1207.4684, version 4, 2014.)

[CEM+15]   M.B. Cohen, S. Elder, C. Musco, C. Musco, and M. Persu. Dimensionality reduction for k-means clustering and low rank approximation. In *Proc. 47th ACM Sympos. Theory Comput.*, pages, 163–172, 2015.

[Cha00]   B. Chazelle. *The Discrepancy Method.* Cambridge University Press, 2000.

[Cha06]   T.M. Chan. Faster core-set constructions and data-stream algorithms in fixed dimensions. *Comput. Geom.*, 35:20–35, 2006.

[Che09]   K. Chen. On coresets for k-median and k-means clustering in metric and Euclidean spaces and their applications. *SIAM J. Comput.*, 39:923–947, 2009.

[CHW12]   K.L. Clarkson, E. Hazan, and D.P. Woodruff. Sublinear optimization for machine learning. *J. ACM*, 59:23, 2012.

[Cla10]   K.L. Clarkson. Coresets, sparse greedy approximation, and the Frank-Wolfe algorithm. *ACM Trans. Algorithms*, 6:4, 2010.

[CLM+15]   M.B. Cohen, Y.T. Lee, C. Musco, C. Musco, R. Peng, and A. Sidford. Uniform sampling for matrix approximation. In *Proc. Conf. Innovations Theor. Comp. Sci.*, pages 181–190, ACM Press, 2015.

[CM96]   B. Chazelle and J. Matoušek. On linear-time deterministic algorithms for optimization problems in fixed dimensions. *J. Algorithms*, 21:579–597, 1996.

[CM05]   G. Cormode and S. Muthukrishnan. An improved data stream summary: The count-min sketch and its applications. *J. Algorithms*, 55:58–75, 2005.

[CGHJ11]   G. Cormode, M. Garofalakis, P.J. Haas, C. Jermaine. Synopses for massive data: Samples, histograms, wavelets, sketches. *Found. Trends Databases*, 4:1–294, 2011.

[CV07]   K.L. Clarkson and K. Varadarajan. Improved approximation algorithms for geometric set cover. *Discrete Comput. Geom.*, 37:43–58, 2007.

[CW17]   K.L. Clarkson and D.P. Woodruff. Low rank approximation and regression in input sparsity time. *J. ACM*, 63:54, 2017.

[CW15a]   K.L. Clarkson and D.P. Woodruff. Input sparsity and hardness for robust subspace approximation. In *Proc. 47th IEEE Sympos. Found. Comp. Sci.*, pages 310–329, 2015.

[CW15b]   K.L. Clarkson and D.P. Woodruff. Sketching for M-estimators: A unified approach to robust regression. In *Proc. ACM-SIAM Sympos. Discrete Algorithms*, pages 921–939, 2015.

[DF03]   M. Durand and P. Flajolet. Loglog counting of large cardinalities. In *Proc. 11th European Sympos. Algorithms*, vol. 2832 of *LNCS*, pages 605–617, Springer, Berlin, 2003.

[DFK+04]   P. Drineas, A. Frieze, R. Kannan, S. Vempala, and V. Vinay. Clustering large graphs via the singular value decomposition. *Machine Learning*, 56:9–33, 2004.

[DG03]   S. Dasgupta and A. Gupta. An elementary proof of a theorem of Johnson and Lindenstrauss. *Random Structures Algorithms*, 22:60–65, 2003.

[DKM06]   P. Drineas, R. Kannan, and M.W. Mahoney. Fast Monte Carlo algorithms for matrices II: Computing a low-rank approximation to a matrix. *SIAM J. Comput.*, 36:158–183, 2006.

[DLOM02]   E.D. Demaine, A. Lopez-Ortiz, and J.I. Munro. Frequency estimation of internet packet streams with limited space. In *Proc. 10th European Sympos. Algorithms*, vol. 2461 of *LNCS*, pages 348–360, Springer, Berlin, 2002.

[DMI+12]   P. Drineas, M. Magdon-Ismail, M.W. Mahoney, and D.P. Woodruff. Fast approximation of matrix coherence and statistical leverage. *J. Machine Learning Research*, 13:3441–3472, 2012.

[DMM08]   P. Drineas, M.W. Mahoney, and S. Muthukrishnan. Relative-error CUR matrix decompositions. *SIAM J. Matrix Analysis Appl.*, 30:844–881, 2008.

[DV06]   A. Deshpande and S. Vempala. Adaptive sampling and fast low-rank matrix approximation. In *Proc. 10th Workshop Randomization and Computation*, vol. 4110 of *LNCS*, pages 292–303, Springer, Berlin, 2006.

[FFK11]   D. Feldman, M. Faulkner, and A. Krause. Scalable training of mixture models via coresets. In *Proc. Neural Information Processing Systems*, pages 2142–2150, 2011.

[FKV04]   A. Frieze, R. Kannan, and S. Vempala. Fast Monte-Carlo algorithms for finding low-rank approximations. *J. ACM*, 51:1025–1041, 2004.

[FL11]   D. Feldman and M. Langberg. A unified framework for approximating and clustering data. In *Proc. 43rd ACM Sympos. Theory Comput.*, pages 569–578, 2011.

[FM85]   P. Flajolet and G.N. Martin. Probabilistic counting algorithms for data base applications. *J. Comp. Syst. Sci.*, 31:182–209, 1985.

[FMS07]   D. Feldman, M. Monemizadeh, and C. Sohler. A PTAS for k-means clustering based on weak coresets. In *Proc. 23rd Sympos. Comput. Geom.*, pages 11–18, ACM Press, 2007.

[FO15]   D. Felber and R. Ostrovsky. A randomized online quantile summary in $O(\frac{1}{\epsilon} \log \frac{1}{\epsilon})$ words. In *Proc. 19th Workshop Randomization and Computation*, vol. 40 of LIPIcs, pages 775–785, Schloss Dagstuhl, 2015.

[FS12]   D. Feldman and L.J. Schulman. Data reduction for weighted and outlier-resistant clustering. In *Proc. 23rd ACM-SIAM Sympos. Discrete Algorithms*, pages 1343–1354, 2012.

[FSS13]   D. Feldman, M. Schmidt, and C. Sohler. Turning big data into tiny data: Constant-size coresets for k-means, PCA, and projective clustering. In *Proc 24th ACM-SIAM Sympos. Discrete Algorithms*, pages 1434–1453, 2013.

[FT15]   D. Feldman and T. Tassa. More constraints, smaller coresets: Constrained matrix approximation of sparse big data. In *Proc. 21st ACM Sympos. Knowledge Discovery Data Mining*, pages 249–258, 2015.

[FVR16]   D. Feldman, M. Volkov, and D. Rus. Dimensionality reduction of massive sparse datasets using coresets. In *Proc. 29th Adv. Neural Inf. Process. Syst.*, pages 2766–2774, 2016.

[FW56]   M. Frank and P. Wolfe. An algorithm for quadratic programming. *Naval Research Logistics Quarterly*, 3:95–110, 1956.

[GI10]   A.C. Gilbert and P. Indyk. Sparse recovery using sparse matrices. In *Proc. IEEE*, 98:937–947, 2010.

[GJ09]   B. Gärtner and M. Jaggi. Coresets for polytope distance. In *Proc. 25th Sympos. Comput. Geom.*, pages 33–42, ACM Press, 2009.

[GK01]   M. Greenwald and S. Khanna. Space-efficient online computation of quantile summaries. In *Proc. ACM Conf. Management Data*, pages 58–66, 2001.

[GLPW16]   M. Ghashami, E. Liberty, J.M. Phillips, and D.P. Woodruff. Frequent directions: Simple and deterministic matrix sketching. *SIAM J. Comput.*, 45:1762–1792, 2016.

[Gon85]   T.F. Gonzalez. Clustering to minimize the maximum intercluster distance. *Theoret. Comput. Sci.*, 38:293–306, 1985.

[GP14]   M. Ghashami and J.M. Phillips. Relative errors for deterministic low-rank matrix approximations. In *Proc. 25th ACM-SIAM Sympos. Discrete Algorithms*, pages 707–717, 2014.

[Han16]   S. Hanneke. The optimal sample complexity of PAC learning. *J. Machine Learning Res.*, 17:115, 2016.

[Hoc97]   D. Hochbaum. *Approximation Algorithms for NP-Hard Problems*. PWS Publishing Company, 1997.

[HLPW16]   L. Huang, J. Li, J.M. Phillips, and H. Wang. ε-Kernel coresets for stochastic points. In *Proc. 24th European Sympos. Algorithms*, vol. 57 of *LIPIcs*, article 50, Schloss Dagstuhl, 2016.

[HP04a]   S. Har-Peled. Clustering motion. *Discrete Comput. Geom.*, 31:545–565, 2004.

[HP04b]   S. Har-Peled. No coreset, no cry. In *Proc. 24th Conf. Found. Software Tech. Theor. Comp. Sci.*, vol. 3821 of *LNCS*, pages 107–115, Springer, Berlin, 2004.

[HP06]   S. Har-Peled. Coresets for discrete integration and clustering. In *Proc Found. Software Tech. Theor. Comp. Sci.*, vol. 4337 of *LNCS*, pages 33–44, Springer, Berlin, 2006.

[HP11]   S. Har-Peled. *Geometric Approximation Algorithms*. AMS, Providence, 2011.

[HPK07]   S. Har-Peled and A. Kushal. Smaller coresets for k-median and k-mean clustering. *Discrete Comput. Geom.*, 37:3–19, 2007.

[HPM04]   S. Har-Peled and S. Mazumdar. Coresets for k-means and k-median clustering and their applications. In *Proc. 36th ACM Sympos. Theory Comput.*, pages 291–300, 2004.

[HPS11]   S. Har-Peled and M. Sharir. Relative (p, ε)-approximations in geometry. *Discrete Comput. Geom.*, 45:462–496, 2011.

[HPV04]   S. Har-Peled and K. Varadarajan. High-dimensional shape fitting in linear time. *Discrete Comput. Geom.*, 32:269–288, 2004.

[HPW04]   S. Har-Peled and Y. Wang. Shape fitting with outliers. *SIAM J. Comput.*, 33:269–285, 2004.

[HW87]   D. Haussler and E. Welzl. Epsilon nets and simplex range queries. *Discrete Comput. Geom.*, 2:127–151, 1987.

[IKP14]   P. Indyk, M. Kapralov, and E. Price. (Nearly) Space-optimal sparse Fourier transform. In *Proc. 25th ACM-SIAM Sympos. Discrete Algorithms*, pages 480–499, 2014.

[JKPV11]   S. Joshi, R.V. Kommaraju, J.M. Phillips, and S. Venkatasubramanian. Comparing distributions and shapes using the kernel distance. In *Proc. 27th Sympos. Comput. Geom.*, pages 47–56, ACM Press, 2011.

[JL84]   W.B. Johnson and J. Lindenstrauss. Extensions of Lipschitz mappings into a Hilbert space. *Contemp. Math.*, 26:189–206, 1984.

[KLL16]   Z. Karnin, K. Lang, and E. Liberty. Optimal quantile approximation in streams. In *Proc. 57th IEEE Sympos. Found. Comp. Sci.*, pages 71–76, 2016.

[KSP03]   R.M. Karp, S. Shenker, and C.H. Papadimitriou. A simple algorithm for finding frequent elements in streams and bags. *ACM Trans. Database Syst.*, 28:51–55, 2003.

[Lar14]   K.G. Larsen. On range searching in the group model and combinatorial discrepancy. *SIAM J. Comput.*, 43:673–686, 2014.

[Lib13]   E. Liberty. Simple and deterministic matrix sketching. In *Proc. 19th ACM SIGKDD Conf. Knowledge Discovery Data Mining*, pages 581–588, 2013.

[LLS01]   Y. Li, P. Long, and A. Srinivasan. Improved bounds on the samples complexity of learning. *J. Comp. Syst. Sci.*, 62:516–527, 2001.

[LS10]   M. Langberg and L.J. Schulman. Universal ε-approximators for integrals. In *Proc. 21st ACM-SIAM Sympos. Discrete Algorithms*, pages 598–607, 2010.

[MAA06]   A. Metwally, D. Agrawal, and A. Abbadi. An integrated efficient solution for computing frequent and top-k elements in data streams. *ACM Trans. Database Syst.*, 31:1095–1133, 2006.

[Mah11]   M.W. Mahoney. Randomized algorithms for matrices and data. *Found. Trends Machine Learning*, 3:123–224, 2011.

[Mat08]   J. Matoušek. On variants of the Johnson-Lindenstrauss lemma. *Random Structures Algorithms*, 33:142–156, 2008.

[Mat10]   J. Matoušek. *Geometric Discrepancy; An Illustrated Guide*, 2nd printing. Springer, Heidelberg, 2010.

[MG82]   J. Misra and D. Gries. Finding repeated elements. *Sci. Comp. Prog.*, 2:143–152, 1982.

[MK08]   R.M. McCutchen and S. Khuller. Streaming algorithms for k-center clustering with outliers and with anonymity. In *Proc. 11th Workshop Approx. Algorithms*, vol. 5171 of *LNCS*, pages 165–178, Springer, Berlin, 2008.

[MNT15]   J. Matoušek, A. Nikolov, and K. Talwar. Factorization norms and hereditary discrepancy. Preprint, `arXiv:1408.1376v2`, version 2, 2015.

[MSF14]   A. Munteanu, C. Sohler, and D. Feldman. Smallest enclosing ball for probabilistic data. In *Proc. 30th Sympos. Comput. Geom.*, page 214, ACM Press, 2014.

[MSW90]   J. Matoušek, R. Seidel, and E. Welzl. How to net a lot with a little: Small ε-nets for disks and halfspaces. In *Proc. 6th Sympos. Comput. Geom.*, pages 16–22, ACM

Press, 1990. Corrected version available at http://kam.mff.cuni.cz/~matousek/enets3.ps.gz, 2000.

[MT83]   N. Megiddo and A. Tamir. Finding least-distance lines. *SIAM J. Algebraic Discrete Methods*, 4:207–211, 1983.

[Mut05]   S. Muthukrishnan. Data streams: Algorithms and applications. *Found. Trends Theor. Comp. Sci.*, 1:117–236, 2005.

[Nao12]   A. Naor. Sparse quadratic forms and their geometric applications (after Batson, Spielman and Srivastava). *Asterisque*, 348:189–217, 2012.

[NN13]   J. Nelson and H.L. Nguyen. OSNAP: Faster numerical linear algebra algorithms via sparser subspace embeddings. In *Proc. 54th IEEE Sympos. Found. Comp. Sci.*, pages 117–126, 2013.

[Phi13]   J.M. Phillips. ε-Samples for kernels. In *Proc. 24th ACM-SIAM Sympos. Discrete Algorithms*, pages 1622–1632, 2013.

[PR08]   E. Pyrga and S. Ray. New existence proofs for ε-nets. In *Proc. 24th Sympos. Comput. Geom.*, pages 199–207, ACM Press, 2008.

[PT13]   J. Pach and G. Tardos. Tight lower bounds for the size of epsilon-nets. *J. Amer. Math. Soc.*, 26:645–658, 2013.

[PZ15]   J.M. Phillips and Y. Zheng. Subsampling in smooth range spaces. In *Proc. 26th Conf. Algorithmic Learning Theory*, vol. 9355 of *LNCS*, pages 224–238, Springer, Berlin, 2015.

[Sar06]   T. Sarlós. Improved approximation algorithms for large matrices via random projections. In *Proc. 47th IEEE Sympos. Found. Comp. Sci.*, pages 143–152, 2006.

[Sri10]   N. Srivastava. *Spectral Sparsification and Restricted Invertibility*. PhD thesis, Yale University, New Haven, 2010.

[STZ06]   S. Suri, C.D. Tóth, and Y. Zhou. Range counting over multidimensional data streams. *Discrete Comput. Geom.*, 36:633–655, 2006.

[TKC05]   I.W. Tsang, J.T. Kwok, and P.-M. Cheung. Core vector machines: Fast SVM training on very large data sets. *J. Machine Learning Res.*, 6:363–392, 2005.

[VC71]   V.N. Vapnik and A.Y. Chervonenkis. On the uniform convergence of relative frequencies of events to their probabilities. *Theory Probab. Appl.*, 16:264–280, 1971.

[VX12]   K. Varadarajan and X. Xiao. On the sensitivity of shape fitting problems. In *Proc. Conf. Found. Software Tech. Theor. Comp. Sci.*, vol. 18 of LIPIcs, pages 486–497, Schloss Dagstuhl, 2012.

[Woo14]   D.P. Woodruff. Sketching as a tool for numerical linear algebra. *Found. Trends Theor. Comp. Sci.*, 10:1–157, 2014.

[WZ13]   D.P. Woodruff and Q. Zhang. Subspace embeddings and lp regression using exponential random variables. In *Proc. Conference on Learning Theory*, 2013.

[YAPV08]   H. Yu, P.K. Agarwal, R. Poreddy, and K. Varadarajan. Practical methods for shape fitting and kinetic data structures using coresets. *Algorithmica*, 52:378–402, 2008.

[ZZ11]   H. Zarrabi-Zadeh. An almost space-optimal streaming algorithm for coresets in fixed dimensions. *Algorithmica*, 60:46–59, 2011.